Advances in Intelligent Systems and Computing

Volume 1292

The series "Advances in Intelligent Systems and Computing" contains publications on theory, applications, and design methods of Intelligent Systems and Intelligent Computing. Virtually all disciplines such as engineering, natural sciences, computer and information science, ICT, economics, business, e-commerce, environment, healthcare, life science are covered. The list of topics spans all the areas of modern intelligent systems and computing such as: computational intelligence, soft computing including neural networks, fuzzy systems, evolutionary computing and the fusion of these paradigms, social intelligence, ambient intelligence, computational neuroscience, artificial life, virtual worlds and society, cognitive science and systems, Perception and Vision, DNA and immune based systems, self-organizing and adaptive systems, e-Learning and teaching, human-centered and human-centric computing, recommender systems, intelligent control, robotics and mechatronics including human-machine teaming, knowledge-based paradigms, learning paradigms, machine ethics, intelligent data analysis, knowledge management, intelligent agents, intelligent decision making and support, intelligent network security, trust management, interactive entertainment, Web intelligence and multimedia.

The publications within "Advances in Intelligent Systems and Computing" are primarily proceedings of important conferences, symposia and congresses. They cover significant recent developments in the field, both of a foundational and applicable character. An important characteristic feature of the series is the short publication time and world-wide distribution. This permits a rapid and broad dissemination of research results.

Indexed by DBLP, EI Compendex, INSPEC, WTI Frankfurt eG, zbMATH, Japanese Science and Technology Agency (JST), SCImago.

All books published in the series are submitted for consideration in Web of Science.

More information about this series at http://www.springer.com/series/11156

Sheng-Lung Peng · Rong-Xia Hao · Souvik Pal
Editors

Proceedings of First International Conference on Mathematical Modeling and Computational Science

ICMMCS 2020

Springer

Editors
Sheng-Lung Peng
Department of Creative Technologies
and Product Design
National Taipei University of Business
Taoyuan, Taiwan

Rong-Xia Hao
Department of Mathematics
Beijing Jiaotong University
Beijing, China

Souvik Pal
Department of Computer Science
and Engineering
Global Institute of Management
and Technology
Krishnanagar, West Bengal, India

ISSN 2194-5357 ISSN 2194-5365 (electronic)
Advances in Intelligent Systems and Computing
ISBN 978-981-33-4388-7 ISBN 978-981-33-4389-4 (eBook)
https://doi.org/10.1007/978-981-33-4389-4

This Springer imprint is published by the registered company Springer Nature Singapore Pte Ltd.
The registered company address is: 152 Beach Road, #21-01/04 Gateway East, Singapore 189721, Singapore

Organizing Committee and Key Members

Conference Committee Members

Conference General Chair

Sheng-Lung Peng, National Taipei University of Business, Taiwan

Program Conveners

Hanaa Hachimi, Secretary General of Sultan Moulay Slimane University, USMS of Beni Mellal, Morocco
G. Suseendran, Vels Institute of Science, Technology, and Advanced Studies, Chennai, India

Conference Organizing Chairs

K. R. Karthikeyan, Caledonian College of Engineering, Sultanate of Oman
Souvik Pal, Global Institute of Management and Technology, India

Program Co-conveners

Duraisamy Balaganesh, Lincoln University College, Malaysia
Ndolane Sene, Cheikh Anta Diop University, West Africa

Program Chairs

D. Akila, Vels Institute of Science, Technology, and Advanced Studies, India
Thulasyammal Ramiah Pillai, Taylor's University, Malaysia
V. Umadevi, New Prince Sri Bhavani Arts and Science College, India

International Advisory Board Members

Srinath Doss, Botho University, Botswana
Anirban Das, University of Engineering and Management, India
Debashis De, Maulana Abul Kalam Azad University of Technology, India
Mario Jose Diván, National University of La Pampa, Argentina
Ahmed A. Elnger, Beni-Suef University, Egypt
Limei Lin, Fujian Normal University, China
Sanjeevikumar Padmanaban, Aalborg University, Denmark
K. Rohini, Vels Institute of Science, Technology, and Advanced Studies, Chennai, India
Anand Nayyar, Duy Tan University, Vietnam
Bikramjit Sarkar, JIS College of Engineering, India
Kusum Yadav, University of Hail, Kingdom of Saudi Arabia
Noor Zaman, Taylor's University, Malaysia

Technical Chairs

V. R. Elangovan, A. M. Jain College, India
Ashish Mishra, Gyan Ganga Institute of Technology and Science, India
P. Sumathi, C. Kandaswami Naidu College for Men, India

Technical Program Committee Members

Jafar Ahmad Abed Alzubi, Al-Balqa Applied University, Salt, Jordan
R. Anandan, Vels Institute of Science, Technology, and Advanced Studies, Chennai, India
R. Amutha, Tiruvalluvar University College Arts and Science, Tamil Nadu, India
J. Deny, Kalasalingam Academy of Research and Education, Srivilliputtur, India
Sayed Ameeddnuddin Irfan, Universiti Teknologi Petronas, Malaysia
S. Elavarasi, New Prince Shri Bhvani arts and Science College, Chennai, India
S. Lavanya, Bharathi Women's College, Chennai, India

K. R. Karthikeyan, Caledonian College of Engineering, Sultanet of Oman
V. R. Elangovan, A. M. Jain College, India
S. Gopinathan, University of Madras, Chennai, India
Cheng-KuanLin, Fuzhou University, China
Dac-Nhuong Le, Haiphong University, Haiphong, Vietnam
M. Manikandan, Actuarial Science and Applied Statistics, Malaysia
Mohammed Kaicer, Ibn Tofail University, Morocco
K. Kavitha, Mother Teresa University, India
Jyoti Mishra, Gyan Ganga Institute of Technology and Science, India
B. Mahalaksmi, Vels Institute of Science, Technology, and Advanced Studies, Chennai, India
Anand Paul, Kyungpook National University, South Korea
G. Palani, Dr. Ambedkar Government Arts College, Chennai, India
T. Gunasekaran, Veltech Rangarajan and Dr. Sagunthala R&D Institute of Science and Technology, India
T. Nathiya, New Prince Shri Bhvani Arts and Science College, Chennai, India
T. Nagarathinam, MASS College of Arts and Science, Thanjavur, India
M. Thiyagaraj, Nazareth College of Arts and Science, Chennai, India
E. Ramaraj, Alagappa University, India
Suresh Rasappan, VelTech Rangarajan and Dr. Sagunthala R&D Institute of Science and Technology, India
D. Senthil, Nazareth College of Arts and Science, Chennai, India
Abhishek Singh, Banasthali Vidhyapith, India

Invited Speakers

Prof. (Dr.) Amiya Bhaumik, Vice Chancellor and CEO, Lincoln University College, Malaysia
Prof. Hanaa Hachimi, Associate Professor, Applied Mathematics and Computer Science, Secretary General of Sultan Moulay Slimane University, USMS of Beni Mellal, Morocco
Prof. Cheng-Kuan Lin, College of Mathematics and Computer Science, Fuzhou University, China
Prof. E. Chandrasekaran, Mathematics Department, VelTech Rangarajan and Dr. Sagunthala R&D Institute of Science and Technology, India
Dr. G. Suseedran, Department of Information Technology, Vels Institute of Science, Technology, and Advanced Studies, India
Dr. D. Balaganesh, Dean, Faculty of Computer Science and Multimedia, Lincoln University College, Malaysia
Dr. Noor Zaman, School of Computing and Information Technology, Taylor's University, Selangor, Malaysia
Prof. D. Akila, Department of Information Technology, Vels Institute of Science, Technology, and Advanced Studies, India
Dr. Hamed Daei Kasmaei, Department of Mathematics and Statistics, Islamic Azad University, Central Tehran Branch, Tehran, Iran

Preface and Acknowledgements

The main aim of this proceedings book is to bring together leading academic scientists, researchers, and research scholars to exchange and share their experiences and research results on all aspects of intelligent ecosystems, data sciences, and mathematics. ICMMCS 2020 is a multidisciplinary conference organized with the objective of bringing together academic scientists, professors, research scholars, and students working in various fields of engineering and technology. On ICMMCS 2020, you will have the opportunity to meet some of the world's leading researchers, to learn about some innovative research ideas and developments around the world, and to become familiar with emerging trends in science–technology. The conference will provide the authors, research scholars, and listeners with opportunities for national and international collaboration and networking among universities and institutions for promoting research and developing the technologies globally. This conference aims to promote translation of basic research into institutional and industrial research and convert applied investigation into real-time application.

ICMMCS 2020 is hosted by Lincoln University College, Malaysia. ICMMCS 2020 has been held during August 14 and 15, 2020, in online mode (Zoom platform). The conference brought together researchers from all regions around the world working on a variety of fields and provided a stimulating forum for them to exchange ideas and report on their researches. The proceeding of ICMMCS 2020 consists of 60 best-selected papers which were submitted to the conferences and peer reviewed by conference committee members and international reviewers. The presenters have presented through a virtual screen. Many distinguished scholars and eminent speakers have joined from different countries like India, Malaysia, Bangladesh, Iraq, Iran, China, Vietnam, Taiwan, and Morocco to share their knowledge and experience and to explore better ways of educating our future leaders. This conference became a platform to share the knowledge domain among different countries research culture. The main and foremost pillar of any academic conference is the authors and the researchers. So, we are thankful to the authors for choosing this conference platform to present their works in this pandemic situation.

We sincerely thank the almighty for supporting and standing at all times with us, whether it is good or tough times, and given ways to concede us. Starting from the call for papers till the finalization of chapters, all the team members have given their contributions amicably, which is a positive sign of significant team works. The editors and conference organizers sincerely thank all the members of Springer, especially Mr. Aninda Bose, for providing constructive inputs and allowing an opportunity to finalize this conference proceedings. We also thank Prof. William Achauer and Prof. Anil Chandy for their support. We are also grateful to Ms. Sharmila Mary Panner Selvam, Project Coordinator, Springer, for her cooperation. We thank all the reviewers who hail from different places in and around the globe shared their support and stand firm toward quality chapter submission in this pandemic situation.

Finally, we would like to wish you have good success in your presentations and social networking. Your strong supports are critical to the success of this conference. We hope that the participants not only enjoyed the technical program in conference but also found eminent speakers and delegates in the virtual platform. Wishing you a fruitful and enjoyable ICMMCS 2020.

Taoyuan, Taiwan — Sheng-Lung Peng
Beijing, China — Rong-Xia Hao
Krishnanagar, India — Souvik Pal

Contents

About the Editors

Sheng-Lung Peng is a Professor and the director (head) of the Department of Creative Technologies and Product Design, National Taipei University of Business, Taiwan. He received the Ph.D. degree in Computer Science from the National Tsing Hua University, Taiwan. He is an honorary Professor of Beijing Information Science and Technology University, China, and a visiting Professor of Ningxia Institute of Science and Technology, China. He is also an adjunct Professor of Mandsaur University, India. Dr. Peng has edited several special issues at journals, such as *Soft Computing*, *Journal of Internet Technology*, *Journal of Real-Time Image Processing*, *International Journal of Knowledge and System Science*, *MDPI Algorithms*, and so on. His research interests are in designing and analyzing algorithms for Bioinformatics, Combinatorics, Data Mining, and Networks areas in which he has published over 100 research papers.

Rong-Xia Hao is a professor and Ph.D. student supervisor of the department of Mathematics at Beijing Jiaotong University, Beijing, P. R. China. She received the Diploma and Ph.D. degrees from Beijing Jiaotong University in 2002. From 1998 to 2006, she was an associate professor. Since 2006, she was a professor at Department of Mathematics, Beijing Jiaotong University. She received Beijing Jiaotong University ZhiJin Foundation Outstanding Youth Teaching Award in 2007 and the First Prize of 2008 Excellent Paper Awarded by Beijing Operations Research Society. She is supported by National Natural Science Foundation of China for several times. She published over 80 international journal papers. Her research interests include graph theory, interconnection network and fault tolerant computing. She visited Georgia Institute of Technology for one year in 2013 and visited West Virginia University for half of a year in 2006 as a visiting scholar. She is a mathematical reviews reviewer, and an associate Editor of the *International Journal of Computer Mathematics Computer Systems Theory*.

Souvik Pal, Ph.D., is currently associated as an Associate Professor and Head of the Computer Science and Engineering Department at the Global Institute of Management and Technology, West Bengal, India. Prior to that, he was associated with Brainware University, Kolkata, India; JIS College of Engineering, Nadia; Elitte College of Engineering, Kolkata; and Nalanda Institute of Technology, Bhubaneswar, India. Dr. Pal received his B.Tech., M.Tech., and Ph.D. degrees in the field of Computer Science and Engineering. He has more than a decade of academic experience. He is author or co-editor of 12 books from reputed publishers, including Elsevier, Springer, CRC Press, and Wiley, and he holds three patents. He is serving as a series editor for "Advances in Learning Analytics for Intelligent Cloud-IoT Systems", published by Scrivener Publishing (Scopus-indexed) and "Internet of Things: Data-Centric Intelligent Computing, Informatics, and Communication", published CRC Press, Taylor & Francis Group, USA. Dr. Pal has published a number of research papers in Scopus/SCI-indexed international journals and conferences. He is the organizing chair of RICE 2019, Vietnam; RICE 2020 Vietnam; ICICIT 2019, Tunisia. He has been invited as a keynote speaker at ICICCT 2019, Turkey, and ICTIDS 2019, Malaysia. His professional activities include roles as associate editor and editorial board member for more than 100+ international journals and conferences of high repute and impact. His research area includes cloud computing, big data, internet of things, wireless sensor network, and data analytics. He is a member of many professional organizations, including MIEEE; MCSI; MCSTA/ACM, USA; MIAENG, Hong Kong; MIRED, USA; MACEEE, New Delhi; MIACSIT, Singapore; and MAASCIT, USA.

Maximum Matching in Koch Snowflake and Sierpinski Triangle

P. Tharaniya and G. Jayalalitha

Abstract The goal of this paper presents the method of finding the maximum matching cardinality in the fractal graph. A fractal is uneven or irregular sets of well-defined structure that has to be broken up into small pieces, the property of having each piece analogous or identical to the overall structure at random ranges. Fractal graph is an excellent construction of well-defined objects. It provides a general frame work-study of regular self-similarity sets and natural phenomena of the fractal graphs. Matching is the collection of non-touching edges in the graph. Maximum matching is the largest possible collection of non-touching edges in the graph. This paper is used to determine the constant ratio of maximum matching cardinality in all the iteration of self-similarity Fractal Graph. This paper is used to find the constant ratio between the number of edges in each iteration of the Fractal Graph and the Maximum Cardinality number of Maximum Matching. This provides general formulae for all the iteration of Koch snowflake and Sierpinski triangle. By using mathematical induction method, it finds the maximum matching constitutive number in all the iterations of them.

Keywords Graph · Matching · Maximum matching · Fractals · Mathematical induction

AMS Classification Key 05C · 05C70 · 28A80 · 51N3

1 Introduction

Graph theory is the study of the graphs which can be made by vertices or dots or points which are joined by edges or sinks or lines or curves [1]. It takes an essential

P. Tharaniya (✉) · G. Jayalalitha
Research Scholar, Department of Mathematics, VELS Institute of Science, Technology and Advanced Studies, Chennai, India
e-mail: tharamanisharmi@gmail.com

G. Jayalalitha
e-mail: g.jayalalithamaths.sbs@velsuniv.ac.in

S.-L. Peng et al. (eds.), *Proceedings of First International Conference on Mathematical Modeling and Computational Science*, Advances in Intelligent Systems and Computing 1292, https://doi.org/10.1007/978-981-33-4389-4_1

part of nature. The graph can be implemented from nature. It implies more basis ways which are used to mathematical structures. It is used to solve many difficult practical problems. It acts as an essential role in many fields like physical, medical, social, chemical and informative system. Fractal is one of the mathematically defined genesis structures or appearances. The special case of the fractal is it has the same characteristics in each iteration. It has a simple and repetitive defined fine structure at approximate scales. Set and functions are not enough to describe sufficiently smooth irregular objects. Mathematician discovered fractal graph which is the smooth self-similarity object. It has a simple and recursive defined fine structure at arbitrary scales. Despite that it has been divided or spitted or enlarged or minimized, it has the following same properties and similar mathematically defined structure. That means, it contains self-similarity [2] in each iteration. Recently, the paper entitled maximum matching in fractal graph [3] obtained the maximum matching constitutive number of Von Koch curve and star fractal graph motivated and connected the above result. This paper is used to obtain the Maximum Matching constitutive number [4] of Koch Snowflake and Sierpenski Triangle [5].

2 Koch Snowflake

Koch snowflake is one of the earliest fractals having been described. It is the next level of the von Koch curve. It has been invented by a Swedish Mathematician Hedge Von Koch [6]. He dissatisfied Karl Weierstrass abstract which is the opponent of similarity mathematical object and gives more geometric definition of self-similarity [7] which is called Koch curve, Koch snowflake and so on. Structure of the Koch snowflake is very interesting. It starts with an equilateral triangle in the initial iteration. In the first iteration, it can be implemented into large. Each side of the triangle split into three segments. The middle segment can be moved into outward and form a triangle. It shows that each side of the triangle spit into four lines. This process is continued in the upcoming iteration. In each iteration, edges of the Koch snowflake increase in the common ratio. In each iteration, Koch snowflake has $3(4^n)$ number of edges where n is increased by one and starts at zero.

3 Sierpinski Triangle

It is the world-famous fractal graph in classical geometry [8]. It consists of full of triangles inside. It consists of overall shape of an equilateral triangle which has each side of the triangle that has an equal length [9]. It has self-similarity. It is used in the structure of jewels and building construction [10]. Hedge Von Koch is the father of the Sierpinski triangle.

Sierpinski triangle can be started with an equilateral triangle. In each side of the triangle, fix the center point. Now joining the center points of the triangle, it may form an equilateral triangle [11]. It can be covered ¾ of the original triangle. It can be removed from the original triangle. Each side decreased 1/3 of the original length. Repeat the same procedure for every line to form the new structure, and repeating forever, got a new structure in turn. It is shown in Fig. 7. Its diameter is infinite and its area is finite. By using Theorem 2, maximum matching cardinality [12] of the Sierpinski triangle can be calculated.

4 Maximum Matching in Koch Snowflake

Matching [13] is the set of independent non-touching edges that do not share a common vertex. Maximum matching is the matching that has the largest collection of non-touching edges. Maximum matching constitutive number or cardinality is the number of collections of non-touching edges. The number of edges in the graph and number of edges in maximum matching is denoted by $N(G)$ and $V(G)$. Here Maximum Matching Cardinality is calculated for all the iterations of Koch Snowflake in Fig. 3 by using the mathematical induction method [14]. It follows by an Algorithm 1.

Algorithm 1

Step 1: Analyze the theorem for initial iteration of the fractal.

Step 2: Assume the theorem for $n = k$.

Step 3: To find the maximum matching cardinality for $n = k + 1$.

- Here, the image is divided into regular meshes.
- To calculate the matching cardinality for each image.
- Applying the concept of permutation, merge the divided images [15].
- To get the maximum matching cardinality for $n = k + 1$.

Step 4: The theorem is proved for all n.

Theorem 1 *Koch snowflake consists of $t(4^n)$ number of edges in each iteration. Then, it has $(2t)4^{n-1}$ number of edges in the maximum matching in the corresponding iteration where t is the number of edges in the equilateral triangle in the initial iteration.*

Proof This theorem can be proved by an induction method.

Step 1:

In the initial iteration, it consists of an equilateral triangle. It has only 3 edges.

It is shown in Fig. 1. Obviously, it is an equilateral triangle; any single edge is the maximum matching cardinality.

Number of edges in Koch snowflake $N(G_0) = 3(4^0) = 3$.

Maximum matching number $V(G_0) = (2 * 3)4^{0-1} = 1$ (consider integer only).

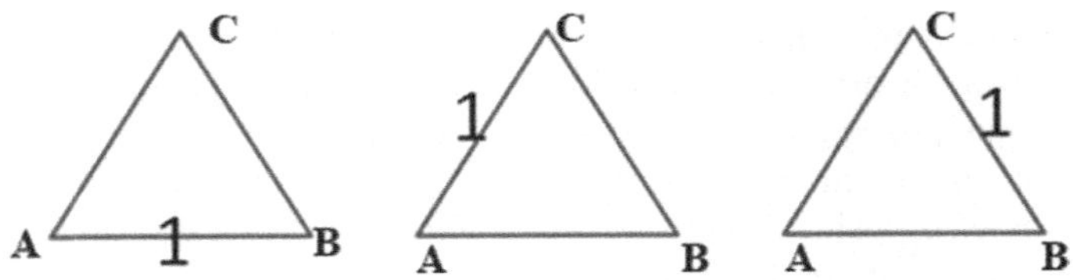

Fig. 1 Initial iteration of Koch snowflake G_0

Fig. 2 First iteration of Koch snowflake G_1

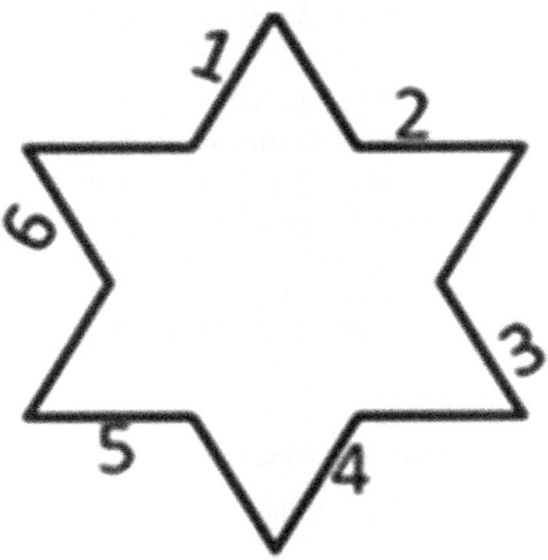

Step 2:

In the first iteration of Koch snowflake, it consists of 12 edges.

It is shown in Fig. 2.

Here $t = 3$ and $n = 1$. Number of edges in the graph $N(G_1) = t\ 4^n = 3 * 4 = 12$.

Total number of edges in maximum matching in the first iteration $V(G_1) = (2t)\ 4^{n-1} = (2*3) * 4^{1-1} = 6$.

Step 3:

Let us assume that this theorem is true for all $n = k$.

For Koch snowflake consisting of $t(4^k)$, edges have $(2t)4^{k-1}$ number of edges in maximum matching.

Step 4:

To prove that the theorem is true for all $n = k + 1$.

In Koch snowflake in $(k + 1)$, iteration has a number of edges t 4^{k+1}.

It can be divided into two components $t\ 4^k$ and 4.

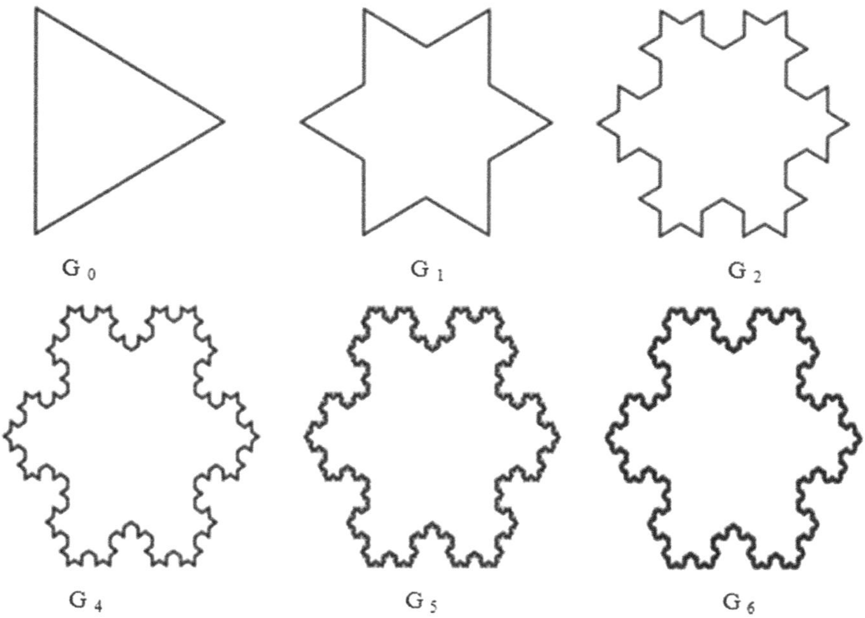

Fig. 3 Different iterations of Koch snowflake

In the first component t 4^k edges of Koch Snowflake has $(2t)\ 4^{k-1}$ edges of Maximum Matching and 4 edges from the second component connected with the first component by permutation method.

Number of edges in maximum matching $= (2t)\ 4^{k-1} * 4 = (2t)4^k$.

The theorem is proved for all n.

Induction method is proved.

A various iteration of Koch snowflake is shown in Fig. 3.

Theorem 2 *Sierpinski triangle has* 3^n *a number of edges in each iteration. It has the common ratio between the number of edges and the cardinality number of maximum matching in each iteration. Then in the first two initial iterations, the Sierpinski triangle has* 3^{n-1} *edges of maximum matching constitutive number* $(1 \leq n \leq 2)$ *and the next level of iterations, and it has* $7{*}3^{n-3}$ *edges of maximum matching constitutive number where* $n \geq 3$.

Proof This theorem can be proved in two steps. In the first step, analyze about initial iteration where $(1 \leq \boldsymbol{n} \leq 2)$.

Step 1:

When $n = 1$, first iteration of the Sierpinski triangle is shown in Fig. 4.

It consists of $3^n = 3$ edges and $3^{n-1} = 3^0 = 1$ edges of maximum matching cardinality.

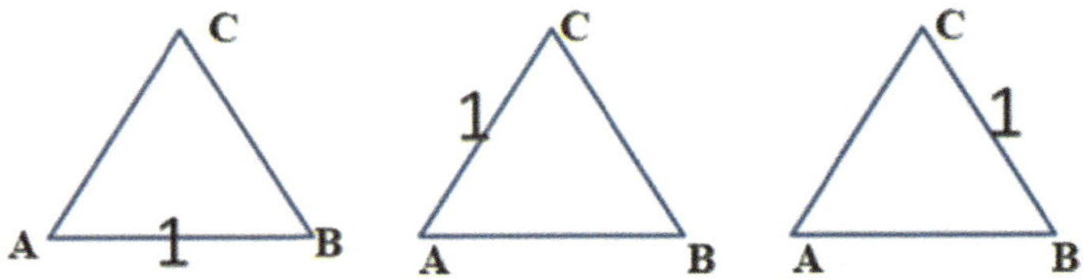

Fig. 4 Maximum matching in first iteration of Sierpinski triangle I_1

Fig. 5 Maximum matching in second iteration of Sierpinski triangle I_2

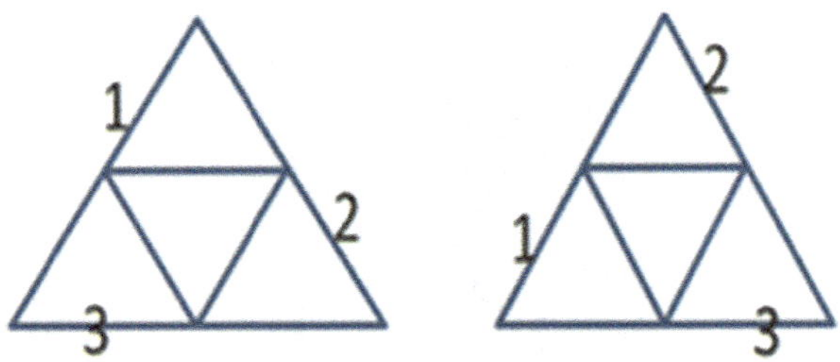

Here in the initial iteration, start with an equilateral triangle. It has only three edges.

It is a closed-loop structure [12]. It implies any one of the edges which is considered as maximum matching cardinality.

Here, maximum matching cardinality $V(G) = 1$.

Step 2:

When $n = 2$, in the second iteration Sierpinski triangle has $3^n = 3^2$ edges and $3^{n-1} = 3^1$ maximum matching cardinality. Here, only three edges are possible to maximum matching set. It is shown in Fig. 5.

Step 3:

After two iterations, the common ratio between the number of edges and maximum matching cardinality has differed.

Hereafter, Sierpinski triangle implies 3^n edges of the fractal graph have $7*3^{n-3}$ maximum matching cardinality.

It can be proved by using mathematical induction method.

Case (i):

In the third iteration, Sierpinski triangle is very interesting. Let $n = 3$.

This Sierpinski triangle has $3^n = 3^3 = 27$ edges.

It has $7*3^{n-3} = 7*3^{3-3} = 7$ number of edges in maximum matching.

It is shown in Fig. 6.

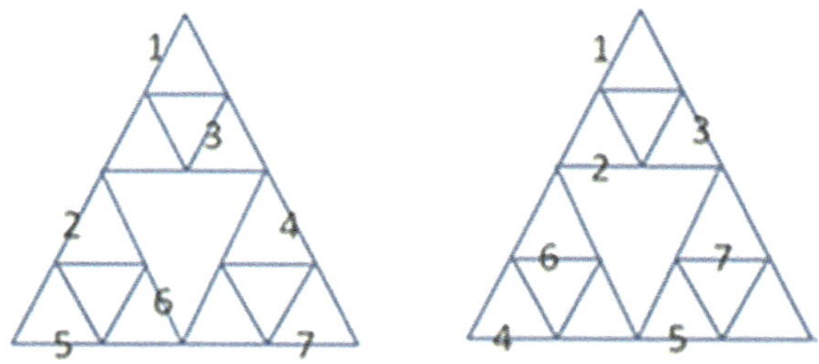

Fig. 6 Maximum matching in third iteration of Sierpinski triangle I_3

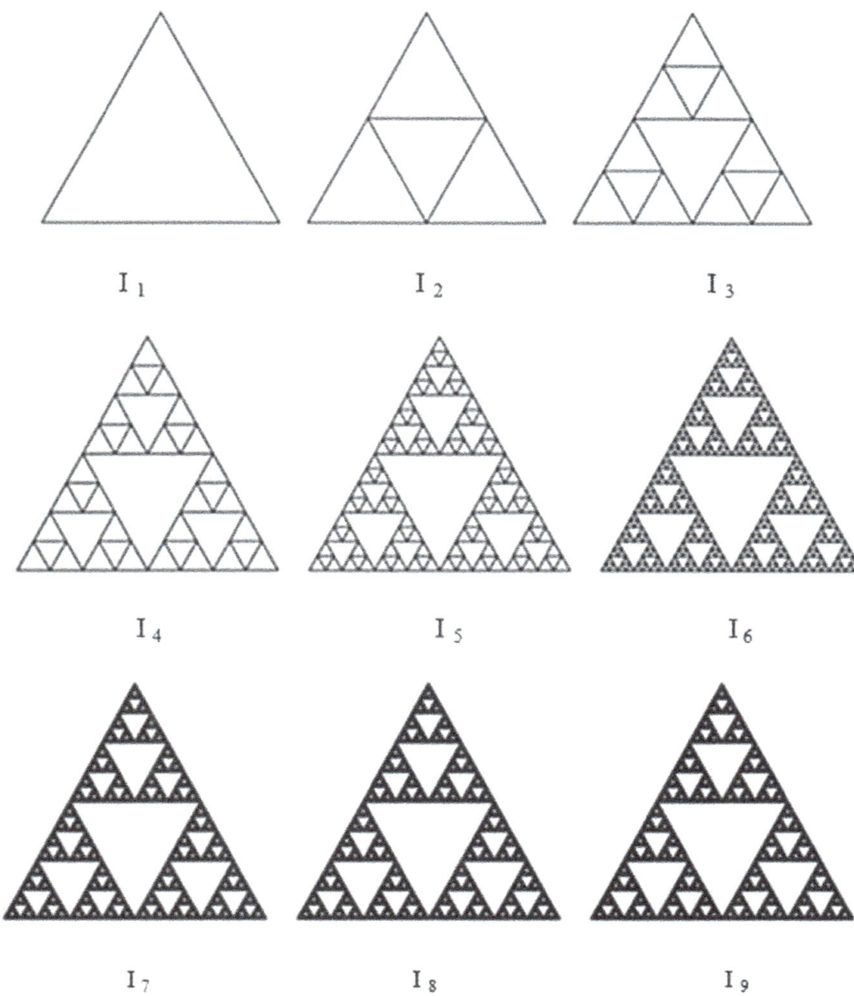

Fig. 7 Different iteration of Sierpinski triangle

Case (ii):

Let us assume that the theorem is true for all $n = k$.

Assume that in kth iteration Sierpinski triangle has 3^k edges and it has $7{*}3^{k-3}$ number edges in maximum matching.

Case (iii):

To prove this theorem is true for $n = k + 1$, i.e., to prove that in this Sierpinski triangle has 3^{k+1} edges and $7{*}3^{k-2}$ edges of a maximum matching set.

In this, Sierpinski triangle has 3^{k+1} edges. It can be divided into two components where one has 3^k edges and other has 3 edges. Assumption of the case (ii) implies the first component of 3^k has $7{*}3^{k-3}$ edges of maximum matching and by the way of permutation method, the second component of three edges can be added to the first component. In this way, permutation implies $7{*}3^{k-2}$ edges of maximum matching cardinality. This theorem is proved for all n.

Various iterations of the Sierpinski triangle are shown in Fig. 7.

Note:

1. In each iteration of Sierpinski triangle, number of edges as well as a number of vertices can be increased.
2. Each iteration number of triangles has to be increased.
3. Way of permutation method is applied for the selection of edges in maximum matching.
4. The number of edges of the Sierpinski triangle consists of only power of 3 values.
5. The diameter of the Sierpinski triangle is increased in the upcoming iteration.
6. The structure of the Sierpinski triangle plays an essential role in designing the new structure of jewels.

5 Conclusion and Future Work

In this paper, maximum matching is applied in self-similarity fractal graph. It is deeply noted it implies a constant ratio between the maximum matching cardinality and the number of edges of each iteration of the fractal graph. The relation and constant ratio between maximum matching cardinality and the edges of the self-similarity fractal graph in the Koch snowflake and Sierpinski triangle are implemented by Theorems 1 and 2. This paper used the concept of coloring of graphs, domination of graphs, creation of new fair ornament, finding injured cells for injecting drugs in medical science, placing stones in jewels, plumping work, designing fractal antennas and so on. This concept is very useful for the architecture field, structures of the building, networking, finding influential people in a social network such as media, Facebook, Instagram, WhatsApp, Google Map and so on.

In the upcoming paper, maximal matching will be analyzed in various types of the graph as well as fractal graph. It will show the role of 'cardinality of maximal matching' in day-to-day real-life problems. The importance of matching in fractals can be explained in various engineering fields like computer science, electrical and biomedical. In the future work, the application of matching in fractals will be explained in medical filed, computer designing, coloring and domination.

References

1. Deo, N., (2017). *Graph theory with applications to engineering and computer science*. Courier Dover Publications.
2. Falconer, K., (2004). *Fractal Geometry: Mathematical Foundations and Applications*. Wiley.
3. Tharaniya, P., & Jayalalitha, G. (2019). Maximum matching in fractal graph. *The International Journal of Analytical and Experimental Modal Analysis, XI*(X), 138–143.
4. Gabow, H. N. (1976). An efficient implementation of Edmonds' algorithm for maximum matching on graphs. *Journal of the ACM (JACM), 23*(2), 221–234.
5. Blackledge, J. M., Evans, A. K., & Turner, M. J., (2002). *Fractal Geometry: Mathematical Methods, Algorithms, Applications*. Elsevier.
6. Mandelbrot, B. (1982). *The Fractal Geometry of Nature*. New York: WH Freeman and Co.
7. Kominek, J. (1997). Advances in fractal compression for multimedia applications. *Multimedia Systems, 5*(4), 255–270.
8. Gyárfás, A., & Lehel, J. (1988). On-line and first fit colourings of graphs. *Journal of Graph Theory, 12*(2), 217–227.
9. Nguyen, H. T., Walker, C. L., & Walker, E. A, (2018). *A First Course in Fuzzy Logic*. CRC.
10. Bollobás, B., (2013). *Modern Graph Theory*. Springer Science & Business Media.
11. Wang, N., (2017). A Framework of Transforming Vertex Deletion Algorithm to Edge Deletion Algorithm (Doctoral dissertation, University of Cincinnati).
12. Trinajstic, N., (2018). *Chemical graph theory*. Routledge.
13. West, D. B. (2001). *Introduction to Graph Theory*. Upper Saddle River: Prentice Hall.
14. Jungnickel, D. (2005). *Graphs, Networks and Algorithms*. Berlin: Springer.
15. Zaslavsky, T., (2012). A mathematical bibliography of signed and gain graphs and allied areas. *The Electronic Journal of Combinatorics*, DS8-Dec.

A Study of Stochastic Ecological Model with Prey Harvesting as a Tool of Disease Control

G. M. Vijaya Lakshmi, Suresh Rasappan, and Pugalarasu Rajan

Abstract This study explores the hierarchical dynamics of a predator model in which the population of prey is comprised of two classes: susceptible and infected (SI) prey including the non-selective prey harvesting. The system's points of equilibrium are found. The model's stability properties were tested by both locally and globally. Next, we have implemented the stochastic perturbations and indicated that the deterministic model of stochastic perturbations is robust. The stochastic perturbations effects on the dynamics were discussed via simulations. The deterministic method results in a stochastic system with an effect of atmospheric random noise were presented. That is, we have explored the SI prey–predator model's stability incorporating prey harvesting by introducing stochastic perturbations that were not previously studied. Finally, some numerical examples and diagrams help our complex dynamics of our model using time series diagrams and phase portraits which supports our theoretical results. Parallel to this, we presented the time series diagram and phase portraits for the stochastic perturbation system.

Keywords SI prey–predator system · Harvesting · Lyapunov's function · Stochastic differential equation · Analyzing stability

1 Introduction

There was abundant interest in the analysis of the complex relationship between prey and predator. In the study of dynamics of prey–predator interactions, the very first step

G. M. Vijaya Lakshmi · S. Rasappan (✉)
Department of Mathematics, Vel Tech Rangarajan Dr Sagunthala R&D Institute of Science and Technology, Avadi, Tamil Nadu 600062, India
e-mail: mrpsuresh83@gmail.com

G. M. Vijaya Lakshmi
e-mail: vijidayal05@gmail.com

P. Rajan
Department of Information Technology, University of Technology and Applied Sciences - Ibri, Ibri, Sultanate of Oman
e-mail: r.pugalarasu@ibrict.edu.om

S.-L. Peng et al. (eds.), *Proceedings of First International Conference on Mathematical Modeling and Computational Science*, Advances in Intelligent Systems and Computing 1292, https://doi.org/10.1007/978-981-33-4389-4_2

is the classical Lotka–Volterra model. Due to its universal nature and significance, the dynamic oscillations of the prey-predator model continued to be one of mathematical ecology's most influential subjects. Plenty of authors [1–4] have worked with the diseased prey on this system and are seeking stability. In the references of Bairagi et al. [5] and Kar et al. [6], we have studied that the harvesting strategy is one of the main issues in environmental, eco-epidemiological, political, etc. According to Peterson et al. [7], the predators only consume the sick prey, since the sick victims were less aggressive and incredibly quick to kill. The prey–predator stability analysis with stochastic disruptions in many ecological systems is a difficult topic where we have been examined in several references ([8–10]). In the work of Srinivasan et al. [11], we studied that the determinist system with the influence of external noise resulted in a stochastic process where input parameters oscillate around its mean values. In this work, we have explored the SI prey–predator model's stability incorporating prey harvesting and we also examined this model by introducing stochastic perturbations through Lyapunov function that were not previously studied. We have evaluated the model's local and global stability. Finally, we portrayed several numerical simulations and have given comparative analysis between deterministic and stochastic system.

2 The Mathematical Model for Prey-Predator

Michaelis–Menten–Holling's revised prey–predator model with dynamics of defenseless infected prey

$$\begin{aligned}
\dot{Y}_1 &= rY_1\left(1 - \frac{Y_1}{\kappa}\right) - PY_1Y_2 \\
\dot{Y}_2 &= PY_1Y_2 - \frac{\gamma Y_2Y_3}{Y_3 + \gamma\beta Y_2} \\
\dot{Y}_3 &= \frac{e\gamma Y_2Y_3}{Y_3 + \gamma\beta Y_2} - dY_3
\end{aligned} \tag{1}$$

where Y_1 and Y_2 are susceptible (S) to prey and infected (I), Y_3—predator, r—the rate of growth, κ—carrying ability, P—the rate of incidence, γ—the maximum level of predator attack, β—processing time, e—efficiency of transition, and d—predator's fatality rate. Furthermore, $\frac{\gamma Y_2Y_3}{Y_3+\gamma\beta Y_2}$ and $\frac{e\gamma Y_2Y_3}{Y_3+\gamma\beta Y_2}$ are Holling's functional response and numerical response due to Michaelis–Menten.

2.1 *Modified Model*

Now, predator–prey dynamics involving harvesting in prey in an indiscriminate manner is given below:

$$\dot{Y}_1 = rY_1\left(1 - \frac{Y_1}{\kappa}\right) - PY_1Y_2 - Eq_1Y_1$$
$$\dot{Y}_2 = PY_1Y_2 - \frac{\gamma Y_2Y_3}{Y_3 + \gamma\beta Y_2} - Eq_2Y_2$$
$$\dot{Y}_3 = \frac{e\gamma Y_2Y_3}{Y_3 + \gamma\beta Y_2} - dY_3 \tag{2}$$

where q_1 and q_2 be S and I prey catchability coefficients, respectively, and E—collective outer action for the harvesting of vulnerable preys in an indiscriminate manner by outside harvesters (not by predators). Thus, Eq_1Y_1 and Eq_2Y_2 represent the capture of vulnerable and infected species of prey. Also q_1—the prone prey catchability coefficient may be lower than q_2—the infected prey catchability coefficient because the diseased prey is worse, less involved than their stable counterpart; and for a similar effort E.

Let's reduce the number of parameters:

$$S = \frac{Y_1}{\kappa}, I = \frac{Y_2}{\kappa}, Y = \frac{Y_3}{\gamma\beta\kappa}, t = r\tau$$

and we are expected to

$$\eta = \frac{P\kappa}{r}, a_1 = \frac{\gamma}{r}, a_2 = \frac{e}{r\beta}, a_3 = \frac{d}{r}, h_1 = \frac{Eq_1}{r}, h_2 = \frac{Eq_2}{r}$$

Hence, the reduced prey–predator model undergoing combined harvesting with prey infection:

$$\dot{S} = S(1 - S) - \eta SI - h_1S$$
$$\dot{I} = \eta SI - a_1\frac{IY}{Y + I} - h_2I$$
$$\dot{Y} = a_2\frac{IY}{Y + I} - a_3I \tag{3}$$

3 The Study of the Deterministic Model

3.1 Boundedness of the Model

Now the below theorem determines the system (3) uniform boundaries.

Theorem 1 Every solution of the system (3) with the initial positive conditions $(S(0), Y(0), I(0))$ is bounded uniformly.

Proof Let $S_0 > 0$, $I_0 > 0$ and $Y_0 > 0$ be any system's solution with positive initial conditions.

We are now defining the function:

$$\Upsilon(t) = S(t) + Y(t) + I(t) \tag{4}$$

Implies the derivative of (4) gives

$$\dot{\Upsilon}(t) = \dot{S}(t) + \dot{I}(t) + \dot{Y}(t) \tag{5}$$

$$= S(1-S) - \eta SI - h_1 S + \eta SI - a_1 \frac{IY}{Y+I} - h_2 I + a_2 \frac{IY}{Y+I} - a_3 Y \tag{6}$$

$$\frac{d\Upsilon}{dt} \le S - h_1 S - h_2 I - a_3 Y \tag{7}$$

$$\frac{d\Upsilon}{dt} \le S - \theta W \tag{8}$$

where $\theta = \min(1, h_1, h_2, a_3)$, which gives:

$$\frac{d\Upsilon}{dt} + \theta\Upsilon \le S \tag{9}$$

Since

$$\frac{d\Upsilon}{dt} \le S(1-S) \tag{10}$$

We have a basic theorem of comparison, by using that

$$\lim_{t\to\infty} \sup S(t) \le \Lambda,$$

where $\Lambda = \max\{S(0), 1\}$, which gives

$$\frac{d\Upsilon}{dt} + \theta\Upsilon \le \Lambda \tag{11}$$

$$\frac{d\Upsilon}{dt} + \theta\Upsilon = \Lambda (\text{we can say}) \tag{12}$$

Now we use inequality theory in differential equation (Birkoff & Rota, 1982)

$$0 \le \Upsilon(S, I, Y) \le \frac{\Lambda}{\theta} + \frac{\Upsilon(S_0, I_0, Y_0)}{e^{\theta t}} \tag{13}$$

as $t \to \infty$, here we get

$$0 \le \Upsilon(S, I, Y) \le \frac{\Lambda}{\theta} \tag{14}$$

Therefore, every solutions of the system (3) join in the region B where,

$$\mathrm{B} = \left\{(S, Y, I) \in \mathbb{R}^3_+ : 0 \le \Upsilon \le \frac{\Lambda}{\theta} + \varepsilon, \forall \varepsilon > 0\right\} \tag{15}$$

That concludes the theorem.

3.2 *Analysis of Equilibrium and Local Stability*

Let us discuss the boundary and interior equilibrium point.

(a) The Trivial equilibrium state $E_T(0, 0, 0)$.
(b) The Boundary equilibrium state $E_{B_1} = (1 - \alpha_1, 0, 0)$.
(c) The Boundary equilibrium state $E_{B_2} = \left(\bar{S}, \bar{I}, 0\right)$ where:

$$\bar{S} = \frac{h_2}{\eta}; \bar{I} = \frac{1}{\eta^2}(\eta - \eta h_1 - h_2)$$

(d) The inner state of equilibrium $E^* = (S^*, I^*, Y^*)$ here $S^* = \frac{1}{\eta}\left(h_2 + \frac{a_3 a_1}{a_2}\right)$, $I^* = \frac{1}{\eta^2 a_2}(\eta a_3 - a_3 a_1 - h_1 \eta a_3 - h_2 a_3)$ and $Y^* = \frac{a_3}{\eta^2 a_3 (a_2 - a_3)}(\eta a_3 - a_3 a_1 - h_1 \eta a_3 - h_2 a_3)$.

Theorem 2 *If $\eta < \frac{h_2}{1-h_1}$ then E_T is stable. If $a_2 > a_3$ then E_{B_2} is unstable. E_T is unstable, can be easily verified. E_{B_1} is stable if $\eta < \frac{h_2}{1-h_1}$. E_{B_2} is unstable if $a_2 > a_3$. E^* asymptotically stable at the local level. Now, we can illustrate that E^* is asymptotically stable globally whenever it persists.*

3.3 *The Analysis of Stability Globally*

Theorem 3 The inward equilibrium point E^* is asymptotically stable globally if:

$$a_2 \delta_3 - a_1 \delta_2 = 0, \tag{16}$$

$$\eta = \frac{1}{S - S^*}, \tag{17}$$

$$\delta_1 - \delta_2(I - I^*) = 0, \tag{18}$$

and

$$\delta_2(I - I^*) + \delta_3(Y - Y^*)^2 = 0 \tag{19}$$

Proof Classify the function of Lyapunov:

$$\begin{aligned} \nu(S, I, Y) = \delta_1\left(S - S^* - S^* \ln \frac{S}{S^*}\right) + \delta_2\left(I - I^* - I^* \ln \frac{I}{I^*}\right) \\ + \delta_3\left(Y - Y^* - Y^* \ln \frac{Y}{Y^*}\right) \end{aligned} \tag{20}$$

where δ_i, $i = 1, 2, 3$ are positive constants.

Apparently, in the region B, ν is a positive definite function apart from on E^*, where it is zero.

By system (3), the rate of change of ν becomes,

$$\begin{aligned} \dot{v}(S, I, Y) = \delta_1(S - S^*)\{1 - S - \eta I - h_1\} \\ + \delta_2(I - I^*)\left\{\eta S - a_1\frac{Y}{Y + I} - h_2\right\} \\ + \delta_3(Y - Y^*)\left\{a_2\frac{I}{Y + I} - a_3\right\} \end{aligned} \tag{21}$$

$$\begin{aligned} = -\delta_1(S - S^*)^2 - \delta_1\eta(S - S^*)(I - I^*) + \delta_2\eta(S - S^*)(I - I^*) \\ -a_1\delta_2\frac{(I - I^*)(Y - Y^*)}{(I - I^*) + (Y - Y^*)} + a_2\delta_3\frac{(I - I^*)(Y - Y^*)}{(I - I^*) + (Y - Y^*)} \end{aligned} \tag{22}$$

Now choosing (16) to (19), we get

$$\frac{dv}{dt} = -\delta_1(S - S^*)^2 - \delta_2(I - I^*)^2 - \delta_3(Y - Y^*)^2 \tag{23}$$

and $\dot{v}$ is thus negative.

Hence, the largest dynamical set where $\dot{v} = 0$ is the point of equilibrium, by Laselle's concept of invariance. E^* asymptotically stable globally.

4 The Stochastic Model

Now we enable the stochastic disturbances of the S^*, I^*, Y^* are proportional to the distances variables in the positive equilibrium E^* along with their values.

Thus, the model (3) is.

$$dS = [S(1 - S) - \eta SI - h_1 S]dt + \varsigma_1(S - S^*)dg_t^1,$$
$$dI = \left[\eta SI - a_1 \frac{IY}{Y + I} - h_2 I\right]dt + \varsigma_2(I - I^*)dg_t^2,$$
$$dY = \left[a_2 \frac{IY}{Y + I} - a_3 Y\right]dt + \varsigma_3(Y - Y^*)dg_t^3. \tag{24}$$

where ς_i and $g_t^i = g_i(t), i = 1, 2, 3$ are real constants and Gaussian white noise. process in three dimensions [6].

$$\mathrm{E}[g_i(t)] = 0, \ \mathrm{i} = 1, 2, 3 \tag{25}$$

$$\mathrm{E}\left[g_i(t)g_j(t')\right] = \gamma_{ij}\gamma(t - t'), i, j = 1, 2, 3 \tag{26}$$

where γ_{ij} and γ are Kronecker and Dirac functions, respectively.

Therefore, $g(t)$ is the Gaussian white noise process with the properties (25) and (26). We will consider (24) as the Ito stochastic differential system.

4.1 The Analysis of Stochastic Model

This study is primarily intended to represent the system dynamics around the point E^* of internal equilibrium. This is why, here we have linearized the model with the following noise process

$$\upsilon_1 = S - S^*; \upsilon_2 = I - I^*; \upsilon_3 = Y - Y^* \tag{27}$$

$$\Rightarrow S = \upsilon_1 + S^*; I = \upsilon_2 + I^*; Y = \upsilon_3 + Y^*$$

Then the linearized form of stochastic differential equation around E^* emerge

$$d\upsilon(t) = f_1(\upsilon(t))dt + f_2(\upsilon(t))dg(t) \tag{28}$$

where $\upsilon(t) = (\upsilon_1(t), \upsilon_2(t), \upsilon_3(t))^T$ and

$$f_1(\upsilon(t)) = \begin{pmatrix} 1 - 2S^* - \eta I^* - & \eta S^* & 0 \\ \eta I^* & \eta S^* - h & 0 \\ 0 & 0 & -a_3 \end{pmatrix} \upsilon(t) \tag{29}$$

$$f_2(\upsilon(t)) = \begin{pmatrix} \varsigma_1\upsilon_1 & 0 & 0 \\ 0 & \varsigma_2\upsilon_2 & 0 \\ 0 & 0 & \varsigma_3\upsilon_3 \end{pmatrix} \tag{30}$$

In (22), the E^* corresponds to the $\upsilon(t) = 0$, which is a trivial solution.

Let $\mathbb{S}$ be the set $\mathbb{S} = (t \geq t_0)X\Re^n, t_0 \in \Re^+$. Hence, $\nu \in C_2^0(\mathbb{S})$ is a twice-continuously differentiable function and a continuous function about t.

Next, the stochastic differential model by using Ito process with the reference to [9]

$$\mathfrak{M}\nu(t,\upsilon) = \frac{\partial\nu(t,\upsilon)}{\partial t} + f_1^T(\upsilon(t))\frac{\partial\nu(t,\upsilon)}{\partial\upsilon} + \frac{1}{2}\text{trace}\left[f_1^T(\upsilon(t))\frac{\partial^2\nu(t,\upsilon)}{\partial t^2}f_2(\upsilon(t))\right] \tag{31}$$

where T—transposition,

$$\frac{\partial\nu}{\partial\upsilon} = \left(\frac{\partial\nu}{\partial\upsilon_1}, \frac{\partial\nu}{\partial\upsilon_2}, \frac{\partial\nu}{\partial\upsilon_3}\right)^{\mathrm{T}}, \tag{32}$$

$$\frac{\partial^2\nu(t,u)}{\partial\upsilon^2} = c\left(\frac{\partial^2\nu}{\partial\upsilon_j\partial\upsilon_i}\right), i, j = 1, 2, 3 \tag{33}$$

$c =$ Columnwise.

Theorem 4 Assume there is a function $\dot{\upsilon}(t,\upsilon)C_2^0(\mathbb{S})$ satisfying the inequalities.

$$\nu(t,\upsilon) \leq \Bbbk_2|\upsilon|^\rho, \tag{34}$$

$$\mathfrak{M}\nu(t,\upsilon) \leq -\Bbbk_3|\upsilon|^\rho, \Bbbk_i > 0, \rho > 0 \tag{35}$$

The trivial solution of (24) is then likely to be globally asymptotically stable [3].

Theorem 5 The null solution of (24) is mean-square stable asymptotically if.

$$\varsigma_1^2 < 2(1 - 2S^* - \eta I^* - h_1), \varsigma_2^2 < 2(\eta S^* - h_2), \varsigma_3^2 < 2a_3.$$

Proof Consider now the role of Lyapunov.

$$V(u) = \frac{1}{2}\left[g_1u_1^2 + g_2u_2^2 + g_3u_3^2\right] \tag{36}$$

It is easy to check that inequalities (34) hold with $\rho = 2$.

Now Ito's process (31) becomes.

$$\mathfrak{M}\nu(t,\upsilon) = g_1\left[(1 - 2S^* - \eta I^* - h_1)\upsilon_1 - \eta S^*\upsilon_2\right]\upsilon_1 + g_2\left[\eta I^*\upsilon_1 + (\eta S^* - h_2)\upsilon_2\right]$$

$$+\frac{1}{2}\text{trace}\left[f_1^T(\upsilon(t))\frac{\partial\nu(t,\upsilon)}{\partial t}f_2(\upsilon(t))\right]\upsilon_2+g_3[-a_3\upsilon_3]\upsilon_3 \tag{37}$$

where

$$\frac{\partial^2\nu}{\partial\upsilon^2}=\begin{pmatrix} g_1 & 0 & 0\\ 0 & g_2 & 0\\ 0 & 0 & g_3\end{pmatrix} \tag{38}$$

and

$$f_1^T(\upsilon(t))\frac{\partial^2\nu(t,\upsilon)}{\partial t^2}f_2(\upsilon(t))=\begin{pmatrix} g_1\varsigma_1^2\upsilon_1^2 & 0 & 0\\ 0 & g_2\varsigma_2^2\upsilon_2^2 & 0\\ 0 & 0 & g_3\varsigma_3^2\upsilon_3^2\end{pmatrix} \tag{39}$$

With

$$\frac{1}{2}\text{trace}\left[f_1^T(\upsilon(t))\frac{\partial^2\nu(t,\upsilon)}{\partial t^2}f_2(\upsilon(t))\right]==\frac{1}{2}[g_1\varsigma_1^2\upsilon_1^2+g_2\varsigma_2^2\upsilon_2^2+g_3\varsigma_3^2\upsilon_3^2] \tag{40}$$

if we choose $S^*g_1=I^*g_2$ in (37) along (40) we get

$$\begin{aligned}\mathfrak{M}\nu(t,\upsilon)=&-g_1\left[h_1+2S^*+\eta I^*-1-\frac{1}{2}\varsigma_1^2\right]\upsilon_1^2\\ &-g_2\left[h_2-\eta S^*-\frac{1}{2}\varsigma_2^2\right]\upsilon_2^2\\ &-g_3\left[a_3-\frac{1}{2}\varsigma_3^2\right]\upsilon_3^2\end{aligned} \tag{41}$$

This proved the theorem in which the function is negative definite.

5 Simulations of Numerical Data

Let the initial values be $(S, I, Y) = (0.9, 0.8, 0.7)$. First, we present the graph for the deterministic model (3).

Figure 1a, b, c exhibits the variance of the specific population to time t for parameters $(a_3, a_1, a_2, \eta, h_1, h_2) = (3.555, 0.25, 0.25, 4.5555, 0.001, 0.9)$ and reaches stability.

Now let us see the numerical simulations for the stochastic system (24).

Figure 2a displays population oscillations in a random environment with reduced

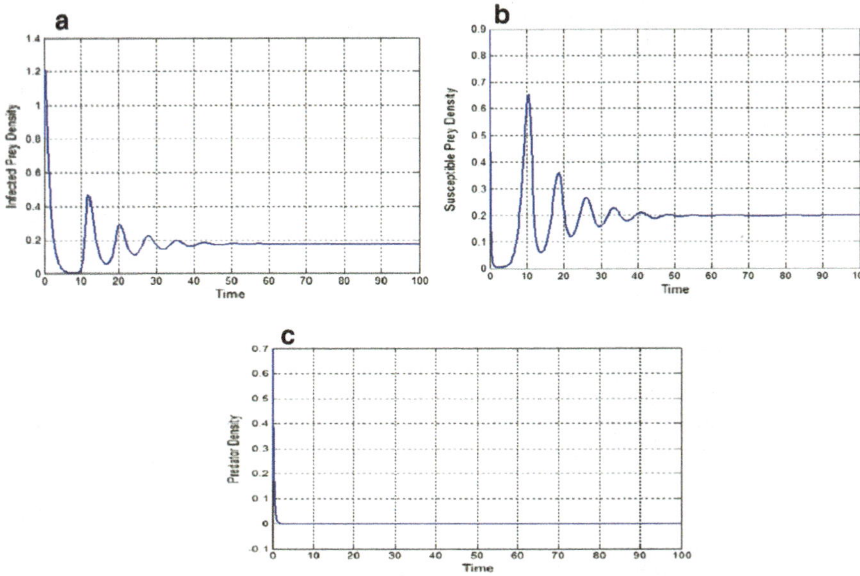

Fig. 1 Population Density

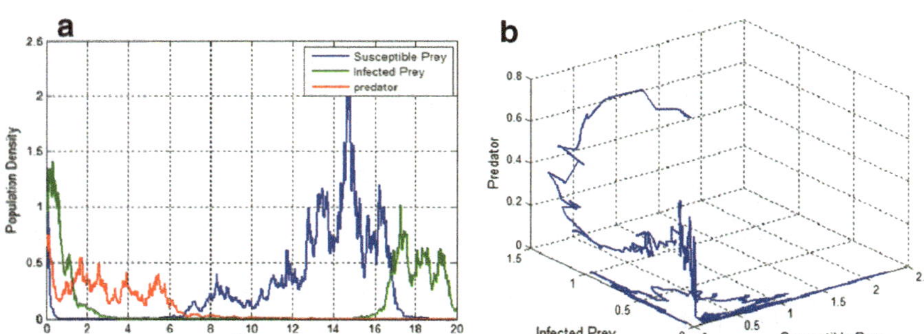

Fig. 2 Species oscillation with random noise

intensity when $(\boldsymbol{a}_3, \boldsymbol{a}_1, \boldsymbol{a}_2, \eta, \boldsymbol{h}_1, \boldsymbol{h}_2) = (0.25, 0.25, 0.25, 4.5555, 0.001, 0.9)$, and Fig. 2b shows the portrait of the concerned phase.

Figure 3a shows the population forms chaos for a stochastic system with random noise, and Fig. 3b shows the concern chaotic phase portrait with increased intensity for the parameters $(\boldsymbol{a}_3, \boldsymbol{a}_1, \boldsymbol{a}_2, \eta, \boldsymbol{h}_1, \boldsymbol{h}_2) = (3.555, 0.25, 0.25, 4.5555, 0.001, 0.9)$. The density of population in Fig. 1a, b, c reaches stability; but in Fig. 3a, for the same parameter values, the system behaves chaotically. Therefore, it is concluded that the system without random noise will approach the stable point more quickly than the stability of the system with random noise.

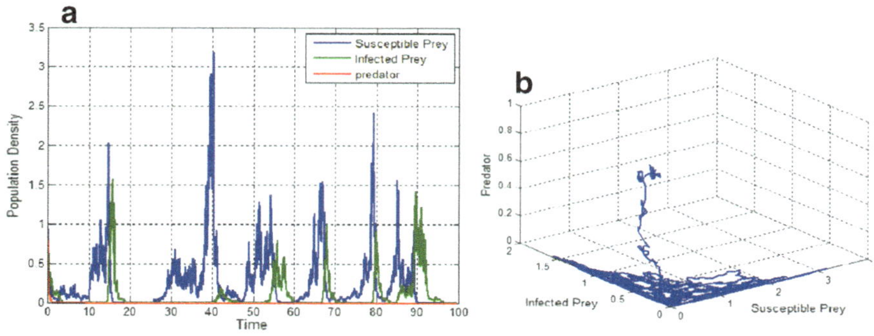

Fig. 3 Sensitive dependence on initial conditions

6 Conclusion

In this work, both analytically and numerically have been analyzed as a computational model of vulnerable prey and predator interactions with Holling's operational response and prey harvesting. We have worked out on the stability analysis on deterministic and on the stochastic system through Ito's process by constructing Lyapunov function. Finally, numerical simulations were done which are supporting our results and the comparative analysis of the diagrams between deterministic and stochastic system are also given.

References

1. Johri, A., Trivedi, N., Sisodiya, A., Sing, B., & Jain, S. (2012). Study of a prey-predator model with diseased prey. *International Journal of Contemporary Mathematical Sciences, 7*(10), 489–498.
2. Mukherjee, D. (2003). Stability analysis of a stochastic model for prey-predator system with disease in the prey. *Nonlinear Analysis: Modelling and Control, 8*(2), 83–92.
3. Rao, V. L. G. M., Rasappan, S., Murugesan, R., & Srinivasan, V. (2015). A prey predator model with vulnerable infected prey consisting of non-linear feedback. *Applied Mathematical Sciences, 9*(42), 2091–2102.
4. Wuhaib, S. A., & Hasan, Y. A. (2012). A Prey Predator Model. *Applied Mathematical Sciences, 6*(107), 5333–5348.
5. Bairagi, N., Chaudhuri, S., & Chattopadhyay, J. (2009). Harvesting as a disease control measure in an eco-epidemiological system–a theoretical study. *Mathematical Biosciences, 217*(2), 134–144.
6. Kar, T. K., & Matsuda, H. (2007). Global dynamics and controllability of a harvested prey–predator system with Holling type III functional response. *Nonlinear Analysis: Hybrid Systems, 1*(1), 59–67.
7. Peterson, R. O., & Page, R. E. (1987). Wolf density as a predictor of predation rate. Swedish Wildlife Research (Sweden).
8. Cai, G. Q., & Lin, Y. K. (2007). Stochastic analysis of predator–prey type ecosystems. *Ecological Complexity, 4*(4), 242–249.

9. Gikhman, I. I., & Skorokhod, A. V. (2004). *The Theory of Stochastic Process I*. Berlin Heidelberg: Springer-Verlag.
10. Das, K., Srinivas, M. N., Srinivas, M. A. S., & Gazi, N. H. (2012). Chaotic dynamics of a three species prey–predator competition model with bionomic harvesting due to delayed environmental noise as external driving force. *Comptes Rendus Biologies, 335*(8), 503–513.
11. Srinivas, M., Reddy, K., Das, K., Sabarmathi, A., & Das, P. (2014). Stochastic effects on an ecosystem with predation, commensalism, mutualism and neutralism. *Journal of Advanced Research in Dynamic Control Systems, 6*, 62–78.

Eccentric Distance Sum in Chemical Compounds Graph Using Domination

M. Raji and G. Jayalalitha

Abstract Any graphical representation of combined points and connecting lines is called a graph. Railway networks, mobile communication networks and any other traffic connections are more often examples. The representation of an atomic structure by a graph, which is connected, undirected and their vertices represent atoms and their edges are bondings, is an important application of graph theory field called chemical graph theory. Chemical graph theory is often used through the interpretation of chemical structures into numerical graph invariants. Graph invariant is a property of the graph that is preserved by isomorphism. Eccentric distance sum is a novel graph invariant which can be used in chemical compounds structure, some drug designs and molecular documentation. Eccentric distance sum presents a huge potential for structure relationships. This paper shows eccentric distance sum in the graph of chemical compounds that depends on minimum dominating distance matrix and then computes total eccentricity of linear benzenoid system (LBS) for h hexagons where $h \leq 5$.

Keywords Distance in graphs · Minimum dominating distance matrix · Eccentric distance sum

1 Introduction

Let G be molecular graph where atoms and bonds are specified by vertices and edges, respectively [1]. Graph theory may be employed through the translation of chemical structures into characteristic polynomial, matrix, sequence and numerical graph invariants [2–5]. Gupta, Singh and Madan established eccentric distance sum [6]. Throughout this paper, graphs are taken into consideration as simple, connected,

M. Raji (✉) · G. Jayalalitha
Department of Mathematics, VELS Institute of Science, Technology and Advanced Studies, Chennai, India
e-mail: rajialagumurugan@gmail.com

G. Jayalalitha
e-mail: g.jayalalithamaths.sbs@velsuniv.ac.in

S.-L. Peng et al. (eds.), *Proceedings of First International Conference on Mathematical Modeling and Computational Science*, Advances in Intelligent Systems and Computing 1292, https://doi.org/10.1007/978-981-33-4389-4_3

finite and undirected graphs. Graph theory is one among the foremost developing branches of recent mathematics and computer applications. The theory of domination has been an important part of research activity in graph theory. Domination in graphs has applications to many fields like engineering, basic sciences, life sciences, etc. For a graph G and a subset D of the vertex set $V(G)$, the set of vertices in G is in D or adjacent to a vertex in D. Then D is called a dominating set. The details about dominating set and minimum dominating distance matrix of G can be found in [7–10]. The eccentricity of a vertex v, $e(v)$ is the maximum distance from v to any other vertex [11]. Total eccentricity of the graph G, TE (G), is defined as the summation of eccentricities of all vertices of graph G [12]. Eccentric distance sum EDS (G) can be stated as the total number of the product of eccentricity and distance sum of each vertex in a graph G [6].

2 Eccentric Distance Sum in Graph of Chemical Compounds

Here it describes the computation of eccentric distance sum in the graph of chemical compounds like benzene, naphthalene and anthracene using minimum dominating distance matrix.

2.1 *Molecular Graph of Benzene*

The structure of benzene consists of six vertices. Its vertices are of degree 2. So all vertices have even degree. The maximum distance of each vertex to another vertex is 3 (Fig. 1 and Table 1).

Fig. 1 Benzene

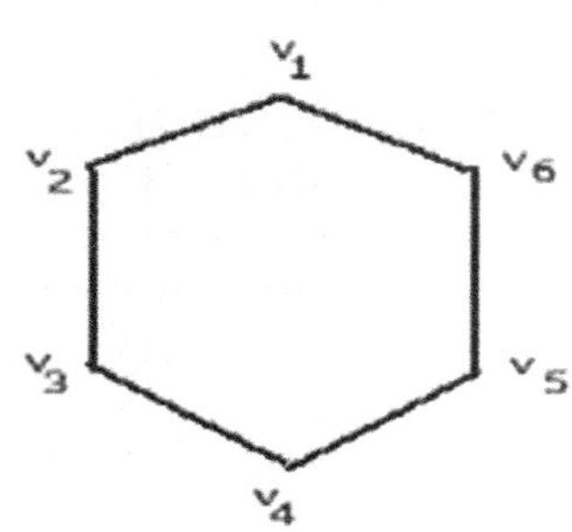

Table 1 Eccentricity of molecular graph of benzene

$e(v_i)$ $(i = 1–6)$	3	3	3	3	3	3

TE (Benzene) = 3 + 3 + 3 + 3 + 3 + 3 = 18.
EDS (Benzene) = 27 + 27 + 30 + 27 + 27 + 30 = 168.

Minimum Dominating Distance Matrix (6 × 6) of Benzene

$$\begin{pmatrix} 0 & 1 & 2 & 3 & 2 & 1 \\ 1 & 0 & 1 & 2 & 3 & 2 \\ 2 & 1 & 1 & 1 & 2 & 3 \\ 3 & 2 & 1 & 0 & 1 & 2 \\ 2 & 3 & 2 & 1 & 0 & 1 \\ 1 & 2 & 3 & 2 & 1 & 1 \end{pmatrix}$$

2.2 *Molecular Graph of Naphthalene*

The structure of naphthalene consists of ten vertices. Its vertices have degree 2 and 3. In this, eight vertices are of degree 2 and two vertices are of degree 3 (Fig. 2 and Table 2).

Minimum Dominating Distance Matrix (10 × 10) of Naphthalene

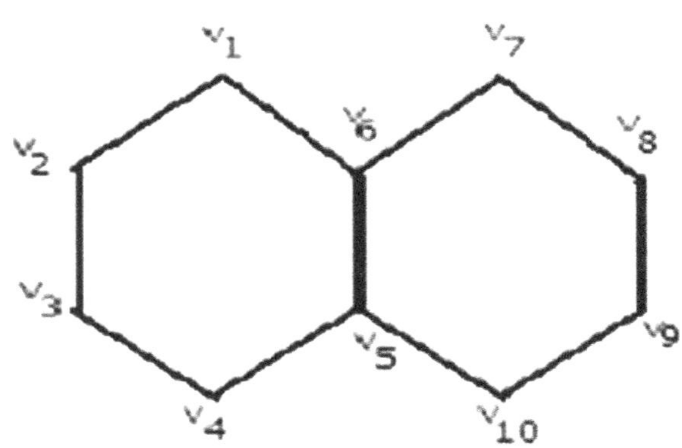

Fig. 2 Naphthalene

Table 2 Eccentricity of molecular graph of naphthalene

$e(v_1)$	$e(v_2)$	$e(v_3)$	$e(v_4)$	$e(v_5)$	$e(v_6)$	$e(v_7)$	$e(v_8)$	$e(v_9)$	$e(v_{10})$
4	5	5	4	3	3	4	5	5	4

TE (Naphthalene) = 4 + 5 + 5 + 4 + 3 + 3 + 4 + 5 + 5 + 4 = 42
EDS (Naphthalene) = 84 + 125 + 130 + 84 + 51 + 54 + 84 + 125 + 130 + 84 = 951

$$\begin{pmatrix} 0&1&2&3&2&1&2&3&4&3 \\ 1&0&1&2&3&2&3&4&5&4 \\ 2&1&1&1&2&3&4&5&4&3 \\ 3&2&1&0&1&2&3&4&3&2 \\ 2&3&2&1&0&1&2&3&2&1 \\ 1&2&3&2&1&1&1&2&3&2 \\ 2&3&4&3&2&1&0&1&2&3 \\ 3&4&5&4&3&2&1&0&1&2 \\ 4&5&4&3&2&3&2&1&1&1 \\ 3&4&3&2&1&2&3&2&1&0 \end{pmatrix}$$

2.3 Molecular Graph of Anthracene

The structure of anthracene consists of 14 vertices. Its vertices have degree 2 and 3. Here, ten vertices are of degree 2 and four vertices are of degree 3 (Fig. 3 and Table 3).

Minimum Dominating Distance Matrix (14 × 14) of Anthracene

Fig. 3 Anthracene

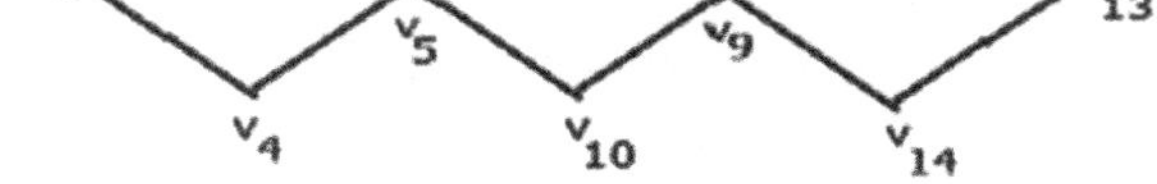

Table 3 Eccentricity of molecular graph of Anthracene

$e(v_1)$	$e(v_2)$	$e(v_3)$	$e(v_4)$	$e(v_5)$	$e(v_6)$	$e(v_7)$	$e(v_8)$	$e(v_9)$	$e(v_{10})$	$e(v_{11})$	$e(v_{12})$	$e(v_{13})$	$e(v_{14})$
6	7	7	6	5	5	4	5	5	4	6	7	7	6

TE (Anthracene) $= 6 + 7 + 7 + 6 + 5 + 5 + 4 + 5 + 5 + 4 + 6 + 7 + 7 + 6 = 80$

EDS (Anthracene) $= \{246 + 343 + 350 + 246 + 165 + 170 + 132 + 165 + 170 + 132 + 246 + 350 + 343 + 246\} = 3304$

$$\begin{pmatrix}
0&1&2&3&2&1&2&3&4&3&4&5&6&5\\
1&0&1&2&3&2&3&4&5&4&5&6&7&6\\
2&1&1&1&2&3&4&5&4&3&6&7&6&5\\
3&2&1&0&1&2&3&4&3&2&5&6&5&4\\
2&3&2&1&0&1&2&3&2&1&4&5&4&3\\
1&2&3&2&1&1&1&2&3&2&3&4&5&4\\
2&3&4&3&2&1&0&1&2&3&2&3&4&3\\
3&4&5&4&3&2&1&0&1&2&1&2&3&2\\
4&5&4&3&2&3&2&1&1&1&2&3&2&1\\
3&4&3&2&1&2&3&2&1&0&3&4&3&2\\
4&5&6&5&4&3&2&3&2&3&0&1&2&3\\
5&6&7&6&5&4&3&2&3&4&1&1&1&2\\
6&7&6&5&4&5&4&3&2&3&2&1&0&1\\
5&6&5&4&3&4&3&2&1&2&3&2&1&0
\end{pmatrix}$$

2.4 Molecular Graph of Tetracene

The structure of tetracene consists of 18 vertices. Here, 12 vertices are of degree 2 and six vertices are of degree 3 (Fig. 4 and Table 4).

Minimum Dominating Distance Matrix (18 x 18) of Tetracene

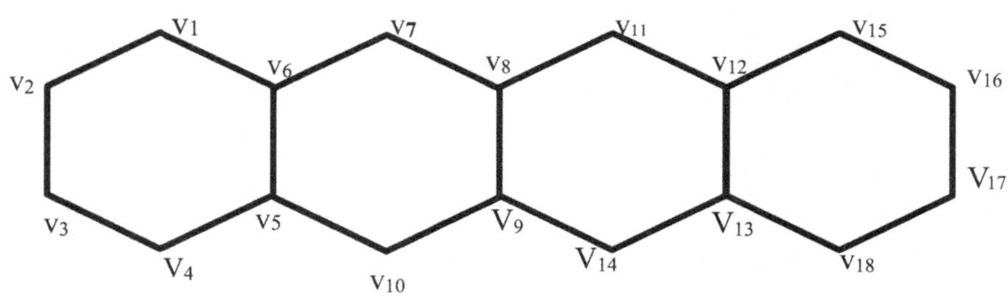

Fig. 4 Tetracene

Table 4 Eccentricity of molecular graph of Tetracene

$e(v_1)$	8
$e(v_2)$	9
$e(v_3)$	9
$e(v_4)$	8
$e(v_5)$	7
$e(v_6)$	7
$e(v_7)$	6
$e(v_8)$	5
$e(v_9)$	5
$e(v_{10})$	6
$e(v_{11})$	6
$e(v_{12})$	7
$e(v_{13})$	7
$e(v_{14})$	6
$e(v_{15})$	8
$e(v_{16})$	9
$e(v_{17})$	9
$e(v_{18})$	8

TE (Tetracene) = 8 + 9 + 9 + 8 + 7 + 7 + 6 + 5 + 5 + 6 + 6 + 7 + 7 + 6 + 8 + 9 + 9 + 8 = 130
EDS (Tetracene) = 8514

$$\begin{pmatrix}
0&1&2&3&2&1&2&3&4&3&4&5&6&5&6&7&8&7\\
1&0&1&2&3&2&3&4&5&4&5&6&7&6&7&8&9&8\\
2&1&1&1&2&3&4&5&4&3&6&7&6&5&8&9&8&7\\
3&2&1&0&1&2&3&4&3&2&5&6&5&4&7&8&7&6\\
2&3&2&1&0&1&2&3&2&1&4&5&4&3&6&7&6&5\\
1&2&3&2&1&1&1&2&3&2&3&4&5&4&5&6&7&6\\
2&3&4&3&2&1&0&1&2&3&2&3&4&3&4&5&6&5\\
3&4&5&4&3&2&1&0&1&2&1&2&3&2&3&4&5&4\\
4&5&4&3&2&3&2&1&1&1&2&3&2&1&4&5&4&3\\
3&4&3&2&1&2&3&2&1&0&3&4&3&2&5&6&5&4\\
4&5&6&5&4&3&2&1&2&3&0&1&2&3&2&3&4&3\\
5&6&7&6&5&4&3&2&3&4&1&1&1&2&1&2&3&2\\
6&7&6&5&4&5&4&3&2&3&2&1&0&1&2&3&2&1\\
5&6&5&4&3&4&3&2&1&2&3&2&1&0&3&4&3&2\\
6&7&8&7&6&5&4&3&4&5&2&1&2&3&0&1&2&3\\
7&8&9&8&7&6&5&4&5&6&3&2&3&4&1&0&1&2\\
8&9&8&7&6&7&6&5&4&5&4&3&2&3&2&1&1&1\\
7&8&7&6&5&6&7&4&3&4&3&2&1&2&3&2&1&0
\end{pmatrix}$$

2.5 Molecular Graph of Pentacene

The structure of tetracene consists of 22 vertices. Here 12 vertices are of degree 2 and ten vertices are of degree 3 (Fig. 5 and Table 5).

Minimum Dominating Distance Matrix (22 × 22) of Pentacene

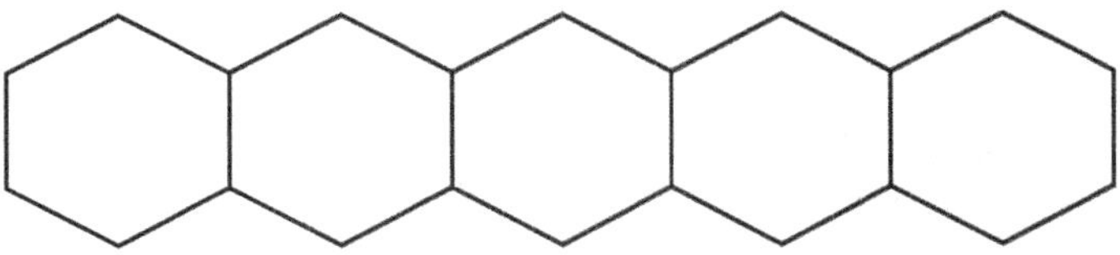

Fig. 5 Pentacene

Table 5 Eccentricity of molecular graph of Pentacene

$e(v_1)$	10
$e(v_2)$	11
$e(v_3)$	11
$e(v_4)$	10
$e(v_5)$	9
$e(v_6)$	9
$e(v_7)$	8
$e(v_8)$	7
$e(v_9)$	7
$e(v_{10})$	8
$e(v_{11})$	6
$e(v_{12})$	7
$e(v_{13})$	7
$e(v_{14})$	6
$e(v_{15})$	8
$e(v_{16})$	9
$e(v_{17})$	9
$e(v_{18})$	8
$e(v_{19})$	10
$e(v_{20})$	11
$e(v_{21})$	11
$e(v_{22})$	10

TE (Pentacene) = $\{10 + 11 + 11 + 10 + 9 + 9 + 8 + 7 + 7 + 8 + 6 + 7 + 7 + 6 + 8 + 9 + 9 + 8 + 10 + 11 + 11 + 10\} = 192$
EDS (Pentacene) = 18,280

$$
\begin{pmatrix}
0 & 1 & 2 & 3 & 2 & 1 & 2 & 3 & 4 & 3 & 4 & 5 & 6 & 5 & 6 & 7 & 8 & 7 & 8 & 9 & 10 & 9 \\
1 & 0 & 1 & 2 & 3 & 2 & 3 & 4 & 5 & 4 & 5 & 6 & 7 & 6 & 7 & 8 & 9 & 8 & 9 & 10 & 11 & 10 \\
2 & 1 & 1 & 1 & 2 & 3 & 4 & 5 & 4 & 3 & 6 & 7 & 6 & 5 & 8 & 9 & 8 & 7 & 10 & 11 & 10 & 9 \\
3 & 2 & 1 & 0 & 1 & 2 & 3 & 4 & 3 & 2 & 5 & 6 & 5 & 4 & 7 & 8 & 7 & 6 & 9 & 10 & 9 & 8 \\
2 & 3 & 2 & 1 & 0 & 1 & 2 & 3 & 2 & 1 & 4 & 5 & 4 & 3 & 6 & 7 & 6 & 5 & 8 & 9 & 8 & 7 \\
1 & 2 & 3 & 2 & 1 & 1 & 1 & 2 & 3 & 2 & 3 & 4 & 5 & 4 & 5 & 6 & 7 & 6 & 7 & 8 & 9 & 8 \\
2 & 3 & 4 & 3 & 2 & 1 & 0 & 1 & 2 & 3 & 2 & 3 & 4 & 3 & 4 & 5 & 6 & 5 & 6 & 7 & 8 & 7 \\
3 & 4 & 5 & 4 & 3 & 2 & 1 & 0 & 1 & 2 & 1 & 2 & 3 & 2 & 3 & 4 & 5 & 4 & 5 & 6 & 7 & 6 \\
4 & 5 & 4 & 3 & 2 & 3 & 2 & 1 & 1 & 1 & 2 & 3 & 2 & 1 & 4 & 5 & 4 & 3 & 6 & 7 & 6 & 5 \\
3 & 4 & 3 & 2 & 1 & 2 & 3 & 2 & 1 & 0 & 3 & 4 & 3 & 2 & 5 & 6 & 5 & 4 & 7 & 8 & 7 & 6 \\
4 & 5 & 6 & 5 & 4 & 3 & 2 & 1 & 2 & 3 & 0 & 1 & 2 & 3 & 2 & 3 & 4 & 3 & 4 & 5 & 6 & 5 \\
5 & 6 & 7 & 6 & 5 & 4 & 3 & 2 & 3 & 4 & 1 & 1 & 1 & 2 & 1 & 2 & 3 & 2 & 3 & 4 & 5 & 4 \\
6 & 7 & 6 & 5 & 4 & 5 & 4 & 3 & 2 & 3 & 2 & 1 & 0 & 1 & 2 & 3 & 2 & 1 & 4 & 3 & 4 & 3 \\
5 & 6 & 5 & 4 & 3 & 4 & 5 & 2 & 1 & 2 & 3 & 2 & 1 & 0 & 3 & 4 & 3 & 2 & 5 & 6 & 5 & 4 \\
6 & 7 & 8 & 7 & 6 & 5 & 4 & 3 & 4 & 3 & 2 & 1 & 2 & 3 & 0 & 1 & 2 & 3 & 2 & 3 & 4 & 3 \\
7 & 8 & 9 & 8 & 7 & 6 & 5 & 4 & 5 & 4 & 3 & 2 & 3 & 4 & 1 & 0 & 1 & 2 & 1 & 2 & 3 & 2 \\
8 & 9 & 8 & 7 & 6 & 7 & 6 & 5 & 4 & 5 & 4 & 3 & 2 & 3 & 2 & 1 & 1 & 1 & 2 & 3 & 2 & 1 \\
7 & 8 & 7 & 6 & 5 & 6 & 7 & 4 & 3 & 4 & 3 & 2 & 1 & 2 & 3 & 2 & 1 & 0 & 3 & 4 & 3 & 2 \\
8 & 9 & 10 & 9 & 8 & 7 & 6 & 5 & 6 & 7 & 4 & 3 & 4 & 5 & 2 & 1 & 2 & 3 & 0 & 1 & 2 & 3 \\
9 & 10 & 11 & 10 & 9 & 8 & 7 & 6 & 7 & 8 & 5 & 4 & 3 & 6 & 3 & 2 & 3 & 4 & 1 & 1 & 1 & 2 \\
10 & 11 & 10 & 9 & 8 & 9 & 8 & 7 & 6 & 7 & 6 & 5 & 4 & 5 & 4 & 3 & 2 & 3 & 2 & 1 & 0 & 1 \\
9 & 10 & 9 & 8 & 7 & 8 & 7 & 6 & 5 & 6 & 5 & 4 & 3 & 4 & 3 & 2 & 1 & 2 & 3 & 2 & 1 & 0
\end{pmatrix}
$$

Problem: For the Molecular Graph of Linear Benzenoid system, The Total Eccentricty of Linear Benzenoid System (LBS) for $h \leq 5$ hexagons can be defined as TE (LBS(h)) = $(1/12)*(h^4 - 14h^3 + 143\text{h}^2 - 58h + 144)$.

Solution

Consider linear benzenoid system LBS (h) with h hexagons.

Using the definition of total eccentricity and using the method of computations (2.1, 2.2, 2.3, 2.4, 2.5).

Total eccentricity of the molecular graph of benzene = 18.

Total eccentricity of the molecular graph of naphthalene = 42.

Total eccentricity of the molecular graph of anthracene = 80.

Total eccentricity of the molecular graph of tetracene = 130.

Total eccentricity of the molecular graph of pentacene = 192.

Let us take x as hexagons and y as the total eccentricity of the molecular graph of linear benzenoid system LBS(h) with the corresponding hexagons, respectively.

We form forward difference table (Table 6).

Forward difference table

x	y	Δy	Δy^2	Δy^3	Δy^4
1	18				
2	42	24			
3	80	38	14		
4	130	50	12	-2	
5	192	62	12	0	2

From the above table, using Newton's forward interpolation formula [13], we get, $y = (1/12)*(x^4 - 14x^3 + 143x^2 - 58x + 144)$.

Now the molecular graph of benzene, naphthalene, anthracene, tetracene and pentacene contains 1, 2, 3, 4 and 5 hexagons, respectively.

Here we can consider hexagon $h = x$.

Molecular graph of linear benzenoid system contains h hexagons.

So, we can conclude that the total eccentricty of linear benzenoid system (LBS) for h hexagons is TE (LBS(h)) $= (1/12)*(h^4 - 14h^3 + 143h^2 - 58h + 144)$ where $h \leq 5$.

3 Conclusion

In this paper, eccentric distance sum in the graph of chemical compounds that depends on minimum dominating distance matrix is shown and then total eccentricity of linear benzenoid system (LBS) for h hexagons where $h \leq 5$ is computed. Aim of this paper is to present the importance of graph theoretical ideas in chemical structure for researches. This paper obtains a new approach associated with the molecular graph using domination concepts.

References

1. Trinajstic, N. (1983). *Chemical Graph Theory* (2nd ed.). CRC.
2. Raut, N. K., Gaikwad, P. D., & Ipper, S. N. (2017). The vertex-degree of a molecular graph. *International Journal of Innovative Research in Science, Engineering and Technology, 6*(10), 19531–19533.

3. Hayat, S., Ahmad, S., Umair, H. M., & Wang, S. (2018). Distance property of chemical graphs. *Hacettepe Journal of Mathematics and Statistics, 47*(5), 1071–1093.
4. Vukicevic, D., & Trinajstic, N. (2004). Wiener indices of benzenoid graphs. *Bulletin of the Chemists and Technologists of Macedonia, 23*(2), 113–129.
5. Klavzar, S., & Gutman, I. (1996). A comparison of the Schultz topological index with the Wiener Index. *Journal of Chemical Information and Computer Science, 36,* 1001–1003.
6. Gupta, S., Singh, M., & Madan, A. K. (2002). Eccentric distance sum: A novel graph invariant for predicting biological and physical properties. *Journal of Mathematical Analysis and Applications, 275,* 386–401.
7. Bondy, J. A., & Murty, U. S. R. (1982). *Graph Theory with Applications.* Elsevier Science Publishing Co., Inc.
8. Deo, N. (**1974**). *Graph Theory With Applications To Engineering And Computer Science.* PHI Learning Pvt. Ltd. ISBN: 9788120301450.
9. Jayalalitha, G., & Raji, M. (2019). Schultz polynomial, modified Schultz polynomial and indices of molecular graph of anthracene based on domination. *International Journal of Research in Advent Technology, 7*(1), 136–140.
10. Rajesh Kanna, M. R., Dharmendra, B. N., & Sridhara, G. (2014). Minimum dominating distance energy of a graph. *Journal of the Indonesia Mathematical Society, 20*(1), 19–29.
11. Arumugam, S. (2014). *Invitation to Graph Theory.* Scitech Publications (India) Pvt Limited. ISBN: 139788187328469.
12. Tang, Y., & Zhou, B. (2012). On average eccentricity. *Match-Communications in Mathematical and in Computer Chemistry, 67,* 405–423.
13. K. Subramani, A. Santha, (2012). *Numerical Methods.* Scitech Publications (India) Pvt. Limited. ISBN: 9788183713641.

Analysis of Sierpinski Triangle Based on Fuzzy Triangular Numbers and Dihedral Group

T. Sudha and G. Jayalalitha

Abstract Fractals are indefinitely complex patterns such as self-similar across at different scales; for example, Sierpinski triangle is a fractal. This paper analysed in the Sierpinski triangle. It is considered as equilateral triangles such as 1 unit, k unit and $k + 1$ unit. Each iteration is divided as [(0, 1/4, ½, ..., 1)], [(0, k/4, k/2..., k)], [(0, (k + 1)/4, (k + 1)/2,... (k + 1))], so on. It analysed this triangle which satisfies fuzzy triangular numbers and the number of the theoretical aspect of fuzzy triangular numbers (FTNs) in self-similarity set of fractal set (Sierpinski triangle) and some arithmetic operations of α-ut and discussed that this triangle satisfied the centroid and median of the normal triangle. Multiplication of fuzzy triangular numbers α-cuts is explained graphically. It also analysed that these smaller equilateral triangles form a group, and this group satisfies the property of dihedral group.

Keywords Fuzzy numbers · Fractal · Sierpinski triangle · Fuzzy triangular numbers · Centroid of the triangle · Dihedral group

AMS Classification 2000 · 03E72 · 28A80 · 20D15

1 Introduction

In earlier times, there was no mathematical concept to describe the uncertain (vagueness) situation. Fuzzy numbers are defined in uncertainty situation, and it is playing a vital role in many applications, but the main problem in the development of the application is the computational complexity. Hence, more attention is needed to simply arithmetic computation with fuzzy numbers. By restricting the fuzzy number to triangular fuzzy numbers, addition and subtraction become simpler [1].

T. Sudha (✉) · G. Jayalalitha
Department of Mathematics, VELS Institute of Science, Technology and Advanced Studies, Chennai, India
e-mail: psudha.thakur@gmail.com

G. Jayalalitha
e-mail: g.jayalalithamaths.sbs@velsuniv.ac.in

S.-L. Peng et al. (eds.), *Proceedings of First International Conference on Mathematical Modeling and Computational Science*, Advances in Intelligent Systems and Computing 1292, https://doi.org/10.1007/978-981-33-4389-4_4

1.1 Fuzzy Number

Fuzzy number, which is the extension of real numbers, has its properties which can be related to the theory of numbers [2]. It is widely used in engineering applications because of their suitability for representing uncertain information [3].

1.2 Fractals

The word fractal was coined by Mandelbrot; *fractus* means broken, to describe the objects that were too irregular to fit into a traditional geometrical setting [4, 5]. Many fractals have characteristics of self-similarity that it made up parts to resemble the whole in some way. The few examples of fractals are such as Cantor sets, Von Koch curve, Menger sponge, dragon curve, Julia sets and Mandelbrot sets [6].

1.3 Sierpinski Triangle

In 1915, Waclaw Sierpinski described the triangle which is a self-similar structure, and it occurs at different levels of iterations or magnifications; it is named as Sierpinski triangle. The Sierpinski triangle is a fractal, and it satisfies all the properties of a fractal [4]. For the triangular area, with each iteration, the side of the inside triangle reduces by a factor of 2. The numbers of these little triangles, on the other, had increased not by 4 but by a factor of 3. The dimension of self-similar object is then (log 3/log 2) = 1.58 approximately [7].

1.4 Dihedral Group

The dihedral groups play a significant role in the group theory, while the dihedral groups are originally produced from the symmetries of regular polygons, which together from surfaces and planes. A regular polygon which has rotations and reflections forms a *d*ihedral group, and the dihedral group for n-polygon is denoted by $D_{2n.}$ [8, 9].

Fuzzy triangular number as a tool for student assessment is explained in this paper [10]. New operation on triangular fuzzy number for solving fuzzy linear programming problem is discussed [3]. It explained about triangular approximation of fuzzy numbers—a new approach [1], and they all discussed with randomly chosen triangular fuzzy numbers. In Sect. 2, operations of fuzzy triangular numbers with graphical representation are discussed using Sierpinski triangle with side $n = 1$, k and $k + 1$ unit and also discussed dihedral group in fuzzy triangular number triangle.

2 Operations on Fuzzy Triangular Numbers

2.1 *Iteration of Equilateral Triangle*

Figure 1 represents the iteration of the Sierpinski triangle from stage 0 to 3.

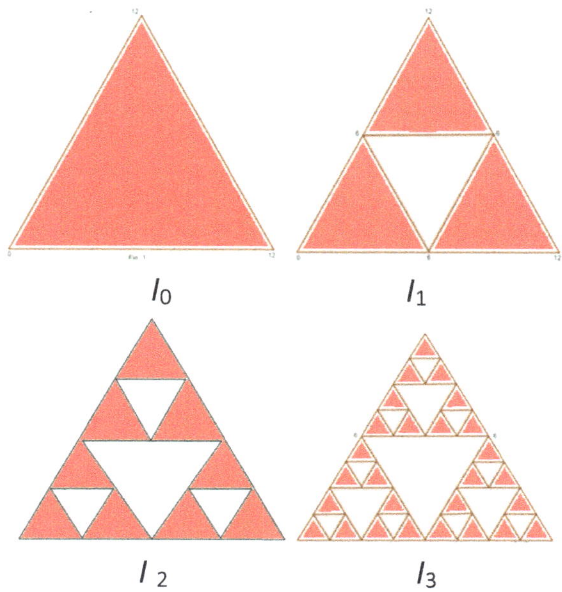

Fig. 1 Iteration of Sierpinski triangle

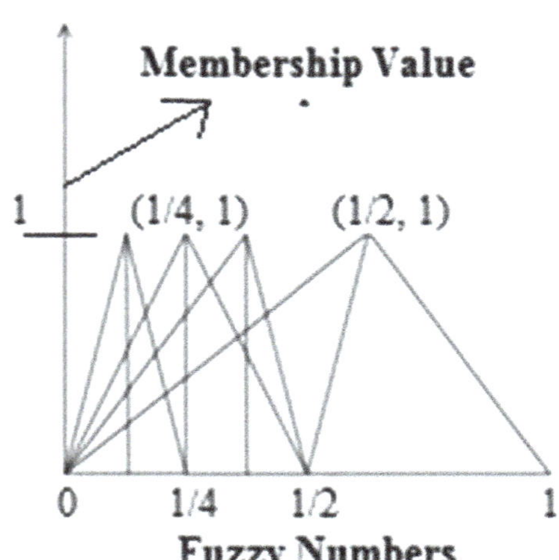

Fig. 2 (1 unit)

Figures 2, 3 and 4 represent fuzzy triangular numbers obtained from an *equilateral triangle* of side 1 unit, k unit and $k + 1$ unit, respectively.

Based on Fig. 1, the iteration of the Sierpinski triangle of side 1 unit, k units and $k + 1$ unit is explained in Tables 1, 2 and 3, respectively.

Fig. 3 (k unit)

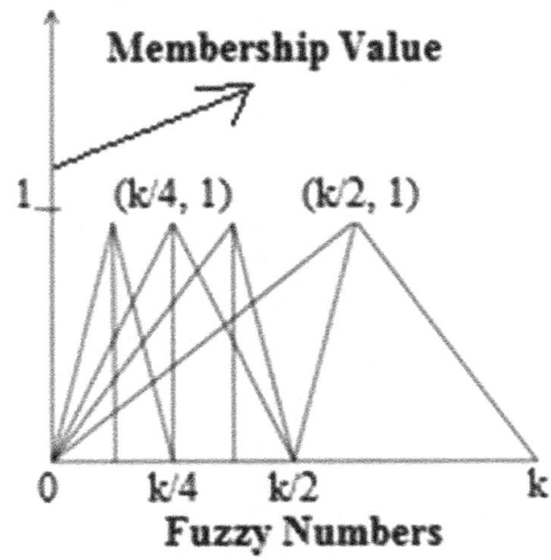

Fig. 4 ($k + 1$ unit)

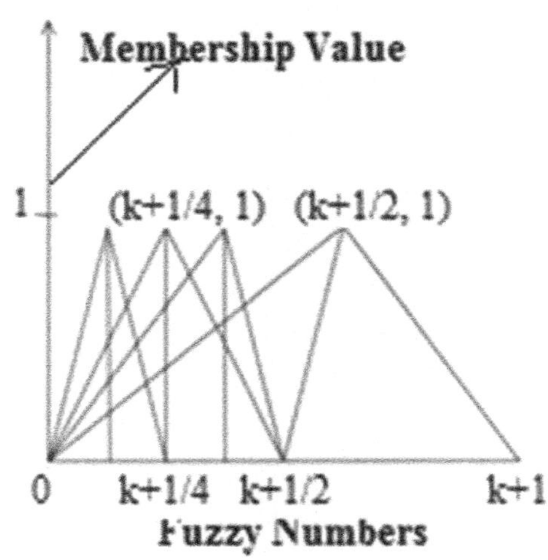

Table 1 Iteration of Sierpinski triangle of side 1 unit

Iteration number	Scaling	TFNs
I_1	$\frac{1}{2}$	$(0, \frac{1}{2}, 1)$
I_2	$\frac{1}{4}$	$(0, \frac{1}{4}, \frac{1}{2}), (\frac{1}{4}, \frac{1}{2}, 1), (0, \frac{1}{8}, \frac{1}{4}),$
I_3	$\frac{1}{8}$	...
I_n	$\left(\frac{1}{2}\right)^n$	n–numbers

Table 2 Iteration of Sierpinski triangle of side K unit

Iteration number	Scaling	TFNs
I_1	$\frac{1}{2}$	$(0, \frac{K}{2}, k)$
I_2	$\frac{1}{4}$	$(0, \frac{k}{4}, \frac{k}{2}), (\frac{k}{4}, \frac{k}{2}, k), (0, \frac{k}{8}, \frac{k}{4}),$
I_3	$\frac{1}{8}$	...
I_n	$\left(\frac{1}{2}\right)^n$	n–numbers

Table 3 Iteration of Sierpinski triangle of side $K + 1$ unit

Iteration number	Scaling	TFNs
I_1	$\frac{1}{2}$	$(0, \frac{k+1}{2}, k+1)$
I_2	$\frac{1}{4}$	$(0, \frac{k+1}{4}, \frac{k+1}{2}), (\frac{k+1}{4}, \frac{k+1}{2}, k+1), (0, \frac{k+1}{8}, \frac{k+1}{4}),.$
I_3	$\frac{1}{8}$	...
I_n	$\left(\frac{1}{2}\right)^n$	n–numbers

2.2 *FTN in* α-cut Operations

By using the definition of [11], let a, b and c be real numbers with $a < b < c$. Then, the TFN $\grave{A} = (a, b, c)$ is a fuzzy number with membership function

$$m(\Phi) = \begin{cases} \frac{\Phi - a}{b - a}, & a \le \Phi \le b \\ \frac{c - \Phi}{c - b}, & b \le \Phi \le c \\ 0, & \Phi < a,\ \Phi > c \end{cases}$$

By using the definition of [10, 12], the α-cuts $\grave{A}_\alpha$ of a TFN $\grave{A} = (a, b, c)$, $\alpha \in [0, 1]$, then

$$\grave{A}_\alpha = [\grave{A}_l\alpha,\ \grave{A}_r\alpha] = [(a + \alpha(b - a)),\ (c - \alpha(c - b))]$$

α-cut operation and from Fig. 3 and from Table 2, $G\ (0, k/2, k)$

$$G_\alpha = [a_1\alpha,\ a_3\,\alpha] = (k\,\alpha,\ -\frac{k}{2}\alpha + k) \tag{1}$$

Operation by α-cut $F\ (0, \frac{k}{4}, \frac{k}{2})$

$$F_\alpha\ = [a_1\alpha,\ a_3\,\alpha]\ = (\frac{k}{4}\alpha,\ -\frac{k}{4}\alpha\ +\ \frac{k}{2}) \tag{2}$$

Approximation for multiplication from Eqs. (1) and (2)

$$G_\alpha(\cdot)\ F_\alpha\ = [(\frac{k}{2}\alpha \cdot \frac{k}{4}\alpha),\ (-\frac{k}{2}\alpha + k \cdot -\frac{k}{4}\alpha + \frac{k}{2})] \tag{3}$$

$$\alpha\ =\ 0\ =\ (0,\ k^2/2)$$
$$\alpha\ =\ 1\ =\ (k^2/8,\ k^2/8)$$

$$G_\alpha(\cdot)\ F_\alpha\ =\ (0,\ k^2/2,\ k^2/8)$$

From Fig. 4 and from Table 3 and based on definition TFN α-cut

$$A\,(0,\ \frac{k+1}{2},\ k+1)$$

$$A_\alpha\ =\ [a_1\alpha,\ a_3\alpha]\ =\ [\{(k+1)\,\alpha,\ -\frac{k+1}{2}\,\alpha + k + 1\}] \tag{4}$$

Operation by α-cut $B\ (0,\ \frac{k+1}{4},\ \frac{k+1}{2})$

$$B_\alpha\ =\ [a_1\alpha,\ a_3\alpha]\ =\ [(\frac{k+1}{2})\,\alpha,\ (-\frac{k+1}{4})\alpha + \frac{k+1}{2})] \tag{5}$$

Approximation for multiplication from Eqs. (4) and (5)

$$A_\alpha.B_\alpha = [\{(k+1)\alpha \cdot (\frac{k+1}{2})\alpha\},\ \{(-\frac{k+1}{2}\alpha + k + 1) \cdot (-\frac{k+1}{4}\alpha + \frac{k+1}{2})\}]$$
$$\alpha = 0 = (0,\ \frac{(k+1)2}{2})$$
$$\alpha = 1 = \{\frac{(k+1)2}{2},\ \frac{(k+1)2}{8}\}$$

$$A_\alpha\ (.)\ B_\alpha = [0,\ \frac{(k+1)2}{2},\ \frac{(k+1)2}{8}] \tag{6}$$

Therefore multiplication of two TFN is Triangular Fuzzy Number but it is not always. The lines connecting the end points are parabolic and triangular form in actual product and standard approximation, respectively.

2.3 Graphical Representation of Multiplication of FTN

Figures 5 and 6 represent a graphical representation of side$k$ units and $k + 1$ units, respectively.

Consider two fuzzy numbers $U = (0, k/2, k)$, $\tilde{U} = (0, k/4, k/2)$, and from Eqs. (1), (2) and (3), multiplication of two fuzzy numbers [1] based on this lemma their respective α*-cut will be* $U(\alpha) = \frac{k^2\alpha}{8}$, $\tilde{U}(\alpha) = (k^2\,\alpha^2 - 4k^2\alpha + 4k^2)\,/\,8$, $v(r) = \frac{k^2}{16} + \frac{k^2 r}{16}$, $v\,{*}(r) = \frac{5k^2}{16} - \frac{3k^2 r}{2}$.

Consider two fuzzy numbers $(0, \frac{k+1}{2}, k + 1)$, $(0, \frac{k+1}{4}, \frac{k+1}{2})$, and from the Eqs. (4), (5) and (6) and the lemma [1], α-cut will be

$$U(\alpha) = [\frac{(k+1)^2\alpha^2]}{8},\ \tilde{U}(\alpha) = \frac{(k+1)^2\alpha^2]}{8} - \frac{(k+1)^2\alpha]}{2} + \frac{(k+1)^2]}{2},$$
$$v(r) = \frac{(k+1)^2]}{16} + \frac{(k+1)^2 r]}{4},\ v^*(r) = 5\frac{(k+1)^2]}{16} - 3\frac{(k+1)^2 r]}{8}$$

Fig. 5 α-cut diagram for k

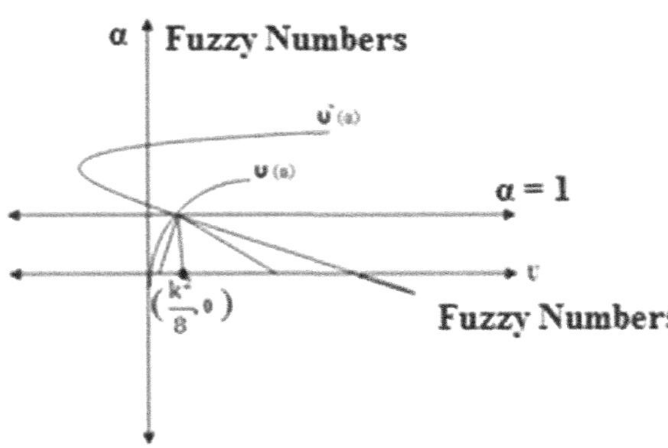

Fig. 6 α-cut diagram of $k + 1$

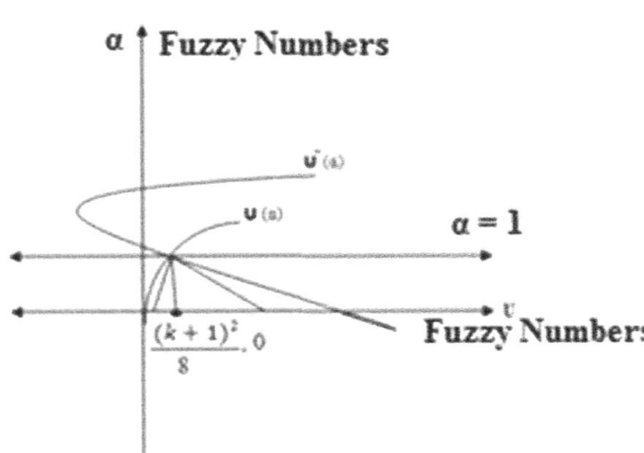

2.4 Calculating Centroid and Median of FTN Triangle of Side k, k + 1 Units

In Fig. 7, Let ABC is a Fuzzy triangle obtained from FTN of side k unit and G is the centroid of the triangle ABC of side k units with the vertices $A(0,0)$, $B(k/2, 1)$, and $C(k, 0)$ and $M(k/2, 0)$ is the midpoint of A and C, and $N(3\,k/4, 1/2)$ is the midpoint of B and C. Based on the definition TFN and from Fig. 6, Equation of the straight line $A(0, 0)$ and $N((3\,k)/4, 1/2)$ on which AN lies in

$$2x - 3ky = 0 \tag{7}$$

From Fig. 7, based on the definition TFN, Equation of B ($k/2$, 1) and M ($k/2$,0)lies in

$$2x + 0y = k \tag{8}$$

The linear system of Eqs. (7), and (8) has a consistent and has unique solution determining the coordinates of the triangle COG in Fig. 7 and the centroid of the triangle after observing the following

$$D = \begin{vmatrix} 2 & -3k \\ 2 & 0 \end{vmatrix} = 0 + 6k = 6k \tag{9}$$

$$D_x = \begin{vmatrix} 0 & -3k \\ K & 0 \end{vmatrix} = 3k^2 \tag{10}$$

$$Dy = \begin{vmatrix} 2 & 0 \\ 2 & k \end{vmatrix} = 2k \tag{11}$$

Therefore, $x = k/2$, $y = 1/3$. Solving Eqs. (7) and (8), based on the definition [10, 13].

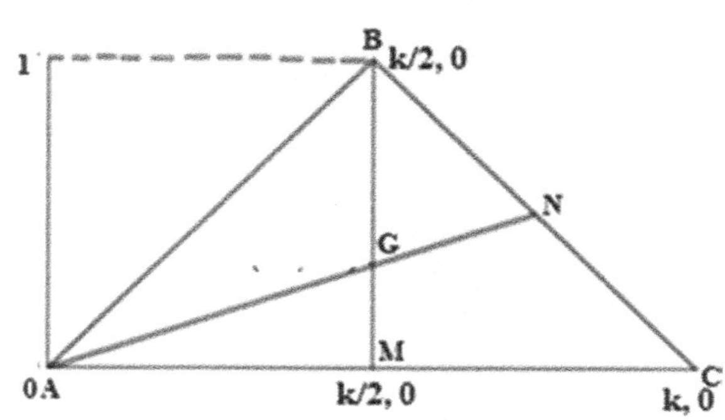

Fig. 7 Fuzzy triangular numbers triangle with centroid and median

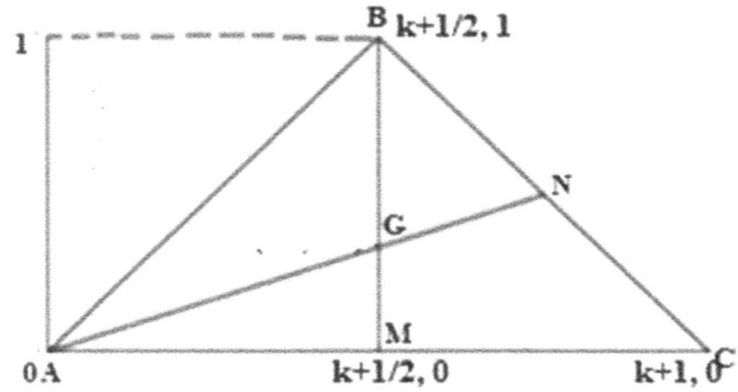

Fig. 8 Fuzzy triangular numbers triangle with centroid and median

[$x = (a + b + c)/3$, $y = 1/3$]. The coordinates of centroid of the triangle G are ($k/2$, 1/3) which is equal to $x = \frac{(0+k+k/2)}{3}$, $y = \frac{(1+0+0)}{3}$, i.e. ($k/2$, 1/3). Similarly, for $k + 1$ units.

In Fig. 8, G represents the centroid of the fuzzy triangular numbers triangle of side, $k + 1$units with the vertices $A(0, 0)$, $B(k + 1/2, 1)$, and $C(k + 1,0)$ and $M(\frac{k+1}{2}$, 0) is the midpoint of A and C, and $N((3k + 3)/4, 1/2)$ is the midpoint of B and C. $x = \frac{(0+\frac{k+1}{2}+(k+1))}{3}$, $y = \frac{(1+0+0)}{3}$, i.e. ($\frac{k+1}{2}$, 1/3).

2.5 Dihedral Group in FTN

From Figs. 2, 3 and 4, the FTNs triangle obtained after the iteration it has six symmetries i.e. three rotations and three reflections; asymmetry can interchange some of the sides and vertices. Figure 9 represents rotations and reflections of the triangle which is obtained from the equilateral triangle after each iteration. (Assume that the sides are 1, 2, 3, respectively) ®, ȣrepresents reflections, rotations, respectively.

Each rotation has binary composite functions because every image has a unique pre-image. This can be represented by matrix also $\begin{bmatrix} 1\ 2\ 3 \\ 3\ 1\ 2 \end{bmatrix}$.

Table 4 clearly shows that it satisfies the group property of **Closure:** If $®_1$ and $®_2$ are in the group, then $®_1 \circ ®_2$are also in the group. **Associativity:** If $ȣ_1$, $ȣ_2$ *and* $ȣ_3$are in the group, then (($ȣ_1 \circ ȣ_2$) $\circ$ $ȣ_3$ = $ȣ_1 \circ$ ($ȣ_2 \circ ȣ_3$)) **Identity:** There is an element of the group such that a $\circ$ *e* = e e $\circ$ a = a, for all a belongs the group, here $e = ®_0$. **Inverse:** For any element a of the group, there is an element a^{-1} such that a $\circ$ a^{-1} = eand a^{-1} $\circ$ a = e, and every element has inverse but not a commutative because $ȣ_3 \circ ®_2 \neq ®_2 \circ ȣ_3$(rotation and reflection will not commute in general); therefore, it is a non-abelian group.

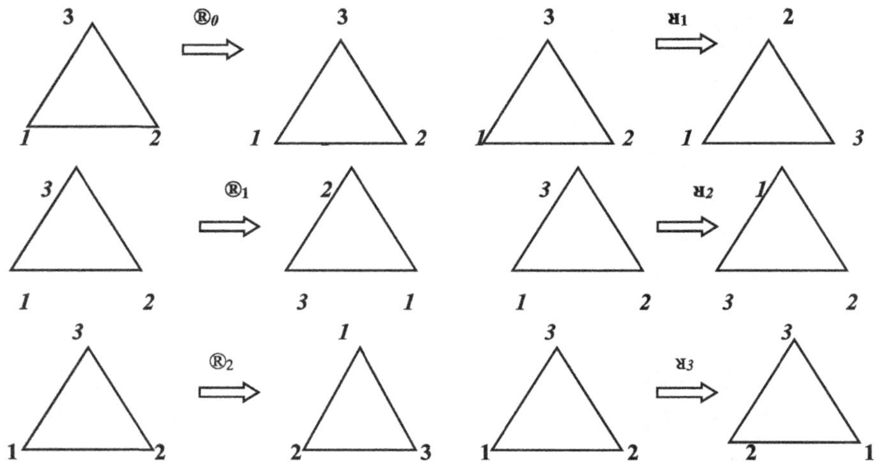

Fig. 9 Rotations and reflections of FTNs triangle

Table 4 Cayley table of usual multiplication

o	®$_0$	®$_1$	®$_2$	ꓤ$_1$	ꓤ$_2$	ꓤ$_3$
®$_0$	®$_0$	®$_1$	®$_2$	ꓤ$_1$	ꓤ$_2$	ꓤ$_3$
®$_1$	®$_1$	®$_2$	®$_0$	ꓤ$_3$	ꓤ$_1$	ꓤ$_2$
®$_2$	®$_2$	®$_0$	®$_1$	ꓤ$_2$	ꓤ$_3$	ꓤ$_1$
ꓤ$_1$	ꓤ$_1$	ꓤ$_2$	ꓤ$_3$	®$_0$	®$_1$	®$_2$
ꓤ$_2$	ꓤ$_2$	ꓤ$_3$	ꓤ$_1$	®$_2$	®$_0$	®$_1$
ꓤ$_3$	ꓤ$_3$	ꓤ$_1$	ꓤ$_2$	®$_1$	®$_2$	®$_0$

3 Conclusion

Multiplication of fuzzy triangular numbers α-cuts is discussed in self-similarity of Sierpinski triangle. This triangle satisfied fractal property. From this, it concludes that fuzzy triangular numbers triangle satisfied FTN properties for all value of n where n belongs to natural number, and multiplication of two fuzzy triangular numbers is approximation of fuzzy triangular numbers. Median and centroid derived from the triangle formed by fuzzy triangular numbers triangle are same as a normal triangle, and it has six symmetries (rotations and reflections); therefore, it forms dihedral group of non-abelian. This dihedral group plays an important role in group theory, chemistry (the group is used in the classification of molecules and crystals) and geometry.

This research article aims is that proposed and bridged some new ideas in mathematics with fuzzy concept and also developing and strengthening the number-theoretical aspects of fuzzy numbers in future.

References

1. Senthilkumar, L. S. (2017). Triangular approximation of fuzzy numbers—a new approach. *International Journal of Pure and Applied Mathematics, 113*(13), 115–121.
2. Clement Joe Anand, M. (2017). Theory of triangular fuzzy number. In *Proceedings of National Conference on Advanced Trends in Mathematics*. ISBN: 978-93-85126-14-7.
3. Nagoor Gani, A., & Mohamed Assarudeen, S. N. (2012). A new operation on triangular fuzzy number for solving fuzzy linear programming problem. *Applied Mathematical Sciences, 6*(11), 525–532.
4. Mandelbrot, B. B., & Freeman, W. H. (1982). *The Fractal Geometry of Nature*. San Francisco.
5. Yu Z.-G. (2005). Visualization and fractal analysis of biological sequences. *Bioinformatics Technologies*, 313–351.
6. Jayalalitha, G., & Sudha, T. (2019). Analysis of diabetes based on fuzzy fractals. *International Journal of Recent Technology and Engineering*, *7*(6S2), 653–658. ISSN: 2277-3878.
7. Nithya, R., Kamali, R., & Jayalalitha, G. (2017). Ramsey numbers in Sierpinski triangle. *International Journal of Pure and Applied Mathematics, 116*(4).
8. Al-Hasanat, B. N., & Almatroud, O. A. (2013). Dihedral groups of order 2^{m+1}. *International Journal of Applied Mathematics, 26*(1), 1–7.
9. Oshman, Y., & Markley, F. L. (1999). Spacecraft Altitude/rate estimation using vector-aided GPS observations. *IEEE Transactions on Aerospace and Electronic System, 35*(3), 1019–1032.
10. Voskoglou, M. G. (2015). An application of triangular fuzzy numbers to learning assessment. *Journal of Physical Sciences, 20,* 63–79.
11. Voskoglou, M. G. (2017). An application of triangular fuzzy numbers for assessing the results of iterative learning. *International Journal of Applications of Fuzzy Sets and Artificial Intelligence, 7,* 59–72.
12. Voskoglou, M. G. (2017). An application of triangular fuzzy numbers to assessment of human skills. *International Journal of Fuzzy System Applications, 6*(3), 59–73.
13. Voskoglou, M. G. (2018). An application of triangular fuzzy numbers to analogical reasoning. *International Journal of Quantitative Research in Education, 4*(3).
14. Voskoglou, M. G. (2015). An application of triangular fuzzy numbers to learning assessment. *Journal of Physical Scences, 20,* 63–79.

Near-Optimal Cable Layout Design of a Wind Farm Using Genetic Algorithm

Chakib El Mokhi and Adnane Addaim

Abstract The wind energy is the fast-growing source of electrical energy. The total new wind power installed worldwide increased by the rate of 9.6% in 2018, reaching the capacity of 591 GW. Therefore, it is very important to optimize the production efficiency in wind farms and to keep the power losses as low as possible. This can be achieved through an optimal cable layout design. This paper deals with the optimal cable routing of a power collection system in a large-scale wind farm using genetic algorithm (GA) combination with multiple traveling salesmen problem (MTSP). The main objective is to reduce the total cable length and thereby the power losses and the investment costs. To check the reliability of the suggested optimization, this methodology is carried out on existing wind farms. The obtained results show an improvement in minimizing cable costs and length using the GA.

Keywords Wind farm cable layout · Radial collector topology · Metaheuristic optimization · Genetic algorithm · Traveling salesman problem

1 Introduction

In recent decades, the use of wind energy as a free source of carbon has increased rapidly. Wind farms are gaining increasing importance as an electricity source that counteracts climate change. In 2018, the growth of wind energy also continued internationally. Worldwide, 51.3 GW of wind power was recently installed. The total energy output rose by 9.6% and now reaches 591 GW [1]. Figure 1 shows the full global wind generation capacity up to 2018. However, developing and building a wind farm (WF) is a complex process and has to go through several steps. The method involves selecting the property correctly with the required wind profile, optimizing the location of each turbine to reduce the wake effect, optimizing the

C. El Mokhi (✉) · A. Addaim
Ibn Tofail University, Kenitra, Morocco
e-mail: chakib.elmokhi@uit.ac.ma

A. Addaim
e-mail: adnane.addaim@uit.ac.ma

S.-L. Peng et al. (eds.), *Proceedings of First International Conference on Mathematical Modeling and Computational Science*, Advances in Intelligent Systems and Computing 1292, https://doi.org/10.1007/978-981-33-4389-4_5

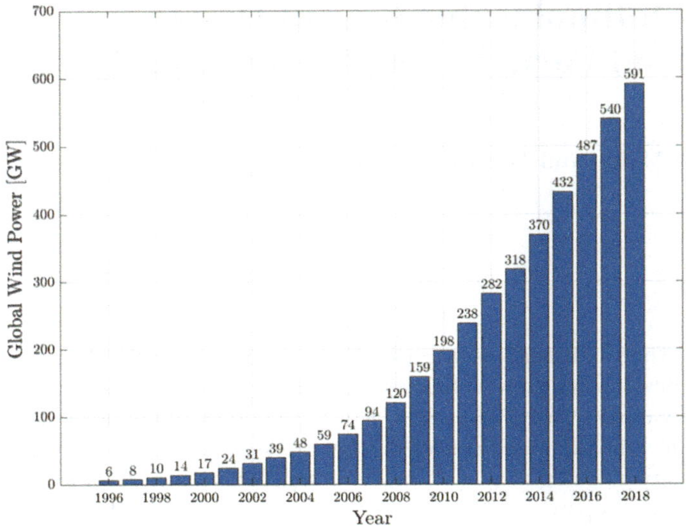

Fig. 1 Global wind generation capacity

locations of the required substations, optimizing the interface cable of the electricity collection system connected by the turbines with each other and with the substation and the generated electricity collection system. The costs of electrical infrastructure in offshore wind farms in particular are very high, which make them uncompetitive with onshore wind farms. Electrical maintenance costs in offshore wind farms are estimated at 15–30% of overall investment costs [2].

A special area has emerged that optimizes the routing of cables to reduce these investment costs by minimizing the total length of the installed cable. As a result, the corresponding power losses on the cables can also be avoided. Famous works in this area, which optimized this issue by using mixed integer linear programming (MILP) are [2] and [3]. Nevertheless, in this paper, we will solve the cable routing issue of the power collection system, which is considered as a multiple traveling salesmen problem (mTSP), by using a metaheuristic algorithm for the optimization.

A large-scale WF can contain hundreds of installed wind turbine (WT) generators interconnected through a medium-voltage network or a power collection system, usually 33 kV. In this way, the collected energy is transferred to the onshore main network over a transmission line. The collection system begins with every WT that has its transformer. Additionally, the collection system consists of several connecting cables or feeders. These feeders, which collect the energy generated by the wind turbines (WTs), are linked to the transformer substation. Finally, the transformer in the substation increases the voltage to a high voltage level for the transmission to the onshore main grid. Figure 2 shows a radial structure of an offshore wind farm (OWF).

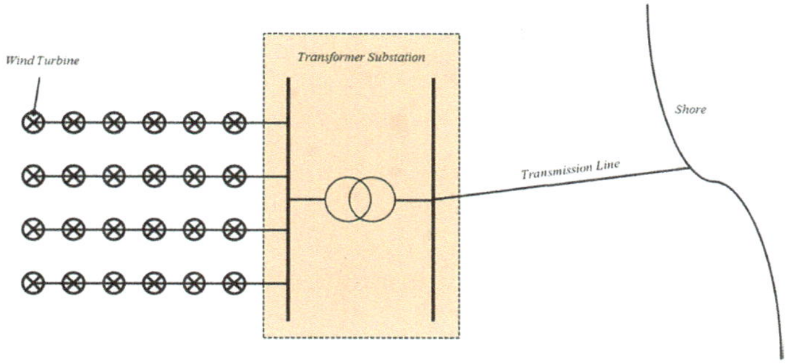

Fig. 2 Radial topology of an offshore wind farm

The paper is divided into the following sections. Section 2 introduces the GA and mTSP used to optimize interface cable routing. Section 3 presents the results of the optimization of electricity collection systems of three real existing OWFs. The conclusions of this work and possible future works are presented in the last section.

2 Optimization Method

2.1 Multiple Traveling Salesmen Problem

The mTSP is a special case of the popular traveling salesman problem (TSP), which is an NP-hard problem [4–6]. Accordingly, mTSP is more difficult than the TSP. The TSP is a typical problem of combinatorial optimization [7] and is generally defined as: Upon visiting n nodes, a salesman $m = 1$ begins heading away from every node, the node depot, and returns so that every node visits once. The goal is to find as short a minimum time as possible. Meanwhile, the mTSP involves searching for a series of routes for $m > 1$ salesmen to deliver a set of, e.g., cities given $n > m$ nodes, so that each node is attended by precisely one salesman except the common point of departure (depot). The routes take a form of loops. The main goal is to reduce the sum of travelled distances by all the salesmen. One wind farm with radial topology for our case, another type of mTSP must be considered, the so-called non-returning multiple traveling salesmen problem (nrmTSP), every single node has to be attended once only by only one salesman without returning to the common starting point. The substation in an OWF is viewed as a depot or as a starting node and the WTs are considered as cities. Although much attention has been paid to the TSP since the 1920s, research on the mTSP is insufficient and the need for research on the nrmTSP is larger. Figure 3 illustrates the differences between TSP, mTSP and nrmTSP. The optimization problem can be described as this [8]:

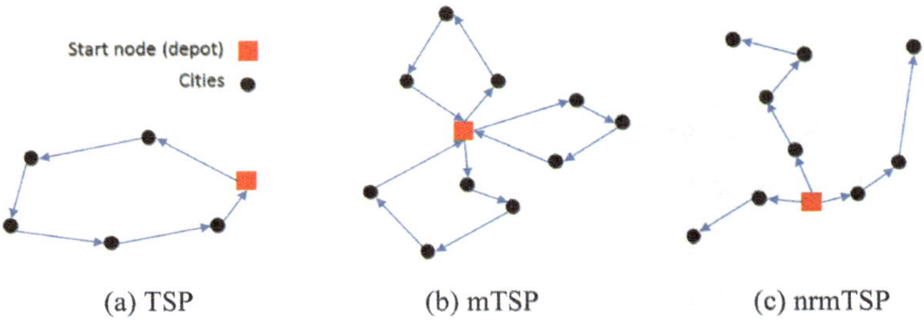

Fig. 3 Illustration of basic TSP, mTSP with returning and nrmTSP without returning to the depot

$$\min_{\substack{\varrho \\ \{\alpha_{h_j}\}}} \left\{ \sum_{i=1}^{n-1} d_{\varrho_i,\varrho_{i+1}} - \sum_{j=1}^{m-1} \alpha_{h_j} d_{\varrho_{h_j},\varrho_{h_j+1}} \right\} \tag{1}$$

$$\text{s.t.}\forall j, \sum_{h_j} \alpha_{h_j} = 1, \alpha_{h_j} \in \{0, 1\}, 1 \leq h_j \leq n - 1 \tag{2}$$

where n is the quantity of wind turbines (cities) and m is the quantity of cable routes (salesmen). The locations of the WTs are denoted by $\{x_i : x_i \in \mathbb{R}^2; 1 \leq i \leq n\}$.

We take $\alpha \in \mathbb{R}^2$ to indicate the position of the substation. The square-euclidean distance between two WTs i and j is indicated by $d_{i,j} = x_i - x_{j_2}^2$. A tour (cable route) is defined by the sequence $(\varrho_1, \ldots, \varrho_n)$, where $\varrho_i \in \{1, \ldots, n\}$ defines the ith WT's index in a cable route; notice that $\varrho_i \neq \varrho_j$ when $i \neq j$ and $1 \leq i, j \leq n$ and that $\varrho_{n+1} = \varrho_1$ [8].

The first part of the cost function in Eq. (1) above accords to reducing the overall distance between the adjacent WTs, while the second part accords to eliminating $m - 1$ edges, thus giving growth to m disjoint routes. The factor $\alpha_{hj} = 1$, if the edge according to the WTs $(\varrho_{h_j}; \varrho_{h_j+1})$ is removed [8].

2.2 *Genetic Algorithm*

The concept of the GA (more generally: evolutionary algorithms) was proposed by Holland in 1975 to describe adaptive systems [9]. The vocabulary used is directly modeling evolution theory and genetics [10] on it. So here we mean people (potential solutions), populations, genes (variables), chromosomes, parents, offspring, reproductions, etc. The points of the search space are then the individuals of a population and the function to be optimized corresponds to their adaptation [11]. These algorithms evolve a population iteratively. Some individuals reproduce, others mutate or even leave and only better-readjusted individuals are purported to withstand. The genetic inheritance of the generations must allow the population to be better and better readjusted and therefore to better respond to the optimization criterion. The flowchart of the GA for cable routing using the nrmTSP is shown in Fig. 4.

The most important parameters that define a GA are consequently the size of the population, the probabilities associated with mutation and crossover as well as the elitism factor. For a particular fitness ch_i chromosome $f(ch_i)$, the probability of its selection in the new population of size N is [12, 13]:

$$P_s(ch_i) = \frac{f(ch_i)}{\sum_{j=1}^{N} f(ch_j)} \tag{3}$$

Crossover is the main operator acting on the population of parents. It creates new individuals by exchanging information between chromosomes through their combination. The current population is divided into two sub-populations of the same size and each pair formed by a member from each sub-population participates in a crossover with a probability often greater than 0.5.

If the crossing took place between two-parent chromosomes (ch_1 and ch_2), a position is drawn randomly from each of the parents. We then exchange the two-terminal sub-chains of each of the chromosomes, which produce two children ch_1' and ch_2' as shown in the figure (crossover). Figure 5 illustrates the crossover. By selecting 2 genes $ch_1(i)$ and $ch_2(i)$ in each of the parents at the same position i, we can then define two new chromosomes $ch_1'(i)$ and $ch_2'(i)$ by a linear combination [7, 12]:

$$\begin{cases} ch_1'(i) = \gamma \times ch_1(i) + (1 - \gamma) \times ch_2(i) \\ ch_2'(i) = (1 - \gamma) \times ch_1(i) + \gamma \times ch_2(i) \end{cases} \tag{4}$$

where γ is a random weighting parameter which generally takes its values in the interval $[-0.5; 1.5]$ [12, 13].

For the mutation, the calculation of the new value of a gene is a little more complex. The x_k gene undergoes significant modifications during the first generations, then gradually decreasing as we progress in the optimization process.

For a generation t, we draw a binary value which will decide whether the change should be positive or negative. The new x_k' value of the x_k gene is given by:

$$x_k' = \begin{cases} x_k + \Delta\left(t, x_k^{max} - x_k\right) \text{if rand} = 0 \\ x_k - \Delta\left(t, x_k - x_k^{min}\right) \text{if rand} = 1 \end{cases} \tag{5}$$

where $\Delta(t; y)$ is a function which defines the difference between the new value and the initial value at generation t and rand is a randomly generated number, which takes the values 0 or 1.

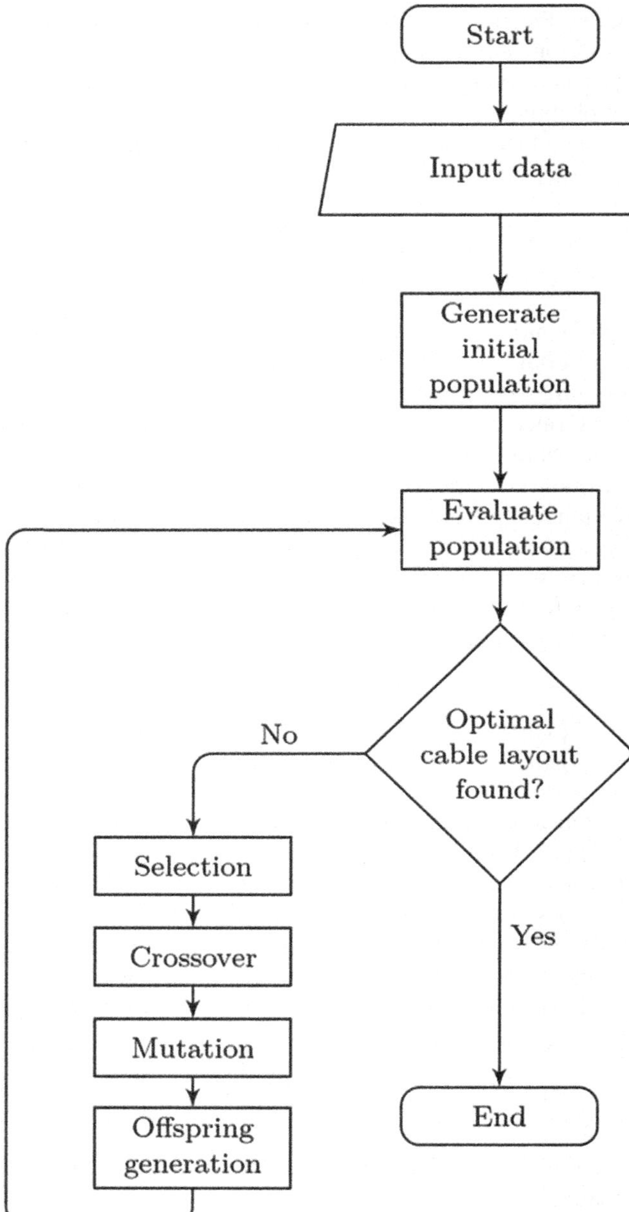

Fig. 4 Flowchart of the genetic algorithm

Fig. 5 Crossover

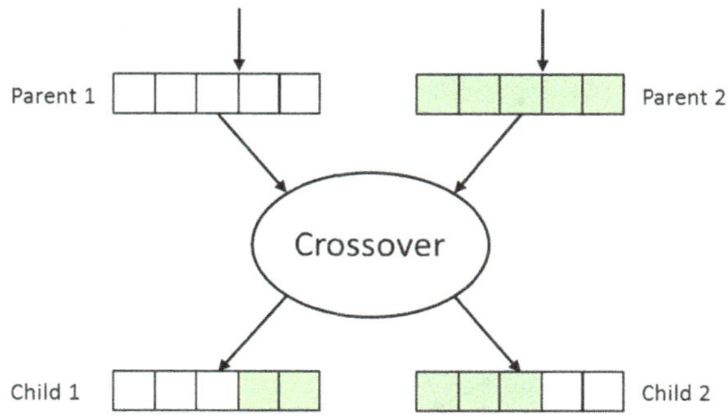

3 Optimization of the Cable Layout

3.1 Westermost Rough

Westermost Rough is an OWF in the United Kingdom situated in the North Sea. It is located 8 km off the coast of Holderness. The wind farm has 35 wind turbines, each with a capacity of 6 MW.

Accordingly, an approximate capacity generation of 210 MW. The annual energy production is around 800 GWh. The OWF has been in operation since May 2015. The total length of the cable for the power collection system is 40.13 km. After using an optimization based on the GA and nrmTSP algorithms, we were able to reduce the total length of the cable by 0.85 km.

That means 2.12% savings. We considered that 6 feeders collect the generated energy and transfer it to the substation. Each feeder may have at least 5 WTs. Figure 6

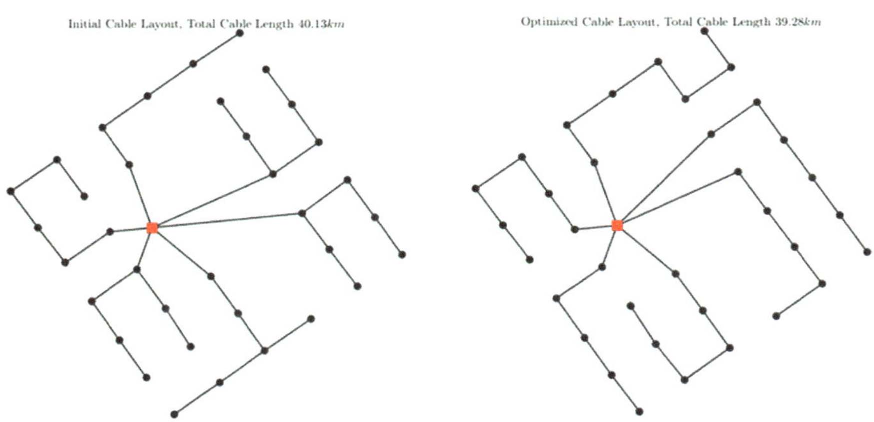

Fig. 6 Initial and optimized inter-array cable routing of Westermost Rough WF

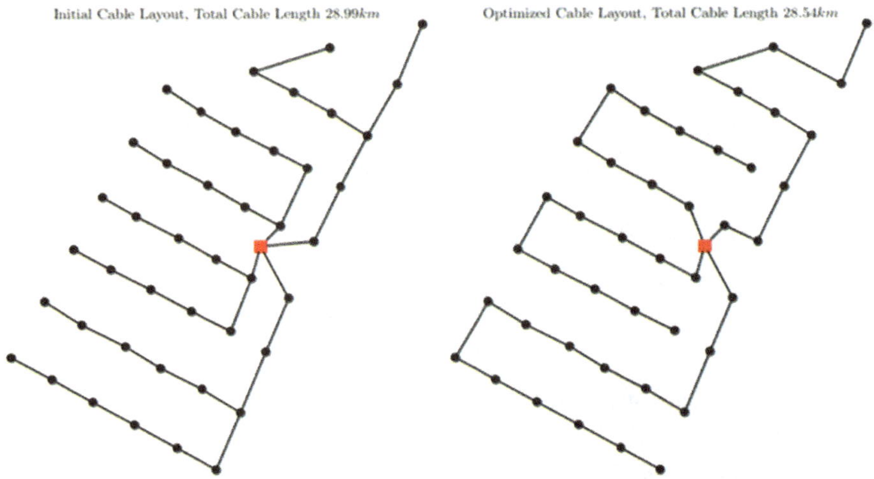

Fig. 7 Initial and optimized inter-array cable routing of Luchterduinen WF

shows the starting position and the optimized cable layout for the Westermost Rough WF.

3.2 *Luchterduinen*

The Luchterduinen offshore wind farm is about 23 km from the Dutch coast. The forecast annual energy production is at 530 GW. Commissioning took place in September 2015. The OWF contains 43 wind turbines, each with 3 MW capacity. The total capacity of the OWF is therefore 129 MW. By using the optimization algorithms, we get a total length of the cable for the power collection system of 28.54 km instead of the initial cable layout 28.99 km. The savings in cable length are 0.45 km, corresponding to 1.55%. When carrying out the optimization, the number of feeders was respected as in the initial situation. Figure 7 shows both the initial and optimized cable routing for the Luchterduinen OWF.

3.3 *Horns Rev 1*

The Horns Rev 1 was built in the North Sea opposite the Danish coast and has been operative since 2002. It has 80 wind turbines. The wind turbine capacity is 2 MW and therefore the offshore wind farm capacity is 160 MW. The wind farm's annual energy production is at 600 GWh. The topology of the power collection system is a string structure, and the total length of the installed cable is 58.50 km. In the application

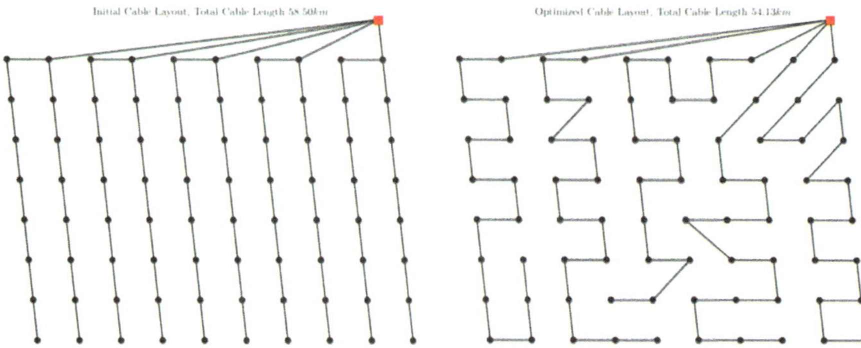

Fig. 8 Initial and optimized inter-array cable routing of Horns Rev 1 WF

Table 1 Results summary for the Horns Rev 1 OWF

Cross-sectional area (mm^2)	Initial cable length (km)	Costs €	Optimized cable length (km)	Costs €	Δ Costs €
3 × 150	42.06	15,982,800	39.39	14,968,200	−1,014,600
3 × 400	16.44	7,891,200	14.74	7,075,200	−816,000
Total	**58.50**	**23,874,000**	**54.13**	**22,043,400**	**−1,830,600**

of the proposed algorithm (GA and nrmTSP), we take into account that each feeder connects 16 WTs and finally connects them to the substation outside the turbine array, making a total of 5 feeders.

The optimized cable routing of the power collection system results in a total cable length of 54.13 km. The saved cable length is 4.37 km, which means 7.47%. Figure 8 shows the initial cable layout of the wind farm and the new cable routing after applying the proposed optimization.

In Horns, Rev 1 are two types of three-core cables installed; the first inter-array cable with a cross-section area of 3×150 mm^2, and the second one, the export cable to the offshore substation with a cross-sectional area of 3×400 mm^2. We estimate that the prices for the two cable types are 130 €/m and 230 €/m, respectively, whereby the installation costs about 250 €/m irrespective of cable type. The economic savings for the cable layout installation after applying the GA-nrmTSP optimization are 7.67%. Table 1 shows more details about the cable length and the economic savings.

4 Conclusions and Future Work

This article aims to optimize the inter-array cable architecture for offshore wind farms with a combination of the genetic algorithm and the problem of not returning many traveling salesmen. In general, the cable length has a direct impact on the power losses and the investment costs for the electrical infrastructure within an OWF. The combined GA-nrmTSP algorithm was tested on real existing OWF and shows interesting and satisfactory results in terms of cable length savings, and these economic advantages being supported by the results of the case study Horns Rev 1. The reliability is therefore confirmed. This work, however, has no limitations and can be extended to optimizing the proper placement of the wind turbines and substations. Also, an optimization for the appropriate selection of the cable cross-section can be added. In conjunction with these options for extending the work, an analysis of the cable power loss can also be implemented.

References

1. Fraunhofer Institute for Energy Economics and Energy System Technology (Fraunhofer IEE) (https://windmonitor.iee.fraunhofer.de/windmonitor_de/index.html)
2. Fischetti, M., & Pisinger, D. (2018). Optimizing wind farm cable routing considering power losses. *European Journal of Operational Research, 270*(3), 917–930.
3. Wedzik, A. (2014). The optimization of cable layout design in wind farm internal networks. *Acta Energetica Power Engineering Quarterly, 20*(3), 144–149.
4. Kencana, E. N., Harini, I., & Mayuliana, K., (Eds.), (2017). The performance of ant system in solving multi traveling salesmen problem. In *Proceedings of the 4th Information Systems International Conference 2017* (pp. 46–52), November 6–8. Bali, Indonesia.
5. Larki, H., & Yousefikhoshbakht, M. (2014). Solving the multiple traveling salesman problem by a novel meta-heuristic algorithm. *Journal of Optimization in Industrial Engineering, 7*(16), 55–63.
6. Hou, M., & Liu, D. (2012). A novel method for solving the multiple traveling salesmen problem with multiple depots. *Chinese Science Bulletin, 57,* 1886–1892.
7. Pérez-Jiménez, M., (Eds.), (2014). *Bio-Inspired Computing—Theories and Applications.* Springer Science and Business Media LLC.
8. Baranwal, M., Roehl, B., Salapaka, S. M., (Eds.), (2017). Multiple traveling salesmen and related problems: A maximum-entropy principle based approach. In *Proceedings of the 2017 American Control Conference (ACC)* (pp. 3944–3949), May 24–26. Seattle, WA, USA.
9. Gass, S. I., Fu, M. C., (Eds.), (2013). *Encyclopedia of Operations Research and Management Science.* Springer Edition.
10. Zeinebou, Z., Benabdelhafid, A., (2014). Comparing the effectiveness of different metaheuristics for optimization of flows of containers in a containers terminal. In *Proceedings of the 2014 International Conference on Advanced Logistics and Transport* (pp.235–240), May 1–2; Hammamet, Tunisia.
11. Kramer, O., (Ed.), (2017). *Genetic Algorithm Essentials.* Springer International Publishing.
12. Hachimi, H., (Ed.), (2013). *Hybridations D'algorithmes Métaheuristiques en Optimisation Globale et Leurs Applications.* HAL Archives.
13. Hachimi, H., El Hami, A., & Ellaia, R. (2012). A new hybrid genetic algorithm and particle swarm optimization. *Key Engineering Materials, 498,* 115–125.

Inventory Cost Estimation Using Brownian Motion

K. Suganthi and G. Jayalalitha

Abstract This paper describes the EOQ Inventory control of production model in partial backlogging which is evaluated through the Inventory-level variation. In this paper, two methods are explained: (i) Inventory different cost with shortage, back order quantity, maximum Inventory, number of order and total cost are found numerically. These parameters are obtained based on the assumption of Inventory carrying cost (m) and the shortage cost (M). (ii) EOQ cost in without shortage. These two methods can be applied to a small company, which includes sales or delivering any category of product; there is no way to stay away from the necessity to hold an Inventory to make sure timely delivery and avoid no-stock conditions. Conversely, holding Inventory does build extra costs for a company. These costs can happen in several ways, based on the features and size of the company. This helps the company Inventory cost reduce to make effective use of the working capital. These parameters of Inventory costs are estimated by Brownian motion. It is observed to be minimized. The computational results are obtained from numerical examples. The results are shown in Brownian motion. This outcomes because of fluctuations. It is incrementing. Consequently, it is points of increases and random walk. The results are irregular and follows the Brownian motion, and it is Fractals.

Keywords Inventory model · Economic order quantity · Back order quantity · Brownian motion · Fractals

AMS Classification 2000 · 90B05 · 90B99 · 60J65

K. Suganthi (✉) · G. Jayalalitha
Pallavaram, Chennai, Tamil Nadu 600117, India
e-mail: suganthi19122015@gmail.com

G. Jayalalitha
e-mail: g.jayalalithamaths.sbs@velsuniv.ac.in

S.-L. Peng et al. (eds.), *Proceedings of First International Conference on Mathematical Modeling and Computational Science*, Advances in Intelligent Systems and Computing 1292, https://doi.org/10.1007/978-981-33-4389-4_6

1 Introduction

1.1 Inventory Model

Inventory is an amount of stock owned and stored by means of a market place that is intended both for resale or as raw materials and additives used in producing items that the commercial enterprise sells [1]. Inventory manipulate or Inventory control can be broadly defined because the interest of checking a store's Inventory [2], However, a more targeted definition takes under consideration the greater technology-based, methodical practice of no longer only verifying a commercial enterprise stock but additionally focusing on the numerous related facets of Inventory control within an business enterprise to fulfill the demand placed upon that commercial enterprise economically. Other Inventory of stock manipulate include supply chain management, manufacturing manipulates monetary flexibility, and consumer delight [3]. Deciding the ordering diagram to reduce the whole Inventory price. The average sized companies, demand is undefined and hard to predict. This paper takes endless model consists in of financial order quantity, period order extent [1].

1.2 Economic Order Quantity

The EOQ is amount designed to assist corporations to not over- or below-stock their inventories and decrease their capital investments on the products that they are selling [4]. The cost of ordering a stock touches down with growth in ordering in bulk. However, as the seller needs to grow the size of the stock, the wearing fees additionally grow [5]. The EOQ is precisely the factor that optimizes both of those charges that is price of ordering and the carrying price which can be inversely associated [6].

1.3 Back Order Quantity

Back orders constitute buy orders made to the supplier for products, which can be already out of stock from a gives vicinity being served. Back ordering is the procedure of selling stock the employer doesn't have accessible. Back ordering takes region only while the call for is captured in a proper way [7]. For instance, in a retail shop, maximum clients could without a doubt circulate on while facing an out-of-shelf situation, without reporting the lacking product to the store. Back orders represent specific challenges in terms of stock optimization, as back ordered measure is typically related to a degree of urgency coming from the patron [8].

1.4 Brownian Motion

The irregular movement of tiny particles suspended in a fluid moved on exceptionally unpredictable path. In 1827, Robert Brown was resulting of bombardment of the particles through molecules of the medium [9].

1.5 Fractals

The Fractals was coined through Benoit Mandelbrot in 1975. It is continuous patterns. Infinitely complicated patterns. It is complicated types which might be self-similar across whole lot of scales. They are formed by means of replicating method and over in an ongoing loop [10].

2 Examples. Inventory Different Cost with Shortage

2.1 Mathematical Model

The cause of keeping the Inventory price of Inventory is to use the available capital perfectly so that Inventory price in keeping with item of material minimum as favor. Hence, the objective of controlling the Inventories is to permit the materials to place and order at right time with the proper supply at proper place to buy right amount. Occasionally it so takes place that the material may not be possible, when needed or when the demand happen in such work has to be stopped until the procurement of the goods, which can also lead miss the supply dates or not on time production. The cost of products will cannot be predicted, that is company is losing of the customers. Here the stock out situation, if the company goods more materials, Inventory carrying cost increases and to take care of Inventory Cost. In the method of balancing the interrelated value that is order quantity, back order quantity, the highest Inventory held in a cycle length, the amount of order/year and total cost. The following assumptions are below:

- By right Inventory cost control.
- The available capital may be used precisely.
- By avoiding the useless consumption on Inventory.
- Proper Inventory will make sure the supply of the specified material in required amount with the minimum Inventory cost.
- Causes of surplus goods may be absolutely prevented.
- To sufficiently deliver products to the consumer or good enough of amount of raw substances to the manufacturing department in order that the state of affairs of inventory out may be reduced.

2.2 Notations

The Mathematical model is generated on the basis of the following notation.

$Q*$ Order quantity.
D The demand rate.
$S*$ Back order quantity.
I Inventory cycle from m to M before backlogging.
$M*$ Shortage cost or back order cost.
$m*$ Inventory carrying cost.
I_m^* The highest Inventory held in a cycle length.
N The amount of order/year.

2.3 Inventory Model in Order Quantity

Order quantity in an Inventory purchasing denotes that a product has been ordered and is waiting for voucher.

The Inventory order quantity can be as follows

$$Q = \sqrt{\frac{2DC_0(m^* + M^*)}{M^* m^*}} \tag{1}$$

$m*$ is Inventory carrying costs. Since D is demand and C_0 is order cost.

2.4 Back Order Quantity

A back order represents the quantity of stock, which the company's consumers have ordered but have not received yet as it is not available in stock currently.

Back order quantity is given by

$$\text{Back Order Quantity} = \frac{\text{Order Quantity} \times \text{Inventory Carrying Cost}}{\text{Inventory Carrying Cost} + \text{Back Order Cost}}$$

$$S = \frac{Qm^*}{m^* + M^*} \tag{2}$$

Since $M* =$ Shortage Cost.

2.5 The Highest Inventory Held in a Cycle is Now

Maximum Inventory = Order Quantity – Back Order Quantity

$$Im^* = Q * -S* \tag{3}$$

2.6 Number Order/Year

The Number of Orders = Demand/Order amount

$$N = \frac{D}{Q} \tag{4}$$

2.7 Total Cost

The optimal total expected cost is evaluated using the expected cost per cycle is given by:

Annual Expense = Annual Order Quantity Expense + Annual Carrying Expense + Annual Back Order in the System.

$$T_c = \frac{DC_0}{Q} + \frac{(Q - S)^2 m^*}{2Q} + \frac{S^2 M^*}{2Q} \tag{5}$$

2.8 Numerical Example

Let $D = 1000$, $C_0 =$ Rs 200/order and $m* = 7$/unit/year, $M* = 70$ per unit/year. The optimal solution are $Q* = 250.71$, $S = 22.79$, $I_{m*} = 227.92$, $N = 3.99$, $T_c = 1595.45$.

Figures 1, 2, 3, 4 and 5 explains the Inventory cost in Brownian motion.

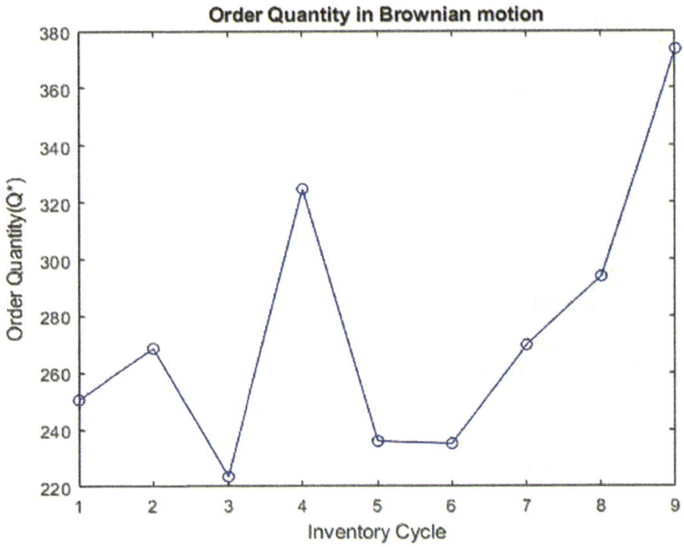

Fig. 1 Order quantity in Brownian motion

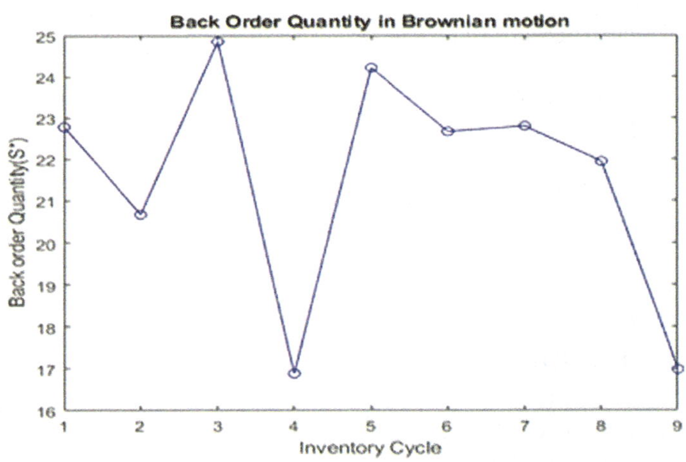

Fig. 2 Back order quantity in Brownian motion

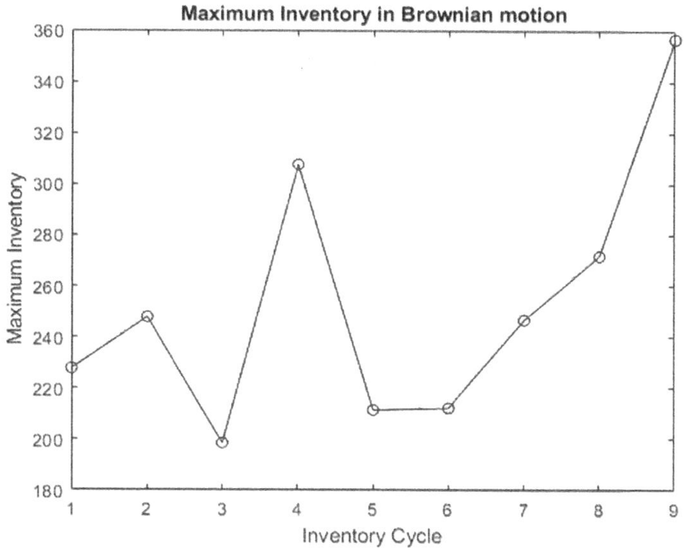

Fig. 3 Maximum inventory in Brownian motion

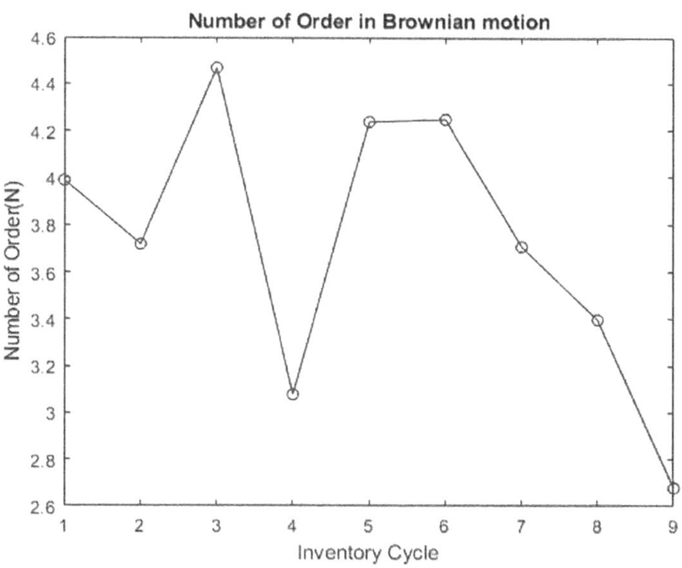

Fig. 4 Number of orders in Brownian motion

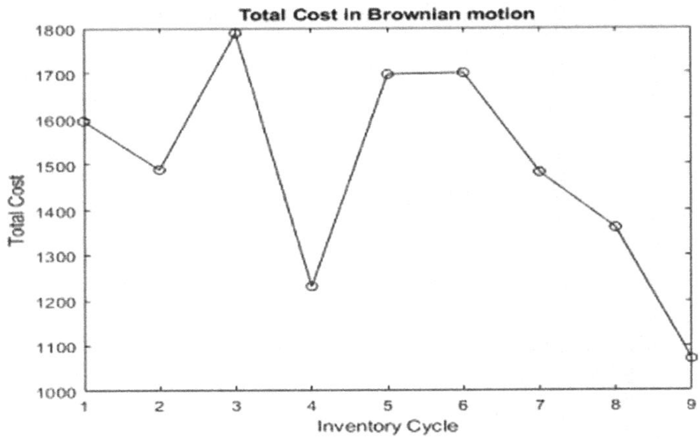

Fig. 5 Total cost in Brownian motion

3 Examples. EOQ Cost in Without Shortage

3.1 Mathematical Model

EOQ is a necessary Inventory management procedure that demonstrates the amount of a stock to minimize the total cost of both conduct of Inventory (carrying cost) and ordering processing (ordering cost).

Considering manufacturing problems, this amount Manufactured in keeping with batch lot size, fixed set up cost per batch. Since method for economic lot length q_1 per cycle of single product is derived so as to reduce the total average variable price per unit time.

3.2 Assumptions and Notations

The following assumptions were made in order to formulate the model.

D = Demand for item.
I_c = Inventory carrying cost.
P = Unit cost.
c_1 = Ordering cost.
C = Average Inventory.
C_0 = Optimal ordering cost.
q_1 = Optimal order quantity.
R = Demand rate.
T_{oc} = Total ordering cost.

$T_{\text{IC}} =$ Total inventory cost.
$t_0 =$ Optimal order period.

(I)

$$\text{Economic order quantity} = \sqrt{\frac{2c_1 \times D}{\text{p} \times \text{c}}} \tag{6}$$

$$\text{Optimal Inventory cost} = \sqrt{2 \times \text{c} \times p \times c_1 \times D} \tag{7}$$

(II)

$$\text{Optimal order time} = \sqrt{\frac{2 \times C_1}{C \times R}} \tag{8}$$

(III) Optimal number of orders $= \frac{D}{q_0}$.

The number of days' supply per most efficient order is obtained by $d = \frac{365}{N_0}$.

(IV)

$$\text{Optimal ordering cost is} C_0 = \sqrt{2 \times c \times c_1 \times D} \tag{9}$$

(V) Inventory carrying cost is given by:

$$\frac{q_1}{2} \times p \times c_1 \tag{10}$$

Numerical Examples

Parameters 1: Economic Order Quantity $D = 2000$ units, $p =$ Re 2/-, $C_0 = 30\% = 0.3$, ordering cost is 13.50 per order, The Optimal solution is $q_1 = 300$, $C_0 = 180$,, $I_c = 90$, $T_{O_c=90,}$ $T_{I_c} = 180$, $N = 6.6$, $t_0 = 54.75$.

Parameters 2: Economic Order Quantity $D = 2000$ units, $p =$ Re 2/-, $C_0 = 40\%$. $= 0.4$. ordering cost is 14.50 per order, The Optimal solution is $q_1 = 269.26$, $C_0 = 215.41$, $I_c = 107.70$, $T_{O_c=107.70,}$ $T_{I_c} = 215.4$, $N = 7.4$, $t_0 = 47.45$.

Parameters 3: Economic Order Quantity $D = 2000$ units, $p =$ Re 2/-, $C_0 = 50\%. = 0.5$. ordering cost is 15.50 per order, The Optimal solution is $q_1 = 249$, $C_0 = 248.99$,, $I_c = 124.50$, $T_{O_c=124.50,}$ $T_{I_c} = 249$, $N = 8.3$, $t_0 = 43.8$.

Parameters 4: Economic Order Quantity $D = 2000$ units, $p =$ Re 2/-, $C_0 = 60\%$. $= 0.6$. ordering cost is 16.50 per order, The Optimal solution is $q_1 = 234.52$, $C_0 = 281.42$,, $I_c = 140.71$, $T_{O_c=140.71,}$ $T_{I_c} = 281.42$, N $= 8.5$, $t_0 = 36.5$.

Figure 6 shown below explains the EOQ cost in Brownian motion.

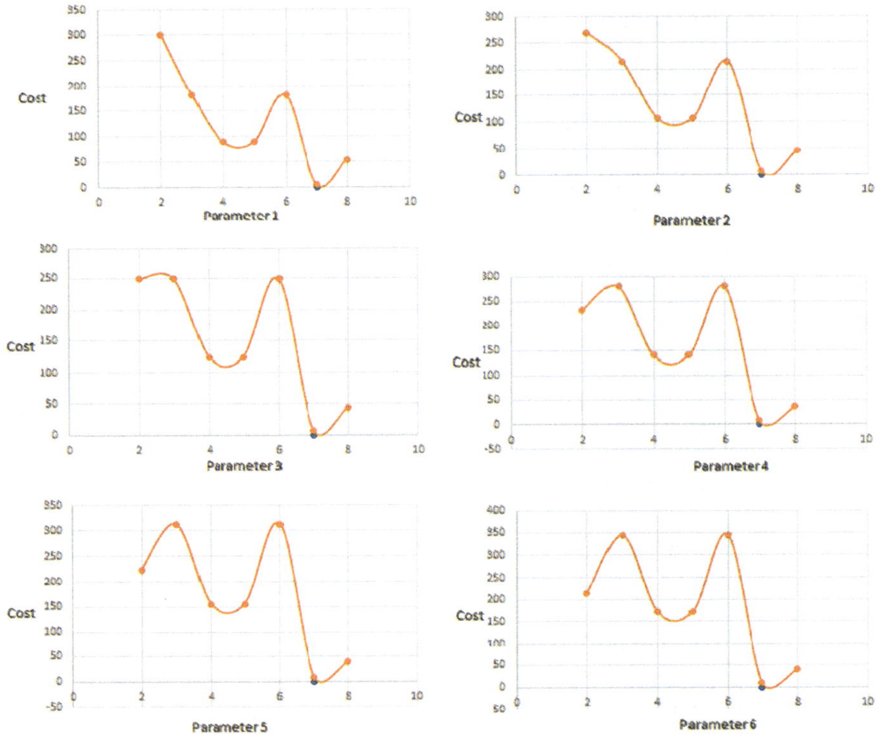

Fig. 6 EOQ cost in Brownian motion

3.3 *Brownian Motion*

A Brownian curve is described to be a set of random variables of time which have the following conditions:

(i) For every $t > 0, h > 0$, the increments $X(t+h) - X(t)$ are approximately normal with mean 0 and variance h.
(ii) The distributions $X(t+h) - X(t)$,$0 < t_1 < t_2 < \ldots\ldots < t_n$ are independent increments.
(iii) With probability 1, $X(0) = 0$ and $X(t)$ is a continuous function of t.
(iv) If $0 \leq t_1 \leq t_2 \leq \ldots. \leq t_{2m,}$ the increments $X(t_2) - X(t_1) \ldots\ldots. X(t_{2m}) - X(t_{2m-1})$ are independent.

Since $X(t)$ is regular with mean 0 and variance t and $t_1, t_2 \ldots\ldots t_n$ are independent random variables.

From (i) and (ii) that $X(t)$ is independent regularly distributed with mean 0 and variance for every t.

The increments of X are stationary, that is $X(t+h) - X(t)$ has distribution independent of t.

For example, $X(t_2) - X(t_1)$ and $X(t_3) - X(t_2)$ are independent and regular with mean 0 and variance $t_2 - t_1$ and $t_3 - t_2$ respectively, the sum $X(t_3) - X(t_1)$ is essential normal with mean 0 and variance $t_3 - t_1$.

The technique to generate a Brownian motion curve is to move alongside one factor at a time in which every factor is displaced vertically upward and downward.

3.4 *Points of Increase*

A real-valued function f has a global factor of growth within the interval (a, b) if there may be aspect $t_0 \in f(a, b)$ Such that

$$f(t) \leq f(t_0) \text{ For all } t \in (a, t_0) \text{ and} f(t_0) \leq f(t) \text{for all} t \in (t_0, b). \quad (11)$$

We say t_0 is a neighborhood factor of growth if it is a global factor of growth in few intervals. Here we show that the no increase phenomenon holds for arbitrary symmetric random walks and might as a result be considered as a combinatorial outcome of fluctuations in random sums.

Theorem 1 *For any random walk* $\{s_j\}$ *on the line,*

$$P(s_0, s_1, s_2 \ldots s_n \text{has a factor of growth}) \leq \leq 2\frac{\sum_{k=0}^{n} p_k p_{n-k}}{\sum_{k=0}^{\frac{n}{2}} p_k^2} \quad (12)$$

Proof The idea is easy. The predicted wide variety of factors increase is the numerator, and for the reason that there may be at least one such point, the anticipated variety is bounded underneath via the denominator; the ratio of these expectations bounds the desired probability [10]. To supply this out, denote by way of using $I_n(k)$ the function that k is a factor of growth for,

$$s_0, s_1, s_2 \ldots\ldots s_n \text{and by} F_n(k) = \frac{I_n(k)}{U_{i=0}^{k-1} I_n(i)} \quad (13)$$

The function that k is the first such factor $\{s_k$biggest among $s_0, s_1, s_2 \ldots\ldots\ldots\ldots s_n\}$ and $s_k$smallest among $s_0, s_1, s_2 \ldots\ldots\ldots\ldots s_n\}$ are independent, and.

Therefore

$$P(I_n(k) = p_k p_{n-k} \quad (14)$$

Notice that if S_j is minimal among $s_j \ldots s_n$, then any factor of increases for $s_0, s_1, s_2 \ldots s_j$ is routinely a factor of increase for $s_0, s_1, s_2 \ldots s_n$. Therefore, for $j \leq k$, $F_n(j) \cap I_n(k)$ is same to

$$F_j(j) \cap \{s_j \le s_i \le s_k \forall i \in [j,k]\} \cap \{s_k = \min(s_k, \ldots s_n\} \tag{15}$$

The three functions on the right-hand side aspect are independent, as they contain disjoint units of summands; the second of these functions is of the kind into taken consideration in Theorem *we have* $p_n^2 \le \mathbb{P}(0 \le s_j \le s_n \text{for all} 1 \le j \le n) \le p_{\frac{n}{2}}^2$. *Proof* the two events are independent, since A depends most effective only on $X_1, \ldots X_{\lfloor[\frac{n}{2}]\rfloor}$ and B relies upon simplest at the remaining $X_{[\frac{n}{2}]+1} \ldots X_n$ therefore,

$$A = \left\{0 \le s_j \text{for all} j \le \frac{n}{2}\right\} \text{and } B = \left\{s_j \le s_n \text{for all} j \ge \frac{n}{2}\right\} \tag{16}$$

$$\mathbb{P}(0 \le s_j \le s_n forall 0 < j < n) \le \mathbb{P}(A \cap B) = \mathbb{P}(A \cap B) = \mathbb{P}(A)\mathbb{P}(B) \le p_{\frac{n}{2}}^2 \tag{17}$$

which proves the upper bound.

Thus, $\mathbb{P}(F_n(j) \cap I_n(k) \ge \mathbb{P}(F_j(j)p_{k-j}^2 p_{n-k} \ge p_{k-j}^2 \mathbb{P}(F_j(j)\mathbb{P}(s_j \text{ is minimal among } s_j \ldots s_n\}$.

Since $p_{n-k} \ge p_{n-j}$. *Here the 2 occasions at the right are independent and their intersection is exactly* $F_n(j)$. *Consequently,*

$$\mathbb{P}(F_n(j) \cap I_n(k)) \ge p_{k-j}^2 \mathbb{P}(F_n(j)). \tag{18}$$

Decomposing the function $I_n(k)$ *in step with the primary factor of will increase gives*

$$\begin{aligned} \sum_{k=0}^{n} p_{n-k}p_{n-k} &= \sum_{k=0}^{n} \mathbb{P}(I_n(k)) = \sum_{k=0}^{n}\sum_{k=0}^{n} p_{n-k}p_{n-k} \\ &\ge \sum_{j=0}^{n}\sum_{k=j}^{j=[\frac{n}{2}]} p_{k-j}^2 \mathbb{P}(F_n(j)) = \sum_{j=0}^{\frac{n}{2}} p_i^2 \end{aligned} \tag{19}$$

This yields an upper bound at the probability. The *point of growth* by time $\frac{n}{2}$; but this random walk has a factor of growth at time k if and simplest if the "reversed" walk. Thus, doubling the upper bound proves the theorem.

4 Results

From Table 1, the optimal order quantity cost decreases, when the back order rate rises. Similarly, when the optimal order quantity increases, the back order rate decreases. These two processes happen based on the Inventory carrying cost and shortage cost. The inventory cycle values are taken in random manner. Hence, a

Table 1 Computational results of Inventory

Inventory cycle	Q^*	S^*	I_{m^*}	N	T_c
$m^* = 7$ $M^* = 70$	250.71	22.79	227.92	3.99	1595.45
$m^* = 6$ $M^* = 72$	268.74	20.67	248.07	3.72	1488.42
$m^* = 9$ $M^* = 72$	223.61	24.85	198.76	4.47	1788.85
$m^* = 4$ $M^* = 73$	324.78	16.87	307.91	3.08	1231.62
$m^* = 8$ $M^* = 70$	236.04	24.21	211.8	4.24	1697.12
$m^* = 8$ $M^* = 75$	235.23	22.67	212.56	4.25	1700.46
$m^* = 6$ $M^* = 65$	269.85	22.80	247.05	3.71	1482.29
$m^* = 5$ $M^* = 62$	294.03	21.94	272.09	3.40	1360.42
$m^* = 3$ $M^* = 63$	373.74	16.99	356.75	2.68	1070.26

better association should be created with consumers and increase the consumer's faithfulness so as to attract consumers choice for back orders. A back order is an order for a good or service that cannot be back order must follow to avoid. m^* is Inventory carrying costs. These costs are involved to carry the products. This expresses that one should take safety actions to maintain the whole expense of the system under control by sustaining the worth of Inventory level inconsistency. The five parameters such as order quantity, back order quantity, maximum Inventory, number of order and total cost are simulated using MATLAB and are found to be Brownian motion. The order quantity is plotted as shown in Fig. 1. The plot follows a Brownian motion. The order quantity ranges between 220 and 380. All the graphs plotted below are obtained from the values present in Table 1. The order quantity is 250.71 but also build a back order of 22.79 units, the highest Inventory maintained in a period is 227.92. The number of order is 3.99. Every cycle is for 0.250 years out for that Inventory cycle is 0.227 years and the shortage period is 0.22, then total cost (Tc) = 1595.45. Total cost is plotted as shown in Fig. 5. All the values are obtained from Table 1. These figures are obtained from Inventory cycle values (m^* Inventory carrying cost and M^* shortage cost). Brownian motions are observed in all the figures.

Table 2 Optimal EOQ in different cost

Parameters	q_1	C_0	I_c	T_{o_c}	T_{I_c}	N	t_0
1	300	180	90	90	180	6.6	54.75
2	269.26	215.41	107.70	107.70	215.4	7.4	47.45
3	249	248.99	124.50	124.50	249	8.03	43.8
4	234.52	281.42	140.71	140.71	281.42	8.5	36.5
5	223.61	313.05	156.53	156.53	313.06	8.9	40.15
6	215.06	344.09	172.05	172.05	344.1	9.3	40.15

5 Conclusion

The main objective of the article is to suggest as new description of product– EOQ Inventory control type, and additionally concentrated on identifying good representations for the functioning attributes. A mixed back orders and proper sales Inventory model are created, in which the manufacturing level is monitored at various time periods. The costs assumed are Inventory holding, back order cost, order quantity and total costs. These costs are found as minimized by using Brownian motion. This will decrease the necessity of manpower and will reduce the labor expenses also. The results in the numerical examples express how administration movement can improve the effectiveness of the Inventory procedure (e.g., to minimize the system expense). Its consequences are proven in Brownian motion. This outcomes because of fluctuations. It is incrementing. Hence, it is points of increase. The consequences are irregular and follows the Brownian motion, and it is Fractals.

References

1. Stopkova, M., Stopka, O., & L'uptak, V. (2019). Inventory model design by implementing new parameters into the deterministic model objective function to streamline effectiveness indicators of the inventory management. *Sustainability, 11,* 4175.
2. Samak-Kulkarni, S. M., & Rajhans, N. R. (2013). Determination of optimum inventory model for minimizing total inventory cost. *Procedia Engineering, 51,* 803–809.
3. Gholami, A., & Mirzazadeh, A. (2018). An inventory model with controllable lead time and ordering cost, log-normal-distributed demand and gamma available capacity. 5, 1469182.
4. Kazemia, Nima, Abdul-Rashida, Salwa Hanim, Ghazillaa, Raja Ariffin Raja (2016). Economic order quantity models for items with imperfect quality and emission considerations. *International Journal of Systems Science.*
5. Dhodi, M. H. (2019). The effect of information technology on inventory management for the manufacturing companies in Mogadishu. *European Journal of Logistics, Purchasing and Supply Chain Management*, 6(3). ISSN 2054-0930.
6. Pandeya, R. K., Singhb, S. R., Vaisha, B., & Tayalb, S. (2017). An EOQ model with quantity incentive strategy for deteriorating items and partial backlogging. *Uncertain Supply Chain Management, 5,* 135–142.

7. Zhang, R.-Q., Wu, Y.-L., Fang, W.-G., & Zhou, W.-H. Inventory model with partial backordering when backordered consumers delay purchase after stockout-restoration (Vol. 2016). Article ID 642540.
8. Kurdhi, N. A., & Doewes, R. I. (2019). Periodic review Inventory policy with variable ordering cost, lead time, and backorder rate. *Scientific Technology, 41*(1), 1–11.
9. Melda, O., Dai, J. G., & Vate, J. V. (2018). Impulse control of Brownian motion: The constrained average cost case. *Operation Research, 56*(3), 618–629.
10. Bishop, C. J., & Peres, Y. (2017). Fractals in Probability and Analysis.
11. Kupta, P. K., & Hira, D. S. (2007). Operations Research.

Applying Negative Association Rule Utilizing Unsupervised Machine Learning Models

Saddam Raheem Salih and Amaal Mohi Nassief

Abstract Association rules are described as the relationships between objects in the database, and these rules are broadly investigated in the analysis of shopping baskets, to examine the influence of one item on the another in the same transaction. In this paper, a novel technique is suggested that can be the ability to mine thrilling negative association rules between items in the transaction database, by extracting knowledge from that database depending on the purchased quantities. Two clustering techniques are examined in this paper, which are the density-based spatial clustering and K-means of applications with noise (DBSCAN) techniques. The outcomes of these methods are compared to the outcomes of extracting negative association rules without any domain knowledge. The utilization of DBSCAN clustering technique has given better negative association rule mining outcomes of 40,862 rules, with a rate of 0.209% support and 91.840% confidence, when examining on a real-life transaction database. Mining negative association rules depends on the domain knowledge extracted utilizing the K-means clustering technique that has 17,801 rules with a rate of 0.189% support and 85.835% confidence, while mining negative rules without any domain knowledge results in 90,655 rules with rate support of 0.122% and rate confidence of 99.365%, utilizing the same database.

Keywords Association rule mining · DBSCAN clustering · Negative association rules · Unsupervised machine learning

1 Introduction

The rapid growth of computer usage in different fields, according to the high performance and accuracy provided by these computers, leads to enormous data being

---

S. R. Salih (✉)
Department of Information Technologies, Directorate General of Education in Diyala, Baghdad, Iraq
e-mail: saddam_rhs72@yahoo.com

A. M. Nassief
Department of Mathematics, College of Science, University of Diyala, Baghdad, Iraq
e-mail: amalmuhi@sciences.uodiyala.edu.iq

S.-L. Peng et al. (eds.), *Proceedings of First International Conference on Mathematical Modeling and Computational Science*, Advances in Intelligent Systems and Computing 1292, https://doi.org/10.1007/978-981-33-4389-4_7

collected for these applications. Some of these data include very valuable information that may assist improving the services provided in these application fields. Extracting such information is accomplished using data mining techniques, which are based on machine learning. Machine learning models may be split into two main classes, supervised and unsupervised, depending on the required knowledge from the database. Supervised machine learning requires prior data labeling, so that these techniques detect patterns and relations between the attributes' values of the objects in the database and label given for them. Unsupervised machine learning, on the other hand, requires no data labeling, as it extracts knowledge by investigating relationships between objects and frequent patterns of their values in the database. One of the largely utilized data mining models is the association rule [1, 2]. Association rules are another unsupervised data mining technique, which investigates relationships between objects and attributes without the need for data labeling. Association rule mining techniques are widely used to analyze transaction databases, where relationships that join items in a similar transaction are studied. The association rule is defined as the relationships among items in the same shopping basket, where the existence of one item in that basket has a direct influence on the chance that the other item is found in the same basket. These relations may be divided into two categories, positive and negative association rules [3, 4]. Two strength measures are used to describe the strength of an association rule, which can be the support and confidence. Support describes how dependable is the extracted association rule, while confidence describes the possibility of this rule to occur. Support is measured based on the display of the objects of that rule in the database, while confidence represents the part of the transaction where this rule is true to the full transaction that includes those items. Thresholds are set for support and confidence to avoid mining weakened rules. Many clustering algorithms are used to distribute objects in a database into more homogenous groups. K-means algorithm is utilized in database clustering [5]. Another popular clustering method is "DBSCAN." This method can automatically detect the optimal number of the cluster for every different database using a predefined setting, which is the max dimension among objects in the similar cluster and the min number of objects in a cluster. Using these settings, the algorithm starts clustering from a random object and then combines all the objects that are closer to that object than the maximum distance set for the algorithm. Then, another object is selected, which is not clustered yet, to start creating a new cluster. Eventually, all objects are distributed on clusters, using a different number of clusters for each database [6].

2 Proposed Method

The proposed negative association rule mining method utilizes machine learning techniques to obtain the knowledge in negative association rule in the database. Thus, it is important to use a technique that has the capability of computing the optimized number of clusters. The elbow method is one of the extensively used methods, which has the required capability. This method computes the summation of dimension among objects of the database and the centroid of the cluster that this centroid belongs to, for the clustering results of one cluster up to a predefined max number of clusters. Then, the rate where the change in the summation dramatically drops, this value is considered as the optimal number of clusters. As the created graph using this method looks like a human elbow, where the point at the joint of the elbow is selected as the optimized number of clusters, this method is named after that point as the elbow method. The less number of cluster imposes huge loss in the extracted domain knowledge, while the higher number of clusters require more processing with no significant addition to the domain knowledge extracted by clustering the database.

A Novel Quantity-Based on Negative Association Rule learning: Domain knowledge, extracted using clustering, is used for negative association rule mining from the item sets that have positive association rule to assure relationship in objects that are being considered for negative association rule. Each item in the transaction database, with support greater than the min allowable support, is replaced with a subset of items depending on the clustering results. Negative association rules between items, or item sets, which have a positive association rule joining them, are investigated between every subset of items in the antecedent and subsets of items. So that, if the purchased quantities of item χ are clustered into two clusters, so are the quantities of item Υ.

The positive association rule $\chi \rightarrow \Upsilon$ exists in the extracted positive association rules. The support values of all possible combinations from χ and Υ are computed, so that the support values for these candidate negative association rules $\chi_1 \rightarrow \Upsilon_1$, $\chi_1 \rightarrow \Upsilon_2$, $\chi_2 \rightarrow \Upsilon_1$, $\chi_2 \rightarrow \Upsilon_2$, $\chi \rightarrow \Upsilon_1$, $\chi \rightarrow \Upsilon_2$, $\chi_1 \rightarrow \Upsilon$, and $\chi_2 \rightarrow \Upsilon$ are computed, and these rules with support greater than the min allowable support are extracted as the resulting negative association rules. The algorithm used for quantity-based negative association rule mining is presented in Fig. 1.

3 Experiment Results

For examining the proposed negative association rule mining methods, three experiments are conducted using a real-life online store transaction database acquired from the University of California, Irvine (UCI) repository [7]. The database consists of 541,909, which include information of 25,295 transactions for 3925 products, tuples characterized by eight attributes, which are invoice ID, stock code, invoice date, description, customer ID, country, quantity, and unit price. A sample of the database used in the experiments is shown in Table 1.

The experiments are conducted by using different groups of the purchased quantities; the first experiment uses each purchased quantity solely, while the second and third experiments use different clustering techniques.

Experiment A: In this experiment, the negative association rules are extracted by investigating each purchased quantity of an antecedent item with every quantity in the

```
Input minimum support(minSup),Input minimum negative support (minNSup),
Input minimum confidence (minConf),Input minimum negative confidence (minNConf)
F = Apriori (I, minSup)
For each x in F
Foreach y in F
If x≠y
C_xy = |χ ∪ Υ|\|χ|
end if
end for
end for
Q = {χ → Υ|C_xy ≥ minConF}.
For each q in Q
S = cluster(F)
NF = Apriori(S_x ∪ S_y, minNSup)
Foreach m in NF
For each n in NF
If m≠n
NC_mn = |m ∪ ¬n|\|m |
end if
end for
end for
end for each
N= {M ⇒ ¬N|NC_mn ≥ minNConf}.
Return N
```

Fig. 1 Mining quantity-based negative association rules

consequent item of the items, and item sets, of the positive association rule, extorted from the database. The summary of the experiment is shown in Table.2.

Experiment B: In this experiment, the purchased quantities are clustered using the K-means clustering method. As this method cannot automatically set the number of clusters, the elbow method is utilized for selecting the optimized number of clusters. Antecedent and consequent objects in positive association rule mining are then investigated using those clusters, which are concluded based on the histogram of the purchased quality. The summary of the negative association rule mining extracted depends on the same positive association rules, extracted in Experiment B as shown in Table 2.

Experiment C: This experiment uses the DBSCAN method to cluster the purchased quantities in the database. This method can automatically cluster the values in the

Table 1 Sample of the database used in tests

Invoice No.	Stock code	Invoice date	Description	Customer ID	Country	Quantity	Unit price
536365	85123A	12.1.2010 8:26	WHITE HANGING HEART T-LIGHT HOLDER	17850	UK	6	2.55
536365	71053	12.1.2010 8:26	WHITE METAL LANTERN	17850	UK	6	3.39
536365	84406B	12.1.2010 8:26	CREAM CUPID HEARTS COAT HANGER	17850	UK	8	2.75
536365	84029G	12.1.2010 8:26	KNITTED UNION FLAG HOT WATER BOTTLE	17850	UK	6	3.39
536365	84029E	12.1.2010 8:26	RED WOOLLY HOTTIE WHITE HEART	17850	UK	6	3.39
536365	22752	12.1.2010 8:26	SET 7 BABUSHKA NESTING BOXES	17850	UK	2	7.65
536365	21730	12.1.2010 8:26	GLASS STAR FROSTED T-LIGHT HOLDER	17850	UK	6	4.25
536366	22633	12.1.2010 8:28	HAND WARMER UNION JACK	17850	UK	6	1.85
536366	22632	12.1.2010 8:28	HAND WARMER RED POLKA DOT	17850	UK	6	1.85
536367	84879	12.1.2010 8:34	ASSORTED COLOUR BIRD ORNAMENT	13047	UK	32	1.69
536367	22745	12.1.2010 8:34	POPPY'S PLAYHOUSE BEDROOM	13047	UK	6	2.1

(continued)

Table 1 (continued)

Invoice No.	Stock code	Invoice date	Description	Customer ID	Country	Quantity	Unit price
536367	22748	12.1.2010 8:34	POPPY'S PLAYHOUSE KITCHEN	13047	UK	6	2.1
536367	22749	12.1.2010 8:34	FELTCRAFT PRINCESS CHARLOTTE DOLL	13047	UK	8	3.75
536370	21731	12.1.2010 8:45	RED TOADSTOOL LED NIGHT LIGHT	12583	France	24	1.65
536370	22900	12.1.2010 8:45	SET 2 TEA TOWELS I LOVE LONDON	12583	France	24	2.95
536370	21913	12.1.2010 8:45	VINTAGE SEASIDE JIGSAW PUZZLES	12583	France	12	3.75
536370	22540	12.1.2010 8:45	MINI JIGSAW CIRCUS PARADE	12583	France	24	0.42
536370	22544	12.1.2010 8:45	MINI JIGSAW SPACEBOY	12583	France	24	0.42
536370	22492	12.1.2010 8:45	MINI PAINT SET VINTAGE	12583	France	36	0.65

Table 2 Summary of negative association rule mining using different techniques

Experiment (clustering)	Number of rules	Exec. time	Support (%)			Confidence (%)		
			Min	Max	Rate	Min	Max	Rate
A (none)	9.0655	1932.74	0.10	0.13	0.122	66.67	100	99.365
B (K-means)	1.7801	353.57	0.10	0.44	0.189	31.69	100	85.835
C (DBSCAN)	40,862	783.30	0.10	0.40	0.209	22.54	100	91.840

histogram depending on predefined values, without the need to provide the number of clusters for each histogram of the purchased items. The summary of the negative association rule mining extracted depends on the clusters generated by a DBSCAN method as shown in Table 2.

Discussion: The results of Experiment A illustrate negative association rule mining without any region knowledge. Such negative association rule mining is considered to produce un-interesting rules, by comparing the results of this experiment to the other experiment conducted in this paper, as shown in Table 2. The average support of the negative association rules extracted without any region knowledge is less than the other rules extracted using region knowledge.

The domain knowledge, in an earlier study [8–17], is provided manually to the model, to utilize this knowledge for mining negative association rule from positive rules, to ensure the extraction of interesting rules. In this paper, the domain knowledge is extracted automatically using data clustering, which is an unsupervised machine learning technique. The use of such knowledge improves the overall strength of the extracted negative association rules, as shown in experiments B and C. Although the K-means clustering method is very popular and is used in many applications [18–23], it cannot automatically delimit the optimize several clusters to utilize for a certain database. This drawback increases the complexity of the algorithm required to cluster the database, where it is important to measure the distortion in the clusters by clustering the database into a range of the number of clusters and measure the distortion per every clustering operation. Based on this distortion, the elbow method is utilized to delimit the optimized several clusters and then cluster the database using the K-means method using that number clusters. The execution time of this method is less than the execution time required by the mining method based on DBSCAN clustering. But the value of mined rules is less than the value of rules extracted based on the DBSCAN clustering method. Thus, the average time consumed per each rule extraction based on the K-means clustering is 198.63 ms, while it is 191.70 ms per each negative association rule extracted based on the clusters of the DBSCAN clustering method. Moreover, the execution times measured are given in Table 2. Such behaviour is caused by the need multiple passes over the database being clustered using the K-means method. In the first pass, the database is clustered into the different values of clusters to select the optimized several clusters, utilizing the "elbow" method, while in a DBSCAN method, the optimized several clusters are set by the method as the objects of the database are being distributed into the clusters.

The DBSCAN clustering method, on the other hand, can automatically detect the optimized several clusters, based on predefined specifications of the required cluster. This reduces the complexity of clustering a database and is also noticed to produce an improved domain knowledge, compared to the use of K-means clustering method based on the number of clusters determined by the "elbow" method. The distribution of the percentage number of rules in the confidence level for the DBSCAN method also shows the high superiority of this method over the K-means method, where the bigger portion of extracted rules is in the higher levels of confidence. The support of the negative association rules extracted using the K-means method based on the number of clusters computed using the "elbow" method shows close average support of the takeout negative association rules, compared to the average support of the rule takeout based on the DBSCAN clustering results. But the average confidence of these rules is significantly lower than those extracted using the domain knowledge provided by the DBSCAN clustering method. This illustrated that the domain knowledge extracted using the DBSCAN clustering method is more valuable than that gained from the K-means clustering method, especially when the number of clusters is concluded automatically.

4 Conclusion

In this paper, a novel negative association rule technique is suggested, which relies on domain knowledge extracted automatically from the database, using the unsupervised data mining technique, clustering. Two clustering methods are tested in this paper, which are the K-means and DBSCAN clustering methods. The average support for the negative association rules extracted without domain knowledge is 0.122%, while it is 0.189 and 0.209% for the K-means and DBSCAN domain knowledge extraction. The average confidence, on the other hand, is 99.365% without the domain knowledge extraction and 85.835 and 91.840% when domain knowledge extracted using K-means and DBSCAN cluster methods, consequentially, is used.

References

1. Buczak, A. L., & Guven, E. (2016). A survey of data mining and machine learning methods for cyber security intrusion detection. *IEEE Communications Surveys and Tutorials, 18*(2), 1153–1176.
2. Holzinger, A., & Jurisica, I. (2014). Knowledge discovery and data mining in biomedical informatics: the future is in integrative, interactive machine learning solutions.
3. Michael, H., & Radoslaw, K. (2017). Visualizing association rules in hierarchical groups. *Journal of Business Economics, 87*(3), 317–335.
4. Brodić, D., & Amelio, A. (2018). Association rule mining for the usability of the CAPTCHA interfaces: A new study of multimedia systems. *Multimedia Systems*, (2).
5. Sai Krishna, T. V., Yesu Babu, A., & Kiran Kumar, R. (2018). Determination of optimal clusters for a non-hierarchical clustering paradigm means algorithm.
6. Rodriguez, A., & Laio, A. (2014). Clustering by fast search and find of density peaks. *Science, 344*(6191), 1492–1496.
7. Chen, D., Sain, S. L., & Guo, K. (2012). Data mining for the online retail industry: A case study of RFM model-based customer segmentation using data mining. *Journal of Database Marketing and Customer Strategy Management, 19*(3), 197–208.
8. Ramasubbareddy, B., Govardhan, A., & Ramamohanreddy, A. (2010). Mining positive and negative association rules. In *2010 5th International Conference on Computer Science and Education (ICCSE)* (pp. 1403–1406).
9. Cornelis, C., Yan, P., Zhang, X., & Chen, G. (2006). Mining positive and negative association rules from large databases. In *2006 IEEE Conference on Cybernetics and Intelligent Systems* (pp. 1–6).
10. Tsiligaridis, J. (2013). Mining positive and negative association rules. In *The International Conference on E-Technologies and Business on the Web (EBW2013)* (pp. 110–114).
11. Wu, X., Zhang, C., & Zhang, S. (2004). Efficient mining of both positive and negative association rules. *ACM Transactions on Information Systems (TOIS), 22*(3), 381–405.
12. Savasere, A., Omiecinski, E., & Navathe, S. (1998). Mining for strong negative associations in a large database of customer transactions. In *Proceedings of the 14th International Conference on Data Engineering 1998* (pp. 494–502).

13. Yuan, X., Buckles, B. P., Yuan, Z., & Zhang, J. (2002). Mining negative association rules. In *Computers and Communications, 2002. Proceedings. ISCC 2002. Seventh International Symposium on* (pp. 623–628).
14. Mahmood, S., Shahbaz, M., & Guergachi, A. (2014). Negative and positive association rules mining from text using frequent and infrequent itemsets. *The Scientific World Journal*, 2014
15. Martin, D., Rosete, A., Alcala-Fdez, J., & Herrera, F. (2014). A new multiobjective evolutionary algorithm for mining a reduced set of interesting positive and negative quantitative association rules. *IEEE Transactions on Evolutionary Computation, 18*(1), 54–69.
16. Rani, B. K., Srinivas, K., Reddy, B. R., & Govardhan, A. (2011). Mining negative association rules. *International Journal of Engineering and Technology, 3*(2), 100–105.
17. Chen, C., & Wang, D. (2016). Research on association rules mining base on positive and negative items of FP-tree.
18. Kodinariya, T. M., & Makwana, P. R. (2013). Review on determining number of cluster in K-means clustering. *International Journal, 1*(6), 90–95.
19. Bholowalia, P., & Kumar, A. (2014). EBK-means: A clustering technique based on elbow method and K-means in WSN. *International Journal of Computer Applications*, 105(9).
20. Cohen, M. B., Elder, S., Musco, C., Musco, C., & Persu, M. (2015). Dimensionality reduction for K-means clustering and low rank approximation. In *Proceedings of the Forty-Seventh Annual ACM Symposium on Theory of Computing* (pp. 163–172).
21. Celebi, M. E., Kingravi, H. A., & Vela, P. A. (2013). A comparative study of efficient initialization methods for the K-means clustering algorithm. *Expert Systems with Applications, 40*(1), 200–210.
22. Chévez, P., Barbero, D., Martini, I., & Discoli, C. (2017). Application of the K-means clustering method for the detection and analysis of areas of homogeneous residential electricity consumption at the Great La Plata region, Buenos Aires, Argentina. *Sustainable Cities and Society, 32,* 115–129.
23. Zhang, Y., Tangwongsan, K., & Tirthapura, S. (2017). Streaming algorithms for K-means clustering with fast queries. arXiv preprint: arXiv:1701.03826.

Integrated Network of Pearson Correlation Coefficients, Kendall Tau, Spearman Rho and Its Impact on Disease, Health Indicators and Mortality Ratings in Babil Province

Asmaa Shaker Ashoor and Ali Abdul Karim Kazem Naji

Abstract The present study explains the methods of parametric and nonparametric correlation coefficients represented by Pearson correlation coefficient, Kendall correlation coefficient Tau, Spearman correlation coefficient Rho with moral correlation coefficient statistical tests applied to some data and health indicators for births by age groups under the age of six years. In this study, we examine the methods of correlation coefficients, which are the parameters of the correlation coefficients and their application to some data, health indicators for births and deaths according to the classification of age groups below age Six years and some classifications of diseases in the province of Babylon, has been inferred from the results of the search on the existence of links at different levels reflected the evolution in the level of health service provided in the province of Babylon, which requires work to do more to raise the level of health in the province.

Keywords Pearson correlation · Kendall tau · Spearman rho · Classification · Statistical analysis

1 Introduction

In scientific life, we face situations involving two variables and more. Someone has to see if there is a relationship between these variables, what forms this relationship and also how to predict one of these two variables. If we learn one of these two variables in the case of our knowledge of the other variable since the focus on data description would be a more accurate description of phenomena and it will be realistic if we could see the strength and form of this relationship and therefore can be used to estimate the values of certain phenomena (variables) to learn the values of other phenomena

A. S. Ashoor (✉) · A. A. K. K. Naji
College Basic of Education, University of Babylon, Hillah, Iraq
e-mail: asmaashaker218@gmail.com

A. A. K. K. Naji
e-mail: ali.babylon1982@gmail.com

S.-L. Peng et al. (eds.), *Proceedings of First International Conference on Mathematical Modeling and Computational Science*, Advances in Intelligent Systems and Computing 1292, https://doi.org/10.1007/978-981-33-4389-4_8

where the relationship between variables may be strong or weak, direct or productive and to examine the relationship of power and direction is called a commit [1, 2].

2 Research Methodology

2.1 Aims of Research

This study aims to identify the concept of correlation and its types by addressing the scientific and non-scientific correlation through the study and application of correlation types (Pearson correlation coefficient, Spearman Rho correlation coefficient, Kendall Tau correlation coefficient) and try to apply some of the indicators in the field of health in the province of Babylon.

2.2 The Importance of Study

The importance of studying the correlation is because it is the basis for the study and analysis of logical relationships between variables because of their great importance to ensure a logical effect relationship with the possibility of conducting statistical tests to discover whether there were statistically significant differences in conceptual correlation coefficient between variables.

2.3 The Problem of Study

The problem of the study is to calculate the correlation between some health variables related to births and deaths and some diseases in Babylon governorate and the extent, strength, type and effect of the correlation between different bilateral variables.

3 Correlation

3.1 Correlation

It is a relationship between two variables representing each particular phenomenon so that if one changes in a particular direction (increase or decrease) the other changes in the same direction, so the correlation will be between positive correlation and a negative one. If the change occurred in the opposite direction and if the increase in the first variable offsets by a decrease in the second variable or vice versa, then the correlation between them is either a positive or a negative one [3, 4].

3.2 Types of Correlation

On the basis of the number of variables involved in the correlation calculation, the following types can be identified.

First: The binary correlation: It includes two types:

a. Simple correlation: It looks at the relationship between two variables only, without taking into account other variables that may be associated with them.
b. The partial correlation is the one that looks at the relationship between two variables among several variables assuming the stability of the other variables, for example, collecting intelligence correlation probably due to the correlation of the intelligence and achievement with a third variable like age or living conditions of the individual. To specify the relationship or correlation between intelligence and learning achievement, we resort to assuming the steady influence of age on each.

Second: Multi-correlation: It is the one who looks at the relationship between a variable on one side and a set of variables on the other and is called a multi-correlation first class.

3.3 Correlation Strength and Direction

It is worth mentioning that the correlation coefficient between two variables takes values between -1, $+1$ and the absence of a relationship or correlation between two variables, the value of the correlation coefficient calculated zero value or very close to zero, and the correlation coefficient is usually denoted by the symbol (r) and notes that whether the link between variable Yen, weak and strong takes one of the two directions, either positive or negative. A positive correlation means that the variation between the variables is one direction that one values, variables increase accelerate in another variable and shortages in one variable values accompanied by shortages of other variable values such as the correlation between educational achievement and intelligence if the correlation is negative as it is between investment and interest rates, the increase in the value of a variable is accompanied by a lack of other variable pairs.

4 Theoretical Background [5, 6]

The methods of calculating the correlation coefficient and its conditions and tests will be discussed as follows.

4.1 Pearson's Correlation Coefficient

One of the parameter tests uses the Pearson correlation coefficient to measure the linear correlation strength between two variables. The correlation is completely positive when the value $r = 1$ means that one of the two variables increases the other variable and may be completely inverse when the value $r = 1$ means that the increase of one of the variables leads to the reduction of the other variable and vice versa and may take values between it and the requirements of correlation coefficient Pearson to meet the following conditions:

- The measurement scale for the variables must be a parameter or ratio.
- The distribution of variables must be a natural distribution or close to it.
- The two variables must be linear.
- There should not be any outliers between the behaviors of the data.
- The Pearson correlation coefficient formula is:

$$r_{xy} = \frac{n\sum x_i y_i - \sum x_i \sum y_i}{\sqrt{n\sum x_i^2 - (\sum x_i)^2}\sqrt{n\sum y_i^2 - (\sum y_i)^2}}$$

For pairs of variables that have a two-dimensional normal distribution, the distribution of samples for a particular Pearson correlation function follows Student's t distribution with degrees of freedom in 2. Specifically, the basic variables have a two-variable medical distribution. The variable will be explained below:

$$t = r\sqrt{\frac{n-2}{1-r^2}}$$

It is a statistic test for a Pearson correlation coefficient in which the significance of the correlation coefficient and its statistical significance are examined.

4.2 Spearman Correlation Coefficient

It is used to find the correlation coefficient between two variables and qualitative data that cannot be measured depending on rank. Quantitative variables may be descriptive or either ordinal or both descriptive ordinal and the test of nonparametric tests that are not based on the parameters of the distribution. The formula for the correlation coefficient for Spearman is described in the following relationship.

$$r_s = 1 - \frac{6\sum d_i^2}{n\left(n^2 - 1\right)}$$

where (d) represents the difference between the two variables.

4.3 *Kendall's Tao Coefficient*

Kendall's rank correlation coefficient is a nonparametric test which measures the strength of dependency between two variables. If we take into account the variables x and y, where each sample size represents (n), we know that the total number of pairs with (x) is n (n-1)/2. The following formula is used to calculate the value of Kendall's rank correlation:

$$\tau = \frac{n_c - n_d}{0.5n(n-1)}$$

(number of discordant n_d and number of concordant where n_c).

5 Scientific Side

Statistical analysis was analyzed in Babil health data through the calculation of correlation types (nonparametric and parametric, Pearson correlation, Spearman correlation coefficient Rho, Kendall correlation coefficient Tau) between study variables described in Tables 1 and 2 with the comparison between results and drew out the best values by the statistical measurements. The correlation coefficients were tested according to the t-test to show the significance of the correlation coefficients and their statistical significance. The analysis includes two sides. The analysis of death data is classified by age group and type of disease.

- The number of deaths is less than one year for bacterial poisoning.
- 3–1-year deaths of pneumonia and the inability to breathe.
- The number of deaths 6–3 years for pneumonia and the inability to breathe.

The second side included analyzing the variables related to births and deaths by age group and included the following variables:

- The number of deaths in Babylon province.
- The number of deaths under the age of 6 years.
- The number of deaths under the age of 1 year.

Table 1 Classification of deaths by diseases and age groups

6–3 years	3–1 years	Less than a year	Year
Pneumonia and the inability to breathe	Pneumonia and the inability to breathe	Bacterial poisoning	
21	19	376	2010
27	36	383	2011
21	22	337	2012
29	139	286	2013
19	36	248	2014
28	39	201	2015
38	33	183	2016
39	36	291	2017

Table 2 Births and deaths (by age group) in Babil Province

No. of deaths 6–3	No. of deaths 3–1	No. of deaths under 1	No. of deaths under 6	No. of births	Year
100	311	2327	2738	60,517	2010
121	311	2287	2719	61,741	2011
188	441	1083	1712	61,887	2012
202	470	1139	1811	62,532	2013
173	383	1198	1754	66,544	2014
115	318	1203	1636	62,119	2015
106	298	1213	1617	57,774	2016
157	274	1208	1639	53,449	2017

- The number of deaths under 1 year.
- The number of deaths from 3–1 years.
- The number of deaths at the age of 6–3 years.

Data were collected from the Ministry of Health, Babil Province. After the classification of data by the researcher, the data are presented in Table 2.

5.1 Analysis of Death Data by Age Group and Type of Disease

The analysis was performed by calculating the following correlation coefficients (Pearson, Spearman and Kendall Tau correlation). The results are presented in Table 3.

Table 3 Pearson correlation coefficient and its statistical tests

		Year	Less than one year Bacterial poisoning	3–1 pneumonia and the inability to breathe	3–1 pneumonia and the inability to breathe
Year	Pearson correlation	1	-0.794-*	0.039	0.739*
	Sig. (2-tailed)		0.019	0.927	0.036
Less than one year Bacterial poisoning	Pearson correlation	-0.794-*	1	-0.115-	-0.431-
	Sig. (2-tailed)	0.019		0.785	0.287
3–1 pneumonia and the inability to breathe	Pearson correlation	0.039	-0.115-	1	0.150
	Sig. (2-tailed)	0.927	0.785		0.722
6–3 pneumonia and the inability to breathe	Pearson correlation	0.739*	-0.431-	0.150	1
	Sig. (2-tailed)	0.036	0.287	0.722	

There is an inverse correlation between the variable and the variable bacterial poisoning disease category less than a year, which reached (-0.794-*) and conceptual correlation to that test values are conceptual. 019 amounted to less than 0.05, the calculated correlation points to a lower number of deaths for such evidence of improved health care over the years, the relationship between pneumonia and the inability to breathe for the age group of 6–3 years with years variable was (.*739) and high value close to 1 and significant and also a reference to the statistical significance of a positive sign denoting a direct correlation relationship which means that the number of deaths to pneumonia, the inability to breathe for the age group of 6–3 years from its relationship with time and no significant relationship between the time variable, lung disease and respiratory deficit for the age group of 3–1 years, and the positive, low and nonsignificant correlation relationship of (- 0.431) between the variable of number of deaths for bacterial poisoning disease (less than one year) Lung and impotence for the 6–3 age group, the remaining relationships were insignificant and the correlation coefficient of Pearson decreased indicating insignificance and no effect on each other variable which is a reference to care decrease for this age group.

The results of the Kendall Tau correlation coefficient confirmed the results of the Pearson molecular test with a slight difference, where there was an inverse relationship between the variable of years and the variable bacterial disease of the class less than one year. The correlation was significant because the test values were 0.26 and 0.05, from which the calculated correlation refers to the low death rate for this group, which is evidence of the improvement of health care provided over the years.

The relationship between pneumonia and the inability to breathe for the age group of 3–6 years with (years) variable (0.546) average value and Pearson correlation coefficient less than the correlation and it is not statistically significant effect and is a positive sign denoting a direct correlation relationship means that deaths number of pneumonia and the inability to breathe for the age group of 3–6 years from its relationship with a time variable and less coefficient of Pearson and is referring to the decrease of care for this age group. The lack of correlation relationship between time variable, pneumonia variable and the inability to breathe for 1–3 years, and inverse, low and nonmoral correlation was (-0.265) between a variable of deaths of bacterial poisoning disease (less than a year) and a variable of number of pneumonia deaths and the inability to breathe for the age group 3–6 years, and the remaining of relationships is nonmoral. The calculated decreased value of the correlation coefficient of Kendall Tau indicates its immateriality in effect (Table 4).

Spearman correlation coefficient results confirmed Rho and Kendall Tau that parametric Pearson correlation coefficient results so that there is an inverse correlation relationship between the year variable and the variable of bacterial poisoning disease category less than a year (0.738*) and moral correlation to that test values (0.037). It is moral because it is less than 0.05. The calculated correlation indicates low numbers of deaths in this category which is an evidence of healthcare improvement appointed within the passage of years. The relationship between pneumonia and the inability to breathe for the age group of 3–6 years with the year variable (0.671) is a high value on the center and also a moral reference to statistical significance and is of a positive

Table 4 Kendall Tau correlation coefficient and its statistical tests

		Year	Less than one year Bacterial poisoning	3–1 years Pneumonia and inability to breathe	6–3 years Pneumonia and inability to breathe
Kendall's tau_b Year	Correlation coefficient	1.000	-0.643-*	0.265	0.546
	Sig. (2-tailed)		0.026	0.373	0.061
Less than one year Bacterial poisoning	Correlation coefficient	-0.643-*	1.000	-0.265-	-0.182-
	Sig. (2-tailed)	0.026		0.373	0.533
1–3 years Pneumonia and inability to breathe	Correlation coefficient	0.265	-0.265-	1.000	0.308
	Sig. (2-tailed)	0.373	0.373		0.305
6–3 Years Pneumonia and inability to breathe	Correlation coefficient	0.546	-0.182-	0.308	1.000
	Sig. (2-tailed)	0.061	0.533	0.305	

Table 5 Spearman Rho correlation coefficients and their statistical tests

		Year	-Less than a year	3–1 years Pneumonia variable and the inability to breathe	6–3 years Pneumonia variable and the inability to breathe
Spearman's Rho Year	Correlation coefficient	1.000	−0.738-*	0.366	0.671
	Sig. (2-tailed)		0.037	0.373	0.069
Less than a year Bacterial poisoning	Correlation coefficient	−0.738-*	1.000	−0.342-	−0.323-
	Sig. (2-tailed)	0.037		0.408	0.435
3–1 years Pneumonia variable and the inability to breathe	Correlation coefficient	0.366	−0.342-	1.000	0.356
	Sig. (2-tailed)	0.373	0.408		0.387
6–3 years Pneumonia variable and the inability to breathe	Correlation coefficient	0.671	-0.323-	0.356	1.000
	Sig. (2-tailed)	0.069	0.435	0.387	

signal which indicates the presence of the correlation function. It means that the number of deaths of pneumonia and deficit breathing to the age group of 6–3 years from its relationship with time variable is a reference to caring decrease for this age group. The lack of moral relationship between a time variable, pneumonia variable and the inability to breathe for 1–3 years is a correlation relationship and low morale (−0.323) between a death variable of bacterial poisoning disease (less than a year) and several death variables, the number of deaths to pneumonia and the inability to breathe for the 3–6 years which indicates a low, inverse value and insignificant relations. The rest of the relationship is insignificant with a significant decrease in the correlation coefficient value of Spearman Rho which indicates its insignificance (Table 5).

5.2 *Analysis of Death and Birth Data by Age Group*

The analysis was performed by calculating the following correlation coefficients (Pearson, Spearman and Kendall Tau correlation). The results are presented in Table 6.

Table 6 Pearson correlation coefficient and its statistical tests

		Year	Number of birth	No. of deaths under 6	No. of deaths under 1	No. of deaths under 3–1	No. of deaths under 6–3
Year	Pearson correlation	1	-0.491-	−0.814-*	−0.708-*	−0.316-	0.056
	Sig. (2-tailed)		0.217	0.014	0.049	0.446	0.896
Number of birth	Pearson correlation	−0.491-	1	0.135	0.023	0.592	0.269
	Sig. (2-tailed)	0.217		0.749	0.957	0.122	0.520
Number of deaths under 6	Pearson correlation	−0.814-*	0.135	1	0.981**	−0.224-	−0.438-
	Sig. (2-tailed)	0.014	0.749		0.000	0.594	0.277
Number of deaths 1	Pearson correlation	−0.708-*	0.023	0.981**	1	-0.404-	-0.590-
	Sig. (2-tailed)	0.049	0.957	0.000		0.321	0.124
Number of deaths under 1–3	Pearson correlation	−0.316-	0.592	−0.224-	−0.404-	1	0.821*
	Sig. (2-tailed)	0.446	0.122	0.594	0.321		0.012
Number of deaths under 3–6	Pearson correlation	0.056	0.269	−0.438-	−0.590-	0.821*	1
	Sig. (2-tailed)	0.896	0.520	0.277	0.124	0.012	

Parametric Pearson correlation coefficient results confirmed the presence of correlation relationship with most variables that are variables of several births, the number of deaths under 6 years, the number of deaths under 1 year, number of deaths (3–1) and its relation with a time variable. The biggest inverse relationship with a large number of deaths under 6 years (−0.814-*.). This decrease is significant, effective, statistically significant and indicates the improvement of health status in Babil for the decrease of overall deaths for the age group of fewer than 6 years. The existence of a relationship between time variable and the number of deaths of 3–6 years because the correlation value is near zero (0.056) and insignificant value, whereas a variable of several births, the largest total correlation (0.592), with a positive reference with a variable of deaths number of 3–1 followed by a variable of several deaths of 6–3 And low values for other variables.

We infer that the increase in the number of births is accompanied by a positive change in the number of deaths with the age category of 3–1 which requires interest to the ages of children after birth and providing health care after birth to achieve a reduction in numbers of deaths for this age. Several death variables under one year are associated with a strong correlation with death variable less than 6 years in a positive and significant form with a correlation coefficient of (0.981**). It is close to positive numbers (1) from which we show that the main factor that increases the number of deaths less than 6 years is the number of deaths in the age group 3–1 years, and the number of deaths less than 6 years is correlated to the inverse relationship with the number of deaths in the age group 6–3 years (438). The decrease in total deaths less than 6 years is due to a decrease in deaths in the 6–3 age group, close to the center. The variable number of deaths less than 1 year is associated with an inverse correlation between the variable of 3–1 years and the number of death variable of 6–3 where the correlation coefficients are (−0.404-) and (−0.590-), respectively. The test shows statistical insignificance. The number of deaths less than 3–1 years is correlated to a positive and strong relationship with the variable of death rate for the age group of 6–3 years and with a correlation of (0.821*). This is a strong and positive correlation because it is close to a number (1) which indicates that the rising of deaths in the (3–1) years age group leads to an increase in deaths in the age group of (6–3) which requires the government and the health sector to take care of the provision of treatment to achieve the reduction of deaths of these age groups to develop the health reality in the province. Kendall's Tau correlation coefficient is calculated as in Table 7 where the previous results of Pearson prove the results of a test, Spearman Rho correlation coefficient is also calculated in Table 8 as the results also confirmed the findings of the previous analysis with a light change in values of correlation coefficients.

6 Conclusions and Recommendations

1. There is an inverse correlation relationship between the years variable and the variable of bacterial poisoning disease category less than a year (−0.794-*), and the calculated correlation points to a decreasing number of deaths for such evidence of improving health care over the years.
2. The relationship between pneumonia and respiratory deficit to 6–3-year correlation is a positive relationship which means that the number of deaths of pneumonia and respiratory deficit to 6–3 years age group is increasing its relationship with time and variable is a reference to reduced care for this age group.
3. There is no correlation relationship between a time variable and variable pneumonia and the respiratory deficit for 3–1 years, low and positive correlation between the number of deaths of bacterial poisoning disease (less than one year) and the variable of deaths of pneumonia and respiratory deficit for the 6–3 age group. The residual relationship was insignificant, and the correlation coefficient of Pearson, Kendall Tau and Spearman Rho decreased, indicating that it is not important and does not affect one another.
4. Parametric Pearson correlation coefficient results in Kendall Tau and Spearman Rho on direct correlation relationship with most of the variables of the number

Table 7 Kendall Tau correlation coefficient and its statistical tests

Correlations

		Year	No. of birth	No. of deaths under 6	No. of deaths under 1	No. of deaths 1–3 year	No. of deaths 3–6 year
Kendall's tau_b Year	Correlation coefficient	1.000	−0.071-	−0.714-*	0.000	−0.327-	0.000
	Sig. (2-tailed)		0.805	0.013	1.000	0.262	1.000
No. of birth	Correlation coefficient	−0.071-	1.000	0.071	−0.357-	0.764**	0.357
	Sig. (2-tailed)	0.805		0.805	0.216	0.009	0.216
No. of deaths under 6	Correlation coefficient	−0.714-*	0.071	1.000	0.143	0.182	0.143
	Sig. (2-tailed)	0.013	0.805		0.621	0.533	0.621
No. of deaths under 1	Correlation coefficient	0.000	−0.357-	0.143	1.000	−0.546-	−0.714-*
	Sig. (2-tailed)	1.000	0.216	0.621		0.061	0.013
No. of deaths 1–3 year	Correlation coefficient	-0.327-	0.764**	0.182	−0.546-	1.000	0.546
	Sig. (2-tailed)	0.262	0.009	0.533	0.061		0.061
No. of deaths 3–6 year	Correlation coefficient	0.000	0.357	0.143	−0.714-*	0.546	1.000
	Sig. (2-tailed)	1.000	0.216	0.621	0.013	0.061	

of births, the number of deaths less than 6 years, the number of deaths less than 1 year, the number of deaths (3–1) from its relation with time variable and the biggest inverse relationship with a larger number of deaths less than 6 years, and this decrease indicates better health status in Babil to the decrease of overall deaths.

5. The variable of the number of births, the largest total correlation (0.592) with a positive signal with a variable number of deaths of 3–1 followed by a variable number of deaths of 6–3 and low values for other variables; we infer that the increase in the number of births accompanied by positive changes in deaths in the age group of 3–1 which requires care to the supply of continuing health care after birth to achieve a reduction in numbers of deaths for this age group.
6. The variable of the number of deaths less than one year is associated with a strong correlation with death variable under 6 years in a positive and significant relationship with a correlation coefficient (**0.981), and we infer that the main

Table 8 Spearman Rho correlation coefficients and statistical tests

		Year	No. of births	No. of deaths under 6	No. of deaths under 1	No. of deaths under 1–3	No. of deaths under 3–6
Spearman's Rho Year	Correlation coefficient	1.000	−0.238-	−0.857-**	−0.190-	−0.395-	0.048
	Sig. (2-tailed)		0.570	0.007	0.651	0.333	0.911
No. of births	Correlation coefficient	−0.238-	1.000	0.214	−0.619-	0.874**	0.548
	Sig. (2-tailed)	0.570		0.610	0.102	0.005	0.160
No. of deaths under 6	Correlation coefficient	−0.857-**	0.214	1.000	0.262	0.275	0.071
	Sig. (2-tailed)	0.007	0.610		0.531	0.509	0.867
No. of deaths under 1	Correlation coefficient	−0.190-	−0.619-	0.262	1.000	−0.755-*	−0.857-**
	Sig. (2-tailed)	0.651	0.102	0.531		0.031	0.007
No. of deaths under 1–3	Correlation coefficient	−0.395-	0.874**	0.275	−0.755-*	1.000	0.683
	Sig. (2-tailed)	0.333	0.005	0.509	0.031		0.062
No. of deaths under 3–6	Correlation coefficient	0.048	0.548	0.071	-0.857-**	0.683	1.000
	Sig. (2-tailed)	0.911	0.160	0.867	0.007	0.062	

factor that increases the number of deaths under 6 years is the number of deaths in the age group (3–1) years.

7. The decrease in total deaths under 6 years is due to a decrease in deaths in the 6–3 age group which is close to the center.
8. The number of deaths under 3–1 years associated with a positive and strong relationship with a variable of a number of deaths for age group 6–3 years and the total correlation was (*0.821) positive and strong because it was close to number 1 that indicates the rising number of deaths for age group 3–1 leads to death rising in the age group of 6–3 which requires the government and health sector care to provide treatment to achieve a reduction in deaths of these age groups to develop the health situation in the province.
9. The need to work in subsequent studies to examine the correlated and partial relations of different diseases in Babil Province and other disease categories.

References

1. Sami, A. A., & Mohammed, Y. H. (2011). Methodology of Scientific Research Concept, Methods, Analysis and Writing. University of Baghdad.
2. Farid, K., Abu Zayna & others, (2007). Methods of Scientific Research Statistics in the Scientific Research. Amman, Dar Al-Masirah for Publishing, Distribution and Printing, 1, (2007).
3. Wajih M., Jassim & others, (1998). Methods of scientific research in the field of physical education (p. 89). Press of the Ministry of Higher Education, Baghdad.
4. Tamimi, R., Fadel, (2014). Principles of applied statistics (for administrative, financial, accounting, economic and statistical). Office of Al-Hadia for Statistical Analysis and Information Technology, Baghdad.
5. Al-Rawi, K. M., (1987). Introduction to regression analysis. Directorate of Dar al-Kutub for Printing and Publishing, University of Mosul, 1987.
6. Al-Mashhadani, M. H., Dulaimi, Muhammad M. A., (1985). Methods of Statistics (Records and Time Series). Baghdad University Press

The Estimate Function of Schrödinger Operator

Yahea Hashem Saleem and Hadeel Ali Hassen Shubber

Abstract In this paper, as a basic step in arriving at a self-adjoint operator proof, we must find an estimate function and a continuation of our previous work in proof of convergence and smoothness of a function $\Psi(t, x)$. We structure the function $\xi(t, x)$, and we show the function $\xi(t, x)$ in $L^2(\mathbb{R}^n, dx)$ and get the relevance among this function and the function $\Psi(t, x)$ by using the characteristics of the first function. Generally previous function for any smooth potential $W(x)$ and we have the same results with the function $\xi(t, x)$. The advantages of $\xi(t, x)$ are investigated. Finally, we evidence that the function $\xi^{(W)}(t, x)$ converges to $\xi(t, x)$ in $L^2(\mathbb{R}^n, dx, dV)$ as $r \to \infty$.

Keywords Schrödinger operator · Electric potential · Magnetic potential · Infinitely differentiable

1 Introduction

The equation of the self-adjoint operator is major in the quantum machine (the Dirac–von Neumann formulation of quantum mechanics, physical observables such as position, momentum, and angular momentum).

Ismagilov [1, 2] showed the Schrödinger operator (denoted by Sch.O.) is essentially self-adjoint (denoted by E.S.A.). After this study, Simon [6] investigated the Sch.O. $-\Delta + q$ on $L^2(\mathbb{R}^m)$, where q is an operator of multiplication by a real-valued measurable function, $q(x)$, on $\mathbb{R}^n$. Many studies interested Essentiality self -adjoint operator refer to [3–5, 7, 9, 10].

Y. H. Saleem (✉) · H. A. H. Shubber
University of Basrah, Basrah, Iraq
e-mail: yahea_h@mail.ru

H. A. H. Shubber
e-mail: hadeelali2007@yahoo.com

S.-L. Peng et al. (eds.), *Proceedings of First International Conference on Mathematical Modeling and Computational Science*, Advances in Intelligent Systems and Computing 1292, https://doi.org/10.1007/978-981-33-4389-4_9

In this article, we define function $\xi(t, x)$ as

$$\xi(t,x) = \frac{\varphi(x)}{\sqrt{t}\Gamma\left(\frac{1}{2}\right)} + e^{ts}(2\pi)^{-n} \int_{-\infty}^{\infty} e^{it\sigma} \frac{\Psi\left(\frac{1}{s+i\sigma}, x\right)\varphi(x) - \varphi(x)}{\sqrt{s+i\sigma}} d\sigma,$$

we study its properties for this function and investigate the relationship with the function $\Psi(t, x)$ of the variable t which is defined as [7]:

$$\Psi(t,x) = \int d\mu_x^t(w)\left\{\exp\left[-i\int_0^t b(\omega(s))d\omega - \frac{1}{2}i\int_0^t \operatorname{div} b(\omega(s))ds - \int_0^t V(\omega(s))ds\right]\right\}\varphi(\omega(t)) \tag{1}$$

for the electromagnetic Schrödinger operator

$$H = \sum_{j=1}^{n} \frac{1}{2}\left(i\partial_j + b_j(x)\right)^2 + V(x),$$

in $L^2(\mathbb{R}^n)$ where $b_j(x)$, $j = 1, 2, \ldots n$ and $V(x)$ are real-valued functions on $\mathbb{R}^n$, $V \in L^1_{loc}(\mathbb{R}^n)$, $b \in C^2(\mathbb{R}^n)$, $\partial_j = \frac{\partial}{\partial x_j}$, and $i = \sqrt{-1}$.

2 The Main Results

We investigate the function $\xi(t, x)$ and their relationship with $\Psi(t, x)$.

Proposition 2.1 *Let $V \in L^2(\mathbb{R}^n)$, $\varphi \in C_0^\infty$. Hence, $\xi(t, x)$ is found and has the characteristics*

i. $\xi(t,x) \in L^2\left(\mathbb{R}^n, dx\right)\forall t, s > 0$

ii. $\int_0^\infty e^{-ts}\xi(t,x)_{L^2(\mathbb{R}^n,dx)}dt < +\infty$ *for all* $t, s > 0$

iii. $\xi(t, x)$ *is finite* ($\xi(t, x) = 0$, *if* $d(x, \operatorname{supp}\varphi) > 2\sqrt{t}$)

iv. $\frac{1}{\sqrt{s}}\Psi\left(\frac{1}{s}, x\right) = \int_0^\infty e^{-ts}\xi(t,x)dt$

v. $\xi(t,x) = \frac{\varphi(x)}{\sqrt{t}\Gamma(\frac{1}{2})} + e^{ts}(2\pi)^{-n} \int\limits_{-\infty}^{\infty} e^{it\sigma} \frac{\Psi(\frac{1}{s+i\sigma},x)\varphi(x)-\varphi(x)}{\sqrt{s+i\sigma}} d\sigma$, *$\Gamma(x)$ is Γ-function,* $\mathrm{Re}\sqrt{s+i\sigma} > 0$.

Proof From (1.1), we have

$$\Psi_\nu(t,x) = \frac{(-1)^\nu}{\nu!} \sum_{j=0}^{\nu} \binom{\nu}{j} (-i)^j b(\omega(s))^j_{L^2[0,t]} \int\limits_0^t ds_1 \int\limits_{s_1}^t ds_2 \dots \int\limits_{s_{\nu-j-1}}^t ds_{\nu-j} \int\limits_{\mathbb{R}^1} \dots$$

$$\int\limits_{\mathbb{R}^1} p(x,x_1,s_1)\dots p\big(x_{\nu-j-1},x_{\nu-j},s_{\nu-j}-s_{\nu-j-1}\big)p\big(x_{\nu-j},y,t-s_{\nu-j}\big)$$

$$\left(-\frac{i}{2}\mathrm{div}\, b(x_1) - V(x_1)\right)\left(-\frac{i}{2}\mathrm{div}\, b(x_2) - V(x_2)\right)\dots$$

$$\left(-\frac{i}{2}\mathrm{div}\, b\big(x_{\nu-j}\big) - V\big(x_{\nu-j}\big)\right)\varphi(y)dx_1dx_2\dots dx_{\nu-j}dy$$

We have

$$\Psi(t,x) = \sum_{\nu=0}^{\infty} \Psi_\nu(t,x)$$

Now we can write the function

$$\Psi_\nu(t,x) = \frac{(-1)^\nu}{\nu!} \sum_{j=0}^{\nu} \binom{\nu}{j} (-i)^j b(\omega(s))^j_{L^2[0,t]} \int\limits_0^t ds_1 \int\limits_{s_1}^t ds_2 \dots \int\limits_{s_{\nu-j-1}}^t ds_{\nu-j} \int\limits_{\mathbb{R}^1}$$

$$\dots \int\limits_{\mathbb{R}^1} p(x,x_1,s_1)p\big(x_{\nu-j-1},x_{\nu-j},s_{\nu-j}-s_{\nu-j-1}\big)p\big(x_{\nu-j},y,t-s_{\nu-j}\big)$$

$$\left(-\frac{i}{2}\mathrm{div} b(x_1) - V(x_1)\right)\left(-\frac{i}{2}\mathrm{div} b(x_2) - V(x_2)\right)\dots$$

$$\left(-\frac{i}{2}\mathrm{div} b\big(x_{\nu-j}\big) - V\big(x_{\nu-j}\big)\right)$$

$$\varphi(y)dx_1dx_2\dots dx_{\nu-j}dy \tag{2}$$

where $p(x,y,t) = (2\pi t)^{-\frac{1}{2}} \exp\left(\frac{-(x-y)^2}{2t}\right)$. By Fourier transformation

$$\widehat{\Psi_\nu}(t,\lambda) = \frac{(-1)^\nu}{\nu!} \sum_{j=0}^{\nu} \binom{\nu}{j} (-i)^j t^{\nu-j} b(\omega(s))^j_{L^2[0,t]}$$

$$\times \int\limits_{\sigma_1+\sigma_2+cldots+\sigma_{\nu-j}<1} \ldots \int d\sigma_1 \ldots d\sigma_{\nu-j} \int\limits_{\mathbb{R}^1}$$
$$\times \int\limits_{\mathbb{R}^1} d\lambda_1 d\lambda_2 \ldots d\lambda_{\nu-j} e^{-t\sigma_1\lambda^2} e^{-t\sigma_2\lambda_1^2} \ldots e^{-t(1-\sigma_1-\sigma_2-\ldots-\sigma_{\nu-j})\lambda_{\nu-j}^2}$$
$$\times \Big(-\frac{i}{2}\widehat{\operatorname{div}} b(\lambda-\lambda_1) - \hat{V}(\lambda-\lambda_1)\Big)\Big(-\frac{i}{2}\widehat{\operatorname{div}} b(\lambda_1-\lambda_2) - \hat{V}(\lambda_1-\lambda_2)\Big)\ldots$$
$$\times \Big(-\frac{i}{2}\widehat{\operatorname{div}} b(\lambda_{\nu-j-1}-\lambda_{\nu-j}) - \hat{V}(\lambda_{\nu-j-1}-\lambda_{\nu-j})\Big)\hat{\varphi}(\lambda_{\nu-j}).$$

We put $t = \frac{1}{s}$ and use the Laplace transformation; then, we have

$$\frac{1}{\sqrt{s}}\widehat{\Psi_\nu}\Big(\frac{1}{s},\lambda\Big) = \frac{(-1)^\nu}{\nu!}\sum_{j=0}^{\nu}\binom{\nu}{j}(-i)^j \frac{1}{s^{\frac{\nu-j}{2}}} b(\omega(s))^j_{L^2\left[0,\frac{1}{s}\right]}$$
$$\int\limits_{\sigma_1+\sigma_2+\ldots+\sigma_{\nu-j}<1} \ldots \int d\sigma_1 \ldots d\sigma_{\nu-j} \int_0^\infty e^{-s\tau} d\tau \int\limits_{\mathbb{R}} \ldots \int\limits_{\mathbb{R}^1} d\lambda_1 d\lambda_2 \ldots d\lambda_{\nu-j}$$
$$\left(\frac{\cos(2\sqrt{\sigma_1\tau_1}\lambda}{\sqrt{\pi\tau_1}}\right) * \left(\frac{\cos(2\sqrt{\sigma_2\tau_2}\lambda_1}{\sqrt{\pi\tau_2}}\right) * \cdots * \left(\frac{\cos(2\sqrt{\sigma_{\nu-j}\tau_{\nu-j}}\lambda_{\nu-j-1}}{\sqrt{\pi\tau_{\nu-j}}}\right) *$$
$$\left(\frac{\cos(2\sqrt{(1-\sigma_1-\sigma_2-\cdots-\sigma_{\nu-j})\tau_{\nu-j+1}}\lambda_{\nu-j}}{\sqrt{\pi\tau_{\nu-j+1}}}\right)\Big(-\frac{i}{2}\widehat{\operatorname{div}}\, b(\lambda-\lambda_1) - \widehat{V}(\lambda-\lambda_1)\Big)$$
$$\Big(-\frac{i}{2}\widehat{\operatorname{div}}\, b(\lambda_1-\lambda_2) - \widehat{V}(\lambda_1-\lambda_2)\Big)\ldots\Big(-\frac{i}{2}\widehat{\operatorname{div}}\, b(\lambda_{\nu-j-1}-\lambda_{\nu-j})$$
$$-\widehat{V}(\lambda_{\nu-j-1}-\lambda_{\nu-j})\Big)\widehat{\varphi}(\lambda_{\nu-j})$$

where $f * g$ is convolution. If $\Delta_q f(x) = 2^{-1}(f(x+q) + f(x-q))$, thus

$$\mathcal{F}\Delta_q\mathcal{F}^{-1} f(x) = \mathcal{F}\Big(\mathcal{F}^{-1}\Big(\frac{1}{2}f(x+q)\Big) + \mathcal{F}^{-1}\Big(\frac{1}{2}f(x-q)\Big)\Big) = \Delta_q f(x)$$

Therefore,

$$\frac{1}{\sqrt{s}}\Psi_\nu\Big(\frac{1}{s},x\Big) = \frac{(-1)^\nu}{\nu!}\sum_{j=0}^{\nu}\binom{\nu}{j}(-i)^j \frac{1}{s^{\frac{\nu-j}{2}}\pi^{\frac{\nu-j+1}{2}}} b(\omega(s))^j_{L^2[0,\frac{1}{s}]}$$
$$\int\limits_{\sigma_1+\sigma_2+\ldots+\sigma_{\nu-j}\le 1} \ldots \int d\sigma_1 \ldots d\sigma_{\nu-j} \int_0^\infty e^{-s\tau} d\tau \int\limits_{\tau_1+\tau_2+\ldots+\tau_{\nu-j}\le\tau} \ldots$$
$$\int d\tau_1 \ldots d\tau_{\nu-j} \frac{1}{\sqrt{\tau_1}}\cdot\frac{1}{\sqrt{\tau_2}} \cdots \frac{1}{\sqrt{\tau-\tau_1-\tau_2-\ldots-\tau_{\nu-j}}} \Delta_{2\sqrt{\sigma_1\tau_1}}$$

$$\left(-\frac{i}{2}\text{div}\, b(\tau_1) - V(\tau_1)\right) \ldots \Delta_{2\sqrt{\sigma_{\nu-j}\tau_{\nu-j}}}$$
$$\left(-\frac{i}{2}\text{div}\, b\big(\tau_{\nu-j}\big) - V\big(\tau_{\nu-j}\big)\right)\Delta_{2\sqrt{(1-\sigma_1-\sigma_2-\ldots-\sigma_{\nu-j})(\tau-\tau_1-\tau_2-\ldots-\tau_{\nu-j})}}\varphi(x).$$

We note that by Cauchy inequality we have

$$\left(\sum_{k=1}^{\nu-j} 2\sqrt{\sigma_k\tau_k}+2\sqrt{(1-\sigma_1-\sigma_2-\cdots-\sigma_{\nu-j})}\sqrt{(\tau-\tau_1-\tau_2-\cdots-\tau_{\nu-j})}\right)^2$$
$$\leq \sum_{k=1}^{\nu-j} 4\sigma_k\tau_k + 4\big(1-(\sigma_1+\sigma_2+\cdots+\sigma_{\nu-j})\big)\big(\tau-(\tau_1+\tau_2+\cdots+\tau_{\nu-j})\big) \leq 4\tau.$$

We get

$$\sum_{k=1}^{\nu-j} 2\sqrt{\sigma_k\tau_k} + 2\sqrt{(1-\sigma_1-\sigma_2-\cdots-\sigma_{\nu-j})}\sqrt{(\tau-\tau_1-\tau_2-\cdots-\tau_{\nu-j})} \leq 2\sqrt{\tau}.$$

So,

$$\Delta_{2\sqrt{\sigma_1\tau_1}}\left(-\frac{i}{2}\widehat{\text{div}\, b}(\tau_1) - \widehat{V}(\tau_1)\right) \ldots \Delta_{2\sqrt{\sigma_{\nu-j}\tau_{\nu-j}}}\left(-\frac{i}{2}\widehat{\text{div}\, b}\big(\tau_{\nu-j}\big) - \widehat{V}\big(\tau_{\nu-j}\big)\right)$$
$$\Delta_{2\sqrt{(1-\sigma_1-\sigma_2-\cdots-\sigma_{\nu-j})(\tau-\tau_1-\tau_2-\cdots-\tau_{\nu-j})}}\varphi(x) = 0$$

If $\mathrm{d}(x, \mathrm{supp}\varphi) > 2\sqrt{\tau}$.
Thus,

$$\sum_{\nu=0}^{\infty} \frac{1}{\sqrt{s}}\Psi_\nu\left(\frac{1}{s}, x\right) = \sum_{\nu=0}^{\infty} \frac{(-1)^\nu}{\nu!} \sum_{j=0}^{\nu} \binom{\nu}{j} (-i)^j \frac{1}{s^{\frac{\nu-j}{2}}\pi^{\frac{\nu-j+1}{2}}} b(\omega(s))^j_{L^2[0,\frac{1}{s}]}$$
$$\int\limits_{\sigma_1+\sigma_2+\ldots+\sigma_{\nu-j}\leq 1} \ldots \int \mathrm{d}\sigma_1 \ldots \int \mathrm{d}\sigma_{\nu-j} \int_0^\infty \mathrm{e}^{-s\tau}\mathrm{d}\tau$$
$$\int\limits_{\tau_1+\tau_2+\ldots+\tau_{\nu-j}\leq\tau} \ldots \int \mathrm{d}\tau_1 \ldots \mathrm{d}\tau_{\nu-j} \frac{1}{\sqrt{\tau_1}} \cdot \frac{1}{\sqrt{\tau_2}}$$
$$\ldots \frac{1}{\sqrt{\tau-\tau_1-\tau_2-\cdots-\tau_{\nu-j}}}$$
$$\Delta_{2\sqrt{\sigma_1\tau_1}}\left(-\frac{i}{2}\widehat{\text{div}\, b}(\tau_1) - V(\tau_1)\right) \ldots \Delta_{2\sqrt{\sigma_{\nu-j}\tau_{\nu-j}}} \Delta_{2\sqrt{(1-\sigma_1-\sigma_2-\ldots-\sigma_{\nu-j})(\tau-\tau_1-\tau_2-\ldots-\tau_{\nu-j})}}\varphi(x) \tag{3}$$

Thus, we have

$$\frac{1}{\sqrt{s}}\Psi\left(\frac{1}{s},x\right)\varphi(x)=\int_{0}^{\infty}\mathrm{e}^{-s\tau}\xi(\tau,x)\mathrm{d}\tau$$

By using Proposition (1.1) in [7], the function $\xi(t,x)$ satisfied the properties (i) and (ii).

Moreover, we can compose (iii) by replacing s by $s+i\sigma$ in Proposition (2.1) in [7]. Hence, we have

$$\frac{1}{\sqrt{s+i\sigma}}\left(\Psi\left(\frac{1}{s+i\sigma},x\right)-\varphi(x)\right)=\int_{0}^{\infty}\mathrm{e}^{-s\tau-i\sigma\tau}\left(\xi(\tau,x)-\frac{1}{\Gamma(\frac{1}{2})\sqrt{\tau}}\varphi(x)\right)\mathrm{d}\tau$$

$$\frac{1}{\sqrt{s+i\sigma}}\Psi\left(\frac{1}{s+i\sigma},x\right)-\varphi(x)_{L^2(\mathbb{R}^n,dx)}\le\frac{1}{|s+i\sigma|^{-3/2}}\mathrm{const}$$

Hence, by inverse Fourier transformation

$$\xi(t,x)=\frac{1}{\Gamma(\frac{1}{2})\sqrt{t}}\varphi(x)+\frac{\mathrm{e}^{s\tau}}{(2\pi)^n}\int_{-\infty}^{\infty}\mathrm{e}^{i\sigma t}\frac{1}{\sqrt{s+i\sigma}}\left(\Psi\left(\frac{1}{s+i\sigma},x\right)-\varphi(x)\right)d\sigma$$

By the Proposition (2.1), we can easily prove the next proposition.

Proposition 2.2 *Suppose that* W(x) *is a smooth function with compact support, suppose* $\varphi\in C_0^\infty(\mathbb{R}^n)$, *let*

$$\Psi(t,x)=\int\mathrm{d}\mu_x^t(w)\left\{\exp\left[-i\int_0^t b(\omega(s))\mathrm{d}\omega\right.\right.$$
$$\left.\left.-\frac{1}{2}i\int_0^t \mathrm{div}b(\omega(s))\mathrm{d}s-\int_0^t W(\omega(s))\mathrm{d}s\right]\right\}\varphi(\omega(t))$$

Then, $\exists$ a family $\xi^{(W)}(t,x)\ni$

i. $\xi^{(W)}(t,x)\in L^2(\mathbb{R}^n,\mathrm{d}x)$,

ii. $\int_{-\infty}^{\infty}\mathrm{e}^{-ts}\xi^{(W)}(t,x)_{L^2(\mathbb{R}^n,\mathrm{d}x)}\mathrm{d}t\langle+\infty$ for any$s,t\rangle 0$

iii. $\xi(t,x)$ is finite $\xi(t,x)=0$, if $\mathrm{d}(x,\mathrm{supp}\,\varphi)>2\sqrt{t}$

iv. $\frac{1}{\sqrt{s}}\Psi^{(W)}\left(\frac{1}{s}, x\right) = \int\limits_0^\infty \mathrm{e}^{-ts}\xi^{(W)}(t, x)\mathrm{d}t$

v. $\xi^{(W)}(t, x) = \frac{\varphi(x)}{\sqrt{t}\Gamma\left({}^1\!/_2\right)} + \mathrm{e}^{ts}(2\pi)^{-n}\int\limits_{-\infty}^{\infty} \mathrm{e}^{it\sigma}\frac{\Psi^{(w)}\left(\frac{1}{s+i\sigma}, x\right) - \varphi(x)}{\sqrt{s+i\sigma}}\mathrm{d}\sigma,$

where $\Psi^{(W)}(t + is, x)$ is an analytic expansion of $\Psi(t, x)$ into a domain $\mathrm{Re}(t + is) > 0$ and an account of a quadrate root is possessed in a right-hand half-plane for any s> 0.

Suppose $V(x)$ have the form

$$v_{j-1,m}\big(x - (a_{j-1,m1}, a_{j-1,m2}, \ldots .a_{j-1,mn})\big)\xi_{j-1,m} + v_{j,m}\big(x - (a_{j,m1}, a_{j,m2}, \ldots, a_{j,mn})\big)\xi_{j,m}$$

For $r \in Z$, put $v_{j,m}(x) = 0$ for $|m| > r$. Thus, the random potential conformable to $v_{j,m}$ is a smooth finite function of x. Stand for W and stratify to the proposition.

We secure random functions $\Psi^{(W)}(t, x)$ and $\xi^{(W)}(t, x)$ together the advantages (i)–(v). From the outcomes of Proposition (2.1) and Lemma (2.2) in [7], we acquire

$$\Psi^{(W)}(t, x)\,\text{converge}\,\Psi(t, x)\text{ in }L^2\big(\mathbb{R}^n, \mathrm{d}x, \mathrm{d}V\big)\text{if } r \to \infty.$$

converge $\Psi(t, x)\mathrm{in}L^2(\mathbb{R}^n, \mathrm{d}x, \mathrm{d}V)\mathrm{if}\ r \to \infty$.

Proposition 2.3 *For $t = \tau + i\gamma$, the assessment holds:*

$$\Psi(t, x) - \varphi(x) + tH\varphi(x) - \frac{t^2}{2!}H^2\varphi(x) + \cdots (-1)^{m+1}\frac{t^m}{m!}H^m\varphi(x)_{L^2(\mathbb{R}^n, \mathrm{d}x, \mathrm{d}V)} \le C_{m+1}\frac{t^{m+1}}{m!} \quad (4)$$

where C_{m+1} is a constant.

Proof As it was shown

$$\frac{\partial^m}{\partial t^m}\int\limits_{\mathbb{R}^n} E\left(\Psi(t, x)\sum_{k,l} C_{k,l}h_k(x)\theta_l(V)\right)\mathrm{d}x = (-1)^m\int\limits_{\mathbb{R}^n} E\left(\widetilde{\Psi}_m(t, x)\sum_{k,l} C_{k,l}h_k(x)\theta_l(V)\right)\mathrm{d}x, \quad (5)$$

where $\widetilde{\Psi}_m(t, x) \equiv H^m\varphi$. By Taylor's formula,

$$\int_{\mathbb{R}^n} E\left(\Psi(t,x)\sum_{k,l} C_{k,l}h_k(x)\theta_l(V)\right)\mathrm{d}x = \sum_{j=0}^{n} \frac{(t-t_1)^j}{j!}\cdot\frac{\partial^j}{\partial t^j}$$

$$\int_{\mathbb{R}^n} E(\Psi(t,x)\sum_{k,l} C_{k,l}h_k(x)\theta_l(V))\,\mathrm{d}x|_{t=t_1} + \int_{t_1}^{t}\mathrm{d}\tau_1\int_{t_1}^{\tau_1}\mathrm{d}\tau_2\ldots\int_{t_1}^{\tau_n}\mathrm{d}\tau_{n+1}\frac{\partial^{n+1}}{\partial t^{n+1}}$$

$$\int_{\mathbb{R}^n} E\left(\Psi(t,x)\sum_{k,l} C_{k,l}h_k(x)\theta_l(V)\right)\mathrm{d}x|_{t=\tau_n} \tag{6}$$

substitute, by formula (6), the derivations into (5). According to: [8]

$$\frac{\partial\Psi}{\partial t} = \sum_{j=1}^{n}\frac{1}{2}\left(i\partial_j + b_j(x)\right)^2\Psi - V\Psi$$

A .e. V, we acquire

$$\lim_{t_1\to 0}\widetilde{\Psi}_m(t_1,x) = H^m\varphi(x).$$

From (6),

$$\int_{\mathbb{R}^n} E\left((\Psi(t,x) - \sum_{j=0}^{m}\frac{(-t)^j}{j!}H^j\varphi(x))\sum_{k,l} C_{k,l}h_k(x)\theta_l(V)\right)\mathrm{d}x$$

$$= (-1)^{m+1}\int_0^t d\tau_1\int_0^{\tau_1} d\tau_2\ldots\int_0^{\tau_m} d\tau_{m+1}\int_{R^n} E\left(\widetilde{\Psi}_{m+1}(t,x)\times\sum_{k,l} C_{k,l}h_k(x)\theta_l(V)\right)\mathrm{d}x \tag{7}$$

by the estimates,

$$\left|\int_{R^n} E\left(\Psi(t,x)\sum_{k,l} C_{k,l}h_k(x)\theta_l(V)\right)\mathrm{d}x\right| \le \text{const.}\sum_{k,l} C_{k,l}h_k(x)\theta_l(V)_{L^2(\mathbb{R}^n,dx,dV)}. \tag{8}$$

Now (4) follows from (7) and (8). ■

We return to $\xi^{(W)}(t,x)$ and (v). By the proposition,

$$\frac{1}{|\sqrt{s+i\sigma}|}\Psi^{(w)}\left(\frac{1}{s+i\sigma},x\right) - \varphi(x)_{L^2(\mathbb{R}^n,dx,dV)} \le C_1\frac{1}{|s+i\sigma|^{3/2}}, \tag{9}$$

$$\frac{1}{\left|\sqrt{s+i\sigma}\right|}\Psi\left(\frac{1}{s+i\sigma},x\right)-\varphi(x)_{L^2(\mathbb{R}^n,dx,dV)}\leq C_1\frac{1}{|s+i\sigma|^{3/2}} \quad (10)$$

Since (9) does not depend on r, then by (v) of Proposition (2.3) that $\xi^{(W)}(t,x)$ have limits in $L^2(\mathbb{R}^n,\mathrm{d}x,\mathrm{d}V)$, and denote by $\xi(t,x)$. In addition,

$$\xi(t,x)=\frac{\varphi(x)}{\sqrt{t}\Gamma\left(1/2\right)}+\mathrm{e}^{ts}(2\pi)^{-n}\int\limits_{-\infty}^{\infty}\mathrm{e}^{it\sigma}\frac{\Psi\left(\frac{1}{s+i\sigma},x\right)-\varphi(x)}{\sqrt{s+i\sigma}}\mathrm{d}\sigma. \quad (11)$$

We note for $\xi(t,x)$ the following:

i. $\xi(t,x)\in L^2(\mathbb{R}^n,\mathrm{d}x,\mathrm{d}V)$.

ii. $\int\limits_0^{\infty}\mathrm{e}^{-ts}\xi(t,x)_{L^2(\mathbb{R}^n,\mathrm{d}x,\mathrm{d}V)}\mathrm{d}t\langle+\infty$ for any $s,t\rangle 0$

iii. $\xi(t,x)$ is finite $\xi(t,x)=0$, if $\mathrm{d}(x,\operatorname{supp}\varphi)>2\sqrt{t}$

iv. $\frac{1}{\sqrt{s}}\Psi\left(\frac{1}{s},x\right)=\int\limits_{-\infty}^{\infty}\mathrm{e}^{-ts}\xi(t,x)\mathrm{d}t,$

which is convergent in $L^2(\mathbb{R}^n,\mathrm{d}x,\mathrm{d}V)$.

These advantages can be found by the limit the conformable advantages of $\xi^{(W)}(t,x)$.

By depending on scalar product in $L^2(\mathbb{R}^n,dx,dV)$ of (11) and $\sum\limits_{k.l}C_{k,l}h_k(x)\theta_l(V)$, thus we acquire a. e. V, the $\xi(t,x)$ is smooth and

$$H\xi(t,x)=\xi^{(1)}(t,x)$$

where $\xi^{(1)}(t,x)$ is found by (11) and $\varphi(x)$ is displaced by $H\varphi(x)$. This pursues from $H\Psi(t,x)=\Psi^{(1)}(t,x)$. Suppose that evidences our proposition.

From (11) and the results in [8],

$$\left|E\int\limits_{\mathbb{R}^n}\Psi(t,x)\sum_{k,l=1}^{p}C_{k,l}\sum_{j=1}^{n}2^{-1}\left(i\partial_j+b_j(x)\right)^2h_k(x)\theta_l(V)\mathrm{d}x\right|$$
$$\leq\text{const}\sum_{k,l=1}^{p}C_{k,l}h_k(x)\theta_l(V)_{L^2(\mathbb{R}^n,dx,dV)}$$

we get

$$E\left(\int_{\mathbb{R}^n} \xi(t,x)\sum_{k,l} C_{k,l}\left(\sum_{j=1}^{n}\frac{1}{2}(i\partial_j + b_j(x))^2\right)h_k(x)\theta_l(V)\mathrm{d}x\right)$$

$$= \frac{1}{\sqrt{t}\Gamma\left({}^1/_2\right)}\int_{\mathbb{R}^n}\left(\sum_{j=1}^{n}\frac{1}{2}(i\partial_j + b_j(x))^2\right)\varphi(x)E\left(\sum_{k,l} C_{k,l}h_k(x)\theta_l(V)\right)\mathrm{d}x$$

$$+ (2\pi)^{-n}\mathrm{e}^{ts}\int_{-\infty}^{\infty}\mathrm{e}^{it\sigma}\frac{1}{\sqrt{s+i\sigma}}E\left[\int_{\mathbb{R}^n}\mathrm{d}x\left(\left(\sum_{j=1}^{n}\frac{1}{2}(i\partial_j + b_j(x))^2\,\Psi\left(\frac{1}{s+i\sigma}, x\right)\right.\right.\right.$$

$$\left.\left.- \left(\sum_{j=1}^{n}\frac{1}{2}(i\partial_j + b_j(x))^2\varphi(x)\right)\sum_{k,l} C_{k,l}h_k(x)\theta_l(V)\right]\mathrm{d}\sigma$$

$$\left(\sum_{j=1}^{n}\frac{1}{2}(i\partial_j + b_j(x))^2\,\Psi\left(\frac{1}{s+i\sigma}, x\right) - \left(\sum_{j=1}^{n}\frac{1}{2}(i\partial_j + b_j(x))^2\varphi(x)\right.\right. \tag{12}$$

$$= H\varphi(x) - H\Psi\left(\frac{1}{s+i\sigma}, x\right) + V\Psi\left(\left(\frac{1}{s+i\sigma}, x\right) - \varphi(x)\right)$$

$$= \widetilde{\varphi}_1(x) - \widetilde{\Psi}_1\left(\frac{1}{s+i\sigma}, x\right) + V\left(\Psi\left(\frac{1}{s+i\sigma}, x\right) - \varphi(x)\right), \tag{13}$$

where $\widetilde{\varphi}_1 = H\varphi(x)$ and $\widetilde{\Psi}_1(t,x)$ corresponds to $\widetilde{\varphi}_1(x)$.

Note that the Proposition (2.2) remains satisfied for $\widetilde{\Psi}_1$ and $\widetilde{\varphi}_1$. In particular,

$$\widetilde{\Psi}_1 - \widetilde{\varphi}_1{}_{L^2(\mathbb{R}^n,\mathrm{d}x,\mathrm{d}V)} \leq \mathrm{const}\frac{1}{|s+i\sigma|}. \tag{14}$$

If we multiply the integrand in (5) by $V(x)$, then (4) remains satisfied

$$V\left(\Psi(t,x) - \sum_{j=0}^{n}(-1)^{j+1}\frac{(t+1)^j}{j!}\varphi(x)\right)_{L^2(\mathbb{R}^n,\mathrm{d}x,\mathrm{d}V)} \leq C_{n+1}\frac{t^{n+1}}{n!}, \tag{15}$$

where C_{n+1} is a constant.

$$V\left(\Psi\left(\frac{1}{s+i\sigma}, x\right) - \varphi(x)\right)_{L^2(\mathbb{R}^n,\mathrm{d}x,\mathrm{d}V)} \leq \frac{\mathrm{const}}{|s+i\sigma|} \tag{16}$$

substitute (13)–(16) into (11). After the Fourier transform, we get

$$\left|E\left(\int_{\mathbb{R}^n}\hat{\xi}(t,q)\left(\frac{-n}{2}|q|^2 - \frac{1}{2}\sum_{j=1}^{n} i\partial_j\hat{b}_j(q) - \frac{1}{2}\sum_{j=1}^{n} i\hat{b}_j(q)|q| + \frac{1}{2}\sum_{j=1}^{n}\hat{b}_j^2(q)\right)\right.\right.$$

$$\sum_{k,l} C_{k,l}\hat{h}_k(q)\theta_l(V)\Bigg)\mathrm{d}q\Bigg| \leq \mathrm{const}\left(\frac{1}{\sqrt{t}} + B_s e^{ts}\right)$$

$$\sum_{k,l} C_{k,l}h_k(x)\theta_l(V)_{L^2(\mathbb{R}^n,\mathrm{d}x,\mathrm{d}V)}, \tag{17}$$

where A, B_s are constants. The estimate (17) implies $\xi(t, x) \in W_1$ a. e. V, also, $\xi^2_{W_1}$ an integrable of V. Further, we can return above by the same method as in [8]. Thus, we get the stated assertion.

3 Conclusion

W construct the function $\xi(t, x)$ and note that the function$\xi^{(W)}(t, x)$ converges to $\xi(t, x)$ in $L^2(R^n, \mathrm{d}x, \mathrm{d}V)$ as $r \to \infty$. So, the same properties of the $\Psi(t, x)$ are satisfied.

References

1. Ismagilov, R. S. (1962). On conditions of self-adjointness of differential operators of a higher order. *Dokl Acad Nauk SSSR, 142,* 1239–1242.
2. Milatovic, O. (2018). Self-adjointness, m-accretivity, and separability for perturbations of Laplacian and Bi-Laplacian on Riemannian manifolds. *Integral Equations, Operator Theory* pp 90–112.
3. Nenciua G, Nenciua I (2017) Drift-diffusion equations on domains in Rd: essential self-adjointness and stochastic completeness. *Journal of Functional Analysis* 2616–2654.
4. Portmann F, Sok J, Solovej JP (2018) Self-adjointness and spectral properties of dirac operators with magnetic links. *Journal de Mathématiques Pures et Appliquées* 114–157.
5. Simon, B. (1973). Essential self-adjointness of Schrödinger operators with singular potentials: a generalized Kalf-Walter—Schmincke theorem, arch. *Rational Mechanics and Analysis, 52,* 44–48.
6. Treust, L. L., Bonafos, T. O. (2018). Self-adjointness of dirac operators with in_nite mass boundary conditions in sectors, Annales Henri Poincar.
7. Saleem, Y. H., & Hassen, H. A. (2019). Convergence of Schrödinger Operator with electromagnetic potential. *Science International Lahore, 31*(2), 303–308.
8. Saleem, Y. H., Shubber, H. A.(2019. The smoothness of Schrödinger operator with electro-magnetic potential. *European Journal of Pure and Application Mathmatics 12*(4), 1441–1454.
9. Saleem, Y. H., Shubber, H. A. (2019). Essential self-adjointness of the Schrödinger operator with electromagnetic potential. *Journal of Interdisciplinary Mathamatics 22*(8), 1537–1542.
10. Saleem, Y. H., Goldstein, M. S. (1989). Self-adjointness of Schrödinger operators for one class of strongly oscillating potential, tez.dokl., Skoly-seminara, aktualnye voprosy kompleksnogo analiza. Tashkent, pp. 152 [Russian].

IoT Implementation at Global Enterprises for Progressive Human Resource Practices

H. M. Moyeenudin and R. Anandan

Abstract The human resource division has skillfully improved to identify the employee performance to meet the employer's commitment through Internet of things (IoT) in advanced multinational establishments, which encounters various jobs at the working environment. The majority of multinational companies is focused on identifying the right tools to improve their recruitment and training processes that deals with the organizational culture, commitment of employee, and overseeing the performance of an employee by IoT. To locate the specific aptitudes of individuals regarding the company's policies, rules and regulation, the human resource management (HRM) recruits employee using varied channels with the aid of IoT, and the virtual platforms support the HRM for recruitment and training. This paper is focused on the implementation of IoT in human resource management for effective recruitment and training process. Secondly, it highlights the purpose and quantity-wise usage for human resource (HR) practices to determine the attendance, feedbacks, appraisals, and performance of an employee.

Keywords IoT · Human resource · Employee training · Human resource practice

1 Introduction

The IoT in HRM supports the organization by providing quick methods of recruitment and examination process. The employee is also scrutinized after recruitment, during the time of training process and further at regular work schedules through IOT virtual platforms [1], which alludes the combination of significant HR information from various sources, the performance of hierarchical and workforce investigation on this

H. M. Moyeenudin (✉)
School of Hotel & Catering Management, Vels Institute of Science, Technology and Advanced Studies, Pallavaram, Chennai, India
e-mail: moyeenudin@velsuniv.org

R. Anandan
Department of Computer Science & Engineering, Vels Institute of Science, Technology and Advanced Studies, Pallavaram, Chennai, India
e-mail: anandan.se@velsuniv.ac.in

S.-L. Peng et al. (eds.), *Proceedings of First International Conference on Mathematical Modeling and Computational Science*, Advances in Intelligent Systems and Computing 1292, https://doi.org/10.1007/978-981-33-4389-4_12

gathered information, and at last, the collected information support to make the choices for better association and execution of work [2]. The recruitment process is designed with the convenience and accessibility of employer and employee. The prepared structure of human resource department of an organization can keep up the HR activities in a sequence with authoritative objectives and procedures with IoT [3]. This is to its most noteworthy bit of leeway, HR interviews and tests are demonstrated to be noteworthy for an organization or an institution, that will play a key role with IoT in HRM. With the assistance of prescient and prescriptive approaches connected to HR practices [4], the investigations can foresee issues and openings concerning HR policies, helping organizations proactively to settle on ideal basic leadership. The various ongoing approaches have appeared a solid connection between the degree and level to which they study and utilize their money-related execution [5]. It is obvious from these investigations with IoT associations generously in front of their associates in their utilization of information and examination are probably going to be considerably ahead in money-related incentives [6].

1.1 Materials and Methods

This research is conducted by using qualitative and quantitative methods, over collecting the data using measurable information in an explicit questionnaire for the human resource department of multinational companies utilizing IoT platforms in HR practices. Around 200 companies are visited with questionnaires and collected the data from the HR managers and executives to identify the implementation of IoT in these enterprises and only 180 finalized feedback form are used to make this analysis. This study was carried out from November 2019 to December 2019 at the global enterprises of Chennai. The Statistical Package for Social Sciences (SPSS 24 version) is used to scrutinize the association between the variables using correlation and linear regression to appraise the dependent and independent variables with their relationship.

2 Results Analysis

2.1 Correlations

Table 1 denotes the relationship between the variables are significant by having strongest correlation by value $r = 0.925^{**}$ with a significant value $p = 0.000$, for the recruitment methods and online recruitment, thus this show in present scenario majority of recruitment methods is carried out by online which is influenced by IoT. Applications for drawing inability, specialist execution examination, candidate following an appraisal, venture the executives, inside the administration, and so on

Table 1 Correlation between human resource practices with IoT

		Online recruitment	Traditional recruitment	Training internet	Training lecture	Recruitment methods	Payroll
Online recruitment	P C	1	−0.054	0.895**	−0.252**	0.925**	0.807**
	Sig		0.472	0.000	0.001	0.000	0.000
	N	180	180	180	180	180	180
Traditional recruitment	P C	−0.054	1	-0.041	-0.078	-0.041	0.019
	Sig	0.472		0.582	0.301	0.581	0.801
	N	180	180	180	180	180	180
Training by internet	PC	0.895**	−0.041	1	−0.327**	0.844**	0.755**
	Sig	0.000	0.582		0.000	0.000	0.000
	N	180	180	180	180	180	180
Training by lecture	P C	−0.252**	−0.078	−0.327**	1	−0.272**	−0.203**
	Sig	0.001	0.301	0.000		0.000	0.006
	N	180	180	180	180	180	180
Recruitment methods	P C	0.925**	−0.041	0.844**	−0.272**	1	0.707**
	Sig	0.000	0.581	0.000	0.000		0.000
	N	180	180	180	180	180	180
Payroll	P C	0.807**	0.019	0.755**	−0.203**	0.707**	1
	Sig	0.000	0.801	0.000	0.006	0.000	
	N	180	180	180	180	180	180

**Correlation is significant at the 0.01 level (2-tailed)

are upgrading better basic leadership with the utilization of machines as a device and associate [7]. There is a stronger correlation with a value $r = 0.895^{**}$ and a significant value $p = 0.000$ for the variables training the employees using the Internet with online recruitment aided by IoT, as many global enterprises prefer to train the applicant to finalize their employees and the majority of multinational organization and large institutions will move to the cloud connected with ERP to make their procedures feasible in training on areas, and as the association between the variables recruitment methods and online recruitment has a strong r value $= 0.844^{**}$ with a significant value $p = 0.000$ for the variables recruitment methods with training by Internet, there is a quickly developing movement toward more information-driven, smart computerized associations, and increasingly independent/scientific workers. This is expanding the requirement for apparatuses that incorporate the ERP in addition to cloud technology, representative interchanges, and help-work area cooperations. Self-administration and representative experience stages with ERP are the foundation of works related to an administration which focuses on disentangling, cost announcing, time following, and pretty much every other HR work. ERP applications are changing the idea of work and the structure of the work environment [8]. Thus, the salaries of employees are confirmed through online after recruitment with the relationship between the variables payroll and online recruitment by value $r = 0.807^{**}$ with a significant p-value $= 0.000$. This Pearson's correlation is significant at the 0.01 level.

2.2 Linear Regression

This Table 2 determines the coefficients with a value R square = 0.881 that shows around 88% of variation with std. error of the estimated value 0.223 on predicting the mean of Wi-Fi usage in HR practices, recruitment, and training using IoT-enabled facilities like training the employees using electronic gadgets connected with Internet, also on recruitment of employees. The market for enlistment gadgets is thriving with headway, as HR needs instruments to help find people with the right limits and learning aptitudes, not just particular or abstract limits. High volume enlistment is being automated, and talented work enrolling is being changed by freely discharging gadgets, selection of the official's systems, and improved assessments [9]. Similarly called appropriated record advancement that is on a very basic level, a database keeps a reliably creating once-over of records with IoT.

The relationship of dependent variable "Recruitment methods" is analyzed with an independent variable to know the results on variation, thus Table 3 ANOVA outcome displays the linear regression in comparison with traditional recruitment methods and IoT assisted online training and recruitment methods. Associations are modernizing in relationship with IoT learning headways that are progressively appearing in the principle of a procedure, and self-facilitated commitment is increasing young experts

Table 2 Model summary of linear regression

Model summary				
Model	R	R square	Adjusted R square	Std. error of the estimate
1	0.939[a]	0.881	0.877	0.223

[a]Predictors: (Constant), Wi-Fi, training by lecture, traditional recruitment, payroll, training by internet, online recruitment

Table 3 ANOVA on recruitment methods

ANOVA						
Model		Sum of squares	df	Mean square	F	Sig
1	Regression	64.162	6	10.694	214.179	0.000[b]
	Residual	8.638	173	0.050		
	Total	72.800	179			

[a]Dependent variable: recruitment methods
[b]Predictors: (Constant), Wi-Fi, training by lecture, traditional recruitment, payroll, training by internet, online recruitment

Table 4 Variation of coefficients on recruitment methods

	Coefficients	Unstandardized coefficients		Standardized coefficients		
Model		B	Std. Er	Beta	t	Sig
1	(Constant)	−7.166	0.556		−12.884	0.000
	Online recruitment	0.110	0.008	1.182	14.653	0.000
	Traditional recruitment	−0.001	0.004	−0.005	−0.185	0.853
	Training by Internet	0.003	0.005	0.030	0.482	0.630
	Training by lecture	0.004	0.007	0.017	0.576	0.566
	Payroll	−0.124	0.043	−0.129	−2.868	0.005
	Wi-Fi	0.503	0.096	0.231	5.225	0.000

[a]Dependent variable: recruitment methods

by using video and another new kind of scaled downscale learning stages including PC produced reality, extended reality, experience stages, modernized learning the board structures, and IoT-based systems for learning and getting ready. It empowers mechanized information to be scattered at this point not copied. In HR, potential used for IoT consolidates mechanized procedure of the board, disentangles affirmation issues, increases straightforwardness, improves abroad electronic fund, and robotizes routine and data overpowering strategies, digital security, and safety [10].

In Table 4, the dependent variable recruiting methods is predicted from an independent variable; there is a stronger relationship between online recruitment from dependent variable recruitment methods with a value ($b = 0.110, p = 0.000$) whereas in another side there is a negative association from traditional recruitment with dependent variable recruitment methods by a value ($b = -0.001, p = 0.853$) that denotes the high amount of variation. The digitalization in HR practices with IoT has skillfully stimulated the complex through employee's participation toward the operational meetings of work atmosphere. The IoT-based training has shown relationship to recruitment methods with a value ($b = 0.003, p = 0.630$). The global environment is touching their prominence on the advancement of technology with the dimensions of corporate companies with an interest in the supervision of results through the HR representatives of an enterprise [11]. The results of linear regression also state that the role of Wi-Fi plays a major role in training their employees as the value shows ($b = 0.503, p = 0.000$). The traditional recruitment shows a negative association with recruitment methods. Thus, these results prove a greater influence of IoT implementation in human resource practices.

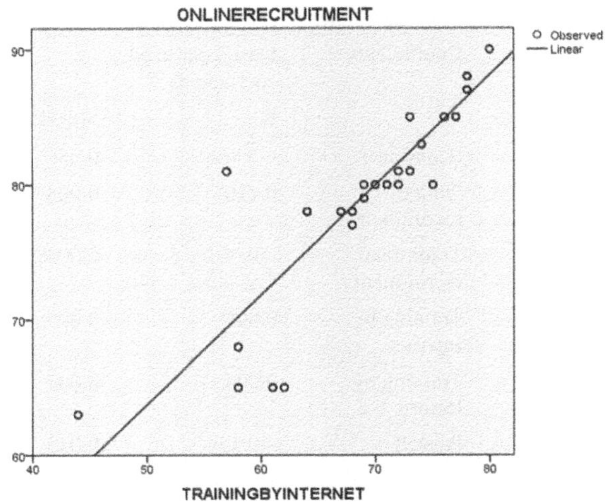

Fig. 1 Online recruitment and training by internet

2.3 Linear Regression Curve with Scattered Plot

In Fig. 1, the influence of IoT is demonstrated with the scattered plots graph of linear regression for online recruitment and training by Internets, more than 90% of human resource division of global enterprise follows online recruitment methods similarly in training methods HR division utilizes the Internet for training their employees, about 60–85% displayed in scattered plot graph of this study. This kind of methodology is changing how HR experts enroll applicants with innovation driving the charge, with the understanding that representatives must be genuinely and physically well to perform at their most extreme levels. In Fig. 2, the linear regression graph shows the number of traditional recruitment methods and training by lecture methods utilized in HR departments of multinational enterprises, as it displays that the traditional HR practices are applied in HR divisions for about 20–40%. This is applicable for a few areas and positions, simultaneously the traditional method of training the employees by lecture was about 20–30% (Fig. 4).

Figure 3 ascertains a utilization of Wi-Fi facility in HR practices at multinational companies, and it shows the majority of HR departments use Wi-Fi for training and development purposes. The recruitment procedure is completed by intimation to candidates through e-mail intimation. The company's HR will have the hands-on knowledge to handle electronic gadgets like worker health applications with weary input apparatuses, profitability devices, and present-day specialized devices aided by Wi-Fi also like some applications that help in understanding and support the devices through improved delegate's significant experience. Widely in HR departments, cloud-based training is conducted with the facility of Wi-Fi that is noticeable with this graph. If the human resource executives can investigate the organization's information cautiously using Wi-Fi during cloud training and recruitment, at that point they can foresee about the requirements of the staff that help with improving

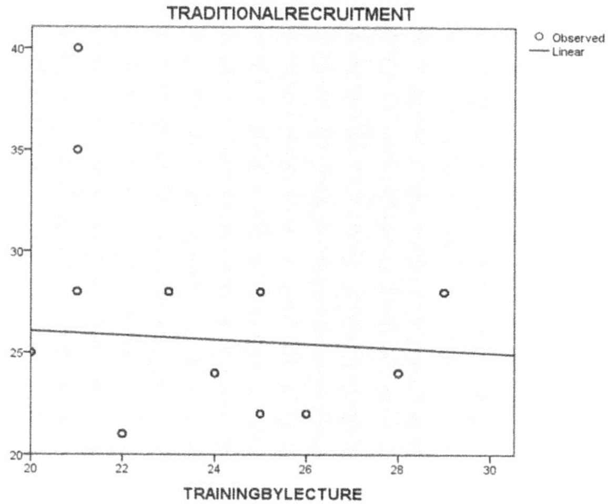

Fig. 2 Traditional recruitment and training by lecture

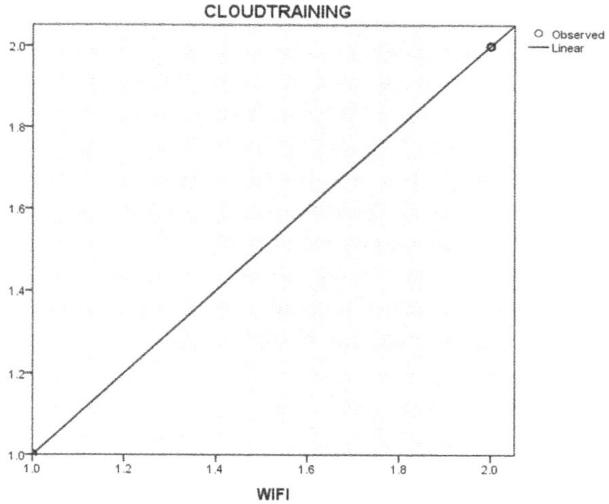

Fig. 3 Wi-Fi with e-mail

worker's vocation with beneficial office climate for easy implementation of the HR procedure and practices, at the general quantity analysis for employees in developing as organizations [12].

3 Conclusion

The human resource departments are experiencing an intense change in recent years, by the expansion IoT associations for advanced recruitment methods, which is proved

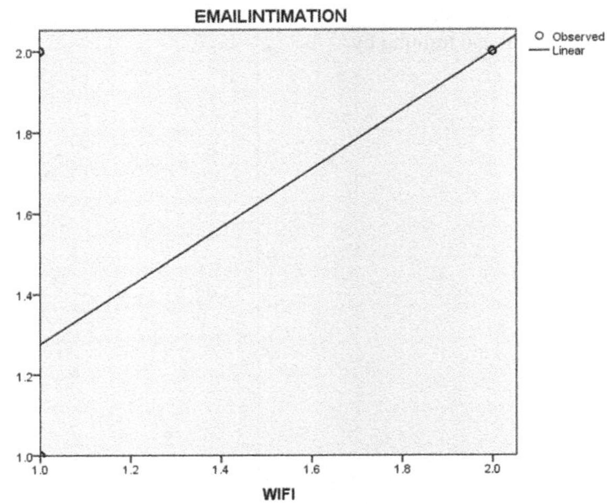

Fig. 4 Cloud training with Wi-Fi

by the regression for online recruitment with the result ($b = 0.110$, $p = 0.000$) in multinational companies. Thus, usage of IoT during training their candidates to appraise and confirm their appointment is confirmed by correlation with a value ($r = 0.895^{**}$, $p = 0.000$). The thought is to concentrate on the requirements of the employees, need for training to strengthen their prosperity, and so on by utilizing more updated innovations through IOT [1]. In future, the sensors will play a key role in HR departments by utilizing the IoT for attendance, health of an employee and also it will support payroll, training for making appraisals by the regular analysis on the performance of an employee, and this research proves that IoT systems are used at multinational companies for efficient recruitment and other human resource practices like training their employees.

References

1. Santoro, G., Vrontis, D., Thrassou, A., & Dezi, L. (2018). The internet of things: Building a knowledge management system for open innovation and knowledge management capacity. *Technological Forecasting & Social Change, 136*, 347–354.
2. Garbarino-Alberti, H. (2013). IT governance and human resources management: A framework for SMEs. *International Journal of Human Capital and Information Technology Professionals (IJHCITP), 4*(3), 40–57. https://doi.org/10.4018/jhcitp.2013070104.
3. Brueller, N., Carmeli, A., Gideon, Markman, D. (2018). Linking merger and acquisition strategies to postmerger integration: A configurational perspective of human resource management. *Journal of Management, 44*(5), 1793–1818.
4. Dash, D., Farooq, R., Panda, J. S., Sandhyavani, K. V. (2019). Internet of things (IoT): The new paradigm of HRM and skill development in the fourth industrial revolution (Industry 4.0). *IUP Journal of Information Technology, 15*(4), 7–30.

5. Bresciania, S., Ferraris, A., & Del Giudice, M. (2018). The management of organizational ambidexterity through alliances in a new context of analysis: Internet of things (IoT) smart city projects. *Technological Forecasting and Social Change, 136,* 331–338.
6. Streumer, J. N., & Kommers, A. M. (2002). Developments in the emerging human resource development discipline. *International Journal of Human Resources Development and Management, 2*(1), 1–16.
7. Gupta, S., Kumar, S., Singh, S., Foropon, C., & Chandra, C. (2018). Role of cloud ERP on the performance of an organization. *International Journal of Logistics Management, 29*(2), 659–675. https://doi.org/10.1108/IJLM-07-2017-0192.
8. Wang, K. H., Chen, C. M., Fang, W., et al. (2018). On the security of a new ultra-lightweight authentication protocol in IoT environment for RFID tags. *he Journal of Supercomputing, 74*(65), 65–70. https://doi.org/10.1007/s11227-017-2105-8.
9. Kapoor, B., & Kabra, Y. (2014). Current and future trends in human resources analytics adoption. *Journal of Cases on Information Technology, 16*(1), 1–10.
10. Gloet, M., & Terziovski, M. (2004). Exploring the relationship between knowledge management practices and innovation performance. *Journal of Manufacturing Technology Management, 15*(5), 402–409.
11. Strohmeier, S. (2018). Smart HRM—A delphi study on the application and consequences of the internet of things in human resource management. *The International Journal of Human Resource Management*, 1–30. https://doi.org/10.1080/09585192.2018.1443963
12. Hollenbeck, J. R., & Jamieson, B. (2015). Human capital, social capital, and social network analysis: Implications for strategic human resource management. *Academy of Management Perspectives, 29*(3), 370–385.

An Empirical Study of Heart Disease Prediction System Using Various Machine Learning Classification Algorithms

V. Poornima and D. Gladis

Abstract In the world, cardiac disease is one of the leading causes to death. Every year, 17.3 million people approximately die based on cardiac disease worldwide. Heart disease prediction system plays a significant role in health care utilization such as identifying individual expenses and disease risks for the patients. Early prediction of heart disease can prevent disease and can help lower the risk of heart diseases. Data mining (DM) techniques have a convincing role in the healthcare industry to empower health systems. This investigation inspects the early prediction of HD based on numerous DM techniques. Here, for experimentation, the dataset of the American Heart Association (AHA) is solicited to analyze the performance of this study and it was estimated with the six classifier evaluation metrics. The system was implemented using Weka tool and three different test cases were explored such as training set, Percentage Split (PS), and tenfold Cross-Validation (CV). The experiments verify that the random tree method could achieve a prediction accuracy of 100% with zero mean absolute error, which is greater than the other machine learning classification algorithms.

Keywords Data mining · Machine learning classification algorithms · Cardiovascular diseases · American health association dataset · Weka tool

1 Introduction

Nowadays, heart disease is the dominant reason for deaths far and wide. World Health Organization (WHO) estimated 17.5 million people deceased from cardiac diseases worldwide in 2012 [1]. Heart-related diseases are responsible for 31% of all global deaths according to the report of WHO [2]. Prediction of HD can reduce

V. Poornima (✉)
New Prince Shri Bhavani Arts and Science College, Chennai, India
e-mail: poornimasudhaagar@yahoo.com

D. Gladis
Bharathi Women's College, Chennai, India
e-mail: gladischristopher@gmail.com

S.-L. Peng et al. (eds.), *Proceedings of First International Conference on Mathematical Modeling and Computational Science*, Advances in Intelligent Systems and Computing 1292, https://doi.org/10.1007/978-981-33-4389-4_13

the calamitous rate of a human. Information technology plays a crucial role in the health field [3]. The quality of the service is to diagnose the disease accurately and yielding effective treatment to patients. Prediction of heart failure on the right time is hard from the point of medical professionals [4]. Data mining techniques are most frequently used in extracting crucial and knowledgeable information from the medical field [5]. Due to unavailability of diagnostic tools and inadequate of medical practitioners, diagnosis and treatment of HD are very thorny. To save humans life, the DM techniques have manifested for early prediction of HD with greater meticulous [6]. Naive Bayes is the kernel of the multifarious machine learning and DM methods [7].

This paper attempts to present the detailed study about the twelve various Machine Learning (ML) techniques such as Bayesian net, Naive Bayes, Multi Layer Perceptron (MLP), Stochastic Gradient Descent (SGD), AdaBoost, K-Nearest Neighbors (KNN), Random Forest, Random Tree, J48, Logistic Model Tree (LMT), Decision Table, and Logit Boost, and their performance were analyzed to predict the heart disease.

The AHA dataset is employed to these 12 ML classification algorithms and their accuracies are estimated. To examine the outcome of various algorithms, three different evaluation methods were implemented such as training set, tenfold CV, and percentage split. In training dataset, the random tree method achieved a prediction accuracy of 100% with zero Mean Absolute Error (MAE), which is greater than the other machine learning classification algorithms. Tenfold cross-validation method concluded that Naïve Bayes achieved an accuracy of 86% with 0.1819 MAE. In percentage split test case, the Bayesian net method obtained 94% accuracy with 0.1224 MAE using Weka 3.8 tool.

2 Literature Review

Over the years, innumerable tasks have been executed related to cardiovascular diseases using different DM algorithms. Subhadra et al. [8] had initiated the heart disease diagnostic system by use of multilayer perceptron neural network with back-propagation. Based on the outcomes, it gives 93.39% accuracy in heart disease prediction. Banu et al. [9] highly suggested the HD prediction using fuzzy c-means classifier with DM techniques. Based on the result, better performance is given by the proposed method compared to other classification algorithms. Seema et al. [10] insisted that predicting chronic illness by exploiting the historical health data by using SVM, ANN, decision tree, and Naïve Bayes. As a result, the maximum accuracy rate is given by SVM, whereas for diabetes the highest accuracy is given by Naïve Bayes.

Mujawar et al. [11] anticipated the prediction of the disease using modified Naïve Bayes and k-means classifier. This forecaster has 93% accurateness in the prediction of heart disease and 89% accurateness in such cases where it identified that a patient does not has heart disease. Kumari et al. [12] have reduced the difficulty in detection of circulatory disease using support vector machine (SVM), ANN, RIPPER classifier, and decision tree algorithm. The result showed that SVM gives a better result for prediction. Sridhar et al. [13] proposed an HD prediction system based on the ML algorithm. From their results, this study showed that decision tree performed better than Naïve Bayes. Dineshgar et al. [14] surveyed on the data mining methods and built a model of intelligent HD prediction system which gave the diagnosis of HD using the historical database.

From the related studies, it is observed that numerous ML classification algorithms are recurrently exploited in different studies to predict HD. The rest of this study is systemized as follows: segment 3 proposed a prediction of heart disease using the different classifier, segment 4 summarizes the outcome of HD prediction using various classifiers, and segment 5 concludes the paper.

3 Machine Learning Algorithms for Predicting Heart Disease

Machine learning (ML) is a kind of artificial intelligence that extracts patterns out of raw data by utilizing a method or an algorithm. The primary attention of machine learning is to allow computer systems to learn from experience without being explicitly programmed or human intervention [15]. Machine learning intends to construct the machine more prosperous, efficient, and reliable. The machine learning tool is the doctor's brain and knowledge in the healthcare system. Machine learning can be categorized as two different types (Fig. 1).

This analysis intends to effectively predict cardiovascular diseases from patient dataset. In this research work, the twelve different classification methods are used to predict the HD and also analyze the execution of these classification techniques in the American Heart Association (AHA) dataset. The problem with the heart-related disease is that there are many kinds of Risk Factors (RF) involved such as age, blood cholesterol, usage of cigarette, blood pressure, person's fitness, and stress. Understanding and categorizing each RF based on its importance is a tedious task. The dataset of 50 people was collected from the American Heart Association [16]. The dataset was comprised of twelve important RFs. The dataset was composed of twelve important RFs which were sex, age, blood pressure, family history, blood cholesterol, smoking habit, alcohol consumption, diabetes, physical inactivity, poor diet, and obesity. This objective of this study specified that whether the people had a risk of HD or not.

4 Results and Discussions

The execution of the ML model is measured by the evaluation metrics. In this study, some of the performance metrics are discussed such as true positive (TP), precision, recall, F1 measures, and ROC curve.

- Precision is the ratio between the number of true positive and the sum of a true positive and false positive.

$$\text{Precision} = \frac{\text{TP}}{\text{TP} + \text{FP}} \tag{1}$$

- The recall is the ratio between the number of true positive and the sum of a true positive and false negative.

$$\text{Recall} = \frac{\text{TP}}{\text{TP} + \text{FN}} \tag{2}$$

- An excellent system should have a good precision as well as a high recall. So ideally, F1 measures combine both these aspects in one single metric.

$$F1\text{Score} = \frac{(2 * \text{precision} * \text{Recall})}{(\text{Precision} + \text{Recall})} \tag{3}$$

- True Positive Rate (TPR) is the total number of accurate predictions that were positive and is defined as follows:

$$\text{TPR} = \frac{\text{TP}}{\text{TP} + \text{FN}} \tag{4}$$

- False Positive Rate (FPR) is the total number of inaccurate predictions that were positive and is defined as follows:

$$\text{FPR} = \frac{\text{FP}}{\text{FP} + \text{TN}} \tag{5}$$

- Accuracy is the ratio of the total number of correct predictions.

$$\text{Accuracy} = \frac{\text{TP} + \text{TN}}{\text{TP} + \text{FP} + \text{FN} + \text{TN}} \tag{6}$$

- ROC curve can be used as an evaluation metric to estimate the efficacy of the models.
- Confusion Matrix (CM) used to evaluate the accuracy of the problem.
- Absolute error is the amount of error and Mean Absolute Error (MAE) is the average of all absolute errors.

The dataset of AHA has been converted as CSV file, and it has been sent to the Weka tool to predict the HD using different machine learning classification algorithms. To compare the performance of various algorithms, three different evaluation methods which are training set, tenfold CV, and PS (66%) were analyzed in Tables 1, 2, and 3, respectively.

Based on the experiment performed, the above table concluded that the MLP, KNN, and random forest achieved 100% accuracy with least MAE values and random tree method achieved 100% accuracy with zero mean absolute error using the training dataset, and the graph of various classifiers using training dataset is shown in Fig. 2.

CM of random tree
x y ← classified as
8 0 | x = Yes
0 32 | y = No
In Table 2, the tenfold cross-validation test case concluded that the Naïve Bayes method obtained 86% accuracy with 0.1819 error value and the graph of the various classifier using tenfold validation is shown in Fig. 3.

CM of Naïve Bayes
x y ← classified as.
15 3 | x = Yes.
4 28 | y = No.
Based on the report of the Weka tool, Table 3 concluded that the Bayes net method obtained 94% accuracy with 0.1224 error values using the percentage split test case.

Graph of various machine learning algorithm using percentage split is shown in Fig. 4.
CM for Bayes net
x y ← classified as
3 1 | x = Yes.
0 13 | y = No.

The evaluation had been estimated in three different ways with WEKA tool for analyzing the presence of HD. The accuracy results obtained were random tree with 100% accuracy using the training set test case, Naïve Bayes with 86% accuracy using tenfold CV, and Bayes net with 94% using PS test case.

5 Conclusions

The exploration of this work attempted to study comparative performances of different machine learning algorithms in HD prediction using Weka tool. Various classifiers were compared regarding their accuracies, error functions, and building times using three different test case such as training dataset, CV and PS. Results

Table 1 Comparison of various classification algorithms using training dataset

S.no.	Classifiers	Correctly identified instances	Correctly classified	MAE	TP	Precision	Recall	F-score	ROC curve
1	Bayes net	46	92	0.1148	0.92	0.92	0.92	0.92	0.96
2	MLP	50	100	0.0093	1	1	1	1	1
3	AdaBoost	46	92	0.1247	0.92	0.925	0.92	0.92	0.983
4	KNN	50	100	0.0192	1	1	1	1	1
5	J48	44	88	0.1794	0.88	0.91	0.88	0.882	0.91
6	Random forest	50	100	0.0927	1	1	1	1	1
7	*Random tree*	*50*	*100*	*0*	*1*	*1*	*1*	*1*	*1*
8	Naïve Bayes	44	88	0.1347	0.88	0.885	0.88	0.881	0.957
9	LMT	48	96	0.1119	0.96	0.96	0.96	0.96	0.983
10	LogitBoost	49	98	0.0979	0.98	0.981	0.98	0.98	0.993
11	Decision table	46	92	0.2266	0.92	0.92	0.92	0.92	0.978
12	SGD	49	98	0.02	0.98	0.981	0.98	0.98	0.984

Table 2 Comparison of various classification algorithms using tenfold cross-validation techniques

S. no.	Classifiers	Correctly identified instances	Correctly classified (%)	MAE	TP	Precision	Recall	F-score	ROC curve
1	Bayes net	42	84	0.1888	0.84	0.846	0.84	0.842	0.91
2	MLP	41	82	0.1809	0.82	0.818	0.82	0.819	0.887
3	AdaBoost	41	82	0.2068	0.82	0.818	0.82	0.816	0.871
4	KNN	41	82	0.1834	0.82	0.818	0.82	0.819	0.821
5	J48	36	72	0.2925	0.72	0.728	0.72	0.723	0.714
6	Random forest	40	80	0.2763	0.80	0.797	0.80	0.797	0.880
7	Random tree	39	78	0.2267	0.78	0.778	0.78	0.779	0.751
8	*Naïve Bayes*	*43*	*86*	*0.1819*	*0.86*	*0.862*	*0.86*	*0.861*	*0.918*
9	LMT	41	82	0.2233	0.82	0.818	0.82	0.816	0.885
10	LogitBoost	40	80	0.2136	0.8	0.789	0.8	0.793	0.885
11	Decision table	35	70	0.3637	0.7	0.725	0.7	0.706	0.731
12	SGD	40	80	0.2	0.8	0.797	0.8	0.797	0.771

Table 3 Comparison of various classification algorithms using percentage split

S. no.	Classifiers	Correctly identified instances	Correctly classified (%)	MAE	TP	Precision	Recall	F-score	ROC curve
1	*Bayes net*	*16*	*94*	*0.1224*	*0.941*	*0.945*	*0.941*	*0.938*	*0.923*
2	MLP	14	82	0.1729	0.824	0.842	0.824	0.83	0.904
3	AdaBoost	15	88	0.1857	0.882	0.898	0.882	0.867	0.894
4	KNN	15	88	0.1384	0.882	0.882	0.882	0.882	0.846
5	J48	14	82	0.2059	0.824	0.899	0.824	0.836	0.885
6	Random forest	13	76	0.2547	0.765	0.813	0.765	0.778	0.942
7	Random tree	13	76	0.2353	0.765	0.882	0.765	0.783	0.588
8	Naïve Bayes	15	88	0.2004	0.882	0.882	0.882	0.882	0.865
9	LMT	14	82	0.3215	0.82	0.899	0.824	0.836	0.942
10	LogitBoost	14	82	0.2226	0.824	0.842	0.824	0.83	0.885
11	Decision table	13	76	0.3456	0.765	0.882	0.765	0.783	0.846
12	SGD	13	76	0.2353	0.765	0.882	0.765	0.783	0.846

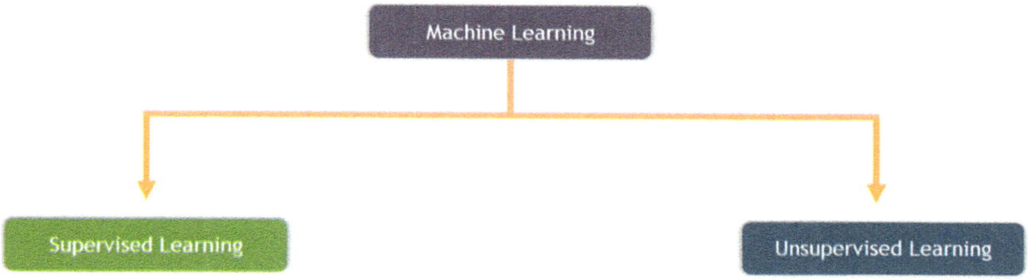

Fig. 1 Types of machine learning algorithms

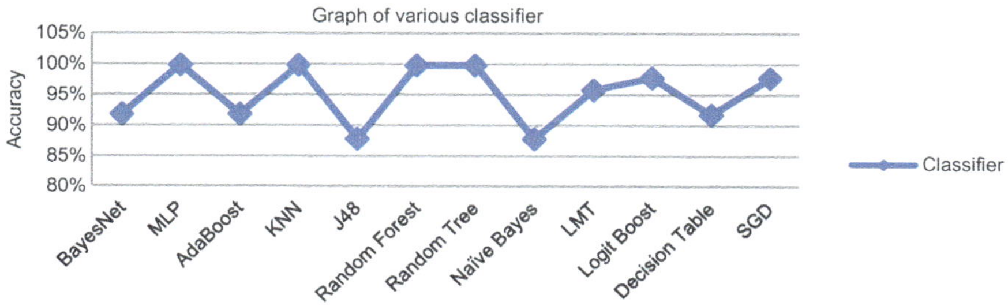

Fig. 2 Accuracy graph of 12 machine learning algorithm using the training dataset

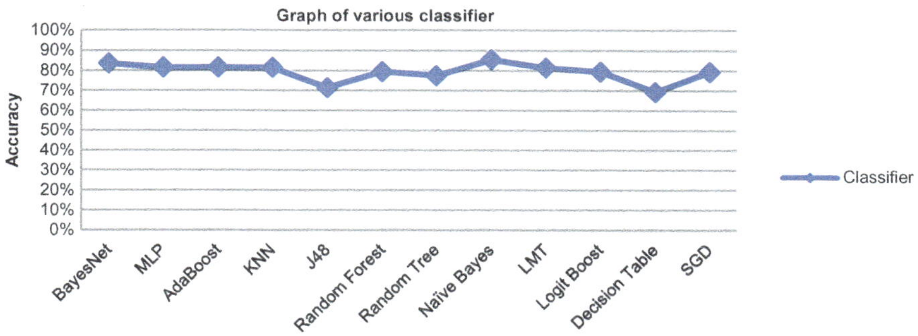

Fig. 3 Graph of 12 machine learning algorithm using tenfold validation methods

showed that the random tree method achieved average prediction 100% accuracy with zero mean absolute error in the training dataset, Naïve Bayes achieved an accuracy of 86% with 0.1819 mean absolute error rate in tenfold cross-validation method, and Bayes net obtained an accuracy of 94% with 0.1224 mean absolute error in percentage split test case. The accuracy of the prediction of HD can be further upgraded by creating different combinations of machine learning algorithms.

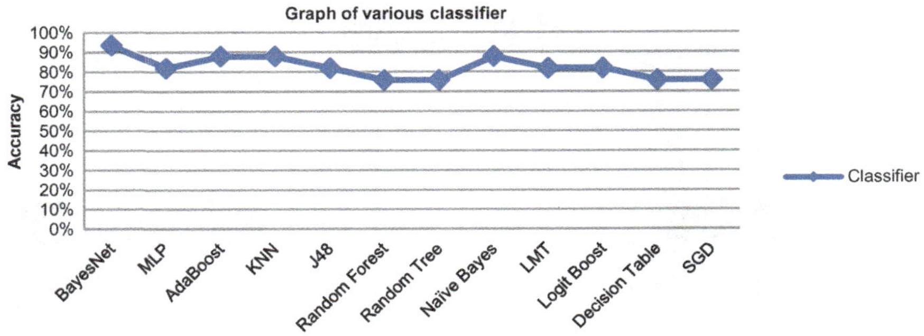

Fig. 4 Graph of 12 machine learning algorithm using percentage split

References

1. World Health Organization. (2016). Hearts: technical package for cardiovascular disease management in primary health care.
2. Ramalingam, V. V., Dandapath, A., & Raja, M. K. (2018). Heart disease prediction using machine learning techniques: a survey. *International Journal of Engineering & Technology, 7*(2.8), 684–687.
3. Devi, M. R. (2016). Analysis of various data mining techniques to predict diabetes mellitus. *International Journal of Applied Engineering Research, 11*(1), 727–730.
4. Alotaibi, F. S. (2019). Implementation of machine learning model to predict heart failure disease. *(IJACSA) International Journal of Advanced Computer Science and Applications, 10*(6).
5. Deekshatulu, B. L., & Chandra, P. (2013). Classification of heart disease using k-nearest neighbor and genetic algorithm. *Procedia Technology, 10,* 85–94.
6. Yahaya, L., Oye, N. D., & Garba, E. J. (2020). A comprehensive review on heart disease prediction using data mining and machine learning techniques. *American Journal of Artificial Intelligence, 4*(1), 20–29.
7. MacLennan, J., Tang, Z., & Crivat, B. (2011). *Data mining with Microsoft SQL server 2008.* Wiley.
8. Subhadra, K., & Vikas, B. (2019). Neural network based intelligent system for predicting heart disease. *International Journal of Innovation Technology and Exploring Engineering (IJITEE), 8*(5), 484–487.
9. Banu, G. R., & Jamala, J. H. (2015). Heart attack prediction using data mining technique. *International Journal of Modern Trends in Engineering and Research, 2*(5), 428–432.
10. Deepika, K., & Seema, S. (2016, July). Predictive analytics to prevent and control chronic diseases. In *2016 2nd International Conference on Applied and Theoretical Computing and Communication Technology (iCATccT)* (pp. 381–386). IEEE.
11. Mujawar, S. H., & Devale, P. R. (2015). Prediction of heart disease using modified K-means and by using Naive Bayes. *International Journal of Innovative Research in Computer and Communication Engineering (An ISO 3297: 2007 Certified Organization), 3*(10), 10265–10273.
12. Kumari, M., & Godara, S. (2011, June). Comparative study of data mining classification methods in cardiovascular disease prediction 1. *IJCST, 2*(2). ISSN: 2229-4333(Print) | ISSN: 0976-8491(Online)
13. Sridhar, A., & Kapardhi, A. (2018). Predicting heart disease using machine learning algorithm. *International Research Journal of Engineering and Technology, 6*(4), 36–38.
14. Dineshgar, G. P., & Singh, L. (2016). A review on DATA mining for heart disease prediction. *International Journal of Advanced Research on Electronics and Communication Engineering (IJARECE), 5*(2), 462–466.

15. https://www.tutorialspoint.com/machine_learning_with_python/index.htm
16. American Heart Association. (2014). Atherosclerosis https://www.heart.org/HEARTORG/Conditions/Cholesterol/WhyCholesterolMatters. *Atherosclerosis_UCM_305564_Article. jsp (18 Maret 2014).*

FPGA Implementation of Scaled "Quasi-Cyclic LDPC" Decoder Using High-Level Synthesis

Sarah Alaa Tamkeen and Ahmed A. Hamad

Abstract The increased demands for quality, like high throughput, low-latency, wide coverage, energy consumption, cost and reliable connections in mobile services, multimedia, and data transmission impose the use of advance technical requirements for the next fifth-generation (5G) new radio (NR). One of the most crucial parts in the physical layer of the new generation is the error correction coding technique. Three schemes, namely "Turbo codes", "low-density parity check (LDPC)", and "polar codes" are potentially considered as the candidate codes for both data and control channels. The competition is evaluated in terms of error correction capability, computational complexity, and flexibility. The parallelism, flexibility, and high processing speed of "field-programmable gate array (FPGA)" make it preferable in prototyping and implementation of different codes. In this paper, design, implementation, experimental verification, and validation of the quasi-cyclic LDPC encoder and decoder are proposed on the AWGN channel using Kintex-7 FPGA. Codewords are constructed and simulated for different length and rates. The performance and architectural complexity in terms of the throughput and silicon area are presented based on Vivado high-level synthesis (HLS). The paper also studies the effect of applying scaling factors to alleviate the overestimation of the "extrinsic information" that exchanges during the iterative decoding process.

Keywords Low-density parity check (LDPC) · Error correction codes · Minimum-sum (MS) · Field-programmable gate array (FPGA) · High-level synthesis (HLS)

S. A. Tamkeen (✉) · A. A. Hamad
Department of Electrical Engineering, Specialty of Communication Engineering, Babylon University, Babel, Iraq
e-mail: saraalaatamkeen1992@gmail.com

A. A. Hamad
e-mail: ahmedbabel@gmail.com

S.-L. Peng et al. (eds.), *Proceedings of First International Conference on Mathematical Modeling and Computational Science*, Advances in Intelligent Systems and Computing 1292, https://doi.org/10.1007/978-981-33-4389-4_14

1 Introduction

"Low-density parity-check codes" and its associated iterative "message-passing" decoding algorithms were first produced by Gallager in his 1962 Ph.D. work [1] and rediscovered again in 1995 by McKay and Neal [2]. Owens to the remarkable error correction performance (specially for large codeword length) and comparable decoding complexity (due to their highly structured parity-check matrices and iterative decoding) compared with the existing error correcting codes, "LDPC" codes have been adopted in several data communication and digital storage applications including LAN [3], WiMax [4], DVB-S2 [5], CCSDS [6], and ITU G.hn [7]. "Turbo codes" and "LDPC" codes are competing to reserve a place into the data channels within the new 5G NR standard. They all seem to have the same opportunity to win the race. The decision to use one of these codes must be based on a comprehensive examination to know that it meets the 5G requirements, namely the latency, throughput, resource requirements, error correction capability, consumed energy, bandwidth flexibility, and efficiency [8]. This imposes the physical implementation of these codes. One of the easiest approaches to accomplish this task is the use of FPGA. Adopting this trend will shorten the time required for prototyping and processing. Because of the probabilistic nature to the bit error rate (BER) performance test, it may take days on computer while on FPGA it can take hours [9]. This is mainly due to the ability of parallel processing that is characterized by FPGA.

Although the register-transfer level (RTL) model of design capture is completely compatible with ASIC design, it is analogous to assembly language programming. High-level synthesis makes it possible to accomplish the required design at abstracted level away from the implementation details, which consumes development time and effort. In this work, "HLS" tool is utilized to transforms a C code into a "RTL" implementation that can synthesize into the FPGA. In that way, the developers can avoid the specific knowledge of VHDL or Verilog RTL details to the designed circuits. Trying to take the advantage of these HLS characteristics, the source, irregular LDPC encoder, decoder, and the AWGN channel that this work describe is going to be implemented in C++ utilizing the Xilinx Vivado High-Level Synthesis (HLS) version 2018.3 over Digilent Genesys 2 board.

All the necessary communication system parts for testing the performance of irregular "LDPC" code are implemented, and we focus mainly on the design and implementation of "irregular LDPC" decoder. The structure of "parity check matrices (PCMs)" presented in the "IEEE 802.11n" standard is adopted [3].

2 "Quasi-Cyclic LDPC" Codes

The "parity check matrix (PCM)" of "QC" codes is based on a prototype matrix of elements that each represents a square submatrix P_i of size Z × Z. These submatrices are either null submatrices or cyclically shifting the columns of Z × Z identity matrix to the right by i elements according to the corresponding shifting value in the prototype matrix [10]. For instance, let us assume that P_0 is given by the following 2×2 identity matrix

$$P_0 = \begin{bmatrix} 1 & 0 \\ 0 & 1 \end{bmatrix} \quad (1)$$

and that the x × y prototype matrix H_p is given by

$$H_p = \begin{bmatrix} 0 & - & 1 & 0 \\ 1 & 0 & 1 & 1 \\ 0 & 1 & - & 0 \end{bmatrix} \quad (2)$$

Vacant entries in H_p denote Z × Z null submatrices. Therefore, the actual PCM $H = \left[H_{ji}\right]$ can be written as

$$H = \begin{bmatrix} 1\,0 & 0\,0 & 0\,1 & 1\,0 \\ 0\,1 & 0\,0 & 1\,0 & 0\,1 \\ 0\,1 & 1\,0 & 0\,1 & 0\,0 \\ 1\,0 & 0\,1 & 1\,0 & 0\,0 \\ 1\,0 & 0\,1 & 0\,0 & 1\,0 \\ 0\,1 & 1\,0 & 0\,0 & 0\,1 \end{bmatrix} \quad (3)$$

The semi-structure of the PCM of QC LDPC facilitates its memory addressing in hardware implementation. A shallow degradation in error correction performance should be accepted as a sacrifice for the benefits obtained by adopting this technique.

3 Minimum-Sum Decoding Algorithm (MSA)

Based on the messages properties that deal with, whether hard (binary) or soft channel information (log-likelihood ratio, LLR), there are mainly two general belief propagation (BP) decoding schemes for LDPC, namely bit-flipping (BF) algorithm and sum-product (SP) [11], 12 or its suboptimal algorithm, the min-sum (MS) [12]. The decoding idea is based on exchanging the extrinsic information iteratively between the decoder parts to increase the belief about the values of codeword bits. The passing of messages is carried out over the nodes of factor graph (or "Tanner graph") [13]. Factor graph is a graphical representation of the PCM H. It consists of two groups of nodes, the one that represents the columns of H, namely N variable or bit nodes (BNs), and M "check nodes" (CNs) that belong to the rows of H. This is depicted in Fig. 1 for the PCM of Eq. 3.

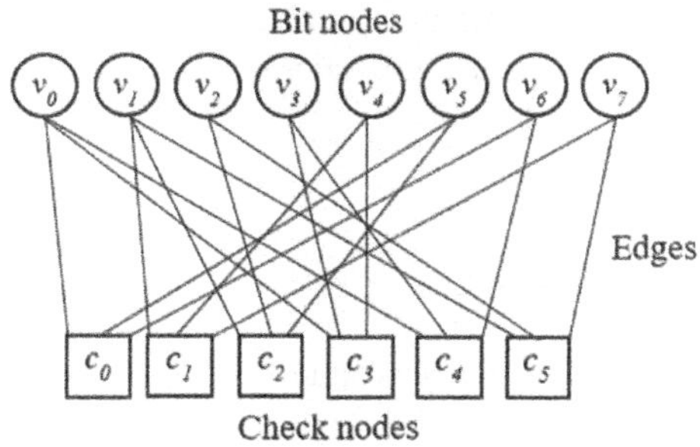

Fig. 1 A factor graph for the PCM of Eq. 3 headings

Let us define the group of adjacent "check nodes" to the "bit node" i is $\mathbb{C}_i = \{j{:}H_{ji} = 1\}$. While the group of adjacent "bit nodes" of the "check node" j is $\mathbb{R}_j = \{i{:}H_{ji} = 1\}$. On the other hand, the group of bits that involved in "check node" j except for bit i is denoted by $\mathbb{R}_{j/i}$. Here, the notation $\mathbb{C}_{i/j}$ refers to the group of checks in which bit i involved excluding check j. For each iteration I, the decoder estimates the following log-likelihood values:

- λ_{ji}^{I}: The LLR information that is transmitted from "check node" j to "bit node" i and recognized as "extrinsic information".
- γ_{ji}^{I}: The "LLR" information that is passed from "bit node" i to "check node" j.
- γ_n^i: The a-posteriori "LLR" of bit i.

The sum-product decoding algorithm can be outlined by the following steps.

Step 1: Set the following parameters:

- $\gamma_{ji}^{I} = r_i$ for each i and j (where r_i is the received symbol corresponds to the codeword bit i).
- the maximum number of iteration $I_{\max}$.
- iteration counter $I = 1$.
- Failure counter $f = 0$.

Step 2: Horizontal update: For each i, j with $= H_{ji} 1$, compute

$$\lambda_{ji}^{I} = 2\tanh^{-1}\left(\prod_{i' \in \mathbb{R}_{j/i}} \tanh\left(\frac{\gamma_{ji'}^{I-1}}{2}\right)\right) \quad (4)$$

Step 3: The a-posteriori LLR: For each j, i with $= H_{ji} 1$, compute

$$\gamma_i^{I} = r_i + \sum_{j \in \mathbb{C}_i} \lambda_{ji}^{I} \quad (5)$$

Step 4: Hard decision: denote the decoded codeword bits as

$$\hat{c}_n = \begin{cases} 0 if \gamma_i^I \geq 0 \\ 1 if \gamma_i^I < 0 \end{cases} \tag{6}$$

Step 5: Calculate syndrome s: As a stopping criteria, if the syndrome $s = H\hat{c}^T = 0$ or the iteration reaches its maximum limit $I = I_{\max}$ jump to step 7, otherwise go to step 6.

Step 6: Vertical update:

$$\gamma_{ji}^I = r_i + \sum_{j' \in \mathbb{C}_{i/j}} \lambda_{j'i}^I \tag{7}$$

set $I = I + 1$ and go back to step 2.

Step 7: Announce that $\hat{c}$ is the estimated codeword. If $H\hat{c}^T \neq 0$ increment the failure counter $f = f + 1$.

Own to the high complexity in implementing Eq. 4, SP is rarely used in practical applications and therefore, its simplified version, MS is usually adopted. It can be easily verified that the effective terms γ_{ji}^I in Eq. 4 is the one with smallest values. Hence, it is possible to approximate Eq. 4 to be

$$\lambda_{ji}^I = \prod_{i' \in \mathbb{R}_{j/i}} sgn\left(\gamma_{ji'}^{I-1}\right) . \min_{i' \in \mathbb{R}_{j/i}} \gamma_{ji'}^{I-1} \tag{8}$$

It is clearly evident that the hardware resources and latency needed to implement Eq. 8 is much efficient than Eq. 4. The reduction in decoding complexity obtained by using MS algorithm is on the account of a loss in bit error rate performance (0.5–1 dB). This reduction is mainly due to the over estimation of extrinsic information passing between *BNs* and *CNs*. Many literatures proposed different remedies to mitigate this degradation [14–20]. Perhaps the most classical approaches are the ones based on normalized and/or offset MS algorithm. Due to its low computational complexity (simple multiplier), the non-adaptive algorithm in [20] is used, in which, a constant scaling factor is applied to the *LLRs* information that interchanged between the nodes of the factor graph.

The modified *LLR* values λ'^I_{ji} and γ'^I_{ji} are given by the following equations:

$$\lambda'^I_{ji} = a^I \lambda_{ji}^I \tag{9}$$

$$\gamma'^I_{ji} = b^I \gamma_{ji}^I \tag{10}$$

In [20], the estimated scaling values are adaptive according to the code and channel characteristics. Meanwhile, it declares that using an average value of 0.899 for both scaling factors a^I and b^I for different code and channel gives a good performance.

4 Proposed Hardware MS Decoder Design

Owen to the semi-structure and sparseness of PCM, the hardware implementation of QC-LDPC decoder can be significantly simplified. A resource efficient fully or partially parallel decoder architecture is particularly suited for "QC-LDPC" [21]. Due to the resources and memory size limitations, it is impractical to store every element in PCM especially for large codeword length. Instead of that, the addresses of nonzero entries are stored in submatrices. In "QC-LDPC" codes, the size of these submatrices depends on the structure of "PCM".

In this work, a flooding schedule in a layered orientation is proposed in processing the BNs and CNs messages. The algorithm comprises of splitting the "PCM" in horizontal and vertical layers. A horizontal layer of PCM is the group of Z successive rows which correspond to certain row within the prototype matrix H_p. In a similar manner, the vertical layer is defined as the group of Z successive columns.

The architecture of irregular LDPC decoder consists of the following main units:

1. Processing nodes: comprises x = N/Z processing bit nodes units PBNs and y = M/Z processing check node units PCNs which are responsible for computing the bit to check and check to bit messages respectively.
2. Bit routing networks (BRNs): there are N/Z routing units corresponds to each bit processing node. Each network unit contains the addresses of all "check nodes" connected to each "bit node" in the bit processing node. Usually dedicated ROM memories are used to store this information.
3. Check routing networks (CRNs): there are M/Z routing units corresponds to each check processing node. Each network unit contains the addresses of all "bit nodes" connected to each "check node" in the check processing node.
4. Memory blocks: required to store both the received "LLRs" and the bit to check and check to bit messages. According to the targeted parallelism, throughput and available resources a decision must be made on applying single-port, multi-port, and distributed block RAM (BRAM) memory.

Given that $N = 8$, $M = 6$, and $Z = 2$, the BRNs for the code presented by the PCM in Eq. 3 are:

$$BRN0 = \begin{bmatrix} 0\ 1 \\ 3\ 2 \\ 4\ 5 \end{bmatrix}, BRN1 = \begin{bmatrix} 2\ 3 \\ 5\ 4 \end{bmatrix}, BRN2 = \begin{bmatrix} 1\ 0 \\ 3\ 2 \end{bmatrix} \text{ and } BRN3 = \begin{bmatrix} 0\ 1 \\ 4\ 5 \end{bmatrix}$$

On the other hand, the CRNs are

$$CRN0 = \begin{bmatrix} 0\ 5\ 6 \\ 1\ 4\ 7 \end{bmatrix}, CRN1 = \begin{bmatrix} 1\ 2\ 5 \\ 0\ 3\ 4 \end{bmatrix} \text{ and } CRN2 = \begin{bmatrix} 0\ 3\ 6 \\ 1\ 2\ 7 \end{bmatrix}$$

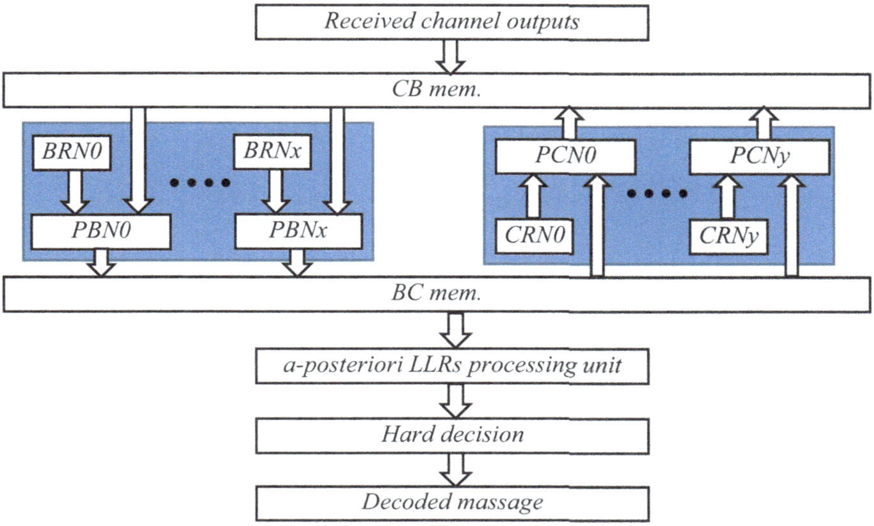

Fig. 2 General block diagram of proposed irregular LDPC decoder

Figure 2 depicts a general block diagram of the proposed "LDPC" decoder. Since each PBN and PCN is working independently, high degrees of parallelism can be achieved.

5 Results

To explore the degree of parallelization that can be achieved using the proposed decoder, we suggest two codes, first one is the (8,2) small-scale LDPC example given by Eq. 3. The second one is rate ½ (1296,648) irregular LDPC code that is produced by IEEE 802.11n.

Vivado HLS facilitates the process of applying different level of pipelining, data flow, and unrolling processes. Table 1 shows details on latency, instantiation interval

Table 1 Latency and utilization estimate for (8,2) LDPC decoder

	Solution1	Solution2
Latency (CLK)	1008	14
Throughput	1008	1
BRAM_18K	4	0
DSP48E	5	30
FF	3892	5961
LUT	7716	6854

Table 2 Latency for (1296,648) MS irregular QC-LDPC decoder with 12 iterations

	Min	Max
Float point	340,692	5,548,937
S float point	354,624	5,716,121
D float point	61,783	701,494
S D float point	62,277	707,444
S D fixed32 point	40,802	449,720

(throughput), and different FPGA resources (LUT, FF, BRAM, and DSP processor) for two syntheses solutions with different optimization directive. Solution 1 uses the default HLS directive which is arranged to achieve a compromise between throughput and resources. On the other hand, in solution 2, different parallelism directives are applied such as partitioning all memories and unrolling all looping iterations. It is evident that by using the proposed decoder architecture one can achieve a throughput of one cycle per decoding iteration process with no significant sacrifice in resources utilization.

For the (1296,648) irregular LDPC code, similar results are achieved. It is worth to mention that not all pipelining directives are applicable as in solution 2 of the (8,6) code. This is due to the resource limitations of the FPGA device. Table 2 shows the latency and resources utilization estimate of this code for different data precision, scaling, and optimization directive.

Different hardware tests for the irregular, rate ½ (1296,648) QC-LDPC are carried out over AWGN channel code to examine the effects of scaling and arbitrary precision on bit error rate performance. All tests are implemented over the Genesys 2 development platform board based on the Kintex-7 FPGA from Xilinx. The schematic diagram of the test circuit that used in this work is shown in Fig. 3. It consists of two types of IP blocks; the ones that generated by the HLS (like AWGN channel, LDPC encoder, and decoder) and the IPs provided by the Vivado (like Microblaze, timer, AXI interface, DDR memory, clock manager, etc.).

Figure 4a, b depicts the BER performance and probability of decoding failure of the (1296,648) QC-LDPC code where the LLRs are given in float format. The figures reveal the effect of scaling on the BER and probability of frailer. Here, the letter D

Table 3 Utilization estimate for (1296,648) MS irregular QC-LDPC decoder with12 iteration

	BRAM_18K % of 890	DSP48E % of 840	FF % of 407,600	LUT % of 203,800
Float point	7	0	3	18
S float point	7	0	3	18
D float point	22	2	12	37
S D float point	22	3	11	35
S D fixed32 point	22	25	7	35

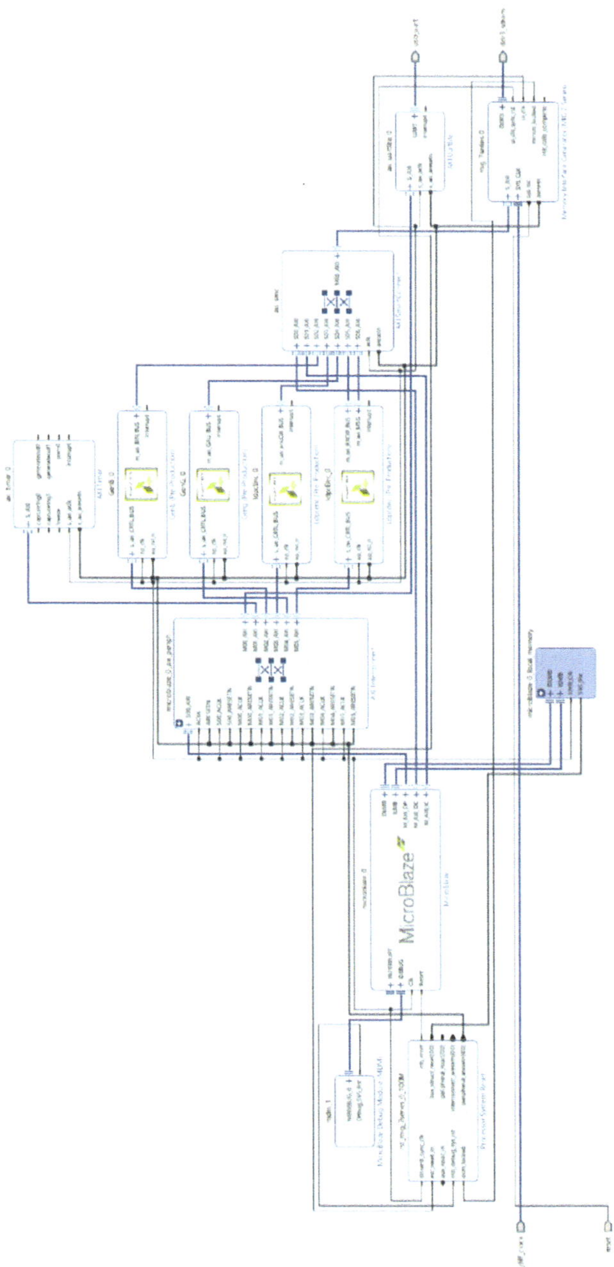

Fig. 3 Schematic diagram of the tested system

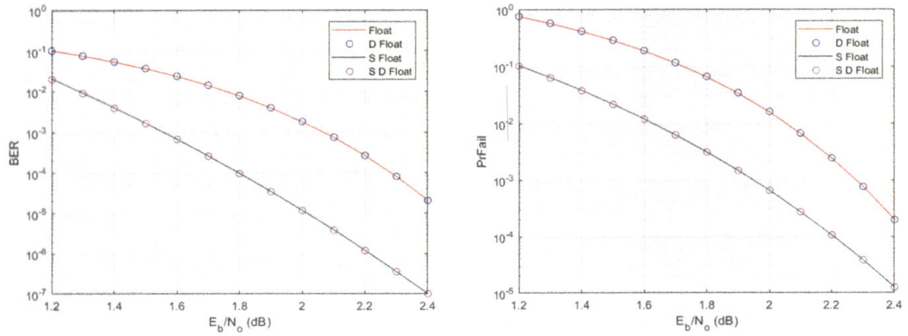

(a) (b)

Fig. 4 **a** BER and **b** probability of decoding failure of (1296,648) irregular QC-LDPC code with maximum iterations = 100

refers to the systems with optimization directive and the letter S refers to the systems using scaling factor.

The effects of scaling on the average number of iterations for the (1296,648) code is given by Fig. 5a. The figure shows also that using HLS directive like partitioning of arrays and loop unrolling has no adverse effect on code performance while it reduces the processing latency significantly as depicted in Fig. 5b.

Figure 6a–c reveals that using arbitrary precision instead of float format, provided by HLS, can reduce the processing time significantly but conserving other parameters

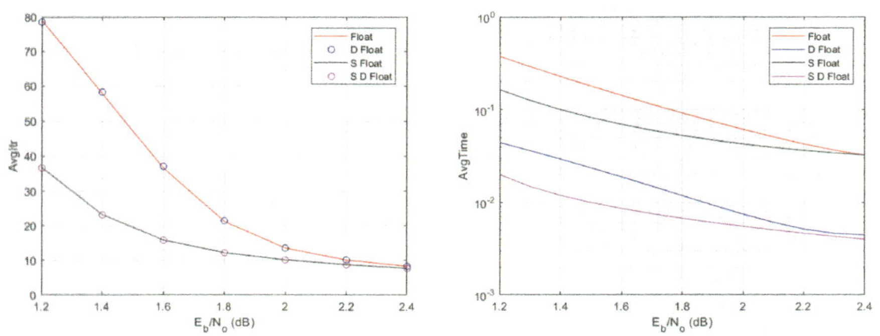

(a) (b)

Fig. 5 **a** Average number of iterations and **b** average estimated decoding time of (1296,648) irregular QC-LDPC code with maximum iterations = 100

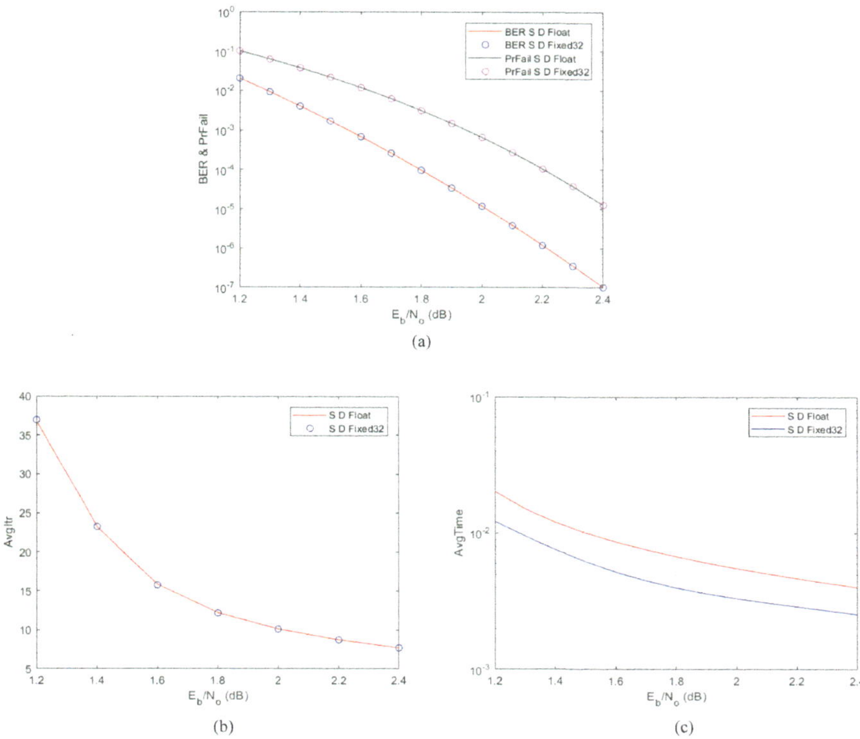

Fig. 6 **a** BER and probability of decoding failure, **b** average number of iterations, **c** average estimated decoding time of (1296,648) irregular QC-LDPC code with maximum iterations = 100

6 Conclusion

In this study, a fully and partially parallel "QC-LDPC" decoders are designed based on high abstract C++ language using Vivado HLS. Utilizing the optimization directive provided by HLS, the same program can be used to generate HDL code to represent both fully or partially parallel decoder. It is proved that by utilizing the semi-structure that the QC-LDPC characterized by, a decoder with processing latency of 1 cycle is feasible conditioned on availability of enough resources in the FPGA device. It is shown that multiplication of extrinsic information by scaling factor improves the BER performance by 0.5 dB and approximately divides the probability of failure by a factor of 30. It is shown that a significant reduction in number of iteration and processing latency is obtained when scaling is applied. The results show that using ap-fixed format instead of float to represent the exchanged LLRs between check and bit nodes can give the same performance with reduced processing latency.

References

1. Gallager, R. G. (1963). *Low Density Parity-Check Codes*. Cambridge, MA: MIT Press.
2. MacKay, D., & Neal, R. (1995, October) Good codes based on very sparse matrices. In: *Proceedings of 5th IMA Conference Cryptography and Coding* (pp. 100–111).
3. IEEE 802.11n. (2006). *Wireless LAN Medium Access Control and Physical Layer Specifications: Enhancement for Higher Throughput*, IEEE P802.11n/D1.0.
4. IEEE 802.16e. (2005). *Air Interface for Fixed and Mobile Broad- band Wireless Access Systems*, IEEE P802.16e/D12 Draft.
5. European Telecommunications Standards Institute (ET-SI). (2005). *Digital Video Broadcasting (DVB) Second Generation Framing Structure for Broadband Satellite Applications*, EN 302 307 v1.1.1.
6. CCSDS. (2011). CCSDS 131.0-B-2 Recommendation for Space Data System Standards; TM Synchronization and Channel Coding.
7. Oksman, V., & Galli, S. (2009). G.hn: The new ITU-T home networking standard. *IEEE Communications Magazine, 47*(10), 138–145.
8. Hailes, P., Xu, L., Maunder, R. G., Al-Hashimi, B. M., & Hanzo, L. (2016). A survey of FPGA-based LDPC decoders. *IEEE Communications Surveys and Tutorials, 18*(2), 1098–1122.
9. Cai, Y., Jeon, S., Mai, K., & Kumar, B. V. K. V. (2009). Highly parallel FPGA emulation for LDPC error floor characterization in perpendicular magnetic recording channel. *IEEE Transactions on Magnetics, 45*(10), 3761–3764.
10. Fossorier, M. P. C. (2004). Quasi-cyclic low-density parity-check codes from circulant permutation matrices. *IEEE Transaction on Information Theory, 50*(8), 1788–1793.
11. Wiberg, N. (1996). *Codes and Decoding on General Graphs*. Ph.D. thesis, Likoping University, Sweden.
12. Fossorier, M. P. C., Mihaljevic, M., & Lmai, H. (1999, May). Reduced complexity iterative decoding of low-density parity check codes based on belief propagation. *IEEE Transactions on Communications, 47*(5), 673–680.
13. Tanner, R. (1981). A recursive approach to low complexity codes. *IEEE Transactions on Information Theory, 27*(5), 533–547.
14. Chen, J., & Fossorier, M. (2002). Near optimum universal belief propagation based decoding of low-density parity check codes. *IEEE Transactions on Communications, 50*(3), 406–414.
15. Chen, J., & Fossorier, M. (2002). Density evolution for two improved bp-based decoding algorithms of LDPC codes. *IEEE Communications Letters, 6*(5), 208–210.
16. Chandrasetty, V. A., & Aziz, S. M. (2012). An area efficient LDPC decoder using a reduced complexity min-sum algorithm. *Integrated VLSI Journal, 45*(2), 141–148.
17. Chandrasetty, V. A., & Aziz, S. M. (2010). FPGA implementation of high performance LDPC decoder using modified 2-bit min-sum algorithm. In: *Proceedings of the 2nd International Conference on Computer Research and Development* (pp. 881–885).
18. Shafiullah, D., Islam, M., Faisal, M., & Rahman, I. Optimized min-sum decoding algorithm for low density pc codes. In: 2012 *14th International Conference on Advanced Communication Technology (ICACT)* (pp. 475–480).
19. Xu, M., Wu, J., Zhang, M. (2010). A modified offset min-sum decoding algorithm for LDPC codes. In: *2010 3rd IEEE International Conference on Computer Science and Information Technology (ICCSIT)* (pp. 19–22).
20. Hamad, A. A. (2013). Estimation of Two-Dimensional Correction Factors for Min-Sum Decoding of Regular LDPC Code. *Scientific Research Magazine, Wireless Engineering and Technology, 4*(4), 181–187. https://doi.org/10.4236/wet.2013.44027.
21. Shimizu, K., Ishikawa, T., Togawa, N., Ikenaga, T., & Goto, S. (2005, October 2–5). Partially-parallel LDPC decoder based on high efficiency message-passing algorithm. In: *Proceedings of the IEEE International Conference on Computer Design: VLSI in Computers and Processors* (pp. 503–510).

Automatic Road Enforcement Detection System Using Intelligent Robust Fuzzy Trained Stochastic Neural Network

M. D. Vimalapriya and S. Vignesh Baalaji

Abstract The intelligent transport system helps to manage the transport systems and transport network and provide better service to the user. Among the various ITS involvements, the most precise is the road enforcement process, monitoring the vehicle activities with the help of a camera to predict the illegal activities on the road. Several automatic systems have been created to predict the disobeying movements on the road. But the traditional system consumes more time and increases computation complexity along with accuracy issues. For overcoming the above problems, in this work, intelligent, robust fuzzy trained stochastic neural network has been designed and developed. Here, the system analyzes the incoming vehicle information by using the stochastic transfer function, and the obtained results are examined using the fuzzy program, which helps to decide on the disobeying activities such as over speed, accidents, level crossing, and so on. The discussed intelligent transport system is developed using MATLAB. During the implementation process, the San Diego Freeway Data Hourly dataset from the Intelligent Transportation Systems Research Data Exchange is used, and the excellence of the system is evaluated with the help of the performance metrics such as error rate (0.022) precision (99.08%), recall (99.06%), and accuracy (99.34%).

Keywords Intelligent transport system · Disobeying activities · Intelligent robust fuzzy trained stochastic neural network · Fuzzy program · And safety system dataset

---

M. D. Vimalapriya (✉)
Women's Christian College, Chennai, India
e-mail: prshvi17375@gmail.com

S. Vignesh Baalaji
Vellore Institute of Technology, Chennai, India
e-mail: vigneshbaalajis@gmail.com

S.-L. Peng et al. (eds.), *Proceedings of First International Conference on Mathematical Modeling and Computational Science*, Advances in Intelligent Systems and Computing 1292, https://doi.org/10.1007/978-981-33-4389-4_15

1 Introduction

Intelligent transport system is one of the effective applications that provide valuable and pioneering services [1, 1] to the transport process and traffic management process. ITS technologies are widely used in emergency services such as accident issues, violation of traffic rules, over-speed, and so on. The ITS system is not only used in the road transport management process, but is also used in the infrastructure analysis, traffic management process, vehicle, user analysis, and mobility management process. The traffic information is collected through CCTV camera that includes the traffic lights, vehicle information, emergency details [2], etc., To manage the difficulties, practical work intelligent, robust fuzzy trained stochastic neural network has been considered. The rest of the manuscript is arranged as follows: Sect. 2 analyzes a few research works of the authors regarding the intelligent transport systems. Section 3 discusses the robust fuzzy trained stochastic neural network related to the ITS for predicting the disobeying activities; Sect. 4 examines the efficiency of the robust fuzzy trained stochastic neural network related to the ITS for predicting the disobeying activities and concludes in Sect. 5.

2 Related Works

We discuss a few authors working concepts of the intelligent transport system for creating an effective roadside communication process [3] making the smart transport system using a neural network to predict the route and destination. Pamuła et al. [4] developing the traffic forecasting system by utilizing the traffic data. Gallo et al. [5] introducing single-layer feedforward neural network-based traffic flow detection of the road traffic sensor data. Bai et al. [6] detecting the short-term traffic flow in the urban network using deep learning neural network structure. Qu et al. [7] creating the system to predict the real-time freeway rear-end crash potentiation using a multi-layer perceptron with a support vector machine that has been analyzed by the author. This process collects data from the Interstate 894 Milwaukee, USA, which was processed by a support vector machine to predict the relationship between the rear-end crash occurrence and respective traffic conditions. Further, the SVM approach was applied to create a useful training model that helps to predict the rear-end crashes successfully. According to the above discussion, the researchers' opinions and discussions on intelligent transport systems were created with the help of machine learning techniques, from which the proposed work has a specific research goal, which was completed using the intelligent machine learning algorithm. The aim of this research incorporates the introduction of an optimized learning algorithm, to minimize the misclassification error rate using a fuzzy-based decision-making process. We also focus to minimize the computational complexity involved in the recognition of disobedient activities. We are creating a system to support large volumes of traffic datasets with minimum deviation and maximum accuracy. The detailed working process of the robust fuzzy trained stochastic neural network is discussed in Sect. 3.

3 Robust Fuzzy Trained Stochastic Neural Network-Based Intelligent Transport System Creation for Disobeying Activities Prediction

3.1 Materials and Methods

This section discusses the robust fuzzy trained stochastic neural network related to the intelligent transport system for predicting the disobeying activities. The introduced system uses the San Diego Freeway Data Hourly dataset [8] from the Intelligent Transportation Systems Research Data Exchange that details to predict the road transport system's disobedience activities effectively. The San Diego environment has 3000 traffic detectors and 1250 lane miles of 1–5 San Diego, which continuously report the traffic information for every 30 s. In addition to this, data are collected with the help of GPS that collects the vehicle's longitude and latitude information; also, individual vehicle time, weather information, and other traffic details that are collected are processed using the robust fuzzy trained stochastic neural network. Furthermore, the detailed description of the work is explained in the following section.

3.2 Robust Fuzzy Trained Stochastic Neural Network-Based Intelligent Transport System Creation for Disobeying Activities Detection Process

The introduced robust fuzzy trained stochastic neural network-based intelligent transport system process works according to the three essential processes as listed as follows: pretraining module, classification module, and transport decision-making model.

3.3 Pretraining Module

The first step of the work is the pretraining module, and in this model, the collected transport data are considered as X, and the relevant output value is represented as Y. The pretraining process is done using the deep learning algorithm [9], which extracts the information from the data, which is used with the help of unlabeled data. As discussed earlier, the X is the input to the deep learning approach. The deep belief network has a process where the traffic data is analyzed according to the restricted Boltzmann machine energy-based model [10]. Initially, the energy value of the model is identified, and the Gaussian noise value is used to represent the traffic data. The trained data collected traffic data-related condition probability, and energy function (u, f) are computed as follows,

$$-\log P(u, f) \propto E(u, f; \theta) = \sum_{i=1}^{|U|} \frac{(u_i - a_i)^2}{2\sigma_i^2} - \sum_{j=1}^{|F|} b_j f_j - \sum_{i=1}^{|U|} \sum_{j=1}^{|F|} \frac{u_i}{\sigma_i} f_i w_{ij} \quad (1)$$

In Eq. (1) visible unit is denoted as u and the hidden unit of the deep belief network is represented as f.

From Eq. (1), θ is computed as follows,

$$\theta = (w, a, b) \quad (2)$$

From the above-defined Eq. (2), θ is denoted as a restricted Boltzmann machine parameter.

w_{ij} is the weight value of the deep network of layer i–j.

a_i and b_j is denoted as a bias value of the network.

According to Eqs. (1 and 2), the energy function of traffic data is estimated, and the respective conditional probability value is computed for (u, f) as follows.

$$\rho(f_j | u; \theta) = \mathrm{sigm}\left(\sum_{i=1}^{|U|} w_{ij} u_i + b_j\right) \quad (3)$$

$$\rho(u_j | h; \theta) = N\left(a_j + \sigma_i \sum_{j=1}^{|H|} w_{ij} f_i, \sigma_i^2\right) \quad (4)$$

In Eq. (4), $N\left(a_j + \sigma_i \sum_{j=1}^{|H|} w_{ij} f_i, \sigma_i^2\right)$ is represented as the Gaussian distribution. Then, the sign is denoted based on the sigmoid function [11], estimated as follows.

$$\mathrm{sigm}(x) = \frac{1}{(1 + e^{-x})} \quad (5)$$

With the help of the computation, the distribution by u is estimated as follows,

$$p(u) = \sum_f p(f|\theta) p(u|f, \theta) \quad (6)$$

The above process is completed in every deep belief neural network in a layer by layer. Here, every layer uses the visible traffic data into the hidden data. The output of hidden data is transmitted to the next hidden layer; during this process, hidden layer output is treated as a visible data vector. During this process, network weight and bias values are continuously examined at the time of the classification process. The effective generation of energy function and distribution probability value determines the efficiency of the traffic data training process and the sigmoid activation function that provide the proper learning concept to the data effectively.

Then, the new incoming data arrived at the system based on the classification process, which is completed using the stochastic neural network, as is explained in the following section.

3.4 Classification Module

The second step of the work is done in the classification module, which is performed using a stochastic neural network approach [12]. It is the type of neural network that is used to classify the incoming traffic data using the neuron stochastic transfer function. During the classification process network, the handling of the local minima problem has been computed. The system uses the Boltzmann machine while classifying the incoming traffic data effectively for the classification model. The stochastic network uses the probability function [13] in a dynamic manner to predict the exact output of the given input. Let us consider the network, which has several states such as s, and each state has respective energy values which are represented as $E_1, E_2, \ldots E_s$. From the initialization, the probability value of one state energy to next state energy value is represented as follows,

$$p_{ij} = \frac{1}{1 + e^{(E_j - E_i)/T}} \tag{7}$$

The computed probability value effectively works on the given traffic systems and reaches to the stable probabilistic state as defined as Boltzmann distribution. The traffic system must reach the thermal equilibrium, which is computed from the given state s, energy value E, and respective probability value.

$$p_i = \frac{e^{-E_i/T}}{Z} \tag{8}$$

In Eq. (8), Z is named as the normalized factor that is estimated as follows,

$$Z = \sum_{i=1}^{m} -E_i/T \tag{9}$$

The computed Boltzmann distribution [14] helps to deal with the optimization problem and benefits to predict the specific transportation activities effectively. After computing the Boltzmann distribution, it has been used for learning and classifying the given traffic inputs. Let α be the input unit index value and β be the hidden unit's index value. Based on the index value, let us consider m as the input unit value, k is the hidden unit, and then the network is structured in terms of $m + k$ units. Based on the network index value and input units, the network probability value is estimated as follows,

$$P_{\alpha} = \sum_{\beta} P_{\alpha\beta} \tag{10}$$

In Eq. (10), $P_{\alpha\beta}$ is represented at the combine network state which is computed from the hidden unit's state summation value, as rewritten as follows,

$$P_{\alpha} = \frac{1}{Z}\sum_{\beta} \exp(-\gamma E_{\alpha\beta}) \tag{11}$$

$$Z = \sum_{\alpha,\beta} \exp(-\gamma E_{\alpha\beta}) \tag{12}$$

In Eqs. (11 and 12), $\gamma = 1/T$, and the computed Boltzmann distribution value has been analyzed using the network energy value as estimated as follows.

$$E_{\alpha\beta} = -\frac{1}{2}\sum_{i=1}^{n}\sum_{j=1}^{n} w_{ij} x_i^{\alpha\beta} x_j^{\alpha,\beta} \tag{13}$$

Based on the above process, the output is computed from the incoming traffic data. Besides, the actual value is compared with the expected value using the distance value. The computation is performed with training and computed output value, which is done as follows.

$$D = \sum_{\alpha} R_{\alpha} \log \frac{R_{\alpha}}{P_{\alpha}} \tag{14}$$

The computed distance value (D) must be 0 when the estimated output value is R_{α} that is equal to the actual output value P_{α}. Else, the network consumes error value, based on the network weight, the value must be updated, which is shown and mathematically computed as follows.

$$\Delta w_{ij} = -\eta \frac{\partial D}{\partial w_{ij}} = \eta \sum_{\alpha} \frac{R_{\alpha}}{P_{\alpha}} \frac{\partial P_{\alpha}}{\partial w_{ij}} \tag{15}$$

$$\frac{\partial P_\alpha}{\partial w_{ij}} = \frac{\gamma \sum_\beta \exp(-\gamma E_{\alpha,\beta}) x_i^{\alpha\beta} x_j^{\alpha,\beta}}{Z} \tag{16}$$

$$= -\frac{\gamma\left(\sum_\beta \exp(-\gamma E_{\alpha,\beta})\right)\sum_{\lambda\mu} \exp(-\gamma E_{\lambda\mu}) x_i^{\lambda\mu} x_j^{\lambda\mu}}{Z} \tag{17}$$

$$= \gamma\left(\sum\nolimits_\beta x_i^{\alpha\beta} x_j^{\alpha,\beta} P_{\alpha\beta} - \sum_\alpha R_\alpha \langle x_i, x_j \rangle_{free}\right) \tag{18}$$

In the above derivation, the expected value of inputs in *i* and *j* state value is denoted as $\langle x_i, x_j \rangle_{\text{free}}$.

From the results, the updating weight value is consolidated as follows.

$$\Delta w_{ij} = \eta\gamma\left(\frac{R_\alpha}{P_\alpha}\sum\nolimits_\beta x_i^{\alpha\beta} x_j^{\alpha,\beta} P_{\alpha\beta} - \sum_\alpha R_\alpha \langle x_i, x_j \rangle_{\text{free}}\right) \tag{19}$$

Based on this process, the network weights are continuously updated to get the output value for the given traffic input. Finally, the computed output value is fed into the decision-making process to know the exact outcome of the given input. The decision-making process is done using the robust fuzzy rules which have been analyzed using the traffic data in different cases, as listed as the worst and best case. The detailed decision making of the intelligent transport system is explained in the following section.

3.5 Decision-Making Module

The final stage of the work is a decision-making module, which is done with the help of a robust fuzzy approach. The fuzzy approach can resolve the mathematical optimization problems under uncertainty. It predicts the obedience of the driving process for the given traffic input value, based on the features like the driving speed, and direction. Then, the optimal value of the system is obtained using the objective function, in which the traffic criteria are analyzed continuously, and the obedience-related information is estimated effectively according to the target value. The computed value is compared with the training value if it is the same, the target having 0 value, then the driver follows the traffic rules, as well as their obedience activities are predicted. Thus, the introduced robust fuzzy trained stochastic neural network-related for creating the intelligent transport system is successfully developed to predict the disobedience activities while driving. The efficiency of the presented system uses a smart transportation management system for the evaluation using experimental analysis, which is discussed in Sect. 4.

4 Results and Discussion

This section discusses the efficiency of the robust fuzzy trained stochastic neural network (RFTSNN) related to the intelligent transport system for predicting the disobedience activities. During this process, the dataset consists of several volumes of traffic data, which was discussed in Sect. 3, materials, and method section.

The gathered information was processed, as discussed in section. The excellence of the system was implemented using the MATLAB tool to evaluate the proficiency of the system. Here, the collected data is divided into two parts, such as training (70%) and testing (30%) dataset based on the worst and best case. According to the different period, the activities of people on the transport system were continuously monitored with the help of the future transportation process. During the training and testing process, the system consumes error rate that reduces the efficiency of the system. So, the intelligent transport system deviation value is estimated using the absolute error value (AEV), mean absolute error value (MAEV), and root mean square error value (RMSE). These error values are computed as follows

$$\text{Absoluteerrorvalue} = \left|\frac{\widehat{y}_i - y_i}{y_i}\right| * 100 \tag{20}$$

$$\text{Meanabsoluteerrorvalue} = \frac{1}{n}.\sum_{i=1}^{n}\left|\frac{\widehat{y}_i - y_i}{y_i}\right| * 100 \tag{21}$$

$$\text{Rootmeansquareerrorvalue} = \left[\frac{1}{n}\sum_{i=1}^{n}\left(\left|y_i - \widehat{y}_i\right|\right)^2\right]^{1/2} \tag{22}$$

In the above Eqs. (20, 21, and 22), y_i represents the actual traffic data and $\widehat{y}_i$ is computed traffic data. According to the discussion, the computed error value is shown in Table 1.

According to the above Table 1, the introduced robust fuzzy trained stochastic neural network (RFTSNN) approach attains a minimum error value which is computed from the Eqs. (20–22). The practical computation of the distance value (D) is estimated in terms of $\frac{R_a}{P_a}$. The computed value helps to find the deviation between the actual and predicted value. The effective utilization of the pretraining model with the Boltzmann machine model improves the overall training rate that minimizes the deviation effectively. The proposed system ensures optimized results while comparing the computed and actual value. According to the analysis, the respective graphical structure is shown in Fig. 1.

From Fig. 1a–d, introduced a robust fuzzy trained stochastic neural network (RFTSNN) ensures minimum error rate on different cases with a different period. The transportation information is examined in terms of the early morning, afternoon, and office hours. Even though the system uses a different time period of time, the system consumes a minimum deviation of expected and predicted traffic value. The

Table 1 Error value—SFNN, SVM, MLP, DNN, RFTSNN

Methods	Period	Best case			Worst case		
		AEV	MAEV	RMSE	AEV	MAEV	RMSE
Single-layer feedforward neural network (SFNN)	Earlier morning (EM)	0.124	0.18	0.12	0.18	0.12	0.09
	Afternoon hour (AH)	0.189	0.112	0.09	0.123	0.098	0.089
	Off hour (OH)	0.176	0.167	0.12	0.176	0.15	0.092
Support vector with MLP (SVM + MLP)	Earlier morning (EM)	0.098	0.18	0.092	0.12	0.09	0.10
	Afternoon hour (AH)	0.10	0.102	0.087	0.0879	0.0839	0.0927
	Off hour (OH)	0.12	0.098	0.0923	0.0923	0.0867	0.0934
Deep learning neural networks (DNN)	Earlier morning (EM)	0.098	0.0868	0.093	0.0967	0.0952	0.0912
	Afternoon hour (AH)	0.087	0.084	0.092	0.089	0.0912	0.093
	Off Hour (OH)	0.067	0.0723	0.0789	0.0724	0.0823	0.0845
Robust fuzzy trained stochastic neural network (RFTSNN)	Earlier morning (EM)	0.0089	0.0933	0.0067	0.057	0.0062	0.00734
	Afternoon hour (AH)	0.0645	0.0067	0.0061	0.0068	0.0075	0.0068
	Off hour (OH)	0.0072	0.0873	0.0073	0.0082	0.0083	0.0081

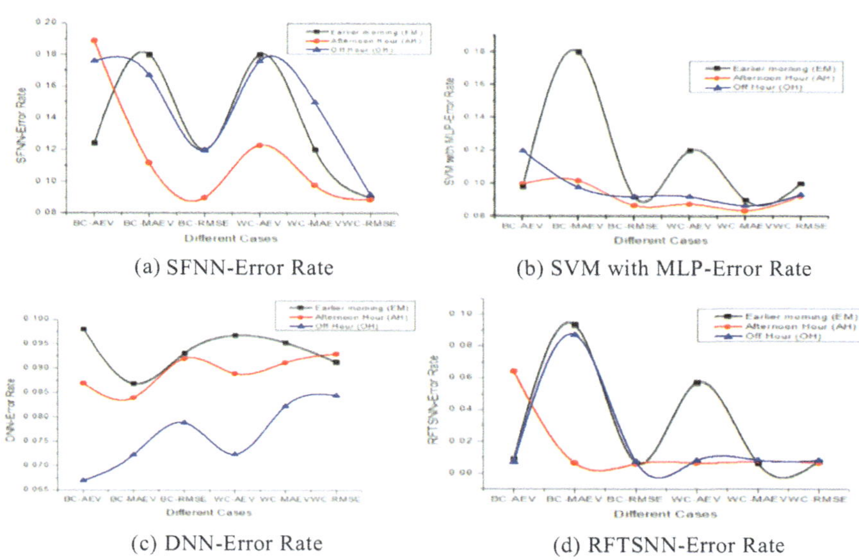

(a) SFNN-Error Rate

(b) SVM with MLP-Error Rate

(c) DNN-Error Rate

(d) RFTSNN-Error Rate

Fig. 1 Error rate

effective utilization of the pretraining module has the weight initialization, weight updating and learning concept which improve the overall transport-related activities effectively compared to the other classifiers such as single-layer feedforward neural network (SFNN), support vector with MLP (SVM + MLP), and deep learning neural network (DNN). From the analysis, the RFTSNN approach attains a minimum error rate (0.022) compared to other approaches such as SFNN (0.133), SVM + MLP (0.097), and DNN (0.086). Due to the minimum deviation of predicted and computed value, the selection and chosen traffic data has maximum accuracy, which is measured in terms of using precision, recall, and accuracy metrics. Then, the obtained value is shown in Table 2.

According to the above Table 2, it is depicted that the introduced robust fuzzy trained stochastic neural network (RFTSNN) approach has the maximum traffic

Table 2 Precision, recall, and accuracy

Methods	Period	Best case			Worst case		
		Precision	Recall	Accuracy	Precision	Recall	Accuracy
Single-layer feedforward neural network (SFNN)	Earlier morning (EM)	95.6	97.24	96.42	96.89	97.23	97.06
	Afternoon hour (AH)	96.72	97.928	97.324	97.92	98.01	97.965
	Off hour (OH)	96.29	97.24	96.765	98.08	97.38	97.73
Support vector with MLP (SVM + MLP)	Earlier morning (EM)	97.89	98.23	98.06	98.24	98.45	98.345
	Afternoon hour (AH)	98.07	98.31	98.19	97.79	98.24	98.015
	Off hour (OH)	97.98	98.21	98.095	98.34	98.14	98.24
Deep learning neural networks (DNN)	Earlier morning (EM)	98.24	98.35	98.295	98.21	97.97	98.09
	Afternoon hour (AH)	98.52	98.21	98.365	98.44	98.23	98.335
	Off hour (OH)	98.14	98.31	98.225	98.32	98.65	98.485
Robust fuzzy trained stochastic neural network (RFTSNN)	Earlier morning (EM)	98.89	98.69	98.79	98.92	99.098	99.009
	Afternoon hour (AH)	98.92	99.02	98.97	99.03	98.98	99.005
	Off hour (OH)	99.45	99.31	99.38	99.24	99.41	99.325

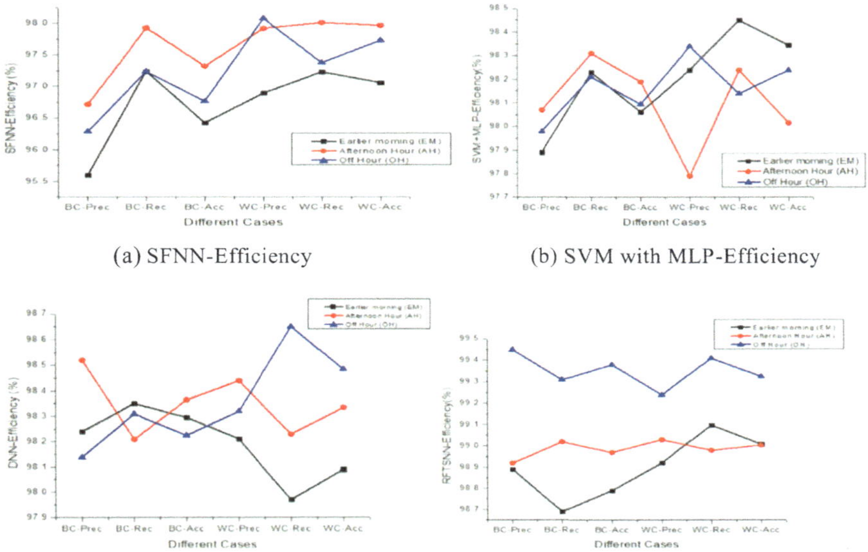

(a) SFNN-Efficiency (b) SVM with MLP-Efficiency

(c) DNN-Efficiency (d) RFTSNN-Efficiency

Fig. 2 Efficiency

activities detection accuracy which is computed from Eqs. 7–19 process. The effective utilization of vector units and hidden units in each layer improves the overall classification accuracy. The excellence of the system is determined using a different period, in which the system ensures the optimized results which classify the incoming traffic details. According to the analysis, the respective graphical analysis is shown in Fig. 2.

From Fig. 2a–d, introduced robust fuzzy trained stochastic neural network (RFTSNN) ensures maximum recognition accuracy on different cases with a different period. The transportation information is examined in terms of the early morning, afternoon, and office hours. Even though the system uses a different period of time, the system identifies the traffic activities with maximum recognition accuracy; it successfully analyzes the incoming traffic data with the pretrained module data, which has been used to improve the overall recognition accuracy. Thus, the introduced system-related transport-related activities are effective when compared to the other classifiers such as single-layer feedforward neural network (SFNN), support vector with MLP (SVM + MLP), and deep learning neural network (DNN). Here, the different period has been analyzed with the method that effectively recognizes the traffic activities around 99.3% of accuracy on worst case and 99.41% of best case on traffic data examination process. Thus, the introduced system effectively predicts traffic-related transport activities with a minimum error rate and maximum recognition accuracy.

5 Conclusion

Thus, the paper analyzes the robust fuzzy trained stochastic neural network related to creating the intelligent transport system for predicting the disobeying activities. Based on the previous discussions, the fuzzy decision is handled to predict the rules and changes of the driver behavior effectively. Then, the efficiency of the system is developed using experimental analysis. The introduced system recognizes the traffic-related activities which were detected with 99.41% for worst case and 99.3% for the best case. In the future, the intelligent transport system is improved using the optimized classifier.

References

1. Khalid, A. S., Al-Khateeb, Jaiz, A. Y., Johari, Wajdi, F. A. (2008). Dynamic traffic light sequence, Science Publications. *Journal of Computer Science, 4*(7), 517–524. https://doi.org/10.3844/jcssp.2008.517.524.
2. Habtemichael, F. G., & Cetin, M. (2016). Short-term traffic flow rate forecasting based on identifying similar traffic patterns. *Transportation Research Part C: Emerging Technologies, 66*, 61–78. https://doi.org/10.1016/j.trc.2015.08.017.
3. Miklušćák, T., Gregor, M., Janota, A. (2012). Using neural networks for route and destination prediction in intelligent transport systems. In Mikulski, J. (Eds.), *Telematics in the transport environment. TST 2012. Communications in computer and information science* (Vol. 329). Berlin, Heidelberg, Springer.
4. Pamuła, T. (2012). Traffic flow analysis based on the real data using neural networks. In Mikulski, J. (Eds.), *Telematics in the transport environment. TST 2012. Communications in computer and information science* (Vol. 329). Berlin, Heidelberg, Springer
5. Gallo, M., & De Luca, G. (2018). Spatial extension of road traffic sensor data with artificial neural networks. *Sensors, 18*, 2640. https://doi.org/10.3390/s18082640.
6. Bai, J., & Chen, Y. (2019). A deep neural network based on classification of traffic volume for short-term forecasting. *Mathematical Problems in Engineering*, Article ID 6318094, 10 pages. https://doi.org/10.1155/2019/6318094.
7. Qu, X., Wang, W., Wang, W., et al. (2017). Real-time rear-end crash potential prediction on freeways. *Journal Central South University, 24*, 2664. https://doi.org/10.1007/s11771-017-3679-2.
8. https://www.its-rde.net/index.php/rdedataenvironment/10012
9. Zhou, Y., Li, Y. P., & Huang, G. H. (2015, December 15). A robust possibilistic mixed-integer programming method for planning municipal electric power systems. *International Journal of Electrical Power & Energy Systems, 73*, 757–772. https://doi.org/10.1016/j.ijepes.2015.06.009.
10. Cao, B., Li, J., Wu, J., & Zhang, B. (2013). Restricted Boltzmann Machine with Adaptive Local Hidden Units. In: Lee M., Hirose A., Hou ZG., Kil R.M. (eds) *Neural Information Processing. ICONIP 2013. Lecture Notes in Computer Science* (Vol. 8227). Berlin, Heidelberg: Springer.
11. Ebert, T., Bänfer, O., & Nelles, O. (2010). Multilayer perceptron network with modified sigmoid activation functions. In: Wang, F. L., Deng, H., Gao, Y., & Lei, J. (Eds.), *Artificial Intelligence and Computational Intelligence. AICI 2010. Lecture Notes in Computer Science* (Vol. 6319). Berlin, Heidelberg: Springer.
12. Li, Y., & Zhu, Q. (2019). Stability analysis for discrete-time stochastic fuzzy neural networks with mixed delays. *Mathematical Problems in Engineering*, Article ID 8529053, 13 pages. https://doi.org/10.1155/2019/8529053.

13. Zhu, Q., & Li, X. (2012). Exponential and almost sure exponential stability of stochastic fuzzy delayed Cohen–Grossberg neural networks. *Fuzzy Sets and Systems, 203*(2), 74–94. View at Publisher, View at Google Scholar, View at MathSciNet
14. Wang, N., Melchior, J., & Wiskott, L. (2012). An analysis of Gaussian-binary restricted Boltzmann machines for natural images. In Verleysen, M. (Ed.), *European symposium on artificial neural networks, computational intelligence and machine learning (ESANN)* (pp. 287–292). d-side publications: Evere.

Human Facial Expression Based Video Retrieval with Query Video Using EBCOT and MLP

S. Jeyalaksshmi, D. Akila, D. Padmapriya, G. Suseendran, and Souvik Pal

Abstract We can find a massive demand for systems that retrieve video with a query as image or video since the text-based retrieval system (text, images, video) was obsolete and the content-based video retrieval oriented. For many applications, 'it is essential for retrieving videos from the extensive database with image and video query for content linking and brand monitoring. In this paper, a new video retrieval scheme developed for retrieving videos from an extensive database with a video query. The process uses EBCOT (Embedded Block Coding with Optimization Truncation) coding for feature extraction and MLP (Multilayer Perceptron) for classification and retrieval. The process uses videos with human facial expression as query and related videos with similar facial expression can be retrieved. The process carried out with three phases i)facial expression recognition ii)frame indexing iii)video indexing and retrieving. This methodology does video retrieval process in which video gives as a query for which facial expression has been recognized videos in the database are indexed with the corresponding expression and finally retrieval has been performed thus in this method retrieval scalability has been improved significantly.

Keywords Facial expression · EBCOT · MLP · Video retrieval

S. Jeyalaksshmi (✉) · D. Akila · D. Padmapriya · G. Suseendran
Department of Information Technology, Vels Institute of Science, Technology & Advanced Studies, Chennai, India
e-mail: pravija.lakshmi@gmail.com

D. Akila
e-mail: akiindia@yahoo.com

D. Padmapriya
e-mail: padmapriya.scs@velsuniv.ac.in

G. Suseendran
e-mail: suseendar_1234@yahoo.co.in

S. Pal
Global Institute of Management and Technology, Krishnagar, India
e-mail: souvikpal22@gmail.com

S.-L. Peng et al. (eds.), *Proceedings of First International Conference on Mathematical Modeling and Computational Science*, Advances in Intelligent Systems and Computing 1292, https://doi.org/10.1007/978-981-33-4389-4_16

1 Introduction

In the bleeding edge time, the quick improvement of web, private frameworks and progression of blended media advancements, various modernized pictures moved every day. With the problematic improvement of the amount of cutting edge pictures available on the web and the openness of picture getting devices, for instance, modernized cameras and picture scanners, the degree of automated things are proliferating, like this powerful requesting and chasing wraps up observably significant down generous picture documents [1]. Therefore, many general-reason picture recuperation structures are in creation. There are 3 classes of picture recuperation frameworks: abstract substance-based, content-based completely, and semantic-based completely [2].

In printed content-based structures, the photographs are genuinely remarked on through substance delineation. Content clarification to all delineations truly is unreasonable as a result of immense naming cost and the subjectivity of human acknowledgement. To triumph over the above risks in abstract substance chiefly based photograph recuperation contraption, 'Content-Based Image Retrieval (CBIR)' advanced toward getting included [3].

Feelings are fundamental to people in everyday co-operations, and it uses in regular day to day existence. Feeling acknowledgement has become a significant and intriguing field of study with regards to HCI, Human–Robot Interaction (HRI) etc. [4]. The six essential feelings are, sickening, glad, dread, outrage, miserable and shock. PC designs, programmed driver exhaustion identification, 3D/4D symbols movement in the media outlets, brain research, video and content visit and gaming applications are in remembrance for differing applications. Acknowledgement of feelings from outward appearances utilizing recordings comprises of preprocessing, highlight pulling out and division. Suwa ET has presented outward appearance acknowledgement in 1978 Al. Making an outward appearance acknowledgement framework the central matter of face recognition and arrangement Feature extraction and characterization, picture normalization [5].

The capacity to distinguish changes in outward appearances is exceptionally critical as we can perceive articulation and states of mind of an individual. The primary issue of development an outward framework for face recognition and standardization of picture [6]. Any number of methods perceives outward appearance. This method uses an optical stream to separate fundamental movements.

Outward appearance acknowledgement (FER) framework is in use fundamentally to give excellent human-PC interface so PCs can ready to perceive the feeling of the client and propose tasks in the rebound to the disposition of the client [7]. The human-PC connection can turn out to be increasingly characteristic if PCs can comprehend human feelings and their conduct.

2 Literature Survey

Patel et al. described that With the improvement of sight and sound information types and available transfer speed there is the colossal interest of video recovery frameworks, as clients move from content-based recovery frameworks to content-based recovery frameworks. Determination of removed highlights assumes a significant job in content-based video recovery, paying little heed to video qualities being getting looked [8]. These highlights plan for choosing, ordering and positioning as indicated by their latent capacity enthusiasm to the client. Great highlights determination additionally permits the existence expenses of the recovery procedure to be diminished. This study surveys the intriguing highlights that can be removed from video information for ordering and recovery alongside closeness estimation techniques. We additionally distinguish present research issues in the region of substance based video recovery frameworks.

Padmakala et al. presented a system that retrieves video interactively. This methodology uses three significant phases in namely structure parsing, summarization and Indexing of videos and Relevance Feedback. The input videos are divided into shots, using the shot detection algorithm.

In the video summarization process from every frame, extracted from this are three significant features they are commonly colour, shape and texture. After summing up the video with the feature set, an index table was created with the aid of these features for all the videos matching the query image. A semantic matching distance used for obtaining relevant video by matching with index table created in the matching process. After obtaining relevant video, in the feedback phase, it is given as input to identify whether to identify that it is relevant for other users. In case the video is relevant, it gives some more related videos. A performance measure in terms of precision-recall and f-measure is given [9].

André Araujo et al. proposed a system that compares database videos related to query image with a new architecture for video retrieval by which the retrieval scalability can improve considerably when compared to basic video frame search engine. It is an asymmetric problem comparing an image query with a video database. An asymmetric comparison for fisher vector is in the proposal. At the same time, with query with varying clutter are explored systematically that shows better benefits of the system. A new system proposes that directly compares the image query with a database of videos [10], Fisher vectors constructed for all video segments with a variety of aggregation methodologies. A two order of magnitude compression gain, obtained for a database of lectures videos for the frame-based scheme with better loss performance in retrieval accuracy. Then fisher embedding is combined in the hacking technique for designing of video descriptor by a bloom filter based flexible framework. With three datasets experiments are carried out on a large scale, the system proved as the most efficient retrieval technique and time-efficient.

3 Methodology

With the advancement in video science technology retrieval of video information from an extensive database has been a unique challenge for research scholars and others. Till now, video information retrieved with the help of text describing their names and search through images is on research. Here we are going to propose a new methodology for retrieving videos with human facial expression as a reference. The process approaches by three significant phases (i) facial expression detection on images (ii) frame indexing (iii) video indexing and retrieval. The process is as described in the below sections.

3.1 Video to Frame Separation

Usually, videos are several images streaming at the same interval of the period. This period is for 15/24/30 frames per second. Here images are described to be frames. Still moving images as a stream gives motion which is said to be video. So far, video processing has its first process as frame separation where each image saves as an individual video file in some standard format. With Matlab coding video reader has been used for reading videos, and those are separated.

3.2 Phase-1: Human Facial Expression Recognition

Human facial expression is considered as the emotions expressed by a human being for any situation. There are six basic expressions shown by a human being they are Happy, Sad, Anger, Disgust, Fear, and surprise. For this process, saved frames is in process. Each frame verifies for the presence of a human face using refined Viola-Jones algorithm. Then spectral features of face images are obtained using the EBCOT algorithm [7].

(i) *EBCOT*

Initially, face images detected for the quantization process with Discrete Fourier transform. Then obtained sub-band divides further, which denotes as Code-Blocks as in Fig. 1. Then independent encoding has been carried on each block. With the bit-plane coding method, every bit plane encoded for generating immediate data.

$$V_i[k] = \frac{|S_i[k]|}{\delta_{b_i}} \tag{1}$$

$$V_i[k],\ S_i[k] = \text{quantizedmagitude}$$

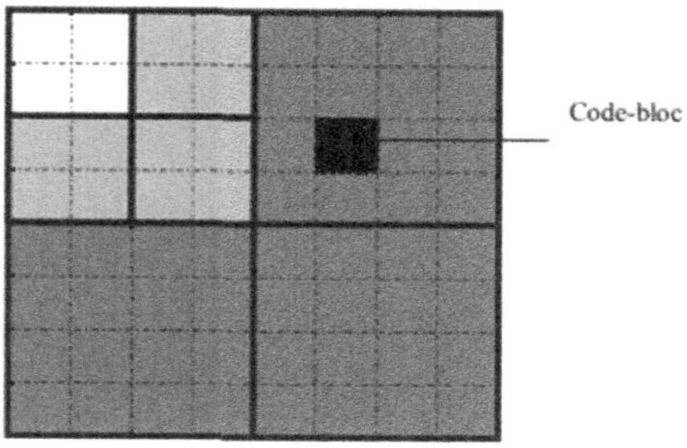

Fig. 1 Partitioning into code-blocks

Wavelet coefficients obtained above are subject to the EBCOT method for encoding with a slice-by-slice basis. Here 3D-IWT are taken as input samples for EBCOT rather than using 2D-IWT coefficients. Then these three-dimensional sub-bands are decomposed to further small sub-bands that encodes independently with EBCOT method [11].

$$X_i = \beta_i^T Y \tag{2}$$

Y Observation in N-dimensional Space.
B orthonormal vector feature set.
X linear feature vector.

Figure 2, shows the EBCOT Architecture. EBCOT receives a set of quantization coefficients within a code block. The coding scheme starts with MSB has driven bit-plane by bit-plane by three coding passes [12].

Feature matrix with unitary transform

$$\mathrm{B} = [\beta 1, \beta 2, \ldots, \beta \mathrm{N}] \tag{3}$$

β_i is an $N \times 1$ column vector.

(ii) *MLP*

Deep learning plays a significant role in Artificial neural network in this modern hardware system. Research scholars have proposed thousands of neural networks.

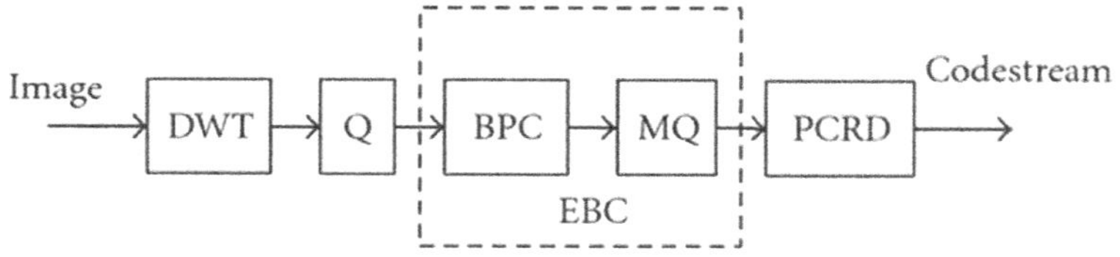

Fig. 2 EBCOT architecture

Fig. 3 MLP sample network

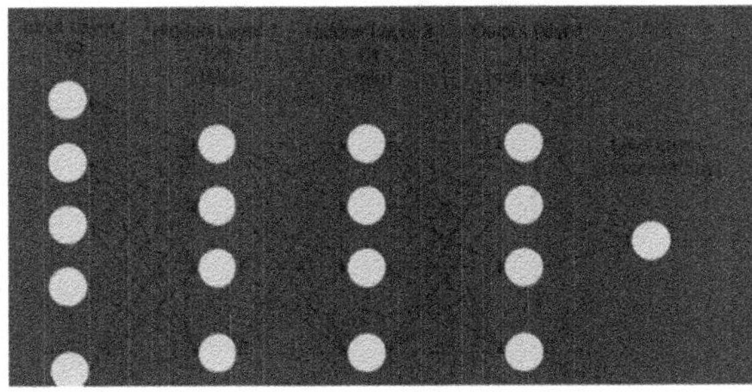

ANN classifies into three categories like Multilayer perceptron, Convolutional Neural Network and Recurrent Neural network.

Here we have used Multilayer perceptron neural network for classification of human facial expression. The network comprises 100 layers of the network.

Inputs sent to stage1 of the neural network. That is a feature vector obtained through the above method feeds as input to the network. Here it has been processed to multiple hidden layers the output obtained from the visible layer. Every emotion represents an index number which feeds as training data. According to that, the network trains with this as reference query images recognize for expression (Fig. 3).

3.3 Phase 2-Frame Indexing

Every frame in the video at this point is analyzed separately for the presence of a human face inside it. Whenever a face image is detected, it is extracted and analyzed for facial expression. And every frame is indexed separately according to the expression detected. Thus feature indices obtained are in the format of a large matrix for videos where vectors are in each row are given with an index number from 1 to 6.

3.4 Phase 3- Video Indexing and Retrieving

Then comes the final process of indexing a complete video according to the significant expression detected by the above process. Still, the above method, every frame in the video with face images indexed with index number according to detected expression, and complete indices analyzed for most redundant expression. The whole video indexed with a single index denoting the significant expression shown by the video in continuous frames. The same process carried out for finding expression on query video as well as videos in the database. Finally, videos with a similar index as the query videos are retrieved.

4 Result and Discussion

Results obtained from the above process has been shows the following performance of the algorithm implemented for facial expression detection and extraction process.

The performance results in Table 1 show EBCOT show better performance compared to other methods like PCA (Principle component analysis) and SPIHT (Set Partition in Hierarchical Tree) by the compression ratio and peak signal to noise ratio (Figs. 4 and 5).

Presence of face image in the above input image has been detected and cropped for reducing the data size when compared to the complete frame (Fig. 6).

As the first part of feature extraction initially, facial features are extracted then followed by defining each feature with a vector formed by the EBCOT method (Fig. 7).

After extraction of face feature, a feature vector formed used for training as well as testing purpose. The images with known face expressions convert to indices. With the indices, the facial expression of the query image is detected (Fig. 8).

By detecting expression of the same feature, the process has been followed for detecting expression of a complete stream of video by finalizing with most redundant expression detected. The same process being carried out for input query also database videos are subjected to, and according to indices videos are retrieved. These video database has taken from some famous movies, serials and some recorded videos. The database used for this process consists of 60 videos and maximum of 5 videos display with a similar expression.

Table 1 Performance result

Parameter	PCA	SPIHT	EBCOT
SE	0.981	0.83	0.64
PSNR	30	48	62
CR %	30	50	74
ER%	50	65	80

CR = Compression Ratio; ER = Efficiency Ratio.

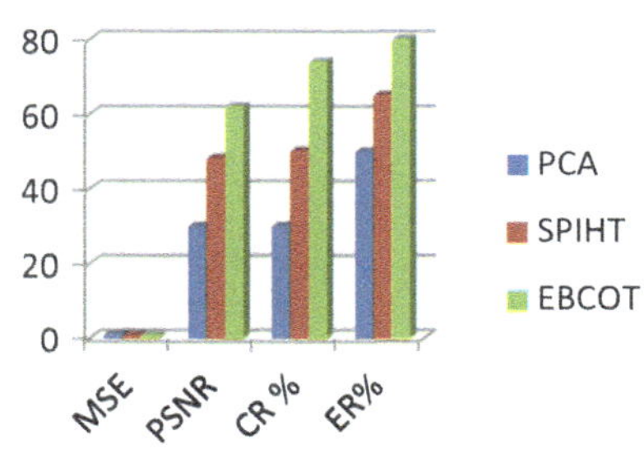

Fig. 4 Performance chart

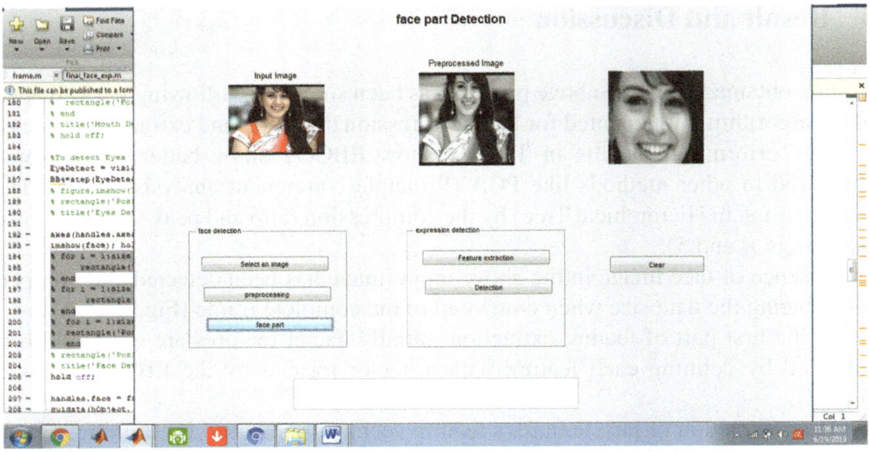

Fig. 5 Face part detection from the input image

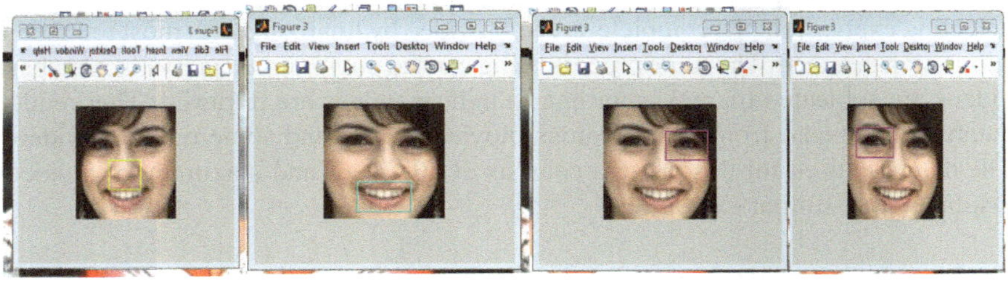

Fig. 6 Nose, mouth, Right eye, left eye detection and extraction

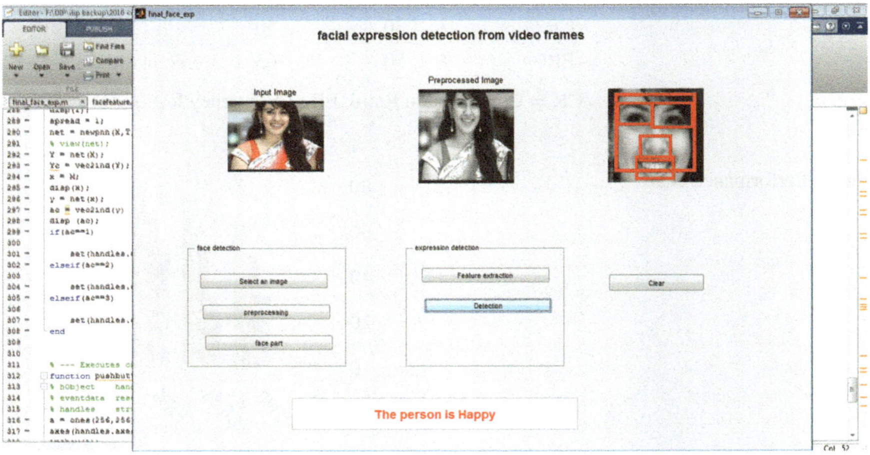

Fig. 7 Expression detection on an image with EBCOT and MLP

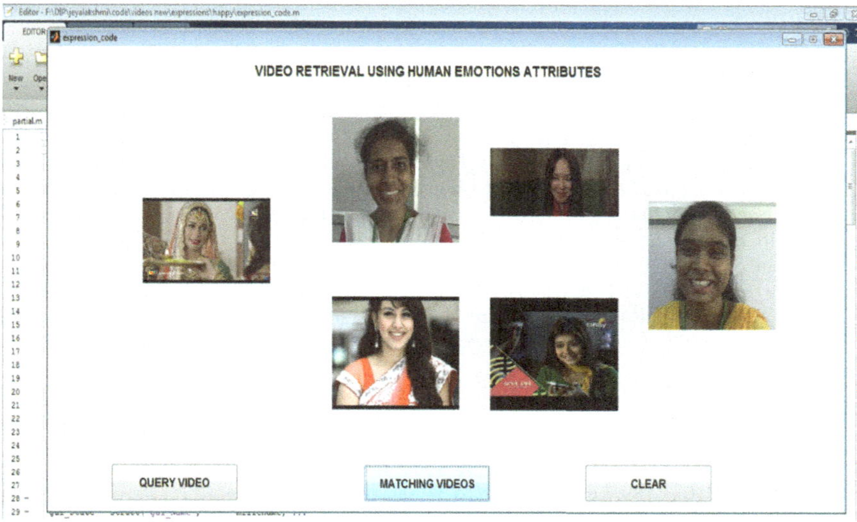

Fig. 8 Retrieved videos for happy expression

5 Conclusion

Thus a new methodology has been implemented for retrieval of video with video as a query. The methodology used shows better performance compared to some basic methods. The videos retrieved with facial expression as a critical feature. Using EBCOT and MLP methodologies facial expressions are recognized. And denoted with an index number and video has been defined with an index number. This number defines the significant expression shown in the video. The same process being carried out for query as well as database video. 60 videos, showing multiple videos are used for the database they are of the length of 30–150 s. Here the use of EBCOT for feature extraction and MLP for classification show a better result.

6 Future Scope

This work could extend for an extensive database of videos with considerable length, which may take an extended period. Feature database with the index for all the videos in the database could be made, which makes the process time-efficient and highly accurate.

References

1. Araujo, A., & Girod, B. (2018). Large-scale video retrieval using image queries. *IEEE Transactions on Circuits and Systems for Video Technology, 28*(6), 1406–1420. https://doi.org/10.1109/tcsvt.2017.2667710.
2. Aly, R. E., & Bayoumi, M. A. (2006). High-speed and low-power IP for embedded block coding with optimized truncation (EBCOT) sub-block in JPEG2000 system implementation. *Journal of VLSI Signal Processing Systems for Signal, Image and Video Technology, 42*(2), 139–148. https://doi.org/10.1007/s11265-005-4179-4.
3. Chauhan, S. (2012). Pattern recognition system using MLP neural networks. *IOSR Journal of Engineering, 02*(05), 990–993. https://doi.org/10.9790/3021-0205990993.
4. Devi, T. Maria, A., & Rajeswari, I. (2016). Hyperspectral band clustering on ebcot pre encoding technique. *Journal of Chemical Pharmaceutical Science (JCPS) Scopus Index Journal, 9*(1), 5.
5. Herrin, V. E., & Gray, G. C. (1996). Decreasing rates of hospitalization for varicella among young adults. *Journal of Infectious Diseases, 174*(4), 835–837. https://doi.org/10.1093/infdis/174.4.835.
6. Huang, Y., Chen, F., Lv, S., & Wang, X. (2019). Facial expression recognition: A survey. *Symmetry, 11*(10), 1189. https://doi.org/10.3390/sym11101189.
7. Jeyalaksshmi, S., & Prasanna, S. (2017). Simultaneous evolutionary neural network based automated video based facial expression analysis. *International Journal of Engineering & Technology, 7*(1.1), 125.https://doi.org/10.14419/ijet.v7i1.1.9211.
8. Ko, B. (2018). A brief review of facial emotion recognition based on visual information. *Sensors, 18*(2), 401. https://doi.org/10.3390/s18020401.
9. Lew, M. S., Sebe, N., & Eakins, J. P. (2002). Challenges of image and video retrieval. In *Lecture notes in computer science* (pp. 1–6). https://doi.org/10.1007/3-540-45479-9_1.
10. Magdin, M., Benko, Ľ, & Koprda, Š. (2019). A case study of facial emotion classification using affdex. *Sensors, 19*(9), 2140. https://doi.org/10.3390/s19092140.
11. Research on Face Expression Recognition. (2019). *International Journal of Innovative Technology and Exploring Engineering, 8*(9S2), 88–91. https://doi.org/10.35940/ijitee.i1017.0789s219.
12. Padmakala, S., & AnandhaMala, G. (2017). Interactive video retrieval using semantic level features and relevant feedback. *International Arabian Journal of Information Technology, 14*(5), 764–773.

Early Diagnosis of Chronic Kidney Disease Using Machine Learning Algorithms with Least Parameters by RFE and Feature Importance Techniques

K. Kavitha and K. Kala

Abstract CKD is a major health problem globally affecting the population severely. Percentage of the prevalence of CKD is increasing day by day all over the world. Early detection of CKD would be very supportive for giving the treatments in the earlier stage and reduce the mortality rate. Most accurate and efficient models have to be implemented for avoiding this risk. Many researchers have undergone research in this area. But still, the prevalence level raised globally. The main intention of this research work is to improvise the diagnosis accuracy by evaluating the best feature selection and establishing prediction model through machine learning techniques. Ranking of each parameter was generated using RFE and the importance of features were identified using statistical values through linear, logistic, decision tree, CART, random forest classifier techniques. The proposed model determined the CKD by applying various classifier algorithms such as SVM-Linear, SVM-Poly, Gaussian NB, K-Neighbors. By applying these machine learning models, SVM-Linear classifier achieved better performance with the diagnosis accuracy 0.925. Hence, it has proven that this proposed model would be pertinent for predicting more complicated diseases.

Keywords Chronic kidney disease (CKD) · Machine learning [ML] · Redundant feature elimination [RFE] · Support vector machine [SVM] · Root mean square error [RMSE]

1 Introduction

Chronic kidney disease (CKD) is a key health concern in every country globally. Once in sixty seconds, the kidney sieve about one-fifth of all of the blood in the

K. Kavitha (✉)
Mother Teresa Women's University, Kodaikanal, Tamilnadu, India
e-mail: kavitha.urc@gmail.com

K. Kala
Nachiappa Swamigal Arts and Science College, Koviloor, Tamilnadu, India
e-mail: kasinathkala1971@yahoo.co.in

S.-L. Peng et al. (eds.), *Proceedings of First International Conference on Mathematical Modeling and Computational Science*, Advances in Intelligent Systems and Computing 1292, https://doi.org/10.1007/978-981-33-4389-4_17

physique. Whenever the blood cut across through the kidneys, they clasp onto things the physique requires and do away with things it does not require. After sieving, half a million liters of blood will be remained, winding up abolishing over one-fourth of urine. CKD means inedible damage to the kidney; i.e., the kidney is not functioning congruously. If the damage worsens, the kidney may halt from working. It does not cleanse the blood correctly. Kidneys must have a conventional flow of blood and undamaged blood vessels to stay healthy and functional presented by Dilli Arasu and Thirumalaiselvi. [12]. A person with diabetes and hypertension can be affected with the vascular health of the kidneys hence leads to the main prospect for both acute and chronic kidney disease, where these two are the types of kidney problems.

Without its proper functioning, a human being may undergo risk and it may challenge their life. A person with this pathological issue may cross many stages from mild damage to complete kidney failure. Moreover, this CKD can be diagnosed and treated by proper medical guidelines. But, there are no symptoms in the early stages for this syndrome. Due to this, a huge percentage of the population suffers from CKD and numerous of them depart this life every year throughout the globe. Universally, on all sides, 850 million individuals were afflicted. Hence, periodic health checkup is the best way to diagnose the problems in the kidney in the earlier stages.

Especially, the tricenarian and quadragenarian should go for a master health checkup recurrently. Not only they should go for a checkup but also they must be treated properly if there are any symptoms for such ailments related to CKD. Furthermore, everyone should aware of health practices to keep themselves healthy. With these practices, one can be prevented from CKD. Dialysis, kidney transplantation, bio-artificial kidneys are some of the boons for a kidney affected person. Nowadays, bio-artificial kidneys play an important role in medicating patients with kidney syndrome. Even though this magical medical world comes across many healthcare facilities, this major concern should be detected with a minimum number of examination and also in a cost-effective manner. So, there comes the main aspiration of this research work to create a proficient machine learning model to detect kidney syndrome with a minimum number of predictors. This research study collected CKD dataset from the UCI repository. The data were collected from hospitals in 2015 which contains 400 samples with 25 predictive variable and one categorical variable (class). This study was conducted in Spyder (Anaconda) which supports all machine learning features.

2 Literature Review

Qin et al. [1] presented that an early diagnose of this disease enables the kidney patients to obtain the treatment in time for avoiding further progression. Machine learning models acted as a major task in the medical field in modern years for providing accurate performance recognition on time. Alloghani et al. [2] proposed that the machine learning models estimates and deducts the relevant information for a specific task and obtains the patient's characteristics. The researcher proved that this machine learning tool can provide an accurate diagnosis of CKD disease. A machine learning tool is rapidly involved in the medical field to record the health-related information electronically and track the details on time. In medicine field, ML has been used to diagnose various diseases. Many researchers have build model using ML algorithms for diagnosing the major risk diseases such as Kate et al. [3], Park et al. [4], and Chen et al. [11] proposed machine learning model for diagnosing kidney diseases, heart disease are diagnosed using machine learning models by Alickovic et al. [5] and Masetic et al. [6], cancer disease diagnosing model was proposed by Patricio et al. [7].

Missing values were dealt with KNN imputation and process the same for each incomplete sample. These models were assumed by various ML algorithms and proved random forest achieved better accuracy to establish the model. To rapidly envisage the major and minor severity of CKD, Xiao et al. [8] Developed predictive model and compared with statistical tools, machine learning techniques and neural network approach. Noia et al. [9] presented a new software tool with ANN, for classifying patient status which leads to diagnose the renal diseases in end-stage (ESRD). The proposed classifiers were examined, assessed, and evaluated using the measures of precision, recall, and F-measure. Microsoft Azure had been used to diagnose CKD, Gunarathne et al. [10] proved that multi-class decision forest achieved a better accuracy of 99.1%.

Feature Reduction method performed to identify the prediction and diagnose the disease with high accuracy with minimum error of 0.11 RMSE.

Many researchers have undergone various researches for effectively detecting kidney disease. But still, it needs more clarity and has to identify the importance of feature selection. Feature selection and importance concept satisfy the needs through redundant frequency elimination concept and feature importance method. Based on the ranking and importance given, the model will be established for detecting the chronic disease.

3 Proposed Methodology

The proposed methodology has the following stages which shown in Fig. 1.

1. Data preprocessing
2. Train and test data
3. Feature selection and importance
4. Predictive model establishment and regression
5. Model assessment.

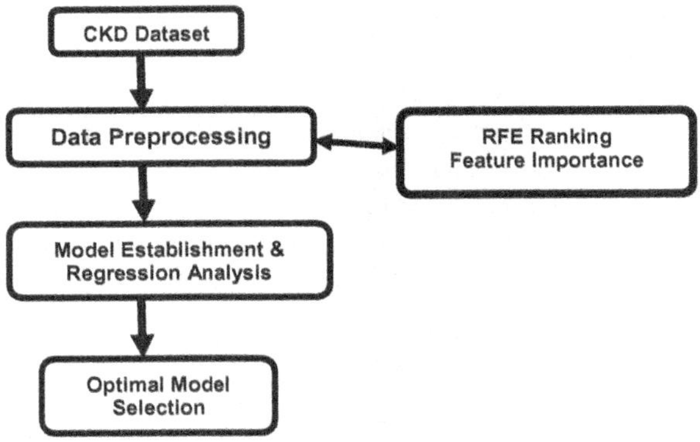

Fig. 1 Proposed framework

3.1 Data Preprocessing

Before fitting machine learning model, data has to be cleaned and formatted. In preprocessing, the selected dataset has many missing values. It has to be replaced by using mean values (numerical data) and mode value (categorical data) for the corresponding parameter. Initially, all the categorical variables were transformed into binary variables.

3.1.1 Data Cleaning

After predetermining the categorical variables, the observed missing values in the selected dataset were replaced by using machine learning techniques (imputer) which processed with Euclidean distance. The highest frequency in the corresponding parameter has been replaced in the missing places. For numerical variable, mean or median can be used to replace the missing values.

3.1.2 Feature Selection

Machine learning offers three stages for selecting the best predictors in the dataset. In the first stage, manipulation and disturbance variable were considered for calculating the correlation between each other and identify the positive, negative correlations using t-test.

In the second stage, the selected features with the target variable have considered for obtaining the correlation values and identify the features which have not relevant to this study. By considering the selected dataset, there was no negative correlation. Finally, RFE method (Redundant Frequency Elimination) applied in this dataset in Table 1 which provided ranking for each parameter that showed the importance and correlation between target variable.

Table 1 Redundant frequency elimination ranking

Variable	RFE rank	Variable	RFE rank
Age	1	Sodium	13
Blood pressure	1	Potassium	14
Specific gravity	1	Hemoglobin	10
Albumin	1	Packed cell volume	6
Sugar	7	White blood cell count	1
Red blood cell	1	Red blood cell count	1
Pus cell	4	Diabetes mellitus	3
Pus cell clumps	1	Coronary artery disease	1
Bacteria	9	Appet	12
Blood glucose random	1	Pedal Edema	8
Blood urea	1	Anemia	5
Serum creatinine	1		

This technique considered the prescribed class for assigning scores to each input features to a predictive model which indicates the importance of the feature and highlights the relevancy with the target variable. Linear, logistic, random forest, decision tree classifier techniques were used to generate the feature importance score. Initially, the research involved entire parameters (26 variables) in the dataset.

After performing the feature selection in Table 1 and feature importance stages in Fig. 2, only 12 parameters have been selected for next stages such as age of targeted, blood pressure range, specific gravity, albumin rate, presence of red blood cell, presence and count of pus cell clumps, blood glucose random, blood urea range, the presence of serum creatinine range, white blood cell count range, red blood cell count range, and coronary artery disease.

3.2 *Training and Test Dataset*

The considered dataset is divided into two sub-datasets based on 80 and 20% ratio. Training data was derived from CKD which contains 320 records and test dataset is of 80 records accordingly.

Features	Scores				
	Linear	Logistic	CART	Decision Tree Classifier	Random Forest
1	0.00000	-0.01679	0.00383	0.01872	0.00235
2	-0.00000	-0.09411	0.00231	0.00667	0.00245
3	-0.00000	-0.01073	0.00282	0.01667	0.00236
4	26.74607	0.10369	0.01735	0.03156	0.01138
5	3.28535	-0.07995	0.00088	0.00853	0.00264
6	86.50811	0.00077	0.40268	0.01555	0.38540
7	0.00000	0.30288	0.00082	0.04221	0.00317
8	93.32225	0.18032	0.52861	0.03633	0.54355
9	0.00000	-0.05088	0.00125	0.05702	0.00337
10	0.00000	-0.00469	0.00217	0.00500	0.00220
11	0.00000	-0.32618	0.00148	0.02415	0.00273
12	12.44483	0.08017	0.00065	0.03584	0.00272
13	0.00000	-0.15793	0.00246	'0.00000	0.00363
14	-0.00000	0.15900	0.00443	0.03626	0.00322
15	0.00000	0.10667	0.00257	0.01350	0.00308
16	-0.00000	0.13172	0.00022	0.04078	0.00170
17	-0.00000	-0.26822	0.00025	0.12367	0.00277
18	0.00000	-0.16714	0.00576	0.10343	0.00357
19	-0.00000	-0.11924	0.00237	0.00834	0.00230
20	-0.00000	-0.00999	0.00244	0.04307	0.00181
21	0.00000	-0.32594	0.00198	0.24777	0.00209
22	-0.00000	-0.10526	0.00124	'0.00000	0.00195
23	0.00000	0.00871	0.00313	0.0080	0.00277

Fig. 2 Feature importance

3.3 *Establishing Predictive Model & Evaluation*

In the model establishment stage, machine learning techniques were used and predictive models were established with the support of ML to predict the accuracy and error rate of the progress of CKD. Model performance assessed by using paired t-test for each sample. Error rates were analyzed by using the following ML techniques such as linear, ElasticNet, Lasso, Ridge, and random forest. These error rates would be very supportive to evaluate the performance of the predicted model. It indicated that the lower error rate value is the more convinced sample for predicting the accuracy.

The extracted feature parameters had removed unrelated parameter for making the efficient model and predicting the optimal accuracy rate. Optimal subset regression technique used to trace the predictors which should be meaningful to identify the predictors. This model technique perceives the performance of all the possible combinations of parameters and chooses the possible samplings. To evaluate the performance of samplings, t-test ($p < 0.05$ assumed the threshold limit) was calculated for each sample. This research study was also investigated the significance of predictor variables.

The coefficients of ElasticNet, Lasso, ridge, random forest, and linear classifier models were utilized as considered importance of predictive factor for identifying the best feature selection. Based on the frequency t-test value, predictors were identified which is positively correlated with process variable for the next stage and removed the parameter from the samplings which are having the highest frequency (significant value). The iteration process repeated continuously until obtained minimum error value for that sample. After performing the feature extraction methods of an optimal subset of predictors regression, the accuracy on the dataset as shown below.

Based on Table 2, it indicated that the error rates need to be reduced in further stages. The probabilistic score of red blood cell is higher than the threshold limit, so far the observed parameter which is extracted from feature importance techniques were removed from the list for further processing. Now it contains 11 parameters such as age, blood pressure range, specific gravity, albumin, presence of pus cell clumps, blood glucose random, range of blood urea, serum creatinine, range of white blood cell count, range of red blood cell count and coronary artery disease. The probabilistic score values have been reduced and shown in Table 3. Optimistic model will be created with the predictive variables for detecting the CKD effectively.

3.4 *Model Assessment*

The proposed model has improved the training dataset and tried to minimize the errors. Initially, model was defined with observed parameters and the dataset were randomly assigned in the train as well as test dataset. Average performance was calculated by using different classifier models such as SVM-Linear, SVM-Poly, Gaussian NB, K-Neighbors. Evaluation indicators used for each model as a confusion matrix, recall, false-positive rate, accuracy precision, etc. Machine learning classifier algorithms were implemented in this model for producing an accurate measure of chronic kidney disease diagnosis on different evaluation measures. This research study carried out in machine learning clustering analysis which is focused on positive false and false negative values. In this study, Python and Spyder were used to build and evaluate the accuracy metrics from ML classifiers.

3.5 *Performance Evaluation*

Symptoms and medical tests information are mostly taken to diagnose the disease. Parameters of the diagnose tests were carried out with specific disease constraints. The best probability measure would be chosen based on the attributes. Figure 3 represents the accuracy score for machine learning classifiers. The results of applying SVM-linear, poly, Gaussian NB, and K-Neighbors classifier algorithms were implemented and compared with the accuracy values. The research findings highlighted that the SVM-Linear regression algorithm provides better accuracy (93% accuracy, 87% precision, 93% recall) for diagnosing the early stage of CKD as shown in Table 4. The accuracy comparison between the classifiers is as shown in Fig. 3. It is proven that the SVM-Linear classifier provides better accuracy (93%) than others. Gaussian and K-neighbor provide the next better accuracy which is very closer to each other.

Table 2 Correlation coefficient for the first stage

Feature Selection: Regression Score				**Predictors: OLS Regression Results**		
Dep. Variable:	**y**			**R-squared:**	**0.703**	
Model:	**OLS**			**Adj. R-squared:**	**0.692**	
Method:	**Least Squares**			**F-Statistic:**	**70.12**	
Date: Tue, 14 Apr 2020				**Prob(F-Statistic):**	**4.92e-93**	
Time:	**14:54:02**			**Log-Likelihood:**	**-34.932**	
No. Observations:	**400**			**AIC:**	**97.86**	
Df Residuals:	**386**			**BIC:**	**153.7**	
Df Model:	**13**					
Covariance Type:	**nonrobust**					

	coef	**std err**	**t**	**P> |t|**	**[0.025**	**0.975]**
const	34.2543	3.222	10.632	0.000	27.920	40.589
x1	-0.0009	0.000	-7.134	0.000	-0.001	-0.001
x2	0.0012	0.001	1.413	0.158	-0.000	0.003
x3	0.0029	0.001	2.726	0.007	0.001	0.005
x4	-32.7833	3.180	-10.310	0.000	-39.035	-26.532
x5	0.0918	0.014	6.394	0.000	0.064	0.120
x6	-0.0120	0.044	-0.270	0.788	-0.099	0.075
x7	-0.0636	0.049	-1.290	0.198	-0.160	0.033
x8	0.0005	0.000	2.707	0.007	0.000	0.001
x9	-0.0004	0.000	-0.990	0.323	-0.001	0.000
x10	0.0056	0.003	1.861	0.063	-0.000	0.011
x11	3.994e-06	5.58e-06	0.715	0.475	-6.98e-06	1.5e-05
x12	-0.1191	0.021	5.742	0.000	-0.160	-0.078
x13	0.0291	0.053	0.553	0.581	-0.074	0.133

Omnibus: **9.925**		**Durbin-Watson:**	**1.223**
Prob(Omnibus):	**0.007**	**Jarque-Bera (JB):**	**8.640**
Skew:	**0.290**	**Prob(JB):**	**0.0133**
Kurtosis:	**2.573**	**Cond. No.**	**2.96e+06**

Table 3 Correlation_coefficient for the second stage

Feature Selection: Regression Score **Predictors: OLS Regression Results**

Dep. Variable:	y	R-squared:	0.671
Model:	OLS	Adj. R-squared:	0.661
Method:	Least Squares	F-Statistic:	65.77
Date:	Tue, 14 Apr 2020	Prob(F-Statistic):	1.27e-85
Time:	14:54:46	Log-Likelihood:	-55.069
No. Observations:	400	AIC:	136.1
Df Residuals:	387	BIC:	188.0
Df Model:	12		
Covariance Type:	nonrobust		

	coef	std err	t	P> |t|	[0.025	0.975]
const	38.1367	3.323	11.476	0.000	31.603	44.670
x1	-0.0010	0.000	-7.141	0.000	-0.001	-0.001
x2	0.0007	0.001	0.828	0.408	-0.001	0.002
x3	0.0027	0.001	2.483	0.013	0.001	0.005
x4	-36.4033	3.286	-11.078	0.000	-42.864	-29.942
x5	-0.0880	0.045	1.962	0.050	-0.176	0.000
x6	0.0165	0.050	0.330	0.742	-0.082	0.115
x7	0.007	0.000	3.473	0.001	0.000	0.001
x8	6.572e-05	0.000	0.170	0.865	-0.001	0.001
x9	0.0053	0.003	1.701	0.090	-0.001	0.011
x10	8.278e-06	5.82e-06	1.422	0.156	-3.17e-06	1.97e-05
x11	-0.1393	0.022	-6.470	0.000	-0.182	-0.097
x12	0.0507	0.055	0.919	0.358	-0.058	0.159

Omnibus:	13.568	Durbin-Watson:	1.131
Prob(Omnibus):	0.001	Jarque-Bera (JB):	12.991
Skew:	0.395	Prob(JB):	0.00151
Kurtosis:	2.604	Cond. No.	2.91e+06

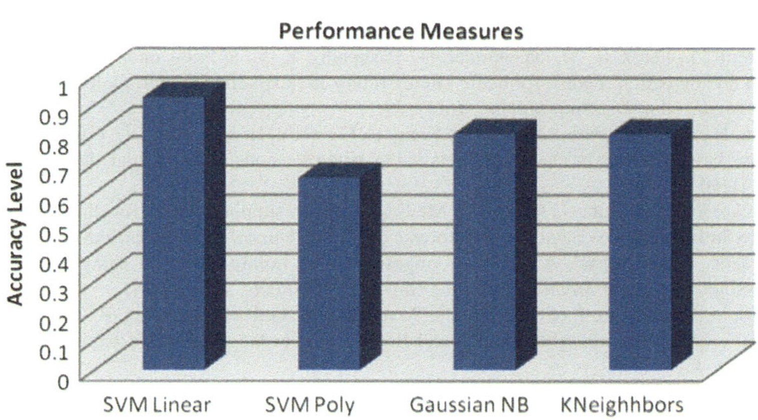

Fig. 3 Performance measure of ML classifiers

Table 4 Regression score for ML classifier algorithms

Model	Accuracy	Precision	Recall	F1-score
SVM-linear	0.93	0.87	0.93	0.90
SVM-poly	0.65	0.00	0.00	0.00
Gaussian NB	0.89	0.77	0.96	0.86
K-neighbor	0.80	0.65	0.93	0.76

4 Conclusion and Future Reference

The research work analyzed the CKD dataset using machine learning algorithms. This research study examined the feature selection and feature importance techniques for selection of best features of diagnosing the kidney disease at an earlier stage. Further machine learning models like Lasso, Ridge, linear, ElasticNet, random forest were established to ensure the model as perfect. Finally, the proposed model was assessed by using SVM, Gaussian NB, and K-Neighbors classifier models. The accuracy score has proven that SVM-Linear regression model could be the better model to predict the accuracy. In future, the large number of datasets will be considered and randomly allocated to train the proposed model and expand the performance metrics level which enables the detection of the severity of the disease.

References

1. Qin, J., Chen, L., Liu, Y., Liu, C., Feng, C., & Chen, B. (2019). A machine learning methodology for diagnosing chronic kidney disease. *IEEE Access, 8,* 20991–21002.
2. Alloghani, M., Al-Jumeily, D., Baker, T., Hussain, A., Mustafina, J., & Aljaaf, A. J. (2018, August). Applications of machine learning techniques for software engineering learning and early prediction of students' performance. In *International Conference on Soft Computing in Data Science* (pp. 246–258). Springer, Singapore.
3. Kate, R. J., Perez, R. M., Mazumdar, D., Pasupathy, K. S., & Nilakantan, V. (2016). Prediction and detection models for acute kidney injury in hospitalized older adults. *BMC Medical Informatics and Decision Making, 16*(1), 39.
4. Park, N., Kang, E., Park, M., Lee, H., Kang, H. G., Yoon, H. J., & Kang, U. (2018). Predicting acute kidney injury in cancer patients using heterogeneous and irregular data. *PLoS ONE, 13*(7), e0199839.
5. Alickovic, E., & Subasi, A. (2016). Medical decision support system for diagnosis of heart arrhythmia using DWT and random forests classifier. *Journal of Medical Systems, 40*(4), 108.
6. Masetic, Z., & Subasi, A. (2016). Congestive heart failure detection using random forest classifier. *Computer Methods and Programs in Biomedicine, 130,* 54–64.
7. Patrício, M., Pereira, J., Crisóstomo, J., Matafome, P., Gomes, M., Seiça, R., & Caramelo, F. (2018). Using resistin, glucose, age and BMI to predict the presence of breast cancer. *BMC Cancer, 18*(1), 29.
8. Xiao, J., Ding, R., Xu, X., Guan, H., Feng, X., Sun, T., & Ye, Z. (2019). Comparison and development of machine learning tools in the prediction of chronic kidney disease progression. *Journal of Translational Medicine, 17*(1), 119.

9. Di Noia, T., Ostuni, V. C., Pesce, F., Binetti, G., Naso, D., Schena, F. P., & Di Sciascio, E. (2013). An end stage kidney disease predictor based on an artificial neural networks ensemble. *Expert Systems with Applications, 40*(11), 4438–4445.
10. Gunarathne, W. H. S. D., Perera, K. D. M., & Kahandawaarachchi, K. A. D. C. P. (2017, October). Performance evaluation on machine learning classification techniques for disease classification and forecasting through data analytics for chronic kidney disease (CKD). In *2017 IEEE 17th International Conference on Bioinformatics and Bioengineering (BIBE)* (pp. 291–296). IEEE.
11. Chen, Z., Zhang, X., & Zhang, Z. (2016, June). Clinical risk assessment of patients with chronic kidney disease by using clinical data and multivariate models. *International Urology and Nephrology, 48*(12), 2069–2075.
12. Arasu, D., & Thirumalaiselvi, R. (2017). Review of chronic kidney disease based on data mining techniques. *International Journal of Applied Engineering Research, 12*(23), 13498–13505.

Mixing Sumudu Transform and a Domain Decomposition Method for Solving a Space–Time Fractional Derivative of Groundwater Flow Equation

Hassan Zaidan Mjthap and Saad Naji AL-Azzawi

Abstract In this work, the aim is to discuss a fractional model for the groundwater flow equation; as groundwater is not static, it flows into the groundwater layer and its flow can be described using the fractional differential equation with the initial and boundary conditions associated with it, and this model is derived using the law of conservation of mass and generalization of the Darcy law with taking into account the flow of water to the piezometric head as a function of the derivative of the fractional arrangement. To solve the groundwater model equation, there are many different techniques such as the Frobenius method and the finite element method. In this work, we present an approximate solution to space–time fractional derivative of the groundwater flow equation, by applying Sumudu transformation and the Adomain decomposition method (ADM). The proposed method is validated by providing two examples and simulations for these examples to complete the paper. Illustrative examples show that the proposed method can be used for similar problems and is very effective. The fractional derivative is Caputo derivative in this paper, and this problem has great importance in petroleum technology.

Keywords Groundwater model · Fractional differential equations · Caputo derivative · Sumudu transforms · Adomain decomposition method

1 Introduction

An important equation to access water sources is the groundwater model because the lack of rivers in many countries has made a second source of water is necessary, which is the main source of water because of the lack of rivers and is useful for agricultural

H. Z. Mjthap (✉)
Department of Mathematics, College of Education for Pure Science, Ibn Al-Haitham, University of Baghdad, Baghdad, Iraq
e-mail: hassanalkinani1978@gmail.com

S. N. AL-Azzawi
Department of Mathematics, College of Science for Women, University of Baghdad, Baghdad, Iraq
e-mail: saadnaji58@gmail.com

S.-L. Peng et al. (eds.), *Proceedings of First International Conference on Mathematical Modeling and Computational Science*, Advances in Intelligent Systems and Computing 1292, https://doi.org/10.1007/978-981-33-4389-4_18

fields and others. Therefore, groundwater management requires predictability the groundwater movement system and human activities on this water and to know the changes imposed by nature. Over the past 20 years, some disciplines in groundwater hydrology have seen technological breakthroughs in groundwater systems and the use of distributed and inevitable simulation models to analyse dissolved water flow. Isotope measurements are used to help explain and determine groundwater flow pathways and their interactions with surface water and leakage as another major technological growth area in applying isotopic analysis to groundwater hydrology [1–4].

In this research, we will focus on one of the mathematical transformations that are used in many fields of application such as computer science, physics, applied mathematics, and engineering which is Sumudu transformation to solve fractional differential equations. In differential equations, this transformation is useful because turning it into equations can be easier to solve, and also, the introduction of boundary conditions into fractional differential equations presents extra challenges. So we will handle these difficulties by use [5], Adomian decomposition method with Sumudu transformation since it can handle with the initial conditions and the fractional space operator only.

2 Basic Concepts

2.1 Caputo Derivative

Definition 1 [6, 7]: Let $\alpha > 0$, the Caputo derivative of order α, of the function $f(t)$, is defined by,

$$D^{\alpha} f(t) = \frac{1}{\Gamma(n-\alpha)} \int\limits_{0}^{x} \frac{f^{(n)}(\tau)}{(t-\tau)^{\alpha+1-n}} \mathrm{d}\tau \tag{1}$$

where $\Gamma(\cdot)$ is the gamma function, and $n - 1 < \alpha < n, n \in \mathbb{N}$. It has linear property:

$$D^{\alpha}(k_1 g_1(t) + k_1 g_1(t)) = k_1 D^{\alpha} g_1(t) + k_2 D^{\alpha} g_2(t),$$

where k_1 and k_2 are constants.

Remark 1 [8]: Let $(0 < \alpha \leq 1)$. From the Definition 1, we can get the following formula

$$D^{\alpha} t^{\gamma} = \begin{cases} 0 & \gamma = 0 \\ \frac{\Gamma(\gamma+1)}{\Gamma(\gamma+1-\alpha)} t^{\gamma-\alpha} & \gamma = 1, 2, 3 \ldots \end{cases} \tag{2}$$

2.2 The Sumudu Transformation

In 1993, Watugala gave this definition [9].

Definition 2 Let

$$A = \left\{ f(t); \exists \tau_{1,} \tau_2 > 0.\text{such that}, |f(x)| < Me^{\frac{|t|}{\tau_j}}, if t \in (-1)^j \times [0, \infty) \right\}.$$

Then Sumudu transform of $f \in A$, denoted by $F(u) = \sim[f(t)]$ is defined as:

$$F(u) = \sim[f(t)] = \int_0^{\infty} f(ut)e^{-t}dt, u \in (\tau_{1,}, \tau_2) \tag{3}$$

where u is a parameter it is independent of t and it may be real or complex. Sumudu transform is related to Laplace transformation by the relation [10], L $\left(\frac{1}{u}\right) = uG(u)$ and $G\left(\frac{1}{s}\right) = s\text{L}(s)$ where $L(u)$ and G(s) are Laplace transform and the Sumudu of the function$f(t)$, respectively, for more details see [11].

Theorem 1 [12]: *The Sumudu transform of the fractional derivative using the Caputo idea of the function $f(t)$ is given by:*

$$\mathbb{S}\left[D^{\alpha} f(t)\right] = \frac{F(u)}{u^{\alpha}} - \sum_{k=0}^{n-1} \frac{f^{(k)}(0)}{u^{\alpha-k}}, n-1 < \alpha \le n, n \in \mathbb{N}. \tag{4}$$

where $F(u) = \sim[f(t)]$.

3 The Mathematical Model

To study natural phenomena uses many different mathematical models, Seyam [13] for solve model of groundwater use the artificial neural networks, the seepage losses through the canals study by Uchdadiya and Patel [14], Willis to describe the mathematical model to analyse the potential impacts associated with sea-level rise, US Geological Survey are used [2]. Many researchers are studying and solving the typical equation of shallow water as part of the studies as with groundwater [4]. This paper, solving and study the equations, governs the flow of groundwater. It is derivative by using the principle of the Darcy law and the continuity low without recourse is as follows [1]:

$$S_0 \partial_t \psi(r, t) = \frac{K}{r^{n-1}} \partial_r \left[r^{n-1} \partial_r \psi(r, t) \right]. \tag{5}$$

Subject to

$$\psi(r, 0) = ar,$$

$$\lim_{r\to\infty} \psi(r, t) = ar,$$

$$Q = \frac{2\pi^{n/2}}{\Gamma(n/2)} r_b^{n-1} K d^{3-n} \partial_r \psi(r_b, t). \tag{6}$$

where S_0, K the specified storage capacity and hydraulic conductivity, respectively, and they are constant and scalar quantities, n is the radial dimension for $n = 1, 2$ or 3. And $\psi(r, 0) = ar$ implies that the initial water level in the aquifer is a before the pumping test begins or in the aquifer, the primary hydraulic head with a positive gradient a, is a linear function of space, $\lim_{r\to\infty} \psi(r, t) = ar$.

It implies that during the pumping test, the water level isn't influenced by the well for a very long distance; $Q = \frac{2\pi^{n/2}}{\Gamma(n/2)} r_b^{n-1} K d^{3-n} \partial_r \psi(r_b, t)$, this means the hydraulic conductivity is commensurate with the pumping rate. Thus, Q is the rate of discharge of the well, with the radius r_b and d groundwater layer thickness from a good tap. Recent research suggests that the geometry of parallel fractures of bedding is influenced by groundwater flow, a feature that cannot be equated (5) and can become unworkable in these broken aquifers. Parker [1] provided a model for eliminating this problem where it was applied. For hydraulic tests and analysis, those of wells in the aquifers where the information provided in the case if the flow dimensions are incorrect fractures.

In the mathematical model $\partial_r \psi$, it is replaced by complementary fractional derivatives (the Weyl-fractional derivatives) of order α, the piezometric head for the sake of candour incorporate the potential impact of the flow. This provides to the flow equation governing of the classical equation a generalized form (5):

$$S_0 \partial_t \psi(r, t) = \frac{K}{r^{n-1}} \partial_r \left[r^{n-1} \partial_r^{\alpha} \psi(r, t) \right]. \tag{7}$$

This parameter is a physical parameter, which appears in the binary integral equation, to describe the flow through geological formations [13], for (6) use the same transformation gives the boundary condition more general form for at the well [13]:

$$Q = (-1)^{m+1} \frac{2\pi^{n/2}}{\Gamma(n/2)\Gamma(\rho)} r_b^{n-1} K d^{3-n+\alpha} \partial_r^{\alpha} \psi(r_b, t). \tag{8}$$

The initial condition described in (6), with relationships (7) and (8) represents for a solution which exists a complete set of equations.

For solve ground water model equation, there are many different techniques such as Frobenius method and finite element method, [15, 16]. In this work, we will use

Sumudu transform and the Adomain decomposition method (ADM) to solve the equation of the fractional groundwater model.

4 The Solution of the Generalized Flow Equation

We consider the Caputo derivation of order α where $\alpha \in \mathbb{R}$ and $0 < \alpha \leq 1$, for groundwater flow, governed.

Example 1: Consider the following time fractional derivatives groundwater flow equation

$$K\partial_r^2\psi(r,t) + \frac{K}{r}\partial_r\psi(r,t) - S_0\partial_t^\alpha\psi(r,t) = 0 \tag{9}$$

Subject Eq. (6). The method used here depends on applying Sumudu transform on (9) to get

$$S[\psi(r,t)] = \psi(r,0) + \nu^\alpha S\left[\frac{K}{S_0}\partial_r^2\psi(r,t) + \frac{K}{S_0 r}\partial_r\psi(r,t)\right]. \tag{10}$$

Take inverse operator, we get

$$\psi(r,t) = \psi(r,0) + S^{-1}\left[\nu^\alpha S[\frac{K}{S_0}\partial_r^2\psi(r,t) + \frac{K}{S_0 r}\partial_r\psi(r,t)]\right]. \tag{11}$$

Assumed series solution by using Adomian decomposition method [5, 17], for (11) to be

$$\psi(r,t) = \sum_{i=0}^{\infty}\psi_i(r,t), \tag{12}$$

The components $\psi_i(r,t)$ are frequently selected. Into both sides of (11) replace by (12) gives

$$\sum_{i=0}^{\infty}\psi_i(r,t), = \psi(r,0) + S^{-1}\left[\nu^\alpha S[\frac{K}{S_0}\partial_r^2\sum_{i=0}^{\infty}\psi_i(r,t), + \frac{K}{S_0 r}\partial_r\sum_{i=0}^{\infty}\psi_i(r,t),]\right]. \tag{13}$$

After lysis method, recursive relationships are presented as

$$\psi_0(r,t) = ar,$$

$$\psi_{i+1}(r,t) = S^{-1}\left[v^{\alpha} S[\frac{K}{S_0}\partial_r^2\psi_i(r,t) + \frac{K}{S_0 r}\partial_r\psi_i(r,t)]\right] i \geq 1. \tag{14}$$

Note that component $\psi_0(r,t)$ is defined so the remaining components $i \geq 1$ are determined using the preconditions.

$$\psi_0(r,t) = \psi(r,0) = ar,$$

$$\begin{aligned}
&\psi_0(r,t) = \psi(r,0) = ar,\\
&\psi_1(r,t) = S^{-1}\left[v^{\alpha} S\left[\frac{K}{S_0}\partial_r^2\psi_0(r,t) + \frac{K}{S_0 r}\partial_r\psi_0(r,t)\right]\right]\\
&= S^{-1}\left[v^{\alpha} S\left[\frac{aK}{S_0 r}\right]\right]\\
&= S^{-1}\left[v^{\alpha} S\left[\frac{aK}{S_0 r}\right]\right] = \frac{aKt^{\alpha}}{rS_0\Gamma(1+\alpha)}
\end{aligned}$$

$$\begin{aligned}
\psi_2(r,t) &= S^{-1}\left[v^{\alpha} S\left[\frac{K}{S_0}\partial_r^2\psi_1(r,t) + \frac{K}{S_0 r}\partial_r\psi_1(r,t)\right]\right]\\
&= S^{-1}\left[v^{\alpha} S\left[\frac{K}{S_0}\left(\frac{2aKt^{\alpha}}{r^3 S_0\Gamma(1+\alpha)}\right) + \frac{K}{S_0 r}\left(\frac{-aKt^{\alpha}}{r^2 S_0\Gamma(1+\alpha)}\right)\right]\right]\\
&= S^{-1}\left[v^{\alpha} S\left[\frac{aK^2 t^{\alpha}}{r^3 S_0^2\Gamma(1+\alpha)}\right]\right]\\
&= S^{-1}\left[\frac{aK^2 v^{2\alpha}}{r^3 S_0^2}\right] = \frac{aK^2 t^{2\alpha}}{r^3 S_0^2\Gamma(1+2\alpha)}.
\end{aligned}$$

$$\begin{aligned}
\psi_2(r,t) &= S^{-1}\left[v^{\alpha} S\left[\frac{K}{S_0}\partial_r^2\psi_1(r,t) + \frac{K}{S_0 r}\partial_r\psi_1(r,t)\right]\right]\\
&= S^{-1}\left[v^{\alpha} S\left[\frac{K}{S_0}\left(\frac{2aKt^{\alpha}}{r^3 S_0\Gamma(1+\alpha)}\right) + \frac{K}{S_0 r}\left(\frac{-aKt^{\alpha}}{r^2 S_0\Gamma(1+\alpha)}\right)\right]\right]\\
&= S^{-1}\left[v^{\alpha} S\left[\frac{aK^2 t^{\alpha}}{r^3 S_0^2\Gamma(1+\alpha)}\right]\right]\\
&= S^{-1}\left[\frac{aK^2 v^{2\alpha}}{r^3 S_0^2}\right] = \frac{aK^2 t^{2\alpha}}{r^3 S_0^2\Gamma(1+2\alpha)}.
\end{aligned} \tag{15}$$

To obtained $\psi(r,t)$ in this example, we evaluated four components of Eq. (12) to give the following expansion

$$\psi(r,t) = ar + \frac{aKt^{\alpha}}{rS_0\Gamma(1+\alpha)} + \frac{aK^2 t^{2\alpha}}{r^3 S_0^2\Gamma(1+2\alpha)} + \frac{9aK^3 v^{3\alpha}}{r^5 S_0^3\Gamma(1+3\alpha)} \tag{16}$$

And we can find a constant value (a) by applying the boundary condition on (16), to get

$$a = \frac{Q}{2\pi r_b K d f(r_b, t)}, \tag{17}$$

where

$$f(r_b, t) = 1 - \frac{K t^{\alpha}}{r_b^2 S_0 \Gamma(\alpha + 1)} + \frac{3K t^{2\alpha}}{r_b^4 S_0^2 \Gamma(2\alpha + 1)} + \dots \tag{18}$$

Example 2: Consider the following space and time fractional derivatives groundwater equation:

$$S_0 \partial_t \psi(r, t) = \frac{K}{r} \partial_r^{\alpha} \psi(r, t) + K \partial_r^{\mu} \psi(r, t), 0 < \alpha \le 1 < \mu \le 2. \tag{19}$$

Subject to Eq. (6) The gradient equals zero, by supposing the water level before pumping is the same throughout the aquifer, for $t = 0$, the fractional change is assumed to be constant in the drag meaning $\partial_r^{\alpha} \psi(r, 0) =$ constant. Apply Sumudu transform on Eq. (19) is as follows:

$$S[\partial_t \psi(r, t)] = S\left[\frac{K}{r S_0} \partial_r^{\alpha} \psi(r, t) + \frac{K}{S_0} \partial_r^{\mu} \psi(r, t)\right],$$

$$v^{-1}(S[\psi(r, t)] - \psi(r, 0)) = S\left[\frac{K}{r S_0} \partial_r^{\alpha} \psi(r, t) + \frac{K}{S_0} \partial_r^{\mu} \psi(r, t)\right],$$

So

$$S[\psi(r, t)] = \psi(r, 0) + v S\left[\frac{K}{r S_0} \partial_r^{\alpha} \psi(r, t) + \frac{K}{S_0} \partial_r^{\mu} \psi(r, t)\right]. \tag{20}$$

Applying the inverse operator on Eq. (20) to get

$$\psi(r, t) = \psi(r, 0) + S^{-1}\left[v S\left[\frac{K}{r S_0} \partial_r^{\alpha} \psi(r, t) + \frac{K}{S_0} \partial_r^{\mu} \psi(r, t)\right]\right]. \tag{21}$$

Into both sides of (21) replace by (12) gives

$$\sum_{i=0}^{\infty} \psi_i(r, t) = \psi(r, 0)$$
$$+ S^{-1}\left[v S\left[\frac{K}{r S_0} \partial_r^{\alpha}\left(\sum_{i=0}^{\infty} \psi_i(r, t)\right) + \frac{K}{S_0} \partial_r^{\mu}\left(\sum_{i=0}^{\infty} \psi_i(r, t)\right)\right]\right]. \tag{22}$$

The recursive relationships are presented as

$$\psi_0(r,t) = ar. \tag{23}$$

$$\psi_{i+1}(r,t) = S^{-1}\left[vS\left[\frac{K}{rS_0}\partial_r^{\alpha}\psi_i(r,t) + \frac{K}{S_0}\partial_r^{\mu}\psi_i(r,t)\right]\right], i \geq 1. \tag{24}$$

The equations below obtained by using that $\partial_r^{\alpha}\psi(r,0) = \text{constant} = a$ with (24)

$$\begin{aligned}
\psi_1(r,t) &= S^{-1}\left[vS\left[\frac{K}{rS_0}\partial_r^{\alpha}\psi_0(r,t) + \frac{K}{S_0}\partial_r^{\mu}\psi_0(r,t)\right]\right] \\
&= S^{-1}\left[vS[\frac{aK}{S_0 r}]\right] \\
&= S^{-1}\left[\frac{aKv}{S_0 r}\right] = \frac{aKt}{rS_0} \\
\psi_2(r,t) &= \frac{aK^2t^2}{2S_0^2}\left(\frac{\Gamma(1+\mu)}{r^{1+\mu}} + \frac{\Gamma(1+\alpha)}{\alpha r^{2+\alpha}}\right), \\
\psi_3(r,t) &= \frac{aK^3t^3}{6S_0^3}\left(\frac{\Gamma(1+2\mu)}{r^{1+2\mu}} + \frac{\Gamma(1+2\alpha)}{\alpha r^{2+2\alpha}}\right).
\end{aligned} \tag{25}$$

Equations (23) and (25) together give the following expansion which is the approximate solution for this problem

$$\begin{aligned}
\psi(r,t) = ar + \frac{aKt}{rS_0} + \frac{aK^2t^2}{2S_0^2}\left(\frac{\Gamma(1+\mu)}{r^{1+\mu}} + \frac{\Gamma(1+\alpha)}{\alpha r^{2+\alpha}}\right) \\
+ \frac{aK^3t^3}{6S_0^3}\left(\frac{\Gamma(1+2\mu)}{r^{1+2\mu}} + \frac{\Gamma(1+2\alpha)}{\alpha r^{2+2\alpha}}\right).
\end{aligned} \tag{26}$$

And we can find a constant value (a) by applying the boundary condition on (26), to get

$$a = \frac{Q}{2\pi r_b K d^{\alpha} f(r_b,t)}. \tag{27}$$

where

$$f(r_b,t) = 1 + \frac{Kt\Gamma(1+\alpha)}{S_{0r_b^{1+\alpha}}} + \frac{K^2t^2\Gamma(1+\mu+\alpha)}{2S_0^2 r_b^{1+\alpha+\mu}} + \frac{K^2t^2\Gamma(2+2\alpha)}{2(1+\alpha)S_0^2 r_b^{2+2\alpha}}$$

Fig. 1 Solution of (9) (drawdown)

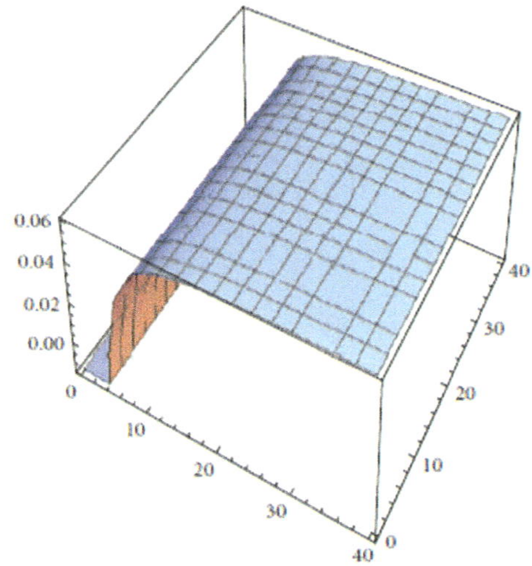

$$+\frac{K^3t^3}{6S_0^3}\left(\frac{\Gamma(1+2\mu+\alpha)}{r_b^{1+2\mu+\alpha}}+\frac{\Gamma(2+3\alpha)}{(1+\alpha)(1+2\alpha)r_b^{2+3\alpha}}\right) \tag{28}$$

For different values of different parameters, the solutions of (9) and (19) are graphically illustrated in Figs. 1, 2, 3, 4, 5, 6, 7, 8 and 9. And these graphs show during the pumping test, the behaviour of the drawdown is as follows

Fig. 2 Solution of (9) (piezometric head)

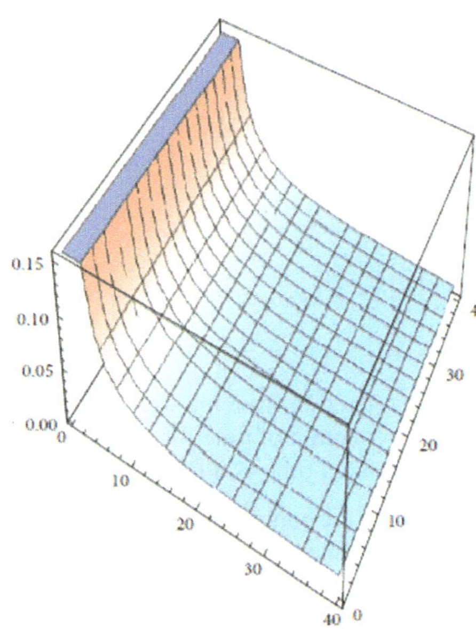

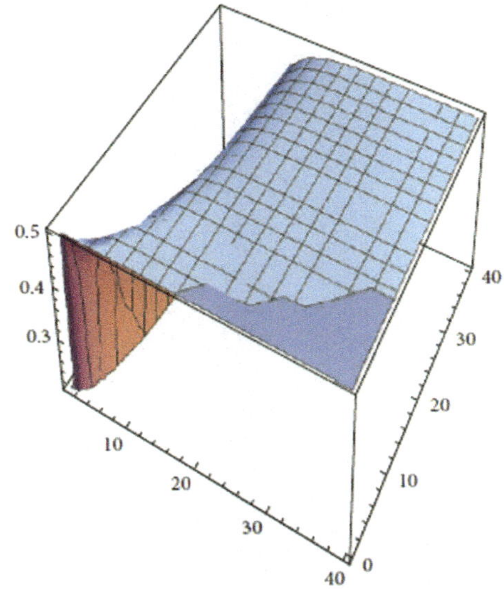

Fig. 3 Solution of (19) (drawdown)

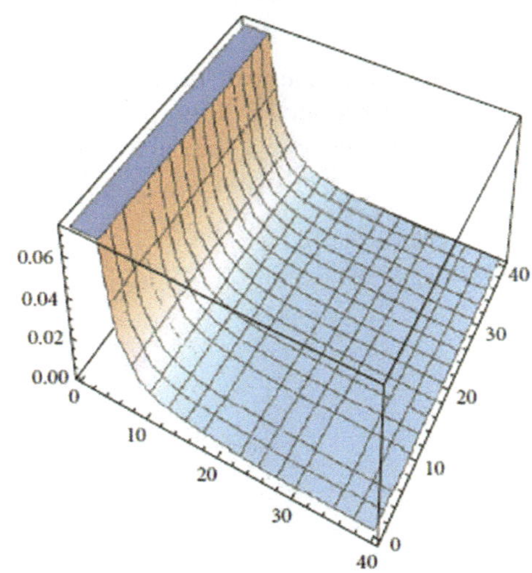

Fig. 4 Solution of (19) (piezometric head)

i. As a function from the well to the perception point of time and space
ii. As a function of time for a fixed distance from the well
iii. For a fixed time as a function of space.

And they are compared with the following solution proposed by Parker for infinite aquifer with line source [1].

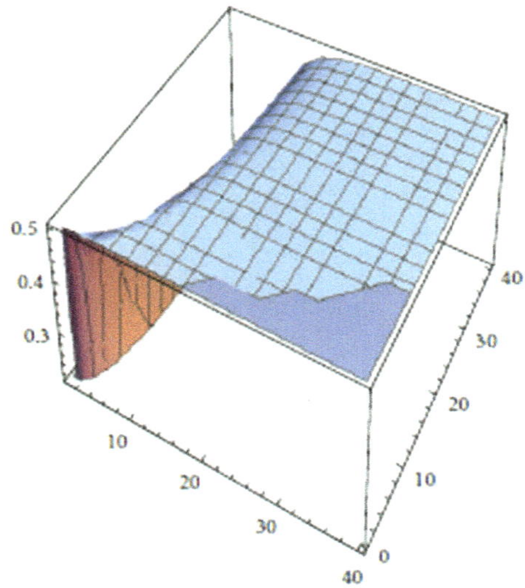

Fig. 5 Baker's solution (drawdown)

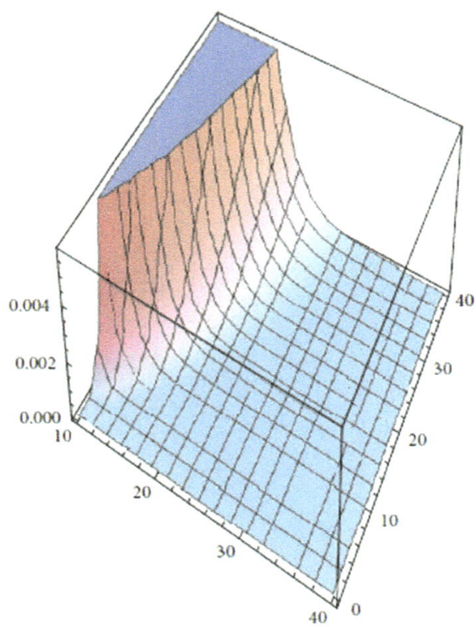

Fig. 6 Baker's solution (piezometric head)

$$\psi(r,t) = \frac{Qr^{2(1-\eta/2)}}{4\pi^{(\eta/2)}Kd^{1-\eta}}\Gamma\left[\frac{\eta}{2}-1, \frac{r^3 S_0}{4Kt}\right]. \tag{29}$$

where

Fig. 7 Solution as function of base of (9)

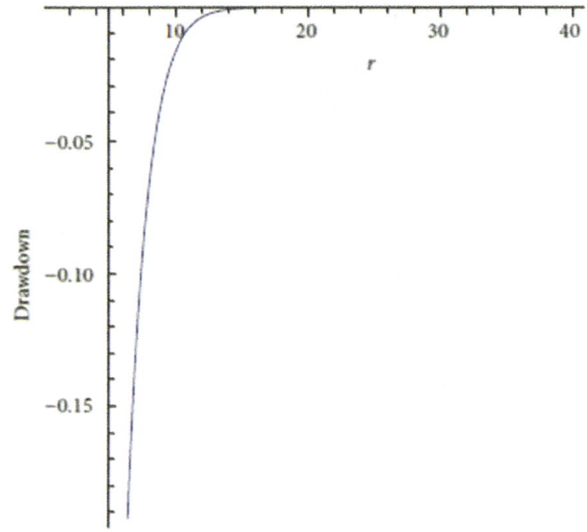

Fig. 8 Solution as function of base of (19)

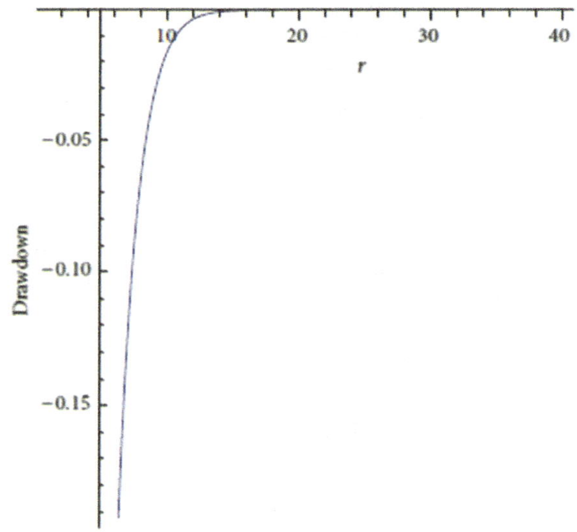

Γ The gamma function.

η The special dimension of the flow, which is equal to the special dimension, n is an integer.

And the other quantities all have the same meaning as before. Table 1 shows the theoretical values of the parameters of the aquifer and the rate of discharge used in numerical simulation.

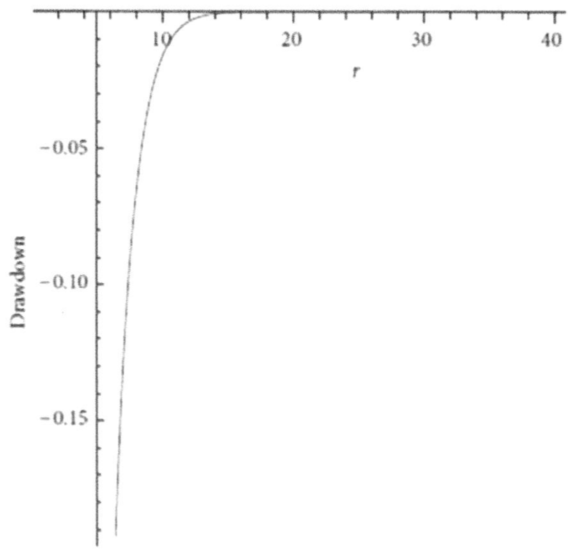

Fig. 9 Baker's solution as function of space

Table 1 Theoretical values

Parameters	Values	Units
S_0	0.75	m^{-1}
Q	10.8	m^3s^{-1}
K	0.65	ms^{-1}
d	40	m
r_b	0.15	m
$\psi_0(r,t)$	0	m
α	0.005	None
μ	1.005	None
η	0.012	None

Remark 2: The main reason to compare these results obtained by the proposed method (STADM) with Barker's solution is to establish a possible relationship between the parameter fractal previously presented by Barker and the fractional order of the derivative.

5 Conclusions

In this paper, the fractional model of groundwater flow equation based on the generalization of the classical Darcy law was studied using the concept of complementary fractional-order derivatives of Weyl and Caputo fractional derivatives. The suggested method is employed to compute the approximate solution to the proposed problem. The construction of method based on using the Sumudu transform and domain decomposition method (ADM) is presented. Illustrative examples show that there is a relationship between the fractional order of the derivative and the non-integral dimension of the flow by comparing these results obtained via the suggested method with that of Barker's fractal radial flow model and also show that the suggested method can be used for similar problems, and it is very effective.

References

1. Barker, J. A. (1988). A generalized radial flow model for hydraulic tests in fractured rock. *Water Resources Research, 24*(10), 1796–1804.
2. Botha, J. F., Verwey, J. P., Van der Voort, I., Vivier, J. J. P., Buys, J., Colliston, W. P., & Loock, J. C. (1998). Karoo aquifers: Their geology, geometry and physical properties. *Water Research Commission, 487*(1), 98.
3. Black, J. H., Barber, J. A., & Noy, D. J. Cross hole investigations: the method, theory and analysis of cross hole sinusoidal pressure tests in fissured rocks. In *Strip a projects internal reports 86-03*. Stockholm, Sweden: SKB.
4. Van Tonder, G. J., Botha, J. F., Chiang, W. H., Kunstmann, H., & Xu, Y. (2001). Estimation of the sustainable yields of boreholes in fractured rock formations. *Journal of Hydrology, 241*(1–2), 70–90.
5. Mjthap, H. Z., & Al-Azzawi, S. N. (2019). Mixing Sumudu transform and Adomain decomposition method for solving Riccati equation of variable fractional order. *Journal of Interdisciplinary Mathematics, 22*(8), 1559–1563.
6. Al-Shammari, A. G. N., AL-Hussein, W. R. A., & AL-Safi, M. G. (2018). A new approximate solution for the Telegraph equation of space-fractional order derivative by using Sumudu method. *Iraqi Journal of Science, 59*(3A), 1301–1311.
7. Gill, V., Modi, K., & Singh, Y. (2018). Analytic solutions of fractional differential equation associated with RLC electrical circuit. *Journal of Statistics and Management Systems, 21*(4), 575–582.
8. Ibrahim, R. W., & Darus, M. (2014). On a new solution of fractional differential equation using complex transform in the unit disk. *Mathematical and Computational Applications, 19*(2), 152–160.
9. Watugala, G. (1993). Sumudu transform: A new integral transform to solve differential equations and control engineering problems. *Integrated Education, 24*(1), 35–43.
10. Belgacem, F. B. M. (2006). Introducing and analyzing deeper Sumudu properties. *Nonlinear Studies, 13*(1), 23.
11. Mjthap, H. Z., & Al-Azzawi, S. N. (2019). Some properties of Sumudu transformation. *Journal of Interdisciplinary Mathematics, 22*(8), 1543–1547.
12. Belgacem, F. B. M., Karaballi, A. A., & Kalla, S. L. (2003). Analytical investigations of the Sumudu transform and applications to integral production equations. *Mathematical Problems in Engineering, 2003*(3), 103–118.
13. Cloot, A., & Botha, J. F. (2006). A generalised groundwater flow equation using the concept of non-integer order derivatives. *Water SA, 32*(1), 1–7.
14. Uchdadiya, K. D., & Patel, J. N. (2014). Seepage losses through unlined and lined canals. *International Journal of Advances in Applied Mathematics and Mechanics, 2*(2), 88–91.
15. Neuringer, J. L. (1978). The Frobenius method for complex roots of the indicial equation. *International Journal of Mathematical Education in Science and Technology, 9*(1), 71–77.

16. Oldham, K. B., & Spanier, J. (1974). The fractional calculus, vol. 111 of *Mathematics in science and engineering*.
17. Adomian, G. (1988). A review of the decomposition method in applied mathematics. *Journal of Mathematical Analysis and Applications, 135*(2), 501–544.

A Meta-Analysis Review of DeLone and McLean Information System Success Model

Barakat Ibrahim, Salih Glood, Seror Bahloos, and Sajad Abd

Abstract DeLone and McLean introduced and released a new model of IS success two decades ago. Their model has been used through several studies in different countries. The relationships between the success model constructs have varying outcomes in prior studies. Therefore, there is a need to conduct a review of previous studies on the IS success model with different contexts. In this paper, we conducted a meta-analysis review among empirical studies in terms of country, year and results in each study. By aggregating the results of 52 empirical studies that tested relationships within the IS success model at the individual level of analysis. Findings showed that different findings occur in various countries with different cultural contexts at the same theory and constructs. Therefore, due to the inconsistency of the findings among the constructs' relationships of the D&M success model, it is necessary to examine the constructs in different environments, such as unstable environments (conflicts and violence).

Keywords A meta-analysis · DeLone and McLean model · IS success model · Civil conflicts

1 Introduction

Information systems (ISS) considered a backbone of developing the communities. Introducing information systems in all respects the life (such as economic, social,

B. Ibrahim · S. Bahloos · S. Abd
Al-Muthanna University, Al-Muthanna, Iraq
e-mail: brakat.s@mu.edu.iq

S. Bahloos
e-mail: seror.manea@mu.edu.iq

S. Abd
e-mail: sajad.abd88@mu.edu.iq

S. Glood (✉) · S. Bahloos · S. Abd
Thi-Qar University, Thi-Qar, Iraq
e-mail: salihglood@utq.edu.iq

S.-L. Peng et al. (eds.), *Proceedings of First International Conference on Mathematical Modeling and Computational Science*, Advances in Intelligent Systems and Computing 1292, https://doi.org/10.1007/978-981-33-4389-4_19

and healthcare), it makes has an important role in public and private sectors. From another side, although many information systems were implemented in different contexts and different countries, there is a huge rate of these information systems is failed [1] which lead to spend much money and time. Therefore, there is a necessary need to examine the influencing constructs of IS success as a critical issue.

In 1992, DeLone and McLean (D&M) [2] proposed six different constructs of IS success in their model that encompasses information quality, system quality, user satisfaction, use, individual impacts, and organizational impacts. The model specified the different relationships between those constructs of success; but, at the same time, alerted researchers that the model needed "*further development and validation*" [2]. In the empirical literature, the various relationships in the IS success model have found varying degrees of support. Some studies found strong build correlations, while others found correlations that were either weak or not significant. Therefore, this study conducted a meta-analysis review among empirical studies in terms of country, year, and results and by aggregating their findings from 52 empirical research, which analyzed relationships at the individual level of the IS success model.

2 DeLone and McLean Success Model

In 1992, D&M revised the previous literature published in 1981–1987 to develop the IS success taxonomy. This taxonomy was based on the modification of Mason's communications model proposed in 1945 by Shannon and Weaver. In Shannon and Weaver model was determined the technical level, semantic level, and effectiveness level as information levels. Similarly, Mason modified and adapted this model, extending the degree of effectiveness level to receiving information, influencing the recipient, and influencing the system. In their study, D&M divided all constructs into program performance categories by mapping every degree of Mason's effectiveness to an aspect of IS performance. This yielded six IS output constructs: Device Quality (SYQ) and Information Quality (IQ) were close, respectively, to the technological and semantic communication levels. While the other four constructs are use (U) related to Mason's "*receipt of information.*" User satisfaction (US) and individual impact were related to the "*information's influence on the recipient.*" Organizational impact was related to the "*influence of the information on the system.*" Figure 1 shows

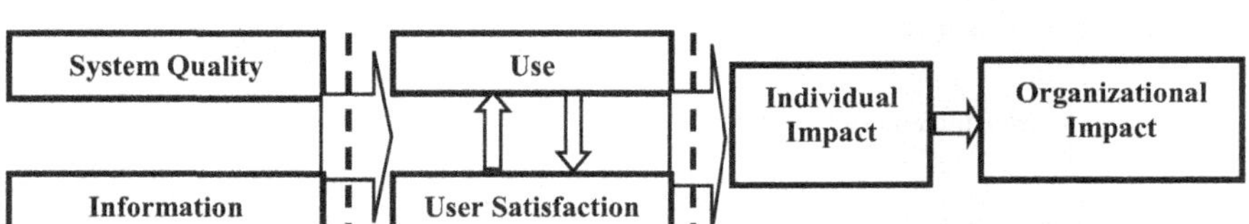

Fig. 1 Power DeLone and McLean's model of IS success. (*Source* DeLone and McLean, 1992, p. 87)

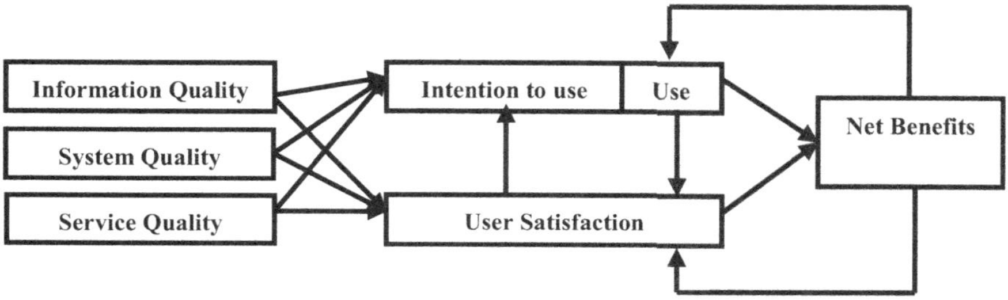

Fig. 2 DeLone and McLean's updated IS success model. (*Source* DeLone and McLean, 2003, p. 24)

the DeLone and McLean's model of IS success.

D&M updated their model in 2003 to fix certain shortcomings of their original template. A major change to the revised model was the introduction of service quality (SQ) as an additional dimension of IS performance [3]; this was implemented because IS's increasing nature warranted an assessment of service quality when evaluating IS performance. Another changing by Seddon and Kiew [4], they replaced individual and organizational impacts as separate constructs with net benefits (NB) construct. This move implies that IS would impact at any level of research (such as workgroups, markets, and societies) other than individuals and organizations. Another enhancement in the D&M updated model by adding "intention to use" construct. This result according to DeLone and McLean [5], they stated that "if intention to use as a measure, then increased user satisfaction would lead to a higher intention to use, which would subsequently affect use." The updated model is shown in Fig. 2.

3 Research Methodology

A meta-analysis defined as an organized method to summarize and interpret a choose sets of empirical studies in a specific domain by systematic procedures. A meta-analytical review considered as a useful technique to conciliate between conflicting results extricated from previous studies in a particular domain. This study as a review study of DeLone and McLean model and focused to study the relationships of DeLone and McLean model in terms of supported or not supported relationships in previous studies. Therefore, this study follows the same way of meta-analysis review through the search for empirical studies literature of IS success model.

In this study, authors try to identify as many previous studies as possible that examine of D&M model. Therefore, multiple full-text searches in online databases were performed for papers published between 1992 and 2020 years. This yielded nearly five hundred manuscripts, the authors included several requirements in their meta-analysis after obtaining 500 manuscripts of searching online databases. They excluded any manuscript if it was: a literature review or purely theoretical; not had quantitative empirical data; did not test any of our hypotheses; also, if it was tested one or more of hypotheses, but did not have enough data. After finishing this step, 150 studies only requiring extra examination. Additionally, when looking at those studies, the authors defined further criteria for their meta-analysis. Firstly, the authors determined the studies using the same dataset and kept only the studies that tested most of our hypotheses. Second, we found that some work centered on IS performance at the organization's level while others analyzed it at the individual level. The authors therefore decided to exclude the studies which tested the hypotheses using a non-individual unit of analysis to ensure the results were interpretable since most of the studies based on IS success at the individual level. Finally, in investigating the retain studies, we find several studies evaluating other models (e.g., TAM, TPB or task-technology fit). Consequently, the authors have ruled out any work using a separate study methodology from the D&M methodology. Although research using alternative models were included if the authors referenced one of the D&M papers for the use of one or more constructs in their analysis. Therefore, using those additional criteria results in a tally of 53 samples from 52 studies. Reviewing those studies according to relations between D&M updated model's constructs would be explained in the next sections.

4 Reviewing the Relationships Between D&M Updated Model's Constructs

4.1 The Relationship Between IQ and IU/U

Information quality (IQ) has played a vital role in the measurement of IS success. In their study, Bento and Costa [6] asserted that IQ dimension as a core component for the valuation of IS success that determines the quality of information's produced by IS. The interrelationships between IQ and IU/U have been gaining diversified support in the perspective of IS success. Several research studies investigated the IQ and IU/U relationship, as shown in Table 1.

4.2 The Relationship Between SYQ and IU/U

System quality (SYQ) referred to the overall quality of the system's performance. According to Bento and Costa [6] asserted that SYQ considered an important construct to evaluate IS success. In the previous literature, the relationship between SYQ and IU/U has achieved varied support as shown in Table 2.

Table 1 Summary of studies that tested the relationship between IQ and IU/U

Relationship	Sources	Year	Country	Result
IQ → IU/U	McGill et al	2003	Australia	Not supported
	Iivari	2005	Finland	Not supported
	Petter and Mclean,	2009	US	Supported
	Bento and Costa	2013	Portugal	Supported
	Cho et al	2015	South Korea	Supported
	Sirsat S. and Sirsat M	2016	India	Supported
	Ojo	2017	Nigeria	Supported
	Yakubu and Dasuki	2018	Nigeria	Supported
	Al-Fraihat et al	2020	United Kingdom	Not supported
	Shim and Jo	2020	South Korea	Supported

Table 2 Summary of studies that tested the relationship between SYQ and IU/U

Relationship	Sources	Year	Country	Result
SYQ → IU/U	Straub et al.	1995	US	Not supported
	Lucas and Spitler	1999	US	Not supported
	Rai et al.	2002	US	Supported
	McGill et al.	2003	Australia	Not supported
	Iivari	2005	Finland	Not supported
	Kositanurit et al.	2006	Canada	Supported
	Wang and Liao	2008	Taiwan	Not supported
	Alshibly	2014	Jordan	Supported
	Cho et al	2015	South Korea	Supported
	Sirsat and Sirsat	2016	India	Supported
	Ojo	2017	Nigeria	Supported
	Yakubu and Dasuki	2018	Nigeria	Supported
	Al-Fraihat et al.	2020	United Kingdom	Supported
	Shim and Jo	2020	South Korea	Not supported

4.3 The Relationship Between SQ and IU/U

Glood et al. [1] describe that SQ intends to evaluate the service received through a comparative analysis of expected and perceived services. Other researchers, Pitt et al. [3] stated that SQ is believed to be a significant measure of IS success. According to the previous literature, SQ has achieved an important role in the assessment of ISs success. The interaction between SQ and IU/U has collected mixed support in IS success context, as shown in Table 3.

Table 3 Summary of studies that tested the relationship between SQ and IU/U

Relationship	Sources	Year	Country	Result
SQ → IU/U	Kositanurit et al.	2006	Canada	Not supported
	Wang and Liao	2008	Taiwan	Supported
	Chatterjee et al.	2009	US	Supported
	Alshibly	2014	Jordan	Supported
	Cho et al.	2015	South Korea	Not supported
	Sirsat and Sirsat	2016	India	Supported
	Ojo	2017	Nigeria	Supported
	Yakubu and Dasuki	2018	Nigeria	Not supported
	Al-Fraihat et al.	2020	United Kingdom	Not supported
	Shim and Jo	2020	South Korea	supported

4.4 *The Relationship Between IQ and US*

Prior studies have given significant support to the relationship between IQ and US [7]. The IQ-US partnership has gained diverse support at the individual analytics level. Several types of research investigated the relationship between IQ and the US, as shown in Table 4.

Table 4 Summary of studies that tested the relationship between IQ and the US

Relationship	Sources	Year	Country	Result
IQ → US	Seddon and Kiew	1996	US	Supported
	Kim et al.	2004	Singapore	Supported
	Wixom and Todd	2005	US	Supported
	Teo et al.	2008	Singapore	Not supported
	Zhou	2013	china	Not supported
	Alshibly	2014	Jordan	Supported
	Sirsat and Sirsat	2016	India	Supported
	Yakubu and Dasuki	2018	Nigeria	Not supported
	Al-Mamary et al.	2019	Yemen	Supported
	Al-Fraihat et al.	2020	United Kingdom	Supported
	Shim and Jo	2020	South Korea	Not supported

Table 5 Summary of studies that tested the relationship between SYQ and US

Relationship	Sources	Year	Country	Result
SYQ → US	Seddon and Kiew	1996	US	Supported
	Rai et al.	2002	US	Supported
	Wixom and Todd	2005	US	Supported
	Leclercq	2007	Italia	Not supported
	Wang and Liao	2008	Taiwan	Supported
	Zhou	2013	china	Supported
	Alshibly	2014	Jordan	Supported
	Cho et al.	2015	South Korea	Supported
	Sirsat and Sirsat	2016	India	Not supported
	Yakubu and Dasuki	2018	Nigeria	Not supported
	Al-Mamary et al.	2019	Yemen	Supported
	Al-Fraihat et al.	2020	United Kingdom	Supported
	Shim and Jo	2020	South Korea	Not supported

4.5 The Relationship Between SYQ and US

To success, any information system, DeLone and McLean [2, 5], as well as Petter et al. [8], asserted that SYQ and US are meaningfully associated with each other. Various researches have scrutinized the relationship between SYQ and US as shown in Table 5.

4.6 The Relationship Between SQ and US

To measure the SQ, numerous techniques have used in prior studies to observe the relation between SQ and US [9]. Petter et al. [8] explored that this relation among SQ and US has mixed support in most of the prior studies. Table 6 shows the summaries of prior studies' results between SQ and US.

4.7 The Relationship Between U and US

Several researchers, who cited or used IS Success model, confirmed that U and US constructs performed an essential role for improved grasping the success of information systems [8]. DeLone and McLean [2] proposed that the use of IS might influence users' satisfaction, and vice versa. In particular, they mentioned that "use can affect the degree of user satisfaction—positively or negatively, as well as the reverse being true" (p. 83). DeLone and McLean [5] declare that " 'use' must precede

Table 6 Summary of studies that tested the relationship between SQ and US

Relationship	Sources	Year	Country	Result
SQ → US	Devaraj et al.	2002	Indiana	Not supported
	Chiu et al.	2007	Taiwan	Not supported
	Teo et al.	2008	Singapore	Not supported
	Zhou	2013	china	Not supported
	Alshibly	2014	Jordan	Supported
	Sirsat and Sirsat	2016	India	Supported
	Yakubu and Dasuki	2018	Nigeria	Supported
	Nugroho and Prasetyo	2018	Indonesia	Supported
	Al-Mamary et al.	2019	Yemen	Not supported
	Al-Fraihat et al.	2020	United Kingdom	Supported
	Shim and Jo	2020	South Korea	Not supported

Table 7 Summary of studies that tested the relationship between U and US

Relationship	Sources	Year	Country	Result
U → US	Seddon and Kiew	1996	US	Not supported
	Rai et al.	2002	US	Supported
	Wang and Liao	2008	Taiwan	Supported
	Alshibly	2014	Jordan	Supported
	Sirsat and Sirsat	2016	India	Supported
	Tam and Oliveira	2016	Portugal	Supported
	Ojo	2017	Nigeria	Not Supported
	Keikhosrokiani et a.l	2018	Malaysia	Supported

'user satisfaction' in a process sense, but a positive experience with 'use' will lead to greater 'user satisfaction' in a causal sense." Similarly, "increased 'user satisfaction' will lead to increased use." Table 7 shows the summaries of prior studies' results U and US.

4.8 The Relationship Between US and IU

User satisfaction (US) is considered among one of the major constructs of IS Success. US has been known as the feeling of aggregate experience, either pleasure or displeasure, that resulted from cumulating the entire expected benefits associated with interaction and usage of information system. Past literature indicates that the success of information systems is largely dependent on the US. Thus, the US is an important determinant of IS success. Previous studies have explored the linkage between the

Table 8 Summary of studies that tested the relationship between US and IU

Relationship	Sources	Year	Country	Result
US → IU	Collopy	1996	US	Not supported
	Ang and Soh	1997	Korea	Not supported
	Rai et al.	2002	US	Supported
	Kim et al.	2004	Singapore	Supported
	Chiu et al.	2007	Taiwan	Supported
	Alshibly	2014	Jordan	Supported
	Tam and Oliveira	2016	Portugal	Supported
	Keikhosrokiani et al.	2018	Malaysia	Supported
	Yakubu and Dasuki	2018	Nigeria	Not supported
	Shim and Jo	2020	South Korea	Supported

satisfaction of users and their intention to use in the context of IS success. However, few studies revealed contrasting results [10]. Table 8 shows the summaries of prior studies' results US and IU.

4.9 *The Relationship Between U and NB*

DeLone and McLean [5] have argued that U is a relevant measure of IS success where the use of the information system is optional. In the previous research, U has achieved a vital role in the measurement of ISs success. The relationship between U of system and NB has performed mixed support in the IS success context, as shown in Table 9.

Table 9 Summary of studies that tested the relationship between U and NB

Relationship	Sources	Year	Country	Result
U → NB	Seddon and Kiew	1996	US	Supported
	Lucas and Spitler	1999	US	Not supported
	McGill et al.	2003	Australia	Not supported
	Wang and Liao	2008	Taiwan	Supported
	Alshibly	2014	Jordan	Supported
	Ojo	2017	Nigeria	Supported
	Nugroho and Prasetyo	2018	Indonesia	Supported
	Shim and Jo	2020	South Korea	Supported
	Al-Fraihat et al.	2020	United Kingdom	Supported

Table 10 Summary of studies that tested the relationship between US and NB

Relationship	Sources	Year	Country	Result
US → NB	Yuthas and Young	1998	China	Not supported
	Rai et al.	2002	US	Supported
	McGill et al.	2003	Australia	Supported
	Cho et al.	2015	South Korea	Supported
	Sirsat and Sirsat	2016	India	Supported
	Al-Mamary et al.	2019	Yemen	Not supported
	Al-Fraihat et al.	2020	United Kingdom	Supported
	Shim and Jo	2020	South Korea	Not supported

4.10 The Relationship Between US and NB

Seddon and Kiew [4] reported that the user hopes to receive certain benefits from using the information system. The extent of the US depends, therefore, on the degree to which the program conforms to the perceptions of the user that are associated with the information system. Many of the prior studies have explored that US is directly associated with system-based benefits such as [9–11]. However, in contrast to the aforementioned findings, Yuthas and Young [12] found a weak correlation between US and performance in decision-making. Table 10 shows the summaries of prior studies' results between US and NB.

4.11 The Feedback Relationship Between NB and Both U and US

The original model of IS success has been revised and modified by DeLone and McLean in 2003. They specified the weaknesses points and made some amendments in the model. The main modification is integrating the two constructs "individual impacts" and "organizational impacts" into one construct called "net benefits (NB)." Bento and Costa [6] reported that NB as the most construct was played a significant role in the updated D&M model, allowing for an effective understanding of the balance of positive/negative impacts of IS. In the previous literature, DeLone and McLean [5] confirmed that NB as important determinant employed to measure the effective success of an IS. The correlations between NB, US, and U of a system have acquired mixed support of IS success. The summary of prior studies' results between NB, US and U can be seen in Table 11 and Table 12, respectively.

Table 11 Summary of studies that tested the relationship between NB and U

Relationship	Sources	Year	Country	Result
NB → U	Straub et al.	1995	US	Not supported
	Lucas and Spitler	1999	US	Not supported
	Wixom and Todd	2005	US	Supported
	Chatterjee et al.	2009	US	Supported
	Bento and Costa	2013	Portugal	Supported
	Alshibly	2014	Jordan	Supported
	Sirsat and Sirsat	2016	India	Supported

Table 12 Summary of studies that tested the relationship between NB and US

Relationship	Sources	Year	Country	Result
NB → US	Seddon and Kiew	1996	US	Supported
	Rai et al.	2002	US	Supported
	Bharati and Chaudhury	2006	US	Supported
	Alshibly	2014	Jordan	Supported
	Sirsat and Sirsat	2016	India	Supported

5 Result and Discussion

The finding from the aforementioned discussion showed that the use of the updated model's D&M and its constructs' relationships in previous researches have conflicting results at different nations and different cultural background. For instance, the relationship between IQ and IU/U has been supported on some studies such as [6, 11, 13, 13–15] while this relation has not supported by Iivari [7] and Al-Fraihat et al. [16]. Additionally, the correlation between SYQ and IU/U considered as a significant relationship in some studies such as [11, 13, 16, 17]; while some studies such as [15] detected a non-significant correlation between SYQ and IU/U.

The interaction between SQ and IU/U has been supported in some studies such as [13, 15, 18] while few researchers such as [9, 14, 16] revealed in their studies that there is a non-significant correlation between SQ and IU/U. Additionally, the relationship between IQ and the US has been supported in several studies, such as [6, 8, 16, 19, 20] while Zhou [21] and Yakubu and Dasuki [14] did not detect a significant relationship between IQ and US. Moreover, the relationship between SYQ and US has been supported by Rai et al. [11], Cho et al. [19], Ojo [17], Al-Mamary et al. [20], and Shim and Jo [15], whereas Sirsat and Sirsat [18] and Yakubu and Dasuki [14] detected a non-significant correlation between SYQ and US.

The relationship between SQ and US has a strong correlation in the previous studies such as [13, 15, 17, 19] while the same relationship has not been supported by Zhou [21] and Al-Mamary et al. [20]. Several researchers such as [7, 13, 22] highlighted that there was a strong association between U and US, while Seddon and

Kiew [4] and Ojo [17] in their studies detected that there was no relationship between U and US. In the same way, Rai et al. [11], Alshibly [13], Tam and Oliveira [22], and Shim and Jo [15] revealed in their studies that there is a significant relationship between US and IU of the IS while Ang and Soh [10] and Yakubu and Dasuki [14] asserted that there was a non-significant correlation between US and IU.

In the same way, the relationship between U and NB was considered to be strongly significant by Alshibly [13], Ojo [17], Shim and Jo [15] while some studies such as [7, 19] established that there is no significant association between U and NB. Furthermore, the relationship between US and NB has been supported in some studies such as [13, 16, 18]; however, Yuthas and Young [12], Ojo [17], Al-Mamary et al. [20], and Shim and Jo [15] explored that there is a weak relationship between US and NB in their study.

The feedback relation between NB and U has been supported by Bento and Costa [6] and Sirsat and Sirsat [18] while the same feedback relationship not supported by Lucas and Spitler [60]. Also, the relation between NB and US has been strongly supported by some researchers such as [11, 13, 18].

The findings from the aforementioned discussion of all the relationships between variables of DeLone and McLean IS success model showed that most of the prior studies were examined the success of DeLone and McLean model have different outcomes in different countries with different cultural backgrounds. Its also showed that most of the prior studies conducted their studies in stable environments. But they are not taking considered the unstable environments and the role of civil conflicts and violence of IS success. Therefore, there is a need to examine DeLone and McLean IS success model in a different environment (such as unstable environment).

6 Civil Conflicts

Civil conflict (CC) and violence have been recorded to be alarmingly high in the most recent times, especially in developing nation such as Iraq, Syria, Libya, Yemen, India, Afghanistan, Sri Lanka, and Pakistan. Rising the numbers of civil conflicts are related to continuous struggle and fight for political superiority to protect political leaders' interests and objectives [23].

Political instability is distinguished as a central challenge facing IS/IT services in developing states. This obstacle or challenge may have a significant impact to adopt and use of the new technology by publics, which leads to generating net benefits of technology usage. Sanginga et al. [24] asserted the presence of a positive relationship between civil conflicts and the adoption of technology. Partridge [25] affirmed that the level of civil conflicts in a state highlights its citizens' intention to use new technology. The findings by Partridge [25] show that the states involved in civil conflicts invested most of their resources and activities on winning the war; therefore, adoption civilian technologies has a lesser degree. Also, Khan [26] affirmed that the relation between civil conflicts and technology adoption and utilization by the public is significant.

However, Barki and Hartwick [27] claimed that the potential effects of civil conflicts on the IS adoption have not been adequately investigated. The influencing of civil conflicts on peoples' behavior of using government services must be further investigated as a crucial issue [26].

Civil conflicts might create several potential effects of society, such as psychosocial and environmental impacts. As well as, civil conflicts and violence generate substantial impacts on citizens' psychosocial life at large. Through looking at prior literature, there very few studies addressed the effect of civil conflicts on technology adoption.

Deduce from above, most of the results among constructs of the updated D&M model are conflicting, despite most of the researchers' studies were conducted in a stable environment. Therefore, there is a need to investigate constructs' relations of model's D&M in a different environment such as an unstable environment.

7 Conclusion

The finding from the study confirms that the use of the updated model's D&M and its constructs in previous researches has conflicts results at different nations and different cultural background. Therefore, there is a need to investigate the constructs' relationships of the updated D&M model in a different environment (such as an unstable environment). Also, there is a lack of studies in D&M updated model to evaluate IS success in an unstable environment. In a priority, evaluation of information systems among users is an important sign in the Middle East, especially in unstable environments such as Syria, Yemen, and Iraq. It will contribute to highlighting the key elements that may affect of continuing to use of any new IS initiatives through non-normal circumstances, thus overcome of the drawbacks that face to successful any IS projects in these countries.

References

1. Glood, S. H., Osman, W. R. S., & Nadzir, M. M. (2016). The effect of civil conflicts and net benefits on m-government success of developing countries: A case study of Iraq. *Journal of Theoretical and Applied Information Technology.*, *88*(3), 541–552.
2. DeLone, W. H., & McLean, E. R. (1992). Information systems success: The quest for the dependent variable. *Information Systems Research.*, *3*(1), 60–95.
3. Pitt, L., Watson, R., & Kavan, C. (1995). Service quality: A measure of information systems effectiveness. *MIS Quarterly.*, *19*(2), 173–187.
4. Seddon, P., & Kiew, M. (1996). A partial test and development of Delone and Mclean's model of is success. *Australasian Journal of Information Systems.*, *4*(1), 90–109.
5. DeLone, W. H., & McLean, E. (2003). The DeLone and McLean model of information systems success: A ten-year updated. *Journal of Management Information Systems.*, *19*(4), 9–30.

6. Bento, F., & Costa, C. J. (2013). ERP measure success model; a new perspective. In *Proceedings of the 2013 International Conference on Information Systems and Design of Communication - ISDOC '13* (p. 16).
7. Iivari, J. (2005). An empirical test of the DeLone-McLean model of information system success. *ACM SIGMIS Database, 36*(2), 8–27.
8. Petter, S., DeLone, W., & McLean, E. (2008). Measuring information systems success: Models, dimensions, measures, and interrelationships. *European Journal of Information Systems., 17*(3), 236–263.
9. Halawi, L., & McCarthy, R. (2007). Measuring faculty perceptions of blackboard using the technology acceptance model. *Issues in Information Systems, 8*(2), 160.
10. Ang, J., & Soh, P. H. (1997). User information satisfaction, job satisfaction and computer background: An exploratory study. *Information & Management, 32*(5), 255–266.
11. Rai, A., Lang, S., & Welker, R. (2002). Assessing the validity of IS success models: An empirical test and theoretical analysis. *Information Systems Research, 13*(1), 50–69.
12. Yuthas, K., & Young, S. T. (1998). Material matters: Assessing the effectiveness of materials management IS. *Information & Management, 33*(3), 115–124.
13. Alshibly, H. H. (2014). Evaluating E-HRM success : A validation of the information systems success model. *International Journal of Human Resource Studies, 4*(3), 107–124.
14. Yakubu, N., & Dasuki, S. (2018). Measuring e-learning success in developing countries: applying the updated DeLone and McLean model. *Journal of Information Technology Education: Research.*
15. Shim, M., & Jo, H. S. (2020). What quality factors matter in enhancing the perceived benefits of online health information sites? Application of the updated DeLone and McLean information systems success model. *International Journal of Medical Informatics, 137*(1), 104093.
16. Al-Fraihat, D., Joy, M., Masa'deh, R., & Sinclair, J. (2020). Evaluating E-learning systems success: An empirical study. *Computers in Human Behavior, 102*, 67–86.
17. Ojo, A. I. (2017). Validation of the Delone and Mclean information systems success model. *Healthcare Informatics Research, 23*(1), 60–66.
18. Sirsat, S. S., & Sirsat, M. S. (2016). A validation of the Delone and Mclean model on the educational information system of the Maharashtra state (India). *International Journal of Education and Learning Systems, 1*, 9–18.
19. Cho, K. W., Bae, S. K., Ryu, J. H., Kim, K. N., An, C. H., & Chae, Y. M. (2015). Performance evaluation of public hospital information systems by the information system success model. *Healthcare Informatics Research, 21*(1), 43–48.
20. Al-Mamary, Y. H. S., Al-Nashmi, M. M., Shamsuddin, A., & Abdulrab, M. (2019). Development of an integrated model for successful adoption of management information systems in yemeni telecommunication organizations. *International Journal of Scientific and Technology Research, 8*(11), 3912–3939.
21. Zhou, T. (2013). An empirical examination of continuance intention of mobile payment services. *Decision Support Systems, 54*(2), 1085–1091.
22. Tam, C., & Oliveira, T. (2016). Understanding the impact of m-banking on individual performance: DeLone & McLean and TTF perspective. *Computers in Human Behavior, 61*, 233–244.
23. HIIK. (2008). *Conflict barometer*.
24. Sanginga, P. C., Kamugisha, R. N., & Martin, A. M. (2007). Conflicts management, social capital and adoption of agroforestry technologies: Empirical findings from the highlands of southwestern Uganda. *Agroforestry Systems, 69*(1), 67–76.
25. Partridge, H. (2007). Redefining the digital divide: Attitudes do matter! *Proceedings of the American Society for Information Science and Technology, 44*(1), 1–13.
26. Khan, G. (2011). Essays on electronic government in developing countries: A socio-technical perspective. Korea Advanced Institute of Science and Technology.
27. Barki, H., & Hartwick, J. (2001). Interpersonal conflict and its management in information system development. *MIS Quarterly*, 195–228.

Γ-n-Derivations on Semigroup Ideals on Prime Γ-Near-Rings

Hiba A. Ahmed and Abdulrahman H. Majeed

Abstract The main aim of this paper is to define the Γ-n-derivation, study and investigate some results of ΓΓ-n-derivation on prime Γ-near-ring G and K are the non-zero ideal G-semigroup, which makes G a commutative ring. The concept of Γ-diverse authors has generalized derivations in various ways. The concept of Γ-n-derivation, permuting derivation, and rig Γ-n-derivation has already been introducing in Γ-near-rings. We investigated commutativity of zero symmetric right near-ring satisfying certain identities involving Γ-derivation. In this paper, we have extended the result in Γ-derivation and semigroup ideal on a zero symmetric left close-ring and show that G satisfies some Γ-n-derivations and ideal for semigroups, then G is a commutative ring

Keywords Prime · Γ-near-ring · Semi-group ideal · Γ-n-derivation

1 Introduction

Throughout this paper, A Γ-near-ring is a triple $(G, +, \Gamma)$, where (i) $(G, +)$ isia (not necessarily abelian) group; (ii) Γ is a binary operation non-emptiest on G such that for each $\gamma \in \Gamma$, $(G +, \gamma)$ is a left near-ring (iii) $s\gamma(r\mu c) = (s\gamma r)\mu c$, for all $s, r, c \in G$ and $\gamma, \mu \in \Gamma$ [1–4]. And G will denote a zero–symmetric left Γ-near-ring with multiplicative center $Z(G)$. For a Γ-near-ring G, the set $G_0 = \{s \in G\text{: } 0\rho s = 0, \forall\ \rho \in \Gamma\}$ is called zero symmetric part of G. If $G = G_0$, then G is called zero-symmetric [4–7]. A Γ-near-ring G is said to be prime Γ-inear-ringi if $s\Gamma G\Gamma r = 0$ impliess $= 0$orr $= 0$, forever $s, r \in G$ and it is said to be semi prime if $s\Gamma G\Gamma s = 0$ implies $s = 0$ for every $s \in G$ [1, 3, 8, 9]. A non-empty subset K of G is called semi-group ideal if $K\Gamma G \subset K$ and $G\Gamma K \subset K$ [10]. The other commutators are: $[s, r]_\rho = s\rho r - r\rho s$ and $(s, r) = s + r - s - r$ denote the additive-group commutator [7,

H. A. Ahmed (✉) · A. H. Majeed
Department of Mathematics, College of Science, University of Baghdad, Baghdad, Iraq
e-mail: hiba.abdulla84@yahoo.com

A. H. Majeed
e-mail: ahmajeed6@yahoo.com

S.-L. Peng et al. (eds.), *Proceedings of First International Conference on Mathematical Modeling and Computational Science*, Advances in Intelligent Systems and Computing 1292, https://doi.org/10.1007/978-981-33-4389-4_20

11, 12]. Γ-near-ring G is called commutative if $(G, +)$ is abelian, and torsion free if $2s = 0$ implies $s = 0$ [13–16].

An additive mapping $h{:}G \rightarrow G$ is called a Γ-derivation on G satisfying $h(s\rho r) = h(s)\rho r + s\rho h(r)$, for every $s, r \in G$ and $\rho\Gamma$ or equivalently $h(s\rho r) = s\rho h(r) + h(s)\rho r$ [4, 12].

An additive mapping $h{:}G \times G \times \cdots \times G \rightarrow G$ is said to be Γ-n-derivation if the relations

$$\begin{aligned} h\big(x_1\gamma x_1', x_2, \ldots, x_n\big) &= h(x_1, x_2, \ldots, x_n)\gamma x_1' + x_1\gamma h\big(x_1', x_2, \ldots, x_n\big) \\ h\big(x_1, x_2\gamma x_2', \ldots, x_n\big) &= h(x_1, x_2, \ldots, x_n)\gamma x_2' + x_2\gamma h\big(x_1, x_2', \ldots, x_n\big) \\ &\vdots \\ h\big(x_1, x_2, \ldots, x_n\gamma x_n'\big) &= h(x_1, x_2, \ldots, x_n)\gamma x_n' + x_n\gamma h\big(x_1, x_2, \ldots, x_n'\big) \end{aligned}$$

Hold for all $x_1, x_1', x_2, x_2', \ldots, x_n, x_n' \in G$ If in addition h is a permuting map, then h is called a Γ-n-derivation of G.

Some recent results on getting closer to a deal with prime near-rings commutativity admitting suitably restricted n-derivations. For further details on prime near-ring, we refer the reader to [1, 2, 6, 7, 14–16].

The purpose of this paper is to study and generalize some results of [10, 11, 13, 17] on commutativity of prime Γ-near-ring and semi-group ideal on which admits suitably constrained Γ-n-derivations.

2 The Main Results

We start with the following lemmas in this section, which are important to the creation of the proofs of our main results.

Lemma 2.1 *Let G be a Γ-near-ring, then h is a Γ-n-derivation of G if and only if* $h(s_1\gamma s'_1, s_2, \ldots, s_n) = s_1\gamma h(s'_1, s_2, \ldots, s_n) + h(s_1, s_2, \ldots, s_n)\gamma s'_1$, for all $s_1, s_1', s_2, \ldots, s_n \in G$ *and* $\gamma \in \Gamma$.

Proof We have

$$\begin{aligned} h(s_1\gamma(s'_1 + s'_1), s_2, \ldots, s_n) &= h(s_1, s_2, \ldots, s_n)\gamma(s'_1 + s'_1) \\ &\quad + s_1\gamma h(s'_1 + s'_1, s_2, \ldots s_n) \\ &= h(s_1, s_2, \ldots, s_n)\gamma s'_1 + h(s_1, s_2, \ldots, s_n)\gamma s'_1 \\ &\quad + s_1\gamma h(s'_1, s_2, \ldots, s_n) + s_1\gamma h(s'_1, s_2, \ldots, s_n) \end{aligned}$$

And

$$\begin{aligned} h(s_1\gamma s'_1 + s_1\gamma s'_1, s_2, \ldots, s_n) &= h(s_1\gamma s'_1, s_2, \ldots, s_n) + h(s_1\gamma s'_1, s_2, \ldots, s_n) \\ &= h(s_1, s_2, \ldots s_n)\gamma s'_1 + s_1\gamma h(s'_1, s_2, \ldots, s_n) \\ &\quad + h(s_1, s_2, \ldots, s_n)\gamma s'_1 s_1\gamma h(s'_1, s_2, \ldots, s_n). \end{aligned}$$

Combining above two equalities, we obtain that

$$\begin{aligned} h(s_1, s_2, \ldots, s_n)\gamma s'_1 + s_1\gamma h(s'_1, s_2, \ldots, s_n) &= s_1\gamma h(s'_1, s_2, \ldots, s_n) \\ &\quad + h(s_1, s_2, \ldots, s_n)\gamma s'_1. \end{aligned}$$

Therefore, $h(s_1\gamma s'_1, s_2, \ldots, s_n) = s_1\gamma h(s'_1, s_2, \ldots, s_n) + h(s_1, s_2, \ldots, s_n)\gamma s'_1$.

Converse can be proved in a similar way.

Lemma 2.2 *Let G be a Γ-near-ring and h be Γ-n-derivation of G, then for every* $s_1, s'_1, \ldots, s_n, r \in G$.

(i) $\{h(s_1, s_2, \ldots, s_n)\gamma s'_1 + s_1\gamma h(s'_1, s_2, \ldots, s_n)\}\beta r = h(s_1, s_2 \ldots, s_n)\gamma s'_1\beta r + n_1\gamma h(s'_1, s_2, \ldots, s_n)\beta r$

(ii) $\{s_1\gamma h(s'_1, s_2, \ldots, s_n) + h(s_1, s_2, \ldots, s_n)\gamma s_1'\}\beta r = s_1\gamma h(s'_1, s_2, \ldots, s_n)\beta r + h(s_1, s_2, \ldots, s_n)\gamma s_1'\beta r$

Proof

(i) For all $s_1, s_1', r, s_2, \ldots, s_n \in G$

$$\begin{aligned} h((s_1\gamma s'_1)\beta r, s_2, \ldots, s_n) &= h(s_1\gamma s'_1, s_2, \ldots, s_n)\beta r + (s_1\gamma s'_1)\beta h(r, s_2, \ldots, s_n) \\ &= \{h(s_1, s_2, \ldots, s_n)\gamma s'_1 + s_1\gamma h(s'_1, s_2, \ldots, s_n)\}\beta r \\ &\quad + (s_1\gamma s'_1)\beta h(r, s_2, \ldots, s_n). \end{aligned}$$

Also

$$\begin{aligned} h(s_1\gamma(s'_1\beta r), s_2, \ldots, s_n) &= h(s_1, s_2, \ldots, s_n)\gamma s'_1\beta r \\ &\quad + s_1\gamma h(s'_1\beta r, s_2, \ldots, s_n) \\ &= h(s_1, s_2, \ldots, s_n)\gamma s'_1\beta r \\ &\quad + s_1\gamma\{h(s'_1, s_2, \ldots, s_n)\beta r + s'_1\beta h(r, s_2, \ldots, s_n)\} \\ &= h(s_1, s_2, \ldots, s_n)\gamma s'_1\beta r \\ &\quad + s_1\gamma h(s'_1, s_2, \ldots, s_n)\beta r \\ &\quad + s_1\gamma s'_1\beta h(r, s_2, \ldots, s_n). \end{aligned}$$

Combining the above two relations, we get

$$\{h(s_1, s_2, \ldots, s_n)\gamma s\prime_1 + s_1\gamma h(s\prime_1, s_2, \ldots, s_n)\}\beta r = h(s_1, s_2, \ldots, s_n)\gamma s\prime_1\beta r + s_1\gamma h(s\prime_1, s_2, \ldots, s_n)\beta r.$$

Lemma 2.3 [6] *Let G be a prime Γ-near-ring.*

(i) *If K is a nonzero semi-group ideal of G and s is an element of G such that* $s\Gamma K = 0$ *(*$K\Gamma s = 0$*), then* $s = 0$*, where* $s \in G$.
(ii) *If K be a nonzero semi-group ideal of G and s is an element of G which centralizes K, then* $s \in Z$.
(iii) *if K is a nonzero semigroup ideal of G. If* $s, r \in G$ *and* $s\Gamma K\Gamma r = \{0\}$*, then* $r = 0$ *or* $s = 0$.

Lemma 2.4 *Let h be an Γ-n-derivation of a Γ-near-ring, then* $h(Z, G, \ldots, G) \subseteq Z$.

Proof If $u \in Z$ then $h(u\gamma s_1, s_2, \ldots, s_n) = h(s_1\gamma u, s_2, \ldots, s_n)$, $\forall\, s_1, s_2, \ldots, s_n \in G$ and $\gamma \in \Gamma$.

Therefore, $h(u, s_2, \ldots, s_n)\gamma s_1 + u\gamma h\,(s_1, s_2, \ldots, s_n) = s_1\gamma h(u, s_2, \ldots, s_n) + h(s_1, s_2, \ldots, s_n)\gamma u$.

Since $u \in Z$, we get $h(u, s_2, \ldots, s_n)\gamma s_1 = s_1\gamma h(u, s_2, \ldots, s_n)$, $\forall\, u \in Z, s_1, s_2, \ldots, s_n \in G$ and $\gamma \in \Gamma$. This implies that $h(Z, G, \ldots, G) \subseteq Z$.

Lemma 2.5 [16] *Let G be a prime Γ-near-ring. If Z comprises a semi-group ideal nonzero left (resp. right), then G is a commutative ring.*

Let G be a prime Γ-near-ring admitting a nonzero Γ-n-derivation h of G and $K_1, K_2, \ldots, K_n$ *is nonzero semi-group ideals of G.*

(i) *If* $x \in G$ and $h(K_1, K_2, \ldots, K_n)\gamma\, x = \{0\}$*, then* $x = 0$.
(ii) If $x \in G$ *and* $x\,\gamma\, h(K_1, K_2, \ldots, K_n) = \{0\}$*, then* $x = 0$.

Proof

(i) By our hypothesis, we have

$$h(s_1, s_2, \ldots, s_n)\gamma x = 0, \forall s_1 \in K_1, s_2 \in K_2, \ldots, s_n \in K_n \text{ and } \gamma \in \Gamma. \quad (1)$$

Putting $r_1\, \beta\, s_1$ in place of s_1, where $r_1 \in G$, in Eq. (1) we get

$h(r_1\, \beta s_1, s_2, \ldots, s_n)\gamma x = 0, \forall s_1 \in K_1, s_2 \in K_2, \ldots, s_n \in K_n, r_1 \in G$ and $\gamma, \beta \in \Gamma$.

By defining property of h and Lemma (2.2), we get

$r_1\, \beta\, h(u_1, u_2, \ldots, u_n)\gamma\, x + h(r_1, s_2, \ldots, s_n)\beta\, s_1\, \gamma\, x = 0$. Using Eq. (1) in previous equation, we get $h(r_1, s_2, \ldots, s_n)\beta\, s_1\, \gamma\, x = 0$. Therefore,

$h(r_1, s_2, \ldots, s_n)\Gamma K_1\Gamma x = \{0\}, \forall\, s_2 \in K_2, \ldots, s_n \in K_n, r_1 \in G$.

Since K_1 is a semi-group ideal and by using Lemma (2.3)(iii), we conclude that either $h(r_1, s_2, \ldots, s_n) = 0, \forall s_2 \in K_2, \ldots, s_n \in K_n, r_1 \in G$ or $x = 0$.

Our claim is that $h(r_1, s_2, \ldots, s_n) \neq 0$, for some $r_1 \in G$, $s_2 \in K_2, \ldots, s_n \in K_n$.

For otherwise, if $h(r_1, s_2, \ldots, s_n) = 0$. Putting $r_2\beta s_2 \in K_2$ in place of s_2, where $r_2 \in G$, in last equation and using it again, we get $h(r_1, r_2, s_3, \ldots, s_n)\beta s_2 = 0$.

Now replacing s_2 again by $w\lambda s_2$, where $w \in G$, in preceding equation, we obtain that

$$h(r_1, r_2, s_3, \ldots, s_n)\beta w\lambda s_2 = 0, \forall s_2 \in K_2, s_3 \in K_3, \ldots, s_n \in K_n, r_1, r_2, w \in G \text{ and } \beta, \lambda \in \Gamma.$$

which implies $h(r_1, r_2, s_3, \ldots, s_n)\Gamma G\Gamma K_2 = \{0\}$. As $K_2 \neq \{0\}$, primeness of G yields that $h(r_1, r_2, s_3, \ldots, s_n) = 0$. Preceding inductively as before, we conclude that

$h(r_1, r_2, \ldots, r_n) = 0$.This shows that $h(G, G, \ldots, G) = \{0\}$.

Leading to a contradiction as h is a non zero Γ-n-derivation. Therefore, our claim is

Correct, and now, we conclude that $x = 0$.

(ii) It can be proved similarly.

Lemma 2.6 *Let G be a prime Γ-near-ring, $K_1, K_2, \ldots, K_n$ is nonzero semi-group ideals of G and let us h be a nonzero Γ-n-derivation of G and $t\gamma k\beta u = t\beta k\gamma u$ for every t, k, u ∈ K and $\gamma, \beta \in \Gamma$. If $h(K_1, K_2, \ldots, K_n) \subseteq Z$, then G is a commutative ring.*

Proof Suppose for all $s_1 \in K_1$, $s_2 \in K_2, \ldots, s_n \in K_n$. $h(s_1, s_2, \ldots, s_n) \in Z$.

Replacing s_1 by $s_1\, \gamma s_1'$ in the previous equation, we get

$h(s_1\, \gamma s_1', s_2, \ldots, s_n) = h(s_1, s_2, \ldots, s_n)\gamma + s_1' s_1 \gamma h(s_1', s_2, \ldots, s_n) \in Z$.

Hence, $\{h(s_1, s_2, \ldots, s_n)\gamma s_1' + s_1\, \gamma\, h(s_1', s_2, \ldots, s_n)\}\beta s_1' = s_1'\beta\{h(s_1, s_2, \ldots, s_n)\gamma s_1' + s_1\, \gamma\, h(s_1', s_2, \ldots, s_n)\}$, $\forall s_1, s_1' \in K_1$, $s_2 \in K_2, \ldots, s_n \in K_n$ and $\gamma, \beta \in \Gamma$.

Using the hypothesis and Lemma (2.2), we get

$s_1\, \gamma\, h(s_1', s_2, \ldots, s_n)\beta s_1' = s_1'\beta\, s_1\, \gamma\, h(s_1', s_2, \ldots, s_n)$. Hence,

$[s_1', s_1]_\gamma\, \beta\, h(s_1', s_2, \ldots, s_n) = 0$, $\forall s_1 \in K_1$, $s_2 \in K_2, \ldots, s_n \in K_n$ and $\in\, \gamma, \beta\Gamma$.

Then, by Lemma (2.6), we get $[s_1', s_1]_\gamma = 0$, for all $s_1 \in s_1' K_1$. It means that s_1' centralizes K_1. Thereby, Lemma (2.3)(ii), we get $\in\, s_1{}'Z$. Now, we conclude that $K_1 \subseteq Z$, and by Lemma (2.4), we get we find that G is a commutative ring.

Corollary 2.8 *Let G be a prime Γ-near-ring admitting a nonzero Γ-n-derivation h of G. If $h(G, G, \ldots, G) \subseteq Z(G)$, then G is a commutative ring.*

Theorem 2.9 *Let G be a prime Γ-near-ring having a nonzero Γ-n-derivation h. If $K_1, K_2, \ldots, K_n$ is nonzero semi-group ideals of G and $t\gamma\, k\beta\, u = t\beta\, k\gamma\, u$, for every t, k, u ∈ K and $\gamma, \beta \in \Gamma$ such that* $h([s, r]_\gamma, s_2, \ldots, s_n) = 0$, for all $s, r \in K_1$, $s_2 \in K_2, \ldots, s_n \in K_n$ *and* $\gamma \in \Gamma$, *then G is a commutative ring.*

Proof Suppose that

$$h([s, r]_\gamma, s_2, \ldots, s_n) = 0, \forall s, r \in K_1, s_2 \in K_2, \ldots, s_n \in K_n \text{and} \gamma \in \Gamma. \quad (2)$$

Replacing r by $s\ \beta\ r$ in Eq. (2), we get

$h([s, s\ \beta\ r]_\gamma, r_2, \ldots, r_n) = 0, \forall s, r \in K_1, s_2 \in K_2, \ldots, s_n \in K_n$ and $\in \gamma, \beta\Gamma$.

which gives us $h(s, s_2, \ldots, s_n)\beta[s, r]_\gamma + s\ \beta\ h([s, r]_\gamma, s_2, \ldots, s_n) = 0$.

By hypothesis, we get $h(s, s_2, \ldots, s_n)\beta[s, r]_\gamma = 0$ that means

$$h(s, s_2, \ldots, s_n)\beta s\gamma r = h(s, s_2, \ldots, s_n)\beta r\gamma s, \forall s, r \in K_1, s_2 \in K_2, \ldots, s_n \in K_n \text{ and } \gamma, \beta \in \Gamma \quad (3)$$

Replacing r by $r\mu u$, where $u \in G$, in Eq. (3) and using it again, we get

$h(s, s_2, \ldots, s_n)\Gamma K_1\Gamma[s, u]_\gamma = \{0\}$. Since K_1 is a semi-group ideal and by Lemma (2.3), we get for each $s \in K_1$, either $s \in Z$ or $h(s, s_2, \ldots, s_n) = 0$. But using Lemma (2.4), we get

$h(s, s_2, \ldots, s_n) \in Z$, for all $s \in K_1$, $s_2 \in K_2, \ldots, s_n \in K_n$. Thus, $h(K_1, K_2, \ldots, K_n) \subseteq Z$.

Now by using Lemma (2.6), we find that G is a commutative ring.

Corrollary 2.10 Let *G be a prime* Γ*-near-ring admitting a nonzero* Γ *-n-derivation h. If* $h([s, r]_\gamma, s_2, \ldots, s_n) = 0$ *for all* $s, r, s_2, \ldots, s_n \in G$ *and* $\gamma \in \Gamma$, *then G is a commutative ring.*

Theorem 2.11 *Let G be a prime* Γ*-near-ring having a nonzero* Γ*-n-derivation h. If* $K_1, K_2, \ldots, K_n$ *is nonzero semi-group ideals of G and* $t\ \gamma\ k\ \beta\ u = t\ \beta\ k\ \gamma\ u$, *, for every* $t, k, u \in K$ *and* $\gamma, \beta \in \Gamma$ *such that* $h([s, r]_\gamma, s_2, \ldots, s_n) = [s, r]_\gamma$, *for all* $s, r \in K_1$, $s_2 \in K_2, \ldots, s_n \in K_n$ *and* $\gamma \in \Gamma$, *then G is a commutative ring.*

Proof Suppose that

$$h([s, r]_\gamma, s_2, \ldots, s_n) = [s, r]_\gamma, \forall s, r \in K_1, s_2 \in K_2, \ldots, s_n \in K_n \text{ and } \gamma \in \Gamma. \quad (4)$$

Replacing r by $r\ \beta\ s$ in Eq. (4), we get
$h([s, r\beta s]_\gamma, s_2, \ldots, s_n) = [s, r\ \beta\ s]_\gamma, \forall s, r \in K_1, s_2 \in K_2, \ldots, s_n \in K_n$ and $\in \gamma, \beta\Gamma$.
Then, $h([s, r]_\gamma, r_2, \ldots, r_n)\beta\ s + [s, r]_\gamma\ \beta\ h(s, s_2, \ldots, s_n) = [s, r]_\gamma\ \beta\ s$.
Thus, we get $[s, r]_\gamma\ \beta\ h(s, s_2, \ldots, s_n) = 0$.

$$s\gamma r\beta h(s, s_2, \ldots, s_n) = r\gamma s\beta h(s, s_2, \ldots, s_n),$$
$$\forall s, r \in K_1, s_2 \in K_2, \ldots, s_n \in K_n \text{ and } \gamma, \beta \in \Gamma. \quad (5)$$

Replacing r by $t\ \gamma\ r$, where $t \in G$ in Eq. (5) and using it again, we get

Then, $[s, t]_\gamma\Gamma K_1\Gamma h(s, s_2, \ldots, s_n) = \{0\}$. Since K_1 is a nonzero semi-group ideal and by Lemma (2.3), we get For each $s \in K_1$, either $s \in Z$ or $h(s, s_2, \ldots, s_n) = 0$. Now using Lemma (2.4), we get $h(s, s_2, \ldots, s_n) \in Z$ that means $h(K_1, K_2, \ldots, K_n) \subseteq Z$. By using Lemma (2.6), we find that G is a commutative ring.

Corollary 2.11 *Let G be a prime Γ-near-ring admitting a nonzero Γ -n-derivation h of G. If* $h([s, r]_\gamma, s_2, \ldots, s_n) = [s, r]_\gamma$, *for all* $s, r, s_2, \ldots, s_n \in G$ *and* $\gamma \in \Gamma$, *then G is a commutative ring.*

3 Main Results

Theorem 3.1 *Let G be a prime Γ-near-ring which admits a nonzero Γ-n-derivation h, if* $K_1, K_2, \ldots, K_n$ *is nonzero semigroup ideals of G and* $t\gamma k\beta u = t\beta k\gamma u$, *for every* $t, k, u \in K$ *and* $\gamma, \beta \in \Gamma$. *If* $h([s, r]_\gamma, s_2, \ldots, s_n) = [h(s, s_2, \ldots, s_n), r]_\gamma$, *for all* $s, r \in K_1, s_2 \in K_2, \ldots, s_n \in K_n$ *and* $\gamma \in \Gamma$, *then G is a commutative ring.*

Proof Suppose that

$$h([s, r]_\gamma, s_2, \ldots, s_n) = [h(s, s_2, \ldots, s_n), r]_\gamma,$$
$$\forall s, r \in K_1, s_2 \in K_2, \ldots, s_n \in K_n \text{ and } \gamma \in \Gamma. \tag{6}$$

Take $r = s$ in Eq. (6), we get

$$[h(s, s_2, \ldots, s_n), s]_\gamma = 0, \forall s \in K_1, s_2 \in K_2, \ldots, s_n \in K_n \text{ and } \gamma \in \Gamma \tag{7}$$

Replacing r by $s\beta r$ in Eq. (6), we get

$h([s, s\beta r]_\gamma, s_2, \ldots, s_n) = [h(s, s_2, \ldots, s_n), s\beta r]_\gamma$, $\forall s, r \in K_1, s_2 \in K_2, \ldots, s_n \in K_n$ and $\gamma, \beta \in \Gamma$.

Then, $h(s\beta[s,r]_\gamma, s_2, \ldots, s_n) = [h(s, s_2, \ldots, s_n), s\beta r]_\gamma$. By defining property of h, we get

$h(s, s_2, \ldots, s_n)\beta[s, r]_\gamma + s\beta h([s, r]_\gamma, s_2, \ldots, s_n) = [h(s, s_2, \ldots, s_n), s\beta r]_\gamma$.

By using Eq. (6) again, we get

$h(s, s_2, \ldots, s_n)\beta[s, r]_\gamma + s\beta[h(s, s_2, \ldots, s_n), r]_\gamma = s\beta[h(s, s_2, \ldots, s_n), r]_\gamma + [h(s, s_2, \ldots, s_n), s]_\gamma\beta r$.

Now using Eq. (7) in previous equation gives

$$h(s, s_2, \ldots, s_n)\beta s\gamma r = h(s, s_2, \ldots, s_n)\beta r\gamma s,$$
$$\forall s, r \in K_1, s_2 \in K_2, \ldots, s_n \in K_n \text{ and } \gamma, \beta \in \Gamma. \tag{8}$$

$h(s, s_2, \ldots, s_n)\beta s\gamma r = h(s, s_2, \ldots, s_n)\beta r\gamma s$, $\forall s, r \in K_1, s_2 \in K_2, \ldots, s_n \in K_n$ and $\gamma, \beta \in \Gamma$.

Replace r by $r\mu t$, where $t \in G$, $\mu \in \Gamma$ in Eq. (8) and using it again, we get

$h(s, s_2, \ldots, s_n)\beta\, r\mu[s, t]_\gamma = 0$. Then, $h(s, s_2, \ldots, s_n)\Gamma K_1\Gamma[s, t]_\gamma = \{0\}$. Since K_1 is a non zero semi-group ideal and by Lemma (2.3), we get for each $s \in K_1$, either $s \in Z$ or

$h(s, s_2, \ldots, s_n) = 0$, but $s \in Z$, by using Lemma (2.4), we get $h(s, s_2, \ldots, s_n) \in Z$. Hence, $h(K_1, K_2, \ldots, K_n) \subseteq Z$. By using Lemma (2.6), we find that G is a commutative ring.

Corollary 3.2 *Let G be a prime* Γ*-near-ring which admits a non zero* Γ*-n-derivation h,* $t\gamma k\beta u = t\beta k\gamma u$ *for every t, k, u* $\in K$ *and* $\gamma, \beta \in \Gamma$ *, if* $h([s, r]_\gamma, s_2, \ldots, s_n) = [h(s, s_2, \ldots, s_n), r]_\gamma$ *, for all s, r,* $s_2, \ldots, s_n \in G$ *and* $\gamma \in \Gamma$ *then G is a commutative ring.*

Theorem 3.3 *Let G be a prime* Γ*-near-ring which admits a nonzero* Γ *-n-derivation h, if* $K_1, K_2, \ldots, K_n$ *is nonzero semi-group ideals of G and* $t\gamma k\beta u = t\beta k\gamma u$ *for every t, k, u* $\in K$ *and* $\gamma, \beta \in \Gamma$*, if* $[h(s, s_2, \ldots, s_n), r]_\gamma = [s, r]_\gamma$*, for all s,* $r \in K_1, s_2 \in K_2, \ldots, s_n \in K_n$ *and* $\gamma \in \Gamma$*, this is a commutative ring.*

Proof Suppose that

$$[h(s, s_2, \ldots, s_n), r]_\gamma = [s, r]_\gamma, \text{ for all } s, r \in K_1, s_2 \in K_2, \ldots, s_n \in K_n \text{ and } \gamma \in \Gamma. \tag{9}$$

If we take $r = s$ in Eq. (9), we get $[h(s, s_2, \ldots, s_n), s]_\gamma = 0$, that is

$$h(s, s_2, \ldots, s_n)\gamma s = s\gamma h(s, s_2, \ldots, s_n), \quad \text{for all } s \in K_1, s_2 \in K_2, \ldots, s_n \in K_n \text{ and } \gamma \in \Gamma. \tag{10}$$

Replacing s by $r\beta s$ in Eq. (10) and using it again, we get
$[h(r\beta ss, s_2, \ldots, s_n), r]_\gamma = [r\beta s, y]_\gamma = r\beta[s, r]_\gamma = r\beta[h(s, s_2, \ldots, s_n), r]_\gamma$. Therefore,

$h(r\beta s, s_2, \ldots, s_n)\gamma\ r - r\ \gamma\ h(r\beta s, s_2, \ldots, s_n) = r\beta h(s, s_2, \ldots, s_n)\gamma\ r - r\beta r\gamma h(s, s_2, \ldots, s_n)$.

By using Lemmas (3.1.3) and (3.1.4), the last equation can be rewritten as

$h(r, s_2, \ldots, s_n)\beta s\gamma r = r\gamma h(r, s_2, \ldots, s_n)\beta s$, for all $s, r \in K_1, s_2 \in K_2, \ldots, s_n \in K_n$ and $\gamma, \beta \in \Gamma$.

By using Eq. (10) in previous equation, we have

$$h(r, s_2, \ldots, s_n)\beta s\gamma r = h(r, s_2, \ldots, s_n)\beta r\gamma s, \quad \forall s, r \in K_1, s_2 \in K_2, \ldots, s_n \in K_n \text{ and } \gamma, \beta \in \Gamma. \tag{11}$$

Since Eq. (11) is the same as Eq. (8), in Theorem (3.1) arguing as in the proof of Theorem (3.1), we find that G is a commutative ring.

Corollary 3.4 *Let G be a prime* Γ*-near-ring which admits a nonzero* Γ*-n-derivation h, if* $[h(s, s_2, \ldots, s_n), r]_\gamma = [s, r]_\gamma$*, for all s, r,* $s_2, \ldots, s_n \in G$ *and* $\gamma \in \Gamma$*, then G is a commutative ring.*

Theorem 3.5 *Let G be a 2-torsion free prime* Γ*-near-ring and h be a nonzero* Γ*-n-derivation of G. Let* $K_1, K_2, \ldots, K_n$ *be nonzero semigroup ideals of G and* $t\gamma k\beta u = t\beta k\gamma u$ *for every t, k, u* $\in K$*. If* $h([s, r]_\gamma, s_2, \ldots, s_n) \in Z$ *for all s, r* $\in K_1, s_2 \in K_2, \ldots, s_n \in K_n$*,* $\gamma \in \Gamma$*, then G is a commutative ring.*

Proof By hypothesis, we have

$$h([s, r]_\gamma, s_2, \ldots, s_n) \in Z, \forall s, r \in K_1, s_2 \in K_2, \ldots, s_n \in K_n \text{ and } \gamma \in \Gamma. \tag{12}$$

(a) If $Z = \{0\}$, then $h([s, r]_\gamma, s_2, \ldots, s_n) = 0$, for all $s, r \in K_1, s_2 \in K_2, \ldots, s_n \in K_n$ and $\gamma \in \Gamma$. By Lemma (2.9), we conclude that G is a commutative ring.

(b) If $Z \neq \{0\}$, replacingi r by $z\beta r$ in Eq. (12), where $z \in Z$,, we get

$h([s, z\beta r]_\gamma, s_2, \ldots, s_n) = h(z\beta[s, r]_\gamma, s_2, \ldots, s_n) \in Z$ which means that

$hz\beta s, r]_\gamma, s_2, \ldots, s_n)\mu t = t\mu h(z\beta[s, r]_\gamma, s_2, \ldots, s_n$

Using Lemma (2.2) in previous equation, we get

$h(z, s_2, \ldots, s_n)\beta[s, r]_\gamma\mu t + z\,\beta\, h([s, r]_\gamma, s_2, \ldots, s_n)\mu t = t\mu h(z, s_2, \ldots, s_n)\beta[s, r]_\gamma + t\mu z\,\beta\, h([s, r]_\gamma, s_2, \ldots, s_n)$.

Using Eq. (12) and Lemma (2.4) in previous equation, we get

$h(z, s_2, \ldots, s_n)\beta[[s, r]_\gamma, t]_\mu = 0, \forall s, r \in K_1, s_2 \in K_2, \ldots, s_n \in K_n, t \in G, z \in Z$ and $\gamma, \beta, \mu \in \Gamma$.

Hence, we get $h(z, s_2, \ldots, s_n)\Gamma K_1 \Gamma[[s, r]_\gamma, t]_\mu = \{0\}$. Since K_1 is a semi-group ideal and by Lemma (2.3), it yields either $h(Z, K_2, \ldots, K_n) = 0$ or $[[s, r]_\gamma, t]_\mu = 0$.

Assume that

$$[[s, r]_\gamma, t]_\mu = 0, \forall s, r \in K_1, t \in G \text{ and } \gamma, \mu \in \Gamma \tag{13}$$

Replacing r by $s\,\beta\, r$ in Eq. (13) yields $0 = [[s, s\,\beta\, r]_\gamma, t]_\mu = [s\,\beta[s, r]_\gamma, t]_\mu$.

Therefore, $[s, t]_\mu\, \beta[s, r]_\gamma = 0, \forall s, r \in K_1, t \in G$ and $\gamma, \beta, \mu \in \Gamma$.

Then, we get $[s, t]_\mu\, \Gamma K_1 \Gamma[s, r]_\gamma = \{0\}$. Since K_1 is a semi-group ideal and by Lemma (2.3), it implies that either $[s, r]_\gamma = 0$, for all $s, r \in K_1$ or $s \in Z$ for all $s \in K_1$.

If $[s, r]_\gamma = 0, \forall s, r \in K_1$, then by Theorem (2.11), we get $h([s, r]_\gamma, s_2, \ldots, s_n) = 0$.

By using Theorem (2.9), we obtain that G is a commutative ring.

Now assume that $s \in Z, \forall s \in K_1$, then by Lemma (2.4), we obtain

$h(K_1, K_2, \ldots, K_n) \subseteq Z$, by Lemma (2.6), we find that G is a commutative ring.

On the other hand, if $h(Z, K_2, \ldots, K_n) = 0$, then $h(h([s, r]_\gamma, s_2, \ldots, s_n), s_2, \ldots, s_n) = 0$.

Replace r by $s\beta r$ in the previous equation and using it again, we get

$0 = h(h([s, s\beta r]_\gamma, s_2, \ldots, s_n), s_2, \ldots, s_n)$.
$= h(h(s\beta[s, r], s_2, \ldots, s_n), s_2, \ldots, s_n)$.
$= h(h(s, s_2, \ldots, s_n)\beta[s, r]_\gamma + s\beta h([s, r]_\gamma, s_2, \ldots, s_n), s_2, \ldots, s_n)$.
$= h(h(s, s_2, \ldots, s_n)\beta[s, r]_\gamma, s_2, \ldots, s_n) + h(s\beta h([s, r]_\gamma, s_2, \ldots, s_n), s_2, \ldots, s_n)$.
$= h(h(s, s_2, \ldots, s_n), s_2, \ldots, s_n)\beta[s, r]_\gamma + h(s, s_2, \ldots, s_n)\beta h([s, r]_\gamma, s_2, \ldots, s_n) + h(s, s_2, \ldots, s_n)\beta h([s, r]_\gamma, s_2, \ldots, s_n) + s\beta h(h([s, r]_\gamma, s_2, \ldots, s_n), s_2, \ldots, s_n)$.

for all $s, r \in K_1, s_2 \in K_2, \ldots, s_n \in K_n$ and $\gamma, \beta \in \Gamma$.

Hence, we get

$$h(h(s, s_2, \ldots, s_n), s_2, \ldots, s_n)\beta[s , r]_\gamma + 2h(s, s_2, \ldots, s_n)\beta h\big([s, r]_\gamma, s_2, \ldots, s_n\big) = 0 \tag{14}$$

Replace s by $[s_1, r_1]_\mu$ in Eq. (14), where $s_1, r_1 \in K_1$, we get

$2\, h([s_1, r_1]_\mu, s_2, \ldots, s_n)\beta h([[s_1, r_1]_\mu, r]_\gamma, s_2, \ldots, s_n) = 0$. But G is 2-torsionifree, so we obtain

$h([s_1, r_1]_\mu, s_2, \ldots, s_n)\beta\, h([[s_1, r_1]_\mu, r]_\gamma, s_2, \ldots, s_n) = 0$. From Eq. (12), we get $h([s_1, r_1]_\mu, s_2, \ldots, s_n)\Gamma G \Gamma h([[s_1, r_1]_\mu, r]_\gamma, s_2, \ldots, s_n) = \{0\}$. Primeness of G yields that for each $s_1, r_1 \in K_1$, either $h([s_1, r_1]_\mu, s_2, \ldots, s_n) = 0$, $\forall s_2 \in K_2, \ldots, s_n \in K_n$ and $\gamma, \mu \in \Gamma$.

Or, $h([[s_1, r_1]_\mu, r]_\gamma, s_2, \ldots, s_n) = 0$, $\forall r \in K_1$, $s_2 \in K_2, \ldots, s_n \in K_n$ and $\gamma, \mu \in \Gamma$.

If $h([s_1, r_1]_\mu, s_2, \ldots, s_n) = 0$, then by Theorem (2.9), we conclude that G is commutative ring. Now, if $h([[s_1, r_1]_\mu, r]_\gamma, s_2, \ldots, s_n) = 0$, $\forall s_1, r_1 \in K_1$, $s_2 \in K_2, \ldots, s_n \in K_n$ and $\gamma, \mu \in \Gamma$.

We get $0 = h(([s_1, r_1]_\mu\, \gamma\, r - r\, \gamma [s_1, r_1]_\mu), s_2, \ldots, s_n)$

$= h([s_1, r_1]_\mu\, \gamma\, r, s_2, \ldots, s_n) - h(r\, \gamma [s_1, r_1]_\mu, s_2, \ldots, s_n)$.

$= [s_1, r_1]_\mu\, \gamma\, h(r, s_2, \ldots, s_n) + h([s_1, r_1]_\mu, s_2, \ldots, s_n)\gamma\, r - (h(r, s_2, \ldots, s_n)\gamma[s_1, r_1]_\mu - r\, \gamma\, h([s_1 r_1]_\mu, s_2, \ldots, s_n))$.

Using Eq. (12) in the last equation yields

$$[s_1, r_1]_\mu \gamma h(r, s_2, \ldots, s_n) = h(r, s_2, \ldots, s_n)\gamma[s_1, r_1]_\mu, \quad s_1, r_1, r \in K_1, s_2 \in K_2, \ldots, s_n \in K_n \text{ and } \gamma, \mu \in \Gamma. \tag{15}$$

Taking $g\, \beta[s_1, r_1]_\mu$ instead of r, where $g \in K_1$ in Eq. (15) to get

$[s_1, r_1]_\mu\, \gamma\, h(g\, \beta[s_1, r_1]_\mu, s_2, \ldots, s_n)) = h(g\beta[s_1, r_1]_\mu, s_2, \ldots, s_n)\gamma[s_1, r_1]_\mu$.

Using Lemma (2.3)(ii), we get

$[s_1, r_1]_\mu\, \gamma\, h(g, s_2, \ldots, s_n)\beta[s_1, r_1]_\mu + [s_1, r_1]_\mu\, \gamma\, g\, \beta\, h([s_1, r_1]_\mu, s_2, \ldots, s_n) = h(g, s_2, \ldots, s_n)\beta[s_1, r_1]_\mu\, \gamma[s_1, r_1]_\mu + g\, \beta\, h([s_1, r_1]_\mu, s_2, \ldots, s_n)\gamma[s_1, r_1]_\mu$.

Using Eqs. (12) and (15) in previous equation implies that

$h([s_1, r_1]_\mu, s_2, \ldots, s_n)\gamma[s_1, r_1]_\mu\, \beta\, g = h([s_1, r_1]_\mu, s_2, \ldots, s_n)\beta\, g\, \gamma[s_1, r_1]_\mu$.

So we get $h([s_1, r_1]_\mu, s_2, \ldots, s_n)\Gamma K_1 \Gamma[[s_1, r_1]_\mu, g]_\beta = \{0\}$. Since K_1 is a semi-group ideal and by Lemma (2.3), we get for each $s_1, r_1 \in K_1$, we get

either $h([s_1, r_1]_\mu, s_2, \ldots, s_n) = 0$, $\forall s_2 \in K_2, \ldots, s_n \in K_n$ and $\mu \in \Gamma$ or $[[s_1, r_1]_\mu, g]_\beta = 0$.

If $h([s_1, r_1]_\mu, s_2, \ldots, s_n) = 0$, then by Theorem (2.9), we conclude that G is commutative ring or if $[[s_1, r_1]_\mu, g]_\beta = 0$. Replacing r_1 by $s_1\, \gamma\, r_1$ in previous equation yields.

$[s_1, g]_\beta\, \gamma[s_1,r_1]_\mu = 0$. Hence, $[s_1, g]_\beta\, \Gamma K_1 \Gamma[s_1,r_1]_\mu = \{0\}$. Since K_1 is a semi-group ideal and by Lemma (2.3), we get $[s_1, r_1]_\mu = 0$, for all $s_1 r_1 \epsilon K_1$. Then, by Theorem (2.11), we get

$h([s_1,r_1]_\gamma, s_2, \ldots, s_n) = 0$. Now by using Lemma (3.1.9), we obtain that G is a commutative ring.

Corollary 3.6 *Let G be a 2-torsion free prime Γ-near-ring and h is a nonzero Γ-n-derivation of G, if $h([s,r]_\gamma, s_2, \ldots, s_n) \epsilon Z$ for all $s, s_2, \ldots, s_n, r \epsilon G$, then G is a commutative ring.*

Theorem 3.7 *Let G be a prime Γ-near-ring admitting a nonzero Γ-n-derivation of G. Let $K_1, K_2, \ldots, K_n$ be nonzero semi-group ideals of G, if $[h(s, s_2, \ldots, s_n), r]_\gamma \epsilon Z$, for all $s, r \epsilon K_1, s_2 \epsilon K_2, \ldots, s_n \epsilon s_n$, then Gi is a commutative ring.*

Proof Suppose that

$$[h(s, s_2, \ldots, s_n), r]_\gamma \in Z, \forall s, r \in K_1, s_2 \in K_2, \ldots, s_n \in K_n \text{ and } \gamma \in \Gamma. \tag{16}$$

Replacing r by $h(s, s_2, \ldots, s_n)\mu r$ in Eq.

$[h(s, s_2, \ldots, s_n), h(s, s_2, \ldots, s_n)\mu r]_\gamma \epsilon Z, \forall\, s, r \epsilon K_1, s_2 \epsilon K_2, \ldots, s_n \epsilon K_n$ and $\mu, \gamma \epsilon \Gamma$.

which means that $[[h(s, s_2, \ldots, s_n), h(s, s_2, \ldots, s_n)\mu r]_\gamma, t]_\beta = 0$.

Therefore, $[h(s, s_2, \ldots, s_n)\mu[h(s, s_2, \ldots, s_n), r]_\gamma, t]_\beta = 0$. Using Eq. (16) in previous equation implies that

$$[h(s, s_2, \ldots, s_n), t]_\beta \mu [h(s, s_2, \ldots, s_n), r]_\gamma = 0, s, r \in K_1, s_2 \in K_2, \ldots, s_n \in K_n, \; t \in G \text{ and } \mu, \gamma, \beta \in \Gamma. \tag{17}$$

In view of Eqs. (16), (17) obtains that

$[h(s, s_2, \ldots, s_n), r]_\gamma\, \Gamma G \Gamma [h(s, s_2, \ldots, s_n), r]_\gamma = \{0\}$. Primeness of G show that $[h(s, s_2, \ldots, s_n), r]_\gamma = 0$.

It mean $h(s, s_2, \ldots, s_n)\gamma r = r\gamma h(s, s_2, \ldots, s_n)$, for all $s, r \epsilon K_1, s_2 \epsilon K_2, \ldots, s_n \epsilon K_n$ and $\gamma \epsilon \Gamma$. Replacing r by $r\, \gamma\, y$ where $y \epsilon G$, in pervious equation and using it again implies that

$r\, \gamma[h(s, s_2, \ldots, s_n), y]_\gamma = 0$. Thus, $K_1 \Gamma[h(s, s_2, \ldots, s_n), y]_\gamma = \{0\}$.

Using Lemma (2.2) implies that $[h(s, s_2, \ldots, s_n), y]_\gamma = 0$. Hence, $h(K_1, K_2, \ldots, K_n) \subseteq Z$.

By Lemma (2.6), we conclude that G is a commutative ring.

Corollary 3.8 *Let G be a prime Γ-near-ring and h be a nonzero Γ-n-derivation of G, if $[h(s_1, s_2, \ldots, s_n), r]_\gamma \epsilon Z$ for all $s_1, s_2, \ldots, s_n, r \epsilon G$, then G is a commutative ring.*

Theorem 3.9 *Let G be a prime Γ-near-ring and h a no nonzero Γ -n-derivation on G such that $K_1, K_2, \ldots, K_n$ be nonzero semi-group ideals of G and $t\gamma k\beta u = t\beta k\gamma u$, for every $t, k, u \in K$ and $\gamma, \beta \epsilon \Gamma$. If $h((s \circ r)_\gamma, s_2, \ldots, s_n) = 0$, for all $s,r \in K_1, s_2 \in K_2, \ldots, s_n \in K_n$ and $\gamma \in \Gamma$, then G is commutative ring.*

Proof We have

$$h((s_1 \circ r_1)_\gamma, s_2, \ldots, s_n) = 0, \forall s_1, r_1 \in K_1, s_2 \in K_2, \ldots, s_n \in K_n \text{ and } \gamma \in \Gamma \quad (18)$$

Substituting $s_1 \beta r_1$ for r_1 in Eq. (18) and using it again, we obtain

$h(s_1, s_2, \ldots, s_n)\beta(s_1 \circ r_1)_\gamma = 0, \forall s_1, r_1 \in K_1, s_2 \in K_2, \ldots, s_n \in K_n$ and $\gamma \in \Gamma$.
Therefore,

$$h(s_1, s_2, \ldots, s_n)\beta s_1 \gamma r_1 = h(s_1, s_2, \ldots, s_n)\beta r_1 \gamma s_1 \quad (19)$$

Putting $r_1 \mu z$ for r_1, where $z \in G$, in Eq. (19), we get

$-h(s_1, s_2, \ldots, s_n)\beta s_1 \gamma r_1 \mu z = h(s_1, s_2, \ldots, s_n)\beta r_1 \mu z \gamma s_1$. Using Eq. (19), we obtain

$(h(s_1, s_2, \ldots, s_n)\beta r_1 \gamma s_1)\mu z = h(s_1, s_2, \ldots, s_n)\beta r_1 \mu z \gamma s_1.$

Hence, $h(s_1, s_2, \ldots, s_n)\Gamma K_1 \Gamma [s_1, z]_\gamma = \{0\}$. Since K_1 be a semi-group ideal and by Lemma (2.3) yields that either $s_1 \in Z$ or $0 = h(u_1, u_2, \ldots, u_n), \forall s_2 \in K_2, \ldots, s_n \in K_n$.

If the first case holds $s_1 \in Z$, then by Lemma (2.4), we get $h(s_1, s_2, \ldots, s_n) \in Z$.

And second case implies $h(s_1, s_2, \ldots, s_n) = 0$, that is, $0 = h(s_1, s_2, \ldots, s_n) \in Z$. Combining both the case, we get $h(s_1, s_2, \ldots, s_n) \in Z$ that means, $h(K_1, K_2, \ldots, K_n) \subseteq Z$. Hence, by Lemma (2.6), we find that G is commutative ring.

Corollary 3.10 *Let G be a prime Γ-near-ring and h be a no nonzero Γ -n-derivation on G such that* $h((s \circ r)_\gamma, s_2, \ldots, s_n) = 0$, *for all* $s, r, s_2, \ldots, s_n \in G$ *and* $\in \gamma\ \Gamma$ *then G is commutative ring.*

Theorem 3.11 *Let G be a prime Γ-near-ring admitting a nonzero Γ-n-derivation such that* $h((s_1 \circ r_1)_\gamma, s_2, \ldots, s_n) = \pm(s_1 \circ r_1)_\gamma$ *for all* $s_1, r_1 \in K_1, s_2 \in K_2, \ldots, s_n \in K_n$ *and* $\gamma \in \Gamma$, *where* $K_1, K_2, \ldots, K_n$ *are nonzero semi-group ideals of G then G is a commutative ring.*

Proof We have,

$$h((s_1 \circ r_1)_\gamma, s_2, \ldots, s_n) = \pm(s_1 \circ r_1)_\gamma,$$
$$\forall s_1, r_1 \in K_1, s_2 \in K_2, \ldots, s_n \in K_n \text{ and } \gamma \in \Gamma \quad (20)$$

Substituting $s_1 \beta r_1$ for r_1 in Eq. (20), we obtain

$$h(s_1\beta(s_1 \circ r_1)_\gamma, s_2, \ldots, s_n) = \pm s_1\beta(s_1 \circ r_1)_\gamma$$

Hence, $h(s_1, s_2, \ldots, s_n)\beta(s_1 \circ r_1)_\gamma + s_1 \beta h(s_1 \circ r_1)_\gamma, s_2, \ldots, s_n) = \pm s_1 \beta(s_1 \circ r_1)_\gamma$.

Using hypothesis in last equation implies

$h(s_1, s_2, \ldots, s_n)\beta(s_1 \circ r_1)_\gamma = 0, \forall s_1, r_1 \in K_1, s_2 \in K_2, \ldots, s_n \in K_n$ and $\gamma, \beta \in \Gamma$.

Thus, we get

$$h(s_1, s_2, \ldots, s_n)\beta r_1 \gamma s_1 = -h(s_1, s_2, \ldots, s_n)\beta s_1 \gamma r_1 \quad (21)$$

Equation (21) is identical to theorem Eq. (19) (3.9). Now, we infer in the same Theorem way (3.9), that G is a commutative ring.

Corollary 3.12 *Let G be a prime Γ-near-ring admitting a nonzero Γ-n-derivation such that $h((s_1 \circ r_1)_\gamma, s_2, \ldots, s_n) = \pm(s_1 \circ r_1)_\gamma$ for all $s_1, r_1, s_2, \ldots, s_n \in G$ and $\gamma \in \Gamma$, then G is a commutative ring.*

Theorem 3.11 *Let G be a 2-torsion free prime Γ-near-ring and $K_1, K_2, \ldots, K_n$ be nonzero semi-group ideals of G, then there exists no nonzero Γ-n-derivation h of G and $t\gamma k\beta u = t\beta k\gamma u$ for every $t, k, u \in G$ and $\gamma, \beta \in \Gamma$ such that $(h(s_1, s_2, \ldots, s_n)\circ r)_\gamma = (s_1 \circ r)_\gamma$, for all $s_1, r \in K_1, s_2 \in K_2, \ldots, s_n \in K_n$ and $\gamma \in \Gamma$.*

Proof Suppose that

$$(h(s_1, s_2, \ldots, s_n) \circ r)_\gamma = (s_1 \circ r)_\gamma, \forall s_1, r \in K_1, s_2 \in K_2, \ldots, s_n \in K_n \text{ and } \gamma \in \Gamma. \quad (22)$$

Replacing s_1 by $r\beta s_1$ in Eq. (22) and using it again, we get

$$(h(r\beta s_1, s_2, \ldots, s_n)\circ r)_\gamma = ((r\beta s_1)\circ r)_\gamma = r\beta(s_1 \circ r)_\gamma = r\beta(h(s_1, s_2, \ldots, s_n)\circ r)_\gamma.$$

Hence, $h(r\beta s_1, s_2, \ldots, s_n)\gamma r + r\gamma h(r\beta s_1, s_2, \ldots, s_n) = r\beta(h(s_1, s_2, \ldots, s_n)\gamma r + r\gamma h(s_1, s_2, \ldots, s_n))$.

Using Lemma (2.1) and (2.2) in last equation, we obtain

$r\beta h(s_1, s_2, \ldots, s_n)\gamma r + h(r, s_2, \ldots, s_n)\beta s_1 \gamma r + r\gamma h(r, s_2, \ldots, s_n)\beta s_1 + r\gamma r\beta h(s_1, s_2, \ldots, s_n) = r\beta h(s_1, s_2, \ldots, s_n)\gamma r + r\beta r\gamma h(s_1, s_2, \ldots, s_n), \forall s_1, r \in K_1, s_2 \in K_2, \ldots, s_n \in K_n$ and $\gamma, \beta \in \Gamma$.

Hence, we get $h(r, s_2, \ldots, s_n)\beta s_1 \gamma r = -r\gamma h(r, s_2, \ldots, s_n)\beta s_1$.

Replacing s_1 by $s_1\mu v$ in pervious equation, where $v \in G$, we find that

$h(r, s_2, \ldots, s_n)\beta s_1\mu v\gamma r = -r\gamma h(r, s_2, \ldots, s_n)\beta s_1\mu v = (-r\gamma h(r, s_2, \ldots, s_n)\beta s_1)\mu v = h(r, s_2, \ldots, s_n)\beta s_1 \gamma r\mu v$.

Hence, $h(r, s_2, \ldots, s_n)\beta s_1\mu[v, r]_\gamma = 0$. So, we obtain

$h(r, s_2, \ldots, s_n)\Gamma K_1\Gamma[v, r]_\gamma = \{0\}$. Since K_1 is a semi-group ideal and by Lemma (2.3) yields that either $h(r, s_2, \ldots, s_n) = 0$ or $r \in Z$. But $r \in Z$, by using Lemma (2.4), we get

$h(r, s_2, \ldots, s_n) \in Z$. Hence, $h(K_1, K_2, \ldots, K_n) \subseteq Z$. By using Lemma (2.6), we find that G is a commutative ring.

Corollary 3.14 Let G be a 2-torsioni free prime Γ-near-ring, then there exists no non zero Γ-n-derivation h of G such that $(h(s_1, s_2, \ldots, s_n) \circ r)_\gamma = (s_1 \circ r)_\gamma$, for all $s_1, r, s_2, \ldots, s_n \in G$ and $\gamma \in \Gamma$.

4 Conclusion

In this paper, we investigated commutativity of zero symmetric right near-ring satisfying certain identities involving Γ-n-derivation, it is proved that a prime Γ-near-ring is commutative under some condition on Γ-n-derivation. In this paper, we have extended the finding to an ideal x-n-derivation and semigroup on a zero symmetric near-ring and shown that G is a commutative ring that satisfies some involving x-n-derivations and semigroup ideal and also proved there exists no nonzero Γ-n-derivation under some condition gives and proved that some characterization of a Γ-near-ring admitting Γ-n-derivation on semigroup ideals is a commutative ring.

References

1. Mustafa, A. S. C. I. (2007). Γ−(σ, τ) Derivation on gamma near rings. In *International Mathematical Forum* (Vol. 2, No. 3, pp. 97–102).
2. Beidar, K. I., Fong, Y., & Wang, X. (1996). Posner and Herstein theorems for derivations of 3-prime near-rings. *Communications in Algebra, 24*(5), 1581–1589.
3. Booth, G. L., & Groenewald, N. J. (1991). Equiprime Γ-near-rings: In Memory of Professor Hennie Schutte. *Quaestiones Mathematicae, 14*(4), 411–417.
4. Cho, Y. U., & Jun, Y. B. (2002). Gamma-derivations in prime and semi prime gamma-near rings. *Indian Journal of Pure Applied Mathematics, 33*(10), 1489–1494.
5. Bell, H. E. (1997). On derivations in near-rings. II. *In Near rings, near fields and K-loops* (Hamburg, 1995), *volume 426 of Mathematical Applications* (pp. 191–197). Kluwer Acad. Publ., Dordrecht.
6. Golbasi, O., & Aydin, N. (2004). Results on prime near-ring with (σ, τ)-derivation. *Mathematical Journal of Okayama University, 46,* 1–7.
7. Ravi, R., Bhat, V. K., & Neetu, K. (2009). Commutativity of prime Γ-near rings with Γ-(σ, τ)-derivation. *Mathematics, 25,*165–173.
8. Dey, K. K., & Chandra, A. On derivation in prime gamma near rings. *Journal of Bangladesh Mathematical Society, 32,* 23–28 (2012) (ISSN 1606–3694).
9. Satnarayan, B. (1984). Contributions to near-ring theory. Ph.D. thesis, Nargarjuna University.
10. Madhuchelvi, L. (2016). On generalized derivations in γ-near ring. *International Journal of Mathematical Sciences and Applications, 6*(1). ISSN N0:2230–9888.
11. Ashraf, M., Ali, A., & Ali, S. (2004). (σ, τ)-derivations on prime near rings. *Arciv der Mathematik (Brno), 40*(3), 281–286.
12. Mustafa, U., Mehmet, A. O., & Young, B. J. (2004). On prime gamma near-rings with derivations. *Communications of the Korean Mathematical Society, 19*(3), 427–433.
13. Asraf, M., & Siddeeque, M. A. (2013). On permuting n-derivations in near-rings. *Communications of the Korean Mathematical Society, 28*(4), 97–707.
14. Asraf, M., Siddeeque, M. A., & Parvee, N. (2015). On semigroup ideals and n-derivations in near-rings. *Science Direct Journal of Taibah University for Science, 9,* 126–132.
15. Asraf, M., & Siddeeque, M. A. (2014). On generalized n-derivations in near-rings. *Palestine Journal of Mathematic, 3,* 468–480.
16. Asraf, M., & Siddeeque, M. A. (2013). On permuting n-derivations in near-rings. *Communications of the Korean Mathematical Society, 28*(4), 697–707.
17. Isamiddin, S. R., Dey, K. K., & Paul, A. C. (s2013). Prime gamma near-rings with derivations. *International Journal of Pure and Applied Mathematics, 83*(2), 221–231.

Priority-Based Optimized Scheduling (PBOS) Algorithm for Big Data Applications in Cloud Computing Environments

P. Tamilarasi and D. Akila

Abstract This paper proposes a priority-based optimized scheduling (PBOS) algorithm for cloud-based big data applications. In this algorithm, for each incoming task request, the task size and expected completion time are estimated. At the same time, for each VM on a host, the processing capability and current load are estimated. A priority queue for each task is then created in which the tasks are stored in the ascending order of task size and completion time. Similarly, a priority queue for VMs is also made in which the VMs are stored in the ascending order of current load and processing capacity. Then, the scheduling algorithm intelligently maps each user tasks from task queue to the suitable matching VM from VM queue. The process is continued until both these queue become empty. By experimental results, it is shown that PBOS achieves lesser response time and execution time.

Keywords Big data · Cloud computing · Task scheduling · Priority · Response delay · Execution time

1 Introduction

Big data consists of huge volumes of data produced by various Internet-related applications. It is used to design social media, business applications, etc. The main operations involved in big data processing are extraction, data storage, data distribution, and analysis of data sets [1]. Cloud computing environment yields appropriate architectures for the execution of huge big data applications. But still, it contains various issues and challenges.

P. Tamilarasi (✉)
Department of Computer Science, School of Computing Sciences, Vels Institute of Science, Technology & Advanced Studies (VISTAS), Chennai, India
e-mail: tamilmalu@yahoo.com

D. Akila
Department of Information Technology, School of Computing Sciences, Vels Institute of Science, Technology & Advanced Studies (VISTAS), Chennai, India
e-mail: akiindia@yahoo.com

S.-L. Peng et al. (eds.), *Proceedings of First International Conference on Mathematical Modeling and Computational Science*, Advances in Intelligent Systems and Computing 1292, https://doi.org/10.1007/978-981-33-4389-4_21

Scheduling is the set of rules for maintaining each flow of the tasks which are executed by different cloud resources [2]. Task scheduling in cloud-based big data applications becomes one of the challenges [3]. The principal challenges of task scheduling include: reducing the makespan time, optimizing the task execution time, and reducing the cost of execution [4].

Thus, the main objectives of the task scheduling algorithm are.

- Optimizing the overall task execution time
- Minimizing the response time
- Minimizing the cost of execution.

To meet these objectives, this paper proposes a priority-based optimized scheduling algorithm cloud-based big data.

2 Related Works

Rjoub et al. [5] have proposed a trust-aware scheduling algorithm which has three phases: (i) mist estimation of VMs; (ii) priority estimation of tasks; (iii) mist-aware scheduling. Rarnezani et al. [6] have developed a multi-constraint PSO (MPSO) solution for enhancing task scheduling procedure by reducing the task execution time, transfer time, and execution price. Islam et al. [7] have proposed scheduling algorithms in cloud-based Apache Spark cluster. These algorithms can minimize resource utilization cost and assign priorities for tasks based on their deadlines. Whenever the cluster changes, the scheduling algorithms perform accordingly. Tani et al. [8] have proposed smarter round-robin (SRR) method, which is an enhancement of traditional round-robin scheduling. They have included a smarter layer to the existing algorithm for quick adaptation to various scenarios and for reduction of the execution time. Gandoini et al. [9] have proposed a hybrid scheduling algorithm which combines both dynamic priority and data localization methods. It increases the locality rate and decreases the task finishing time.

3 PBOS Algorithm for Big Data Cloud

3.1 Overview

This paper proposes a PBOS algorithm for big data. For each incoming task request, the task size and expected completion time estimated. Similarly, for each VM on a host, its processing capability, current load are calculated. A priority queue for the task (PT_Q) is created in which the tasks stored in the ascending order of task size and completion time. Similarly, a priority queue for VMs (PV_Q) is created in which the VMs are stored in the ascending order of current load and processing capacity.

The scheduling algorithm intelligently maps each user tasks from PT_Q to the suitable matching VM from PV_Q. The process continued until both these queue become empty.

3.2 System Model

In the system model, a virtualized cloud environment is considered.

Let $H = \{h_1, h_2, \ldots, h_l\}$ be the set of computing hosts.

Let $\{VM_i^j\}$ $j = 1, 2, \ldots, m$, be the set of virtual machines running on host i.

Let TRC_i^j be the total resource capacity of VM_j at host i such that

$$\sum_{j=1}^{||V_i} TRC_i^j < TRC_i$$

where TRC_i is processing capability of ith host.

Let FRC_i^j be the free resource capacity of VM_j at host i.

3.3 Task Execution Time

The task execution time is estimated as follows:

Let $\{T_1, T_2, \ldots, T_n\}$ be the tasks arrived for execution.

Let VM_j, $j = 1, 2, \ldots, m$ be the set of virtual machines running on each host.

Let X_{ik} be a binary variable such that.

$X_{ik} = 1$, if task T_i is assigned to VM_k

$= 0$, otherwise.

Let DE_{ik} be the data size of task T_i assigned VM_k.

Let VM_{mk} be the memory capacity of VM_k.

Let VM_{ck} be the CPU capacity of VM_k.

Then, the task execution time $Texe_k$ of VM_k is given by

$$T_{\text{exe}k} = \sum_{i=1}^{n} x_{ik} * \frac{DE_{ik}}{VM_{mk} * VM_{ck}} \tag{1}$$

3.4 Task Priority Queues Based on Requirements

In this section, the tasks are grouped based on their resource needs and task size. Initially, the tasks are sorted based on their resource requirement metrics such as task size and completion time. Then, a lower and upper threshold values are calculated for the tasks. Here, lower threshold indicates the tasks whose requirements falls below it. Upper threshold indicates the tasks whose requirements fall above it.

The actual grouping of tasks is performed by allocating the tasks whose resource expectations are less than the lower threshold to the low priority queue $PT_Q^{low.}$ The tasks whose resource expectations are greater than the upper threshold value are allocated to the high priority queue PT_Q^{high}. The rest of the values automatically allocated to the medium priority queue PT_Q^{medium}.

Algorithm: Task Grouping

Input:

N: size of the dataset containing the requirements
T: set of task where$t_i \in T$
M = *{task size TS, completion time TC}*: set of metrics required by each task
R_i^{TS} : Task size requirements for the task t_i

R_i^{TC} : Task completion time requirements for the task t_i

L_{th} and U_{th} : Lower and Upper threshold values for resource requirements
Output: QueuesPT_Q^{low}, PT_Q^{medium}, PT_Q^{high}

Procedure Task Grouping

For each $ti \in T$ do

Sort ti with respect to { R_i^{TS} , R_i^{TC} }

if $R_i^{TS} \leq L_{th}$ and $R_i^{TC} \leq L_{th}$

Assign t_i to PT_Q^{low}

Else if $R_i^{TS} \geq U_{th}$ and $R_i^{TC} \geq U_{th}$

Assign $t_i$toPT$_Q^{high}$

Else

Assign t_i to PT_Q^{medium}

End if

End for

End Procedure

3.5 VM Priority Queues

The total workload (TL) of each VM can be predicted using an exponentially weighted moving average (EWMA) of past workloads of all VMs

$$TL(t) = \alpha.EL(t-1) + (1-\alpha).OL(t) \quad (2)$$

where EL and OL are the estimated and observed loads at time t and α is a constant.

Algorithm VM Grouping

Input:
m: Total number of VMs on Host h_i.
VM: set of VMs where $vm_i \in VM$
M = *{Free resource capacity FRC, total load TL}*: a set of metrics for each vm
V_j^{FRC} : Measured FRC values for the vm_j

V_j^{TL} : Measured TL values for the vm_j

VL_{th} and VU_{th} : Lower and Upper threshold values for the metrics
Output: Queues PV_Q^{low}, PV_Q^{medium}, PV_Q^{high}

```
Procedure VM Grouping
For each vm_j ϵ VM
Do
        Sort vm_j with respect to {V_j^FRC , V_j^TL}

        if (V_j^FRC ≤ VL_th or V_j^TL ≥ VU_th)

                Assign vm_j to PV_Q^low

        Else if (V_j^FRC ≥ VU_th or V_j^TL ≤ VL_th)

                Assign vm_j to PV_Q^high
        Else
                Assign vm_j to PV_Q^medium
        End if
End for
End Procedure
```

In this algorithm, VMs are grouped on the basis of free resource capacity and total load.

3.6 *Optimized Task Scheduling*

After assigning priority queues for the tasks and the VMs, the scheduling algorithm intelligently maps each user tasks from the task queue to the suitable matching VM from the VM queue. The process continued until both these queue become empty.

Algorithm - Optimized Task Scheduling

```
//For High priority tasks

For each ti ϵ PT_Q^high
Do
      For each vm_j ϵPV_Q^high
          If (vm_j is not assigned yet)
        Assign t_i to vm_j
                    Remove vm_j from PV_Q^high
          break
         End if
      End For
      If no vm_j found on PV_Q^high
                    Continue
     End if
If( t_i completed)
    Remove t_i from PT_Q^high
End if
End For
//For Medium priority tasks
For each ti ϵ PT_Q^medium
Do
      For each vm_j ϵPV_Q^medium
          If (vm_j is not assigned yet)
        Assign t_i to vm_j
                    Remove vm_j from PV_Q^medium
          break
         End if
      End For
      If no vm_j found on PV_Q^medium
                    Continue
     End if
If( t_i completed)
    Remove t_i from PT_Q^medium
 End if
End For
//For Low priority tasks
For each t_i ϵ PT_Q^low
Do
      For each vm_j ϵPV_Q^low
          If (vm_j is not assigned yet)
        Assign t_i to vm_j
                    Remove vm_j from PV_Q^low
          break
         End if
      End For
      If no vm_j found on PV_Q^low
                    Continue
     End if
If( t_i completed)
    Remove t_i from PT_Q^low
 End if
End For
```

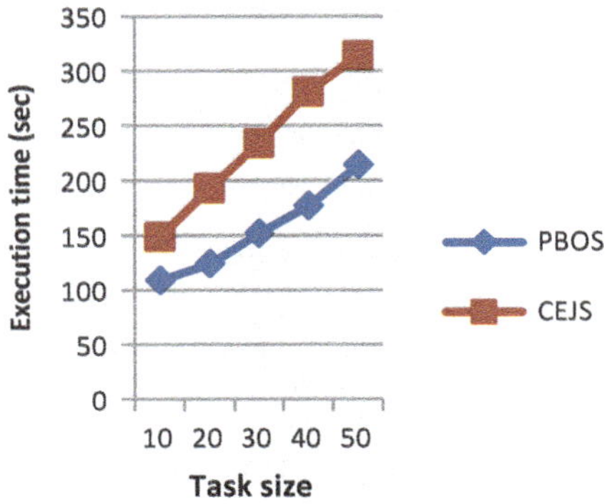

Fig. 1 Execution times for input task sizes

4 Experimental Results

4.1 Results

A workload of 25 GB is fed into Amazon Elastic Compute Cloud (EC2). The PBOS algorithm is implemented on EC2 large instances. The performance of PBOS is compared with the cost-efficient job scheduling (CEJS) algorithm [6], based on the parameters execution time, response delay, and CPU utilization.

4.2 Effect of Task Size

Experiments are conducted by varying the task size from 10 to 50.

Figure 1 shows the task execution times measured for PBOS and CEJS for various task sizes. The figure shows that the execution time becomes least when the number of tasks is less. On the other hand, it becomes the highest when the number of tasks is high. It can be inferred that the execution times of PBOS is 33% lesser than CEJS.

Figure 2 shows the response delay of each task measured for PBOS and CEJS for various task sizes. It can be observed from the figure that the response delay increases linearly when the task size is increased. The results show that the response delay of PBOS is 25% lesser than CEJS.

4.3 Effect of VMs

Experiments are conducted by varying the number of VMs from 2 to 10.

Figure 3 shows the task execution times measured for PBOS and CEJS for number of VMs. It can be observed from the figure that the execution time is lowest when

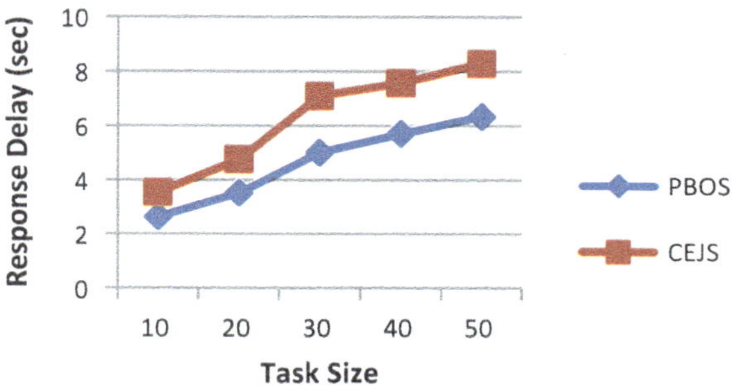

Fig. 2 Response delay for various task sizes

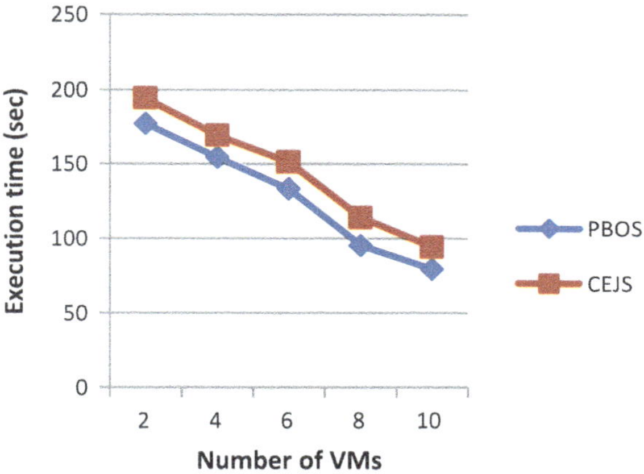

Fig. 3 Execution times for the number of VMs

the number of VMs is maximum, i.e., 10. Similarly, it becomes the highest when the number of VMs is minimum, i.e., 2. It can be inferred that the execution time of PBOS is 12% lesser than CEJS.

Figure 4 shows the response delay measured for PBOS and CEJS for number of VMs. It can be observed from the figure that the delay is least when the number of VMs is maximum, i.e., 10. Similarly, it becomes the highest when the number of VMs is minimum, i.e., 2. It can be inferred that the response delay of PBOS is 13% lesser than CEJS.

5 Conclusion

This paper has proposed PBOS algorithm for cloud-based big data. In this algorithm, a priority queue for each task is then created in which the tasks are stored in the ascending order of task size and completion time. Similarly, a priority queue for VMs is also created in which the VMs are stored in the ascending order of current

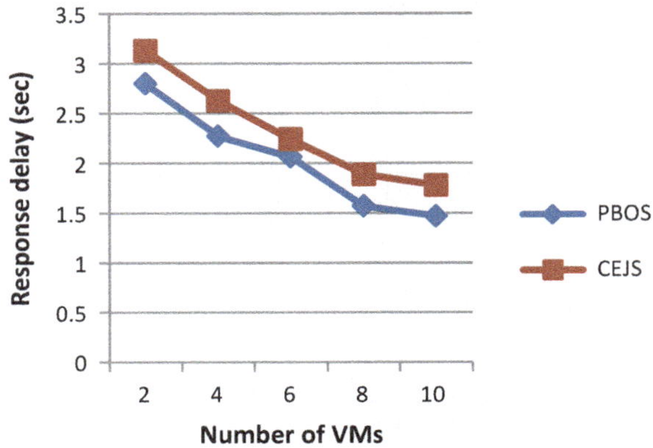

Fig. 4 Response delay for the number of VMs

load and processing capacity. Then, the scheduling algorithm intelligently maps each user tasks from task queue to the suitable matching VM from VM queue. The process is continued until both these queue become empty. By experimental results, it is shown that PBOS achieves lesser response time and execution time, compared to the CEJS.

References

1. Usama, M., Liu, M., & Chen, M. (2017). Job schedulers for big data processing in hadoop environment: Testing real-life schedulers using benchmark programs. *Digital Communications and Networks, 3*(4), 260–273.
2. Mishra, S. K., Satya Manikyam, P., Sahoo, B., Obaidat, M. S., Puthal, D., & Pratama, M. (2017). Response-aware scheduling of big data applications in cloud environments. *Future Technologies Conference (FTC)*, Canada.
3. Tamilarasi, P., & Akila, D. (2020). Task allocation and re-allocation for big data applications in cloud computing environments. In *Intelligent Computing and Innovation on Data Science* (pp. 679–686). Singapore: Springer.
4. Ebrahimi, M., Mohan, A., & Lu, S. (2018, March). Scheduling big data workflows in the cloud under deadline constraints. In *2018 IEEE Fourth International Conference on Big Data Computing Service and Applications (BigDataService)* (pp. 33–40). New York: IEEE.
5. Rjoub, G., Bentahar, J., & Wahab, O. A. (2019). BigTrustScheduling: Trust-aware big data task scheduling approach in cloud computing environments. *Future Generation Computer Systems.*
6. Ramezani, F., Lu, J., & Hussain, F. (2013, December). Task scheduling optimization in cloud computing applying multi-objective particle swarm optimization. In *International Conference on Service-oriented computing* (pp. 237–251). Berlin, Heidelberg: Springer.
7. Islam, M. T., Srirama, S. N., Karunasekera, S., & Buyya, R. (2020). Cost-efficient dynamic scheduling of big data applications in apache spark on cloud. *Journal of Systems and Software, 162,* 110515.
8. EL Amrani, C., & Gibet Tani, H. (2018). Smarter round robin scheduling algorithm for cloud computing and big data. *Journal of Data Mining & Digital Humanities.*
9. Gandomi, A., Reshadi, M., Movaghar, A., & Khademzadeh, A. (2019). HybSMRP: A hybrid scheduling algorithm in Hadoop MapReduce framework. *Journal of Big Data, 6*(1), 106.

Prediction-Based Cost-Efficient Resource Allocation Scheme for Big Data Streams in Cloud Systems

R. Kiruthiga and D. Akila

Abstract Resource allocation (RA) for cloud-based big data systems mainly consists of assigning the tasks to suitable resources. A dynamic unbalanced load in individual virtual machines (VMs) causes the complete inefficient utilization of resources. Apart from this, high execution cost and lower performance are the other issues of RA. A predictive cost-efficient resource allocation (PCRA) scheme for big data applications is developed. In this scheme, the user specifies the number of executors, cores, memory, etc., for the required application. Then, the task accomplishment period of a solicitation is predicted centered on the number of initiators and the specific properties of the application. Then, VMs are determined for the task requests, such that the number of executors is sufficient to minimize the task completion time and total cost. Some idle executors are released to return some resources to the hosts which could be revoked in the future if needed.

Keywords Cloud · Prediction · Cost · Resource · Big data

1 Introduction

Cloud computing is a novel and developing hypothesis in which works are managed by numerous amenities organized on a group of dispersed servers and retrieved through a communal system. This system of hosts and components, together with recognized as the cloud, can bid a noteworthy decrease in calculating and stowage charge sowing to the frugality of scale. Several end operators may access a sole amenity, and several amenities can be organized in a similar data center. The number of data centers joined with the self-enhancing behavior of the diverse amenities organized in them forbids dominant synchronization and brings about several trials

R. Kiruthiga (✉)
Vels Institute of Science, Technology & Advanced Studies, Chennai, India
e-mail: kirthimrk238@gmail.com

D. Akila
Department of Information Technology, Vels Institute of Science, Technology & Advanced Studies, Chennai, India
e-mail: akiindia@yahoo.com

S.-L. Peng et al. (eds.), *Proceedings of First International Conference on Mathematical Modeling and Computational Science*, Advances in Intelligent Systems and Computing 1292, https://doi.org/10.1007/978-981-33-4389-4_22

in the monarchy of source distribution and organization [1]. Resources of computer, storage resources, and the diverse uses can vigorously offer charge as per usage and far ahead can be unconfined if not desirable [2].

One of the most basic recompenses of cloud computing is lesser charges, re-allocation of resources, and isolated availability. To increase the supreme amount of these welfares, the amenities presented regarding resources ought to be allotted accurately to the uses successively in the cloud. The next piece deliberates the consequence of resource division [3].

RA is the procedure of allocating accessible resources to the required cloud users across the Internet. Resource distribution famishes amenities if the distribution is not accomplished exactly. Resource assignment cracks that issue by letting the facility sources to accomplish the resources for every discrete element. Resource allocation strategy (RAS) is all about assimilating cloud source happenings for using and allotting uncommon resources within the boundary of the cloud atmosphere to encounter the requirements of cloud users. It necessitates the kind and quantity of resources required by every use to finish a customer job [4].

Big data is the current publicity in information technology. Technical uses produce a big amount of data that is utilized for detections and examinations. Also, sentiment scrutiny and business data scrutiny are vital for business administration to approve client requirements and gain more incomes. Big data refers to the processes interfered with the collection, storage, distribution, and analysis of massive data sets with various edifices and a supreme proportion of data [12].

Cloud computing can be a relevant result to swarm big data uses, but several encounters need to be answered to utilize the prevailing cloud designs for big data uses [5]. It bids computers as physical or more frequently as virtual machines (VMs). A group of VMs, virtual cluster, is commonly entreated as a podium for operators to run similar or disseminated uses like MapReduce and Dryad applications [6].

2 Related Works

Dai et al. [7] have offered a multi-objective developmental procedure to exchange the presentation, obtainability, and cost of resource request running on cloud. After examining and demonstrating the interweaved associations among these aims, we plan and execute our method on the investigational atmosphere. In conclusion, three groups of try-outs demonstrate that our method can run about 20% quicker than outmoded developmental methods, and can attain about 15% sophisticated presentation than other experiential procedures while saving 4–20% charge.

Jarray et al. [8] have suggested a CG-based one-shot exemplary. For every recognized appeal, an ideal one-shot schmoozing and presenting system is intended to guarantee QoS necessities. Replication outcomes demonstrated that the L-CG-MEC method done meaningfully improved related to the standard techniques from the works.

Moreno-Vozmediano et al. [9] have offered and assessed a new prognostic technique centered on machine learning methods for time sequence predicting. The novel mechanism goals are to precisely forecast the handling load of a disseminated server and evaluate the suitable amount of resources that need to be provisioned to enhance the provision retort period and accomplish the SLA constricted by the operator, while weakening resource over-allocation to lessen energy ingestion and substructure charges. The consequences demonstrate that the suggested exemplary acquires an improved predicting precision than other traditional copies and creates a source allocation closer to the ideal circumstance.

Omotunde et al. [10] have deliberated the chief source distribution policies presently being utilized and associated them using five factors: the capability to evade under-provisioning, the capability to evade overprovisioning, capability to evade resource disputation, the capability to evade resource shortage, and capability to evade source disintegration.

Deepa et al. [11] have defined the current stage of cloud computing investigation by scrutinizing literature and recognizing present study styles. They have investigated the source distribution technique and resolved. It characteristically plans for a high presentation that aids the highest source necessities. After numerous studies, the power ingestion of data center and cloud systems have augmented almost innumerable times. There is an absence of investigation that addresses tasks of handling manifold resources to allot sufficient resources for every workload for enhancing power ingestion.

3 PCRA Scheme

3.1 Overview

This paper proposes PCRA scheme for big data streaming applications. In this scheme, the user specifies the number of executors, cores, memory, etc., for the required application. Then, the mission accomplishment period of a presentation is predicted centered on the number of facilitators and the specific properties of the application. Then, VMs are determined for the task requests, such that the number of executors is sufficient to minimize the task completion time and total cost.

3.2 Constraints of Resource Allocation

Consider a host with N number of worker nodes. Let each worker W_j has C_j CPU cores and M_j total reminiscence. The charge of running every single VM_i is Pr_i per second.

For a given application A, the operator postulates a target (D) and also the number of cores per executor (C_e). Let E be the number of executors required for A, and I be the size of input data associated to A.

Then, the memory required for each executor will be

$$\mathrm{Mr} = C_e * \left(M_j / C_j\right) \tag{1}$$

Cost Constraint

The aggregate rate of a definite application for a user is given by the summation of VMs utilized in this application as given by Eq. (2)

$$\mathrm{Cost} = \sum_{i=1}^{n} (\mathrm{pr}_i) \tag{2}$$

Pr_i is the price of ith VM, and n is the number of utilized VMs.

Then, the constraint for the cost is given by

$$\mathrm{Cost} < \mathrm{Cost}_{\mathrm{min}} \tag{3}$$

where $\mathrm{Cost}_{\mathrm{min}}$ is the minimum bound for cost.

Task Completion Time Constraint

The task completion time (T) is given by

$$T = E * I \tag{4}$$

As shown in Eq. (4), T is obtained from the executor (E) and the feedback of an aggregate petition (I). If the number of input hunks is greater than the number of doers, these input hunks in each executor are viewed as a community. By comparison, having too many executors will result in overheads.

Then, the constraint for task completion time is given by

$$T < D \tag{5}$$

As stated in (5), the task accomplishment time of a designated pattern should encounter the operator deadline D.

Performance Constraint

In Service Level Agreement (SLA), the response time (RT) is frequently utilized to assess the enactment, which can be computed as

$$\mathrm{RT} = \frac{1}{\sum_{i=1}^{n} \frac{1}{rt_i}} - \sum_{i=1}^{\frac{n(n-1)}{2}} (cc_i) \quad (6)$$

where rt_i and cc_i are the RT and communication charge of ith VM.

Then, the constraint for response time is given by

$$\mathrm{RT} < T + t \quad (7)$$

where T is the task completion time and t is the communication time required to respond.

3.3 Objective function for Finding Number of executors

Hence, the objective is to find the number of executors E for A, meeting constraint (3) and (5) and (7).

Find E such that

$$1 < E < E_{\max} \quad (8)$$

where 1 is the lower bound for E and $E_{\max}$ is the upper bound for E which can be given by

$$E_{\max} = N * (C_j / C_e) \quad (9)$$

3.4 Resource Allocation Algorithm

The algorithm presented below explains the steps involved in allocating the resources by satisfying all the constraints.

Algorithm

Input: User Requests U(D,Ce,I)

Output: Number of executors (E)

```
1.    Let E = {1}
2.    For each user request U(D,Ce,I)
3.    Do
4.        Find Cost using (1)
5.        Find T using (3)
6.        Find RT using (5)
7.        Find E_max
8.        If E ≤ E_max , then
9.            If Cost <Cost_min and T < D and RT < (T+t),
then
10.                   Allocate E to U
11.            Else
12.                   Increment E by 1
13.                   If E >E_max, then
14.                         Reject the request U
15.                   Else
16.   Repeat from step (4).
17.                  End if
18.            End if
19.       End if
20.   End For
```

4 Experimental Results

4.1 Results

The proposed PCRA scheme is implemented on Apache Spark. The accomplishment of PCRA is conducted by comparing with the deadline-based resource allocation (DRA) algorithm based on the parameters execution time, response delay, and CPU utilization.

4.2 Effect of Task Size

Experiments are conducted by varying the input tasks from 25 to 125.

Figure 1 shows the task execution times measured for PCRA and DRA for various input tasks. The figure shows that the execution time becomes least when the number of tasks is less. On the other hand, it becomes the highest when the number of tasks is high. It can be inferred that the execution time of PCRA is 21% lesser than RA.

Figure 2 shows the response delay of each task measured for PCRA and DRA for various input tasks. It can be observed from the figure that the response delay increases linearly when the number of tasks is increased. The results show that response delay for each task is 18% less in the case of PCRA when compared to DRA.

Figure 3 shows the CPU utilization for PCRA and DRA for various input tasks. It can be observed from the figure that the utilization decreases gradually when the number of tasks becomes larger. It can be inferred that PCRA has 1.3% higher CPU utilization than DRA.

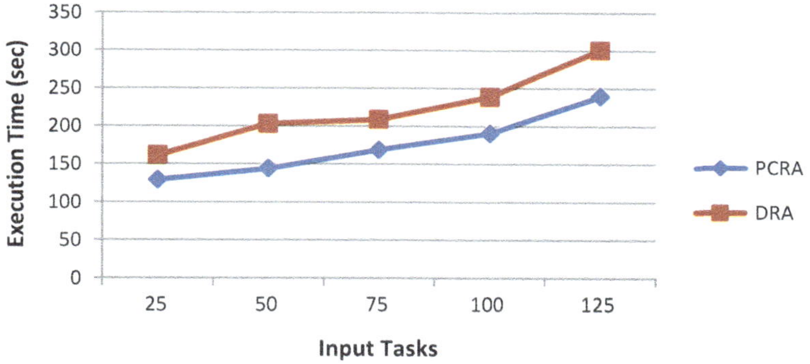

Fig. 1 Execution times for input tasks

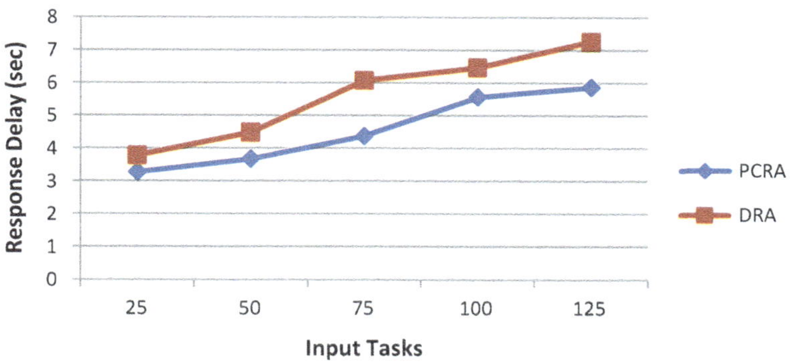

Fig. 2 Response delay for input tasks

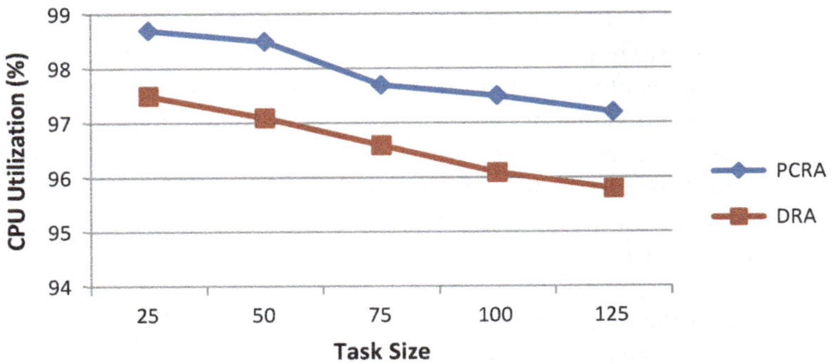

Fig. 3 CPU Utilization for input tasks

4.3 Effect of Deadlines

Experiments are conducted by varying the deadlines from 100 to 500 s.

Figure 4 shows the response delay measured for PCRA and DRA for various deadlines. It can be observed from the figure that the delay is least when the deadline is high, i.e., 500. Similarly, it becomes the highest when the deadline is minimum, i.e., 100. The results show that response delay for the task is 16% less in the case of PCRA when compared to DRA.

Figure 5 shows the CPU utilization for PCRA and DRA for various deadlines. It can be observed from the figure that the utilization increases gradually when the deadline becomes higher. It can be inferred that PCRA achieves 1.2% higher CPU utilization than DRA.

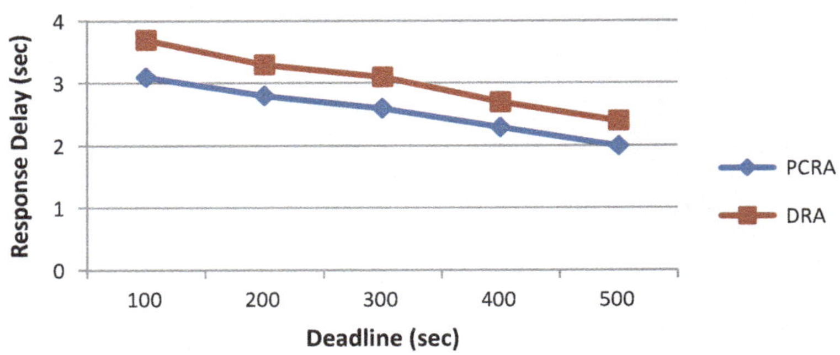

Fig. 4 Response delay for deadlines

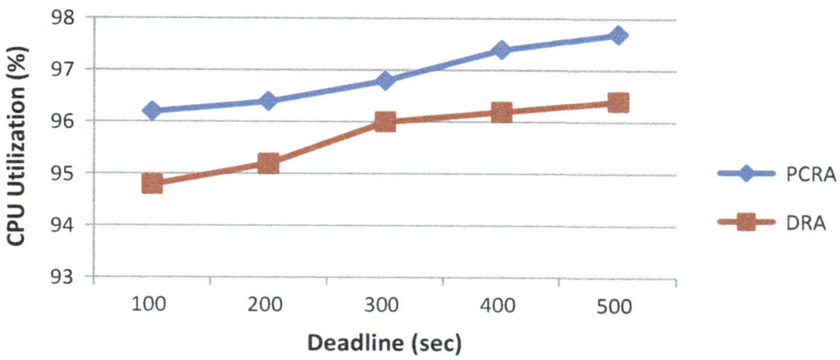

Fig. 5 CPU utilization for deadlines

5 Conclusion

A PCRA scheme has been developed for big data streaming applications. In this scheme, the user specifies the number of executors, cores, memory, etc., for the required application. The completion time of a request is then estimated based on the executors and the application specifications. Then, VMs are determined for the task requests, such that the number of executors is sufficient to minimize the task completion time and total cost. The proposed PCRA scheme is implemented on Apache Spark. The accomplishment of PCRA is conducted comparing with the deadline-based resource allocation (DRA) algorithm based on the parameters execution time, response delay, and CPU utilization. Results show that PCRA achieved lesser execution time and delay with higher CPU utilization.

References

1. Nahir, A., Orda, A., & Raz, D. (2015, May). Resource allocation and management in cloud computing. In *2015 IFIP/IEEE International Symposium on Integrated Network Management (IM)* (pp. 1078-1084). IEEE.
2. Mongia, V., & Kumar, D. (2018). Resource allocation in cloud computing: A review. *International Journal of Computer Sciences and Engineering, 6*(5), 79–84.
3. Vinothina, V., Sridaran, R., & Ganapathi, P. (2012). A survey on resource allocation strategies in cloud computing. *International Journal of Advanced Computer Science and Applications, 3*(6), 97–104.
4. Patel, R., & Patel, S. (2013). Survey on resource allocation strategies in cloud computing. *International Journal of Engineering Research & Technology (IJERT), 2*(2), 1–5.
5. Islam, M. T., & Buyya, R. (2019). Resource management and scheduling for big data applications in cloud computing environments. In *Handbook of Research on Cloud Computing and Big Data Applications in IoT* (pp. 1–23). IGI Global.
6. Priyanka, G., Deshpande, A., Rao, A., Ashik, G., & Kavitha, S. (2017). Cloud infrastructure resource allocation for big data applications. *International Journal of Innovative Research in Computer and Communication Engineering, 5*(11), 16401–16407.

7. Dai, W., Qiu, L., Wu, A., & Qiu, M. (2016). Cloud infrastructure resource allocation for big data applications. *IEEE Transactions on Big Data, 4*(3), 313–324
8. Jarray, A., Karmouch, A., Salazar, J., Elias, J., Mehaoua, A., & Zaman, F. (2017). Efficient resource allocation and dimensioning of media edge clouds infrastructure. *Journal of Cloud Computing, 6*(1), 1–16.
9. Moreno-Vozmediano, R., Montero, R.S., Huedo, E., & Llorente, I. M. (2019). Efficient resource provisioning for elastic Cloud services based on machine learning techniques. *Journal of Cloud Computing, 8*, 1–18.
10. Omotunde, A., & Okolie, S. (2015). Resource allocation in cloud computing – an exposé. *International Journal of Advanced Research in Computer Science and Software Engineering, 5*, 543–8552.
11. Deepa, R., & Dheeba, J. (2019). Resource allocation in cloud computing for energy efficiency. *International Journal of Engineering and Advanced Technology, 8*(6), 1170–1174
12. Islam, M. T., Karunasekera, S., & Buyya, R. (2017, October). dSpark: deadline-based resource allocation for big data applications in Apache Spark. In *2017 IEEE 13th International Conference on E-Science (e-Science)* (pp. 89–98). IEEE.
13. Nahir, A., Orda, A., & Raz, D. (2015, May). Resource allocation and management in cloud computing. In *2015 IFIP/IEEE International Symposium on Integrated Network Management (IM)* (pp. 1078–1084). IEEE

Pythagorean Fuzzy Sets and Its Application in Crop Cultivation

S. G. Karpagavalli and T. Iswarya

Abstract In any decision-making problem, ambiguity is an important factor in finding a solution. Many advanced methods and techniques have been introduced to tackle this unreliable environment. Pythagorean fuzzy sets, a generalization of IFS, is a most flexible tool to handle this situation in decision-making problems. Various factors such as soil type, land type, crop pattern, and investment influence the crop cultivation in a region. This article analyzes the properties of Pythagorean fuzzy sets and Pythagorean fuzzy functions taking the above factors into account and applies the same for the problem of crop cultivation. Finally, a decision-making approach to crop cultivation process is carried out using a hypothetical database using the Pythagorean fuzzy sets and max–min–max operator.

Keywords Crop cultivation · Fuzzy set · Intuitionistic fuzzy set · Pythagorean fuzzy set · Pythagorean fuzzy relations

1 Introduction

Zadeh introduced the fuzzy set theory in which he gave importance of membership value than the non-membership value [15]. The extension of the fuzzy set theory is the intuitionistic fuzzy set (IFS) developed by Atanassov [1–3]. He considered membership, non-membership along with the hesitation index values in his definition [4]. The concept of IFS became a powerful tool in solving many real-life problems like medical diagnosis, career placement, and personal appointments [4–6, 9–12]. The limitations in the above concepts are tackled by Pythagorean fuzzy set proposed by Yager [13, 14]. The Pythagorean fuzzy relations and their composition constructed by Ejegwa [7, 8] are more reliable to find the solutions of the above problems.

S. G. Karpagavalli (✉) · T. Iswarya
Department of Mathematics, Vels Institute of Science, Technology & Advanced Studies, Chennai, India
e-mail: sgkarpa@gmail.com

T. Iswarya
e-mail: ice.rup@gmail.com

S.-L. Peng et al. (eds.), *Proceedings of First International Conference on Mathematical Modeling and Computational Science*, Advances in Intelligent Systems and Computing 1292, https://doi.org/10.1007/978-981-33-4389-4_23

In this paper, Sect. 2 discusses the basic definitions of fuzzy and Pythagorean fuzzy sets, Sect. 3 gives the notions of Pythagorean fuzzy relations a max–min–max rule, Sect. 4 deals with the construction of the problem using composition rule, Sect. 5 shows the numerical illustration to the problem, and remaining sections conclude that crop cultivation can be achieved with the Pythagorean fuzzy relations approach.

2 Preliminaries

Some basic definitions are discussed.

Definition 2.1 Let U be a fixed set. A fuzzy set F in U is defined as set $A = \{\langle p, \lambda_F(p)\rangle / p \in U\}\}$ where the function $\lambda_F(p)$: $U \rightarrow [0, 1]$ defines the degree of membership of the element, $p \in U$.

Definition 2.2 Let U be fixed set. An IFS F in U is a set $A = \{\langle p, \lambda_F(p), \eta_F(p)\rangle \mid p \in U\}$.

where the functions $\lambda_F(p)$: $U \rightarrow [0, 1]$ and $\eta_F(p)$: $U \rightarrow [0, 1]$ are the degree of membership and non-membership, respectively. The hesitation index $\pi_F(p)$ is the degree of non-determinacy of $p \in U$, to the set F and $\pi_F(p) \in [0, 1]$. Thus, $\lambda_F(p) + \eta_F(p) + \pi_F(p) = 1$.

Definition 2.3 Let U be a fixed set. Then a Pythagorean fuzzy set is defined as $F = \{\langle p, \lambda_F(p), \eta_F(p)\rangle \mid p \in U\}$ is a subset of U, and for every $p \in U$: $0 \leq (\lambda_F(p))^2 + (\eta_F(p))^2 \leq 1$. We denote the set of all PFSs over U by PFS(U).

Theorem 2.1 *Let $U = \{u_j\}$ be a non-empty set, for $j = 1, \ldots, n$ and $F \in$ PFS(U). Let $\pi_F(u_j) = 0$. Then,*

(i) $|\lambda_F(u_j)| = \sqrt{|(\eta_F(u_j) + 1)(\eta_F(u_j) - 1)|}$.

(ii) $|\eta_F(u_j)| = \sqrt{|(\lambda_F(u_j) + 1)(\lambda_F(u_j) - 1)|}$.

Proof Suppose that $u_j \in U$ and $F \in$ PFS(U).

Assume that $\pi_F(u_j) = 0$, for $u_j \in U$.

we have,

$$\begin{aligned}
(\lambda_F(u_j))^2 + (\eta_F(u_j))^2 = 1 &\Rightarrow -(\lambda_F(u_j))^2 = (\eta_F(u_j))^2 - 1 \\
&\Rightarrow -(\lambda_F(u_j))^2 = (\eta_F(u_j) + 1)(\eta_F(u_j) - 1) \\
&\Rightarrow \left|(\lambda_F(u_j))^2\right| = |(\eta_F(u_j) + 1)(\eta_F(u_j) - 1)| \\
&\Rightarrow |\lambda_F(u_j)|^2 = |(\eta_F(u_j) + 1)(\eta_F(u_j) - 1)| \\
&\Rightarrow |\lambda_F(u_j)| = \sqrt{|(\eta_F(u_j) + 1)(\eta_F(u_j) - 1)|}.
\end{aligned}$$

Definition 2.5 The complement of a PFS F is defined as follows: Fc $= \{\langle p, \eta_F(p), \lambda_F(p)\rangle | p \in U\}$.

Definition 2.6 The operations union and intersection of two PFSs F and G are defined as.

(i) $F \cup G = \{\langle p, \max(\lambda_F(p), \lambda_G(p)), \min(\eta_F(p), \eta_G(p))\rangle | \; p \in U\}$.
(ii) $F \cap G = \{\langle p, \min(\lambda_F(p), \lambda_G(p)), \max(\eta_F(p), \eta_G(p))\rangle | \; p \in U\}$.

Definition 2.7 Let F be a PFS. Its score is given by the function $sc(F) = (\lambda_F(p))^2 - (\eta_F(p))^2$, where $sc(F) \in [-1, 1]$.

Definition 2.8 The accuracy function of a PFS F is defined as $ac(F) = (\lambda_F(p))^2 + (\eta_F(p))^2$ for $ac(F) \in [0, 1]$.

Theorem 2.2 *Let* $F \in$ PFS(U). *Then*:

(i) $ac(F) = 1 \Leftrightarrow \pi_F(p) = 0$.
(ii) $ac(F) = 0 \Leftrightarrow |\lambda_F(p)| = |n_E(p)|$.

Proof **I Part**.

(i) Suppose $ac(F) = 1$.

Then, $(\lambda_F(u))^2 + (\eta_F(u))^2 = 1$, $(\because \pi_F(u) = 0)$.

II Part.

Assume that $\pi_F(u) = 0$.

Then, it follows that: $(\lambda_E(u))^2 + (\eta_E(u))^2 = 1 \Rightarrow ac(F) = 1$.

(ii) Suppose $ac(F) = 0$. Then, $(\lambda_F(u))^2 = -(\eta_F(u))^2$ or $(\eta_F(u))^2 = -(\lambda_F(u))^2 \Leftrightarrow |\lambda_F(u)|^2 = |\eta_F(u)|^2 \Leftrightarrow |\lambda_F(u)| = |\eta_F(u)|$.

Theorem 2.3 *Let* $F, G \in$ PFS(U). Then, *the following holds*:

(i) $sc(F) = sc(G) \Leftrightarrow F = G$.
(ii) $sc(F) \leq sc(G) \Leftrightarrow F \subseteq G$.
(iii) $sc(F) < sc(G) \Leftrightarrow$ F $\subseteq G$ *and* $F \neq G$.

The proof is obvious.

3 Pythagorean Fuzzy Relation

Basic definitions of Pythagorean fuzzy relations are given.

Definition 3.1 Let U and V be fixed sets. Let f: $U \rightarrow V$ be a function. Let F and G be PFS over U and V. Then,

(i) The image of F is $f(F)$, is a Pythagorean fuzzy set of V defined as

$$\lambda_{f(F)}(q) = \begin{cases} \bigvee_{p \in f^{-1}(q)} \lambda_F(p), & f^{-1}(q) \neq \emptyset \\ 0, & \text{otherwise} \end{cases}$$

and

$$\eta_{f(F)}(q) = \begin{cases} \bigwedge_{p \in f^{-1}(q)} \eta_F(p), & f^{-1}(q) \neq \emptyset \\ 1, & \text{otherwise} \end{cases}$$

for each $q \in V$.

(ii) The inverse image of G is $f^{-1}(G)$, is a Pythagorean fuzzy set of U defined as.
$\lambda_{f-1(G)}(p) = \lambda_G(f(p))$ and
$\eta_{f-1(G)}(p) = \eta_G(f(p))\ \forall\, p \in U$.

Definition 3.2 A Pythagorean fuzzy relation (PFR), R, from U to V is a PFS of U × V given by the membership function, λ_R and non-membership function, η_R. A PFR from U to V is denoted by $R\ (U \to V)$.

Definition 3.3 Let F be a PFS over U. Let R be a PFR from U to V. Then, the composite mapping of $R(U \to V)$ with F is a PFS G of V denoted by $G = R \cdot F$, such that its membership and non-membership functions are defined by the following: $\lambda_G(y) = \bigvee_p(\min[\lambda_F(p), \lambda_R(p, q)])$ and $\eta_G(q) = \bigwedge_p(\max[\eta_F(p), \eta_R(p, q)])\forall\, p \in U$ and $q \in V$, where $\bigvee$ = maximum, $\bigwedge$ = minimum.

Definition 3.4 Let $Q\ (U \to V)$ and $R\ (V \to W)$ be two PFRs.

Then, the (max–min–max) composition $R \cdot Q$ is a PFR from U to W, such that its membership and non-membership functions are defined by the following: $\lambda_{R \cdot Q}(p, r) = \bigvee_q\left(\min\left[\lambda_Q(p, q), \lambda_R(q, r)\right]\right)$ and $\eta_{R \cdot Q}(p, r) = \bigwedge_q\left(\max\left[\eta_Q(p, q), \eta_R(q, r)\right]\right)$, $\forall\, (p, r) \in U \times W$ and $\forall\, q \in V$.

Result 3.1 From Definitions 4.3 and 4.4, the PFR $B = R \bullet Q$ is found by the following:
$B = \lambda_B(q) - \eta_B(p)\pi_B(q), \forall\, q \in V$.

4 Crop Cultivation by Composite Relations of Pythagorean Fuzzy Sets

Let us construct the problem as follows.

Let $P = \{p_1, p_2, \ldots, p_t\}$, $S = \{s_{1,} s_2, \ldots, s_m\}$, $C = \{c_1, c_2, \ldots, c_n\}$ be the collection of states, soils, and crops, respectively.

Let $R_1(P \to S)$ and $R_2(S \to C)$ be two Pythagorean fuzzy relations s.t $R_1 = \{(p, s), \lambda_R(p, s), \eta_R(p, s)/(p, s) \in P \times S\}$ and $R_2 =$

$\{(s, c), \lambda_U(s, c), \eta_U(s, c)/(s, c) \in S \times C\}$, where $\lambda_{R_1}(p, s)$ denotes the degree to which the state p related to the soil s, $\eta_{R_1}(p, s)$ denotes the degree to which the state p not related to the soils.

Similarly, $\lambda_{R_2}(s, c)$ denotes the degree to which the soil s related to the crop c, $\eta_{R_2}(s, c)$ denotes the degree to which the soil s not related to the crops, respectively.

The composite mapping M of R_1 and R_2 is denoted by $R_1 \bullet R_2$ which describes the state in which the states p_i concerning the soil s_j the cultivated crop c_k.

$\lambda_M(p_i, c_k) = \bigvee_{s_j \in S}\{\min[\lambda_{R_1}(p_i, c_j), \lambda_{R_2}(s_j, c_k)]\}$ and $\eta_M(p_i, c_k) = \bigwedge_{s_j \in S}\{\max[\eta_{R_1}(p_i, s_j), \eta_{R_2}(s_j, c_k)]\}$.

5 Case Study

Consider the following factors as sets,

the set of states P is

$$P = \{\text{Assam, Gujarat, Karnataka, Uttar Pradesh, West Bengal}\},$$

the set of soils S is

$$S = \{\text{Alluvial Soil, Black Soil, Red Soil, Desert Soil, Laterite Soil, Mountain Soil}\}.$$

the set of crops C is

$$C = \{\text{Rice, Wheat, Sugarcane, Vegetables, Tea, Coffee}\}$$

The Pythagorean fuzzy relations $R_1(P \to S)$ and R_2 $(S \to C)$ are given hypothetically in Tables 1 and 2. λ_M and η_M function values of the composite relation $M = R_1 \bullet R_2$ are given in Table 3. After finding the hesitation index $(\pi = \sqrt{1 - (u^2 + v^2)})$, we calculate the values of M given in Table 4, respectively.

Table 1 $R_1(P \to S)$ denotes the degree of states to different types of soil

R_1	Alluvial soil	Black soil	Red soil	Desert soil	Laterite soil	Mountain soil
Assam	⟨0.6, 0.1⟩	⟨0.5, 0.4⟩	⟨0.3, 0.4⟩	⟨0.7, 0.2⟩	⟨0.3, 0.4⟩	⟨0.3, 0.4⟩
Gujarat	⟨0.8, 0.1⟩	⟨0.6, 0.1⟩	⟨0.2, 0.8⟩	⟨0.6, 0.1⟩	⟨0.1, 0.6⟩	⟨0.2, 0.8⟩
Karnataka	⟨0.0, 0.8⟩	⟨0.4, 0.4⟩	⟨0.6, 0.1⟩	⟨0.1, 0.7⟩	⟨0.1, 0.8⟩	⟨0.5, 0.4⟩
Uttar Pradesh	⟨0.8, 0.1⟩	⟨0.8, 0.1⟩	⟨0.0, 0.6⟩	⟨0.2, 0.7⟩	⟨0.0, 0.5⟩	⟨0.6, 0.1⟩
West Bengal	⟨0.2, 0.8⟩	⟨0.6, 0.1⟩	⟨0.0, 0.8⟩	⟨0.4, 0.4⟩	⟨0.6, 0.1⟩	⟨0.1, 0.7⟩

Table 2 $R_2(S \to C)$ denotes the degree of soil related to different types of crop

R_2	Rice	Wheat	Sugarcane	Vegetables	Tea	Coffee
Alluvial soil	$\langle 0.1, 0.7\rangle$	$\langle 0.0, 0.9\rangle$	$\langle 0.2, 0.7\rangle$	$\langle 0.8, 0.0\rangle$	$\langle 0.2, 0.8\rangle$	$\langle 0.7, 0.0\rangle$
Black soil	$\langle 0.4, 0.3\rangle$	$\langle 0.7, 0.0\rangle$	$\langle 0.2, 0.6\rangle$	$\langle 0.2, 0.7\rangle$	$\langle 0.2, 0.8\rangle$	$\langle 0.7, 0.0\rangle$
Red soil	$\langle 0.4, 0.0\rangle$	$\langle 0.7, 0.0\rangle$	$\langle 0.3, 0.3\rangle$	$\langle 0.1, 0.7\rangle$	$\langle 0.1, 0.8\rangle$	$\langle 0.2, 0.6\rangle$
Desert soil	$\langle 0.3, 0.5\rangle$	$\langle 0.2, 0.6\rangle$	$\langle 0.6, 0.1\rangle$	$\langle 0.2, 0.4\rangle$	$\langle 0.0, 0.8\rangle$	$\langle 0.2, 0.4\rangle$
Laterite soil	$\langle 0.1, 0.7\rangle$	$\langle 0.1, 0.8\rangle$	$\langle 0.1, 0.9\rangle$	$\langle 0.2, 0.7\rangle$	$\langle 0.0, 0.8\rangle$	$\langle 0.2, 0.7\rangle$
Mountain soil	$\langle 0.7, 0.0\rangle$	$\langle 0.3, 0.3\rangle$	$\langle 0.4, 0.3\rangle$	$\langle 0.7, 0.0\rangle$	$\langle 0.2, 0.6\rangle$	$\langle 0.2, 0.8\rangle$

Table 3 $\lambda_M(p, c)$ and $\eta_M(p, c)$ denote the function values of M

λ_M, η_M	Rice	Wheat	Sugarcane	Vegetables	Tea	Coffee
Assam	$\langle 0.60, 0.05\rangle$	$\langle 0.40, 0.35\rangle$	$\langle 0.50, 0.25\rangle$	$\langle 0.45, 0.45\rangle$	$\langle 0.70, 0.05\rangle$	$\langle 0.40, 0.35\rangle$
Gujarat	$\langle 0.50, 0.25\rangle$	$\langle 0.75, 0.05\rangle$	$\langle 0.70, 0.10\rangle$	$\langle 0.50, 0.25\rangle$	$\langle 0.45, 0.30\rangle$	$\langle 0.50, 0.25\rangle$
Karnataka	$\langle 0.70, 0.10\rangle$	$\langle 0.45, 0.45\rangle$	$\langle 0.55, 0.20\rangle$	$\langle 0.55, 0.20\rangle$	$\langle 0.55, 0.25\rangle$	$\langle 0.70, 0.10\rangle$
Uttar Pradesh	$\langle 0.60, 0.05\rangle$	$\langle 0.60, 0.10\rangle$	$\langle 0.70, 0.10\rangle$	$\langle 0.50, 0.25\rangle$	$\langle 0.50, 0.25\rangle$	$\langle 0.60, 0.05\rangle$
West Bengal	$\langle 0.55, 0.20\rangle$	$\langle 0.45, 0.35\rangle$	$\langle 0.50, 0.25\rangle$	$\langle 0.75, 0.05\rangle$	$\langle 0.75, 0.05\rangle$	$\langle 0.60, 0.05\rangle$

Table 4 $M = \lambda_M - \eta_M \pi_M$ denotes the composition values of R_1 and R_2

M	Rice	Wheat	Sugarcane	Vegetables	Tea	Coffee
Assam	0.5600	0.1035	0.2927	0.1029	0.6644	0.1035
Gujarat	0.3508	0.7170	0.6293	0.2927	0.1977	0.2927
Karnataka	0.6293	0.1029	0.3878	0.3878	0.3508	0.6293
Uttar Pradesh	0.5601	0.5206	0.6293	0.2927	0.2927	0.5601
West Bengal	0.3878	0.1624	0.2927	0.7170	0.7170	0.5601

6 Decisions on the States' Crop Yield

Here, we made the following two decisions on the crop yield:

Row-wise observations: This is based on chances of the state that it can cultivate one or more crop.

Assam can cultivate tea, Gujarat can cultivate wheat, Karnataka can cultivate rice and coffee, Uttar Pradesh can cultivate sugarcane, West Bengal can cultivate tea and vegetables.

Column-wise observations: This decision is centered on the relation that the maximum number of state that can cultivate a crop.

It is noticed that rice can be cultivated in Karnataka, wheat can be cultivated in Gujarat, sugarcane can be cultivated in Gujarat and Uttar Pradesh, vegetables can be cultivated in West Bengal, tea can be cultivated in West Bengal, and coffee can be cultivated in Karnataka.

7 Conclusion

From the above discussion, it is concluded that PFR has vast applications in solving many multi-decision-making problems. Also, it is proved that this approach yields better crop production for economic benefits. Agriculture is the main occupation of our country, yielding a variety of crops in a state in this pandemic, and escalates the country's economy. The same problem can be extended by taking climatic conditions and degree of rainfall as additional input factors and applying the fuzzy neural network to obtain improved results.

References

1. Atanassov, K. T. (1986). Intuitionistic fuzzy sets. *Fuzzy Sets System, 20,* 87–96.
2. Atanassov, K. T. (1989). Geometrical interpretation of the elements of the intuitionistic fuzzy objects. Preprint IM-MFAIS-1–89, Sofia. *International Journal of Bioautomation, 20*(S1), S27–S42 (Reprinted: (2016)).
3. Atanassov, K. T. (2012). *On intuitionistic fuzzy sets theory.* Berlin, Germany: Springer.
4. Davvaz, B., & Sadrabadi, E. H. (2016). An application of intuitionistic fuzzy sets in medicine. *International Journal of Biomathematics, 9*(3), 1650037.
5. De, S. K., Biswas, R., & Roy, A. R. (2001). An application of intuitionistic fuzzy sets in medical diagnosis. *Fuzzy Sets System, 117*(2):209–213.
6. Ejegwa, P. A., & Modom, E. S. (2015). Diagnosis of viral hepatitis using new distance measure of intuitionistic fuzzy sets. *International Journal of Fuzzy Mathematical Archive, 8*(1), 1–7.
7. Ejegwa, P. A. (2015). Intuitionistic fuzzy sets approach in appointment of positions in an organization via max-min-max rule. *Global Journal of Science Frontier Research F: Mathematics & Decision Science, 15*(6), 1–6.
8. Ejegwa, P. A., & Modom, E. S. (2019). Pythagorean fuzzy set and its application in career placements based on academic performance using max–min–max composition. *Complex & Intelligent Systems, 5,* 165–175.
9. Garg, H., & Kumar, K. (2018). An advance study on the similarity measures of intuitionistic fuzzy sets based on the set pair analysis theory and their application in decision making. *Soft Computing, 22*(15), 4959–4970.
10. Garg, H., & Singh, S. (2018). A novel triangular interval type-2 intuitionistic fuzzy set and their aggregation operators. *Iranian Journal of Fuzzy Systems, 15*(5), 69–93.
11. Garg, H., & Kumar, K. (2018). Distance measures for connection number sets based on set pair analysis and its applications to decision making process. *Applied Intelligence, 48*(10), 3346–3359.
12. Szmidt, E., & Kacprzyk, J. (2001). Intuitionistic fuzzy sets in some medical applications. *Note IFS, 7*(4), 58–64.

13. Yager, R. R. (2013). Pythagorean fuzzy subsets. In *Proceedings of the Joint IFSA World Congress NAFIPS Annual Meeting* (pp. 57–61).
14. Yager, R. R. (2014). Pythagorean membership grades in multicriteria decision making. *IIEEE Transactions on Fuzzy Systems, 22*(4), 958–965.
15. Zadeh, L. A. (1965). Fuzzy sets. *Information and Control, 8,* 338–353.

Extraction and Search of Relevant Chemical Documents from the Web

R. Hema

Abstract As the sources of chemical literature are vast, researchers and readers have to spend a great amount of time to obtain specific information. In addition to this, the ambiguity among the names of chemical entities often gives results which are irrelevant to the query. So it is better to provide the researchers and readers with the facility to carry out chemistry-oriented search. Nowadays, domain-specific search engines have become popular because they offer increased accuracy and additional functionality for that domain. In this paper, a domain-specific search application is developed to retrieve the relevant chemical documents from a large collection. This search application has advanced features such as specialized index and re-ranking of documents by chemical entities, functional groups, and phrases as keywords. At present, 85,680 patents and 66,425 chemical abstracts are indexed with 36,984 entities of 23 different types and 16 functional groups.

Keywords Lucene · Specialized index · Extended page rank · Domain-specific search · NDCG

1 Introduction

Nowadays, it is a trend to store the information incorporates, industry, government, and other industries in the form of digital text [1]. The information needs of the scientific community are often more specific than common people. A relevant response to the search problems is the creation of domain-specific search engines. When we want to retrieve a certain type of information, a domain-specific search engine may be a powerful tool to retrieve the specific type of documents [2]. General search engines will have small and more manageable indexes and have a powerful domain-specific search capability. Quick responses to searches for chemical names over large corpora of chemical documents are expected by researchers.

R. Hema (✉)
IDE, University of Madras, Chennai, Tamil Nadu, India
e-mail: hemaramji@gmail.com

S.-L. Peng et al. (eds.), *Proceedings of First International Conference on Mathematical Modeling and Computational Science*, Advances in Intelligent Systems and Computing 1292, https://doi.org/10.1007/978-981-33-4389-4_24

A chemical search engine must identify the occurrences of all instances of chemical entities appearing in text documents and index them to enable fast access. For this purpose, there is a need to specialize the search relevance criteria that well characterize the sub-collections in question. Chemists and other users of chemical search engines may desire to query with a part of a chemical term or full chemical term. To enable fast and efficient search and to support partial or full-term search, indexes are used by search engines.

Chemical search engine ChemXSeer [3] is a tool capable of identifying and extracting the chemical formulae and chemical names from chemical documents and disambiguating the chemical-related terms from general terms. Those disambiguated terms are then used for doing searches. A fast structure and text search was done by ChemSpider [4] from hundreds of data sources. Compared to the existing chemical search engines, our search engine will use both the chemical terms (both partial and exact terms) and functional groups as keywords to search the relevant chemical documents. To build the specialized inverted index, we have used additional fields like chemical entities and functional groups.

Then, the traditional Lucene ranking system is modified by re-ranking the chemical documents to improve the relevance of our chemical search engine as well as the precision rate. The experimental results are evaluated for the relevance of retrieved chemical documents.

Figure 1 shows the architecture of our chemical domain-specific search application. In this architecture, the general search engine is modified into specialized search application operating on a sub-collection and using specialized ranking method relying on additional fields in the selected sub-collection of documents (e.g., chemical entity and functional group).

2 Related Work

Hunting down relevant information from the different acronyms and multiple names that represent chemical compounds is hard enough for an untrained human. Most generic Internet search engines cannot be used for retrieving the chemical compounds. Giles et al. [3] designed ChemXSeer, a chemical search engine that understands chemical information to return relevant documents with high accuracy. This search engine is open source and uses the relevance measure to rank the documents. Sun et al. [5, 6] had developed a search engine to search both chemical entities and chemical formulae.

ChemXSeer [3] is a chemical search engine which extracts chemical entities from chemical documents. Novel indexing is designed to extract chemical formulae and names. In this search engine, they have used the hierarchical conditional random fields (HCRFs) for tagging the chemical entities and an algorithm of independent frequent subsequence mining (IFSM) have been developed to identify the subterms in the chemical entities and the probabilities of occurrence are also evaluated. Finally,

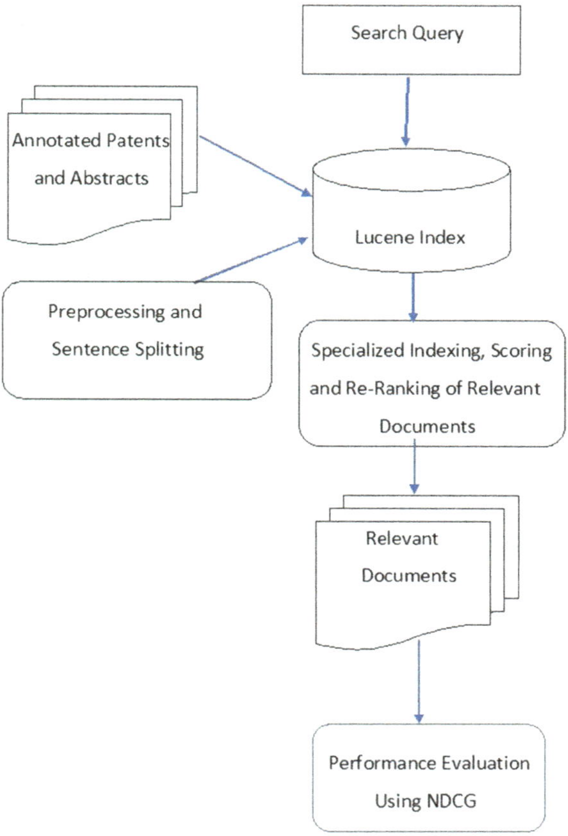

Fig. 1 Architecture of chemical document search application

they have created unique formulae ranking algorithms for formulae search and introduced various query models for chemical name searches.

Sun et al. [5, 6] proposed algorithms for tagging both chemical name and chemical formulae by using conditional random fields (CRFs) and support vector machines (SVMs) and achieved the higher accuracy. This search system indexes the substrings of chemical names. The scoring scheme has been used to rank the retrieved chemical entities based on the vector space model. To rank, the documents, subsequence frequency, and inverse entity frequency were defined.

Often chemists want to query the chemical search engines with chemically related attributes. A chemistry-based search application must identify the occurrences of all instances of chemically related attributes appearing in text documents and index them to enable fast access. Both [4] and [5, 6] are the search engines used chemical names as keywords to search the chemical literature. In this paper, our contribution is that we have developed a domain-specific search application for chemistry which uses the chemical entities, functional groups, and phrases as keywords for searching the relevant chemical documents. To build the specialized inverted index, we have used the additional fields of chemical attributes in Lucene indexing. Then, the traditional Lucene library is modified for our domain-specific requirements in the ranking system by re-ranking the chemical documents to improve the relevance as well as the precision rate.

3 Methodology

3.1 System Description

Chemgateway uses only annotated chemical documents from various resources like gold standard annotated dataset produced by Akhondi et al. [7] and corpora developed by Kolarik et al. [8] for the recognition of IUPAC and IUPAC-like names. Like Friedrich et al. [9] we used the GENIA corpus which is an annotated corpus for bio-text mining and replaced the protein names with IUPAC names from public chemical databases.

There are two broad ways in which to make a search engine specialized [10]. The first is to build or make use of an index that is focused on some particular category of information. The second is to present a search strategy to users that support and implements specialized searches within some category. In this paper, category refers to the chemical terms (both exact match and partial match) and the functional group terms. To provide beneficial access, our search engine has the following advanced features: (i) a specialized inverted index for efficient searching (ii) searching by using chemical entities and functional groups as keywords and (iii) re-ranking of chemical documents for relevant retrieval. The following sections will explain the building of specialized index and a specialized search strategy.

3.2 Building a Specialized Index

When the contents of the search index are restricted to a specific category, then that index becomes a special index. In our paper, we are creating a specialized index using the Lucene library. Lucene builds an index over a collection of documents, and the index is updated when the documents are added and deleted from the collection [11]. Lucene indexes only the terms, and so the search is based only on the terms. A term is a combination of a field name and a token. The fields may be text or non-text fields. Here the text fields consist of field name and token and the non-text fields consist of field name and the corresponding value. Lucene indices consist of several data fields which are connected to each document. This feature makes it very useful for our approach in which we want to store three layers of annotation, namely chemical entity, functional group, and phrase, for each document in the collection. The field-based customized indexes are shown in Fig. 2.

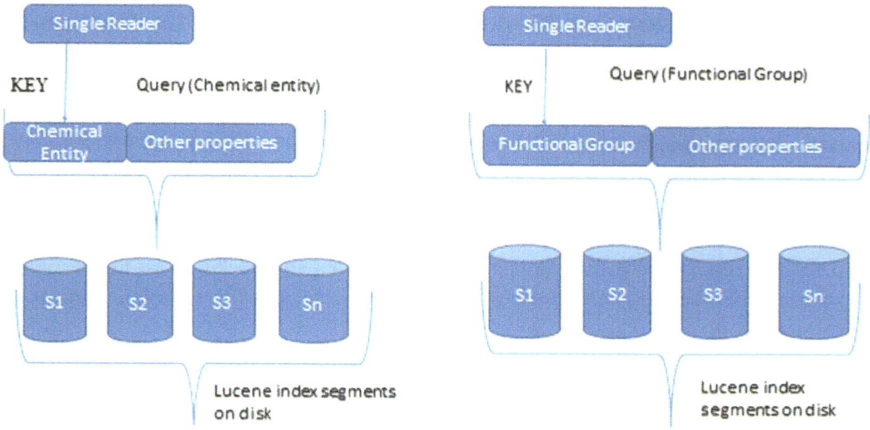

Fig. 2 Customized index for the chemical attributes

To do the specialization, the contents of the search index should be constrained to a particular category. This section explains the way to build such an index via crawling. In our paper, we are creating a specialized index using Lucene library. Lucene builds an index over a collection of documents, and the index is updated when the documents are added and deleted from the collection [11]. Lucene indexes only the terms and so the search is based only on the terms. A term is a combination of a field name and a token. The fields may be text or non-text fields. Here the text fields consist of field name, and token and non-text fields consist of field name and the corresponding value. Lucene indices may include several data fields connected to each document. This feature makes it very useful for our approach in which we want to store two layers of annotation, namely chemical entity and the corresponding functional group, for each document in the collection.

When an index provides a mapping from terms to documents, then that index is called an inverted index. The inverted index gives scores to the search terms. If the searching terms are present in the same document, then that document is said to be relevant. In Table 1, a snippet of entries of our chemical index over a part of the chemical documents is shown as below:

Table 1 A snippet of entries in the inverted index for chemical search engine

Term	Document IDs
text:benzene	2, 7, 18, 21, 26, 35, 64, 76, 120, 149, 234, 289, 356, 390, 521, …
text:methyl	1, 3, 4, 7, 8, 9, 10, 11, 12, 15, 16, 21, 23, 24,25, 27, 29, 30, 33, 34, 35, 36, 37, 38, 41, 43, 46, 49, 51, 52, 58, 62, 63, 64, 67, 68,70, 71, 75, 78, 85, 128, 187, 259, 324, …
functional:alcohol	10, 45, 53, 99, 111, 178, 265, 376, …
functional:aldehyde	9, 34, 56, 78,99, 102, 167, 185, 255, 284, …

In Table 1, the document ids are given by the Lucene library for its an internal reference. These document ids are temporary. Whenever the documents are added or deleted from the index, Lucene will change the document ids of the documents. Conceptually, Lucene carried out the indexing and searching over the fields rather than the documents.

During searching, the field name controls the search over specific fields. For example, for our chemical search engine, the documents can be represented as a sequence of fields: the first field for the IUPAC list of chemical terms and the second field for the functional group name. Different names should be given to the fields. At search time, the user should specify that the searching is for a chemical entity or functional group.

3.3 Boosting

Lucene has the idea of boosting by which we can give importance to certain documents or fields than others [12]. There is a differentiation between search time boosting and index. Lucene has two types of boosting methods. They are: (i) Boosting a term in the index statically and (ii) Boosting a term in the index dynamically.

3.3.1 Boosting a Term in the Index (Static)

@Boost is a property available in Lucene library to boost a term in the static index. This property can be used inside the @Field property or we can specify it on class or method level directly. Here the boost factor is independent of the words in the search query.

3.3.2 Boosting a Term in the Index (Dynamic)

Whenever we want to boost the terms in the search query at runtime, @DynamicBoost is one of the properties of Lucene which helps to boost the terms in query dynamically together with the normal strategy of boosting.

3.3.3 Boosting a Term

Lucene finds the relevant documents based on the tokens or terms. When we want to give importance for a particular term in a search query, we can do this by using the caret, "^", a symbol to boost that particular term with a boost factor (indicates as a number) as a suffix of the term. If the boost factor is high, then the term will be more relevant. By boosting a term, we can control the document relevance. For example, if a search query term is "diethyl ether" and if we know that the term "diethyl" is more relevant, then boosting can be done using the ^ symbol with the boost factor (or) number next to the term. For that, we have to write the query as:

text: "diethyl^4ether"

This will make the chemical documents with the term diethyl appear to be more relevant. We can also boost phrases as in the example:

text: "diethyl ether"^4

3.4 Specialized Searching and Ranking

An advantage of a domain-specific search engine is that the information retrieved is constrained to a particular category, and it provides rich search space. Here, our chemical search engine does not restrict its searches to a subset of the chemical documents available but rather restricts the types of search that can be done.

3.4.1 Search Queries

The input for Lucene search API is a search query, and the query terms are categorized into operators and terms. Again, the terms are classified into single terms and multiple terms or Phrases. A single term is a word like "sodium" or "ethylene." Multiple term is a group of more than one word enclosed within double quotes like "benzoic acid."

Complex queries can be formed by combining multiple terms and boolean operators. Lucene supports the data in the field property. For doing a search using Lucene, we have to specify a field name. If a field name is not specified, then the default field will be taken for search. For our chemical search engine, the default field is a *text* field.

Lucene has its own language to express the search queries. As an example, *methane NOT methyl* is a query that specifies that *methane* should be in the default field and *methyl* should not be in that field. Queries are expressed along with the name of the fields, as in *text:benzene*, which requires the term *benzene* should present in the text field.

3.4.2 Specifying Search Criteria

Lucene supports a wide array of possible searches which includes various modifiers like specifications in term or field, wildcard, searching through ranges, boolean operators and complex queries with the help of logical operators [11]. Sub-queries can also be focused by boosting their relevance using numeric values to the specific parts of a search query. Let us take a look at some example queries for our chemical search engine:

Query 1: *text: 2,7-diazacarbazole*
Response: *The above query will find all chemical documents that contain 2,7-diazacarbazole in the text field*
Query 2: *text: "2-methyl propane" OR text: "2,2-dimethyl propane"*
Response: *This query will find all documents that contain "2 -methyl propane" or "2,2-dimethyl propane"*
Query 3: *text: meth**
Response:*This query will retrieve all documents whether they contain methane or methyl or methanoic and so on*

Range queries are one which allows us to retrieve the relevant documents whose field values fall within the lower and upper bound values. Curly brackets are used for exclusive range queries and square brackets are used for inclusive range queries.

Query 4: *text: (Ethyne To Propanamine)*
Answer: *This query will retrieve all documents whose texts are between Ethyne and Propanamine, but not including Ethyne and Propanamine*

Query 5: *functional: alcohol*
Answer: Find all documents which contain the functional group alcohol in functional field

3.5 Search Scoring and Ranking

The success of a search engine is it's ability to rank the results. The concept of the ranking method is to reorder documents to bring better ones to the top. The scores produced by the baseline engine are too close and did not differentiate heavily between different documents. In our chemical search engine, the standard Lucene ranking was extended to improve results. First, the documents and primary scores were obtained and then the documents were re-ranked by analyzing the query to understand the focus of the query. Once the final results are received, they are displayed to the user.

Lucene library first weights the result documents using term frequency-inverse document frequency (TF-IDF) which is a scoring algorithm. Term frequency defines the importance of the terms in a document, and therefore, the score is very high then the other terms. Inverse document frequency gives importance to the terms which occur only in very few documents. So this measure filters out very common words.

4 Experimental Results

This section presents the findings of our chemical search engine experiments. 10 input documents are taken for testing purpose. Such input documents are preprocessed, and the tf-idf values are then determined for the documents. The results obtained for a question are ordered in descending order, using their ranking. Each document has a score and it indicates that the document's relevancy to the query in searching. Table 2 shows the tf-idf values for the input documents and their ranking.

4.1 *Performance Evaluation*

Evaluation is the key to develop better search engines, and it is also an important aspect of information retrieval. However, it is extremely difficult to find reliable measures to calculate the efficiency of search engines. Our chemical search engine is evaluated using normalized discounted cumulative gain (NDCG) [13]. In the next sections, the metrics and their results are explained.

4.2 *NDCG*

Discounted cumulative gain (DCG): The search results can be used for assessment. DCG method rates the content of the returned documents by calculating the gain in cumulative significance of all the returned documents by rating algorithms. The DCG in a given rank position p is defined as [13]:

Table 2 Ranking using *td-idf* scores

Top "*n*" position	Tf-Idf score	Document Ids
1	0.87	D6
2	0.82	D2
3	0.75	D1
4	0.73	D8
5	0.70	D10
6	0.66	D3
7	0.58	D5
8	0.52	D9
9	0.43	D4
10	0.31	D7

Table 3 NDCG evaluation

Documents ranked by human	Ideal ranking	Tf-Idf Ranking
D4	1	3
D1	2	2
D2	3	5
D7	4	6
D5	5	1
D6	6	4
D3	7	7
D8	8	8
D10	9	10
D9	10	9

$$\mathrm{DCG}_P = \mathrm{rel}_1 + \sum_{i=2}^{p} \frac{\mathrm{rel}_i}{\log_2(i)} \tag{1}$$

where rel_i is the graded relevance of the result at position i.

When the retrieved documents are ordered by human effort, then it is called as the ideal ranking or user ranking. The search results vary depending on the length of the search query. So, only using DCG for evaluating search engine's performance is not a sensible decision.

To enhance this, the cumulative gain at all positions of p should be normalized for all queries with the help of ideal ranking. It is also known as the discounted cumulative benefit ratio of the ranked documents, derived from the tf-idf method to the ideal ranking. The calculation values for nDCG will range between 0.0 and 1.0.

It is also known as the discounted cumulative benefit ratio of the ranked documents, derived from the tf-idf method to the ideal ranking. The calculation values for nDCG will range between 0.0 and 1.0. Here the importance of the document was evaluated by the human experts (called the ideal ranking), and a significant score was given to each document by a graded scale of 10 points where 10 indicated the most important document and 0 indicated the most irrelevant. All the documents are sorted by the relevance scores and summarized in Table 3.

The NDCG value for the results obtained through Tf-Idf algorithm for relevance ranking is 0.99 which is nearer to 1. Therefore, the correlation method of relevance ranking is accurate.

5 Comparison with Existing IR Techniques

We compare our approach with the existing chemical search engines [3, 5, 6] by analyzing the precision of the top 25 returned documents. The precision is defined as the percentage of returned documents which contain the query terms. In ChemXseer paper [3], the index was built with the help of the feature selection. In Sun et al. [5, 6] papers, the size of the index was reduced by using their index pruning method to enhance the search relevance. In this paper, an inverted index is used to index the terms and to map the query terms with the relevant documents. Inverted index means that instead of "what terms does this document contain" you will know "what documents contain this term." This structure speeds up the searching process a lot. In addition to this, if the query terms are not present in the returned documents, then once again the returned documents are scored and re-ranked after analyzing the focus of the query terms. Because of this inverted index and re-ranking approaches, the precision of our approach is better than the existing approaches.

As Sun et al. [6], we also evaluate generic search engines (Google and Yahoo) with the same type of query terms. Since the documents indexed by different search engines are different, the results of general-purpose search engines are not directly comparable with our results. We have not used the traditional Lucene library but we have customized the Lucene library and the index by specifying the field terms. In addition to that, the boosting factor for the query terms will enhance the relevance of the returned results.

From the results, it is observed that the customized Lucene library is having the following advantages. (i) Complete query capabilities like specifying the boolean operators, wildcards, fielded searching along with the query terms. (ii) Portability and (iii) Scalability. But the only disadvantage is that when the user is not familiarized with the querying language, then it will take time for the end-user to learn the query language and to retrieve the documents.

6 Conclusion

This search engine has been evaluated with 1,32,424 documents by using both the chemical entity and the functional group as keywords. We have achieved 0.81 of precision for queries using chemical entities (exact match) and 0.78 precision for queries using chemical entities (partial match) and 0.75 precision for queries using the functional group as keywords. In addition to that, we have evaluated our chemical search engine using NDCG and the calculation and results were discussed. But due to ambiguous chemical entities, it may lead to the retrieval of different documents. Spelling variations need to be addressed to get the relevant documents. So it is a need for an automation tool to rectify these errors.

References

1. Concepts, D. M. (2006). Technique, Jiawei Han and Micheline Kamber. University of Illinois at Urbana-Champaign.
2. McCallumzy, A., Nigamy, K., Renniey, J., & Seymorey, K. (1999). Building domain-specific search engines with machine learning techniques. In *Proceedings of the AAAI Spring Symposium on Intelligent Agents in Cyberspace* (pp. 28–39). *Citeseer*
3. Mitra, P., Giles, C. L., Sun, B., & Liu, Y. (2007, November). Chemxseer: a digital library and data repository for chemical kinetics. In *Proceedings of the ACM first workshop on CyberInfrastructure: information management in eScience* (pp. 7–10).
4. Pence, H. E., & Williams, A. (2010). ChemSpider: An online chemical information resource.
5. Sun, B., Mitra, P., Lee Giles, C., & Mueller, K. T. (2011). Identifying, indexing, and ranking chemical formulae and chemical names in digital documents. *ACM Transactions on Information Systems (TOIS), 29*(2), 1–38.
6. Sun, B., Tan, Q., Mitra, P., & Giles, C. L. (2007, May). Extraction and search of chemical formulae in text documents on the web. In *Proceedings of the 16th international conference on World Wide Web* (pp. 251–260).
7. Akhondi, S. A., Klenner, A. G., Tyrchan, C., Manchala, A. K., Boppana, K., Lowe, D., & Muresan, S. (2014). Annotated chemical patent corpus: A gold standard for text mining. *PLoS ONE, 9*(9), e107477.
8. Kolárik, C., Klinger, R., Friedrich, C. M., Hofmann-Apitius, M., & Fluck, J. (2008). Chemical names: Terminological resources and corpora annotation. In *Workshop on Building and Evaluating Resources for Biomedical Text Mining (6th edition of the Language Resources and Evaluation Conference).*
9. Friedrich, C. M., Revillion, T., Hofmann, M., & Fluck, J. (2006, April). Biomedical and chemical named entity recognition with conditional random fields: The advantage of dictionary features. In *SMBM*
10. Steele, R. J. (2001). Techniques for specialized search engines. In *International Conference on Internet Computing*. CSREA Press.
11. Mohd, M. (2011). Development of Search Engines using Lucene: An Experience. *Procedia-Social and Behavioral Sciences, 18,* 282–286.
12. Gondaliya, T. P., & Joshi, H. D. (2017). Journey of Information Retrieval to Information Retrieval Tools-IR&IRT A Review.
13. Rahayu, S. B., Noah, S. A., & Wardhana, A. A. (2011, June). User-centered evaluation for IR: Ranking annotated document algorithms. In *International Conference on Software Engineering and Computer Systems* (pp. 306–312). Berlin, Heidelberg: Springer.

Convention of Application Service Providers (ASP) to Applied Cloud Computing Circumvents Defeat

D. Karthik and A. Vinayagam

Abstract Organizations have gotten subject to regularly increasing procedures of automated statistics also profligate interchange speeds. Specialists, for example, conjecture that the finish of secluded registering is close by and that in the following decade, most organizations will have made the move to utility processing. To reduce incidentals although, by way of so far accomplishing gradually composite data innovation supervisions, plentiful establishments went towards solicitation specialist organizations. For the reasons that of unfortunate strategies of accomplishment, finished contention, besides underprivileged network availability also, statistics transfer capacity, frequent Active Server Page's disappeared through the website smash. Poles apart Active Server Pages, in any case, who grabbed network directions proposal also unaffected network transmission, remained all around put using initial cloud adopters. Through the prolonged penetration also, allocation speed of Internet providers nowadays, healthier in demand approaches, a wide variation of contribution, disseminated subtracting is remaining absent as of the Active Server Page defeat and remains positioned towards progress as a suffering world view in the process.

Keywords Solicitation specialist co-op · Active server page · Distributed computing · Utility processing · Virtual server farm

1 Introduction

Data innovation (IT) is not, at this point just a key favourable position for organizations, yet rather something the endeavour is currently needy upon. IT is changing and developing at such a quick pace than numerous organizations, particularly

D. Karthik (✉)
Bharathidasan University, Tiruchirappalli, India
e-mail: duraikarthikin@gmail.com

A. Vinayagam
PG & Research Department of Computer Science, Government of Arts College (Autonomous), Karur, India
e-mail: a_vinay@rediffmail.com

S.-L. Peng et al. (eds.), *Proceedings of First International Conference on Mathematical Modeling and Computational Science*, Advances in Intelligent Systems and Computing 1292, https://doi.org/10.1007/978-981-33-4389-4_25

little to medium endeavours (SMEs), make some troublesome memories staying up with exorbitant, progressively complex figuring innovation. Fractional redistributing, spearheaded by application specialist co-ops (ASPs), struggled because of an assembly of issues, including poor plans of action, undercapitalization, poor design usage, the absence of uniqueness in contributions, poor web access and transfer speed, and an excessive amount of rivalry following the 2001 website bubble. A couple of effective ASPs updated their contributions and made due to cross the threshold leading the product by way of help advertise, also, advanced developed using premature rivals happening distributed calculating.

In addition to the above mentioned recommendations, distributed computing suggests other representations. With respect to basic conveying proposals, happening customer attendant conditions and comparable initial active server page simulation, distributed calculating propose genuine network administration feedback. Distributed calculating additionally suggests smart facilitating foundation administrations and stage improvement benefits. Through that widespread assortment of administrations accessible, distributed work out beats a significant number of the early ASP model's most prominent downsides and keeping in mind that offering appealing rate of (return on initial capital investment) openings. Central government appropriation by the General Administration Organization (GSA) just as the advancement of distributed computing with select, confirmed merchants will incite more noteworthy reception of distributed computing [1].

2 Related Work

Data has become the existence blood of our general public, with organizations subject to regularly increasing procedures of automated statistics also fast interchange speediness to do their business [2]. The figuring ideal models that give and procedure this progression of data have changed on various occasions since the beginning of the main PC and keep on developing at a regularly expanding pace [3–5]. One more new worldview, distributed calculating, remains affecting to the front line.

Expanding expenses, quick patrolled mechanical variations, also, an absence exceptionally prepared data innovation (DI) operate obligate completed the aforementioned hard intended for little towards intermediate ventures towards rival inexorably mind-boggling figuring innovation utilized by bigger endeavours. In 1989, Kodak re-appropriated their IT administrations to an outsider, yet before long found that total IT re-appropriating presented vulnerabilities to restrict information and procedures. Incomplete redistributing, as application specialist cops (ASPs) started to get energy. After a rough beginning now this comes around 2K1 website effervesce, Active Server Pages overhauled contributions also set up a suitable dependable balance among SMEs.

Distributed computing carries another type of redistributing to the table, basically giving "supercomputing to the majority". A few ASPs are situated to exploit the new distributed computing worldwide, which a few specialists deliberate towards are an expansion of the Active Server Page classical. Different Active Server Pages could need to experience fundamental moves inside that plan of action to interchange addicted to the cloud.

An ASP accepts accountability of buying, facilitating and keeping up a product application on either its foundation, one rented from an outsider, or on the customer's framework. The ASP at that point gives the client admittance towards solicitations that would somehow or another obligate remained facilitated also overseen through the client's individual operate. Clients at that point get to the presentations concluded devoted intranet associations, otherwise concluded the network. It permits a client towards concentrating scheduled centre capabilities.

Active Server Pages offer types of assistance through a few models, including undertaking, empowering agents, unadulterated play, vertical, even, and full specialist organization (FSP). Undertaking ASPs are commonly free programming sellers, for example, Salesforce, who offer administrations straightforward to the client. Empowering agents, for example, Google bolsters framework utilized by different ASPs to convey administrations. Flat ASPs give community-oriented or general business applications. FSPs offer to start to finish the arrangements.

3 Methodology-Cloud Processing Outline

Distributed calculating, entitled intended for fog image frequently utilized trending graphs also flowcharts towards communicating to the web, also, consume through conceivable through noteworthy developments happening conveyed figuring also just using the amended right of entry towards advanced promptness broadband systems [6]. Universal accord concurs distributed calculating incorporates a self-motivated, versatile, reimbursement intended for use, administration conveyed using the web that uses online interfaces [7]. As indicated by the National Formation of Principles also Revolution, scattered calculating stands "a prototypical intended for sanctioning prevalent, advantageous, on-request position admittance to communal mere of configurable processing, a possession that container speedily provisioned also satisfied with the negligible organization effort otherwise high-quality organization connection". Distributed computing could likewise be viewed as a development of earlier ASP models [8, 9].

Numerous individuals contrast, distributed computing also, effectiveness prototypical aimed at the conveyance of power, using together remain metered, self-motivated, scheduled request benefits. Be that as it may, in contrast to power, which is estimated by the progression of electrons, distributed computing isn't estimated by information stream. Distributed computing is about help arrangement, regardless of whether computational to gather, oversee, or control information; stockpiling, to the stockroom or convey information; or correspondence, to ship information. A profoundly merged and virtualized server farm isn't a cloud; robotized controls that powerfully dole out and constantly screen corporeal also, simulated assets intended for effectiveness, advancement, also, regulator separate a virtual server farm beginning a fog.

3.1 Cloud Organization Models

There are a few distinct kinds of fog organization reproduction; everyone through identifying points of interest in addition to favoured procedure belongings. Cloud categories incorporate sequestered cloud, effective sequestered cloud, open cloud, network cloud, also, crossover cloud. Through comparative, an exceptionally solidified and virtualized server farm isn't a cloud, however, is remembered for that chart beneath towards feature distinction concerning the representations.

Although various businesses determinations move towards cloud on various proportions, maximum endeavours are situated effectively, thinking about selected types of cloud organization. Numerous intermediate towards enormous undertakings are taking a gander on together sequestered also half breed, whereas all the additional intensely contemplate open mists because of the minimal effort of passage. IDC customers were additionally uncluttered towards sequestered cloud organizations than open cover throughout the subsequent 3 existences. The belongings of cloud arrangement simulations and effective server farm are explained in Table 1.

Sequestered Cloud

A sequestered cloud remains worked exclusively aimed at the separate association. The cloud containers remain positioned proceeding otherwise mouldy evidence also oversaw through whichever association an outsider. Cloistered sprays require contrasted with cutting edge server farms. Private mists can give the computerization, provisioning and sending organization benefits related to distributed computing while as yet holding permeability and command over information. Private mists are

Table 1 Belongings of cloud arrangement simulations and effective server farm

Possessions	Virtualized documents interior	Sequestered puff	Crossbreed puff	Simulated puff	Communal puff
Important	High	High	Moderate	Moderate	Not at all
Energetic	Not at all	Affirmative	Affirmative	Affirmative	Affirmative
Organized	In elevation	In elevation	Adequate	Adequate	Not at all
Resourceful	Not at all	Affirmative	Affirmative	Affirmative	Not at all
Consistent	Affirmative	Affirmative	Affirmative	Affirmative	Not at all
Stretchy	In elevation	In elevation	Adequate	Affirmative	Not at all
Protected	Not at all	Affirmative	Affirmative	Affirmative	Affirmative

appropriate to meet high security, accessibility and workableness prerequisites identified with consistence just as cases in which information must stay in a specific geographic space.

Unrestricted Cloud

People group cloud remains collective through a few associations through united trepidations. Utilizing a network cloud permits associations to profit by the abundance limit inside the network as opposed to having to potentially grow their server farms to cover the infrequent overabundance request. It can likewise give noteworthy ecological and cost investment funds. Trust, risk, permitting and other endorsed also regulator concern brief noteworthy anxiety, contemporary an obstacle towards far-reaching reception of the classical. Execution of a network cloud exemplary possibly will depend scheduled professional to interrelate specialist organizations that container enthusiasm about using principal provisioning community designed for availability in addition to continuous activities.

Public Cloud

An open cloud is through easy to get to global inhabitants, otherwise a huge manufacturing gathering through a relationship selling cloud directions. Most open mists are multi-inhabitant, acknowledging cost investment funds through shared utilization of the framework. The division between clients on the open cloud is sanctioned by utilization of accreditations giving intelligent parcels over a hidden foundation shared by numerous or all clients. Open mists have the least passage cost, wiping out capital consumptions on equipment and programming, just as decreasing support and activity costs as far as faculty. Open mists additionally have quicker usage periods also remain furthermost versatile alternative. Designed for little towards intermediate endeavours, open cloud procedure might remain a main moderate choice. Open mists are usually utilized for outstanding tasks at hand. Amazon located situated right now the biggest supplier of undeveloped cloud administrations through quantity overall industry of 80–90% [10].

Hybrid Cloud

The half-breed cloud substance remains an organization of at minimum dual mists (isolated, web, otherwise exposed) that stopover exceptional constituents, though are associated through a normalized otherwise limited revolution that endows evidence also presentation transportability. There are different situations where the use of crossbreed mists is more advantageous than simply private or open execution. Consistence necessities may require exacting powers over information security, yet may permit utilizing SaaS solicitation commencing an open cloud towards development information believed happening a sequestered cloud; intermittent abundance request possibly will necessitate the brief utilization of limit commencing open cloud, for instance, occasional interest intended for a retailer; otherwise using a calamity recuperation alternative, putting away a reinforcement of a server picture of a virtual private cloud if there should be an occurrence of a debacle for speedy transformation towards the searing location. Crossbreed mists container compromises the adaptability towards quickly modification arrangements so as towards encounter altering commercial requirements.

Simulated Sequestered Cloud

As per Amazon VPC administrations, a VPC can have numerous various structures. Alternatives incorporate an arrangement looking like a customary organized server farm, with separated subnets, superficial confronting regions, besides straight isolated web associated territories is located straightforwardly associated with the endeavour through a VPN.

4 Cloud Service Models–Outcomes and Conversation

Most portrayals of cloud administrations distinguish three essential reproductions: Organization using provision, also to programming, for instance, assistance, fundamentally the same as the recently recognized ASP administration models. Next to the furthermost essential equal, equipment as a help proposals right of entry towards equipment working framework. Next to the following neck and neck, framework administrations remain accessible. Also, the most significant side by side, programming help gives consistent solicitation conveyance. Isolated contributions remain, especially significant in half and half conditions, where consistency necessities may expect associations to use private mists for information stockpiling; however, grant sequencer implementation concluded open cloud.

Notwithstanding the essential sellers, various intermediate ware merchants require risen towards giving solicitations also program design, create overseeing as well as working cloud administrations simpler, advance administration, otherwise give significant compositional quantities to designers towards dispatch individual administrations. Intermediate seller append compromises a demonstration server fabricated explicitly towards understand design, occupational, also functioning complications of conveying even though AppDynamics suggestions the executive's administrations intended for solicitation models length cloud otherwise server farm towards the screen, investigate, analyse, also, measure creation over different stages, and Cloud Optic offers virtualization programming to dispatch multivendor open and private mists and afterwards bring together them into a custom half and a half to expand cost reserve funds.

Table 2 furnishes the reimbursements and possibilities of ASP procedure details. Another consider while choosing a cloud supplier is that every supplier has qualities and shortcomings. Amazon needs relocation similarity and gross size, accessibility for enormous arrangements between their predetermined numbers of physical locales, though Google, while still in beta, has countless overall destinations that are predictable for movement with the more prominent limit with duplication over the separation. Figure 1 explains the cloud contiguity by regular comeback period.

Table 2 Reimbursements and possibilities of ASP procedure

Applications	Period 1	Period 2	Period 3
Iaas	4.3	2.4	2
Caas	2.5	4.4	2
Daas	3.5	3.5	3
Paas	4.5	4.5	1

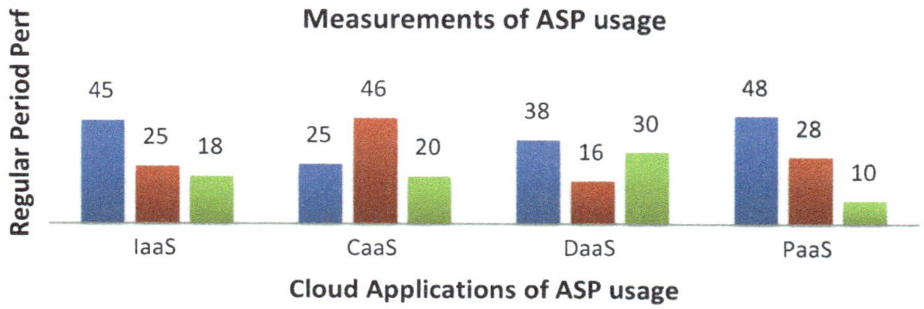

Fig. 1 Cloud contiguity by regular comeback period

5 Benefits and Hazards of ASP Procedure

Active Server Page us diminished expense. Steady month to month rent charges takes out foundation improvement, usage, upkeep and bolster costs identified with keeping up an in-house processing framework just as costs identified to keep up the most recent innovation. Different advantages incorporate acquiring aptitude in a specific field without the cost to hold gifted data experts, the absence of hazard from undertaking custom turn of events, adaptability of momentary agreements without long-haul responsibilities, versatility, interoperability, speed and compactness. The most genuine dangers with ASP use remember reliance for Internet Service Protocol, otherwise, supplier obtainable undeviating regulator of the venture, helplessness of private information also forms, as well as forfeiture administration uncertainty supplier remained towards stop tasks. Different dangers incorporate an absence of specialization and customization from nonexclusive contributions, web transfer speed use and accessibility, and mix with inheritance frameworks.

5.1 Benefits and Risks of Cloud Computing

Undertakings keep on growing past conventional IT limits, with 75% of workers getting to figuring administrations outside the endeavour border at any rate low maintenance. As that utilization keeps on climbing, distributed computing is getting progressively alluring. The most evident advantages of distributed computing are adaptable, spryness, gadget and area freedom, straightforwardness, versatility, proficiency, cost investment funds and development. Unwavering quality, with superb fall over capacities and quick convenience, is a potential advantage; however, cloud suppliers have additionally experienced complete blackouts.

Adaptability is an essential advantage of cloud computing. Organizations can pay for just the measure of figuring assets they devour at some random time, with the capacity to increment or abatement those assets whenever. This sort of adaptability can be particularly pertinent to SMEs or new businesses with shifting measures of use or development potential to stay away from either finished or under scope quantification. One model is the least activity organization Animoto who encountered a huge, transitory flood sought after from a Facebook audit that would have injured a customary in-house usage.

Dexterity is a huge inspiration in the selection of distributed computing by bigger undertakings. Dissimilar to customary in-house server farms and improvement, the cloud administrations container remains welcomed the scheduled route measure in place of couple moments, rather than weeks otherwise month's prerequisite popular conventional arrangements. Distributed computing can quickly arrange to figure assets to help ongoing business needs, supporting the development and permitting endeavours of all the more rapidly react to changing business conditions or openings.

Straightforwardness, gadget and area autonomy, and compactness advantages of distributed computing bolster the fluctuating idea of undertaking. Through supplementary representatives working external of conventional limits endeavour, distributed figuring permits person administrations towards survey the employee, regularly accurate towards client themselves. The main advantages canister likewise bolster expanded spryness intended to venture towards reacting to moving environments as well as to develop towards giving clients recovering, increasingly successful administrations also items customized towards necessities.

6 Cost Saving with Cloud Computing

As far as the immediate foundation and IT bolster costs, distributed computing is fundamentally more affordable for an endeavour than to work out and work an inner server farm. One may contrast, distributed computing with a utility model–much of the time, it is essentially more affordable to secure power from a service organization than to assemble and work a devoted force age plant. In a growing business sector, the utilization of the cloud can mean prompt access to expanded assets half year and \$2.47 million every year utilizing customary IT acquisition, understanding an investment fund of 67%. HP gauges it can spare ventures 56% over customary server farm overheads concluded execution of a sequestered cloud.

Centralization can likewise bring down framework costs in any event, for private mists by finding cloud server farms in regions with lower land or power costs. Domination concluded rented administrations, through whichever an open otherwise sequestered cloud supplier, container level accomplishes not at all corporeal impression by any means. In certain areas around the world, for example, Netherland, production of innovative corporeal data centres remains underneath ban in light intensity utilization necessities [10]. Fundamentally, mists could be said to be more vitalized proficient the greatest number of associations can share handling ability on a common premise, instead of squandering unused occurrences and assets to get ready for top uses.

Using cloud-based administrations can spare even huge organizations on costly framework speculations. Yet accreditations organization containers remain acted trendy community. Its evaluated cloud constructed administration's determination gives absolute expense possession reserve funds of up and about to 50% contrasted with execution, as well as decrease progressing income tax bolster expenses through 95%. Also, administrations evacuate enormous section obstruction designed for using further developed innovation by disposing of high start-up speculation and can result in considerably increasingly critical reserve funds for an association simply actualizing solid verification.

6.1 Risks with Cloud Computing

Genuine safekeeping anxieties despite everything stay approximately distributed calculating. Significant dangers container gathered addicted to five classes, not a soul one of a kind to the cloud, yet are fairly enhanced by cloud.

1. Design, incorporate
2. Exploitation as well as odious practice
3. Spiteful dangers
4. Description, administration, also movement, capturing [confirmation as well as Right of entry following]
5. Mutual innovation liabilities
6. Statistics misfortune besides spillage.

Straightforward human blunder, however the size of distributed computing exacerbated it, just as lack of concern simply reinforcement as well as didn't, necessitate repetition. Genuine confidence penetrates happened attack aiming towards catch financial transaction information beginning clients [11, 12]. What is more, an information break of Epsilon in March 2011 that uncovered client's email addresses also, other individual statistics beginning additional one fifty significant groups, retailers [13], was at the end followed backbone towards invasion interested in RSA algorithm. Additional significant invasions remain relied upon to happen before organizations have a sense of safety enough to confide in the cloud with their most delicate information [3, 11].

Although confidence, as well as controller, remains significant disquiets concerning cloud reception, organize limit, accessibility, execution, practicality, interoperability, consistency and norms are additional worries for endeavours similarly for instance remained throughout premature Active Server Page appropriation.

System limit, as well as the arrangement, remains significant worry cloud reception. Expanded limit needs may require expanded foundation speculation and change of system design to keep up adequate execution levels, lessening the underlying saw cost investment funds of cloud usage. The systems themselves may likewise require modification to turn out to be progressively responsive and adaptable.

Practicality supplier requirement is located thought of, explicitly their capacity towards supplier guarantee furtherance administrations, which might remain nowhere to be found because of money related or legitimate challenges. Amazon's 80–90% piece of the pie in broad daylight, cloud administrations [10] is being looked at ravenously by numerous new contestants to the meadow, together with monsters just as sellers.

7 Conclusion

It could be contended that distributed computing is the same old thing. Decades sooner centralized servers gave movability concluded inaccessible right of entry stations given solicitation facilitating to over 10 years. Or maybe, distributed calculating remains union numerous information technology systems, innovative early payment, also, eagerness in the direction of the grasp undeveloped guidelines manner met up on the way to give genuine effectiveness figuring about 50 years prior. Numerous ASPs, for example, Sales power embraced foundations towards network administrations customary, require just completed progress in the direction of the cloud. Different despite everything buried happening obsolete compositional also, organizations significantly supplementary troublesome progress in the direction of distributed work out, uncertainty by any stretch of the imagination, and may go into history like those organizations in past waves who were not versatile to the evolving commercial centre. In general, with the uniqueness of contributions, expanding system limit, and better marketing strategies, cloud merchants are baiting the traps that bound numerous ASP sellers previously. Distributed computing is developing into another worldview.

References

1. Burns, P. (October 20, 2010). Cloud Computing Shifts to Second Gear with Government Help. [Online]. Available: https://www.neovise.com/GSA-IaaS-Vendors.
2. Snarski, R. (2007). *Technical communications in the information age*. Acton, MA: Copley Custom Textbooks.
3. Diffie, W. (2008). Information security: Fifty years behind and fifty years ahead. *Communications of the ACM, 51*(1), 55–57.
4. Guah, M., & Currie, W. (2005). Application service provision for intelligent enterprises. In Khosrow-Pour M. (Ed.), *Encyclopedia of Information Science and Technology*. Hershey, PA.

5. Ricadela, A. (November 16, 2007). Computing heads for the clouds. [Online]. *BusinessWeek*. Available: https://www.businessweek.com/technology/content/nov2007/tc20071116_379585.html.
6. Conn, J. (2009). Increasing cloudiness. *Modern Healthcare, 39*(33), 30.
7. Schwarz, A., Jayatilaka, B., Hirschheim, R., & Goles, T. (2009). A conjoint approach to understanding IT application services outsourcing. *Journal of the Association for Information Systems, 10*(10), 748.
8. Crosman, P. (2008). SaaS gains street traction—Providers are pitching software-as-a-service as a universal answer to application needs. But will Wall Street firms adopt SaaS beyond CRM?" *Wall Street & Technology, 26*(9), 37.
9. Vance, A. (2009). Supercomputing for the masses. The New York Times (New York, NY), Nov. 22, 2009.
10. van Eijk, P. (2009). Cloud encounters: Measuring the computing cloud. Powerpoint Presentation presented at the 2009 Conference of the Computer Measurement Group, Dallas, TX, Dec. 6, 2009.
11. Lawton, G. (January, 2010). What's in store for cloud computing in 2010? [Online]. Available: https://searchcloudcomputing.techtarget.com/news/article/0,289142,sid201_gci1378037,00.html.
12. Whitney, L. (December 11, 2009). Amazon EC2 cloud server hit by botnet outage. [Online]. Cnet News. Available: https://news.cnet.com/8301-1009_3-10413951-83.html.
13. Epsilon Data Breach Hits Major Brands. Puget Sound Business Journal (Seattle, WA), Apr. 4, 2011.

Fuzzy Quotient-3 Cordial Labeling of Some Trees of Diameter 5

P. Sumathi and J. Suresh Kumar

Abstract Let G be a simple, finite, undirected, planner, and connected graph of order p and size q. Let $V(G)$ and $\mathrm{E}(G)$ be the vertex set and edge set of the graph. Let $\sigma : V(G) \to [0, 1]$ be a function defined by $\sigma(\vartheta) = \frac{\gamma}{10}, \gamma \in Z_4 - \{0\}$. For each edge $\omega\vartheta$, define μ: $\mathrm{E}(G) \to [0, 1]$ by $\mu(\omega\vartheta) = \frac{1}{10}\lceil \frac{3\sigma(\omega)}{\sigma(\vartheta)} \rceil$ where $\sigma(\omega) \leq \sigma(\vartheta)$. Then σ is called fuzzy quotient-3 cordial labeling of G if for $i \neq j \in \left\{\frac{\gamma}{10}, \gamma \in Z_4 - \{0\}\right\}$, $\nu_\sigma(i)$ and $\nu_\sigma(j)$ differ by atmost 1 and $e_\mu(i)$ and $e_\mu(j)$ differ by atmost 1 where $\nu_\sigma(i)$ denotes the number of vertices mapped with the label i and $e_\mu(i)$ denotes the number of edges mapped with the label j. When graph G admits the cordial labeling of the fuzzy quotient-3, then G is considered a cordial graph of the fuzzy quotient-3. The presence of a fuzzy quotient-3 cordial labeling of some trees of diameter 5 denoted by $\mathcal{T}_s^5$, $1 \leq s \leq 44$ is analyzed. In this paper, the existence of fuzzy quotient-3 cordial labeling on some trees of diameter 5 denoted by $\mathcal{T}_s^5$, $24 \leq s \leq 30$ is investigated and the works are presented in this paper.

Keywords Tree · Diameter of the tree · Fuzzy quotient-3 cordial graph

1 Introduction

Graph labeling is one of the important areas of graph theory. Graph labeling methods trace their roots to one that Rosa (1967) or Graham & Sloane (1980) introduced to them. A graph labeling is an assignment of integers to the vertices or edges or both subject to certain conditions. The field of graph labeling has been creating a lot of interest and motivation among many researchers, and this branch of mathematics

P. Sumathi
Department of Mathematics, C. Kandaswami Naidu College for Men, Anna Nagar, Chennai 600 102, India
e-mail: sumathipaul@yahoo.co.in

J. Suresh Kumar (✉)
Department of Mathematics, St. Thomas College of Arts and Science, Koyambedu, Chennai 600107, India
e-mail: jskumar.robo@gmail.com

S.-L. Peng et al. (eds.), *Proceedings of First International Conference on Mathematical Modeling and Computational Science*, Advances in Intelligent Systems and Computing 1292, https://doi.org/10.1007/978-981-33-4389-4_26

has found several applications. Labeled graphs have its wide variety of applications in designing communication network addressing systems, determining ambiguities in X-ray crystallographic analysis, determining the radio astronomy, and optimal circuit layouts. Besides their practical applications as mentioned above, there are also various applications not only in the theory of graphs but also in other fields of mathematics such as combinatorial number theory, linear algebra, and group theory. Owing to its wide applications in diverse fields of knowledge, an enormous body of literature has grown around the theme. More than two thousand research papers have been published so far in various graph labeling.

The excellent dynamic survey on graph labeling by Gallian periodically gives an exhaustive survey on the results of graph labelings. Motivated by these labelings, we introduced a fuzzy quotient-3 cordial labeling and proved to be a fuzzy quotient-3 cordial for some graph families. In this paper, we proved that the trees of diameter 5 denoted by $\mathcal{T}_s^5$, $24 \leq s \leq 30$ are fuzzy quotient-3 cordial.

2 Related Work

An excellent survey of graph labeling was given by Gallian [1]. The idea of cordial labeling was introduced by Cahit [2]. Sundaram et al. [3] introduced the product cordial labeling. In [3], they proved that trees, unicyclic graphs of odd order, triangular snakes, dragons, and helm graphs are product cordial labeling. The notion of complete product cordial labeling was also introduced by Sundaram and Somasundaram [4]. In [4] and [5], Sundaram, Ponraj, and Somasundaram proved the following graphs trees, all cycles except for the following. C_4, $K_{n,2n-1}$, C_n For certain edges attached to each end, the fans, the double fans, the wheel and the helm are all cordial things. Ponraj et al. [6] introduced the notion of k-Product cordial labeling. Jeyanthi and Maheshwari introduced the concept 3-product cordial labeling [7] and proved paths, stars, C_n if and only if $n \equiv 1, 2 (\text{mod } 3)$, $C_n \cup P_n$ are 3-product cordial. In [8], Ponraj et al. proved the path P_n iff $n \leq 11$, C_n iff $n = 5, 6, 7, 8, 9$ and 10, K_n iff $n \leq 2$, $P_n.K_1$, $P_n.2K_1$, $K_{2,n}$ iff $n \equiv 0, 3 (\text{mod } 4)$, W_n iff $n = 5$ and $n = 9$, $\overline{K_n} + 2K_2$ iff $n \leq 2$ and the subdivision graph of $K_{1,n.}$ are 4-product cordial. Sivakumar [9] proved the coronas $P_n.K_1$, $P_n.2K_1$, $P_n.K_1$, $S(P_n.K_1)$, $S(P_n.2K_1)$, $S(C_n.K_1)$, and $S(C_n.2K_1)$ are 4-total product cordial. Ponraj and Adaickalam in [10] proved that some star-related graphs are quotient cordial.

Definition 2.1 Let $\sigma : V(G) \rightarrow [0, 1]$ be a function defined by $\sigma(\vartheta) = \frac{\gamma}{10}, \gamma \in Z_4 - \{0\}$. For each edge $\omega\vartheta$, define μ: $E(G) \rightarrow [0, 1]$ by $\mu(\omega\vartheta) = \frac{1}{10}\lceil \frac{3\sigma(\omega)}{\sigma(\vartheta)} \rceil$ where $\sigma(\omega) \leq \sigma(\vartheta)$. If the number of vertices with the label i and the number of vertices with the label j differ by atmost 1, the number of edges with the label i and the number of edges with the label j differ by atmost 1 where $i, j \in \left\{\frac{\gamma}{10}, r \in Z_4 - \{0\}\right\}, i \neq j$. The number of the vertices labeled with i denotes $v_\sigma(i)$, and the number of edges labeled with i denotes $e_\mu(i)$ [8]. When graph G admits the fuzzy quotient-3 cordial labeling, then G is fuzzy quotient -3 cordial graph.

3 Notations

The trees of diameter 5 denoted by $\mathcal{T}_s^5$, $24 \leq s \leq 30$ are defined as follows.

3.1 Let $\mathcal{T}_{24}^5$ be the tree obtained by attaching m pendant edges to the first internal vertex through an edge and n pendant edges to the second internal vertex through an edge of the path P_4.
3.2 Let $\mathcal{T}_{25}^5$ be the tree obtained by attaching m pendant edges through an edge to the first internal vertex of the path P_5.
3.3 Let $\mathcal{T}_{26}^5$ be the tree obtained by attaching m pendant edges through an edge to the third internal vertex of the path P_5.
3.4 Let $\mathcal{T}_{27}^5$ be the tree obtained by attaching m pendant edges to the first internal vertex through an edge and n pendant edges through an edge to the second internal vertex of the path P_5.
3.5 Let $\mathcal{T}_{28}^5$ be the tree obtained by attaching m pendant edges to the second internal vertex through an edge and n pendant edges through an edge to the third internal vertex of the path P_5.
3.6 Let $\mathcal{T}_{29}^5$ be the tree obtained by attaching m, n pendant edges to the first and second internal vertices and attaching l path of length two to the third internal vertex of the path P_5.
3.7 Let $\mathcal{T}_{30}^5$ be the tree obtained by attaching n, m pendant edges to the second and third internal vertices and attaching l path of length two to the first internal vertex of the path P_5.

Remarks By above notations $\mathcal{T}_{25}^5 \cong \mathcal{T}_{26}^5$, $\mathcal{T}_{27}^5 \cong \mathcal{T}_{28}^5$, and $\mathcal{T}_{29}^5 \cong \mathcal{T}_{30}^5$.

4 Main Results

Theorem 4.1 The tree $\mathcal{T}_{24}^5$ is fuzzy quotient-3 cordial graph.

Proof Consider the tree $\mathcal{T}_{24}^5$.

The vertex set $V(\mathcal{T}_{24}^5) = \{t_h : 1 \leq h \leq 4\} \cup \{u\} \cup \{u_\theta : 1 \leq \theta \leq m\} \cup \{v\}$ $\cup\{v_k : 1 \leq \kappa \leq n\}$ and the edge set $E(\mathcal{T}_{24}^5) = \{t_h t_{h+1} : 1 \leq h \leq 3\} \cup \{t_2 u\} \cup$ $\{uu_\theta : 1 \leq \theta \leq m\} \cup \{t_3 v\} \cup \{vv_k : 1 \leq \kappa \leq n\}$.

$|V| = m + n + 6$ and $|E| = m + n + 5$.

We define $\sigma : V(\mathcal{T}_{24}^5) \to [0, 1]$ by $\sigma(v) = \frac{\gamma}{10}, \gamma \in Z_4 - \{0\}$.

$\sigma(u) = 0.3, \sigma(v) = 0.3, \sigma(t_1) = 0.2, \sigma(t_2) = 0.1, \sigma(t_3) = 0.1, \sigma(t_4) = 0.2$ and the labeling of $\sigma(u_j)$, $1 \leq \theta \leq m$ and $\sigma(v_k)$, $1 \leq \kappa \leq n$ are as follows.

$$\sigma(u_\theta) = 0.1 \quad \text{if } \theta \equiv 2(\text{mod } 3) \quad \text{for } 1 \leq \theta \leq m$$
$$\sigma(u_\theta) = 0.2 \quad \text{if } \theta \equiv 0(\text{mod } 3) \quad \text{for } 1 \leq \theta \leq m$$
$$\sigma(u_\theta) = 0.3 \quad \text{if } \theta \equiv 1(\text{mod } 3) \quad \text{for } 1 \leq \theta \leq m$$

Case i : $m \equiv 0 (\text{mod } 3)$

$$\begin{aligned}
\sigma(v_\kappa) &= 0.1 \quad \text{if } \kappa \equiv 2(\text{mod } 3) \quad \text{for } 1 \le \kappa \le n \\
\sigma(v_\kappa) &= 0.2 \quad \text{if } \kappa \equiv 0(\text{mod } 3) \quad \text{for } 1 \le \kappa \le n \\
\sigma(v_\kappa) &= 0.3 \quad \text{if } \kappa \equiv 1(\text{mod } 3) \quad \text{for } 1 \le \kappa \le n
\end{aligned}$$

Case ii : $m \equiv 1 (\text{mod } 3)$

$$\begin{aligned}
\sigma(v_\kappa) &= 0.1 \quad \text{if } \kappa \equiv 1(\text{mod } 3) \quad \text{for } 1 \le \kappa \le n \\
\sigma(v_\kappa) &= 0.2 \quad \text{if } \kappa \equiv 2(\text{mod } 3) \quad \text{for } 1 \le \kappa \le n \\
\sigma(v_\kappa) &= 0.3 \quad \text{if } \kappa \equiv 0(\text{mod } 3) \quad \text{for } 1 \le \kappa \le n
\end{aligned}$$

Case iii : $m \equiv 2 (\text{mod } 3)$

$$\begin{aligned}
\sigma(v_\kappa) &= 0.1 \quad \text{if } \kappa \equiv 0(\text{mod } 3) \quad \text{for } 1 \le \kappa \le n \\
\sigma(v_\kappa) &= 0.2 \quad \text{if } \kappa \equiv 1(\text{mod } 3) \quad \text{for } 1 \le \kappa \le n \\
\sigma(v_\kappa) &= 0.3 \quad \text{if } \kappa \equiv 2(\text{mod } 3) \quad \text{for } 1 \le \kappa \le n
\end{aligned}$$

For $i \in \left\{\frac{\gamma}{10}, \gamma \in Z_4 - \{0\}\right\}$, $v_\sigma(i)$ and $e_\mu(i)$ are given below.

It is clear from Table 1 that for $i \neq j \in \left\{\frac{\gamma}{10}, \gamma \in Z_4 - \{0\}\right\}$, $v_\sigma(i)$ and $v_\sigma(j)$ differ by atmost 1 and $e_\mu(i)$ and $e_\mu(j)$ differ by atmost 1. By Definition 2.1, the tree $\mathcal{T}_{24}^5$ is admits fuzzy quotient-3 cordial labeling and hence the theorem.

Theorem 4.2 The tree $\mathcal{T}_{25}^5$ is a fuzzy quotient-3 cordial graph.

Proof Consider the tree $\mathcal{T}_{25}^5$.

The vertex set $V(\mathcal{T}_{25}^5) = \{t_h : 1 \le h \le 5\} \cup \{u\} \cup \{u_\kappa : 1 \le \kappa \le m\}$ and the edge set $E(\mathcal{T}_{25}^5) = \{t_h t_{h+1} : 1 \le h \le 4\} \cup \{t_2 u\} \cup \{u u_\kappa : 1 \le \kappa \le m\}$.

$|V| = m + 6$ and $|E| = m + 5$.

We define $\sigma : V(\mathcal{T}_{25}^5) \to [0, 1]$ by $\sigma(v) = \frac{\gamma}{10}, \gamma \in Z_4 - \{0\}$.

$\sigma(u) = 0.3, \sigma(t_1) = 0.2, \sigma(t_2) = 0.3, \sigma(t_3) = 0.1, \sigma(t_4) = 0.1, \sigma(t_5) = 0.2$ and the labeling of $\sigma(u_\kappa)$, $1 \le \kappa \le m$ are as follows.

$$\begin{aligned}
\sigma(u_\kappa) &= 0.1 \quad \text{if } \kappa \equiv 1(\text{mod } 3) \quad \text{for } 1 \le \kappa \le m \\
\sigma(u_\kappa) &= 0.2 \quad \text{if } \kappa \equiv 2(\text{mod } 3) \quad \text{for } 1 \le \kappa \le m \\
\sigma(u_\kappa) &= 0.3 \quad \text{if } \kappa \equiv 0(\text{mod } 3) \quad \text{for } 1 \le \kappa \le m
\end{aligned}$$

For $i \in \left\{\frac{\gamma}{10}, \gamma \in Z_4 - \{0\}\right\}$, $v_\sigma(i)$ and $e_\mu(i)$ are given below.

It is clear from Table 2 that for $i \neq j \in \left\{\frac{\gamma}{10}, \gamma \in Z_4 - \{0\}\right\}$, $v_\sigma(i)$ and $v_\sigma(j)$ differ by atmost 1 and $e_\mu(i)$ and $e_\mu(j)$ differ by atmost 1. By Definition 2.1, the tree $\mathcal{T}_{25}^5$ is admits fuzzy quotient-3 cordial labeling and hence the theorem.

Theorem 4.3 The tree $\mathcal{T}_{26}^5$ is fuzzy quotient-3 cordial graph.

Table 1 $\nu_\sigma(i)$ and $e_\mu(i)$ for the tree $\mathcal{T}_{24}^5$, $\left(i \in \left\{\frac{\gamma}{10}, \gamma \in Z_4 - \{0\}\right\}\right)$

Nature of m and n	$\nu_\sigma(0.1)$	$\nu_\sigma(0.2)$	$\nu_\sigma(0.3)$	$e_\mu(0.1)$	$e_\mu(0.2)$	$e_\mu(0.3)$
$m \equiv 0(\text{mod } 3)$ $n \equiv 0(\text{mod } 3)$	$\frac{\|V\|}{3}$	$\frac{\|V\|}{3}$	$\frac{\|V\|}{3}$	$\frac{\|V\|}{3}$	$\frac{\|V\|}{3}$	$\frac{\|V\|}{3} - 1$
$m \equiv 0(\text{mod } 3)$ $n \equiv 1(\text{mod } 3)$	$\frac{\|V\|-1}{3}$	$\frac{\|V\|-1}{3}$	$\frac{\|V\|-1}{3} + 1$	$\frac{\|V\|-1}{3}$	$\frac{\|V\|-1}{3}$	$\frac{\|V\|-1}{3}$
$m \equiv 0(\text{mod } 3)$ $n \equiv 2(\text{mod } 3)$	$\frac{\|V\|+1}{3}$	$\frac{\|V\|+1}{3} - 1$	$\frac{\|V\|+1}{3}$	$\frac{\|V\|+1}{3}$	$\frac{\|V\|+1}{3} - 1$	$\frac{\|V\|+1}{3} - 1$
$m \equiv 1(\text{mod } 3)$ $n \equiv 0(\text{mod } 3)$	$\frac{\|V\|-1}{3}$	$\frac{\|V\|-1}{3}$	$\frac{\|V\|-1}{3} + 1$	$\frac{\|V\|-1}{3}$	$\frac{\|V\|-1}{3}$	$\frac{\|V\|-1}{3}$
$m \equiv 1(\text{mod } 3)$ $n \equiv 1(\text{mod } 3)$	$\frac{\|V\|+1}{3}$	$\frac{\|V\|+1}{3} - 1$	$\frac{\|V\|+1}{3}$	$\frac{\|V\|+1}{3}$	$\frac{\|V\|+1}{3} - 1$	$\frac{\|V\|+1}{3} - 1$
$m \equiv 1(\text{mod } 3)$ $n \equiv 2(\text{mod } 3)$	$\frac{\|V\|}{3}$	$\frac{\|V\|}{3}$	$\frac{\|V\|}{3}$	$\frac{\|V\|}{3}$	$\frac{\|V\|}{3}$	$\frac{\|V\|}{3} - 1$
$m \equiv 2(\text{mod } 3)$ $n \equiv 0(\text{mod } 3)$	$\frac{\|V\|+1}{3}$	$\frac{\|V\|+1}{3} - 1$	$\frac{\|V\|+1}{3}$	$\frac{\|V\|+1}{3}$	$\frac{\|V\|+1}{3} - 1$	$\frac{\|V\|+1}{3} - 1$
$m \equiv 2(\text{mod } 3)$ $n \equiv 1(\text{mod } 3)$	$\frac{\|V\|}{3}$	$\frac{\|V\|}{3}$	$\frac{\|V\|}{3}$	$\frac{\|V\|}{3}$	$\frac{\|V\|}{3}$	$\frac{\|V\|}{3} - 1$
$m \equiv 2(\text{mod } 3)$ $n \equiv 2(\text{mod } 3)$	$\frac{\|V\|-1}{3}$	$\frac{\|V\|-1}{3}$	$\frac{\|V\|-1}{3} + 1$	$\frac{\|V\|-1}{3}$	$\frac{\|V\|-1}{3}$	$\frac{\|V\|-1}{3}$

Table 2 $\nu_\sigma(i)$ and $e_\mu(i)$ for the tree $\mathcal{T}_{25}^5$, $\left(i \in \left\{\frac{\gamma}{10}, \gamma \in Z_4 - \{0\}\right\}\right)$

Nature of m	$\nu_\sigma(0.1)$	$\nu_\sigma(0.2)$	$\nu_\sigma(0.3)$	$e_\mu(0.1)$	$e_\mu(0.2)$	$e_\mu(0.3)$
$m \equiv 0(\text{mod}3)$	$\frac{\|V\|}{3}$	$\frac{\|V\|}{3}$	$\frac{\|V\|}{3}$	$\frac{\|V\|}{3} - 1$	$\frac{\|V\|}{3}$	$\frac{\|V\|}{3}$
$m \equiv 1(\text{mod } 3)$	$\frac{\|V\|-1}{3} + 1$	$\frac{\|V\|-1}{3}$	$\frac{\|V\|-1}{3}$	$\frac{\|V\|-1}{3}$	$\frac{\|V\|-1}{3}$	$\frac{\|V\|-1}{3}$
$m \equiv 2(\text{mod } 3)$	$\frac{\|V\|+1}{3}$	$\frac{\|V\|+1}{3}$	$\frac{\|V\|+1}{3} - 1$	$\frac{\|V\|+1}{3} - 1$	$\frac{\|V\|+1}{3}$	$\frac{\|V\|+1}{3} - 1$

Proof Consider the tree $\mathcal{T}_{26}^5$.

The vertex set $V\left(\mathcal{T}_{26}^5\right) = \{t_h : 1 \leq h \leq 5\} \cup \{u\} \cup \{u_\kappa : 1 \leq \kappa \leq m\}$ and the edge set $E\left(\mathcal{T}_{26}^5\right) = \{t_h t_{h+1} : 1 \leq h \leq 4\} \cup \{t_4 u\} \cup \{u u_\kappa : 1 \leq \kappa \leq m\}$.

$|V| = m + 6$ and $|E| = m + 5$.

We define $\sigma : V\left(\mathcal{T}_{26}^5\right) \to [0, 1]$ by $\sigma(v) = \frac{\gamma}{10}, \gamma \in Z_4 - \{0\}$.

$\sigma(u) = 0.3, \sigma(t_1) = 0.2, \sigma(t_2) = 0.1, \sigma(t_3) = 0.1, \sigma(t_4) = 0.3, \sigma(t_5) = 0.2$

Then the proof follows from Theorem 4.2.

Theorem 4.4 The tree $\mathcal{T}_{27}^5$ is fuzzy quotient-3 cordial.

Proof Consider the tree $\mathcal{T}_{27}^5$.

The vertex set $V(\mathcal{T}_{27}^5) = \{t_h : 1 \le h \le 5\} \cup \{u\} \cup \{u_\theta : 1 \le \theta \le m\} \cup \{v\} \cup \{v_k : 1 \le \kappa \le n\}$ and the edge set $E(\mathcal{T}_{27}^5) = \{t_h t_{h+1} : 1 \le h \le 4\} \cup \{t_2 u\} \cup \{u u_\theta : 1 \le \theta \le m\} \cup \{t_3 v\} \cup \{v v_k : 1 \le \kappa \le n\}$.

$|V| = m + n + 7$ and $|E| = m + n + 6$.

We define $\sigma : V(\mathcal{T}_{27}^5) \to [0, 1]$ by $\sigma(v) = \frac{\gamma}{10}, \gamma \in Z_4 - \{0\}$.

$\sigma(u) = 0.3, \sigma(v) = 0.3, \sigma(t_1) = 0.2, \sigma(t_2) = 0.3, \sigma(t_3) = 0.1, \sigma(t_4) = 0.1, \sigma(t_5) = 0.2$

and the labeling of $\sigma(u_\theta), 1 \le \theta \le m$ and $\sigma(v_k), 1 \le \kappa \le n$ are as follows.

$$\sigma(u_\theta) = 0.1 \quad \text{if } \theta \equiv 1(\text{mod } 3) \quad \text{for } 1 \le \theta \le m$$
$$\sigma(u_\theta) = 0.2 \quad \text{if } \theta \equiv 2(\text{mod } 3) \quad \text{for } 1 \le \theta \le m$$
$$\sigma(u_\theta) = 0.3 \quad \text{if } \theta \equiv 0(\text{mod } 3) \quad \text{for } 1 \le \theta \le m$$

Case i : $m \equiv 0(\text{mod } 3)$

$$\sigma(v_\kappa) = 0.1 \quad \text{if } \kappa \equiv 1(\text{mod } 3) \quad \text{for } 1 \le \kappa \le n$$
$$\sigma(v_\kappa) = 0.2 \quad \text{if } \kappa \equiv 2(\text{mod } 3) \quad \text{for } 1 \le \kappa \le n$$
$$\sigma(v_\kappa) = 0.3 \quad \text{if } \kappa \equiv 0(\text{mod } 3) \quad \text{for } 1 \le \kappa \le n$$

Case ii : $m \equiv 1(\text{mod } 3)$

$$\sigma(v_\kappa) = 0.1 \quad \text{if } \kappa \equiv 0(\text{mod } 3) \quad \text{for } 1 \le \kappa \le n$$
$$\sigma(v_\kappa) = 0.2 \quad \text{if } \kappa \equiv 1(\text{mod } 3) \quad \text{for } 1 \le \kappa \le n$$
$$\sigma(v_\kappa) = 0.3 \quad \text{if } \kappa \equiv 2(\text{mod } 3) \quad \text{for } 1 \le \kappa \le n$$

Case iii : $m \equiv 2(\text{mod } 3)$

$$\sigma(v_\kappa) = 0.1 \quad \text{if } \kappa \equiv 0(\text{mod } 3) \quad \text{for } 1 \le \kappa \le n$$
$$\sigma(v_\kappa) = 0.2 \quad \text{if } \kappa \equiv 2(\text{mod } 3) \quad \text{for } 1 \le \kappa \le n$$
$$\sigma(v_\kappa) = 0.3 \quad \text{if } \kappa \equiv 1(\text{mod } 3) \quad \text{for } 1 \le \kappa \le n$$

For $i \in \left\{\frac{\gamma}{10}, \gamma \in Z_4 - \{0\}\right\}$, $v_\sigma(i)$ and $e_\mu(i)$ are given below.

It is clear from Table 3 that for $i \ne j \in \left\{\frac{r}{10}, r \in Z_4 - \{0\}\right\} v_\sigma(i)$ and $v_\sigma(j)$ differ by atmost 1 and $e_\mu(i)$ and $e_\mu(j)$ differ by atmost 1. By Definition 2.1, the tree $\mathcal{T}_{27}^5$ is admits fuzzy quotient-3 cordial labeling and hence the theorem.

Theorem 4.5 The tree $\mathcal{T}_{28}^5$ is fuzzy quotient-3 cordial.

Proof Consider the tree $\mathcal{T}_{28}^5$.

The vertex set $V(\mathcal{T}_{28}^5) = \{t_h : 1 \le h \le 5\} \cup \{u\} \cup \{u_\theta : 1 \le \theta \le m\} \cup \{v\} \cup \{v_k : 1 \le \kappa \le n\}$ and the edge set $E(\mathcal{T}_{28}^5) = \{t_h t_{h+1} : 1 \le h \le 4\} \cup \{t_4 u\} \cup \{u u_\theta : 1 \le \theta \le m\} \cup \{t_3 v\} \cup \{v v_k : 1 \le \kappa \le n\}$.

Table 3 $\nu_\sigma(i)$ and $e_\mu(i)$ for the tree $\mathcal{T}_{27}^5$, $(i \in \{\frac{\gamma}{10}, \gamma \in Z_4 - \{0\}\})$

Nature of m and n	$\nu_\sigma(0.1)$	$\nu_\sigma(0.2)$	$\nu_\sigma(0.3)$	$e_\mu(0.1)$	$e_\mu(0.2)$	$e_\mu(0.3)$
$m \equiv 0(\text{mod } 3)$ $n \equiv 0(\text{mod } 3)$	$\frac{\|V\|-1}{3}$	$\frac{\|V\|-1}{3}$	$\frac{\|V\|-1}{3}+1$	$\frac{\|V\|-1}{3}$	$\frac{\|V\|-1}{3}$	$\frac{\|V\|-1}{3}$
$m \equiv 0(\text{mod } 3)$ $n \equiv 1(\text{mod } 3)$	$\frac{\|V\|+1}{3}$	$\frac{\|V\|+1}{3}-1$	$\frac{\|V\|+1}{3}$	$\frac{\|V\|+1}{3}$	$\frac{\|V\|+1}{3}-1$	$\frac{\|V\|+1}{3}-1$
$m \equiv 0(\text{mod } 3)$ $n \equiv 2(\text{mod } 3)$	$\frac{\|V\|}{3}$	$\frac{\|V\|}{3}$	$\frac{\|V\|}{3}$	$\frac{\|V\|}{3}$	$\frac{\|V\|}{3}$	$\frac{\|V\|}{3}-1$
$m \equiv 1(\text{mod } 3)$ $n \equiv 0(\text{mod } 3)$	$\frac{\|V\|+1}{3}$	$\frac{\|V\|+1}{3}-1$	$\frac{\|V\|+1}{3}$	$\frac{\|V\|+1}{3}$	$\frac{\|V\|+1}{3}-1$	$\frac{\|V\|+1}{3}-1$
$m \equiv 1(\text{mod } 3)$ $n \equiv 1(\text{mod } 3)$	$\frac{\|V\|}{3}$	$\frac{\|V\|}{3}$	$\frac{\|V\|}{3}$	$\frac{\|V\|}{3}$	$\frac{\|V\|}{3}$	$\frac{\|V\|}{3}-1$
$m \equiv 1(\text{mod } 3)$ $n \equiv 2(\text{mod } 3)$	$\frac{\|V\|-1}{3}$	$\frac{\|V\|-1}{3}$	$\frac{\|V\|-1}{3}+1$	$\frac{\|V\|-1}{3}$	$\frac{\|V\|-1}{3}$	$\frac{\|V\|-1}{3}$
$m \equiv 2(\text{mod } 3)$ $n \equiv 0(\text{mod } 3)$	$\frac{\|V\|}{3}$	$\frac{\|V\|}{3}$	$\frac{\|V\|}{3}$	$\frac{\|V\|}{3}$	$\frac{\|V\|}{3}$	$\frac{\|V\|}{3}-1$
$m \equiv 2(\text{mod } 3)$ $n \equiv 1(\text{mod } 3)$	$\frac{\|V\|-1}{3}$	$\frac{\|V\|-1}{3}$	$\frac{\|V\|-1}{3}+1$	$\frac{\|V\|-1}{3}$	$\frac{\|V\|-1}{3}$	$\frac{\|V\|-1}{3}$
$m \equiv 2(\text{mod } 3)$ $n \equiv 2(\text{mod } 3)$	$\frac{\|V\|+1}{3}-1$	$\frac{\|V\|+1}{3}$	$\frac{\|V\|+1}{3}$	$\frac{\|V\|+1}{3}-1$	$\frac{\|V\|+1}{3}$	$\frac{\|V\|+1}{3}-1$

$|V| = m + n + 7$ and $|E| = m + n + 6$.

We define $\sigma : V(\mathcal{T}_{28}^5) \to [0, 1]$ by $\sigma(v) = \frac{\gamma}{10}, \gamma \in Z_4 - \{0\}$.

$\sigma(u) = 0.3, \sigma(v) = 0.3, \sigma(t_1) = 0.2, \sigma(t_2) = 0.1, \sigma(t_3) = 0.1, \sigma(t_4) = 0.3, \sigma(t_5) = 0.2$

Then the proof follows from Theorem 4.4.

Theorem 4.6 The tree $\mathcal{T}_{29}^5$ is fuzzy quotient-3 cordial.

Proof Consider the tree $\mathcal{T}_{29}^5$.

The vertex set $V(\mathcal{T}_{29}^5) = \{t_h : 1 \le h \le 5\} \cup \{u_\theta : 1 \le \theta \le m\} \cup \{v_\varepsilon : 1 \le \varepsilon \le n\} \cup \{w_\kappa^1 : 1 \le \kappa \le l\} \cup \{w_\kappa^2, 1 \le \kappa \le l\}$ and the edge set $E(\mathcal{T}_{29}^5) = \{t_h t_{h+1} : 1 \le h \le 4\} \cup \{t_2 u_\theta : 1 \le \theta \le m\} \cup \{t_3 v_\varepsilon : 1 \le \varepsilon \le n\} \cup \{t_4 w_\kappa^1 : 1 \le \kappa \le l\} \cup \{w_\kappa^1 w_\kappa^2 : 1 \le \kappa \le l\}$.

$|V| = m + n + 2l + 5$ and $|E| = m + n + 2l + 4$.

We define $\sigma : V(\mathcal{T}_{29}^5) \to [0, 1]$ by $\sigma(v) = \frac{\gamma}{10}, \gamma \in Z_4 - \{0\}$.

$\sigma(u) = 0.3, \sigma(v) = 0.3, \sigma(t_1) = 0.2, \sigma(t_2) = 0.3, \sigma(t_3) = 0.1, \sigma(t_4) = 0.1, \sigma(t_5) = 0.2$ and the labeling of $\sigma(u_\theta), 1 \le \theta \le m, \sigma(v_\iota), 1 \le \varepsilon \le n, \sigma(w_k^s), 1 \le \kappa \le l, s = 1, 2$ are given below.

$$\sigma(u_\theta) = 0.1 \quad \text{if } \theta \equiv 2(\text{mod } 3) \quad \text{for } 1 \le \theta \le m$$

$$\sigma(u_\theta) = 0.2 \quad \text{if } \theta \equiv 0 (\text{mod } 3) \quad \text{for } 1 \le \theta \le m$$
$$\sigma(u_\theta) = 0.3 \quad \text{if } \theta \equiv 1 (\text{mod } 3) \quad \text{for } 1 \le \theta \le m$$

Case I : $m \equiv 0(\text{mod } 3)$

$$\sigma(v_\varepsilon) = 0.1 \quad \text{if } \varepsilon \equiv 2(\text{mod } 3) \quad \text{for } 1 \le \varepsilon \le n$$
$$\sigma(v_\varepsilon) = 0.2 \quad \text{if } \varepsilon \equiv 0(\text{mod } 3) \quad \text{for } 1 \le \varepsilon \le n$$
$$\sigma(v_\varepsilon) = 0.3 \quad \text{if } \varepsilon \equiv 1(\text{mod } 3) \quad \text{for } 1 \le \varepsilon \le n$$

Subcase i : $n \equiv 0(\text{mod } 3)$

$$\sigma\left(w_\kappa^1\right) = 0.1 \quad \text{if } \kappa \equiv 0, 1(\text{mod } 3) \quad \text{for } 1 \le \kappa \le l$$
$$\sigma\left(w_\kappa^1\right) = 0.3 \quad \text{if } \kappa \equiv 2(\text{mod } 3) \quad \text{for } 1 \le \kappa \le l$$
$$\sigma\left(w_\kappa^2\right) = 0.2 \quad \text{if } \kappa \equiv 0, 2(\text{mod } 3) \quad \text{for } 1 \le \kappa \le l$$
$$\sigma\left(w_\kappa^2\right) = 0.3 \quad \text{if } \kappa \equiv 1(\text{mod } 3) \quad \text{for } 1 \le \kappa \le l$$

Subcase ii : $n \equiv 1(\text{mod } 3)$

$$\sigma\left(w_\kappa^1\right) = 0.1 \quad \text{if } \kappa \equiv 1, 2(\text{mod } 3) \quad \text{for } 1 \le \kappa \le l$$
$$\sigma\left(w_\kappa^1\right) = 0.3 \quad \text{if } \kappa \equiv 0(\text{mod } 3) \quad \text{for } 1 \le \kappa \le l$$
$$\sigma\left(w_\kappa^2\right) = 0.2 \quad \text{if } \kappa \equiv 0, 2(\text{mod } 3) \quad \text{for } 1 \le \kappa \le l$$
$$\sigma\left(w_\kappa^2\right) = 0.3 \quad \text{if } \kappa \equiv 1(\text{mod } 3) \quad \text{for } 1 \le \kappa \le l$$

Subcase iii : $n \equiv 2(\text{mod } 3)$

$$\sigma\left(w_\kappa^1\right) = 0.1 \quad \text{if } \kappa \equiv 0, 2(\text{mod } 3) \quad \text{for } 1 \le \kappa \le l$$
$$\sigma\left(w_\kappa^1\right) = 0.3 \quad \text{if } \kappa \equiv 1(\text{mod } 3) \quad \text{for } 1 \le \kappa \le l$$
$$\sigma\left(w_\kappa^2\right) = 0.2 \quad \text{if } \kappa \equiv 1, 2(\text{mod } 3) \quad \text{for } 1 \le \kappa \le l$$
$$\sigma\left(w_\kappa^2\right) = 0.3 \quad \text{if } \kappa \equiv 0(\text{mod } 3) \quad \text{for } 1 \le \kappa \le l$$

Case II : $m \equiv 1(\text{mod } 3)$

$$\sigma(v_\varepsilon) = 0.1 \quad \text{if } \varepsilon \equiv 2(\text{mod } 3) \quad \text{for } 1 \le \varepsilon \le n$$
$$\sigma(v_\varepsilon) = 0.2 \quad \text{if } \varepsilon \equiv 0(\text{mod } 3) \quad \text{for } 1 \le \varepsilon \le n$$
$$\sigma(v_\varepsilon) = 0.3 \quad \text{if } \varepsilon \equiv 1(\text{mod } 3) \quad \text{for } 1 \le \varepsilon \le n$$

Subcase i : $n \equiv 0(\text{mod } 3)$

$$\sigma\left(w_\kappa^1\right) = 0.1 \quad \text{if } \kappa \equiv 0, 2(\text{mod } 3) \quad \text{for } 1 \le \kappa \le l$$
$$\sigma\left(w_\kappa^1\right) = 0.3 \quad \text{if } \kappa \equiv 1(\text{mod } 3) \quad \text{for } 1 \le \kappa \le l$$

$$\begin{aligned}
\sigma\left(w_\kappa^2\right) &= 0.2 \quad \text{if } \kappa \equiv 1 (\text{mod } 3) \quad \text{for } 1 \le \kappa \le l \\
\sigma\left(w_\kappa^2\right) &= 0.3 \quad \text{if } \kappa \equiv 0, 2 (\text{mod } 3) \quad \text{for } 1 \le \kappa \le l
\end{aligned}$$

Subcase ii : $n \equiv 1(\text{mod } 3)$

$$\begin{aligned}
\sigma\left(w_\kappa^1\right) &= 0.1 \quad \text{if } \kappa \equiv 1, 2 (\text{mod } 3) \quad \text{for } 1 \le \kappa \le l \\
\sigma\left(w_\kappa^1\right) &= 0.3 \quad \text{if } \kappa \equiv 0 (\text{mod } 3) \quad \text{for } 1 \le \kappa \le l \\
\sigma\left(w_\kappa^2\right) &= 0.2 \quad \text{if } \kappa \equiv 1 (\text{mod } 3) \quad \text{for } 1 \le \kappa \le l \\
\sigma\left(w_\kappa^2\right) &= 0.3 \quad \text{if } \kappa \equiv 0, 2 (\text{mod } 3) \quad \text{for } 1 \le \kappa \le l
\end{aligned}$$

Subcase iii : $n \equiv 2(\text{mod } 3)$

$$\begin{aligned}
\sigma\left(w_\kappa^1\right) &= 0.1 \quad \text{if } \kappa \equiv 0, 2 (\text{mod } 3) \quad \text{for } 1 \le \kappa \le l \\
\sigma\left(w_\kappa^1\right) &= 0.3 \quad \text{if } \kappa \equiv 1 (\text{mod } 3) \quad \text{for } 1 \le \kappa \le l \\
\sigma\left(w_\kappa^2\right) &= 0.2 \quad \text{if } \kappa \equiv 1, 2 (\text{mod } 3) \quad \text{for } 1 \le \kappa \le l \\
\sigma\left(w_\kappa^2\right) &= 0.3 \quad \text{if } \kappa \equiv 0 (\text{mod } 3) \quad \text{for } 1 \le \kappa \le l
\end{aligned}$$

Case III : $m \equiv 2(\text{mod } 3)$

$$\begin{aligned}
\sigma(v_\varepsilon) &= 0.1 \quad \text{if } \varepsilon \equiv 0 (\text{mod } 3) \quad \text{for } 1 \le \varepsilon \le n \\
\sigma(v_\varepsilon) &= 0.2 \quad \text{if } \varepsilon \equiv 2 (\text{mod } 3) \quad \text{for } 1 \le \varepsilon \le n \\
\sigma(v_\varepsilon) &= 0.3 \quad \text{if } \varepsilon \equiv 1 (\text{mod } 3) \quad \text{for } 1 \le \varepsilon \le n
\end{aligned}$$

Subcase i : $n \equiv 0(\text{mod } 3)$

$$\begin{aligned}
\sigma\left(w_\kappa^1\right) &= 0.1 \quad \text{if } \kappa \equiv 0, 2 (\text{mod } 3) \quad \text{for } 1 \le \kappa \le l \\
\sigma\left(w_\kappa^1\right) &= 0.3 \quad \text{if } \kappa \equiv 1 (\text{mod } 3) \quad \text{for } 1 \le \kappa \le l \\
\sigma\left(w_\kappa^2\right) &= 0.2 \quad \text{if } \kappa \equiv 1 (\text{mod } 3) \quad \text{for } 1 \le \kappa \le l \\
\sigma\left(w_\kappa^2\right) &= 0.3 \quad \text{if } \kappa \equiv 0, 2 (\text{mod } 3) \quad \text{for } 1 \le \kappa \le l
\end{aligned}$$

Subcase ii : $n \equiv 1(\text{mod } 3)$

$$\begin{aligned}
\sigma\left(w_\kappa^1\right) &= 0.1 \quad \text{if } \kappa \equiv 0, 1 (\text{mod } 3) \quad \text{for } 1 \le \kappa \le l \\
\sigma\left(w_\kappa^1\right) &= 0.3 \quad \text{if } \kappa \equiv 2 (\text{mod } 3) \quad \text{for } 1 \le \kappa \le l \\
\sigma\left(w_\kappa^2\right) &= 0.2 \quad \text{if } \kappa \equiv 1, 2 (\text{mod } 3) \quad \text{for } 1 \le \kappa \le l \\
\sigma\left(w_\kappa^2\right) &= 0.3 \quad \text{if } \kappa \equiv 0 (\text{mod } 3) \quad \text{for } 1 \le \kappa \le l
\end{aligned}$$

Subcase iii : $n \equiv 2(\text{mod } 3)$

$$\sigma\left(w_\kappa^1\right) = 0.1 \quad \text{if } \kappa \equiv 1, 2(\text{mod } 3) \quad \text{for } 1 \le \kappa \le l$$
$$\sigma\left(w_\kappa^1\right) = 0.3 \quad \text{if } \kappa \equiv 0(\text{mod } 3) \quad \text{for } 1 \le \kappa \le l$$
$$\sigma\left(w_\kappa^2\right) = 0.2 \quad \text{if } \kappa \equiv 0, 2(\text{mod } 3) \quad \text{for } 1 \le \kappa \le l$$
$$\sigma\left(w_\kappa^2\right) = 0.3 \quad \text{if } \kappa \equiv 1(\text{mod } 3) \quad \text{for } 1 \le \kappa \le l$$

For $i \in \left\{\frac{\gamma}{10}, \gamma \in Z_4 - \{0\}\right\}$, $v_\sigma(i)$ and $e_\mu(i)$ are given below.

It is clear from Table 4 that for $i \neq j \in \left\{\frac{\gamma}{10}, \gamma \in Z_4 - \{0\}\right\}$, $v_\sigma(i)$ and $v_\sigma(j)$ differ by atmost 1 and $e_\mu(i)$ and $e_\mu(j)$ differ by atmost 1. By Definition 2.1, the tree $\mathcal{T}_{29}^5$ is admits fuzzy quotient-3 cordial labeling and hence the theorem.

Theorem 4.7 The tree $\mathcal{T}_{30}^5$ is fuzzy quotient-3 cordial.

Proof Consider the tree $\mathcal{T}_{30}^5$.

The vertex set $V\left(\mathcal{T}_{30}^5\right) = \{t_h : 1 \le h \le 5\} \cup \{u_\theta : 1 \le \theta \le m\} \cup \{v_\varepsilon : 1 \le \varepsilon \le n\} \cup \left\{w_\kappa^1 : 1 \le \kappa \le l\right\} \cup \left\{w_\kappa^2, 1 \le \kappa \le l\right\}$ and the edge set $E\left(\mathcal{T}_{30}^5\right) = \{t_h t_{h+1} : 1 \le h \le 4\} \cup \{t_4 u_\theta : 1 \le \theta \le m\} \cup \{t_3 v_\varepsilon : 1 \le \varepsilon \le n\} \cup \left\{t_2 w_\kappa^1 : 1 \le \kappa \le l\right\} \cup \left\{w_\kappa^1 w_\kappa^2 : 1 \le \kappa \le l\right\}$ $|V| = m + n + 2l + 5$ and $|E| = m + n + 2l + 4$.

We define $\sigma : V\left(\mathcal{T}_{30}^5\right) \to [0, 1]$ by $\sigma(v) = \frac{\gamma}{10}, \gamma \in Z_4 - \{0\}$.

$\sigma(u) = 0.3$, $\sigma(v) = 0.3$, $\sigma(t_1) = 0.2$, $\sigma(t_2) = 0.1$, $\sigma(t_3) = 0.1$, $\sigma(t_4) = 0.3$, $\sigma(t_5) = 0.2$

Then the proof follows from Theorem 4.6.

5 Conclusion

Fuzzy graph theory is finding an increasing number of application in modeling real-time systems. Many authors discussed various types of labeling and its applications. We have introduced the new concept called fuzzy quotient-3 cordial labeling and proved some families of graphs are fuzzy quotient-3 cordial labeling. As every graph does not admit fuzzy quotient-3 cordial labeling it is very interesting to investigate graphs or graph families which admit fuzzy quotient-3 cordial labeling. The fuzzy quotient-3 cordial labeling for some more family of graphs and application of fuzzy quotient-3 cordial on certain class of graphs shall be our future work.

Table 4 $\nu_\sigma(i)$ and $e_\mu(i)$ for the tree $\mathcal{T}_{24}^5$, $(i \in \{\frac{\gamma}{10}, \gamma \in Z_4 - \{0\}\})$

Nature of m, n and l	$\nu_\sigma(0.1)$	$\nu_\sigma(0.2)$	$\nu_\sigma(0.3)$	$e_\mu(0.1)$	$e_\mu(0.2)$	$e_\mu(0.3)$
$m \equiv 0 (\text{mod } 3)$ $n \equiv 0 (\text{mod } 3)$ $l \equiv 0 (\text{mod } 3)$	$\frac{\lvert V\rvert+1}{3}$	$\frac{\lvert V\rvert+1}{3}$	$\frac{\lvert V\rvert+1}{3} - 1$	$\frac{\lvert V\rvert+1}{3} - 1$	$\frac{\lvert V\rvert+1}{3}$	$\frac{\lvert V\rvert+1}{3} - 1$
$m \equiv 0 (\text{mod } 3)$ $n \equiv 0 (\text{mod } 3)$ $l \equiv 1 (\text{mod } 3)$	$\frac{\lvert V\rvert-1}{3} + 1$	$\frac{\lvert V\rvert-1}{3}$	$\frac{\lvert V\rvert-1}{3}$	$\frac{\lvert V\rvert-1}{3}$	$\frac{\lvert V\rvert-1}{3}$	$\frac{\lvert V\rvert-1}{3}$
$m \equiv 0 (\text{mod } 3)$ $n \equiv 0 (\text{mod } 3)$ $l \equiv 2 (\text{mod } 3)$	$\frac{\lvert V\rvert}{3}$	$\frac{\lvert V\rvert}{3}$	$\frac{\lvert V\rvert}{3}$	$\frac{\lvert V\rvert}{3}$	$\frac{\lvert V\rvert}{3}$	$\frac{\lvert V\rvert}{3} - 1$
$m \equiv 0 (\text{mod } 3)$ $n \equiv 1 (\text{mod } 3)$ $l \equiv 0 (\text{mod } 3)$	$\frac{\lvert V\rvert}{3}$	$\frac{\lvert V\rvert}{3}$	$\frac{\lvert V\rvert}{3}$	$\frac{\lvert V\rvert}{3}$	$\frac{\lvert V\rvert}{3}$	$\frac{\lvert V\rvert}{3} - 1$
$m \equiv 0 (\text{mod } 3)$ $n \equiv 1 (\text{mod } 3)$ $l \equiv 1 (\text{mod } 3)$	$\frac{\lvert V\rvert+1}{3}$	$\frac{\lvert V\rvert+1}{3} - 1$	$\frac{\lvert V\rvert+1}{3}$	$\frac{\lvert V\rvert+1}{3}$	$\frac{\lvert V\rvert+1}{3} - 1$	$\frac{\lvert V\rvert+1}{3} - 1$
$m \equiv 0 (\text{mod } 3)$ $n \equiv 1 (\text{mod } 3)$ $l \equiv 2 (\text{mod } 3)$	$\frac{\lvert V\rvert-1}{3}$	$\frac{\lvert V\rvert-1}{3}$	$\frac{\lvert V\rvert-1}{3} + 1$	$\frac{\lvert V\rvert-1}{3}$	$\frac{\lvert V\rvert-1}{3}$	$\frac{\lvert V\rvert-1}{3}$
$m \equiv 0 (\text{mod } 3)$ $n \equiv 2 (\text{mod } 3)$ $l \equiv 0 (\text{mod } 3)$	$\frac{\lvert V\rvert-1}{3} + 1$	$\frac{\lvert V\rvert-1}{3}$	$\frac{\lvert V\rvert-1}{3}$	$\frac{\lvert V\rvert-1}{3}$	$\frac{\lvert V\rvert-1}{3}$	$\frac{\lvert V\rvert-1}{3}$

(continued)

Table 4 (continued)

Nature of m, n and l	$v_\sigma(0.1)$	$v_\sigma(0.2)$	$v_\sigma(0.3)$	$e_\mu(0.1)$	$e_\mu(0.2)$	$e_\mu(0.3)$
$m \equiv 0 \pmod 3$ $n \equiv 2 \pmod 3$ $l \equiv 1 \pmod 3$	$\frac{\lvert V\rvert}{3}$	$\frac{\lvert V\rvert}{3}$	$\frac{\lvert V\rvert}{3}$	$\frac{\lvert V\rvert}{3}$	$\frac{\lvert V\rvert}{3}$	$\frac{\lvert V\rvert}{3} - 1$
$m \equiv 0 \pmod 3$ $n \equiv 2 \pmod 3$ $l \equiv 2 \pmod 3$	$\frac{\lvert V\rvert+1}{3}$	$\frac{\lvert V\rvert+1}{3}$	$\frac{\lvert V\rvert+1}{3} - 1$	$\frac{\lvert V\rvert+1}{3} - 1$	$\frac{\lvert V\rvert+1}{3}$	$\frac{\lvert V\rvert+1}{3} - 1$
$m \equiv 1 \pmod 3$ $n \equiv 0 \pmod 3$ $l \equiv 0 \pmod 3$	$\frac{\lvert V\rvert}{3}$	$\frac{\lvert V\rvert}{3}$	$\frac{\lvert V\rvert}{3}$	$\frac{\lvert V\rvert}{3} - 1$	$\frac{\lvert V\rvert}{3}$	$\frac{\lvert V\rvert}{3}$
$m \equiv 1 \pmod 3$ $n \equiv 0 \pmod 3$ $l \equiv 1 \pmod 3$	$\frac{\lvert V\rvert+1}{3} - 1$	$\frac{\lvert V\rvert+1}{3}$	$\frac{\lvert V\rvert+1}{3}$	$\frac{\lvert V\rvert+1}{3} - 1$	$\frac{\lvert V\rvert+1}{3}$	$\frac{\lvert V\rvert+1}{3} - 1$
$m \equiv 1 \pmod 3$ $n \equiv 0 \pmod 3$ $l \equiv 2 \pmod 3$	$\frac{\lvert V\rvert-1}{3}$	$\frac{\lvert V\rvert-1}{3}$	$\frac{\lvert V\rvert-1}{3} + 1$	$\frac{\lvert V\rvert-1}{3}$	$\frac{\lvert V\rvert-1}{3}$	$\frac{\lvert V\rvert-1}{3}$
$m \equiv 1 \pmod 3$ $n \equiv 1 \pmod 3$ $l \equiv 0 \pmod 3$	$\frac{\lvert V\rvert-1}{3}$	$\frac{\lvert V\rvert-1}{3}$	$\frac{\lvert V\rvert-1}{3} + 1$	$\frac{\lvert V\rvert-1}{3}$	$\frac{\lvert V\rvert-1}{3}$	$\frac{\lvert V\rvert-1}{3}$
$m \equiv 1 \pmod 3$ $n \equiv 1 \pmod 3$ $l \equiv 1 \pmod 3$	$\frac{\lvert V\rvert}{3}$	$\frac{\lvert V\rvert}{3}$	$\frac{\lvert V\rvert}{3}$	$\frac{\lvert V\rvert}{3} - 1$	$\frac{\lvert V\rvert}{3}$	$\frac{\lvert V\rvert}{3}$

(continued)

Table 4 (continued)

Nature of m, n and l	$v_\sigma(0.1)$	$v_\sigma(0.2)$	$v_\sigma(0.3)$	$e_\mu(0.1)$	$e_\mu(0.2)$	$e_\mu(0.3)$
$m \equiv 1(\text{mod } 3)$ $n \equiv 1(\text{mod } 3)$ $l \equiv 2(\text{mod } 3)$	$\frac{\lvert V\rvert+1}{3}$	$\frac{\lvert V\rvert+1}{3} - 1$	$\frac{\lvert V\rvert+1}{3}$	$\frac{\lvert V\rvert+1}{3} - 1$	$\frac{\lvert V\rvert+1}{3} - 1$	$\frac{\lvert V\rvert+1}{3}$
$m \equiv 1(\text{mod } 3)$ $n \equiv 2(\text{mod } 3)$ $l \equiv 0(\text{mod } 3)$	$\frac{\lvert V\rvert+1}{3}$	$\frac{\lvert V\rvert+1}{3} - 1$	$\frac{\lvert V\rvert+1}{3}$	$\frac{\lvert V\rvert+1}{3} - 1$	$\frac{\lvert V\rvert+1}{3} - 1$	$\frac{\lvert V\rvert+1}{3}$
$m \equiv 1(\text{mod } 3)$ $n \equiv 2(\text{mod } 3)$ $l \equiv 1(\text{mod } 3)$	$\frac{\lvert V\rvert-1}{3}$	$\frac{\lvert V\rvert-1}{3}$	$\frac{\lvert V\rvert-1}{3} + 1$	$\frac{\lvert V\rvert-1}{3}$	$\frac{\lvert V\rvert-1}{3}$	$\frac{\lvert V\rvert-1}{3}$
$m \equiv 1(\text{mod } 3)$ $n \equiv 2(\text{mod } 3)$ $l \equiv 2(\text{mod } 3)$	$\frac{\lvert V\rvert}{3}$	$\frac{\lvert V\rvert}{3}$	$\frac{\lvert V\rvert}{3}$	$\frac{\lvert V\rvert}{3} - 1$	$\frac{\lvert V\rvert}{3}$	$\frac{\lvert V\rvert}{3}$
$m \equiv 2(\text{mod } 3)$ $n \equiv 0(\text{mod}3)$ $l \equiv 0(\text{mod } 3)$	$\frac{\lvert V\rvert-1}{3} + 1$	$\frac{\lvert V\rvert-1}{3}$	$\frac{\lvert V\rvert-1}{3}$	$\frac{\lvert V\rvert-1}{3}$	$\frac{\lvert V\rvert-1}{3}$	$\frac{\lvert V\rvert-1}{3}$
$m \equiv 2(\text{mod } 3)$ $n \equiv 0(\text{mod } 3)$ $l \equiv 1(\text{mod}3)$	$\frac{\lvert V\rvert}{3}$	$\frac{\lvert V\rvert}{3}$	$\frac{\lvert V\rvert}{3}$	$\frac{\lvert V\rvert}{3}$	$\frac{\lvert V\rvert}{3}$	$\frac{\lvert V\rvert}{3} - 1$
$m \equiv 2(\text{mod}3)$ $n \equiv 0(\text{mod } 3)$ $l \equiv 2(\text{mod } 3)$	$\frac{\lvert V\rvert+1}{3}$	$\frac{\lvert V\rvert+1}{3} - 1$	$\frac{\lvert V\rvert+1}{3}$	$\frac{\lvert V\rvert+1}{3}$	$\frac{\lvert V\rvert+1}{3} - 1$	$\frac{\lvert V\rvert+1}{3} - 1$

(continued)

Table 4 (continued)

Nature of m, n and l	$\nu_\sigma(0.1)$	$\nu_\sigma(0.2)$	$\nu_\sigma(0.3)$	$e_\mu(0.1)$	$e_\mu(0.2)$	$e_\mu(0.3)$
$m \equiv 2(\text{mod } 3)$ $n \equiv 1(\text{mod } 3)$ $l \equiv 0(\text{mod } 3)$	$\frac{\lvert V\rvert+1}{3}$	$\frac{\lvert V\rvert+1}{3} - 1$	$\frac{\lvert V\rvert+1}{3}$	$\frac{\lvert V\rvert+1}{3}$	$\frac{\lvert V\rvert+1}{3} - 1$	$\frac{\lvert V\rvert+1}{3} - 1$
$m \equiv 2(\text{mod } 3)$ $n \equiv 1(\text{mod } 3)$ $l \equiv 1(\text{mod } 3)$	$\frac{\lvert V\rvert-1}{3} + 1$	$\frac{\lvert V\rvert-1}{3}$	$\frac{\lvert V\rvert-1}{3}$	$\frac{\lvert V\rvert-1}{3}$	$\frac{\lvert V\rvert-1}{3}$	$\frac{\lvert V\rvert-1}{3}$
$m \equiv 2(\text{mod } 3)$ $n \equiv 1(\text{mod } 3)$ $l \equiv 2(\text{mod } 3)$	$\frac{\lvert V\rvert}{3}$	$\frac{\lvert V\rvert}{3}$	$\frac{\lvert V\rvert}{3}$	$\frac{\lvert V\rvert}{3}$	$\frac{\lvert V\rvert}{3}$	$\frac{\lvert V\rvert}{3} - 1$
$m \equiv 2(\text{mod } 3)$ $n \equiv 2(\text{mod } 3)$ $l \equiv 0(\text{mod } 3)$	$\frac{\lvert V\rvert}{3}$	$\frac{\lvert V\rvert}{3}$	$\frac{\lvert V\rvert}{3}$	$\frac{\lvert V\rvert}{3}$	$\frac{\lvert V\rvert}{3}$	$\frac{\lvert V\rvert}{3} - 1$
$m \equiv 2(\text{mod } 3)$ $n \equiv 2(\text{mod } 3)$ $l \equiv 1(\text{mod } 3)$	$\frac{\lvert V\rvert+1}{3}$	$\frac{\lvert V\rvert+1}{3} - 1$	$\frac{\lvert V\rvert+1}{3}$	$\frac{\lvert V\rvert+1}{3}$	$\frac{\lvert V\rvert+1}{3} - 1$	$\frac{\lvert V\rvert+1}{3} - 1$
$m \equiv 2(\text{mod } 3)$ $n \equiv 2(\text{mod } 3)$ $l \equiv 2(\text{mod } 3)$	$\frac{\lvert V\rvert-1}{3} + 1$	$\frac{\lvert V\rvert-1}{3}$	$\frac{\lvert V\rvert-1}{3}$	$\frac{\lvert V\rvert-1}{3}$	$\frac{\lvert V\rvert-1}{3}$	$\frac{\lvert V\rvert-1}{3}$

References

1. Gallian, J. A. (2014). A dynamic survey of graph labeling. *The Electronic Journal of Combinatorics, 17,* 60–62.
2. Cahit, I. (1987). Cordial graphs-a weaker version of graceful and harmonious graphs. *Arscombinatoria, 23,* 201–207.
3. Sundaram, M., Ponraj, R., & Somasundaram, S. (2004). Product cordial labeling of graphs. *Bulletin of Pure & Applied Sciences- Mathematics & Statistics E, 23,* 155–163.
4. Sundaram, M., Ponraj, R., & Somasundram, S. (2006). Total product cordial labeling of graphs. *Bulletin of Pure & Applied Sciences- Mathematics and Statistics, 25,* 199–203.
5. Sundaram, M., Ponraj, R., & Somasundram, S. (2006). Some results on total product cordial labeling of graphs. *The Journal of the Indian Academy of Mathematics, 28,* 309–320.
6. Ponraj, R., Sivakumar, M., & Sundaram, M. (2012). k-Product cordial labeling of graphs. *International Journal of Contemporary Mathematical Sciences, 7*(15), 733–742.
7. Jeyanthi, P., & Maheswari, A. (2012). 3- Product cordial labelling. *SUT Journal of Mathematics, 48*(12), 231–240.
8. Ponraj, R., Sivakumar, M., & Sundaram, M. (2012). On 4-product cordial graphs. *International Journal of mathematical archive, 7,* 2809–2814.
9. Sivakumar, M. (2016). On 4-total product cordiality of some corona graphs. *IInternational Journal of Mathematical Combinatorics, 3,* 99–106.
10. Ponraj, R., & Adaickalam, M. M. (2015). Quotient cordial labeling of some star related graphs. *The Journal of the Indian Academy of Mathematics, 37*(2), 313–324.

Fuzzy Tri-Magic Labeling of Non-isomorphic Caterpillar of Diameter 5

P. Sumathi and C. Monigeetha

Abstract The graphs considered in this paper are finite, simple, undirected, and non-trivial. Graph theory has a good development in the graph labeling and has a broad range of applications. Let G be a finite, simple, undirected and non-trivial graph. A fuzzy graph is said to admit tri-magic labeling if the number of magic membership values $K_i's$ and $K_j's$ $(1 \leq i, j \leq 3)$ differs by atmost 1 and $\left|K_i - K_j\right| \leq \frac{2}{10^r}$ for $1 \leq i, j \leq 3$, $r \geq 2$. The fuzzy graph which admits a tri-magic labeling is called a fuzzy tri-magic labeling graph. The fuzzy tri-magic labeling graphs are denoted by $\tilde{T}\ m_0 G$. A caterpillar is defined as a tree in which all the vertices are within distance one of a central path. In this paper, we have taken the non-isomorphic caterpillar of diameter 5 and proved that it is fuzzy tri-magic, and we also gave the magic membership values, their corresponding edges, and the number of magic membership values as tables.

Keywords Fuzzy labeling · Magic membership value · Fuzzy tri-magic labeling · Diameter of a graph · Caterpillar graph

1 Introduction

The graphs considered here are finite, simple, undirected, and non-trivial [1]. Graph theory has a good development in the graph labeling and has a broad range of applications [2]. Fuzzy is a newly emerging mathematical framework to exhibit the phenomenon of uncertainty in real-life tribulations [3]. A fuzzy set is defined mathematically, by assigning a value to each possible individual in the universe of discourse, representing its grade or membership which corresponds to the degree

P. Sumathi
Department of Mathematics, C. Kandaswami Naidu College for Men, Anna Nagar, Chennai 600102, India
e-mail: sumathipaul@yahoo.co.in

C. Monigeetha (✉)
Department of Mathematics, Sri Muthukumaran Arts and Science College, Chikkarayapuram, Chennai 600 069, India
e-mail: monimagendra@gmail.com

S.-L. Peng et al. (eds.), *Proceedings of First International Conference on Mathematical Modeling and Computational Science*, Advances in Intelligent Systems and Computing 1292, https://doi.org/10.1007/978-981-33-4389-4_27

to which that individual is similar or compatible with the concept of the fuzzy set. Ameenal Bibi and Devi [4] introduced fuzzy bi-magic labeling; motivated by this labeling, we introduced fuzzy tri-magic labeling and proved some families of graphs are fuzzy tri-magic. In this paper, it is proved that non-isomorphic caterpillar of diameter 5 is fuzzy tri-magic.

Definition 1.1 *Fuzzy graph*

A fuzzy graph $G : (\sigma, \mu)$ is a non-empty set V together with a pair of functions $\sigma : V \rightarrow [0, 1]$ and $\mu : V \times V \rightarrow [0, 1]$, such that for all $u, v \in V$, $\mu(u, v) \leq \sigma(u) \Lambda \sigma(v)$, where σ is the fuzzy vertex set of G and μ is the fuzzy edge set of G, respectively.

Definition 1.2 *Fuzzy Labeling*

Let $G = (V,E)$ be a graph; the fuzzy graph $G : (\sigma, \mu)$ is said to have a fuzzy labeling, if $\sigma : V \rightarrow [0, 1]$ and $\mu : V \times V \rightarrow [0, 1]$ are bijective such that the membership value of edges and vertices is distinct and $\mu(u, v) \leq \sigma(u) \Lambda \sigma(v)$ for all $u, v \in V$ [5].

Definition 1.3 *Magic membership value*

Let $G : (\sigma, \mu)$ be a fuzzy graph; the induced map g: $E(G) \rightarrow [0, 1]$ defined by $g(u, v) = \sigma(u) + \mu(u, v) + \sigma(v)$ is said to have a magic membership value. It is denoted by MMV [6, 7].

Definition 1.4 *Fuzzy tri-magic labeling*

A fuzzy graph is said to admit tri-magic labeling if the MMVs $K_i's$ $(1 \leq \boldsymbol{i} \leq 3)$ are constants where the number of $K_i's$ and $K_j's$ differs by at most 1 and $\left|K_i - K_j\right| \leq \frac{2}{10^r}$ for $1 \leq i, j \leq 3$, $r \geq 2$.

Definition 1.5 *Fuzzy tri-magic labeling graph*

A fuzzy labeling graph which admits a tri-magic labeling is called a fuzzy tri-magic labeling graph. The fuzzy tri-magic labeling graphs are denoted by $\tilde{T}\ m_0G$.

Definition 1.6 *Diameter of a graph*

The maximum distance between the two vertices of a graph is called the diameter of a graph [8].

Definition 1.7 *Caterpillar graph*

A caterpillar graph G is a tree having a chordless path P_t on t vertices, called central path, which contains at least one end point of every edge. Vertices connecting the leaves with the central path are called support vertices [6, 7, 9, 10].

Notation 1.8

Let $\mathscr{C}^{s_1}_{t_1,t_2}$ is a caterpillar graph obtained by attaching m pendant edges and n pendant edges to the path P_6 at the vertices v_{t_1+1} and v_{t_2+1}, respectively.

2 Main Result

Theorem 2.1 *The caterpillar graph $\mathscr{C}_{3,4}^{6}$ of diameter 5 admits fuzzy tri-magic labeling for $m, n \geq 3$.*

Proof Let G be a caterpillar graph $\mathscr{C}_{3,4}^{6}$ of diameter 5. Let m, n be the pendant edges attaching to the path P_6 at the vertices v_3 and v_4, respectively. Let the vertex set and edge set of G be

$$V(G) = \{v_j : 1 \leq j \leq 6\} \cup \{x_j : 1 \leq j \leq m\} \cup \{y_j : 1 \leq j \leq n\}.$$

$$E(G) = \{v_j v_{j+1} : 1 \leq j \leq 5\} \cup \{v_3 x_j : 1 \leq j \leq m\} \cup \{v_4 y_j : 1 \leq j \leq n\}.$$

$$|V(G)| = m + n + 6 \text{ and } |E(G)| = m + n + 5.$$

Let $r \geq 2$ be any positive integer.
We define $\sigma : V \to [0, 1]$ such that

$$\sigma(v_j) = \frac{2m + 2n + 21 - j}{10^r} \quad 1 \leq j \leq 6$$

For $m \equiv 0(mod 3)$

$$\sigma(x_j) = (2m + 2n + 15 - j)\frac{1}{10^r} \quad 1 \leq j \leq \frac{m}{3};$$

$$\sigma(x_j) = (2m + 2n + 14 - j)\frac{1}{10^r} \quad \frac{m}{3} + 1 \leq j \leq \frac{2m}{3};$$

$$\sigma(x_j) = (2m + 2n + 13 - j)\frac{1}{10^r} \quad \frac{2m}{3} + 1 \leq j \leq m$$

For $m \equiv 1(mod 3)$

$$\sigma(x_j) = (2m + 2n + 15 - j)\frac{1}{10^r} \quad 1 \leq j \leq \frac{m-1}{3};$$

$$\sigma(x_j) = (2m + 2n + 14 - j)\frac{1}{10^r} \quad \frac{m-1}{3} + 1 \leq j \leq \frac{2(m-1)}{3};$$

$$\sigma(x_j) = (2m + 2n + 13 - j)\frac{1}{10^r} \quad \frac{2(m-1)}{3} + 1 \leq j \leq m$$

For $m \equiv 2(mod 3)$

$$\sigma(x_j) = (2m + 2n + 15 - j)\frac{1}{10^r} \quad 1 \leq j \leq \frac{m+1}{3};$$

$$\sigma(x_j) = (2m + 2n + 14 - j)\frac{1}{10^r} \quad \frac{m+1}{3} + 1 \le j \le \frac{2m-1}{3};$$

$$\sigma(x_j) = (2m + 2n + 13 - j)\frac{1}{10^r} \quad \frac{2m-1}{3} + 1 \le j \le m$$

For $n \equiv 0(mod3)$ and $m \equiv 0, 1, 2(mod3)$

$$\sigma(y_j) = (m + 2n + 13 - j)\frac{1}{10^r} \quad 1 \le j \le \frac{n}{3};$$

$$\sigma(y_j) = (m + 2n + 12 - j)\frac{1}{10^r} \quad \frac{n}{3} + 1 \le j \le \frac{2n}{3};$$

$$\sigma(y_j) = (m + 2n + 11 - j)\frac{1}{10^r} \quad \frac{2n}{3} + 1 \le j \le n;$$

For $n \equiv 1(mod3)$ and $m \equiv 0, 1, 2(mod3)$

$$\sigma(y_j) = (m + 2n + 13 - j)\frac{1}{10^r} \quad 1 \le j \le \frac{n-1}{3};$$

$$\sigma(y_j) = (m + 2n + 12 - j)\frac{1}{10^r} \quad \frac{n-1}{3} + 1 \le j \le \frac{2(n-1)}{3};$$

$$\sigma(y_j) = (m + 2n + 11 - j)\frac{1}{10^r} \quad \frac{2(n-1)}{3} + 1 \le j \le n;$$

For $n \equiv 2(mod3)$ and $m \equiv 0, 1(mod3)$

$$\sigma(y_j) = (m + 2n + 13 - j)\frac{1}{10^r} \quad 1 \le j \le \frac{n+1}{3};$$

$$\sigma(y_j) = (m + 2n + 12 - j)\frac{1}{10^r} \quad \frac{n+1}{3} + 1 \le j \le \frac{2n-1}{3};$$

$$\sigma(y_j) = (m + 2n + 11 - j)\frac{1}{10^r} \quad \frac{2n-1}{3} + 1 \le j \le n$$

For $n \equiv 2(mod3)$ and $m \equiv 2(mod3)$

$$\sigma(y_j) = (m + 2n + 13 - j)\frac{1}{10^r} \quad 1 \le j \le \frac{n-2}{3};$$

$$\sigma(y_j) = (m + 2n + 12 - j)\frac{1}{10^r} \quad \frac{n-2}{3} + 1 \le j \le \frac{2n-1}{3};$$

$$\sigma(y_j) = (m + 2n + 11 - j)\frac{1}{10^r} \quad \frac{2n-1}{3} + 1 \le j \le n$$

Define $\mu : V \times V \to [0, 1]$ such that

$$\mu(v_3x_j) = (7 + j)\frac{1}{10^r} \quad 1 \le j \le m$$

$$\mu(v_4y_j) = (m + 10 + j)\frac{1}{10^r} \quad 1 \le j \le n$$

$\mu(v_1v_2) = \frac{1}{10^r}$, $\mu(v_2v_3) = \frac{3}{10^r}$, $\mu(v_3v_4) = \frac{4}{10^r}$, $\mu(v_4v_5) = \frac{6}{10^r}$, $\mu(v_5v_6) = \frac{7}{10^r}$.
From different nature of m, n, MMVs $K_i's$, their corresponding edges and number of $K_i's$ $(1 \le i \le 3)$ are given in Table 1.

From Table 1, the maximum difference between the number of $K_i's$ and $K_j's$ is 1 and $|K_i - K_j| \le \frac{2}{10^r}$ for $1 \le i, j \le 3$. Hence, the caterpillar graph $\mathscr{C}^6_{3,4}$ of diameter 5 admits fuzzy tri-magic labeling for $m, n \ge 3$.

Theorem 2.2 *The caterpillar graph $\mathscr{C}^6_{2,5}$ of diameter 5 admits fuzzy tri-magic labeling for $m, n \ge 3$.*

Proof Let G be a caterpillar graph $\mathscr{C}^6_{2,5}$ of diameter 5. Let m, n be the pendant edges attaching to the path P_6 at the vertices v_2 and v_5, respectively. Let the vertex set and edge set of G be

$$|V(G)| = m + n + 6 \text{ and } |E(G)| = m + n + 5.$$

$$V(G) = \{v_j : 1 \le j \le 6\} \cup \{x_j : 1 \le j \le m\} \cup \{y_j : 1 \le j \le n\}.$$

$$E(G) = \{v_jv_{j+1} : 1 \le j \le 5\} \cup \{v_2x_j : 1 \le j \le m\} \cup \{v_5y_j : 1 \le j \le n\}.$$

$V(G) = \{v_j : 1 \le j \le 6\} \cup \{x_j : 1 \le j \le m\} \cup \{y_j : 1 \le j \le n\}$.
Let $r \ge 2$ be any positive integer.
We define $\sigma : V \to [0, 1]$ such that

$$\sigma(v_j) = \frac{2m + 2n + 24 - j}{10^r} \quad 1 \le j \le 6$$

For $m \equiv 0(mod3)$

$$\sigma(x_j) = (2m + 2n + 17 - j)\frac{1}{10^r} \quad 1 \le j \le \frac{m}{3};$$

$$\sigma(x_j) = (2m + 2n + 16 - j)\frac{1}{10^r} \quad \frac{m}{3} + 1 \le j \le \frac{2m}{3};$$

$$\sigma(x_j) = (2m + 2n + 15 - j)\frac{1}{10^r} \quad \frac{2m}{3} + 1 \le j \le m$$

Table 1 MMVs $K_i's$, their corresponding edges and the number of $K_i's$ $(1 \le \boldsymbol{i} \le 3)$

Nature of m, n	Edges	MMV $K_i's$ $1 \le \boldsymbol{i} \le 3$	Number of $K_i's$ $1 \le \boldsymbol{i} \le 3$
$\boldsymbol{m} \equiv 0(\boldsymbol{mod}3)$ $\boldsymbol{n} \equiv 0(\boldsymbol{mod}3)$	$g(v_3x_j)\ 1 \le j \le \frac{m}{3}$ $g(v_1v_2), g(v_2v_3)$ $g(v_4y_j)\ 1 \le j \le \frac{n}{3}$	$(4m + 4n + 40)\frac{1}{10^r}$ for $i = 1$	$\frac{m+n+6}{3}$ for $i = 1$
	$g(v_3x_j)\ \frac{m}{3} + 1 \le j \le \frac{2m}{3}$ $g(v_3v_4), g(v_4v_5)$ $g(v_4y_j)\ \frac{n}{3} + 1 \le j \le \frac{2n}{3}$	$(4m + 4n + 39)\frac{1}{10^r}$ for $i = 2$	$\frac{m+n+6}{3}$ for $i = 2$
	$g(v_3x_j)\ \frac{2m}{3} + 1 \le j \le m$ $g(v_5v_6)$ $g(v_4y_j)\ \frac{2n}{3} + 1 \le j \le n$	$(4m + 4n + 38)\frac{1}{10^r}$ for $i = 3$	$\frac{m+n+3}{3}$ for $i = 3$
$\boldsymbol{m} \equiv 0(\boldsymbol{mod}3)$ $\boldsymbol{n} \equiv 1(\boldsymbol{mod}3)$	$g(v_3x_j)\ 1 \le j \le \frac{m}{3}$ $g(v_1v_2), g(v_2v_3)$ $g(v_4y_j)\ 1 \le j \le \frac{n-1}{3}$	$(4m + 4n + 40)\frac{1}{10^r}$ for $i = 1$	$\frac{m+n+5}{3}$ for $i = 1$
	$g(v_3x_j)\ \frac{m}{3} + 1 \le j \le \frac{2m}{3}$ $g(v_3v_4), g(v_4v_5)$ $g(v_4y_j)\ \frac{n-1}{3} + 1 \le j \le \frac{2(n-1)}{3}$	$(4m + 4n + 39)\frac{1}{10^r}$ for $i = 2$	$\frac{m+n+5}{3}$ for $i = 2$
	$g(v_3x_j)\ \frac{2m}{3} + 1 \le j \le m$ $g(v_5v_6)$ $g(v_4y_j)\ \frac{2(n-1)}{3} + 1 \le j \le n$	$(4m + 4n + 38)\frac{1}{10^r}$ for $i = 3$	$\frac{m+n+5}{3}$ for $i = 3$

(continued)

Table 1 (continued)

Nature of m, n	Edges	MMV $K_i's$ $1 \le \boldsymbol{i} \le 3$	Number of $K_i's$ $1 \le \boldsymbol{i} \le 3$
$\boldsymbol{m} \equiv 0(\boldsymbol{mod}3)$ $\boldsymbol{n} \equiv 2(\boldsymbol{mod}3)$	$g(v_3x_j)$ $1 \le j \le \frac{m}{3}$ $g(v_1v_2), g(v_2v_3)$ $g(v_4y_j)$ $1 \le j \le \frac{n+1}{3}$	$(4m + 4n + 40)\frac{1}{10^r}$ for $i = 1$	$\frac{m+n+7}{3}$ for $i = 1$
	$g(v_3x_j)$ $\frac{m}{3} + 1 \le j \le \frac{2m}{3}$ $g(v_3v_4), g(v_4v_5)$ $g(v_4y_j)$ $\frac{n+1}{3} + 1 \le j \le \frac{2n-1}{3}$	$(4m + 4n + 39)\frac{1}{10^r}$ for $i = 2$	$\frac{m+n+4}{3}$ for $i = 2$
	$g(v_3x_j)$ $\frac{2m}{3} + 1 \le j \le m$ $g(v_5v_6)$ $g(v_4y_j)$ $\frac{2n-1}{3} + 1 \le j \le n$	$(4m + 4n + 38)\frac{1}{10^r}$ for $i = 3$	$\frac{m+n+4}{3}$ for $i = 3$
$\boldsymbol{m} \equiv 1(\boldsymbol{mod}3)$ $\boldsymbol{n} \equiv 0(\boldsymbol{mod}3)$	$g(v_3x_j)$ $1 \le j \le \frac{m-1}{3}$ $g(v_1v_2), g(v_2v_3)$ $g(v_4y_j)$ $1 \le j \le \frac{n}{3}$	$(4m + 4n + 40)\frac{1}{10^r}$ for $i = 1$	$\frac{m+n+5}{3}$ for $i = 1$
	$g(v_3x_j)$ $\frac{m-1}{3} + 1 \le j \le \frac{2(m-1)}{3}$ $g(v_3v_4), g(v_4v_5)$ $g(v_4y_j)$ $\frac{n}{3} + 1 \le j \le \frac{2n}{3}$	$(4m + 4n + 39)\frac{1}{10^r}$ for $i = 2$	$\frac{m+n+5}{3}$ for $i = 2$
	$g(v_3x_j)$ $\frac{2(m-1)}{3} + 1 \le j \le m$ $g(v_5v_6)$ $g(v_4y_j)$ $\frac{2n}{3} + 1 \le j \le n$	$(4m + 4n + 38)\frac{1}{10^r}$ for $i = 3$	$\frac{m+n+5}{3}$ for $i = 3$

(continued)

Table 1 (continued)

Nature of m, n	Edges	MMV $K_i's$ $1 \le \boldsymbol{i} \le 3$	Number of $K_i's$ $1 \le \boldsymbol{i} \le 3$
$\boldsymbol{m} \equiv 1(\boldsymbol{mod}3)$ $\boldsymbol{n} \equiv 1(\boldsymbol{mod}3)$	$g(v_3x_j)\ 1 \le j \le \frac{m-1}{3}$ $g(v_1v_2), g(v_2v_3)$ $g(v_4y_j)\ 1 \le j \le \frac{n-1}{3}$	$(4m+4n+40)\frac{1}{10^r}$ for $i=1$	$\frac{m+n+4}{3}$ for $i=1$
	$g(v_3x_j)\ \frac{m-1}{3}+1 \le j \le \frac{2(m-1)}{3}$ $g(v_3v_4), g(v_4v_5)$ $g(v_4y_j)\ \frac{n-1}{3}+1 \le j \le \frac{2(n-1)}{3}$	$(4m+4n+39)\frac{1}{10^r}$ for $i=2$	$\frac{m+n+4}{3}$ for $i=2$
	$g(v_3x_j)\ \frac{2(m-1)}{3}+1 \le j \le m$ $g(v_5v_6)$ $g(v_4y_j)\ \frac{2(n-1)}{3}+1 \le j \le n$	$(4m+4n+38)\frac{1}{10^r}$ for $i=3$	$\frac{m+n+7}{3}$ for $i=3$
$\boldsymbol{m} \equiv 1(\boldsymbol{mod}3)$ $\boldsymbol{n} \equiv 2(\boldsymbol{mod}3)$	$g(v_3x_j)\ 1 \le j \le \frac{m-1}{3}$ $g(v_1v_2), g(v_2v_3)$ $g(v_4y_j)\ 1 \le j \le \frac{n+1}{3}$	$(4m+4n+40)\frac{1}{10^r}$ for $i=1$	$\frac{m+n+6}{3}$ for $i=1$
	$g(v_3x_j)\ \frac{m-1}{3}+1 \le j \le \frac{2(m-1)}{3}$ $g(v_3v_4), g(v_4v_5)$ $g(v_4y_j)\ \frac{n+1}{3}+1 \le j \le \frac{2n-1}{3}$	$(4m+4n+39)\frac{1}{10^r}$ for $i=2$	$\frac{m+n+3}{3}$ for $i=2$
	$g(v_3x_j)\ \frac{2(m-1)}{3}+1 \le j \le m$ $g(v_5v_6)$ $g(v_4y_j)\ \frac{2n-1}{3}+1 \le j \le n$	$(4m+4n+38)\frac{1}{10^r}$ for $i=3$	$\frac{m+n+6}{3}$ for $i=3$

(continued)

Table 1 (continued)

Nature of m, n	Edges	MMV $K_i's$ $1 \le \boldsymbol{i} \le 3$	Number of $K_i's$ $1 \le \boldsymbol{i} \le 3$
$\boldsymbol{m} \equiv 2(\boldsymbol{mod}3)$ $\boldsymbol{n} \equiv 0(\boldsymbol{mod}3)$	$g(v_3x_j)$ $1 \le j \le \frac{m+1}{3}$ $g(v_1v_2), g(v_2v_3)$ $g(v_4y_j)$ $1 \le j \le \frac{n}{3}$	$(4m + 4n + 40)\frac{1}{10^r}$ for $i = 1$	$\frac{m+n+7}{3}$ for $i = 1$
	$g(v_3x_j)$ $\frac{m+1}{3} + 1 \le j \le \frac{2m-1}{3}$ $g(v_3v_4), g(v_4v_5)$ $g(v_4y_j)$ $\frac{n}{3} + 1 \le j \le \frac{2n}{3}$	$(4m + 4n + 39)\frac{1}{10^r}$ for $i = 2$	$\frac{m+n+4}{3}$ for $i = 2$
	$g(v_3x_j)$ $\frac{2m-1}{3} + 1 \le j \le m$ $g(v_5v_6)$ $g(v_4y_j)$ $\frac{2n}{3} + 1 \le j \le n$	$(4m + 4n + 38)\frac{1}{10^r}$ for $i = 3$	$\frac{m+n+4}{3}$ for $i = 3$
$\boldsymbol{m} \equiv 2(\boldsymbol{mod}3)$ $\boldsymbol{n} \equiv 1(\boldsymbol{mod}3)$	$g(v_3x_j)$ $1 \le j \le \frac{m+1}{3}$ $g(v_1v_2), g(v_2v_3)$ $g(v_4y_j)$ $1 \le j \le \frac{n-1}{3}$	$(4m + 4n + 40)\frac{1}{10^r}$ for $i = 1$	$\frac{m+n+6}{3}$ for $i = 1$
	$g(v_3x_j)$ $\frac{m+1}{3} + 1 \le j \le \frac{2m-1}{3}$ $g(v_3v_4), g(v_4v_5)$ $g(v_4y_j)$ $\frac{n-1}{3} + 1 \le j \le \frac{2(n-1)}{3}$	$(4m + 4n + 39)\frac{1}{10^r}$ for $i = 2$	$\frac{m+n+3}{3}$ for $i = 2$
	$g(v_3x_j)$ $\frac{2m-1}{3} + 1 \le j \le m$ $g(v_5v_6)$ $g(v_4y_j)$ $\frac{2(n-1)}{3} + 1 \le j \le n$	$(4m + 4n + 38)\frac{1}{10^r}$ for $i = 3$	$\frac{m+n+6}{3}$ for $i = 3$

(continued)

Table 1 (continued)

Nature of m, n	Edges	MMV $K_i's$ $1 \leq \boldsymbol{i} \leq 3$	Number of $K_i's$ $1 \leq \boldsymbol{i} \leq 3$
$\boldsymbol{m} \equiv 2(\boldsymbol{mod}3)$ $\boldsymbol{n} \equiv 2(\boldsymbol{mod}3)$	$g(v_3x_j)\ 1 \leq j \leq \frac{m+1}{3}$ $g(v_1v_2), g(v_2v_3)$ $g(v_4y_j)\ 1 \leq j \leq \frac{n-2}{3}$	$(4m+4n+40)\frac{1}{10^r}$ for $i = 1$	$\frac{m+n+5}{3}$ for $i = 1$
	$g(v_3x_j)\ \frac{m+1}{3}+1 \leq j \leq \frac{2m-1}{3}$ $g(v_3v_4), g(v_4v_5)$ $g(v_4y_j)\ \frac{n-2}{3}+1 \leq j \leq \frac{2n-1}{3}$	$(4m+4n+39)\frac{1}{10^r}$ for $i = 2$	$\frac{m+n+5}{3}$ for $i = 2$
	$g(v_3x_j)\ \frac{2m-1}{3}+1 \leq j \leq m$ $g(v_5v_6)$ $g(v_4y_j)\ \frac{2n-1}{3}+1 \leq j \leq n$	$(4m+4n+38)\frac{1}{10^r}$ for $i = 3$	$\frac{m+n+5}{3}$ for $i = 3$

For $m \equiv 1(mod 3)$

$$\sigma(x_j) = (2m + 2n + 17 - j)\frac{1}{10^r} \quad 1 \le j \le \frac{m-1}{3};$$

$$\sigma(x_j) = (2m + 2n + 16 - j)\frac{1}{10^r} \quad \frac{m-1}{3} + 1 \le j \le \frac{2(m-1)}{3};$$

$$\sigma(x_j) = (2m + 2n + 15 - j)\frac{1}{10^r} \quad \frac{2(m-1)}{3} + 1 \le j \le m$$

For $m \equiv 2(mod 3)$

$$\sigma(x_j) = (2m + 2n + 17 - j)\frac{1}{10^r} \quad 1 \le j \le \frac{m+1}{3};$$

$$\sigma(x_j) = (2m + 2n + 16 - j)\frac{1}{10^r} \quad \frac{m+1}{3} + 1 \le j \le \frac{2m-1}{3};$$

$$\sigma(x_j) = (2m + 2n + 15 - j)\frac{1}{10^r} \quad \frac{2m-1}{3} + 1 \le j \le m$$

For $n \equiv 0(mod 3)$ and $m \equiv 0, 1, 2(mod 3)$

$$\sigma(y_j) = (m + 2n + 15 - j)\frac{1}{10^r} \quad 1 \le j \le \frac{n}{3};$$

$$\sigma(y_j) = (m + 2n + 14 - j)\frac{1}{10^r} \quad \frac{n}{3} + 1 \le j \le \frac{2n}{3};$$

$$\sigma(y_j) = (m + 2n + 13 - j)\frac{1}{10^r} \quad \frac{2n}{3} + 1 \le j \le n;$$

For $n \equiv 1(mod 3)$ and $m \equiv 0, 1, 2(mod 3)$

$$\sigma(y_j) = (m + 2n + 15 - j)\frac{1}{10^r} \quad 1 \le j \le \frac{n-1}{3};$$

$$\sigma(y_j) = (m + 2n + 14 - j)\frac{1}{10^r} \quad \frac{n-1}{3} + 1 \le j \le \frac{2(n-1)}{3};$$

$$\sigma(y_j) = (m + 2n + 13 - j)\frac{1}{10^r} \quad \frac{2(n-1)}{3} + 1 \le j \le n;$$

For $n \equiv 2(mod 3)$ and $m \equiv 0, 1(mod 3)$

$$\sigma(y_j) = (m + 2n + 15 - j)\frac{1}{10^r} \quad 1 \le j \le \frac{n+1}{3};$$

$$\sigma(y_j) = (m + 2n + 14 - j)\frac{1}{10^r} \quad \frac{n+1}{3} + 1 \le j \le \frac{2n-1}{3};$$

$$\sigma(y_j) = (m + 2n + 13 - j)\frac{1}{10^r} \quad \frac{2n-1}{3} + 1 \le j \le n$$

For $n \equiv 2(mod3)$ and $m \equiv 2(mod3)$

$$\sigma(y_j) = (m + 2n + 15 - j)\frac{1}{10^r} \quad 1 \le j \le \frac{n-2}{3};$$

$$\sigma(y_j) = (m + 2n + 14 - j)\frac{1}{10^r} \quad \frac{n-2}{3} + 1 \le j \le \frac{2n-1}{3};$$

$$\sigma(y_j) = (m + 2n + 13 - j)\frac{1}{10^r} \quad \frac{2n-1}{3} + 1 \le j \le n$$

Define $\mu : V \times V \to [0, 1]$ such that

$$\mu(v_3 x_j) = (7 + j)\frac{1}{10^r} \quad 1 \le j \le m$$

$$\mu(v_4 y_j) = (m + 10 + j)\frac{1}{10^r} \quad 1 \le j \le n$$

$\mu(v_1v_2) = \frac{1}{10^r}$, $\mu(v_2v_3) = \frac{3}{10^r}$, $\mu(v_3v_4) = \frac{4}{10^r}$, $\mu(v_4v_5) = \frac{6}{10^r}$, $\mu(v_5v_6) = \frac{7}{10^r}$.

From different nature of m, n, MMVs $K_i's$, their corresponding edges and number of $K_i's$ $(1 \le i \le 3)$ are presented in Table 2.

From Table 2, the maximum difference between the number of $K_i's$ and $K_j's$ is 1 and $|K_i - K_j| \le \frac{2}{10^r}$ for $1 \le i, j \le 3$. Hence, the caterpillar graph $\mathscr{C}_{2,5}^6$ of diameter 5 admits fuzzy tri-magic labeling for $m, n \ge 3$.

3 Conclusion

Graph theory is one of the most important branches of discrete mathematics. From graph theory, graph labeling evolved as one of the important areas. A graph labeling is an assignment of integers to the vertices or edges, or both, subject to certain conditions [1]. It is used in many applications like coding theory, X-ray crystallography, radar, astronomy, circuit design, communication network addressing and database management [11]. The fuzzy graph $G : (\sigma, \mu)$ is said to have a fuzzy labeling, if $\sigma : V \to [0, 1]$ and $\mu : V \times V \to [0, 1]$ are bijective such that the membership value of edges and vertices is distinct and $\mu(u, v) \le \sigma(u) \Lambda \sigma(v)$ for all $u, v \in V$ [5]. In the previous papers, we have introduced the concept of fuzzy tri-magic labeling and proved some star-related graphs are fuzzy tri-magic. Also, we have proved that some

Table 2 MMVs $K_i's$, their corresponding edges and number of $K_i's$ $(1 \le \boldsymbol{i} \le 3)$

Nature of m, n	Edges	MMV $K_i's$ $1 \le \boldsymbol{i} \le 3$	Number of $K_i's$ $1 \le \boldsymbol{i} \le 3$
$\boldsymbol{m} \equiv 0(\boldsymbol{mod}3)$ $\boldsymbol{n} \equiv 0(\boldsymbol{mod}3)$	$g(v_2x_j)$ if $1 \le j \le \frac{m}{3}$ $g(v_1v_2), g(v_2v_3)$ $g(v_5y_j)$ if $1 \le j \le \frac{n}{3}$	$(4m + 4n + 46)\frac{1}{10^r}$ for $i = 1$	$\frac{m+n+6}{3}$ for $i = 1$
	$g(v_2x_j)$ if $\frac{m}{3} + 1 \le j \le \frac{2m}{3}$ $g(v_3v_4), g(v_4v_5)$ $g(v_5y_j)$ if $\frac{n}{3} + 1 \le j \le \frac{2n}{3}$	$(4m + 4n + 45)\frac{1}{10^r}$ for $i = 2$	$\frac{m+n+6}{3}$ for $i = 2$
	$g(v_2x_j)$ if $\frac{2m}{3} + 1 \le j \le m$ $g(v_5v_6)$ $g(v_5y_j)$ if $\frac{2n}{3} + 1 \le j \le n$	$(4m + 4n + 44)\frac{1}{10^r}$ for $i = 3$	$\frac{m+n+3}{3}$ for $i = 3$
$\boldsymbol{m} \equiv 0(\boldsymbol{mod}3)$ $\boldsymbol{n} \equiv 1(\boldsymbol{mod}3)$	$g(v_2x_j)$ if $1 \le j \le \frac{m}{3}$ $g(v_1v_2), g(v_2v_3)$ $g(v_5y_j)$ if $1 \le j \le \frac{n-1}{3}$	$(4m + 4n + 46)\frac{1}{10^r}$ for $i = 1$	$\frac{m+n+5}{3}$ for $i = 1$
	$g(v_2x_j)$ if $\frac{m}{3} + 1 \le j \le \frac{2m}{3}$ $g(v_3v_4), g(v_4v_5)$ $g(v_5y_j)$ if $\frac{n-1}{3} + 1 \le j \le \frac{2(n-1)}{3}$	$(4m + 4n + 45)\frac{1}{10^r}$ for $i = 2$	$\frac{m+n+5}{3}$ for $i = 2$
	$g(v_2x_j)$ if $\frac{2m}{3} + 1 \le j \le m$ $g(v_5v_6)$ $g(v_5y_j)$ if $\frac{2(n-1)}{3} + 1 \le j \le n$	$(4m + 4n + 44)\frac{1}{10^r}$ for $i = 3$	$\frac{m+n+5}{3}$ for $i = 3$

(continued)

Table 2 (continued)

Nature of m, n	Edges	MMV $K_i's$ $1 \le \boldsymbol{i} \le 3$	Number of $K_i's$ $1 \le \boldsymbol{i} \le 3$
$\boldsymbol{m} \equiv 0(\boldsymbol{mod}3)$ $\boldsymbol{n} \equiv 2(\boldsymbol{mod}3)$	$g(v_2x_j)$ if $1 \le j \le \frac{m}{3}$ $g(v_1v_2), g(v_2v_3)$ $g(v_5y_j)$ if $1 \le j \le \frac{n+1}{3}$	$(4m+4n+46)\frac{1}{10^r}$ for $i=1$	$\frac{m+n+7}{3}$ for $i=1$
	$g(v_2x_j)$ if $\frac{m}{3}+1 \le j \le \frac{2m}{3}$ $g(v_3v_4), g(v_4v_5)$ $g(v_5y_j)$ if $\frac{n+1}{3}+1 \le j \le \frac{2n-1}{3}$	$(4m+4n+45)\frac{1}{10^r}$ for $i=2$	$\frac{m+n+4}{3}$ for $i=2$
	$g(v_2x_j)$ if $\frac{2m}{3}+1 \le j \le m$ $g(v_5v_6)$ $g(v_5y_j)$ if $\frac{2n-1}{3}+1 \le j \le n$	$(4m+4n+44)\frac{1}{10^r}$ for $i=3$	$\frac{m+n+4}{3}$ for $i=3$
$\boldsymbol{m} \equiv 1(\boldsymbol{mod}3)$ $\boldsymbol{n} \equiv 0(\boldsymbol{mod}3)$	$g(v_2x_j)$ if $1 \le j \le \frac{m-1}{3}$ $g(v_1v_2), g(v_2v_3)$ $g(v_5y_j)$ if $1 \le j \le \frac{n}{3}$	$(4m+4n+46)\frac{1}{10^r}$ for $i=1$	$\frac{m+n+5}{3}$ for $i=1$
	$g(v_2x_j)$ if $\frac{m-1}{3}+1 \le j \le \frac{2(m-1)}{3}$ $g(v_3v_4), g(v_4v_5)$ $g(v_5y_j)$ if $\frac{n}{3}+1 \le j \le \frac{2n}{3}$	$(4m+4n+45)\frac{1}{10^r}$ for $i=2$	$\frac{m+n+5}{3}$ for $i=2$
	$g(v_2x_j)$ if $\frac{2(m-1)}{3}+1 \le j \le m$ $g(v_5v_6)$ $g(v_5y_j)$ if $\frac{2n}{3}+1 \le j \le n$	$(4m+4n+44)\frac{1}{10^r}$ for $i=3$	$\frac{m+n+5}{3}$ for $i=3$

(continued)

Table 2 (continued)

Nature of m, n	Edges	MMV $K_i's$ $1 \le \boldsymbol{i} \le 3$	Number of $K_i's$ $1 \le \boldsymbol{i} \le 3$
$\boldsymbol{m} \equiv 1(\boldsymbol{mod}3)$ $\boldsymbol{n} \equiv 1(\boldsymbol{mod}3)$	$g(v_2x_j)$ if $1 \le j \le \frac{m-1}{3}$ $g(v_1v_2), g(v_2v_3)$ $g(v_5y_j)$ if $1 \le j \le \frac{n-1}{3}$	$(4m+4n+46)\frac{1}{10^r}$ for $i=1$	$\frac{m+n+4}{3}$ for $i=1$
	$g(v_2x_j)$ if $\frac{m-1}{3}+1 \le j \le \frac{2(m-1)}{3}$ $g(v_3v_4), g(v_4v_5)$ $g(v_5y_j)$ if $\frac{n-1}{3}+1 \le j \le \frac{2(n-1)}{3}$	$(4m+4n+45)\frac{1}{10^r}$ for $i=2$	$\frac{m+n+4}{3}$ for $i=2$
	$g(v_2x_j)$ if $\frac{2(m-1)}{3}+1 \le j \le m$ $g(v_5v_6)$ $g(v_5y_j)$ if $\frac{2(n-1)}{3}+1 \le j \le n$	$(4m+4n+44)\frac{1}{10^r}$ for $i=3$	$\frac{m+n+7}{3}$ for $i=3$
$\boldsymbol{m} \equiv 1(\boldsymbol{mod}3)$ $\boldsymbol{n} \equiv 2(\boldsymbol{mod}3)$	$g(v_2x_j)$ if $1 \le j \le \frac{m-1}{3}$ $g(v_1v_2), g(v_2v_3)$ $g(v_5y_j)$ if $1 \le j \le \frac{n+1}{3}$	$(4m+4n+46)\frac{1}{10^r}$ for $i=1$	$\frac{m+n+6}{3}$ for $i=1$
	$g(v_2x_j)$ if $\frac{m-1}{3}+1 \le j \le \frac{2(m-1)}{3}$ $g(v_3v_4), g(v_4v_5)$ $g(v_5y_j)$ if $\frac{n+1}{3}+1 \le j \le \frac{2n-1}{3}$	$(4m+4n+45)\frac{1}{10^r}$ for $i=2$	$\frac{m+n+3}{3}$ for $i=2$
	$g(v_2x_j)$ if $\frac{2(m-1)}{3}+1 \le j \le m$ $g(v_5v_6)$ $g(v_5y_j)$ if $\frac{2n-1}{3}+1 \le j \le n$	$(4m+4n+44)\frac{1}{10^r}$ for $i=3$	$\frac{m+n+6}{3}$ for $i=3$

(continued)

Table 2 (continued)

Nature of m, n	Edges	MMV $K_i^{\prime}s$ $1 \le i \le 3$	Number of $K_i^{\prime}s$ $1 \le i \le 3$
$m \equiv 2(mod 3)$ $n \equiv 0(mod 3)$	$g(v_2x_j)$ if $1 \le j \le \frac{m+1}{3}$ $g(v_1v_2), g(v_2v_3)$ $g(v_5y_j)$ if $1 \le j \le \frac{n}{3}$	$(4m+4n+46)\frac{1}{10^r}$ for $i=1$	$\frac{m+n+7}{3}$ for $i=1$
	$g(v_2x_j)$ if $\frac{m+1}{3}+1 \le j \le \frac{2m-1}{3}$ $g(v_3v_4), g(v_4v_5)$ $g(v_5y_j)$ if $\frac{n}{3}+1 \le j \le \frac{2n}{3}$	$(4m+4n+45)\frac{1}{10^r}$ for $i=2$	$\frac{m+n+4}{3}$ for $i=2$
	$g(v_2x_j)$ if $\frac{2m-1}{3}+1 \le j \le m$ $g(v_5v_6)$ $g(v_5y_j)$ if $\frac{2n}{3}+1 \le j \le n$	$(4m+4n+44)\frac{1}{10^r}$ for $i=3$	$\frac{m+n+4}{3}$ for $i=3$
$m \equiv 2(mod 3)$ $n \equiv 1(mod 3)$	$g(v_2x_j)$ if $1 \le j \le \frac{m+1}{3}$ $g(v_1v_2), g(v_2v_3)$ $g(v_5y_j)$ if $1 \le j \le \frac{n-1}{3}$	$(4m+4n+46)\frac{1}{10^r}$ for $i=1$	$\frac{m+n+6}{3}$ for $i=1$
	$g(v_2x_j)$ if $\frac{m+1}{3}+1 \le j \le \frac{2m-1}{3}$ $g(v_3v_4), g(v_4v_5)$ $g(v_5y_j)$ if $\frac{n-1}{3}+1 \le j \le \frac{2(n-1)}{3}$	$(4m+4n+45)\frac{1}{10^r}$ for $i=2$	$\frac{m+n+3}{3}$ for $i=2$
	$g(v_2x_j)$ if $\frac{2m-1}{3}+1 \le j \le m$ $g(v_5v_6)$ $g(v_5y_j)$ if $\frac{2(n-1)}{3}+1 \le j \le n$	$(4m+4n+44)\frac{1}{10^r}$ for $i=3$	$\frac{m+n+6}{3}$ for $i=3$

(continued)

Table 2 (continued)

Nature of m, n	Edges	MMV $K_i's$ $1 \le \boldsymbol{i} \le 3$	Number of $K_i's$ $1 \le \boldsymbol{i} \le 3$
$\boldsymbol{m} \equiv 2(\boldsymbol{mod}3)$ $\boldsymbol{n} \equiv 2(\boldsymbol{mod}3)$	$g(v_2x_j)$ if $1 \le j \le \frac{m+1}{3}$ $g(v_1v_2), g(v_2v_3)$ $g(v_5y_j)$ if $1 \le j \le \frac{n-2}{3}$	$(4m+4n+46)\frac{1}{10^r}$ for $i=1$	$\frac{m+n+5}{3}$ for $i=1$
	$g(v_2x_j)$ if $\frac{m+1}{3}+1 \le j \le \frac{2m-1}{3}$ $g(v_3v_4), g(v_4v_5)$ $g(v_5y_j)$ if $\frac{n-2}{3}+1 \le j \le \frac{2n-1}{3}$	$(4m+4n+45)\frac{1}{10^r}$ for $i=2$	$\frac{m+n+5}{3}$ for $i=2$
	$g(v_2x_j)$ if $\frac{2m-1}{3}+1 \le j \le m$ $g(v_5v_6)$ $g(v_5y_j)$ if $\frac{2n-1}{3}+1 \le j \le n$	$(4m+4n+44)\frac{1}{10^r}$ for $i=3$	$\frac{m+n+5}{3}$ for $i=3$

ladder graphs and generalized Jahangir graph are fuzzy tri-magic. Then we extended our work and proved that some trees of diameter less than 5 and isomorphic caterpillar of diameter 5 are fuzzy tri-magic [6, 7]. In this paper, we have taken non-isomorphic caterpillar of diameter 5 graphs $\mathscr{C}^6_{3,4}$ and $\mathscr{C}^6_{2,5}$ and shown that these graphs admit fuzzy tri-magic by proving the maximum difference between the number of $K_i's$ and $K_i's$ $(1 \le i \le 3)$ is 1 and $|K_i - K_j| \le \frac{2}{10^r}$ for $1 \le i, j \le 3, r \ge 2$; we also tabulated MMVs $K_i's$, their corresponding edges and the number of $K_i's$ $(1 \le i \le 3)$ for the graph $\mathscr{C}^6_{3,4}$ in Table 1 and also tabulated MMVs $K_i's$, their corresponding edges and the number of $K_i's$ $(1 \le i \le 3)$ for the graph $\mathscr{C}^6_{2,5}$ in Table 2. All the remaining cases of a caterpillar of diameter 5 are also done by us, and they will be published in the upcoming papers.

References

1. Gallian, J. A. (2014). A dynamic survey of graph labeling. *The Electronic Journal of Combinatorics, 17,* 60–62.
2. Harary, F. (2013). *Graph theory*. New Delhi: Narosa Publishing House Pvt. Ltd.
3. Thirusangu, K., & Jeevitha, D. (2018). Fuzzy bimagic and antimagic labeling in star graphs. *International Journal of Mathematics Trends and Technology (IJMTT), Special Issue,* 41–43.
4. Ameenal Bibi, K., & Devi, M. (2017). Fuzzy bi-magic labeling on cycle graph and star graph. *Global Journal of Pure and Applied Mathematics, 13*(6), 1975–1985.
5. Nagoor Gani, A., & Rajalaxmi (a) Subahashini, D. (2014). A note on fuzzy labeling. *International Journal of Fuzzy Mathematical Archive, 4*(2), 2320–3250.
6. Sumathi, P., & Monigeetha, C. (2020). Fuzzy tri-magic labeling of isomorphic caterpillar of diameter 5—Paper 1. *Journal of Engineering, Computing and Architecture, 10,* 182–189.
7. Sumathi, P., & Monigeetha, C. (2020). Fuzzy tri-magic labeling of isomorphic caterpillar of diameter 5—Paper 2. *Journal of Information and Computational Science, 10,* 1463–1481.
8. Hrnčiar, P., & Haviar, A. (2001). All trees of diameter five are graceful. *Discrete Mathematics, 233*(1–3), 133–150.
9. Sumathi, P., & Esther Felicia, R. (2018). Enresdowedness of some caterpillar graphs. *International Journal of Pure and Applied Mathematics, 118,* 2938–2957.
10. Juliana Felix., & Marcia Cappelle. (2016). *1-identifying codes on caterpillar graphs*. Instituto de Informatica INF UFG.
11. Prasanna, N. L., Sravanthi, K., & Sudhakar, N. (2014). Applications of graph labeling in communication networks. *Oriental Journal of Computer Science and Technology, 7*(1), 139–145.

Multi-metaheuristic Features Selection Model for High-Dimensional Biomedical Data

Saba M. Hussain

Abstract This paper proposes a novel multi-metaheuristic features selection model (MHFS) that potentially improves the features selection problems. The gene expression profiles are important biosensor in the clinical diagnosis of cancer types. The challenge is to certain classification techniques for gene expression which has several thousands of features for each exam sample. The proposed system is used the two algorithms to search in parallel for the best number of features that are particle swarm optimization and bat algorithm. Each algorithm is just searching individually the best solutions that they share between them. Experimentally, MHFS's performance has provided a better classification of accurate results with a minimum number of features than both particle swarm optimization and bat algorithm. It achieved an average accuracy rate over ten datasets which is the amount to 93% with selecting 90 features, while the particle swarm optimization gets on 92% with select 252 features, and bat algorithm obtains 86% with 477 features.

Keywords Metaheuristic optimization · Particle swarm optimization · Bat algorithm · Biomedical information · Feature selection

1 Introduction

The gene expression profiles (GEP) are microarray technology, record ten thousands features of the gene in cancer cells versus normal tissues. Therefore, selecting the high important features leads to increase precision in diagnosis [1]. Besides, the reduction of data dimensions and select important features are the necessity to enhance the performance of a machine learning algorithm [2]. Wrapper model is one of power feature selection (FS) techniques that select the optimal feature's base on testing different groups of subsets [3]. Metaheuristic algorithms use to optimize the subset features that feed into the wrapper model. They inspired by natural and have been used the randomness to find an optimal solution [4]. The engine search of metaheuristic

S. M. Hussain (✉)
College of Information Technology, University of Babylon, Babil, Iraq
e-mail: saba.alshebeeb@uobabylon.edu.iq

S.-L. Peng et al. (eds.), *Proceedings of First International Conference on Mathematical Modeling and Computational Science*, Advances in Intelligent Systems and Computing 1292, https://doi.org/10.1007/978-981-33-4389-4_28

based on both exploration the diverse solutions and exploitation best local optimum to find another best one. The particle swarm optimization (PSO) and bat search algorithm (BA) have solved several optimization problems successfully [5–9]. In this paper, the proposed system is applied to select the optimum features of ten GEP dataset.

1.1 Motivation

The gene expression profiles are very sensitive and accurate biosensor to detect the cancer case [1]. The challenge of using this bio-information has ten thousands of feature for each sample [8, 10]. Furthermore, not all features that extracted from cells are interesting and therefore omit the unimportant features that potential increasing examination accurate case. The data which has high undesired features often adversely affects on diagnosis machine performance. Although PSO and BA algorithms have powerful characteristics and succeeded in solving several features selection problems, the algorithms are currently suffering from significantly limitations:

Stagnancy at Local Optimum
PSO and BA find the first local optimum solution and then often lose the initiative to improve the search position. Technically, the one way to generate new candidate solutions in metaheuristic algorithm leads to weak in exploration. Therefore, significantly suffering to overcome local optimum and jump to the best one.

Inaccurate with High-Dimensional Problem
Both PSO and BA are significantly suffering when dealing with the high-dimensional search problem. They fall in local optimum frequently and low convergence rate in search progress.

In persistent Results
The solution engine of PSO and BA is unstable and has very different production solutions especially when dealing with high dimension.

1.2 Contributions

In this paper, we proposed a new hybrid features selection model based on multi-metaheuristic algorithms POS and BA (MHFS-PSO-BA). The proposed model applies both PSO and BA parallel to optimize features selection; meanwhile, algorithms are exchanged the best solutions during the research process. The proposed method inspired by the diverse sources of generating new solution leads to enhance the search performance. It has provided the following achievements:

Un-stagnancy at Local Optimum
The proposed MHFS-PSO-BA after finding the first local optimal solution tries significantly to improve the search position. It generates diverse solutions to promote exploration process.

Accurate with High-Dimensional Problem
The MHFS-PSO-BA performs as a hybrid algorithm to change its manner frequently during the search process. Therefore, it deals efficiently with high-dimensional problems.

Persistent Results
The MHFS-PSO-BA gives roughly the same results when facing high-dimensional problems.

1.3 Paper Organization

The rest of this paper has been organized as follows: Sect. 2 reviews the related works. The bat algorithm and particle swarm optimization are discussed in Sects. 3 and 4, respectively. Section 5 presents features encoding and selection. Section 6 details out the principles of the proposed system. Finally, Sects. 7 and 8, respectively, describe the experimental result and conclusion.

2 Related Works

To optimize feature selection (FS) techniques, several studies have proposed. Most of those studies have applied standard metaheuristic algorithms such as particle swarm optimization (PSO), bee colony algorithm (BCA) and bat algorithm (BA). Researchers in the area of optimization FS often applied the original version of metaheuristic algorithms select optimal features without improving the vulnerabilities in those algorithms.

Amrita et al. [11] proposed a hybrid model based on BA to optimize paraments of support vector machine (SVM) and selected optimal features. Each bat vector in optimization pool consists of two parts: the first two positions allocated to SVM parameters and the rest of the vector represent the mask to select features. The proposed system optimizes simultaneously the SVM parameters and features. Applying the metaheuristic to optimize more than one part of the system increases the bearable the randomization and unstable in final results.

Taha et al. [12] applied naïve Bayes (NB) to guide BA in selecting optimal subset features. They proposed decrees bat velocity when the different past and current position is negative. NB sets bat velocity to adaptive the number of features with the current search process. The feature that corresponding 1 in bat vector is selected and 0 is omitted.

Taha et al. [13] proposed the binary version of particle swarm optimization (BPSO) for optimizing the features selection. The proposed system applies k-nearest neighbours (KNN) as a classifier of the new query. It improves the performance of KNN without developing the PSO. Dhrif et al. [14] developed PSO by modifying the coefficient parameters (c_1, c_2) and weight inertia simultaneously with search progress. The sigmoid function is applied to enhance these parameters. The authors introduced a new encoding technique that keeps the dynamics of the PSO particles by preservation continuous behaviour of their search process.

Search space in [12–14] is limited either 0 or 1 (binary search space) this restricted the search operations at a small range. Generally, the probability stuck at local optimum is increased when metaheuristic algorithms search in binary search space [15].

Ray et al. [16] proposed multi-operator differential evolution (DE) for feature selection. They extended search space by using continuous search space rather than binary. Authors set the threshold for feature selection select features that corresponding gene value greater than the threshold.

3 Bat Algorithm (BA)

The bat algorithm (BA) is an evolutionary search algorithm (EA) which inspired by bat behaviour interceded in [17]. Technically, it generates a new candidate solution ($x_i^{'}$) by adding bat velocity (v_i) to current position x_i. Bat velocity (v_i) is updated dynamically after each optimization iteration [11]. BA exploits the best current solution position (x_*) to modify bat velocity, that promotes the exploration by adding the frequency to a new value of bat velocity ($v_i^{'}$). Mathematically, BA modifies bat position as follows [17]:

$$f_i = f\text{min} + \left(f\text{max} - f\text{min}\right) * \beta \tag{1}$$

where

$f\text{max}$ and $f\text{min}$ are maximum and minimum bat frequency, respectively, that set by a user, β is random within the interval $[-1, 1]$.

$$v_i^{'} = v_i + (x_i - x_*) f_i \tag{2}$$

$$x_i^{'} = x_i + v_i^{'} \tag{3}$$

BA supports the exploration in the search process by compare bat loudness (A_i) and emission pulse rate (r_i) with a random value. It adds random value between [0, 1] to the current bat position if r_i greater than a random value between [0, 1]. The new bat position ($x_i^{'}$) accept as the best solution when satisfying two conditions: the cost function of $x_i^{'}$ better than the cost function of current and value of A_i greater than a random value between [0, 1]. Equations 4 and 5 calculate the A_i and r_i, respectively [17].

$$A_i^{'} = \boldsymbol{\alpha} A_i \tag{4}$$

where $A_i^{'}$ is new bat loudness value and $\boldsymbol{\alpha}$ random value in the interval [1, 2]

$$r_i^{'} = r_i\left[1 - e^{-\gamma t}\right] \tag{5}$$

where $r_i^{'}$ is a new value of emission pulse rate, $\boldsymbol{\gamma}$ is the random value in the interval [0, 1].

4 Particle Swarm Optimization Algorithm (PSO)

The particle swarm optimization (PSO) is an evolutionary search algorithm (EA) inspired by the social behaviour of animals that lives within groups [18]. PSO has been favoured over other EA because it has a simple model with few parameters. The PSO is different from the other EA by the structure of the population (particles) where each particle of PSO has two positions: the current position and the best position found by itself (p_{best}) [8]. It chooses best p_{best} g among particles to be global best position (g_{best}). Technically, the PSO updates particles position by adding the current position and new velocity of particle as shown in Eq 7. The PSO has in calculating particles velocity both exploration and exploitation of search processes. It modifies particles velocity dynamically during the search process. The factors that govern particles movements are inertia component, the cognitive component and social component [14]. Equation 6 calculates the new velocity of particles.

$$v_i^{'} = \underbrace{wv_i}_{\text{inertia component}} + \underbrace{c_1 r_1\left(p_{best_i} - x_i\right)}_{\text{cognitive component}} + \underbrace{c_2 r_2 (g_{best} - x_i)}_{\text{social component}}. \tag{6}$$

where w is weight inertia, c_1, c_2 are coefficient constant, r_1, r_2 random value within interval $-1, +$ 1].

$$x_i^{'} = x_i + v_i^{'} \tag{7}$$

where x_i current particle position. $v_i^{'}$, $x_i^{'}$ new particle velocity and position.

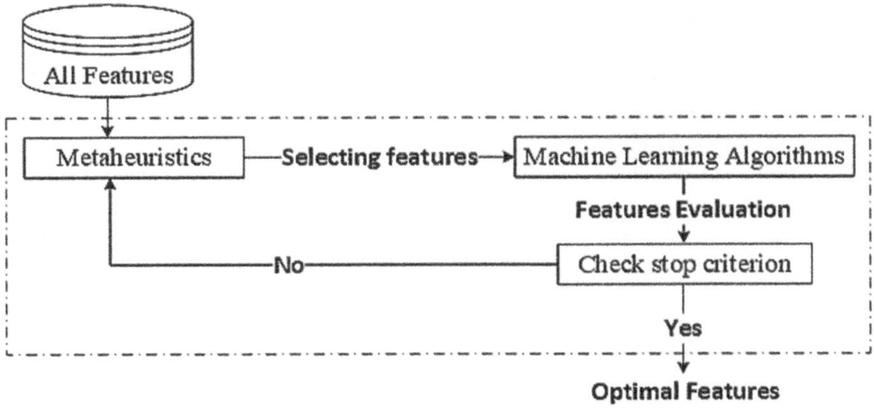

Fig. 1 Features selection based on wrapper model and metaheuristic

5 Features Selection Methodology

In the proposed system and both BA and PSO, the wrapper model is used for applying in the features selection approach. A wrapper model is a feature selection method tests deferent groups of features and then selects the group that satisfying best result [19]. In selecting, features must be based on randomness. The metaheuristic techniques have success in applying randomness for search on an optimal solution. Therefore, it significantly optimizes in selecting features that feed into the wrapper model [20]. Figure 1 illustrates the principles of the wrapper model with metaheuristic for selecting the optimum feature.

Generally, the wrapper model is based on binary feature selection [21]—only features that corresponding 1 in optimization algorithm vector are selected—that restricts the metaheuristic algorithms to search in limited boundaries either 0 or 1. The extending search space is enhanced performance of metaheuristic algorithms [15]. The applied space searching interval range is [−5, +5] in the proposed system and with both PSO and BA. The threshold for feature selection is determined by the user for each experiment [10]. Figure 2 illustrates the features of encoding and selecting.

6 Proposed Multi-metaheuristic Features Selection (MHFS)

In this section, we present our proposed system for optimizing features selection. Technically, it consists of multi-metaheuristic algorithms (PSO and BA) that search parallelly of the best group of features. They exchange the best solution simultaneously during search progress. Enhancement of the exploration in the proposed system

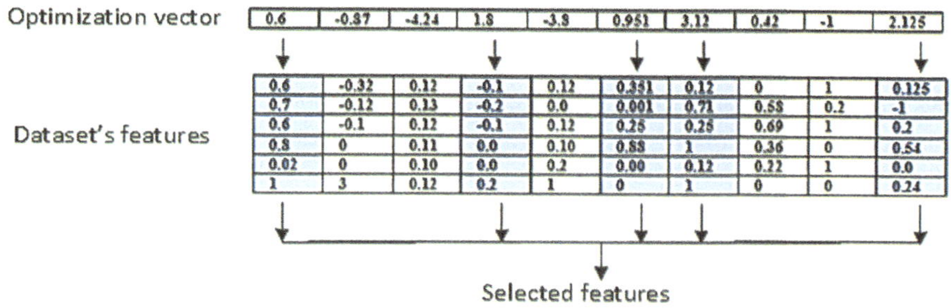

Fig. 2 Example of features encoding and selecting with threshold 0.5

(MHFS) lies in multiple search methodologies to provide divers candidate solutions. It exploits the best solution in each search algorithms (PSO or BA) to enhance the search process of another algorithm. The proposed system consists of four parts initialization, parallel search methodology, solutions aggregate and terminal criteria. Figure 3 shows the main component of MHFS model.

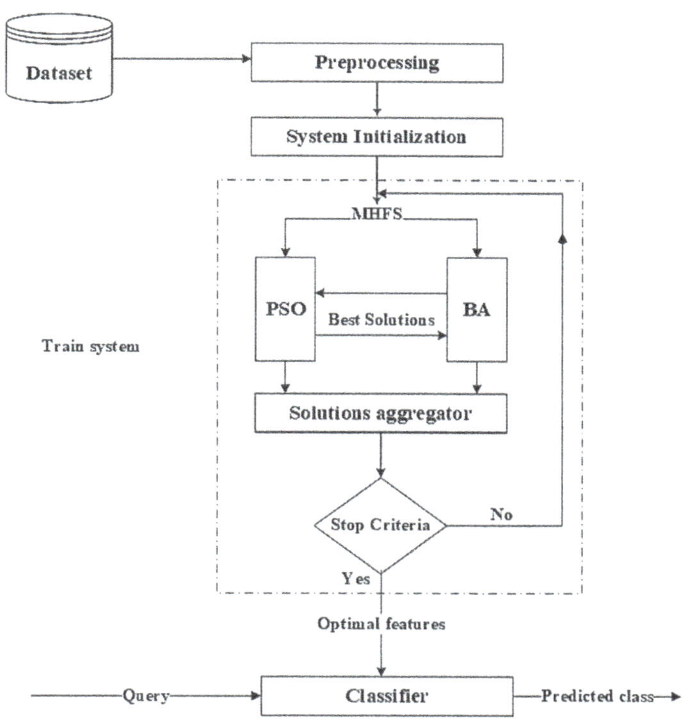

Fig. 3 Proposed MHFS model

6.1 Initialization

The proposed MHFS starts by setting the initial parameters of both PSO and BA. It generates population have random value with zero velocity. The population size is equal in both PSO and BA. The first best local optimum is allocated randomly from the population [22, 23]. Therefore, the proposed system does not exchange the best solutions in the initial step.

6.2 Parallel Search Methodology

In this step, it allows each shared algorithms (PSO and BA) to search in parallel. The best solutions are exchanged periodically during search progress between algorithms. The PSO updates the position of its particles simultaneously with bats position updated by BA.

Let BAS = $\{B_i | i = 1 \ldots m\}$ and PSOS = $\{PS_i | i = 1 \ldots m\}$ be a set of bat and particle swarm population, respectively, B_i and PS_i is solution vector, and f is fitness(cost) function. The algorithm I illustrates the proposed parallel search methodology.

Algorithm I: Proposed Parallel Search Methodology

```
Input: BAS,PSOS
Output: best bat position (Best_B ), best particle position(Best_P)
procedure Parallel Search Methodology (BAS,PSOS)
      B'_i ←  update Bat position (B_i)
      PS'_i ←  update particle position (PS_i)
      if      f(B'_i) > f(Best_B)
                   best bat position ← Best_B
      if   f(B'_i) > f(Best_p)
             best bat position ← Best_p
      if  f(Best_p) > f(Best_B)      Best_p  ← Best_B
                 otherwise    Best_B  ← Best_P
  Return best Best_B, Best_P
```

6.3 Solutions Aggregate

In this section, the proposed system selects that the best solution is funded by BA and PSO. Equation 8 selects best solution ($MHFS_{best}$).

$$MHFS_{best} = \begin{cases} PS_{best_{PSO}}, PS_{best_{PSO}} < B_{best_{BA}} \\ B_{best_{BA}}, B_{best_{BA}} > PS_{best_{PSO}} \end{cases} \tag{8}$$

Finally, MHFS like other EAs stop when satisfying one of the termination criteria, it exceeds maximum iteration, no improvement in search progress, or satisfying object requirement [15]. The proposed system stop when exceeding maximum iteration.

7 Results and Experimental Discussion

In this section, the datasets have been used to evaluate the proposed system and discuss the results that have been got by proposed MHFS PSO-BA, PSO and BA which are presented.

7.1 Dataset Description

The proposed system is tested over ten medical datasets with deferent dimension. They are gathered from two websites [24]. The first one is at [25] consist of five datasets that are 9_tumors, brain_tumor1, cancer, prostate_tumor and DLBCL, and the second one is at [26] that consists of rest datasets in Table 1.

The complexity of data measured parameter C.F/O as C is the number of classes, F is the number of features and O is the number of observations [24]. Technically, a bigger number of features than the number of observations means the dataset has high complexity.

Table 1 Dataset details [24]

Sq.	Dataset name	No. Class	No. Samples	No. Features	C.F/O
1	DLBCL	2	77	5469	142
2	Breast3	3	95	4869	154
3	prostate	2	38	3051	161
4	Lymphoma	3	62	4026	195
5	Prostate_tumor	2	102	10,510	206
6	Adenocarcinoma	2	76	9868	260
7	Lung_cancer	4	203	12,600	310
8	Brain_tumor1	5	90	5920	329
9	Brain_tumor2	4	50	10,367	829
10	9_tumors	9	60	5726	859

7.2 Experimental Result

With the aim of proven an accurate assessment of the proposed features selection technique, the experimental evaluation has considered a wide variety of biomedical dataset features number as shown in Table 1. The K-nearest neighbour (KNN) with set the nearest neighbour (K) by 1 is used as an objective function to features that selected by MHFS PSO-BA, PSO and PSO. In this work, the comparison criteria are best prediction accuracy, high persistent results and better exploration over running thirty times of all algorithm (proposed MHFS PSO-BA, PSO, PSO). The average prediction accuracy of thirty times is calculated as accuracy performance of all compared algorithms, the standard deviation (Std) to measure persistent results of them. The average standard deviation for the number of features that exported during the search process is calculated to determine exploration rate of all MHFS PSO-BA, BA and PSO. Table 2 illustrates the experimental result of accuracy and number of features of MHFS PSO-BA, BA and PSO over run thirty times.

To sum up, the MHFS(PSO-BA) achieved a better accuracy with less number of a selected feature and high accuracy in 80% of datasets compared with other algorithms. Furthermore, it satisfied a markable pre-performance over highest complexity datasets (high C.F/O). Figure 4 shows the persistent results of better than both PSO and BA.

The exchanging best solution between algorithm enhances the proposed system to provide an efficient exploration compared with other algorithms. Figure 5 shows the exploration's rate of the proposed system, PSO and BA.

The proposed MHFS gets the best result with the minimum number of feature. Figure 6 illustrates the percentage of selected features by each algorithm with its accuracy.

Table 2 Accuracy and number of features of MHFS PSO-BA, BA and PSO over run thirty times, 1NN is used as object classification function

Sq.	Dataset name	BA		PSO		MHFS(PSO-BA)	
		Accuracy	No. features	Accuracy	No. features	Accuracy	No. features
1	DLBCL	95.3	687	99.84	360	99.69	184
2	Breast3	64.94	493	78.77	101	79.26	30
3	prostate	99.7	115	100	158	100	42
4	Lymphoma	88	226	93.89	164	94.77	28
5	Prostate_tumor	84.11	395	95	233	94.89	39
6	Adenocarcinoma	91.67	370	96.52	291	97.12	109
7	Lung_cancer	95.25	850	96.72	378	97.35	181
8	Brain_tumor1	92.95	741	96.28	348	95.9	103
9	Brain_tumor2	84.72	200	95.56	293	96.66	89
10	9_tumors	66.67	215	74.36	192	75.13	130

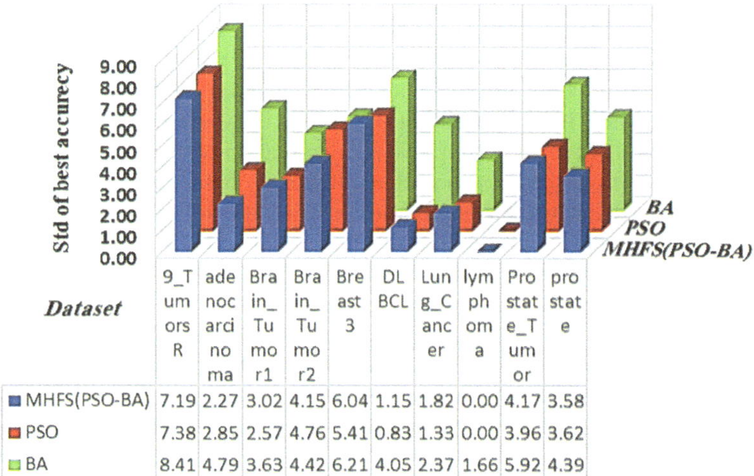

Fig. 4 Standard deviation of best result run MHFS PSO-BA, BA and PSO over thirty times

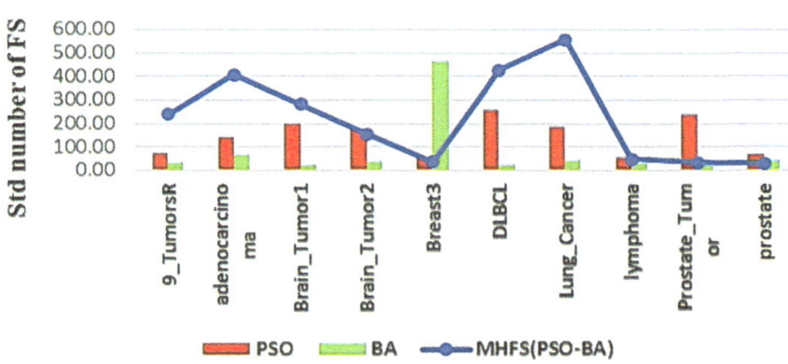

Fig. 5 Exploration rate of the proposed system, PSO and BA

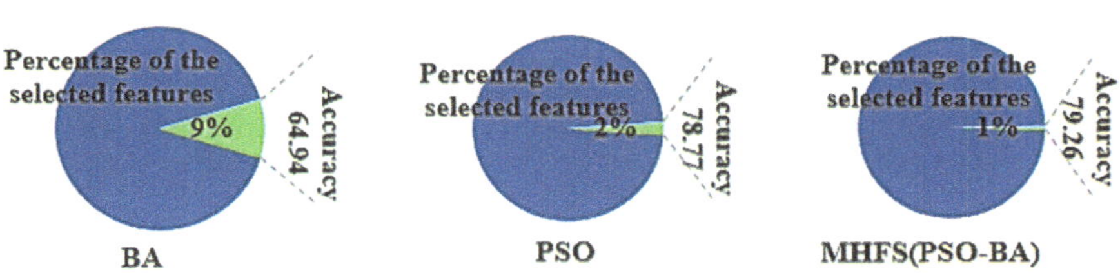

Fig. 6 Accuracy and percentage of selected feature for Breast3 dataset of proposed MHFS, PSO and BA

8 Conclusion and Future Works

In conclusion, EAs are typically suffering from stagnation during the search process as a result of weakness in its exploration. Therefore, increasing the diversity solution in search progress is essential to improve the exploration of these algorithms. In this research, we have proposed a new search methodology based on multiple EAs (PSO and BA) that have enhanced the exploration rate significantly. Our proposed MHFS has achieved better results through applying the PSO and BA to search on optimal solutions parallelly and shared the best solutions between algorithms simultaneous search progress. The proposed system is compared with PSO and BA to select the best group of features that enhance the classification accuracy of biomedical dataset. Experimentally, the proposed MHFS has outflanked other models and demonstrated to be a potential optimization alternative model for features selection. For future work, a proactive random model would be applied to adapt the variable search volume in most optimization techniques.

References

1. Van't Veer, L. J., et al. (2002). Gene expression profiling predicts clinical outcome of breast cancer. *Nature, 415*(6871), 530–536.
2. Ramaswamyreddy, A., Prasad, Shiva, Rangarao, K. V., & Saranya, A. (2019). Efficient datamining model for prediction of chronic kidney disease using wrapper methods. *International Journal of Informatics and Communication Technology, 8*(2), 63.
3. Sánchez-Maroño, N., Alonso-Betanzos, A., & Tombilla-Sanromán, M. (2007). Filter methods for feature selection—A comparative study. *Lecture Notes on Computer Science (including Subser. Lect. Notes Artif. Intell. Lect. Notes Bioinformatics) LNCS, 4881,* 178–187.
4. Lones, M. (2011). *Sean Luke: Essent Metaheur, 12,* 3.
5. Yu, R., An, X., Jin, B., Shi, J., Move, O. A., & Liu, Y. (2018). Particle classification optimization-based BP network for telecommunication customer churn prediction. *Neural Computation and Application, 29*(3), 707–720.
6. Zakaria, M. Z., Mutalib, S., Rahman, S. A., Elias, S. J., & Shahuddin, A. Z. (2019). Solving RFID mobile reader path problem with optimization algorithms. *Indonesian Journal of Electrical Engineering and Computer Science, 13*(3), 1110–1116.
7. Mohd Ali, N. Z., Musirin, I., & Mohamad, H. (2019). Clonal evolutionary particle swarm optimization for congestion management and compensation scheme in power system. *Indonesian Journal of Electrical Engineering and Computer Science, 16*(2), 591–598.
8. Jabor, A. H., & Ali, A. H. (2019). Dual heuristic feature selection based on genetic algorithm and binary particle swarm optimization. *Journal of University of Babylon for Pure and Applied Sciences, 27*(1), 171–183.
9. Topal, A. O., & Altun, O. (2016). A novel meta-heuristic algorithm: dynamic virtual bats algorithm. *Information Science (Ny), 354,* 222–235.
10. Khushaba, R. N., Al-Ani A., & Al-Jumaily, A. (2008). Differential evolution based feature subset selection. In *Proceedings of International Conference on Pattern Recognition.*
11. Amrita, & Ahmed, P. (2014). A hybrid-based feature selection approach for IDS. *Lecture Notes on Electronics and Engineering (LNEE), 284,* 195–211.
12. Taha, A. M., Mustapha, A., Der Chen, S. (2013). Naive Bayes-guided bat algorithm for feature selection. *Science World Journal.*

13. Chuang, L. Y., Chang, H. W., Tu, C. J., & Yang, C. H. (2008). Improved binary PSO for feature selection using gene expression data. *Computational Biology and Chemistry, 32*(1), 29–38.
14. Dhrif, H., Giraldo, L. G. S., Kubat, M., & Wuchty, S. (2019). A stable combinatorial particle swarm optimization for scalable feature selection in gene expression data, pp. 1–13.
15. Al-saeedi, A. H., & Altun, O. (2016). Binary mean-variance mapping optimization algorithm (BMVMO). *J. Appl. Phys. Sci., 2*(2), 42–47.
16. Ray, T., Sarker, R., & Li, X. (2016). Preface. Lect. Notes Comput. Sci. (including Subser. Lect. Notes Artif. Intell. Lect. Notes Bioinformatics). *9592*(v).
17. Yang, X. S. (2013). Bat algorithm: Literature review and applications. *International Journal of Bio-Inspired Computation, 5*(3), 141–149.
18. Eberhart, R., & Kennedy, J. (1995). New optimizer using particle swarm theory. In *Proceedings of International Symposium on Micro Mechanical Human Science* (pp. 39–43).
19. Chandrashekar, G., & Sahin, F. (2014). A survey on feature selection methods. *Computers & Electrical Engineering, 40*(1), 16–28.
20. Zeng, H., & Trussell, H. J. (2006). Feature selection using. *IAENG International Journal of Computer Science, 33*(February), 997–1000.
21. Zhang, Y., Wang, S., Phillips, P., & Ji, G. (2014). Binary PSO with mutation operator for feature selection using decision tree applied to spam detection. *Knowledged Based Systems, 64,* 22–31.
22. Kaveh, A., & Farhoudi, N. (2013). A new optimization method: Dolphin echolocation. *Advanced Engineering Software, 59,* 53–70.
23. Hansen, P., & Mladenović, N. (2018). Variable neighborhood search. *Handbook of Heuristics, 1–2*(1), 759–787.
24. Park, C. H., & Kim, S. B. (2015). Sequential random k-nearest neighbor feature selection for high-dimensional data. *Expert Systems with Applications, 42*(5), 2336–2342.
25. www.gems-system.org.
26. www.ligarto.org/rdiaz/Papers/rfVS/randomForestVarSel.html.

Synchronization of Chaos-Based Recurrent Neural Networks Indicating the Cause of Stochastic Time-Varying Delays

Suresh Rasappan and Sathish Kumar Kumaravel

Abstract The main focus in this article designs a mathematical model to deduct the error of recurrent chaotic neural networks. Employing basic techniques such as the drive response principle is used to evaluate via LMI by different algorithms of convex optimization. LMI procedure along with energy source function (Lyapunov function) is adapted to perform the control strategy. In order to achieve exponential state estimation, the delay contingent feedback mechanism is employed.

Keyword Neural network (NN) · LMI · Lyapunov-Krasovkii · Estimator

1 Introduction

Nowadays, increasing interest has been shown in the emerging applications of artificial neural networks in several diverse branches of knowledge [1–3]. In these models, the characteristics of the equilibrium points are playing a predominant role. As a consequence, wide investigations have been carried out by various researchers in different aspects of NN. Associated time-varying delays and many stability criteria thereof have been obtained [4, 5]. Owing to the exhibition of chaotic behaviour by artificial neural network models, synchronization of chaotic neural networks has become an attractive branch of research. Currently, the researchers are concentrating on state estimation of chaotic NN and coupled delay neural network. By assuming the suitable filters (coupling matrix), the utility of state estimation of chaotic cellular neural network can be examined. When one assumes the irreducibility of the coupling matrix, by applying new technique of Cao and Lu system of equations, the stability is investigated via nonlinear feedback mechanism. In our present work, we provide a new technique to state estimation of two identical coupled chaotic recurrent neural

S. Rasappan (✉) · S. K. Kumaravel
Department of Mathematics, Vel Tech Rangarajan Dr Sagunthala R&D Institute of Science and Technology, Avadi, Chennai, Tamil Nadu 600062, India
e-mail: mrpsuresh83@gmail.com

S. K. Kumaravel
e-mail: k.sathi89@gmail.com

S.-L. Peng et al. (eds.), *Proceedings of First International Conference on Mathematical Modeling and Computational Science*, Advances in Intelligent Systems and Computing 1292, https://doi.org/10.1007/978-981-33-4389-4_29

networks using nonlinear feedback controller by employing the theory of Lyapunov stability [6, 7]. A different algorithm of convex optimization is evaluated with LMI. To employ the above technique, the stability is analysed around slave system.

2 Global State Estimator Result

In accordance with the objective set forth earlier, we deal with the chaotic neural network

$$\mathrm{d}r(t_s) = \left[-Cr(t_s) + Wf(r(t_s)) + W^* f(r(t_s - \phi(t_s))) + J\right]\mathrm{d}t_s \tag{1}$$

where $r(t_s) = (r_1(t_s), r_2(t_s), r_3(t_s), \ldots r_n(t_s))^{\mathrm{T}} \in R^n$ is the state vector of the neural network and W is diagonal matrix with $c_i > 0$, W* is the weight matrix and S the delayed weight matrix with the input vector function $J = [J_1, J_2, \ldots J_n]^{\mathrm{T}}$, the transmission delay ϕ and the activation function $f(r(t_s)) = [f_1(r(t_s)), f_2(r(t_s)), \ldots f_n(r(t_s))]^{\mathrm{T}}$.

The present study has been carried out with two assumptions specified below:

(A1): There exists a positive scalar k_j satisfying $\left\| f_j(r) - f_j(\hat{r}) \right\| \le k_j \left| r - \hat{r} \right|$ for $j = 1, 2, \ldots n$.

(A2): There exists a differential function $(\phi)(t_s) \ge 0$ such as write $(\phi)^* = \max(\phi(t_s))$ and $0 \le \dot{\phi}(t_s) \le \sigma < 1$.

To achieve the synchronization of the system (1), the corresponding controlled slave system has to be

$$\begin{aligned}\mathrm{d}\hat{r}(t_s) &= \left[-C\hat{r}(t_s) + Wf\left(\hat{r}(t_s)\right) + W^* f\left(\hat{r}(t_s - \phi(t_s))\right) + J + u(t_s)\right]\mathrm{d}t_s \\ &\quad + \sum_{i=1}^{n}\left[\sigma_{ij}(t_s, x_c(t_s), x_c(t_s - \phi))\right]\mathrm{d}\omega_j(t_s)\end{aligned} \tag{2}$$

where $\hat{r}(t_s) = \left(\widehat{r_1}(t_s), \widehat{r_2}(t_s), \ldots \widehat{r_n}(t_s)\right)^{\mathrm{T}} \in R^n$ and $u(t_s)$ denotes the driving signal.

For the system furnished by (2), the initial conditions turn out to be $\widehat{r_i}(t_s) = \psi_i(t_s)$. The basic idea present in investigation aims to build a suitable controller $u(t_s)$ where in the synchronization of the coupled neural network takes place [8].

Considering Eqs. (1) and (2), it is seen that the error dynamics system is governed by

$$x_c(t_s) = r(t_s) - \hat{r}(t_s) \text{ and}$$
$$\mathrm{d}x_c(t_s) = \mathrm{d}r(t_s) - \mathrm{d}\hat{r}(t_s)$$

Consequently, it follows that

$$\mathrm{d}x_c(t_s) = \left[-Cx_c(t_s) + Wg(x_c(t_s)) + W^*g(x_c(t_s - \phi(t_s))) - u(t_s)\right]\mathrm{d}t \\ - \sum_{i=1}^{n}\left[\sigma_{ij}(t_s, x_c(t_s), x_c(t_s - \phi))\right]\mathrm{d}\omega_j(t_s) \tag{3}$$

Remark 1 Given constant matrices Θ_1, Θ_2, Θ_3 with $\Theta_1 = \Theta_1^{\mathrm{T}}$ and $0 < \Theta_2 = \Theta_2^{\mathrm{T}}$, then $\Theta_1 + \Theta_3^{\mathrm{T}}\Theta_2^{-2}\Theta_3 < 0$ if and only if

$$\begin{bmatrix} \Theta_1 & \Theta_3^{\mathrm{T}} \\ \Theta_3 & -\Theta_2 \end{bmatrix} \tag{4}$$

Lemma 1 *If Σ_1, Σ_2, Σ_3 are matrices with $\Sigma_3 > 0$, the vector $r(t_s)$ and $\hat{r}(t_s)$ satisfy*

$$2r^T \sum_1^{\mathrm{T}} \sum_2 \hat{r} \le r^{\mathrm{T}} \sum_1^{\mathrm{T}} \sum_3 \sum_1 r + \hat{r}^{\mathrm{T}} \sum_2^{\mathrm{T}} \sum_3^{-1} \sum_2 \hat{r} \tag{5}$$

Lemma 2 *Suppose $r(t_s)$ is an $n \times 1$ vector and $\dot{r}(t_s)$ represents a continuous nonlinear system*

$$\dot{r}(t_s) = g(r(t_s), t_s) \tag{6}$$

Let us associate the Lyapunov function $V(r, t_s)$ imbibed with the following properties:

$$\begin{aligned} (\lambda_1\|r\|)^2 \le{}& V(r, t_s) \le (\lambda_2\|r\|)^2, \forall r, t_s \in R^n \times R \\ \dot{V}(r, t_s) \le{}& -\lambda_3 V(r, t_s) + \lambda_4 \mathrm{e}^{-\alpha t_s}, \forall r, t_s \in R^n \times R \end{aligned} \tag{7}$$

where λ_1, λ_2, λ_3, λ_4 and α are positive scalar constants. The state $r(t_s)$ is observe to be exponentially stable. If the Lyapunov function satisfies the relations (5) and (6), i.e.,

$$\|r(t_s)\| \le \left[\frac{\lambda_2}{\lambda_1}\|r(0)\|^2 \mathrm{e}^{-3\lambda(t_s)} + \frac{\lambda_4}{\lambda_1^2} t\mathrm{e}^{-3\lambda(t_s)}\right]^{\frac{1}{2}} \tag{8}$$

$$\|r(t_s)\| \le \left[\frac{\lambda_2}{\lambda_1}\|r(0)\|^2 \mathrm{e}^{-3\lambda(t_s)} + \frac{\lambda_4}{\lambda_1^2(\lambda_3 - \alpha)}\left(\mathrm{e}^{-\alpha(t_s)} - \mathrm{e}^{-3\lambda(t_s)}\right)\right]^{\frac{1}{2}} \tag{9}$$

Controller design and adaptive synchronization scheme:

Theorem 1 *For the stability of the deriving signal, it has to satisfy $u(t_s)$, where*

$$\gamma P_c x_s(t_s) + \frac{1}{2}\rho \frac{\|x_s(t_s - \phi(t_s))\|^2 \|M_c\|}{P_c \|x_c(t_s)\|} - \frac{1}{2} \frac{K_c^2 x_c^{\mathrm{T}}(t_s) x_c(t_s) P_c^{-1} Q_{c1}^T Q_{c1} x_c t_s}{K_c \|Q_{c1}\| \|x_c(t_s)\|^2 + \in x_c^{-\alpha t_s}} \quad (10)$$

Proof Concerning the error dynamical system (3), one may deal with the Lyapunov function

$$V(t_s) = x_c^{\mathrm{T}}(t_s) P_c x_c(t_s) + \frac{1}{1-\sigma} \int_{t-\phi(t_s)}^{t} g^{\mathrm{T}}(x_c(s)) Q_{c2} g(x_c(s)) \mathrm{d}s \quad (11)$$

Invoking formula

$$\mathrm{d}V(t_s) = LV(t_s)\mathrm{d}t + 2x_c^{\mathrm{T}}(t_s) P_c \sigma(x_c(t_s)) x_c(t_s - \phi(t_s)) \mathrm{d}w(t_s) \quad (12)$$

$$\begin{aligned} LV(t_s) = {} & V_t(x_c(t_s), t_s) \\ & + V_{x_c}(x_c(t_s), t_s)[-K_c x_c(t_s) + W g(x_c(t_s)) \\ & + W^* g(x_c(t_s - \phi(t_s))) - u(t_s)] \\ & + \frac{1}{2} \mathrm{trace}[\sigma(t_s, x_c(t_s)), x_c(t_s - \phi(t_s)) \\ & V_{x_c x_c} \sigma^{\mathrm{T}}(t_s, x_c(t_s)), x_c(t_s - \phi(t_s))] \end{aligned} \quad (13)$$

$$\frac{\partial V(t_s)}{\partial(t_s)} = \frac{1}{1-\sigma} g^{\mathrm{T}}(x_c(t_s)) Q_{c2} g(x_s(t_s)) - g^{\mathrm{T}}(x_c(t_s - \phi)) Q_{c2} g(x_c(t_s - \phi(t_s))) \quad (14)$$

On substitution of Eq. (11) in (13), it is observed that

$$\begin{aligned} LV(t_s) = {} & x_c^{\mathrm{T}}(t_s)\left[-P_c K_c - K_c^{\mathrm{T}} P_c\right] x_s(t_s) + 2x_c^{\mathrm{T}}(t_s) P_c A g(x_c(t_s)) \\ & + 2x_c^{\mathrm{T}}(t_s) P_c B_c g x_c(t_s - \phi(t_s)) - 2\mathrm{e}^{\mathrm{T}}(t_s) P_c u(t_s) \\ & + \frac{1}{1-\sigma} g^{\mathrm{T}}(x_c(t_s)) Q_{c2} g(x_c(t_s)) - g^{\mathrm{T}}(x_c(t_s - \tau(t_s))) Q_{c2} \\ & g(x_c(t_s - \phi(t_s))) + \mathrm{trace}\left[\sigma^{\mathrm{T}}(t_s, x_c(t_s), x_c(t_s - \phi(t_s)))\right. \\ & \left. P_c \sigma(t_s, x_c(t_s), x_c(t_s - \phi(t_s))\right] \end{aligned} \quad (15)$$

Consider the assumption

$$\mathrm{trace}\left[\sigma^{\mathrm{T}}(t_s), x_c(t_s), x_c(t_s - \phi(t_s)) P_c \sigma(t_s, x_c(t_s), x_c(t_s - \phi))\right] \leq Z \quad (16)$$

$$Z = \rho \mathrm{trace}\left[\sigma^{\mathrm{T}}(t_s, x_c(t_s), x_c(t_s - \phi)(\sigma t_s, x_c(t_s), x_c(t_s - \phi(t_s))))\right]$$

One obtains the relation

$$
\begin{aligned}
\frac{\partial V(t_s)}{\partial(t_s)} &= x_c^{\mathrm{T}}(t_s)[\mathrm{T}]x_c(t_s) \\
&+ 2x_c^{\mathrm{T}}(t_s)P_cW_cg(x_c(t_s)) \\
&+ 2x_c^{T}(t_s)P_cW_c^{*}gx_c(t_s-\phi(t_s)) - 2x_c^{\mathrm{T}}(t_s)P_cu(t_s) \\
&+ \frac{1}{1-\sigma}g^{\mathrm{T}}(x_c(t_s))Q_{c2}g(x_c(t_s)) \\
&- g^{\mathrm{T}}(x_c(t_s-\phi(t_s)))Q_{c2}g(x_c(t_s-\phi(t_s))) \\
&+ \rho\big[x_c^{\mathrm{T}}M_c^{\mathrm{T}}M_cx_c(t_s) \\
&+ x_c^{\mathrm{T}}(t_s-\phi(t_s))M_c^{T}M_cx_c(t_s) + x_c^{\mathrm{T}}(t_s-\phi(t_s))\big]
\end{aligned}
\tag{17}
$$

Considering the background of lemma (1) and taking Σ_3 as the identity matrix, it is observed that

$$
\begin{aligned}
2x_c^{\mathrm{T}}(t_s)P_cWg(x_c(t_s)) &\le x_c^{\mathrm{T}}(t_s)P_cWQ_{c1}^{-1}W^{\mathrm{T}}P_cx_c(t_s) \\
&+ g^{\mathrm{T}}(x_c(t_s))Q_{c1}g(x_c(t_s))
\end{aligned}
$$

$$
\begin{aligned}
2x_c^{\mathrm{T}}(t_s)P_cW^{*}g(x_c(t_s-\phi(t_s))) &\le x_c^{\mathrm{T}}(t_s)P_cW^{*}Q_{c2}^{-1}B_c^{\mathrm{T}}P_cx_c(t_s) \\
&+ g^{\mathrm{T}}(x_c(t_s-\tau(t_s)))Q_{c2}g(x_c(t_s-\phi(t_s)))
\end{aligned}
$$

$$
\begin{aligned}
\frac{\partial V(t_s)}{\partial(t_s)} &= x_c^{\mathrm{T}}(t_s)[-P_cK_c - K_c^{\mathrm{T}}P_c + P_cW_cQ_{c1}^{-1}W^{\mathrm{T}}P_c \\
&+ P_cW^{*}Q_{c2}^{-1}B_c^{\mathrm{T}}P_c + M_c^{\mathrm{T}}\rho M_c]x_s(t_s) \\
&+ g^{\mathrm{T}}(x_c(t_s))Q_{c1}g(x_c(t_s)) + \frac{1}{1-\sigma}g^{\mathrm{T}}(x_c(t_s))Q_{c2}g(x_c(t_s)) \\
&+ x_c^{\mathrm{T}}(t_s-\phi(t_s))M_{c1}^{\mathrm{T}}\rho M_{c1}x_c(t_s-\phi(t_s)) - 2x_c^{\mathrm{T}}(t_s)\rho u(t_s)
\end{aligned}
\tag{18}
$$

Referring to the assumption

$$
g^{\mathrm{T}}(x_c(t_s))Q_{c2}g(x_c(t_s)) \le x_c^{\mathrm{T}}(t_s)K_c^{\mathrm{T}}Q_{c2}K_cx_c(t_s) \tag{19}
$$

$$
g^{\mathrm{T}}(x_c(t_s))Q_{c1}g(x_c(t_s)) \le x_c^{\mathrm{T}}(t_s)K_c^{\mathrm{T}}Q_{c1}K_cx_c(t_s) \tag{20}
$$

We get

$$
\begin{aligned}
LV(t_s) = -x_c^{\mathrm{T}}(t_s)\Big[&P_cK_c + K^{c\mathrm{T}}P_c - P_cW^{*}Q_{c1}^{-1}W^{\mathrm{T}}P_c \\
&- P_cW^{*}Q_{c2}^{-1}W^{*\mathrm{T}}P_c - M_c^{\mathrm{T}}\rho M_c - K_c^{\mathrm{T}}Q_{c1}K_c \\
&- \frac{1}{1-\sigma}K_c^{\mathrm{T}}Q_{c2}K_c - \frac{\phi^{*}}{1-\sigma}K_c^{\mathrm{T}}Q_{c2}K_c\Big]x_c(t_s)
\end{aligned}
$$

$$+ \|x_c(t_s - \phi(t_s))\|^2 \rho \|M_{c1}\| - 2x_c^{\mathrm{T}}(t_s)\rho u(t_s)$$
$$- \frac{\phi^*}{1-\sigma} x_c^{\mathrm{T}}(t_s) K_c^{\mathrm{T}} Q_{c2} K_c x_c(t_s) \tag{21}$$

Now, substitution of the nonlinear adaptive feedback control (8) leads to the relation

$$\mathrm{d}V(t_s) = \left[\frac{K_c \|x_c(t_s)\|^2 \|Q_{c1}\| \varepsilon x_c^{-\alpha t_s}}{K_c \|x_c(t_s)\|^2 \|Q_{c1}\| + \varepsilon x_c^{-\alpha t_s}} - \frac{\phi^*}{1-\sigma} x_c^{\mathrm{T}}(t_s) K_c^{\mathrm{T}} Qc_2 K_c x_c(t_s) \right] \mathrm{d}t_s$$
$$+ 2x_c^{\mathrm{T}}(t_s)\sigma(x_c(t_s), x_c(t_s - \tau(t_s)))\mathrm{d}w(t_s) \tag{22}$$

Taking into account the expectation on both sides, one observes that

$$E[\mathrm{d}V(t_s)] = E\left[\frac{K_c \|x_c(t_s)\|^2 \|Q_{c1}\| \varepsilon x_c^{-\alpha t_s}}{K_c \|x_c(t_s)\|^2 \|Q_{c1}\| + \varepsilon x_c^{-\alpha t_s}} - \frac{\phi^*}{1-\sigma} x_c^{\mathrm{T}}(t_s) K_c^{\mathrm{T}} Qc_2 K_c x_c(t_s) \right] \mathrm{d}t_s \Bigg]$$
$$+ E[2x_c^{\mathrm{T}}(t_s)\sigma(x_c(t_s), x_c(t_s - \tau(t_s)))\mathrm{d}w(t_s)] \tag{23}$$

$$\dot{V}(t_s) \le -x_c^{\mathrm{T}}(t_s)\Omega x_c(t_s) + \frac{K_c \|x_c(t_s)\|^2 \|Q_{c1}\| \in x_c^{-\alpha t_c}}{K_c \|x_c(t_s)\|^2 \|Q_{c1}\| + \in x_c^{-\alpha t_c}}$$
$$- \int_{t_s - \phi(t_s)}^{t_c} g^{\mathrm{T}}(x_c(t_s)) Q_{c2} g(x_c(t_s)) \mathrm{d}t_s \tag{24}$$

Employing the inequality

$$0 \le \frac{ab}{a+b} \le a$$

with

$$a = \in x_c^{-\alpha t}, b = K_c \|x_c(t_s)\|^2 \|Q_{c1}\|$$

it is seen that

$$\dot{V} \le -\frac{\lambda(\Omega)}{\lambda(P_c)} V(t_s) + \in x_c^{-\alpha t_s}. \tag{25}$$

By Eq. (6)

$$\dot{V} \le -\lambda_3 V(t_s) + \lambda_4 x_c^{-\alpha t_s}$$

$$\dot{V} \leq -\frac{\lambda(\Omega)}{\lambda(P_c)} + \in x_c^{-\alpha t_s}$$

where

$$\lambda_3 = \frac{\lambda(\Omega)}{\lambda(P_c)}, \lambda_4 = \in$$

Hence, we obtain (24).

Next, we have

$$\Delta\Omega = \begin{bmatrix} -P_c C_c - K_C^{\mathrm{T}} K_C + \frac{1+\phi^*}{1-\sigma} K_C^{\mathrm{T}} Q_{C2} K_C + K_C^{\mathrm{T}} Q_{C1} K_C + M_C^{\mathrm{T}} & P_C & P_C W & P_C W^* \\ P_C & \gamma_1^{-1} & 0 & 0 \\ W^T P_C & 0 & -Q_{C1} & 0 \\ W^{*T} P_C & 0 & 0 & -Q_{C2} \end{bmatrix} < 0 \tag{26}$$

This relationship implies the occurrence of stability exponentially for the proposed system. It follows that global synchronization is achieved by the resulting controlled slave system along with the master system.

3 Numerical Simulation

The chaotic recurrent neural network with stochastic time-varying delay dynamics is assumed for the numerical example. The CRNN equations are given as

$$\begin{aligned} \dot{x}_1(t_c) &= -5x_1(t_c) + 0.8f(x_1(t_c - \tau)) + 0.12f(x_2(t_c - \tau)) \\ &\quad + 0.45 \int_{-\infty}^{t} (K_1(t_c - s) f_1(x_1(s))\, \mathrm{d}s \\ &\quad + 0.1 \int_{-\infty}^{t} (K_2(t_c - s) f_1(x_2(s))\mathrm{d}s + 0.59\dot{x}_1(t_c - h) + 1 \\ \dot{x}_2(t) &= -5x_2(t_c) + 0.8f(x_1(t_c - \tau)) + 0.3f(x_2(t_c - \tau)) \\ &\quad + 0.45 \int_{-\infty}^{t} (K_1(t_c - s) f_2(x_1(s))\, \mathrm{d}s \\ &\quad + 0.1 \int_{-\infty}^{t} (K_2(t_c - s) f_2(x_2(s))\mathrm{d}s + 0.59\dot{x}_2(t_c - h) + 2 \end{aligned} \tag{27}$$

where $\tau = h = 1$ and the stimulated function is described by a nonlinear function $f_i(x) = \tanh(x)$, $i = 1, 2$. The constant matrices are

$$A = \begin{bmatrix} 5 & 0 \\ 0 & 5 \end{bmatrix}, B = \begin{bmatrix} 0.8 & 0.12 \\ 0.8 & 0.3 \end{bmatrix}, C = \begin{bmatrix} 0.45 & 0.1 \\ 0.45 & 0.1 \end{bmatrix}, D = \begin{bmatrix} 0.59 & 0 \\ 0 & 0.59 \end{bmatrix}$$

Solving by generalized eigenvalue problem in Theorem 3.3 by invoking the MATLAB LMI control toolbox, we obtain the solution as

$$\alpha = 1.2109$$

$$P = 10^{-16}\begin{bmatrix} 0.4008 & -0.3193 \\ -0.3193 & 0.2802 \end{bmatrix}$$

$$Q = 10^{-16}\begin{bmatrix} 0.6265 & -0.0966 \\ -0.0966 & 0.0172 \end{bmatrix}$$

$$R = 10^{-17}\begin{bmatrix} 0.6715 & -0.5332 \\ -0.5332 & 0.4752 \end{bmatrix}$$

$$E = 10^{-18}\begin{bmatrix} 0.2956 & 0 \\ 0 & 0.2956 \end{bmatrix}$$

$$K_1 = 10^{-16}\begin{bmatrix} 0.2254 & -0.1832 \\ -01832 & 0.1446 \end{bmatrix}$$

$$K_2 = 10^{-16}\begin{bmatrix} 0.4055 & -0.0599 \\ -0.0599 & 0.0116 \end{bmatrix}$$

$$K_3 = 10^{-17}\begin{bmatrix} 0.6250 & -0.4934 \\ -0.4934 & 0.4522 \end{bmatrix}.$$

4 Conclusion

Chaos-based recurrent neural networks indicating the cause of varying delay are considered in this chapter. The state estimation sufficient conditions are derived under LMI. The global exponential order synchronization is obtained. For this calculation, the new recurrent chaotic neural network models are developed. The threshold function tanh acts as an activation function of the model. The numerical calculations are demonstrated for the derived results. The MATLAB DDE solver and LMI toolbox are used.

References

1. Cui, B., & Lou, X. (2009). Synchronization of chaotic recurrent neural networks with time-varying delays using nonlinear feedback control. *Chaos, Solitons & Fractals, 39*(1), 288–294.
2. Arik, S. (2000). Stability analysis of delayed neural networks. *IEEE Transactions on Circuits and Systems I: Fundamental Theory and Applications, 47*(7), 1089–1092.
3. Forti, M., Nistri, P., & Papini, D. (2005). Global exponential stability and global convergence in finite time of delayed neural networks with infinite gain. *IEEE Transactions on Neural Networks, 16*(6), 1449–1463.
4. Chen, W. H., Lu, X., Guan, Z. H., & Zheng, W. X. (2006). Delay-dependent exponential stability of neural networks with variable delay: An LMI approach. *IEEE Transactions on Circuits and Systems II: Express Briefs, 53*(9), 837–842.
5. Xu, S., Lam, J., Ho, D. W., & Zou, Y. (2005). Delay-dependent exponential stability for a class of neural networks with time delays. *Journal of Computational and Applied Mathematics, 183*(1), 16–28.
6. Liu, Y., Wang, Z., & Liu, X. (2007). Design of exponential state estimators for neural networks with mixed time delays. *Physics Letters A, 364*(5), 401–412.
7. Zhang, H., Xie, Y., Wang, Z., & Zheng, C. (2007). Adaptive synchronization between two different chaotic neural networks with time delay. *IEEE Transactions on Neural Networks, 18*(6), 1841–1845.
8. Balasubramaniam, P., & Vembarasan, V. (2012). Synchronization of recurrent neural networks with mixed time-delays via output coupling with delayed feedback. *Nonlinear Dynamics, 70*(1), 677–691.

Analysis of Observability and Stability of HIV Epidemiology in a Migrant Heterosexual Population

Suresh Rasappan and Regan Murugesan

Abstract The government and various authorized agencies are trying to reduce HIV population day by day. The proposed mathematical model may help to observe the patch of HIV infected or suspected population by nominated agencies. The proposed mathematical model supports the observability and asymptotic stability of the propagation of HIV epidemiology in a migrant heterosexual affected population.

Keywords Stability · HIV · Heterosexual · Lyapunov · AVERT · Observer

1 Introduction

The human immunodeficiency virus (HIV) turned into diagnosed in 1980s. It is a type of lentivirus. It continues to be a noteworthy worldwide health problem [1–3]. According to 2017 statistics, 36.9 million populace live with HIV, around 25% do not know that they have the virus. 940,000 individuals are expired because of AIDS-related illnesses.

The majority of HIV people are located in low-level income countries [4]. In India, people live with HIV are counted as 2.1 million as well as it has the third-largest HIV epidemic around the world. Compared to 2010 statistics, presently there is 27% slow down in HIV epidemic. Labourers involving in sex, sexual intercourse between men (MSM), people who inject drugs (PWID), Hijras/transgender individuals, migrant workers and truck drivers are the major key affected HIV population in India. In particularly 86% of HIV people account for sexual transmission.

NACO [4] is in the place of defining policy and executing programs for hindrance and controlling the HIV pestilence in India. The main motive of the NACO is to decrease the annual rate of new HIV infections with the support of its treatment,

---

S. Rasappan (✉) · R. Murugesan
Department of Mathematics, Vel Tech Rangarajan Dr. Sagunthala R&D Institute of Science and Technology, Avadi, Chennai, Tamil Nadu 600062, India
e-mail: mrpsuresh83@gmail.com

R. Murugesan
e-mail: mreganprof@gmail.com

S.-L. Peng et al. (eds.), *Proceedings of First International Conference on Mathematical Modeling and Computational Science*, Advances in Intelligent Systems and Computing 1292, https://doi.org/10.1007/978-981-33-4389-4_30

awareness on HIV through education, care and give moral support to the people who are at high risk of HIV. This study focuses attention on the prevention of HIV affected people by mathematical analysis. Its main focus is on HIV education, counselling support. Due to continuous monitoring (feedback from two HIV prevention programmes) of key affected people, the annual threshold level may be stabilized. Our results may also support the progress towards 90–90–90 objective in India. In Sect. 2, the observation algorithm and stability concept is discussed. Finally, in Sect. 3, the conclusion is discussed and results presented.

2 An Observation Algorithm and Stability for HIV Epidemiological Model

HIV susceptible, infected male and female population considered a class of epidemiological dynamical system. It is framed by the mathematical equation as follows [5–7]:

$$\dot{x} = Ax + f(t, x). \tag{1}$$

where A represents the constant matrix. The coefficient of matrix A is collected by a number of susceptible females, infected female, susceptible male and infected male by HIV. $f(t, x)$ represents any one of the following situations: it may represent infected male mat with a suspected female, infected male mat with an infected female, infected female mat with suspected male or an infected female mat with an infected male. These are the causes of nonlinearity involvement in mobile heterosexual population dynamics.

The HIV programmes are designed to handle the HIV epidemiology tailored under two main category: one is prevention and another one is treated. The proposed concept is concentrated on prevention.

The prevention of HIV initiatives is very much required to stop HIV transmission among various key affcted population and also the general public. The main objective of the proposed model is

- To lessen the spread of HIV infection or contamination.
- To increase the quantity to response the HIV on a lengthy basis.

"National Adolescent Education Programme" (NAEP), counselling, testing and PPRT programmes, "National Anti Retroviral Treatment Programmes" and setting up of the national council on AIDS are identified as some of the main prevention programmes stretch the HIV and reduction of the spread of HIV infection.

Based on the concept of controlled non-autonomous, epidemiology on HIV prevention is given by the system of equation.

$$\left.\begin{aligned}\dot{y} &= Ay + f(t, y) + Bu(t) + C(t)\\ u(t) &= kz\end{aligned}\right\} \tag{2}$$

where A is a constant matrix. The coefficient of matrix A is collected by several observer ones who take care of a susceptible female, infected female, susceptible male and infected male by HIV. $f(t, y)$ represents the observer who can take care of any one of the following situations: it may represent infected male mat with a suspected female, infected male mat with an infected female, infected female mat with suspected male or an infected female mat with an infected male. These are the cause of then on linearity involvement in mobile heterosexual population dynamics. $u(t)$ is the feedback controller, it is the treatment part or counselling part in the HIV epidemic.

The continuous feedback from various nominated agencies, councillors and active involvement of HIV under treatment peoples are the key factors to design the controller. The controller reduces the HIV awareness or sustains some level of population in society.

Consider the error dynamics,

$$Z(t) = x(t) - y(t)$$

Thus, the observation error between the epidemiological dynamics given in Eqs. (1) and (2) can be derived by

$$\begin{aligned}\dot{z} &= \dot{x} - \dot{y}\\ \dot{z} &= Az + f(t, z) - kBz - C(t)\end{aligned} \tag{3}$$

Theorem 1 *If B is a matrix with the positive elements and $C(t)$ holds on the inequality $f(t, z) - C(t) \leq \sum_{i=1}^{r} \alpha_i z$ in R, then the system of* Eq. (3) *is asymptotically stable in the region R.*

Proof

$$\dot{\theta} = -\frac{1}{\theta}\sum_{i=1}^{r}\frac{\alpha_i^2}{4\beta_i}\left\|\frac{\partial V}{\partial z}\right\|^2, \tag{4}$$

where θ is the parameter with feedback control and β_j are denoted as positive constants. Since the matrix A and B contain an eigenvalue as negative and positive, respectively, then there exists a matrix D such that

$$A^{\mathrm{T}}B + BA = -D$$

Now, consider the equation,

$$V = \sum_{i=1}^{r}\frac{1}{i}\left(z^{\mathrm{T}}Dz\right)^{i}. \tag{5}$$

Now, define a Lyapunov functional as

$$L(t,z)=\sum_i^r \frac{1}{i}\left(z^{\mathrm{T}}Dz\right)^i+\frac{1}{2}\theta^2$$

$$\dot{L}(t,z)=\sum_i^r \frac{1}{i}\left(z^{\mathrm{T}}Dz\right)^{i-1}2z^T D\dot{z}+\theta\dot{\theta}$$

$$=\sum_i^r (z^{\mathrm{T}}Dz)^{i-1}2z^{\mathrm{T}}D[Az+f(t,z)-kBZ-C(t)]+\theta\left[-\frac{1}{\theta}\sum_{i=1}^r \frac{\alpha_i^2}{4\beta_i}\left\|\frac{\partial V}{\partial z}\right\|^2\right]$$

$$=\sum_i^r (z^{\mathrm{T}}Dz)^{i-1}2z^T D[Az]+\sum_i^r (z^{\mathrm{T}}Dz)^{i-1}2z^{\mathrm{T}}D[f(t,z)-C(t)]$$

$$-\sum_i^r (z^{\mathrm{T}}Dz)^{i-1}2z^{\mathrm{T}}D[kBz]-\sum_i^r \frac{\alpha_i^2}{4\beta_i}\left\|\frac{\partial V}{\partial z}\right\|^2$$

$$\le\sum_i^r (z^{\mathrm{T}}Dz)^{i-1}2z^{\mathrm{T}}DAz+\left\|\frac{\partial V}{\partial z}\right\|\alpha_i z^i-\sum_i^r \frac{\alpha_i^2}{4\beta_i}\left\|\frac{\partial V}{\partial z}\right\|^2$$

$$-\sum_i^r (z^{\mathrm{T}}Dz)^{i-1}2z^{\mathrm{T}}D[\mathrm{kBz}]$$

$$\le\sum_i^r (z^{\mathrm{T}}Dz)^{i-1}2z^{\mathrm{T}}DAz+\sum_i^r\left[\frac{\alpha_i^2}{4\beta_i}\left\|\frac{\partial V}{\partial z}\right\|^2+\beta_i z^{2i}\right]-\sum_i^r \frac{\alpha_i^2}{4\beta_i}\left\|\frac{\partial V}{\partial z}\right\|^2$$

$$-\sum_i^r (z^{\mathrm{T}}Dz)^{i-1}2z^{\mathrm{T}}D[\mathrm{kBz}]$$

$$\le\sum_i^r (z^{\mathrm{T}}Dz)^{i-1}2z^{\mathrm{T}}DAz+\sum_i^r\left[\frac{\alpha_i^2}{4\beta_i}\left\|\frac{\partial V}{\partial z}\right\|^2+\beta_i z^{2i}\right]-\sum_i^r \frac{\alpha_i^2}{4\beta_i}\left\|\frac{\partial V}{\partial z}\right\|^2$$

$$-\sum_i^r (z^{\mathrm{T}}Dz)^{i-1}2z^{\mathrm{T}}D[\mathrm{kBz}]$$

$$\le\sum_i^r (z^{\mathrm{T}}Dz)^{i-1}2z^{\mathrm{T}}DAz+\sum_i^r \beta_i\|z\|^{2i}-\sum_i^r (z^T Dz)^{i-1}2z^{\mathrm{T}}D[\mathrm{kBz}]$$

$$\le\lambda_{\min}(DA)\lambda_{\min}^{i-1}(D)\sum_{i=1}^r\|z\|^{2i}+\sum_i^r \beta_i\|z\|^{2i}-\sum_i^r (z^{\mathrm{T}}Dz)^{i-1}2z^{\mathrm{T}}D[\mathrm{kBz}] \quad (6)$$

Here, the parameter β_i values are enough to small, so

$$\beta_i\le\lambda_{\min}(DA)\lambda_{\min}^{i-1}(D) \quad (7)$$

From Eq. (6) and (7), we get

$$\dot{L} \leq 0.$$

Hence, the error system (3) is asymptotically stable based on the stability concept of Lyapunov theory for the non-autonomous system in region R.

The dynamics of the propagation of HIV in a migrant heterosexual affected population is

$$\begin{aligned}
\frac{dx_1^{(r)}}{dt} &= \frac{\mu}{2} - \sum_{j=1}^{r} \beta_{rj} \frac{y_2^{(j)}}{n_M^{(j)}} x_1^{(r)} - \mu x_1^{(r)} \\
\frac{dx_2^{(r)}}{dt} &= \frac{\mu}{2} - \sum_{j=1}^{r} \beta_{rj} \frac{y_2^{(j)}}{n_M^{(j)}} x_1^{(r)} - \mu x_1^{(r)} \\
\frac{dy_1^{(r)}}{dt} &= \frac{\mu}{2} - \sum_{j=1}^{r} \beta_{rj} \frac{x_2^{(j)}}{n_F^{(j)}} y_1^{(r)} - \mu y_1^{(r)} \\
\frac{dy_2^{(r)}}{dt} &= \sum_{j=1}^{r} \beta_{rj} \frac{x_2^{(j)}}{n_F^{(j)}} y_1^{(r)} - (\mu + \gamma) y_2^{(r)}
\end{aligned} \tag{8}$$

where $n_F^{(r)} = x_1^{(r)} + x_2^{(r)}, n_M^{(r)} = y_1^{(r)} + y_2^{(r)}$ and β_{rj} denotes the infection rate of susceptible (males and females) in group "r" by infected individuals from group "j" . The mobility of infection in each group "r" is given by the term $\sum_{j=1}^{k} \beta_{rj} \frac{y_2^{(j)}}{n_M^{(j)}}$ for susceptible females and $\sum_{j=1}^{k} \beta_{rj} \frac{x_2^{(j)}}{n_F^{(j)}}$ for susceptible males. The natural death and death due to AIDS diseases are at the rate of μ and δ, and the individuals who are infected which develop AIDS at the rate of γ and their observation dynamics are

$$\begin{aligned}
\frac{d\hat{x}_1^{(r)}}{dt} &= \frac{\mu}{2} - \sum_{j=1}^{r} \beta_{rj} \frac{\hat{y}_2^{(j)}}{n_M^{(j)}} \hat{x}_1^{(r)} - \mu \hat{x}_1^{(r)} \\
\frac{d\hat{x}_2^{(r)}}{dt} &= \sum_{j=1}^{r} \beta_{rj} \frac{\hat{y}_2^{(j)}}{n_M^{(j)}} \hat{x}_1^{(r)} - (\mu + \gamma) \hat{x}_2^{(r)} \\
\frac{d\hat{y}_1^{(r)}}{dt} &= \frac{d\hat{y}_1^{(r)}}{dt} = \frac{\mu}{2} - \sum_{j=1}^{r} \beta_{rj} \frac{\hat{x}_2^{(j)}}{n_F^{(j)}} \hat{y}_1^{(r)} - \mu \hat{y}_1^{(r)} \\
\frac{d\hat{y}_2^{(r)}}{dt} &= \sum_{j=1}^{r} \beta_{rj} \frac{\hat{x}_2^{(j)}}{n_F^{(j)}} \hat{y}_1^{(r)} - (\mu + \gamma) \hat{y}_2^{(r)}
\end{aligned} \tag{9}$$

Here,

$$A = \begin{bmatrix} \mu & 0 & 0 & 0 \\ 0 & -(\mu+\gamma) & 0 & 0 \\ 0 & 0 & \mu & 0 \\ 0 & 0 & 0 & -(\mu+\gamma) \end{bmatrix}, f(t,x) = \begin{bmatrix} -\sum_{j=1}^{r} \beta_{rj} \frac{y_2^{(j)}}{n_M^{(j)}} x_1^{(r)} \\ \sum_{j=1}^{r} \beta_{rj} \frac{y_2^{(j)}}{n_M^{(j)}} x_1^{(r)} \\ -\sum_{j=1}^{r} \beta_{rj} \frac{x_2^{(j)}}{n_F^{(j)}} y_1^{(r)} \\ \sum_{j=1}^{r} \beta_{rj} \frac{x_2^{(j)}}{n_F^{(j)}} y_1^{(r)} \end{bmatrix},$$

$$B = \begin{bmatrix} 0.01 & 0 & 0 & 0 \\ 0 & 0.01 & 0 & 0 \\ 0 & 0 & 0.01 & 0 \\ 0 & 0 & 0 & 0.01 \end{bmatrix}, C(t) = \begin{bmatrix} \frac{\mu}{2} \\ 0 \\ \frac{\mu}{2} \\ 0 \end{bmatrix}$$

The estimation of state error dynamics may be obtained by

$$z(t) = \hat{x}(t) - x(t);\ z_1 = \hat{x}_1^{(r)} - x_1^{(r)};\ z_2 = \hat{x}_2^{(r)} - x_2^{(r)}$$

$$z_3 = \hat{y}_1^{(r)} - y_1^{(r)};\ z_4 = \hat{y}_2^{(r)} - y_2^{(r)};\ z_4^{(j)} = \hat{y}_2^{(j)} - y_2^{(j)}$$

$$z_5^{(j)} = \hat{x}_2^{(j)} - x_2^{(j)};\ \dot{z}(t) = \dot{\hat{x}}(t) - \dot{x}(t);\ \dot{z}_1 = \dot{\hat{x}}_1^{(r)} - \dot{x}_1^{(r)}$$

$$\dot{z}_2 = \dot{\hat{x}}_2^{(r)} - \dot{x}_2^{(r)};\ \dot{z}_3 = \dot{\hat{y}}_{*-1}^{(r)} - \dot{y}_1^{(r)};\ \dot{z}_4 = \dot{\hat{y}}_2^{(r)} - \dot{y}_2^{(r)}$$

which gives

$$\begin{aligned} \dot{z}_1 &= -\sum_{j=1}^{k} \frac{\beta_{rj}}{n_M^{(j)}} \left[\hat{y}_2^{(j)} z_1 + x_1^{(r)} z_4^{(j)} \right] - \mu z_1 \\ \dot{z}_2 &= \sum_{j=1}^{k} \frac{\beta_{rj}}{n_M^{(j)}} \left[\hat{y}_2^{(j)} z_1 + x_1^{(r)} z_4^{(j)} \right] - (\mu+\gamma) z_2 \\ \dot{z}_3 &= -\sum_{j=1}^{k} \frac{\beta_{rj}}{n_F^{(j)}} \left[\hat{y}_1^{(r)} z_5^{(j)} + x_2^{(j)} z_3 \right] - \mu z_3 \\ \dot{z}_4 &= \sum_{j=1}^{k} \frac{\beta_{rj}}{n_F^{(j)}} \left[\hat{x}_2^{(j)} z_3 + y_1^{(r)} z_5^{(j)} \right] - (\mu+\gamma) z_4 \end{aligned} \quad (10)$$

Now, define the Lyapunov function as

$$V = \sum_{i=1}^{4} \frac{1}{2}\left(z^{\mathrm{T}} D z\right). \tag{11}$$

Now, the derivative of the Eq. (11) is defined by

$$\dot{V} = \sum_{i=1}^{4} \frac{1}{2} z_i \dot{z}_i. \tag{12}$$

The controller u_1, u_2, u_3 and u_4 is designed as

$$\begin{aligned}
u_1 &= \sum_{j=1}^{k} \frac{\beta_{rj}}{n_M^{(j)}}\left[\hat{y}_2^{(j)} z_1 + x_1^{(r)} z_4^{(j)}\right] \\
u_2 &= -\sum_{j=1}^{k} \frac{\beta_{rj}}{n_M^{(j)}}\left[\hat{y}_2^{(j)} z_1 + x_1^{(r)} z_4^{(j)}\right] z_2 \\
u_3 &= \sum_{j=1}^{k} \frac{\beta_{rj}}{n_F^{(j)}}\left[\hat{y}_1^{(r)} z_5^{(j)} + x_2^{(j)} z_3\right] \\
u_4 &= -\sum_{j=1}^{k} \frac{\beta_{rj}}{n_F^{(j)}}\left[\hat{x}_2^{(j)} z_3 + y_1^{(r)} z_5^{(j)}\right]
\end{aligned} \tag{13}$$

which implies that

$$\dot{V} = -\sum_{i=1}^{4} z_i^2. \tag{14}$$

The error system (10) is asymptotically stable by the stability concept of Lyapunov theory for non-autonomous system [8–10]

From these results, the following conclusions are drawn

1. An epidemic will maintain some threshold population occur only if the counselling or any treatment is continuously fed in susceptible female or male.
2. The spread of HIV does not stop when there is a lack of susceptible populace, and at the same time, it stops when there is a lack of infective. Specifically, some people will escape from HIV altogether.

The result (1) respect to the general observations that epidemic tends to build up very much quickly when the density of susceptible is high because of overcrowding and removal rate is low due to inadequate medical care and awareness of HIV.

3 Conclusion

This paper investigated the observation and asymptotic stability of the patch of HIV infected or suspected population. The observation feedback control is designed: this leads to the result as indicated by our mathematical model, continuous observability will lead the society to maintain a pre-decided threshold value of the HIV affected population, and by feedback control, the stability of suspected population of HIV can be maintained below a threshold value. These measures will ensure that there is no expansion of the endangered HIV population.

References

1. Anderson, R. M., Medley, G. F., May, R. M., & Johnson, A. M. (1986). A preliminary study of the transmission dynamics of the human immunodeficiency virus (HIV), the causative agent of AIDS. *Mathematical Medicine and Biology: a Journal of the IMA, 3*(4), 229–263.
2. Afanas'ev, V. N., Kolmanovskij, V. B., & Nosov, V. R. (1996). *Mathematical theory of control systems design*. Dordrecht: Kluwer Academic.
3. Hyman, J. M., & LaForce, T. (2003). Modeling the spread of influenza among cities. In *Bioterrorism: Mathematical modeling applications in homeland security* (pp. 211–236). Society for Industrial and Applied Mathematics.
4. Global information and education on HIV and AIDS. https://www.avert.org/global-hiv-and-aids-statistics.
5. Murugesan, R., & Rasappan, S. (2019). State estimation of a stochastic model for the spread of HIV in a mobile heterosexual population. *Indian Journal of Public Health Research & Development, 10*(6), 1444–1454.
6. Suresh, R., & Sundarapandian, V. (2010). Synchronization of an optical hyper-chaotic system. *International Journal of Computational and Applied Mathematics, 5*(2), 201–209.
7. Sani, A., Kroese, D. P., & Pollett, P. K. (2007). Stochastic models for the spread of HIV in a mobile heterosexual population. *Mathematical Biosciences, 208*(1), 98–124.
8. Mukherjee, D. (2003). Stability analysis of a stochastic model for prey-predator system with disease in the prey. *Nonlinear Analysis: Modelling and Control, 8*(2), 83–92.
9. Ding, Y., Xu, M., & Hu, L. (2008). Asymptotic behavior and stability of a stochastic model for AIDS transmission. *Applied Mathematics and Computation, 204*(1), 99–108.
10. Fang, Y., Wu, H., & Zhu, L. X. (2011). A two-stage estimation method for random coefficient differential equation models with application to longitudinal HIV dynamic data. *Statistica Sinica, 21*(3), 1145.

Proper Lucky Labeling of Graph

T. V. Sateesh Kumar and S. Meenakshi

Abstract The labeling is said to be lucky labeling of the graph if the vertices of the graph are labeled by natural number with satisfying the condition that sum of labels over the adjacent of the vertices in the graph are not equal, and if vertices are isolated vertex, then the sum of the vertex is zero. The least natural number which labeled the graph is the lucky number. The lucky number of graph G is denoted by $\eta(G)$. The labeling is defined as proper labeling if the vertices of the graph are labeled by natural number with fulfilling the condition that label of adjacent vertices is not the same. The labeling is defined as proper lucky labeling if labeling is proper and also lucky. The proper lucky number of graph G is denoted by $\eta_p(G)$. Here, we obtain a proper lucky number for complete bipartite graph $K_{m,n}$, friendship graph F_n and certain book graph such as triangular book graph B_3^t and rectangular book graph B_4^t.

Keywords Proper lucky labeling · Book graph · Triangular book · Rectangular book · Quadrilateral book · Complete bipartite graph · Friendship graph

1 Introduction

In graph theory, graph labeling has a vast area. It has a solid connection among numbers and the structure of the graph. It was introduced by Rosa in 1967 [13, 14]. Labeling is a task to assign the number to a graph vertex or edge or both with the definite condition. Gallian [8] gave a dynamic survey of graph labeling. Several graph labeling was initiated by many researchers in the past decades. The proper labeling was initiated by Karonsik [9] which mean naming adjacent vertices by a different natural number. In several cases, it is fascinating to the condition that the sum of neighbor labeling is not the same together with the label of adjacent also

T. V. Sateesh Kumar · S. Meenakshi (✉)
Department of Mathematics, Vels Institute of Science, Technology and Advanced Studies, Pallavaram, Chennai, India
e-mail: meena.sbs@velsuniv.ac.in

T. V. Sateesh Kumar
e-mail: sateeshkumarsks@gmail.com

S.-L. Peng et al. (eds.), *Proceedings of First International Conference on Mathematical Modeling and Computational Science*, Advances in Intelligent Systems and Computing 1292, https://doi.org/10.1007/978-981-33-4389-4_31

not the same which is called by proper lucky labeling. The graph which fulfills the proper lucky labeling conditions is called a proper lucky graph. It has a very close association with graph coloring.

Lucky labeling for three colourable graphs was studied by Ahadi et al. [1]. Lucky labeling is compared with proper vertex coloring by Akbari et al. [3]. For some proper labeling problems, algorithmic complexity was studied by Dehghan et al. [7]. The recently lucky number of $m \times n$ dimensional bloom graph is derived by Kujur and Chiranjilal [11]. Labeling application is computer networking, clustering, image segmentation, etc., and it is also applicable in astronomy, circuit designing, image process, cryptography and information security. The objective of this paper is to study the proper lucky number of a complete bipartite graph, friendship graph and certain book graph such as triangular book graph and rectangular book graph.

2 Preliminaries

The following are some essential definitions and preliminaries related to this paper.

2.1 Lucky Labeling

Let $f: V(G) \to \mathbb{N}$ be a labeling of the vertices of a graph G by a natural number. Let $S(v)$ indicate the sum of labels over the neighbors of the vertex v in G. If v is an isolated vertex of G, we put $S(v) = 0$. A labeling f is lucky if $S(u) \neq S(v)$ for an adjacent vertex u and v [6].

2.2 Proper Lucky Labeling

A labeling f is proper lucky labeling, if u and v are adjacent in G, then $f(u) \neq f(v)$ & $S(u) \neq S(v)$. The proper lucky number of G is denoted by $\eta_p(G)$, is the least natural number k such that G has proper lucky labeling with $\{1, 2, \ldots, k\}$ as the set of labels [5, 7, 10].

Example Refer Fig. 1, the value mentioned within parenthesis is the sum of labels over neighbor vertices.

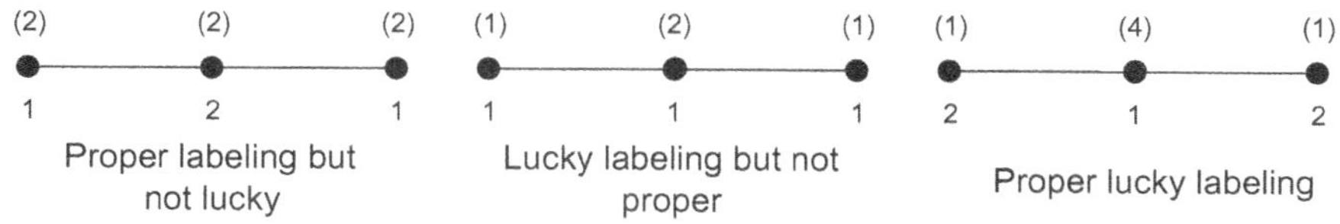

Fig. 1 Illustration for labeling

2.3 *Book Graph*

Arranging the cycles of the same length in a single edge is called a book. The book is denoted by B_n^t where t is a copy and n is the length of the cycle. If $n = 3$, then the book B is said to be a triangular book. If $n = 4$, then the book B is said to be the rectangular (or quadrilateral) book [2, 12, 15].

2.4 *Friendship Graph*

The friendship graph F_n is n copies of C_3 with the common vertex [8, 16].

2.5 *Complete Bipartite Graph*

A complete bipartite graph $K_{m,n}$ is with bipartition $V(G) = A \cup B$ where $A = \{u_1, u_2, u_3, \ldots, u_m\}$ and $B = \{v_1, v, v_3, \ldots, v_m\}$, and all vertices in A are neighbor to all vertices in B, but vertices in the same set are edge-disjoint [4, 15].

3 Main Result

3.1 *Theorem*

The triangular book B_3^t admits proper lucky labeling with $\eta_p\left(B_3^t\right) = 3$.

Proof Let $f : V\left(B_3^t\right) \to \{1, 2, 3\}$ for triangular book B_3^t with common edge $v_1 v_2$ refer Fig. 2a be defined by

$$f(v_i) = \begin{cases} 1 \text{ for } i = 1 \\ 2 \text{ for } i = 2 \\ 3 \text{ otherwise} \end{cases}$$

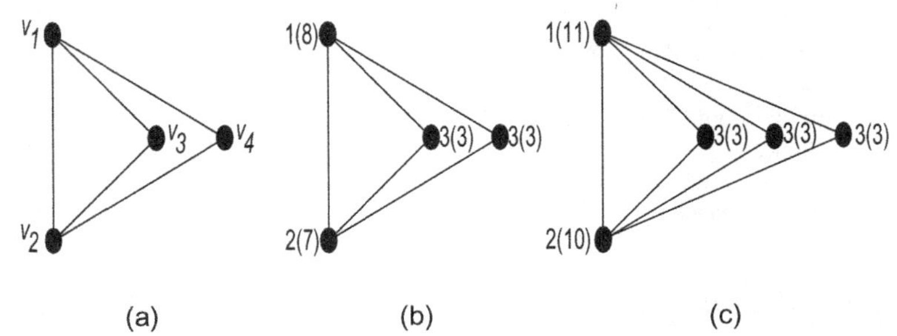

Fig. 2 **a** triangular book B_3^2, **b** labeled B_3^2, **c** labeled B_3^3

Here, we obtain

$$S(v_1) = 3p + 2$$
$$S(v_2) = 3p + 1$$
$$S(v_i) = 3 \text{ for } i > 2$$

This shows $S(v_1) \neq S(v_2) \neq S(v_i)$, i.e., the sum of neighbor vertices is not the same. Therefore, triangular book B_3^t admits proper lucky labeling with $\eta_p(B_3^t) = 3$.

3.2 Illustration

The lucky number 3 of the triangular book with pages 2 and 3 is shown in Fig. 2b, c, respectively.

Algorithm for Proper Lucky Number of Triangular Book B_3^t Input:
A triangular book B_3^t with common edge v_1v_2.

Algorithm

Step 1 Order the vertices from 1 to n.
Step 2 Mark the common vertices as v_1 and v_2.
Step 3 Label the vertex v_1 as 1.
Step 4 Label the vertex v_2 as 2.
Step 5 Label the other vertices as 3. At this stage, neighbor labels are not the same.
Step 6 Add the labels of neighbor vertices. At this stage sum of neighbor, vertices are not the same. The greatest integer labeled 3 is required lucky number.

Output:

$$\eta_p(B_3^t) = 3$$

3.3 Theorem

The rectangular (quadrilateral) book $\boldsymbol{B}_4^t$ admits proper lucky labeling with $\boldsymbol{\eta}_p(\boldsymbol{B}_4^t) = 2$.

Proof Let $\boldsymbol{f} : \boldsymbol{V}(\boldsymbol{B}_4^t) \rightarrow \{1, 2\}$ for rectangular (or quadrilateral) book $\boldsymbol{B}_4^t$ with vertices $\boldsymbol{v}_i$, if $\boldsymbol{i}$ is an odd number, it will be on top of the pages, and if $\boldsymbol{i}$ is even number, it will be on the bottom of the pages refer to Fig. 3a, be defined by.

$$f(v_i) = \begin{cases} 1 \text{ for } i = 1 \text{ or even number} > 2 \\ 2 \text{ for } i = 2 \text{ or odd number} > 1 \end{cases}$$

Here, we obtain

$$S(v_1) = 2p + 2$$
$$S(v_2) = p + 1$$
$$S(v_i) = \begin{cases} 2 \text{ for } i \text{ odd and } > 1 \\ 4 \text{ for } i \text{ even and } > 2 \end{cases}$$

This shows $S(v_1) \neq S(v_2) \neq S(v_i)$, i.e., the sum of adjacent vertices is not the same. Therefore, the rectangular (or quadrilateral) book B_4^t admits proper lucky labeling with $\eta_p(B_4^t) = 2$.

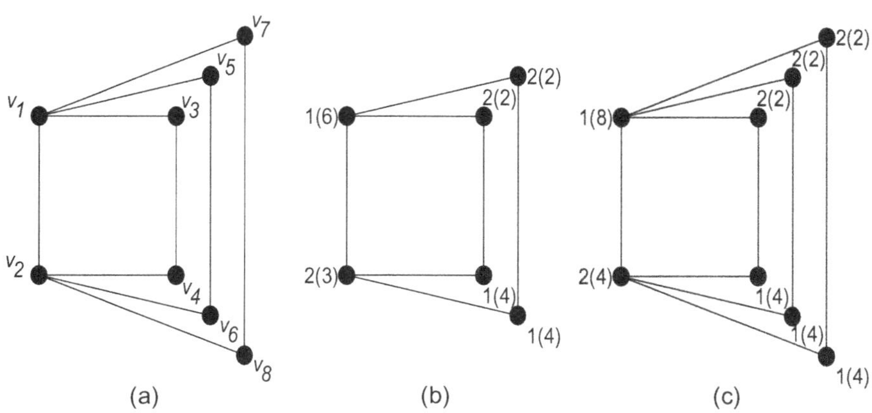

Fig. 3 **a** rectangular (quadrilateral) book $\boldsymbol{B}_4^3$, **b** labeled $\boldsymbol{B}_4^2$, **c** labeled $\boldsymbol{B}_4^3$

3.4 Illustration

The lucky number 2 of the rectangular (or quadrilateral) book with pages 2 and 3 is shown in Fig. 3b, c, respectively.

Algorithm for Proper Lucky Number of Rectangular (Quadrilateral) Book B_4^t
Input:
A rectangular (quadrilateral) book B_4^t.
Algorithm

Step 1 Order the vertices of rectangular book v_i, $i = 1, 2, \ldots n$ by an odd number of i on top of the pages and even number of i on the bottom of the pages.
Step 2 Mark the common vertices as v_1 and v_2.
Step 3 Label the vertex v_i as 1 for $i = 1$ or even number > 2.
Step 4 Label the vertex v_i as 2 for $i = 2$ or odd number > 1. At this stage, neighbor labels are not the same.
Step 5 Add the labels of neighbour vertices. At this stage sum of neighbour, vertices are not the same. The greatest integer labeled 2 is required lucky number.

Output:

$$\eta_p(B_4^t) = 2$$

3.5 Theorem

The friendship graph F_n admits proper lucky labeling with $\eta_p(F_n) = 3$.

Proof Let $f : V(F_n) \to \{1, 2, 3\}$ for friendship graph F_n with vertices v_i, where v_1 be the common vertices of all cycle, and the rest of the vertices takes integer greater than 1 so that each cycle will have an odd number (other than 1) and even numbers refer to Fig. 4a, be defined by.

$$f(v_i) = \begin{cases} 1 \text{ for } i = 1 \\ 2 \text{ for } i \text{ is even} \\ 3 \text{ otherwise} \end{cases}$$

Here, we obtain

$$S(v_1) = 5n$$
$$S(v_i) = \begin{cases} 3 \text{ for } i \text{ odd and } > 1 \\ 4 \qquad \text{for } i \text{ even} \end{cases}$$

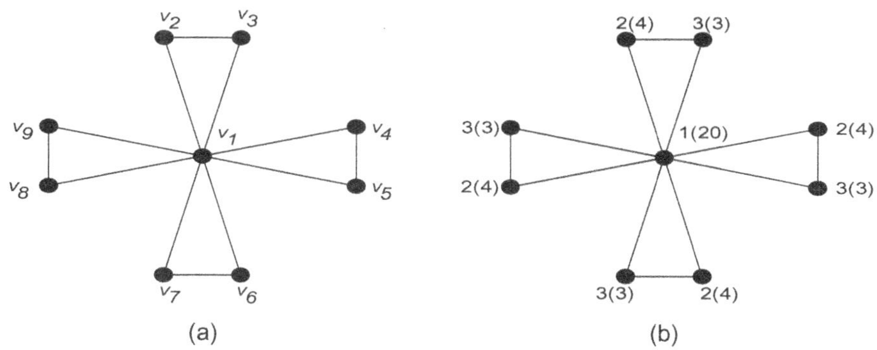

Fig. 4 **a** Friendship graph F_4, **b** labeled F_4

Sum of adjacent vertices is not the same. Therefore, the friendship graph F_n admits proper lucky labeling with $\eta_p(F_n) = 3$.

3.6 *Illustration*

The lucky number 3 of the friendship graph with cycle 4 is shown in Fig. 4b.

Algorithm for Proper Lucky Number of Friendship Graph F_n
Input:

A friendship graph F_n.

Algorithm

Step 1 Order the vertices of friendship $v_i, i = 1, 2, \ldots n$ by v_1 be the common vertices of all cycle and each cycle will have an odd number (other than 1) and even numbers.
Step 2 Mark the common vertex as v_1.
Step 3 Label the vertex v_i as 1 for $i = 1$.
Step 4 Label the vertex v_i as 2 for $i =$ even numbers.
Step 5 Label the vertex v_i as 3 for $i =$ odd number and > 1. At this stage, neighbor labels are not the same.
Step 6 Add the labels of neighbour vertices. At this stage sum of neighbour, vertices are not the same. The greatest integer labelled 3 is required lucky number.

Output:

$$\eta_p(F_n) = 3$$

3.7 Theorem

The complete bipartite graph $K_{m,n}$ admits proper lucky labeling with $\eta_p(K_{m,n}) = 2$.

Proof

Case (i)

When $m = n$. Consider $m = n = 2$, $u \in A$, $v \in B$. Let $f : V(K_{m,n}) \to \{1, 2\}$ for a complete bipartite graph $K_{m,n}$, be defined by

$$f(u) = 1$$
$$f(v) = 2$$

The sum of the label of neighbor vertices are

$$S(u) = 4$$
$$S(v) = 2$$

We know that vertices in the same set are edge-disjoint in the complete bipartite graph. Here, $S(u) \neq S(v)$, i.e., adjacent vertices have no equal neighborhood sum.

In general,

$$S(u) = 2n$$
$$S(v) = m$$

Therefore, it admits proper lucky labeling with $\eta_p(K_{2,2}) = 2$.

Case (ii)

When $m \neq n$. Consider $m = 3$ and $n = 4$, $u \in A$, $v \in B$ Let $f : V(K_{m,n}) \to \{1, 2\}$ for a complete bipartite graph $K_{m,n}$, be defined by

$$f(u) = 1$$
$$f(v) = 2$$

The sum of label of neighbor vertices are

$$S(u) = 8$$
$$S(v) = 3$$

We know that vertices in the same set are edge-disjoint in the complete bipartite graph. Here, $S(u) \neq S(v)$, i.e., adjacent vertices have no equal neighborhood sum.

In general,

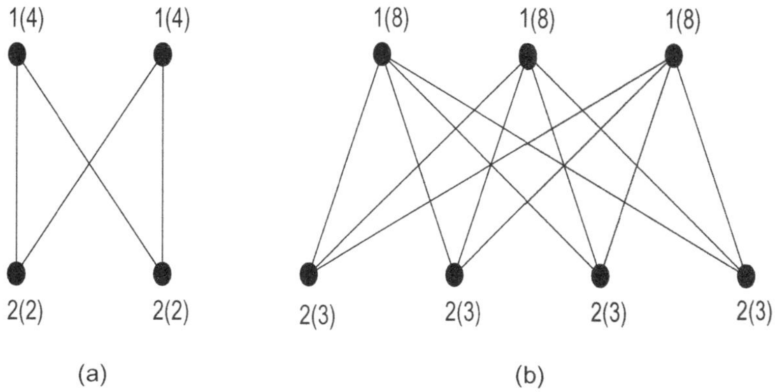

Fig. 5 Labeled **a** $K_{2,2}$, **b** $K_{3,4}$

$$S(u) = 2n$$
$$S(v) = m$$

Therefore, it admits proper lucky labeling with $\eta_p(K_{3,4}) = 2$.

Hence, the complete bipartite graph $K_{m,n}$ admits proper lucky labeling with $\eta_p(K_{m,n}) = 2$.

3.8 Illustration

The lucky number 2 of the complete bipartite graph $K_{2,2}$ and $K_{3,4}$ is shown in Fig. 5a, b, respectively.

Algorithm for proper lucky labeling of $K_{m,n}$.

Input.

A complete bipartite graph $K_{m,n}$ with bipartite A and B, $|A| = m$, $|B| = n$.

Algorithm

Step 1: Mark the vertices of A as v_i where $i = 1, 2, \ldots m$.

Step 2: Mark the vertices of B as v_j where $i = 1, 2, \ldots n$.

Step 3: Label the vertices v_i as 2.

Step 4: Label the vertices v_j as 2. At this stage, neighbor labels are not same.

Step 5: Add the labels of neighbor vertices. At this stage sum of neighbor, vertices are not the same. The greatest integer labeled 2 is required lucky number.

Output

$$\eta_p(K_{m,n}) = 2$$

4 Conclusion

Lucky number for mesh and its derived architectures is computed by Kins at 2017, and identified triangular mesh and identified Sierpinski Gasket graphs are computed by Chiranjilal at 2017. In this paper, we compute the lucky number of some special graph like triangular book graph, rectangular (or quadrilateral) book graph, friendship graph and complete bipartite graph. Observed that a lucky number of a triangular book graph is same as friendship graph since it has cycled with length 3 and a rectangular (or quadrilateral) book graph is same as complete bipartite graph since it was partly bipartite. So, further, we carried this work to various graphs, networks and algorithm can be created for the same.

References

1. Ahadi, A., Dehghan, A., Kazemi, M., & Mollaahmadi, E. (2012). Computation of lucky number of planar graphs is NP-hard. *Information Processing Letters, 112*(4), 109–112. https://doi.org/10.1016/j.ipl.2011.11.002.
2. Aishwarya, A. (2015). Lucky Edge Labeling of certain graphs. *IJIRSET, 4,* 469–471. https://doi.org/10.15680/IJIRSET.2015.0401099.
3. Akbari, S., Ghanbari, M., Manaviyat, R., & Zare, S. (2013). On the lucky choice number of graphs. *Graphs and Combinatorics, 29*(2), 157–163. https://doi.org/10.1007/s00373-011-1112-4.
4. Bondy, J. A., & Murty, U. S. R. (1982). *Graph theory with application.* North Holland.
5. Chiranjilal Kujur, D., Antony Xavier, S., & Raja, ArulAmirtha. (2017). Proper Lucky labeling of K-identified triangular mesh and K-identified Sierpinski Gasket graphs. *International Journal of Scientific and Engineering Research (IJSER), 8*(4), 249–261.
6. Czerwiński, S., Grytczuk, J., & Żelazny, W. (2009). Lucky labelings of graphs. *Information Processing Letters, 109*(18), 1078–1081. https://doi.org/10.1016/j.ipl.2009.05.011.
7. Dehghan, A., Sadeghi, M. R., & Ahadi, A. (2013). Algorithmic complexity of proper labeling problems. *Theoretical Computer Science, 495,* 25–36. https://doi.org/10.1016/j.tcs.2013.05.027.
8. Gallian, J. (2019). A Dynamic survey of graph labeling. The Electronic Journal of Combinatorics DS6
9. Karoński, M., Łuczak, T., & Thomason, A. (2004). Edge weights and vertex colours. *Journal of Combinatorial Theory Series B, 91*(1), 151–157. https://doi.org/10.1016/j.jctb.2003.12.001.
10. Kins Yenoke, R.C. Thivyarathi, & Anthony Xavier, D. (2017). Proper lucky number of mesh and it's derived architectures. *Journal of Computer and Mathematical Sciences (An International Research Journal), 8*(5), pp.187-195.
11. Kujur, C. (2017). Lucky Labeling and Proper Lucky Labeling for Bloom Graph. *IOSR Journal of Mathematics, 13,* 52–59. https://doi.org/10.9790/5728-1302025259.
12. Nellai Murugan, A., & Chitra, R. M. I. A. (2016). Lucky Edge Labeling of Triangular Graphs. *International Journal of Mathematics Trends and Technology (IJMTT), 36,* 116–118. https://doi.org/10.14445/22315373/IJMTT-V36P516.
13. Ramya, M., & Meenakshi, S. (2017). Labeling techniques of some special graphs. *International Journal of Pure and Applied Mathematics, 116*(24), 93–102.

14. Rosa, A. (1966, July). On certain valuations of the vertices of a graph. In *Theory of Graphs (Internat. Symposium)*, Rome (pp. 349–355).
15. Sridevi, R., & Ragavi, S. (2016). Lucky Edge Labeling of Kn and Special Types of Graphs. *International Journal of Mathematics and its Applications*, 4(1-C), pp.125–131.
16. El Deen, M. R. Z. (2019). Edge even graceful labeling of some graphs. *Journal of the Egyptian Mathematical Society, 27*(1), 20. https://doi.org/10.1186/s42787-019-0025-x.

Context Mining for Deep Learning Over Community Health Data for Disease Inference

M. Dhanamalar and K. Kavitha

Abstract Text mining techniques and information retrieval over medical data have always been an area of interest for the bio-information researchers. The researches mostly involve systematically collected health records which are rich in terms of medical terminologies and accurate information regarding symptoms, disease and their treatments, which is not the same case when comes to community data such as the healthcare forums such as webMD, wherein the users posting questions often give very vague descriptions of their symptoms and often miss out critical details regarding the medical conditions. With such inaccurate details, even excellent learning algorithms in the sequential domain such as GRU and LSTM will fail, since they end up learning the irrelevant information from the data and provide inaccurate predictions. We needed a context mining technique to get the best possible information from the available data, such that the words group up to form a medical term used by the doctors to provide better disease inference. To achieve this context mining, we use graph representation of the question–answer and find dense subgraphs to determine the medical term to represent the disease or medical condition described by the user. This does the same job as a restricted Boltzmann machine providing a strong connection between the hidden layer and input layer neurons which are the critical context for prediction and knock the neurons in hidden layers with least context. We have created a comparative study to find an accurate method between the two techniques for context mining above, namely the subgraph mining and Boltzmann machine.

Keywords Context mining · Recurrent neural networks · Boltzmann machine · NLP · Dense sub graph mining

M. Dhanamalar (✉)
Research Scholar, Mother Teresa Women's Universtiy, Kodaikanal, India
e-mail: dhanamca03@gmail.com

K. Kavitha
Assistant Professor, Mother Teresa Women's Universtiy, Kodaikanal, India
e-mail: Kavitha.urc@gmail.com

S.-L. Peng et al. (eds.), *Proceedings of First International Conference on Mathematical Modeling and Computational Science*, Advances in Intelligent Systems and Computing 1292, https://doi.org/10.1007/978-981-33-4389-4_32

1 Introduction

With increasing population and repetitive health issues, the medical service access is getting limited, the ratio of doctors to the patient is very miserable in countries which are less developed and overpopulated, and these issues can be resolved with health assistance as accurate as trained medical practitioners with your mobile phones from anywhere in the world. Most of the medical-assisting bots in today's generation face major hurdle, i.e., the user's lack of medical vocabulary in inputs provided, and our technique of mining context by providing normalised medical terms could be a game-changer in such applications. The reason for such applications of this technique is the source of data, the data from healthcare forum is how an actual user would provide the response to bots, the words which often come together to form a medical term in a medical dictionary are normalized to a signature which accurately described the medical condition.

We represent the question–answer pairs in the graph and search for dense subgraphs which are further subjected to correlation on global mining then compared in SNOMED CT an authenticate online medical dictionary where we convert these mined terms to normalized signature which is trained by our neural network further to provide a medical inference with the disease. These signatures act as our context extracted in the hidden layer from the input layers. This is similar to what a restricted Boltzmann machine does to an auto-encoder wherein the energy-based model which uses a probabilistic function to determine whether the input neuron would make a connection with the hidden layer or not; this is a process of inducing context in the dataset (Fig. 1).

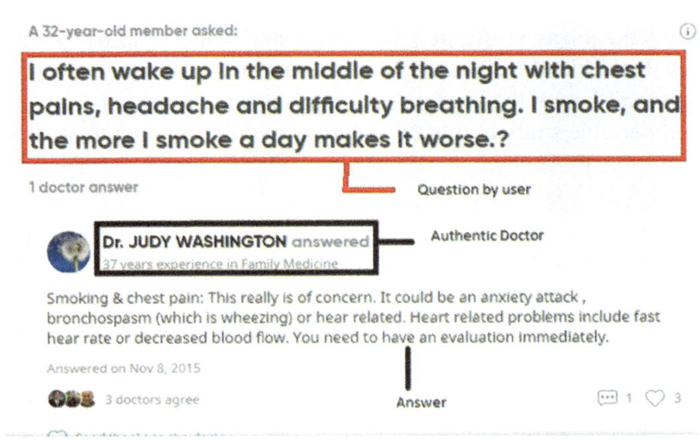

Fig. 1 Example of a QA pair from community-based health services

1.1 Motivation

For any case of medical guidance, the main problem to be solved is the part of identifying the disease based on the symptoms. According to our preliminary study of the datasets, we found the following point being most sought after:

1. Given the diagnosis, what are the inferences
2. Preventive measures for the diagnosed conditions based on the symptoms
3. The medical conditions or disease which can give the signals of given symptoms.

The first two issues can be addressed by using the exact disease types and the possible side effects of the medications taken, which can be solved by fetching relevant information from the healthcare journals and another medical document from authentic resources using various record-matching techniques and information retrieval techniques. But for the third issue, we need to deep dive into complete details of the patient including medical history, allergies and underlying health conditions, also the demographics of the seekers. There are not many efficient techniques to resolve the third type of issue when it comes to the community data, which can change the way the online health care works given a lot of tagging and solutions already available, it's just the correct mapping in context to such a tag is all needed where the simple record matching or plain NLP compelled with information will fail to deliver the desired results. The problem of lack of medical vocabulary in the third issue can be resolved by mining the signature terms in a medical dictionary which can describe the conditions coming together and thus provide a better chance of training the model more accurately.

1.2 Literature Survey

In the work of Doan et al. [1] titled "Recognizing medication-related entities in hospital discharge summaries using support vector machine", provided first of its kind natural language processing approach for entity relation identification in the medical domain; this was part of the 12b2 challenge here; their work had an f-score of 90%, but the dataset provided in i2b2 challenge consists of electronic health records which are rich with medical terms, and they will fail in case of online data since the medical terms used are not very accurate.

On another research by Wang et al. [2] in his work "Aligning temporal data by sentinel events: discovering patterns in electronic health records" which provides an optimized search mechanism based on disease type, although this work has proved to be highly effective in disease inference line, but will fail if data quality doesn't meet the standard and cannot be a robust solution for very large and inconsistent data. Shouman et al. [3] in their work of "Using decision tree for diagnosing heart disease patients" had the first time used a data mining technique for effective classification system, but decision tree is not a very efficient method when comes to building a solution of a very large-scale data such as ours. An integrated machine learning approach to stroke prediction by Khosla et al. [4] changed the dynamics of creating a solution with a special set of SVMs to classify the health issues but even most effective dimensionality reduction techniques cannot help in this case of public data. In 2013 Limsopatham et al. [5] did research on learning to combine representations for medical records search which introduced Information retrieval techniques like a bag of words to healthcare records and this technique may fail if incorrect or very large misleading data are bagged together. Heart disease diagnosis using support vector machine by Ghumbre et al. [6] which was one of the first papers to use advanced ML techniques and functions to establish the relationship between diseases and medical symptoms, but RBF and OLS require proper relationship extractions which isn't available in case of community data.

2 Proposed Methodology Local Mining

As mentioned in the literature survey, there are no methods in machine learning which will be the best fit in case of the web forum data. The main requirement of this research is to provide a way to convert the terms used by the users to words used in medical dictionaries for the apt description of the disease. We need to make the system identify the apt context in medical terms to understand the underlying medical conditions, the demographics and possible side effects for a patient. We have the question and answer pairs, and these answers are provided by medical experts with proper terms and tagged diseases; now we create a graph of these words in questions and their answers, there some n-gram words which often co-occur. The vertex of these graphs is the words in questions and the answers, and the weights for the edge between these vertices are the co-occurrence values which are generated by counting the counts of occurrence of the two words coming together, This helps in removing redundant features coming again and again for different disease say for migraine and brain tumours may have similar symptoms such as headache and vomiting, but when we club the other contexts together while learning using this technique, we get a model with much more accuracy. The occurrence clubbed graph is then subjected to k-clique to get the dense subgraphs which form the signatures which are normalized using medical dictionary to get the features which are stitched to get the sparse learning model.

k-clique

Once the co-occurrence graph is created, we then traverse the graph for k vertices and check if there is an edge, if there is an edge then we use the resulting graph over the complete graph for further similar edges, if they are found the edge weights are summed up and if they exceed a particular threshold, then they constitute our medical signatures and thus will be normalized to build our secondary features. When we build the 3 or 4 clique signatures, we run the algorithm to identify the complete graphs, if they fail to form the complete graphs then we discard them.

The medical signatures are then mapped to the normalized medical terminologies to build the hidden layer neurons which act as the resulting context that we would get when the hidden layer weights are perfectly trained over a huge number of computations.

Boltzmann Machine
We needed to compare our algorithm against traditional feature filtering/selection algorithms, our purpose of graph mining technique is to select the features or phrases that find the most relevance in medical terms, which indirectly knocks of the unnecessary input neurons from the network, such a process of filtering is known as feature selection. When it comes to feature selection, Boltzmann machines have always played a key role in knocking out irrelevant features with the highest precision. We needed to better the results of the Boltzmann machine with our graph-based algorithm for context mining. Boltzmann machine uses an energy-based probabilistic model for determining the state of the hidden neural network nodes; this is expected to remove the irrelevant words from the user questions in our case to present the use with data that could be used to determine the disease. The Boltzmann machine has two layers, the visible layer and the hidden layer; in our case we will provide the questions as the nodes for the visible layer. Instead of training the complete neural network and back-propagating the error to determine the weights in neural networks to predict values, we train network between input and hidden layer, we used Boltzmann's probabilistic functions to reconstruct the visible layer from hidden layer over multiple iterations to get best set of weights for further predictions.

3 Implementation

We first studied various sources of authentic medical forums and their skeleton in web and identified ones with right disease tag and question and answer pairs. We used a beautiful soup library in python and selenium web driver to automatically extract the data into a JSON file. We then processed this JSON file to eliminate the URLs and unnecessary characters from the data, and we then perform an important task of stemming the words, we don't want the tense or words with common root create an issue while training the model and then we passed it against the NLTK noun phrase extractor to build our data source for training. The extracted noun phrases are then converted to medical terms with SNOMED CT.

We used a medical tagger called meta-map to build our signatures, and the purpose of the tool is to produce a term which would represent the signatures mined by applying the k-clique algorithm over the co-occurrence graph. This resulting subgraph would have terms which are just co-occurring; we still don't know if they are accurately representing the medical terms, to make sure that we apply something called the global mining.

Once the signatures are mined from the complete graph, the next process is to make sure that these signatures are pure medical terms which are the task of global mining where we run the signatures over two dictionaries—one PubMed research corpus and other is LDC essay corpus which would return the occurrence in each of them if the co-occurrence is more in PubMed; then the set is considered as the medical signature while if it is more in LDC, then a regular English phrase and thus discarded.

The training of the model was done in two phases—one involving partial dataset where 3500 entries of 85 disease types were used as the data source. The training set was first subjected to subgraph mining for extracting the medical signatures which were stored in separate files. These terms would form the hidden layers. The vector representation of the medical terms and their mapped hidden layers were then trained over 10,000 iterations for appropriate weights against the class values which is the tagged diseases extracted from the forums validated by authentic medical experts in the forums. In the next phase, the complete data was trained for 177 disease types over 10,000 iterations again for complete model training. The data was trained over a simple ad recurrent neural network to study the accuracy of the learning of the model.

We used the representation matrix constructed above with the questions of the healthcare forums and constructed our visible layers, and then we trained our input layer and a randomly generated hidden layer nodes with the Boltzmann ma-chine's energy-based model for optimizing the weights assigned to the hidden neurons, thereby knocking the hidden neurons which doesn't help in reconstructing the input layer successfully does create a shallow network of strongly connected hidden and input layers. The formulae for the computations are given below. The weights are adjusted for the hidden layer by subtracting the Δw_{ij} derived below. This method of computing chained difference of weights is known as Gibbs divergence; this helps to choose the hidden layer nodes which contribute to forming the medical term context with given visible input layer.

$$E(v,h) = -\sum_{i \in \text{Visible}} a_i v_i - \sum_{j \in \text{hidden}} b_i h_i - \sum_{i,j} v_i h_j w_i j \tag{1}$$

$$p(v) = \frac{1}{Z} \sum_{h} e^{-E(v,h)}, \quad \text{where } Z = \sum_{v,h} e^{-E(v,h)} \tag{2}$$

$$p(v) = \frac{\sum_h e^{-E(v,h)}}{\sum_{v,h} e^{-E(v,h)}} \tag{3}$$

$$\Delta w_{ij} = \alpha\left(\langle v_i h_j \rangle_{\text{data}} - \langle v_i h_j \rangle_{\text{model}}\right) \tag{4}$$

4 Experimental Results

In Fig. 2, we have present $E(v, h) = \sum_{i \in \text{visible}}^{n} a_i v_i - \sum_j j \in x^k a^{n-k}$ ed a screenshot of a question–answer pair extracted from everyone healthy and stored in a JSON file with.

its disease as the tag for the question and answer. We then pre-processed the JSON file to remove unnecessary values and URLs to get us the plain text of the question and answer phrases shown in Fig. 3. The pre-processed text files are then matched in SNOMED CT to get the medical terms, and final part of data preparation is to mine the medical signatures using our k-clique algorithm, whose results could be found in Fig. 5. These files with normalized medical signatures will constitute the hidden layer for our sparse neural network for further training (Figs. 4 and 6).

When we trained our model with plain LSTM, the accuracy was pretty average given LSTM's credibility over sequential data since the terms extracted from the question–answer pairs which are straight away inaccurate in themselves have given the wrong context while training LSTM over the data. The LSTM model kept on learning attributes or features which proved a context that may not accurately match the disease over repetitive time-stamp training. The model was then upgraded with the signature mining algorithm which changed the picture by providing top-notch context by selecting the relevant neurons in hidden layers for more accurate results. The results are given in Table 1, and the graph of the same can be seen in the figure.

The feature selection methods have always been a forte of restricted Boltzmann machine which train the neural network over the hidden layer to generate the input layer, thus providing the sets of weights which will knock off the irrelevant neurons in hidden layers thus providing us with a context which can help in better the prediction,


```
    "question": "zirconium dental implants. how common is it  sed now. is there any advantages or benefits over titanium implants
& pros please. thanks.",
    "short_answer": " nDental implants n",
    "answer": "a majority of the dental implants placed are titanium. they are highly s ccessf l with many years  se ; many st di
ch lower in cost ; have many restorative options. zirconia implants are newer fewer st dies on s ccess lesser restorative options
however they can be more aesthetic in certain anterior(front) sit ations. let yo r dentist/oral s rgeon chose what they feel will
best for yo .",
    "tags": [
      "dentistry"
    ],
```


Fig. 2 Extracted data from the webMD

```
1 zirconium dental implants. how common is it used now. is there any advantages or benefits over titanium implants. cons & pros please.
  thanks.
2 a majority of the dental implants placed are titanium. they are highly successful with many years use ; many studies much lower in
  cost ; have many restorative options. zirconia implants are newer fewer studies on success lesser restorative options. however they can
  be more aesthetic in certain anterior(front) situations. let your dentist/oral surgeon chose what they feel will be best for you.
```

Fig. 3 After processing the data

```
1 Academic problem (finding)
2 Educational problem (finding)
3 Basic motor learning problem (finding)
4 Consultation for hearing and/or speech problem (procedure)
```

Fig. 4 Normalized medical features

```
['zirconium', 'implants', 'dentist', 'study']
['zirconium dental implants. how common is it used now. is there any advantages
or benefits over titanium implants. cons & pros please. thanks.', 'a majority of
 the dental implants placed are titanium. they are highly successful with many y
ears use ; many studies much lower in cost ; have many restorative options. zirc
onia implants are newer fewer studies on success lesser restorative options. how
ever they can be more aesthetic in certain anterior(front) situations. let your
dentist/oral surgeon chose what they feel will be best for you.', 'zirconium den
tal implants. how common is it used now. is there any advantages or benefits ove
r titanium implants. cons & pros please. thanks.', 'and the data on zirconia imp
lants is much more limited.']
{('zirconium', 'implants'): 2, ('implants', 'dentist'): 1, ('zirconium', 'dentis
t'): 0, ('implants', 'study'): 0, ('zirconium', 'study'): 0, ('dentist', 'study'
): 0}
{'study': 0, 'dentist': 1, 'implants': 4, 'zirconium': 2}
['study', 'implants', 'dentist', 'zirconium']
[('implants', 'zirconium')]
```

Fig. 5 Signatures mined

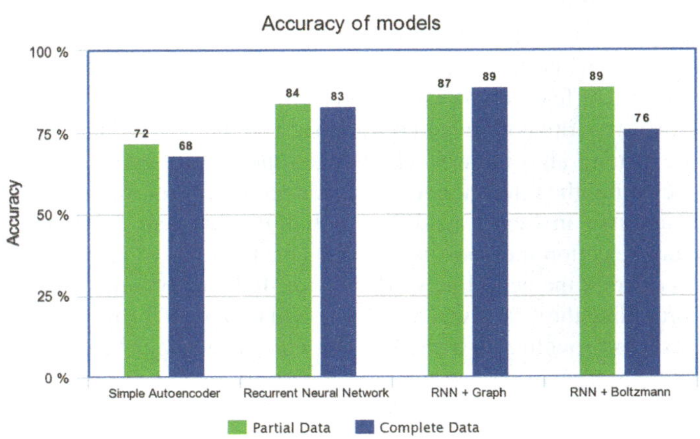

Fig. 6 Graphical comparison of the accuracy

Table 1 Comparison of accuracy of the models

Algorithm/technique	Accuracy on partial data (%)	Accuracy on complete data (%)
Simple autoencoder	72.88	68
Recurrent neural network	83.88	82.78
Recurrent neural network + Graph mining	87.88	88.28
RNN + Boltzmann machine	89.58	76

although this works well over less data when more datasets are to be trained the accuracy is deteriorated since the probability values are no very different for the neurons with correct and incorrect contexts. The complete results can be summarized in the graph.

5 Conclusion

With the Boltzmann machine, although a simple encoder gets the best features to train when we need a mechanism for getting context from a language which is out of context for the machine trained over medical terms, it may not be as efficient, during our experiments when we trained the model with Boltzmann machine instead of the graph mining technique we found the model becoming less accurate with increasing data points or more precisely when many diseases are to be classified. The problem of learning the features of irrelevant context arising in community data cannot be resolved with LSTM but with our methodology, we have achieved the construction of language model to fetch best results out of forum data thus building a context generated by the medical signatures which will help the LSTM model to train hidden neurons much accurately. While the LSTM has its own limitations, on the other hand, the sparse network's powerful context is of no use in case of the huge dataset when the retention rate has to be of the adequate amount. The signature mining had provided a way to bridge the vocabulary gap in the questions asked by the customers which don't have the accurate medical description of the condition. The sparse connections in the network will limit the nodes of the hidden layer from providing inefficient weights to the context of less importance and thus improves the efficiency in learning the required knowledge.

The model can be more accurate if trained with actual medical conversations between a doctor and patient over a chat, thus providing a more enhanced data source which can in turn help produce an accurate model. The model can be clubbed with a medical chatbot to fetch such data and thus retrain to produce the desired results.

References

1. Doan, S., & Xu, H. (2010, August). Recognizing medication-related entities in hospital discharge summaries using support vector machine. In *Proceedings of COLING. International Conference on Computational Linguistics* (Vol. 2010, p. 259). NIH Public Access.
2. Wang, T. D., Plaisant, C., Quinn, A. J., Stanchak, R., Murphy, S., & Shneiderman, B. (2008, April). Aligning temporal data by sentinel events: discovering patterns in electronic health records. In *Proceedings of the SIGCHI conference on Human factors in computing systems* (pp. 457–466).
3. Shouman, M., Turner, T., & Stocker, R. (2011, December). Using a decision tree for diagnosing heart disease patients. In *Proceedings of the Ninth Australasian Data Mining Conference-Volume 121* (pp. 23–30).
4. Khosla, A., Cao, Y., Lin, C. C. Y., Chiu, H. K., Hu, J., & Lee, H. (2010, July). An integrated machine learning approach to stroke prediction. In *Proceedings of the 16th ACM SIGKDD International Conference on Knowledge Discovery and Data Mining* (pp. 183–192).
5. Limsopatham, N., Macdonald, C., & Ounis, I. (2013, July). Learning to combine representations for a medical records search. In *Proceedings of the 36th International ACM SIGIR Conference on Research and Development in Information Retrieval* (pp. 833–836).

6. Guo, D., Duan, G., Yu, Y., Li, Y., Wu, F. X., & Li, M. (2020). A disease inference method based on symptom extraction and bidirectional Long Short Term Memory networks. *Methods, 173,* 75–82.
7. Fox, S., & Duggan, M. (2013). Health online 2013. *Health, 2013,* 1–55.
8. Ji, S., Gu, Q., Weng, H., Liu, Q., Zhou, P., Chen, J., Li, Z., Beyah, R., et al. (2020, April). De-Health: All Your Online Health Information Are Belong to Us. In *2020 IEEE 36th International Conference on Data Engineering (ICDE)* (pp. 1609–1620). IEEE.
9. Zhou, T. C., Lyu, M. R., & King, I. (2012, April). A classification-based approach to question routing in community question answering. In *Proceedings of the 21st International Conference on World Wide Web* (pp. 783–790).
10. Chawla, D. A., Blumm, N. V., Christakis, N., Barabasi, A. L. (2008, October). Predicting individual disease risk based on medical history. In *Proceedings of the 17th ACM Conference on Information and Knowledge Management* (pp. 769–778).
11. Nie, L., Wang, M., Zha, Z., Li, G., & Chua, T.-S. (2011). Multimedia answering: Enriching text QA with media information. In: Proceedings of International ACM SIGIR Conference on Research Development in Information Retrieval, 2011, 695–704.
12. Nie, L., Wang, M., Gao, Y., Zha, Z. J., & Chua, T. S. (2012). Beyond text QA: Multimedia answer generation by harvesting web information. *IEEE Transactions on Multimedia, 15*(2), 426–441.
13. Nie, L., Zhao, Y. L., Wang, X., Shen, J., & Chua, T. S. (2014). Learning to recommend descriptive tags for questions in social forums. *ACM Transactions on Information Systems (TOIS), 32*(1), 1–23.
14. Zhang, D., & Lee, W. S. (2006, August). Extracting key-substring-group features for text classification. In *Proceedings of the 12th ACM SIGKDD International Conference on Knowledge Discovery and Data Mining* (pp. 474–483).
15. Gallé, M. (2013, July). The bag-of-repeats representation of documents. In *Proceedings of the 36th International ACM SIGIR Conference on Research and Dvelopment in information Retrieval* (pp. 1053–1056).

User Behaviour-Based Accurate Detection of Credit Card Frauds

C. Sudha and D. Akila

Abstract The arrival of communication methods, along with online imbursement dealings, is growing every day. Together with this, fiscal scams linked with these dealings are also escalating. Amid several fiscal scams, credit card scam is the utmost common and hazardous one owing to its extensive practice. To perceive these illegal bases of communications, a credit card scam recognition structure is obligatory. In this paper, we propose to design a user behaviour-based accurate detection for credit card frauds. In this technique, user behaviours over a banking website are collected and analysed. The main features related to user behaviour are frequently visited pages, time spent on each page, clicking, etc. From these details, the page reachability and page utility are determined. From this, the normal and abnormal behaviours of users are classified using web Markov skeleton process (WMSP) model.

Keywords WMSP · Credit card · Communication · Markov model

1 Introduction

In the present condition of the world, financial companies develop their commercial facilities available to all by utilizing the exclusive services like credit cards, automated teller machines (ATM), Internet and mobile banking. Due to the fast developments in e-commerce, the usage of credit cards has become an essential component of financial life. A credit card is a payment card issued to customers as a mode of payment [1].

The arrival of communication methods has made an augmentation in e-commerce, along with online imbursement, dealings every day. In conjunction with this fiscal

C. Sudha (✉)
Department of Computer Science, Vels Institute of Science, Technology & Advanced Studies, Chennai, India
e-mail: srisudhasri.kpm@gmail.com

D. Akila
Department of Information Technology, Vels Institute of Science, Technology & Advanced Studies, Chennai, India
e-mail: akiindia@yahoo.com

S.-L. Peng et al. (eds.), *Proceedings of First International Conference on Mathematical Modeling and Computational Science*, Advances in Intelligent Systems and Computing 1292, https://doi.org/10.1007/978-981-33-4389-4_33

scam related to these dealings are also escalating, which consequence in forfeiture of billions of dollars every annum internationally. Amid the several fiscal scams, credit card scam is the utmost ancient, common, and hazardous one owing to its extensive practice on account of the expediency it bids to the client. Likewise, the several sorts of welfares such as cashback, deduction proposals on procurements made at certain shops, and henceforth lure the consumers to utilize credit card as an alternative of cash for their consumptions [2].

Credit card scam is an inclusive tenure for scams devoted concerning an imbursement card as a tainted source of funds in dealings. To perceive these illegal sources of dealings credit card, scam recognition scheme is made use of. An incredible intensification is there in the number of scam dealings happening throughout the universe. Scam dealings are the false dealings made via credit cards deprived of the acquaintance of the actual card owners. Deceitful dealings have been made in private commodities, daily transactions, bill outlays, etc. [3].

The scam is an action in which an individual unswervingly or incidentally utilizes the currency of prey by false dealings and deprived of making him informed about the deal [4]. Perceiving illegal credit card dealings is a tremendously intricate issue, as structures are rarely beneficial if considered independently. While data removal procedure scan sense concealed designs in data, they generally deficiency the ability to create metrics defining the universal edifice shaped by the communications amid the diverse structures [1].

A virtuous fraud detection model should be able to recognize the fraud deal precisely and should ensure that the recognition is feasible in present dealings. Fraud detection schemes can be split into two groups: anomaly and misuse schemes. Anomaly detection schemes fetch regular deals to be skilled and utilize methods to decide new scams. On the other hand, a misuse detection fraud recognition scheme utilizes the branded deal as usual or scam deal to be proficient in the database history. So, this misuse detection scheme belongs to supervised learning and anomaly detection scheme belongs to unsupervised learning [5, 6].

Majority of the works on credit card fraud detection is based on machine learning schemes.

But machine learning techniques involve the following drawbacks:

- Initially, working out superior machine learning replicas necessitates strong core historical data. That is, if we do not have adequate earlier scam and usual dealings, it would be tough to run a machine learning exemplary on it.
- If the stage maintainers did not gather and categorize the data gracefully and appropriately, or even assorted the info on scam dealing with the info on usual ones that is possible to grounds a giant prejudice in the outcomes of the exemplary.
- One more great encounter convoluted in the machine learning method is untrue positives. Therefore, it is the utmost impactful item to advance the exactness of these machine learning replicas.

2 Related Works

This section presents some of the credit card fraud detection schemes using machine learning techniques.

This credit card scam recognition scheme works on neural networks [7], which can perceive up to 80% accurateness with model deal data. But the scheme may not be able to recognize the scam interactions/dealings if the deceiver is a novel customer in the bank, deprived of the early data input, feed the neural network cannot be proficient or organized.

The trapping ensemble classifier centred on the decision tree [8] is a new method for the credit card scam recognition scheme. A common dataset on credit card dealings is made use of for presentation assessment and it has been discovered that the seizing classifier revenue improved consequences, as it is liberated of characteristic values.

In [9], a list from decision theory has been industrialized inside the setting of credit card scams. An efficacy function is utilized to recognize the finest cam method in.

financial terms. They have protracted the confrontational learning exemplary to comprise a hypothetical agenda. Their methods are verified contrary to the results to rehabilitate and not ever tore instruct constantly.

Gabriele Gianini et al. [10] have lectured the succeed opinions: the principles made use of to choose which rubrics to retain functioning in the NRT group are conventionally centred on the past presentation of the distinct rubrics, deliberated in separation. This method disrespects the non-additively of regulation configuration within the group. They suggested utilizing method centred on approximating the distinct regulation influence to the complete group enactment via the Shapley value (SV).

Deng et al. [11–14] have suggested a real-world data focused on recognition example, which can senses cam handlers and improve its recognition capacity routinely. Depending on the novel example, they intended a new scam recognition exemplary, Fraud Judger, to assess customers' deeds on digital imbursement boards and perceive scam customers with less branded data in training. Fraud Judger is able to learn the dormant exemplifications of operators from unlabelled data with the aid of adversarial autoencoder (AAE). Also, Fraud Judger is able to discover new fangled scam designs from unidentified operators by group exploration.

3 Proposed Solution

3.1 Overview

To solve these problems, we propose to design a user behaviour-based accurate detection of credit card frauds. In this technique, the user behaviours over a banking website are collected and analysed. Operator deed comprises the entire activities the operators undertake on a banking website: where and what they connect on, how they roll down a page, where they stagger, and where they ultimately not off the page and leave, etc.

The main features related to user behaviour are frequently visited pages, time spent on each page, clicking, etc. From these details, the page reachability and page utility are determined. For calculating the page reachability and staying time, the server log is considered and the web user's navigation or browsing behaviour is taken into account.

From this, the normal and abnormal behaviours of users are classified using web Markov skeleton process (WMSP) model.

3.2 Web Markov Skeleton Process (WMSP) Model

WMSP is a random and uncertain process which contains a Markov chain as its structure. The main applications areas of WMSP are queueing theory and reliability engineering.

Let WMSP is a random and uncertain process Z. Let X is a Markov chain with state-space S and transition probability matrix P.

Let $X = \{X_1, X2, \ldots, X_n\}$ be the sequence where X_n is a state and $_{n+1} = P(X_{n+1}|X_n),\ n = 0, 1, 2, \ldots$

Let Y is a random and uncertain process, given by $Y = \{Y_1, Y_2, \ldots Y_n\}$ be the sequence.

Let $S_0, S_1, \ldots S_n \subset S$.

Then, Y is determined by the probability function $P(Y_n|S_n)$.

Then, Z is a random and uncertain process on Y given by

$$Z = \boldsymbol{Y}_0 \rightarrow \boldsymbol{Y}_1 \rightarrow \boldsymbol{Y}_n \rightarrow$$

X_n depends on X_{n+1} and Y_n depends on S_n $(n = 1, 2 \ldots)$.

WMSP represents the random walk of a web user and includes the entirely random and uncertain processes used in the analysis of web page importance.

Let $Z = (X, Y)$ be a WMSP on S, where $X = \{X_n\}$ denotes a state and $Y = \{Y_n\}$ denotes the resident time at X_n for each n. Here, S denotes the web pages and Z denotes the random walk on the web pages. Then, the stationary distribution of X $St(X)$ denotes the page reachability and the exponential distribution of mean resident time Y, $e(Y)$ denotes the page utility.

Then, user behaviour metric (U_{m}) is calculated as the product of $St(X)$ and $e(Y)$. (i.e.)

$$U_{\mathrm{m}} = St(X) \cdot e(Y) \tag{1}$$

3.2.1 Page Reachability (PR) Estimation Using Stationary Distribution

Page reachability represents the probability that the user visits each page. It represents the average value that the page provides random surfers during the browsing process.

We compute PR using user browsing behaviour data. A graph is created from the log where nodes represent web pages and edges represent the transition between each web pages viewed by the user.

Let P be a probability matrix of Markov chain such that

$$P = \begin{pmatrix} p_{0,0} & p_{0,1} & p_{o,2} \cdots & p_{0,n} \\ p_{1,0} & p_{1,1} & p_{1,2} & p_{1,n} \\ \cdots & \cdots & \cdots & \cdots \\ p_{n,0} & p_{n,1} & p_{n,2} & p_{n,n} \end{pmatrix} \quad (2)$$

where $P_{i,j}$ represents the transition state from page i to j.

Then, the stationary distribution of P is a probability distribution π such that

$$\pi \cdot P = \pi \quad (3)$$

where $\pi = \{\pi_1, \pi_2, \ldots, \pi_n\}$ can be computed by the power method.

3.2.2 Page Utility Estimation Using Exponential Distribution

Page utility indicates the score of the page assigned by the user in a visit.

The cumulative distribution function $F(x, \lambda)$ is given by

$$F(x, \lambda) = \lambda \cdot e^{-\lambda x} \quad (4)$$

where x denotes the states and λ denotes the staying time at x.

3.3 *Server Log File*

The logging technique is pre-eminent to track back to the offender in the instance of a solicitation scam. For instance, when a cheat may have unlawfully moved currency from someone else account to his account.

Records record info about the actions of solicitations and its operators. Records are beneficial to achieve analytical processes for troubleshooting or criminological investigation. Web solicitations have numerous stages and sorts of records.

Web applications involve application logs. These logs offer a pure image of the actions of a specific operator with the solicitation. It can comprise evidence such as the dealings performed; data arrived into the solicitation, unsuccessful login efforts,

Table 1 Fields of the log file and their description

125.125.125.125 - uche [20/Jul/2008:12:30:45 +0700] "GET /index.html HTTP/1.1" 200 2345

or any other doubtful action with the solicitation. The logs can enclose username, customer IP address, date, time, etc., for every process. Therefore, for an account transmission process, the solicitation log will comprise evidence such as the username, the bank accounts convoluted in the deal, amount shifted and customer IP address. These particulars can be made use of to perceive a scam and trackback to the offender.

For calculating the page reachability and staying time, the application log is considered and the web user's navigation or browsing behaviour is taken into account. Using the log, the number of hits for a web page and the total time length which the web user spent on the particular web page is extracted.

A sample Apache server log file format will be (Table 1).

3.4 Credit Card Fraud Detection Algorithm

Notations	Meaning
U_{id}	User id of U_m
URL_j	URL of jth result U_j
$link_{in}$	Incoming links
$link_{out}$	Outgoing links
P	Transition probability matrix of X
St	Stationary distribution
λ	Staying time
P_{imp}	Page importance
P_{rank}	Page rank
F	Exponential distribution

(continued)

(continued)

Notations	Meaning
1. For each p_i of U	
2. Do	
3. Fetch the user log file from the web server	
4. Search for the URL of p_i	
5. Find out the $\text{link}_{\text{in}}(U_i)$ and $\text{link}_{\text{out}}(U_i)$	
6. Build $P(X)$ corresponding to $\text{link}_{\text{in}}(U_i)$ and $\text{link}_{\text{out}}(U_i)$	
7. Derive St of $P(X)$ using (2)	
8. Calculate λ of U_i	
9. Derive F using (3)	
10. Determine U_{m} using (1)	
11. End For	
12. Sort p_i in descending order of U_{m}	

For a set of pages p_i of user U, the corresponding URL is fetched from the server log file. The number of inlinks and outlinks (i.e. transition from and to this page) is then estimated from the log file, based on which the transition probability matrix P is determined. The stationary distribution of P is derived using Eq. (2).

The staying time of the page λ is also determined from the date/time field of the log file. Then, the exponential distribution F is derived based on λ using Eq. (3). The UB is computed using Eq. (1). This process is repeated for all the pages of U (Fig. 1).

4 Experimental Results

4.1 Datasets

We evaluated the web server logs from a banking organization over a progression of 16 days in late August 2001. The website bids consumers' normal online banking choices like examining a balance, disbursing a bill, transmitting money, etc. The arrangement of the log records is exposed in Table 2. As is normal for such logs, if the customer who send appeals to the web server is behindhand a deputation, then it will be the deputation.

IP address that is documented in the log, the date and time are documented at the determination of one second. The enduring content was almost completely minor image documentations utilized in the page design.

4.2 Results

The WMSP model-based credit card fraud detection system is compared with the k-nearest neighbour (k-NN), support vector machine (SVM) and artificial neural

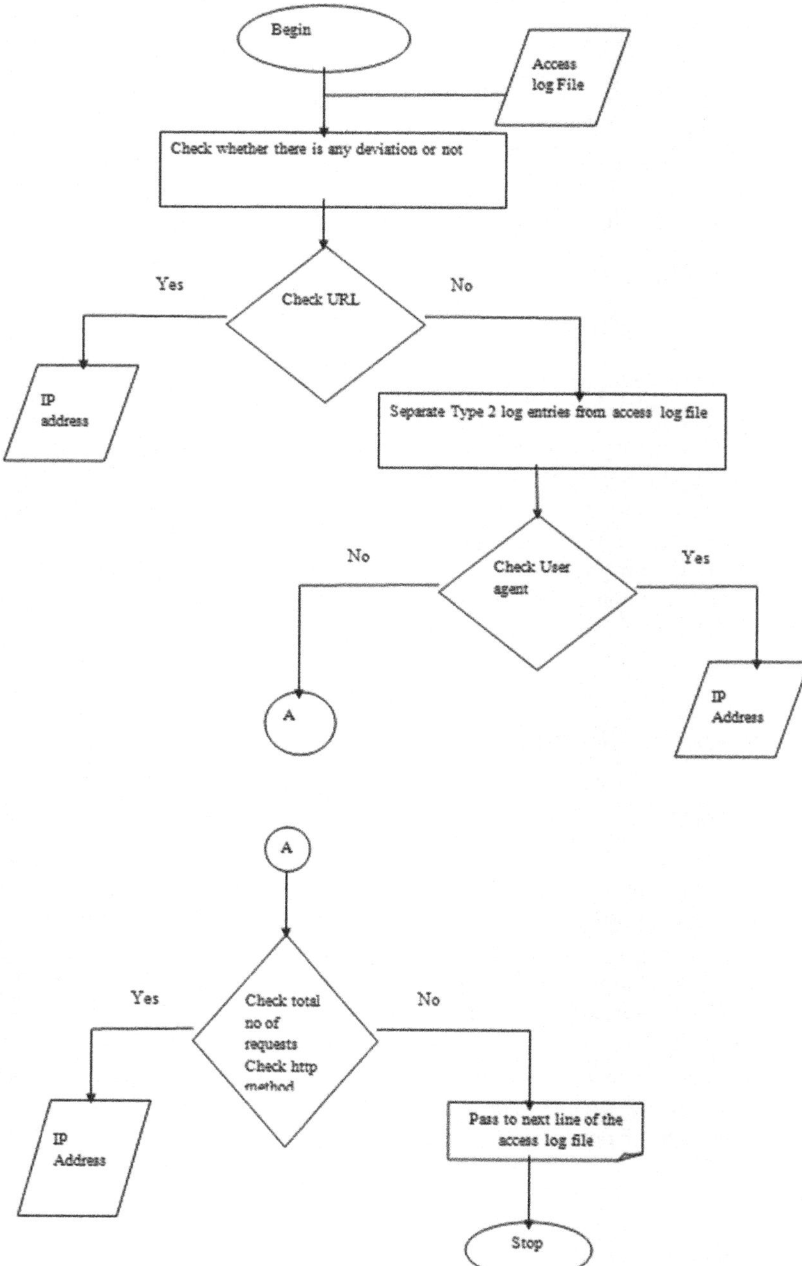

Fig. 1 Fraud detection process

Table 2 Details of log file

IP address	Date and time	Request type	URL	HTTP version	Response code	Content length

network (ANN) algorithms. For these algorithms, the performance comparison is performed concerning the metrics accuracy, sensitivity, specificity and detection rate.

Figure 2 exhibits the performance of the algorithms for accuracy, sensitivity and specificity metrics. As seen from Fig. 3, WMSP attains an accuracy of 0.81, followed by ANN with 0.76, SVM with 0.74% and lastly k-NN with 0.71%. Regarding sensitivity measure, WMSP attains 0.83, followed by ANN with 0.77, SVM with 0.75 and finally k-NN with 0.6. Regarding specificity measure, WMSP attains 0.72%, followed by ANN with 0.68, SVM with 0.63 and finally k-NN with 0.55.

Figure 3 exhibits the detection rate of the algorithms for different numbers of pages ranging from an average of 10 to 50. As seen from Fig. 3, WMSP attains an

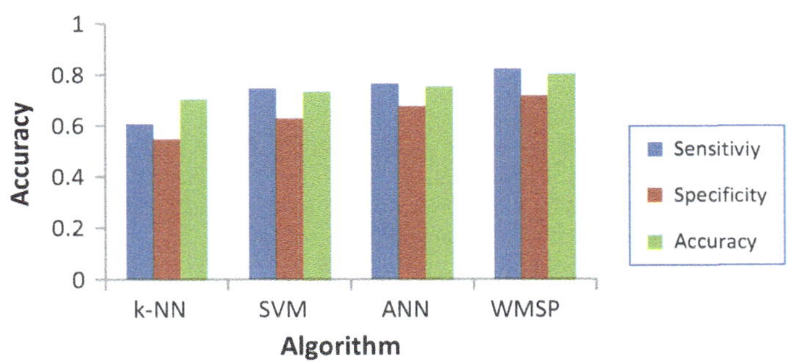

Fig. 2 Performance accuracy of various techniques

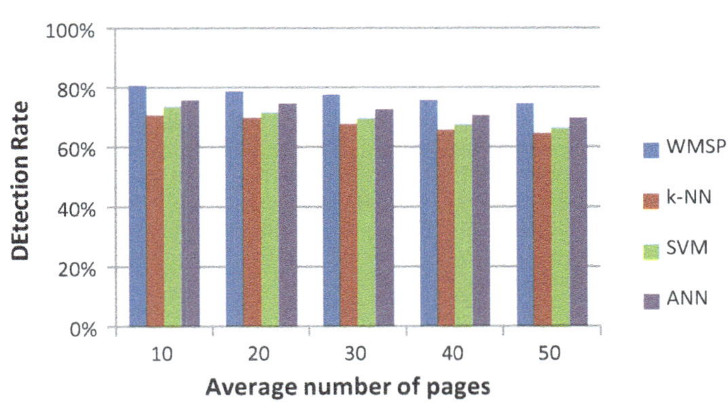

Fig. 3 Performance of detection rate of various techniques

average detection rate of 78%, followed by ANN with 73%, SVM with 70% and lastly k-NN with 68%.

5 Conclusion

In this paper, we propose to design user behaviour-based accurate detection for credit card frauds. In this technique, user behaviours over a banking website are collected and analysed. The main features related to user behaviour are frequently visited pages, time spent on each page, clicking, etc. From these details, the page reachability and page utility are determined. From this, the normal and abnormal behaviours of users are classified using web Markov skeleton process (WMSP) model. By experimental results, it has been shown that the proposed WMSP model attains higher accuracy and detection rate when compared to other algorithms.

References

1. Zanin, M., Romance, M., Moral, S., &Criado, R. (2018). Credit card fraud detection through parenclitic network analysis. *Complexity*.
2. Seeja, K. R., & Zareapoor, M. (2014). Fraud Miner: A novel credit card fraud detection model based on frequent itemset mining. *The Scientific World Journal*.
3. Sailusha, R., Gnaneswar, V., Ramesh, R., &Rao, G. R. (2020, May). Credit card fraud detection using machine learning. In *2020 4th International Conference on Intelligent Computing and Control Systems (ICICCS)* (pp. 1264–1270). IEEE.
4. Saragih, M. G., Chin, J., Setyawasih, R., Nguyen, P. T., & Shankar, K. (2019). machine learning methods for analysis fraud credit card transaction. *International Journal of Engineering and Advanced Technology, 8*.
5. Pumsirirat, A., & Yan, L. (2018). Credit card fraud detection using deep learning based on auto-encoder and restricted boltzmann machine. *International Journal of Advanced Computer Science and Applications, 9*(1), 18–25.
6. Mehndiratta, S., & Gupta, M. K. (2019). *Credit card fraud detection techniques: A review*.
7. Gulati, A., Dubey, P., MdFuzail, C., Norman, J., &Mangayarkarasi, R. (2017, November). Credit card fraud detection using neural network and geolocation. In *IOP Conference Series: Materials Science and Engineering* (Vol. 263, No. 4, p. 042039).
8. Zareapoor, M., & Shamsolmoali, P. (2015). Application of credit card fraud detection: Based on bagging ensemble classifier. *Procedia Computer Science, 48*(2015), 679–685.
9. Cody, T., Adams, S., &Beling, P. A. (2018, April). A utilitarian approach to adversarial learning in credit card fraud detection. In *2018 Systems and Information Engineering Design Symposium (SIEDS)* (pp. 237–242). IEEE.
10. Gianini, G., Fossi, L. G., Mio, C., Caelen, O., Brunie, L., & Damiani, E. (2020). Managing a pool of rules for credit card fraud detection by a Game Theory based approach. *Future Generation Computer Systems, 102,* 549–561.
11. Deng, R., &Ruan, N. (2019). FraudJudger: Real-world data oriented fraud detection on digital payment platforms. arXiv preprint arXiv:1909.02398.
12. Sudha, C., &Akila, D. *Email and IP detection of credit card fraud detection using KNN*.

13. Sudha, C., &Akila, D. (2019). *Detection OFAES algorithm for data security on credit card transaction.*
14. Sudha, C., &Akila, D. (2020). Credit card fraud detection using AES Technic. In *Intelligent computing and innovation on data science* (pp. 91–98). Singapore: Springer.

An Extension of Al-Zughair Integral Transform for Solving Some LPDE'S

Ali Hassan Mohammed, Asraa Obaid Saud, and Ameer Qassim Majde

Abstract There are many types of transformations are used in many applications such as Laplace transform and Fourier transform that use to solve ordinary and partial differential equations. Integral transformation of Al-Zughair is one of the important improvements that plays a significant role in many different physical applications such as transforming the Laplace. Our objective in this paper is to add an extension of Al-Zughair transformation with some properties, constant function theorems and transformations, logarithms functions and other functions also implement the reverse of the extension of Al-Zughair transform and demonstrate how to use it to solve other forms of partial differential equations.

Keywords Al-zughair integral transform · Partial differential equations

1 Introduction

In the twentieth century, general theory was affected by the arrival of set theory in mathematical analysis, with consequences for theorization, including further topological aspects [1].

The new applications are made for the dynamical systems and mathematical quantities as well as relativity theory [2–4]. The Fourier or Laplace transformation methods are established strict mathematical basics that are corresponding to the regenerate practical calculus. Indeed, there are a lot of integration measurements of transformation that involve Hankel transforms, Hilbert transforms, and Stilettoes.

A. H. Mohammed (✉) · A. O. Saud · A. Q. Majde
Department of Mathematics, College of Education of Women, University of Kufa, Najaf, Iraq
e-mail: prof.ali75hassan@gmail.com

A. O. Saud
e-mail: Asraaobaid7@gmail.com

A. Q. Majde
e-mail: ameermath@gmail.com

S.-L. Peng et al. (eds.), *Proceedings of First International Conference on Mathematical Modeling and Computational Science*, Advances in Intelligent Systems and Computing 1292, https://doi.org/10.1007/978-981-33-4389-4_34

These transforms have widely been applied and used to find the solution of problems with initial and boundary value for 'natural differential equations' and other problems in science engineering and mathematics [5–7].

Definition 1.1 An Al-Zughair extension transformation for the function $u(x, f(|lnt|))$ is defined by the following integral: $EZ[u(x, f(|lnx|))] = \int_{e^{-1}}^{e} \frac{(lnt)^{\mu}}{t} u(x, f(|lnt|))dt = v(x, \mu)$.

μ is constant, such that this integral is convergent.

Property 1.2 (Linear property).

The linear property is defined by an extension of the Al-Zughair integral transform as follows:

$$EZ[Au_1(x, f(|lnt|)) \pm Bu_2(x, f(|lnt|))]$$
$$= AEZ[u_1(x, f(|lnt|))] \pm BEZ[u_2(x, f(|lnt|))]$$

where A and B are constants and the functions $u_1(x, f(|lnt|))$ and $u_2(x, f(|lnt|))$ *are defined when* $t \in [e, e^{-1}]$.

Proof

$$EZ[Au_1(x, f(|lnt|)) \pm Bu_2(x, f(|lnt|))]$$
$$= \int_{e^{-1}}^{e} \frac{(lnt)^{\mu}}{t} [Au_1(x, f(|lnt|)) \pm Bu_2(x, f(|lnt|))]dt$$
$$= \int_{e^{-1}}^{e} \frac{(lnt)^{\mu}}{t} Au_1(x, f(|lnt|))dt \pm \int_{e^{-1}}^{e} \frac{(lnt)^{\mu}}{t} Bu_2(x, f(|lnt|))dt$$
$$= A\int_{e^{-1}}^{e} \frac{(lnt)^{\mu}}{t} u_1(x, f(|lnt|))dt \pm B\int_{e^{-1}}^{e} \frac{(lnt)^{\mu}}{t} u_2(x, f(|lnt|))dt$$
$$= AEZ[u_1(x, f(|lnt|))] \pm BEZ[u_2(x, f(|lnt|))]$$

2 An Extension of Al-Zughair Integral Transform for Some Functions

For constant functions, logarithm functions, trigonometric functions, and other functions, an extension of Al-Zughair is given as follows.

1. If $u(x, f(|lnt|)) = 1, x \in \mathbb{R}, t \in [e^{-1}, e], \mu > -1$ then

$$EZ[u(x, f(lnt))] = \frac{2}{\mu + 1}, \quad \mu \text{ is an even number.}$$

Proof

$$EZ[u(x, f(|lnt|))] = \int_{e^{-1}}^{e} \frac{(lnt)^{\mu}}{t} \cdot (1)dt = \left.\frac{(lnt)^{\mu+1}}{\mu+1}\right|_{e^{-1}}^{e}$$

If μ is an even number.

$$= \frac{2}{\mu+1}$$

If μ is an odd number $= 0$.

2. If $u(x, f(|lnt|)) = k, x \in \mathbb{R}, t \in [e^{-1}, e], \mu > -1$ and k is constant, then if μ is an even number

$$EZ[u(x, f(|lnt|))] = \frac{2k}{\mu+1},$$

Proof

$$EZ[u(x, f(|lnt|))] = \int_{e^{-1}}^{e} \frac{(lnt)^{\mu}}{t} \cdot kdt = k\int_{e^{-1}}^{e} \frac{(lnt)^{\mu}}{t}dt = k\left.\frac{(lnt)^{\mu+1}}{\mu+1}\right|_{e^{-1}}^{e}$$

If μ is an even number

$$= \frac{2k}{\mu+1}$$

If μ is an odd number $= 0$.

3. If $u(x, f(|lnt|)) = (lnt)^{n}, x \in \mathbb{R}, t \in [e^{-1}, e], \mu > -(n+1)$, then

$$EZ[u(x, f(|lnt|))] = \frac{2}{\mu+(n+1)}, \quad \text{if } \mu+n \text{ is an even number.}$$

Proof

$$EZ[u(x, f(|lnt|))] = \int_{e^{-1}}^{e} \frac{(lnt)^{\mu}}{t}(lnt)^{n}dt = \int_{e^{-1}}^{e} \frac{(lnt)^{\mu+n}}{t}dt$$

$$= \left.\frac{(lnt)^{\mu+n+1}}{\mu+n+1}\right|_{e^{-1}}^{e}$$

If $\mu+n$ is an even number.

$$= \frac{2}{\mu + (n+1)}$$

If $\mu + n$ is an odd number $= 0$.

4. If $u(x, f(|lnt|)) = \ln|lnt|, x \in \mathbb{R}, t \in [e^{-1}, e], \mu > -1$, then if $\mu + n$ is an even number.

$$EZ[u(x, f(|lnt|))] = \frac{-2}{(\mu+1)^2}$$

Proof

$$EZ[u(x, f(|lnt|))] = \int_{e^{-1}}^{e} \frac{(lnt)^{\mu}}{t} ln|lnt|dt$$

Integrate by parts,
Let $h = ln|lnt| \Longrightarrow dh = t^{-1}(lnt)^{-1}dt$, and $dk = \frac{(lnt)^{\mu}}{t}dt \Rightarrow k = \frac{(lnt)^{\mu+1}}{\mu+1}$

$$\text{so,} \quad \int_{e^{-1}}^{e} \frac{(lnt)^{\mu}}{t} \ln|lnt|dt = \ln lnt . \frac{(lnt)^{\mu+1}}{\mu+1}\Big|_{e^{-1}}^{e} - \frac{1}{\mu+1}\int_{e^{-1}}^{e} \frac{(lnt)^{\mu}}{t}dt$$

If $\mu + n$ is an even number

$$= \frac{-2}{(\mu+1)^2}$$

If $\mu + n$ is an odd number $= 0$.

5. If $u(x, f(|lnt|)) = (\ln|lnt|)^n, x \in \mathbb{R}, t \in [e^{-1}, e], \mu > -1, n = 1, 2, 3, \ldots$, then $EZ[u(x, f(|lnt|))] = \frac{(-1)^n 2 * n!}{(\mu+1)^{n+1}}$; if $\mu + n$ is an even number.

Proof If $n = 1$, and if μ is an odd number; $EZ[u(x, f(|lnt|))] = \frac{-2*1!}{(\mu+1)^2}$.
If μ is an even number $= 0$.
If $n = 2$, and if μ is an even number;

$$EZ[u(x, f(|lnt|))] = \int_{e^{-1}}^{e} \frac{(lnt)^{\mu}}{t} (\ln|lnt|)^2 dt$$

Integrate by parts,
Let $h = (\ln|lnt|)^2 \Rightarrow dh = 2t^{-1}(lnt)^{-1}\ln|lnt|dt$, and, $dk = \frac{(lnt)^{\mu}}{t}dt \Rightarrow k = \frac{(lnt)^{\mu+1}}{\mu+1}$

$$\int_{e^{-1}}^{e} \frac{(lnt)^{\mu}}{t}(\ln|lnt|)^2 dt = (\ln|lnt|)^2 \frac{(lnt)^{\mu+1}}{\mu+1}\Big|_{e^{-1}}^{e} - \frac{2}{\mu+1}\int_{e^{-1}}^{e} \frac{(lnt)^{\mu}}{t}\ln|lnt|dt$$

$$= \frac{2}{\mu+1}.\frac{2}{(\mu+1)^2} = \frac{2*2!}{(\mu+1)^3}$$

If μ is an odd number $= 0$

$$\text{if } \ n = 3 \Rightarrow EZ[u(x, f(|lnt|))] = \int_{e^{-1}}^{e} \frac{(lnt)^{\mu}}{t}(\ln|lnt|)^3 dt$$

Integrate by parts,
Let $h = (\ln|lnt|)^3 \Rightarrow dh = 3t^{-1}(lnt)^{-1}(\ln|lnt|)^2 dt$, and $dk = \frac{(lnt)^{\mu}}{t} \Rightarrow k = \frac{(lnt)^{\mu+1}}{\mu+1}$

$$\int_{e^{-1}}^{e} \frac{(lnt)^{\mu}}{t}(\ln|lnt|)^3 dt = (\ln|lnt|)^3 \frac{(lnt)^{\mu+1}}{\mu+1}\Big|_{e^{-1}}^{e} - \frac{3}{\mu+1}\int_{e^{-1}}^{e} \frac{(lnt)^{\mu}}{t}(\ln|lnt|)^2 dt$$

From the previous step $[n = 2] \rightarrow \left[\int_{e^{-1}}^{e} \frac{(lnt)^{\mu}}{t}(\ln|lnt|)^2 dt = \frac{2*2!}{(\mu+1)^3}\right]$.
If μ is an odd number

$$= \frac{-3}{\mu+1}.\frac{2*2!}{(\mu+1)^3} = \frac{-2*3!}{(\mu+1)^4}$$

If μ is an even number $= 0$

$$\text{If } n = 4 \Rightarrow EZ[u(x, f(|lnt|))] = \int_{e^{-1}}^{e} \frac{(lnt)^{\mu}}{t}(\ln|lnt|)^4 dt$$

Integrate by part,
Let $h = (\ln|lnt|)^4 \Rightarrow dh = 4t^{-1}(lnt)^{-1}(\ln|lnt|)^3 dt$, and $dk = \frac{(lnt)^{\mu}}{t} \Rightarrow k = \frac{(lnt)^{\mu+1}}{\mu+1}$

$$\int_{e^{-1}}^{e} \frac{(lnt)^{\mu}}{t}(\ln|lnt|)^4 dt = (\ln|lnt|)^4 \frac{(lnt)^{\mu+1}}{\mu+1}\Big|_{e^{-1}}^{e} - \frac{4}{\mu+1}\int_{e^{-1}}^{e} \frac{(lnt)^{\mu}}{t}(\ln|lnt|)^3 dt$$

If μ is an even number

$$= \frac{-4}{\mu+1} \cdot \frac{-2*3!}{(\mu+1)^4} = \frac{2*4!}{(\mu+1)^5}$$

If μ is an odd number = 0.
Thus, we obtain:
If $\mu + n$ is an even number

$$EZ[u(x, f(|lnt|))] = \frac{(-1)^n 2*n!}{(\mu+1)^{n+1}}, \quad n = 1, 2, 3, \ldots$$

If μ + n is an odd number = 0.

6. If $u(x, f(|lnt|)) = \sin(aln|lnt|), \quad x \in \mathbb{R}, t \in [e^{-1}, e], \mu > -1, a \in \mathbb{R}$, then if μ + n is an even number

$$EZ[u(x, f(|lnt|))] = \frac{-2a}{(\mu+1)^2 + a^2}$$

Proof

$$EZ[u(x, f(|lnt|))] = \int_{e^{-1}}^{e} \frac{(lnt)^{\mu}}{t} \sin(aln|lnt|)dt$$

$$= \int_{e^{-1}}^{e} \frac{(lnt)^{\mu}}{t} \left(\frac{e^{ialn|lnt|} - e^{-ialn|lnt|}}{2i} \right) dt$$

$$= \int_{e^{-1}}^{e} \frac{(lnt)^{\mu}}{t} \left(\frac{|lnt|^{ia} - |lnt|^{-ia}}{2i} \right) dt$$

$$= \frac{1}{2i} \left(\int_{e^{-1}}^{e} \frac{|lnt|^{\mu+ia}}{t} dx - \int_{e^{-1}}^{e} \frac{|lnt|^{\mu-ia}}{t} dt \right)$$

$$= \frac{1}{2i} \left[\frac{|lnt|^{\mu+ia+1}}{\mu+ia+1} - \frac{|lnt|^{\mu-ia+1}}{\mu-ia+1} \right]_{e^{-1}}^{e}$$

If $\mu + n$ is an even number.

$$= \frac{1}{2i} \left[\left[\frac{1}{\mu+ia+1} + \frac{1}{\mu+ia+1} \right] - \left[\frac{1}{\mu-ia+1} + \frac{1}{\mu-ia+1} \right] \right]$$

$$= \frac{-2a}{(\mu+1)^2 + a^2}$$

If $\mu + n$ is an odd number = 0.

7. If $u(x, f(|lnt|)) = \cos(aln|lnt|)$, $x \in \mathbb{R}, t \in [e^{-1}, e], \mu > -1, a \in \mathbb{R}$, then if $\mu + n$ is an even number

$$EZ[u(x, f(|lnt|))] = \frac{2(\mu+1)}{(\mu+1)^2 + a^2}$$

Proof

$$EZ[u(x, \cos(aln|lnt|))] = \int_{e^{-1}}^{e} \frac{(lnt)^{\mu}}{t} \cos(aln|lnt|)dt$$

$$= \int_{e^{-1}}^{e} \frac{(lnt)^{\mu}}{t} \left(\frac{e^{ialn|lnt|} + e^{-ialn|lnt|}}{2} \right) dt$$

$$= \int_{e^{-1}}^{e} \frac{(lnt)^{\mu}}{t} \left(\frac{|lnt|^{ia} + |lnt|^{-ia}}{2} \right) dt$$

$$= \frac{1}{2} \left(\int_{e^{-1}}^{e} \frac{|lnt|^{\mu+ia}}{t} dt + \int_{e^{-1}}^{e} \frac{|lnt|^{\mu-ia}}{t} dt \right)$$

$$= \frac{1}{2} \left[\frac{|lnt|^{\mu+ia+1}}{\mu+ia+1} + \frac{|lnt|^{\mu-ia+1}}{\mu-ia+1} \right]_{e^{-1}}^{e}$$

If $\mu + n$ is an even number.

$$= \frac{1}{2} \left[\frac{2}{\mu+ia+1} + \frac{2}{\mu-ia+1} \right]$$

$$= \frac{2(\mu+1)}{(\mu+1)^2 + a^2}$$

If $\mu + n$ is an odd number $= 0$.

8. If $u(x, f(|lnt|)) = \sinh(aln|lnt|)$, $x \in \mathbb{R}, t \in [e^{-1}, e]$; $|\mu + 1| > a$, $a \in \mathbb{R}$, then
if $\mu + n$ is an even number

$$EZ[u(x, f(|lnt|))] = \frac{-2a}{(\mu+1)^2 - a^2}$$

Proof

$$EZ[u(x, f(|lnt|))] = \int_{e^{-1}}^{e} \frac{(lnt)^{\mu}}{t} \sinh(aln|lnt|)dt$$

$$= \int_{e^{-1}}^{e} \frac{(lnt)^{\mu}}{t}\left(\frac{e^{aln|lnt|} - e^{-aln|lnt|}}{2}\right)dt$$

$$= \int_{e^{-1}}^{e} \frac{(lnt)^{\mu}}{t}\left(\frac{(lnt)^{a} - (lnt)^{-a}}{2}\right)dt$$

$$= \frac{1}{2}\left(\int_{e^{-1}}^{e} \frac{(lnt)^{\mu+a}}{t}dt - \int_{e^{-1}}^{e} \frac{(lnt)^{\mu-a}}{t}dt\right)$$

$$= \frac{1}{2}\left(\frac{(lnt)^{\mu+a+1}}{\mu+a+1} - \frac{(lnt)^{\mu-a+1}}{\mu-a+1}\right)\Bigg|_{e^{-1}}^{e}$$

If $\mu + n$ is an even number.

$$= \frac{1}{2}\left[\frac{2}{\mu+a+1} - \frac{2}{\mu-a+1}\right] = \frac{-2a}{(\mu+1)^2 - a^2}$$

If $\mu + n$ is an odd number $= 0$.

9. If $u(x, f(|lnt|)) = \cosh(aln|lnt|)$; $x \in \mathbb{R}, t \in \left[e^{-1}, e\right]$, $|\mu + 1| > a$, $a \in \mathbb{R}$, then

if $\mu + n$ is an even number $EZ[u(x, f(|lnt|))] = \frac{2(\mu+1)}{(\mu+1)^2-a^2}$

Proof

$$EZ[u(x, f(|lnt|))] = \int_{e^{-1}}^{e} \frac{(lnt)^{\mu}}{t}\cosh(aln|lnt|)\,dt$$

$$= \int_{e^{-1}}^{e} \frac{(lnt)^{\mu}}{t}\left(\frac{e^{aln|lnt|} + e^{-aln|lnt|}}{2}\right)dt$$

$$= \int_{e^{-1}}^{e} \frac{(lnt)^{\mu}}{t}\left(\frac{(lnt)^{a} + (lnt)^{-a}}{2}\right)dt$$

$$= \frac{1}{2}\left(\int_{e^{-1}}^{e} \frac{(lnt)^{\mu+a}}{t}dt + \int_{e^{-1}}^{e} \frac{(lnt)^{\mu-a}}{t}dt\right)$$

if $\mu + n$ is an even number $= \frac{1}{2}\left[\frac{2}{\mu+a+1} + \frac{2}{\mu-a+1}\right] = \frac{2(\mu+1)}{(\mu+1)^2-a^2}$.
If $\mu + n$ is an odd number $= 0$.

Example 2.1

- $EZ\left[lnx(|lnt|)^5 + x\cos\ln(|lnt|)\right] = \frac{2lnx}{\mu+6} + \frac{2x(\mu+1)}{(\mu+1)^2+1}$
- $EZ\left[3x^2(lnt)^3 + (\ln(|\text{lnt}|))^2\right] = \frac{6x^2}{\mu+4} + \frac{4}{(\mu+1)^3}$
- $EZ\left[\sinh(3\ln(|lnt|)) + (\ln(|lnt|))^7\right] = \frac{-6}{(\mu+1)^2-9} + \frac{(-1)^7*2*7!}{(\mu+1)^8}$
- $EZ\left[(lnt)^{-3} + \cosh(5ln(|lnt|))\right] = \frac{2}{\mu-2} + \frac{2(\mu+1)}{(\mu+1)^2-25}$

Theorem 2.2 *If* $EZ[u(x, f(|lnt|)] = v(x, \mu)$ *and s is constant, then*

$$EZ\left[(lnt)^a u(x, f(|lnt|)\right] = v(x, \mu + a)$$

Proof Since $EZ[u(x, f(|lnt|)] = v(x, \mu)$, then.

$$EZ\left[(lnt)^a u(x, f(|lnt|)\right] = \int_{e^{-1}}^{e} \frac{(lnt)^{\mu}}{t}(lnx)^a u(x, f(|lnt|)\mathrm{d}t$$

$$= \int_{e^{-1}}^{e} \frac{(lnt)^{\mu+a}}{t} u(x, f(|lnt|)\mathrm{d}t = v(x, \mu + a)$$

Example 2.3

- $Ez\left[x(lnt)^{-2}\cosh(6\ln(|lnt|))\right] = \frac{2(\mu-1)x}{(\mu-1)^2-36}$

Definition 2.4 Let $u(x, f(|lnt|)$ be a function where $t \in \left[e^{-1}, e\right]$, $x \in \mathbb{R}$ and.

$EZ[u(x, f(|lnt|)] = v(x, \mu)$, $u(x, f(|lnt|)$ *is* called to be an inverse for an extension Al-Zughair transformation and written as:

$EZ^{-1}[v(x, \mu)] = u(x, f(|lnt|))$ where EZ^{-1} returns the transformation to the original function.

Example 2.5

$$EZ^{-1}\left[\frac{14(lnx)^2}{(\mu + 1)^2 - 49}\right] = -\sinh(7\ln(|\text{lnt}|))(lnx)^2$$

Theorem 2.6

$$If\ EZ^{-1}[v(x, \mu)] = u(x, f(|lnt|)), \quad \text{then } EZ^{-1}[v(x, \mu + a)]$$
$$= (\ln t)^a u(x, f(|lnt|), \quad where\ a\ is\ constant.$$

Proof From Theorem 2.2

$$EZ\left[(lnt)^a u(x, f(|lnt|))\right] = v(x, \mu + a)$$

take EZ^{-1} both sides: $EZ^{-1}[v(x, \mu + a)] = (lnt)^a u(x, f(|lnt|)$.

Example 2.7 • $EZ^{-1}\left[\frac{8\cosh x}{(\mu-3)^2-16}\right] = -(lnt)^{-4}\sinh(4\ln(|lnt|))\cosh x.$

3 The Recent Pattern of LPDEs

Definition 3.1 The equation:

$$\begin{aligned} &a_0(lnt)^m u_t^{(m)}(x, lnt) + a_1(lnt)^{m-1}u_t^{(m-1)}(x, lnt) \\ &\quad + \cdots + a_{m-1}(lnt)u_t(x, lnt) + a_m u(x, lnt) = f(x, t) \end{aligned}$$

where $a_0, a_1 \ldots, a_n$ are constants and $f(x, t)$ is a function of x and t is called **Ali'sequation**.

Note 3.2 If the function $u(x, lnt)$ is defined for $t \in \left[e^{-1}, e\right]$, $x \in \mathbb{R}$ and its derivatives $u_t(x, lnt), u_{tt}(x, lnt), \ldots, u_t^{(m)}(x, lnt)$ are exist then:

1. $EZ[lntu_t(x, |lnt|)] = 2u(x, 1) - (\mu + 1)EZ[u(x, |lnt|)]$, if μ is an even number.
   $= -(\mu + 1)EZ[u(x, |lnt|)]$, if μ is an odd number.

Proof

$$EZ[lntu_t(x, |lnt|)] = \int_{e^{-1}}^{e} \frac{(lnt)^{\mu}}{t} lntu_t(x, |lnt|)dt = \int_{e^{-1}}^{e} \frac{(lnt)^{\mu+1}}{t} u_t(x, |lnt|)dt$$

Let $h = (lnt)^{\mu+1} \Rightarrow dh = (\mu + 1)\frac{(lnt)^{\mu}}{t}dt$, $dv = \frac{u_t(x,|lnt|)}{t}dt \Rightarrow v = u(x, |lnt|)$.

$$\int_{e^{-1}}^{e} \frac{(lnt)^{\mu+1}}{t} u_t(x, |lnt|)\, dt = (lnt)^{\mu+1} \cdot u(x, |lnt|)\Big|_{e^{-1}}^{e} - (\mu + 1)\int_{e^{-1}}^{e} \frac{(lnt)^{\mu}}{t} u(x, |lnt|)dt$$

If μ is an even number

$$\begin{aligned} &= [u(x, 1) + u(x, 1)] - (\mu + 1)EZ[u(x, |lnt|)] \\ &= 2u(x, 1) - (\mu + 1)EZ[u(x, |lnt|)] \end{aligned}$$

If μ is an odd number

$$= -(\mu + 1)EZ[u(x, lnt)]$$

2. $EZ\left[(lnt)^2 u_{tt}(x, |lnt|)\right] = 2u_t(x, 1) + (\mu + 2)(\mu + 1)EZ[u(x, |lnt|)]$, if μ is an odd number.

   $= -2(\mu + 2)u(x, 1) + (\mu + 2)(\mu + 1)EZ[u(x, |lnt|)]$, if μ is an even no.

Proof

$$EZ\left[(lnt)^2 u_{tt}(x,|lnt|)\right] = \int_{e^{-1}}^{e} \frac{(lnt)^{\mu}}{t}(lnt)^2 u_{tt}(x,|lnt|)dt = \int_{e^{-1}}^{e} \frac{(lnt)^{\mu+2}}{t} u_{tt}(x,|lnt|)dt$$

Let $h = (lnt)^{\mu+2} \Rightarrow dh = (\mu+2)\frac{(lnt)^{\mu+1}}{t}dt$, $dv = \frac{u_{tt}(x,|lnt|)}{t}dt \Rightarrow v = u_t(x,|lnt|) \int_{e^{-1}}^{e} \frac{(lnt)^{\mu+2}}{t} u_{tt}(x,|lnt|)dt = (|lnt|)^{\mu+2} u_t(x,|lnt|)\Big|_{e^{-1}}^{e} - (\mu+2)\int_{e^{-1}}^{e} \frac{(lnt)^{\mu+1}}{t} u_t(x,|lnt|)dt.$

If μ is an odd number.

$$= [u_t(x,1) + u_t(x,1)] - (\mu+2)$$

$$\times \left[(lnt)^{\mu+1} u(x,|lnt|) - \int_{e^{-1}}^{e} (\mu+1)\frac{(lnt)^{\mu}}{t} u(|x,lnt|)\, dt\right]$$

$$= 2u_t(x,1) - (\mu+2)[0 - (\mu+1)EZ[u(x,|lnt|)]]$$

$$= 2u_t(x,1) + (\mu+2)(\mu+1)EZ[u(x,|lnt|)]$$

If μ is an even number

$$= 0 - (\mu+2)[u(x,1) + u(x,1)] + (\mu+2)(\mu+1)EZ[u(x,|lnt|)]$$

$$= -2(\mu+2)u(x,1) + (\mu+2)(\mu+1)EZ[u(x,|lnt|)]$$

4 An Extension of Al-Zughair Integral Transform for Solving a New Types of LPDE's

Example 4.1 To solve the differential equation:

$$(\ln x)(lnt)^2 u_{tt}(x,lnt) + (lnx)(lnt)u_t(x,lnt) = \begin{cases} 1 \text{ if } \mu \text{ is an even no.} \\ (\ln(|lnt|))^3 \text{ if } \mu \text{ is an odd no.} \end{cases}$$

$$u(x,1)_{=} -1, \quad u_t(x,1) = -1$$

Take EZ to transform to both sides of above equation to get:

If μ is an even no.

$$EZ\left[(lnx)(lnt)^2 u_{tt}(x,lnt)\right] + EZ[(lnx)(lnt)u_t(x,lnt)] = EZ[1]$$

$$-2lnx(\mu+2)u(x,1)+lnx(\mu+2)(\mu+1)EZ[u(x,|lnt|)]$$
$$+2lnxu(x,1)-lnx(\mu+1)EZ[u(x,|lnt|)]$$
$$=\frac{2}{\mu+1}$$
$$2lnx(\mu+2)+lnx(\mu+2)(\mu+1)EZ[u(x,|lnt|)]$$
$$-2lnx-lnx(\mu+1)EZ[u(x,|lnt|)]=\frac{2}{\mu+1}$$
$$EZ[u(x,|lnt|)]=\frac{2-2lnx-2lnx\mu^2-4lnx\mu}{(\mu+1)^3\text{lnx}}$$

Take EZ^{-1} transform to both sides of the above equation to get:

$$u(x,|lnt|)=EZ^{-1}\left[\frac{\text{A}}{(\mu+1)lnx}+\frac{B}{(\mu+1)^2}+\frac{C}{(\mu+1)^3}\right]$$
$$A=-2\ln x$$
$$2A+\ln xB=-4\ln x$$
$$A+\ln xB+\ln xC=-2\ln x+2$$

Hence,

$$A=-2\text{lnx},\ B=0,\ C=\frac{2}{\text{lnx}}$$
$$u(x,|lnt|)=-1+\frac{1}{2lnx}(\ln(|lnt|))^2$$

If μ is an odd no.
Take *EZ* to transform to both sides of the above equation to get:

$$EZ\left[(\text{lnx})(lnt)^2u_{tt}(x,lnt)\right]+EZ[(lnx)(lnt)u_t(x,lnt)]$$
$$=EZ\left[(\ln(|lnt|))^3\right]$$
$$2lnx\ u_t(x,1)+lnx(\mu+2)(\mu+1)EZ[u(x,|lnt|)]$$
$$-lnx(\mu+1)EZ[u(x,|lnt|)]=\frac{-12}{(\mu+1)^4}$$
$$EZ[u(x,|lnt|)]=\frac{-12+2lnx+2\mu^4lnx+8\mu^3lnx+12\mu^2lnx+8\mu lnx}{(\mu+1)^6lnx}$$

Take EZ^{-1} transform to both sides of above equation to get:

$$u(x,|lnt|)=EZ^{-1}$$
$$\times\left[\frac{A}{(\mu+1)lnx}+\frac{B}{(\mu+1)^2}+\frac{C}{(\mu+1)^3}+\frac{D}{(\mu+1)^4}+\frac{E}{(\mu+1)^5}+\frac{K}{(\mu+1)^6}\right]$$

$$\begin{aligned}
&5A + \ln x B = 2\ln x \\
&10A + 4\ln x B + \ln x C = 8\ln x \\
&10A + 6\ln x B + 3\ln x C + \ln x D = 12\ln x \\
&5A + 4\ln x B + 3\ln x C + 2\ln x D + \ln x E = 8\ln x \\
&A + \ln x B + \ln x C + \ln x D + \ln x E + \ln x K = -12 + 2\ln x
\end{aligned}$$

$$u(x, |lnt|) = -\ \ln(lnt) + \frac{6}{5!lnx}(\ln(lnt))^5$$

Example 4.2 To solve the differential equation:

$$(lnt)^3 u_{ttt}(x, lnt) - 2(lnt)u_t(x, lnt) = \begin{cases} x^3(lnt)^2 \text{if } \mu \text{ is an even no.} \\ \ln(|lnt|) \text{if } \mu \text{ is an odd no.} \end{cases}$$

$$u_{tt}(x, 1) = 1,\ u_t(x, 1) = 0,\ u(x, 1) = 0$$

If μ is a even no.
Take *EZ* to convert the above equation on both sides on obtain:

$$\begin{aligned}
&EZ\left[(lnt)^3 u_{ttt}(x, lnt)\right] - EZ[2(lnt)u_t(x, lnt)] = EZ\left[x^3 \ln(lnt)^2\right] \\
&2u_{tt}(x, 1) + 2(\mu + 3)(\mu + 2)u(x, 1) \\
&\quad - (\mu + 3)(\mu + 2)(\mu + 1)EZ[u(x, |lnt|)] + 4u(x, 1) \\
&\quad + 2(\mu + 1)EZ[u(x, |lnt|)] = \frac{2x^3}{\mu + 3} \\
&EZ[u(x, |lnt|)] = \frac{-2x^3 + 3\mu + 6}{(\mu + 1)^2(\mu + 3)(\mu + 4)}
\end{aligned}$$

Take EZ^{-1} transform to both sides of the above equation to get

$$u(x, |lnt|) = EZ^{-1}\left[\frac{A}{(\mu + 1)} + \frac{B}{(\mu + 1)^2} + \frac{C}{(\mu + 3)} + \frac{D}{(\mu + 4)}\right]$$

$$\begin{aligned}
&A + C + D = 0 \\
&8A + B + 6C + 5D = 0 \\
&19A + 7B + 9C + 7D = 2 \\
&12A + 12B + 4C + 3D = -2x^3 + 6
\end{aligned}$$

Hence,

$$A = \frac{-5x^3 + 4}{18},\ B = \frac{-x^3 + 2}{3},\ C = \frac{-x^3}{2},\ D = \frac{2(x^3 + 1)}{9}$$

$$u(x,|lnt|)=\frac{-5x^3+4}{36}(1)+\frac{-x^3+2}{6}\ln(lnt)-\frac{x^3}{4}(lnt)^2+\frac{\left(x^3+1\right)}{9}(lnt)^3$$

If μ is an odd number.
Take *EZ* to transform the above equation on both sides to get:

$$\begin{aligned}&EZ\left[(lnt)^3u_{ttt}(x,lnt)\right]-EZ[2(lnt)u_t(x,lnt)]=EZ[\ln(lnt)]\\&-2(\mu+3)u_t(x,1)-(\mu+3)(\mu+2)(\mu+1)EZ[u(x,|lnt|)]\\&+2(\mu+1)EZ[u(x,|lnt|)]=\frac{-2}{(\mu+1)^2}\\&EZ[u(x,|lnt|)]=\frac{2}{(\mu+1)^4(\mu+4)}\end{aligned}$$

Take EZ^{-1} transform to both sides of the above equation to get:

$$u(x,|lnt|)=EZ^{-1}\left[\frac{A}{(\mu+1)}+\frac{B}{(\mu+1)^2}+\frac{C}{(\mu+1)^3}+\frac{D}{(\mu+1)^4}+\frac{E}{\mu+4}\right]$$

$$\begin{aligned}&A+E=0\\&7A+B+4E=0\\&15A+6B+C+6E=0\\&13A+9B+5C+D+4E=0\\&4A+4B+4C+4C+4D+E=2\end{aligned}$$

Hence,

$$A=\frac{-2}{81},B=\frac{2}{27},C=\frac{-2}{9},D=\frac{2}{3},E=\frac{2}{81}$$

$$u(x,|lnt|)=\frac{-1}{81}(1)-\frac{2}{27}\ln(lnt)-\frac{1}{18}(\ln(lnt))^2-\frac{1}{18}(\ln(lnt))^3+\frac{1}{81}(\ln t)^3$$

5 Conclusion

In order to solve partial differential equations, we introduced an efficient Al-Zughair-type integral transform called Al-Zughair extension integral transform. We proposed a transformation of its life and the inverse. We provided some useful properties for solving partial differential equations and their applications. We have some essential properties for integral transforming Al-Zughair extension. The transformation can be used to solve, with instance, a new type of partial differential equations. This can easily be extended directly to any class of partial differential equations that are not solved by other transformations such as transforming Laplace.

References

1. Mohammed, A. H., & Habeeb, N. A. (2017). *AL-Zughair transformation and its uses for solving partial differential equations* (M.Sc., thesis). Submitted to the Council of University of Kufa, Faculty of Education for Girls.
2. Grattan-Guiness, I. (Ed.). (2004). *Companion encyclopedia of the history and philosophy of the mathematical sciences* (Vol. 1). London: Routledge
3. Berglund, N. (2001). *Geometrical theory of dynamical systems*. Department of Mathematics, Switzerland, Lecture Notes Winter Semester 2000–2001 (pp. 1–79).
4. Klages, R. (2008). Introduction to dynamical systems. Lecturer Notes MAS424/MTHM021 Queen Mary University, London.
5. Mohammed, A. H., Hussain, Z. M., & Hadi, A. S. (2015). On Al-tememe transform and solving some kinds of ordinary differential equations with initial conditions and without it and some applications in another sciences (M.Sc. thesis). Submitted to the Council of University of Kufa, Faculty of Education for Girls.
6. Zhang, Yu. (2014). solving initial-boundary value problems for local fractional differential equation by local fractional fourier series method. *Hindawi Publishing Corporation Abstract and Applied Analysis, 2014*, 1–5.
7. Erbe, L. (1971). Boundary value problems for ordinary differential equations. *The Rocky Mountain Journal of Mathematics, 1*(4), 709–729.

Solving Realistic Applications of Volterra Integral Equation Using Different Integral Transformations

Ameera N. Alkiffai and Sarah Th. Alaraji

Abstract Many advance scientific problems of biology, chemistry, physics and engineering can represent mathematically in the form of Volterra integral equations. In this paper, we used Sadik transform, Mahgoub transforms, Kamal transforms and Laplace transforms to solve an integral equation and some practical applications of the transformations to solve these kinds of equations are given. The results of given applications show that the linear integral equation has a unique solution despite solving it in several transformations and there is a variety in convolution theorem for each transform.

Keywords Integral equations · Linear volterra integral equation · Integral transforms · Convolution theorem

1 Introduction

An integral equation is an equation in which the universal (x) function is unknown within an integral symbol. The form is the most common type of integral equation in $x(x)$ [1–10]:

$$\eta(x) = F(x) + \lambda \int_{s(x)}^{r(x)} N(x,t)\eta(t)\mathrm{d}t \tag{1.1}$$

In many types, detailed equations appear. The types mainly rely on the integration limits and the kernel of the equation as Fredholm Equations, Volterra Equation which defines as

A. N. Alkiffai (✉) · S. Th. Alaraji
Department of Mathematics, College of Education for Girls, University of Kufa, Najaf, Iraq
e-mail: ameeran.alkiffai@uokufa.edu.iq

S. Th. Alaraji
e-mail: sarahalarrgi@gmail.com

S.-L. Peng et al. (eds.), *Proceedings of First International Conference on Mathematical Modeling and Computational Science*, Advances in Intelligent Systems and Computing 1292, https://doi.org/10.1007/978-981-33-4389-4_35

$$F(x) = \int_{0}^{x} N(x, t)\eta(t)\mathrm{d}t. \tag{1.2}$$

And Volterra–Fredholm Which define as follows

$$\eta(x) = F(x) + \lambda_1 \int_{a}^{x} N_1(x, t)\eta(t)\mathrm{d}t + \lambda_2 \int_{a}^{b} N_2(x, t)\eta(t)\mathrm{d}t \tag{1.3}$$

$$\eta(x, t) = F(x, t) + \lambda \int_{0}^{t} \int_{\Omega} (x, t, \xi, T, \eta(\xi, T))\mathrm{d}\xi d, \quad (x, t) \in \Omega \times [O, T] \tag{1.4}$$

The Volterra Singular Integral Equations are also integral First or Second class equations, respectively:

$$F(x) = \lambda \int_{s(x)}^{r(x)} N(x, t)\eta(t)\mathrm{d}t \tag{1.5}$$

$$\eta(x) = F(x) + \lambda \int_{s(x)}^{r(x)} N(x, t)\eta(t)\mathrm{d}t \tag{1.6}$$

Is called singular when $r(x)$, $s(x)$, or both of the limits of integration is infinite. See also [1, 10].

Some practical applications are provided in this paper to demonstrate the efficiency or effectiveness of the transformations to solve these kinds of integral equations.The results of given applications show that the linear Volterra integral equation has a unique solution despite solving it in several transformations and there is a variety in convolution theorem for each transform.

2 The Transformation for Some Functions

Table 1 show some important transformation for some functions that we need in our results.

Table 1 The transformation for some elementary functions [3–7]

S. No.	$F(t)$	$K\{F(t)\} = G(V)$	$\mu\{F(t)\} = H(V)$	$\delta\{F(t)\} = T(v^{a}, \beta)$	$\ell\{F(t)\} = Y(V)$
1	1	v	1	$\frac{1}{v^{\alpha+\beta}}$	$\frac{1}{v}$
2	t	v^2	$\frac{1}{v}$	$\frac{1}{v^{2\alpha+\beta}}$	$\frac{1}{v^2}$
3	t^2	$2!v^3$	$\frac{2!}{v^2}$	$\frac{2!}{v^{3\alpha+\beta}}$	$\frac{2!}{v^3}$
4	$t^n,\ n \in N$	$n!v^{n+1}$	$\frac{n!}{v^n}$	$\frac{n!}{v^{(n+1)\alpha+\beta}}$	$\frac{n!}{v^{n+1}}$
5	e^{at}	$\frac{v}{1-av}$	$\frac{v}{v-a}$	$\frac{1}{v^{\beta}(v^{\alpha}-a)}$	$\frac{1}{v-a}$
6	sinat	$\frac{av^2}{1+a^2v^2}$	$\frac{av}{v^2+a^2}$	$\frac{a}{v^{\beta}(v^{2\alpha}+a^2)}$	$\frac{a}{v^2+a^2}$
7	cosat	$\frac{v}{1+a^2v^2}$	$\frac{v^2}{v^2+a^2}$	$\frac{v^{\alpha}}{v^{\beta}(v^{2\alpha}+a^2)}$	$\frac{v}{v^2+a^2}$
8	sinhat	$\frac{av^2}{1-a^2v^2}$	$\frac{av}{v^2-a^2}$	$\frac{a}{v^{\beta}(v^{2\alpha}-a^2)}$	$\frac{a}{v^2-a^2}$

3 Applications

3.1 The Population Growth Equation

The population growth (growth of country, or a plant, or a bacterium, or a cell, or an organ, or a species) is governed by the equation [1]:

$$f(x) = bx + \int_0^x N(x-y)f(y)dy \tag{3.1}$$

Consider the number of children born at time $x = 0$ is b(where b is arbitrary constant) and the survival function is x, so we will solve this equation as shown below.

ℓ- Transform

$$f(x) = bx + \int_0^x N(x-y)\cdot f(y)dy.$$

$$\ell(f(x)) = \ell(bx) + \ell\left[\int_0^x N(x-y)\cdot f(y)dy\right]$$

$$(f(x)) = \frac{b}{v^2} + \frac{1!}{v^2}\cdot \ell[f(x)]$$

$$\therefore f(x) = \ell^{-1}\left(\frac{b}{(\upsilon^2 - 1)}\right) = b\ell^{-1}\left(\frac{1}{\upsilon^2 - 1}\right) = b \sinh x$$

3.2 The Storekeeper's Control Problem

The storekeepers want to keep their stores stock of goods is fixed for using to use the storage space in the best case we can formulate this store problem as an integral equation [2], as follows:

$$f(t) = rt^2 + \int_0^t N(t - \tau) f(\tau)\, d\tau \tag{3.2}$$

where

r the number of goods at time t = 0, (r is arbitrary constant);
$N(t)$ the remainder of goods in stock at t;
t the velocity (goods\time) with which new products are bought.

Accordingly, Eq. (3.2) expresses the number of products (goods) in stock which is the second type of Volterra integral equation, so we can solve this equation with some transforms as shown.

A: κ-Transform

$$f(t) = rt^2 + \int_0^t N(t - \tau) f(\tau) d\tau$$

$$\mathrm{K}[f(t)] = r\mathrm{K}\left[t^2\right] + \mathrm{K}\left[\int_0^t N(t - \tau) f(\tau) d\tau\right]$$

$$\mathrm{K}[f(t)] = r2!\upsilon^3 + \upsilon^2\mathrm{K}[f(t)]$$

$$\therefore f(t) = 2rt \cosh t = pt \cosh t \quad (\text{where } p = 2r)$$

B: ℓ-Transform

$$\ell(f(t)) = \ell\left(rt^2\right) + \ell\left[\int_0^t N(t - \tau) f(\tau) d\tau\right]$$

$$\ell[f(t)] = r\frac{2!}{\upsilon^3} + \frac{1}{\upsilon^2}\ell[f(t)]$$

$$\ell[f(t)]\left(1 - \frac{1}{\upsilon^2}\right) = r\frac{2}{\upsilon^3} \to \ell[f(t)]\left(\frac{\upsilon^2 - 1}{\upsilon^2}\right) = \frac{2r}{\upsilon^3}$$

$$\therefore f(t) = 2rt \cosh t = pt \cosh t \ (\text{where } p = 2r).$$

C: δ-Transform

$$f(t) = rt^2 + \int_0^t N(t-\tau)\cdot f(\tau)\mathrm{d}\tau$$

$$\delta[f(t)] = \delta\left[rt^2\right] + \delta\left[\int_0^t N(t-\tau) f(\tau)\mathrm{d}\tau\right]$$

$$= r\cdot\frac{2!}{\upsilon^{3\alpha+\beta}} + \frac{1}{\upsilon^{2\alpha+\beta}}\cdot\upsilon^{\beta}\cdot\delta[f(t)]$$

$$\delta[f(t)]\left(1 - \frac{\upsilon^{\beta}}{\upsilon^{2\alpha+\beta}}\right) = \frac{r2!}{\upsilon^{3\alpha+\beta}}$$

$$\therefore f(t) = 2rt\cosh t = pt\cosh t \quad (\text{where } p = 2r)$$

4 Conclusion

Integral equations are used in many scientific and engineering fields. This kind of equations may be converted from problems with initial and limit value. A transformation is an operational method for solving constant linear integral equations by coefficients. The transform of the function $F(x)$ for $x \geq 0$ exist if it is piecewise continuous and of exponential order. In this work, we solved some of the useful applications of integral equations using famous transforms in order to show that this method is more accurate than the normal one.

References

1. ALKiffai, A. N. (2000). *Solution of inverse problem of integral equations using variatioal technique* (M.Sc. Thesis, University of Babylon).
2. ALKiffai, A. N., & ALShihmani, N. N. (2019). *Variational technique for storekeeper control problem as an integral equation* (M.Sc. Thesis, Kufa University).
3. Aggarwal, S., Chauhan, R., & Sharma, N. (2018). A new application of Kamal Transform for solving linear volterra integral equations. *International journal of latest technology in engineering management and applied science, 7*(4), 132–140.
4. Aggarwal, S., Chauhan, R., & Sharma, N. (2018). Application of Kamal transform for solving linear volterra integral equation of the first kind. *International journal of research in Advent Technolog, 6*(8).
5. Aggarwal, S , Sharma, N., & Chauhan, R. (2019). A new application of mahgoub transform for solving.
6. Linear volterra integral equation. *Asian Resonance,7*(2), (2018).
7. Aggarwal, S., Gupta, A., & Sharma, S. (2019). Application of SadiK transform for handling linear volterra integro—differential equation of second kind. *Universal Review Scientific Information and Technological Board of Sadhana,10*(7).
8. Finan, M. (2010). *Laplace transforms theory, problems and solutions*. New York: Arkansas Tech University.

9. Odibat, Z. M. (2008). Differential transform method for solving Volterra integral equation with separable kernels. *Mathematical and Computer Modelling, 48*(7–8).
10. Schiff, J. (2013). *The laplace transform theory and applications.* New Zealand: Springer Science and Businees Media.
11. Wazwaz, A. (2011). *Linear and nonlinear integral equations methods and applications.* London: Springer Science and Businees Media.

The Generalization of New Integrative Sequence with Parameter s

Ali J. Mohammad and Hadeel O. Muslim

Abstract This paper described a generalization of the integral sequence of form $W_n(f; x)$ depends on a real number $s > -1/2$, denote by $W_{n,s}(f; x)$. This sequence gives a better approximation from the sequence $W_n(f; x)$. The approximation order is improved by using this generalization to $O\left(n^{-(s+1)}\right)$, $s > 0$. Some basic properties of this generalization and the convergence theorem are introduced. Also, we find a recurrence relation for the moment of order r for the generalization and we discuss the Voronovskaja type limit for this approximation. An estimation of the error by $W_{n,s}(f; x)$ is defined and given by the continuity modulus. A number of examples are also given to estimate a test function for the value chosen of $n = 30, 100$ and $\delta_0 = 0.01, 1$ to explain the modification of the approximation by the sequence $W_{n,s}(f; x)$. Finally, we evaluate the error that occurs between the test function and the approximation and shown this by graphs. The numerical results are compared for the values of $s = 0, 1, 2$.

Keywords Simultaneous approximation · Voronovskaja theorem · Degree of approximation · Modulus of continuity

1 Introduction

Given n a positive integer and $f \in C[0, 1]$, the nth order Bernstein polynomial is defined as [3]

$B_n(f; x) = \sum_{i=0}^{n} b_{n,i}(x) f\left(\frac{i}{n}\right)$, where $b_{n,i}(x) = \frac{n!}{i!(n-i)!} x^i (1-x)^{n-i}$, $x \in [0, 1]$.

Pallini defined and studied the generalization of $B_n(f; x)$ given as:

A. J. Mohammad (✉) · H. O. Muslim
University of Basrah, Basrah, Iraq
e-mail: alijasmoh@gmail.com

H. O. Muslim
e-mail: hadeel.oalkh@gmail.com

S.-L. Peng et al. (eds.), *Proceedings of First International Conference on Mathematical Modeling and Computational Science*, Advances in Intelligent Systems and Computing 1292, https://doi.org/10.1007/978-981-33-4389-4_36

$$B_n^s(f;x) = \sum_{i=0}^{n} b_{n,i}(x) f\left(x + \frac{n^{-1}i - x}{n^s}\right) \quad \text{where } s > -\frac{1}{2}. \tag{1}$$

Also, he showed that the sequence $B_n^s(f;x)$ has a batter order of approximation than the sequence $B_n(f;x)$ whenever $s > 0$ [9].

Many researchers introduced and studied some modification sequences of linear and positive operators [1, 2, 4–7].

Recently, Mohammed and Muslim studied the following new integrative operator sequence [8]

$$W_n(f;x) = \int_0^x K_n(t;x) f(t)\mathrm{d}t \tag{2}$$

where

$$K_n(t;x) = \frac{n\cosh(nt)}{(\delta_o + \sinh(nx))}, \tag{3}$$

$n \in N := \{1, 2, \ldots\}, x \in [0, \infty)$ arbitrary but fixed and δ_0 be a positive parameter.

Space $C_\alpha[0,\infty) = \{f \in C[0,\infty) : |f(t)| = O(e^{\alpha t}), \text{for some} \alpha > 0\}$ is normed by $f_{C_\alpha[0,\infty)} = \sup_{t\in[0,x]}|f(t)|e^{-\alpha t}$. We define the generalization of our integrative operators (2) in the same manner as the generalization of $B_n(f;x)$ in (1) as follows

$$W_{n,s}(f;x) = \int_0^x K_n(t;x) f\left(x + \frac{t-x}{n^s}\right)\mathrm{d}t \tag{4}$$

where $K_n(t;x)$ defined in (3), $s \geq -\frac{1}{2}$ and $f \in C_\alpha[0,\infty)$. Note that, if $s = 0$, then we get (2). For practices in this paper, we will take $s = 0, 1, \ldots$.

Firstly, we study the basic properties and the convergence theorem of $W_{n,s}(f;x)$ and find the recurrence relation for the moment of order r for $W_{n,s}(f;x)$. Next, we drive a Voronovskaja asymptotic limit. Also, we consider this approximation error defined by the continuity modulus. Finally, some approximation example of $W_{n,s}(f;x)$ is given and comparison of the results is obtained.

2 Preliminary Lemmas

Lemma 2.1 [8]

For $x \in [0,\infty)$ and $\rho_i(t) = t^i$, we get

(i) $W_n(\rho_0; x) = \frac{\sinh(nx)}{(\delta_0+\sinh(nx))} \to 1$ as $n \to \infty$;
(ii) $W_n(\rho_1; x) = \frac{x\sinh(nx)}{(\delta_0+\sinh(nx))} + \frac{(1-\cosh(nx))}{n(\delta_0+\sinh(nx))} \to x$ as $n \to \infty$;
(iii) $W_n(\rho_2; x) = \frac{x^2\sinh(nx)}{(\delta_0+\sinh(nx))} - \frac{2x\cosh(nx)}{n(\delta_0+\sinh(nx))} + \frac{2\sinh(nx)}{n^2(\delta_0+\sinh(nx))} \to x^2$ as $n \to \infty$.

Lemma 2.2 [8]

For the rth order moment, $V_{n,r}(x)$ for the operator $W_n(f(t); x)$ defined as: $V_{n,r}(x) = W_n\big((t-x)^r; x\big)$, we have

$$V_{n,o}(x) = \frac{\sinh(nx)}{(\delta_o + \sinh(nx))}; \quad V_{n,1}(x) = \frac{1-\cosh(nx)}{n(\delta_o + \sinh(nx))},$$
$$V_{n,r}(x) = \frac{(r-1)r}{n^2} V_{n,r-2}(x) - \frac{(-1)^r r}{n(\delta_o + \sinh(nx))} x^{r-1}, \quad r \geq 2. \tag{5}$$

Lemma 2.3 *For $x \in [0, \infty)$ and $\rho_i(t) = t^i$, we get:*

(i) $W_{n,s}(\rho_0; x) = V_{n,0}(x) \to 1$ as $n \to \infty$;
(ii) $W_{n,s}(\rho_1; x) = xV_{n,0}(x) + n^{-s}V_{n,1}(x) \to x$ as $n \to \infty$;
(iii) $W_{n,s}(\rho_2; x) = x^2V_{n,0}(x) + 2xn^{-s}V_{n,1}(x) + n^{-2s}V_{n,2}(x) \to x^2$ as $n \to \infty$.

Proof The proof follows with use of Lemma 2.2.

Definition 2.1 For $r \in N^{\circ}$, the order moment of order r for $W_{n,s}(f(t); x)$ is defined and denoted as

$$V^s_{n,r}(x) = W_{n,s}\big((t-x)^r; x\big).$$

Lemma 2.4 *We have that*

$$V^s_{n,0}(x) = V_{n,0}(x); \quad V^s_{n,1}(x) = n^{-s}V_{n,1}(x); \quad V^s_{n,2}(x) = n^{-2s}V_{n,2}(x);$$
$$V^s_{n,r}(x) = n^{-rs}V_{n,r}(x),\ r \geq 2 \tag{6}$$

Also, $\forall x \in [0, \infty)$, $V^s_{n,r}(x) = O\big(n^{-r(s+1)}\big)$.

Proof Using Lemmas 2.2 and 2.3.

Lemma 2.5 *For $r \geq 2$, we have:*

$$W_{n,s}\big(t^r; x\big) = x^rV_{n,0}(x) + \frac{r}{n^s}V_{n,1}(x)x^{r-1} + O\big(n^{-2(s+1)}\big).$$

Proof By Lemma 2.3, we can proof of this lemma.

Lemma 2.6 [8]
Let $\delta, \alpha \in \mathbb{R}^+$ and $[a, b] \subset (0, \infty)$. Then, for $\lambda > 0$, we have:

$$\sup_{x\in[a,b]}\left|\int_{x-t\geq\delta} K_n(t;x)e^{\alpha t}\mathrm{d}t\right| = O\left(n^{-\lambda}\right).$$

Lemma 2.7 [10]
Let $p \in N^0$ and f and g are p-times differentiable functions of x, then:

$$\frac{\mathrm{d}^p}{\mathrm{d}x^p}(fg) = \sum_{l=0}^{p}\binom{p}{l}\frac{\mathrm{d}^{p-l}}{\mathrm{d}x^{p-l}}(f)\frac{\mathrm{d}^l}{\mathrm{d}x^l}(g).$$

Lemma 2.8 [8]
For $p \in N$, we have

$$\lim_{n\to\infty}\frac{\sinh(nx)}{(\delta_0+\sinh(nx))} = 1;\quad \lim_{n\to\infty}\frac{\cosh(nx)}{(\delta_0+\sinh(nx))} = 1;$$
$$\lim_{n\to\infty}\frac{\mathrm{d}^p}{\mathrm{d}x^p}\left(\frac{\sinh(nx)}{(\delta_0+\sinh(nx))}\right) = 0;\quad \lim_{n\to\infty}\frac{\mathrm{d}^p}{\mathrm{d}x^p}\left(\frac{\cosh(nx)}{(\delta_0+\sinh(nx))}\right) = 0;$$
$$\lim_{n\to\infty}\frac{\mathrm{d}^p}{\mathrm{d}x^p}\left(\frac{1}{(\delta_0+\sinh(nx))}\right) = 0;$$

Lemma 2.9 [4]
For $F(x) = \int_{a(x)}^{b(x)} f(x,y)\mathrm{d}y$. Then,

$$F'(x) = f(x,b(x))b'(x) - f(x,a(x))a'(x) + \int_{a(x)}^{b(x)} f_x(x,y)\mathrm{d}y.$$

3 Main Results

First, we prove that:

$$W_{n,s}^{(p)}(f(t);x) \to f^{(p)}(x),\qquad \text{as } n\to\infty,\ p\in N. \tag{7}$$

Theorem 3.1 *Suppose that $p \in N$, $f \in C_\alpha[0,\infty)$ and $f^{(p+1)}(x)$ exists at a point $x \in (0,\infty)$, then:*

$$\lim_{n\to\infty} W_{n,s}^{(p)}(f(t);x) \to f^{(p)}(x) \tag{7}$$

Further, if $f^{(p+1)}$ is exist and continuous on $(a-\eta, b+\eta) \subset (0,\infty)$, η positive than (7) uniformly on $[a,b]$.

Proof By Taylor's formula of f, $\varsigma \in (t,x)$, we get:

$$f(t) = \sum_{i=0}^{p} \frac{f^{(i)}(x)}{i!}(t-x)^i + \frac{f^{(p+1)}(\varsigma)}{(p+1)!}(t-x)^{p+1},$$

$$\begin{aligned} W_{n,s}^{(p)}(f(t);x) &= \sum_{i=0}^{p} \frac{f^{(i)}(x)}{i!} W_{n,s}^{(p)}\left((t-x)^i;x\right) \\ &+ \frac{f^{(p+1)}(\varsigma)}{(p+1)!} W_{n,s}^{(p)}\left((t-x)^{p+1};x\right) \\ &\sim= \sum_1 + \sum_{2.} \end{aligned}$$

Hence, using Lemmas 2.5, 2.7, and 2.8, we acquire:

$$\begin{aligned} \sum_1 = \frac{f^{(p)}(x)}{p!} \Bigg\{ & p!\left(\frac{\sinh(nx)}{(\delta_0+\sinh(nx))}\right) + pp!\frac{nx\cosh(nx)}{(\delta_0+\sinh(nx))} \\ & \times \left(1-\frac{\sinh(nx)}{(\delta_0+\sinh(nx))}\right) + \cdots + x^p \frac{d^p}{dx^p}\left(\frac{\sinh(nx)}{(\delta_0+\sinh(nx))}\right) \\ & - \frac{pp!}{n^s}\left(\frac{\sinh(nx)}{(\delta_0+\sinh(nx))} + \frac{\cosh(nx)(1-\cosh(nx))}{(\delta_0+\sinh(nx))^2}\right) \\ & - \cdots - \frac{p}{n^{s+1}} x^{p-1} \frac{d^p}{dx^p}\left(\frac{\cosh(nx)-1}{(\delta_0+\sinh(nx))}\right)\Bigg\} \to f^{(p)}(x) as\ n \to \infty. \end{aligned}$$

By using Lemmas 2.7 and 2.9, we acquire:

$$\begin{aligned} \sum_2 &= \frac{f^{(p+1)}(\varsigma)}{(p+1)!} \frac{\mathrm{d}^p}{\mathrm{d}x^p}\left(\frac{n}{(\delta_0+\sinh(nx))}\right) \int_0^x \cosh(nt)\left(\frac{t-x}{n^s}\right)^{p+1} \mathrm{d}t \\ &+ \frac{f^{(p+1)}(\varsigma)}{(p+1)!} \sum_{l=1}^{p} \binom{p}{l} \frac{\mathrm{d}^{p-l}}{\mathrm{d}x^{p-l}}\left(\frac{n}{(\delta_0+\sinh(nx))}\right) \\ &\times \frac{\mathrm{d}^l}{\mathrm{d}x^l}\left(\int_0^x \cosh(nt)\left(\frac{t-x}{n^s}\right)^{p+1} \mathrm{d}t\right) \\ &= I_1 + I_2 \end{aligned}$$

Then, $I_1 = o(1)$.

Now, from Lemmas 2.7 and 2.9, we obtain

$$|I_2| = M\Bigg\{\Bigg[\frac{np^2}{(\delta_0+\sinh(nx))}\frac{\mathrm{d}^{p-1}}{\mathrm{d}x^{p-1}}(\delta_0+\sinh(nx))^0 + \frac{p^2(p-1)}{2}\frac{n}{(\delta_0+\sinh(nx))^2}\frac{\mathrm{d}^{p-1}}{\mathrm{d}x^{p-1}}(\delta_0+\sinh(nx)) + \cdots + \frac{np}{(\delta_0+\sinh(nx))^p}\frac{\mathrm{d}^{p-1}}{\mathrm{d}x^{p-1}}(\delta_0+\sinh(nx))^{p-1}\Bigg] \times \left(\int_0^x \cosh(nt)\left|-\frac{(p+1)}{n^s}\left(\frac{t-x}{n^s}\right)^p\right|\mathrm{d}t\right) + \cdots + \frac{n}{(\delta_0+\sinh(nx))}\left(\int_0^x \cosh(nt)\left|(-1)^p\frac{(p+1)!}{n^{ps}}\left(\frac{t-x}{n^s}\right)\right|\mathrm{d}t\right)\Bigg\}.$$

Hence, $I_2 = o(1)$. Therefore, $\sum_2 = o(1)$.

The uniformity case can be followed from $\delta(\varepsilon)$ in all steps, in the proof above, i.e., δ is independent of $x \in [a, b]$.

Theorem 3.2 *Suppose that* $p \in N$, $f \in C_\alpha[0,\infty)$ *and* $f^{(p+3)}$ *exists at* $x \in (0,\infty)$, *then:*

$$\lim_{n\to\infty} n^{s+1}\left(W_{n,s}^{(p)}(f(t);x) - f^{(p)}(x)\right) = -f^{(p+1)}(x)$$

Further, if $f^{(p+3)}$ is exist and continuous on $(a-\eta, b+\eta) \subset (0,\infty)$, η is positive than (8) uniformly on $[a, b]$.

Proof Using formulae of Taylor, when $\varsigma \in (t, x)$, we have:

$$f(t) = \sum_{i=0}^{p+2}\frac{f^{(i)}(x)}{i!}(t-x)^i + \frac{f^{(p+3)}(\varsigma)}{(p+3)!}(t-x)^{p+3},$$

$$W_{n,s}^{(p)}(f(t);x) = \sum_{i=0}^{p+2}\frac{f^{(i)}(x)}{i!}W_{n,s}^{(p)}\left((t-x)^i;x\right) + \frac{f^{(p+3)}(\varsigma)}{(p+1)!}W_{n,s}^{(p)}\left((t-x)^{p+3};x\right) \sim= \sum_1 + \sum_2.$$

Using Lemmas 2.5, 2.7, 2.8, and 2.9, we obtain:

$$\sum_1 = \frac{f^{(p)}(x)}{p!}\Bigg\{p!\left(\frac{\sinh(nx)}{(\delta_0+\sinh(nx))}\right) + \frac{pp!nx\cosh(nx)}{(\delta_0+\sinh(nx))}$$

$$\times\left(1-\frac{\sinh(nx)}{(\delta_0+\sinh(nx))}\right)+\cdots+x^p\frac{\mathrm{d}^p}{\mathrm{d}x^p}\left(\frac{\sinh(nx)}{(\delta_0+\sinh(nx))}\right)$$
$$-\frac{pp!}{n^s}\left(\frac{\sinh(nx)}{(\delta_0+\sinh(nx))}+\frac{\cosh(nx)(1-\cosh(nx))}{(\delta_0+\sinh(nx))^2}\right)$$
$$-\frac{p}{n^{s+1}}x^{p-1}\frac{\mathrm{d}^p}{\mathrm{d}x^p}\left(\frac{\cosh(nx)-1}{(\delta_0+\sinh(nx))}\right)\Bigg\}-f^{(p)}(x)$$
$$+\frac{f^{(p+1)}(x)}{(p+1)!}\Bigg\{-\frac{p(p+1)!}{2}\frac{nx^2\cosh(nx)}{(\delta_0+\sinh(nx))}\left(1-\frac{\sinh(nx)}{(\delta_0+\sinh(nx))}\right)$$
$$-\cdots-px^{p+1}\frac{d^p}{dx^p}\left(\frac{\sinh(nx)}{(\delta_0+\sinh(nx))}\right)-\frac{(p+1)!}{n^{s+1}}\left(\frac{\cosh(nx)-1}{(\delta_0+\sinh(nx))}\right)$$
$$+\cdots+\frac{(p+1)(p-1)}{n^{s+1}}x^p\frac{\mathrm{d}^p}{\mathrm{d}x^p}\left(\frac{\cosh(nx)-1}{(\delta_0+\sinh(nx))}\right)\Bigg\}$$
$$+\frac{f^{(p+2)}(x)}{(p+2)!}\Bigg\{\frac{p(p+2)!}{6}\frac{nx^3\cosh(nx)}{(\delta_0+\sinh(nx))}\left(1-\frac{\sinh(nx)}{(\delta_0+\sinh(nx))}\right)$$
$$+\cdots+\frac{p(p+1)}{2}x^{p+2}\frac{\mathrm{d}^p}{\mathrm{d}x^p}\left(\frac{\sinh(nx)}{(\delta_0+\sinh(nx))}\right)$$
$$+\cdots-\frac{p(p-1)(p+2)}{2n^{s+1}}x^{p+1}\frac{\mathrm{d}^p}{\mathrm{d}x^p}\left(\frac{\cosh(nx)-1}{(\delta_0+\sinh(nx))}\right)\Bigg\}.$$

Therefore:

$$\lim_{n\to\infty} n^{s+1}\left(W_{n,s}^{(p)}(f(t);x)-f^{(p)}(x)\right)=-f^{(p+1)}(x).$$

By the same manner in Theorem 3.1, we obtain $\sum_2 = o(1)$.

The uniformity case can be followed from $\delta(\varepsilon)$ in all steps, in the proof above, i.e., δ is independent of $x\in[a,b]$.

Theorem 3.3 *Let* $\in C_\alpha[0,\infty)$, *for some* $\upsilon>0$, $p\le q\le p+2$. *If* $f^{(q)}$ *exists and is continuous on* $(a-\eta,b+\eta)\subset(0,\infty)$, $\eta>0$, *then for sufficiently large* n,

$$W_{n,s}^{(p)}(f;x)-f^{(p)}(x)_{C[a,b]}\le C_1 n^{-(s+1)}\left[\sum_{i=p}^{q} f_{C[a,b]}^{(i)}\right]$$
$$+C_2 n^{-(s+1)}\omega_{f^{(q+1)}}\left(n^{-(s+1)};(a-\eta,b+\eta)\right)+O\left(n^{-2(s+1)}\right),$$

where C_1, C_2 *are constants independent of* f, n.

Proof By Taylors formulas,

$$f(t)=\sum_{i=0}^{q}\frac{f^{(i)}(x)}{i!}(t-x)^i+\chi(t)\frac{f^{(q+1)}(\varsigma)-f^{(q+1)}(x)}{(q+1)!}(t-x)^{q+1}$$
$$+(1-\chi(t))\upsilon(t,x),$$

where $\chi(t)$ the characteristic interval function $(a-\eta, b+\eta)$ and $\varsigma \in (t, x)$.

For $t \in (a-\eta, b+\eta)$ and $x \in (0, \infty)$, we get:

$$f(t) = \sum_{i=0}^{q} \frac{f^{(i)}(x)}{i!}(t-x)^i + \frac{f^{(q+1)}(\upsilon) - f^{(q+1)}(x)}{(q+1)!}(t-x)^{q+1}.$$

For $t \in [0, x)\backslash(a-\eta, b+\eta)$ and $x \in [a, b]$, we define

$$\upsilon(t, x) = f(t) - \sum_{i=0}^{q} \frac{f^{(i)}(x)}{i!}(t-x)^i.$$

$$\frac{\partial^p}{\partial x^p}\upsilon(t, x) = \sum_{i=0}^{q} \frac{f^{(i)}(x)}{i!}(t-x)^i$$

Now,

$$\begin{aligned} W_{n,s}^{(p)}(f(t); x) - f^{(p)}(x) &= \left[\sum_{i=0}^{q} \frac{f^{(i)}(x)}{i!} W_{n,s}^{(p)}\left((t-x)^i; x\right) - f^{(p)}(x)\right] \\ &\quad + W_{n,s}^{(p)}\left(\frac{f^{(q+1)}(\varsigma) - f^{(q+1)}(x)}{(q+1)!}(t-x)^{q+1}\chi(t); x\right) \\ &\quad + W_{n,s}^{(p)}(\upsilon(t, x)(1-\chi(t)); x) \\ &:= \sum_1 + \sum_2 + \sum_3. \end{aligned}$$

By using Lemma 2.5, we get:

$$\left\|\sum_1\right\|_{C[a,b]} \le C_1 n^{-(s+1)}\left[\sum_{i=r}^{q} f_{C[a,b]}^{(i)}\right] + O\left(n^{-2(s+1)}\right), \quad \text{uniformly on } [a, b].$$

To estimate $\sum_{2,}$ we proceed as follows:

$$\begin{aligned} \left|\sum_2\right| &\le W_{n,s}^{(p)}\left(\frac{\left|f^{(q+1)}(\varsigma) - f^{(q+1)}(x)\right|}{(q+1)!}|(t-x)|^{q+1}\chi(t); x\right) \\ &\le \frac{\omega_{f^{(q+1)}}(\delta; (a-\eta, b+\eta))}{q!} \frac{\mathrm{d}^p}{\mathrm{d}x^p}\left[\frac{n}{(\delta_0 + \sinh(nx))}\right. \\ &\quad \left. \times \int_0^x \cosh(nt)\left(\left|\left(\frac{t-x}{n^s}\right)^{q+1}\right| + \delta^{-1}\left|\left(\frac{t-x}{n^s}\right)^{q+2}\right|\right)\mathrm{d}t\right], \quad \delta > 0. \end{aligned}$$

For $\ell = 0, 1, \ldots$ and applying the Lemmas 2.7, 2.8, 2.9, we get:

$$\left|\frac{\mathrm{d}^p}{\mathrm{d}x^p}\left[\frac{n}{(\delta_0+\sinh(nx))}\int_0^x \cosh(nt)\left(\frac{t-x}{n^s}\right)^{\ell+1}\mathrm{d}t\right]\right|$$

$$=\left|\frac{\mathrm{d}^p}{\mathrm{d}x^p}\left(\frac{n}{(\delta_0+\sinh(nx))}\right)\right|\left|\int_0^x \cosh(nt)\left|\frac{t-x}{n^s}\right|^{\ell+1}\mathrm{d}t\right.$$

$$+\sum_{l=1}^{p}\binom{p}{l}\left|\frac{\mathrm{d}^{p-l}}{\mathrm{d}x^{p-l}}\left(\frac{n}{(\delta_0+\sinh(nx))}\right)\right|\left|\frac{\mathrm{d}^l}{\mathrm{d}x^l}\left(\int_0^x \cosh(nt)\left|\frac{t-x}{n^s}\right|^{\ell+1}\mathrm{d}t\right)\right|$$

$$:= J_1 + J_2.$$

Now, we get

$$J_1 = O\left(n^{p-(s+1)(\ell+1)}\right),\quad \text{uniformly on } [a,b]$$

and $J_2 = o(1)$.
Choosing $\delta = n^{-(s+1)}$ and using (9), then, we get

$$\sum_{2C[a,b]} \leq \frac{\omega_{f^{(q+1)}}\left(n^{-(s+1)};(a-\eta,b+\eta)\right)}{(q+1)!}\left[O\left(n^{p-(s+1)(q+1)}\right)\right.$$
$$\left.+\,n^{(s+1)}O\left(n^{p-(s+1)(q+2)}\right)\right],\quad \leq \left(C_2 n^{-(p-(s+1)(q+1))}\right)\omega_{f^{(q+1)}}\left(n^{-(s+1)};(a-\eta,b+\eta)\right).$$

Since $t \in [0,x)\backslash(a-\eta,b+\eta)$, we can choose $\delta > 0$ if $x-t \geq \delta$ for all $x \in [a,b]$.
Thus,

$$\left|\sum_3\right| = \left|\frac{\mathrm{d}^p}{\mathrm{d}x^p}\left(\frac{n}{(\delta_0+\sinh(nx))}\right)\right|\left|\int_{x-t\geq\delta} \cosh(nt)\upsilon(t,x)\mathrm{d}t\right.$$

$$+\sum_{l=1}^{p}\binom{p}{l}\left|\frac{\mathrm{d}^{p-l}}{\mathrm{d}x^{p-l}}\left(\frac{n}{(\delta_0+\sinh(nx))}\right)\right|\left|\frac{\mathrm{d}^l}{\mathrm{d}x^l}\left(\int_{x-t\geq\delta} \cosh(nt)\upsilon(t,x)\mathrm{d}t\right)\right|$$

$$= J_3 + J_4$$

For $x-t \geq \delta$, we can find a constant $C > 0$ such that $|\upsilon(t,x)| \leq Ce^{\alpha t}$ for some $\alpha > 0$ and $\left|\frac{\partial^p}{\partial x^p}\upsilon(t,x)\right| \leq Ce^{\alpha t}$. Using Lemma 2.6, we get:

$$\left|\sum_3\right| = O\left(n^{-\lambda}\right),\quad \lambda > 0 \text{ uniformly on } [a,b]$$

4 Applications

We introduce a numerical example for the generalization sequence of the positive linear operator $W_{n,s}(.;x)$, by using the test function $g_1(t) = \sin(10t)e^{-2t}$, $t \in [0, 2]$. In this sequence, we calculate the error function and compare the results for $W_{n,s}(g_1, x)$ when $s = 0, 1, 2$ for some $n = 30, 100$ and parameter $\delta_0 = 0.01, 1$, respectively.

Example 4.1 For $n = 30, 100$ and $\delta_0 = 0.01, 1$, respectively, the comparison of the sequence $W_{n,s}(f; x)$, $s = 0, 1, 2$ converges to $g_1(x) = \sin(10x)e^{-2x}$ with the error function given as (Figs. 1, 2, 3 and 4):

5 Conclusion

We gave a numerical example to approximate the test function $g_1(x) = \sin(10x)e^{-2x}$ by the sequence $W_{n,s}(.;x)$ for different values of n, s, and δ_0. Also, we calculate the error function $E(x) = \left|W_{n,s}(g_1; x) - g_1(x)\right|$ to explain the approximation and decide what is the better approximation getting by the values n, s, and δ_0. We get that the approximation becomes better whenever s values and when the values of δ_o

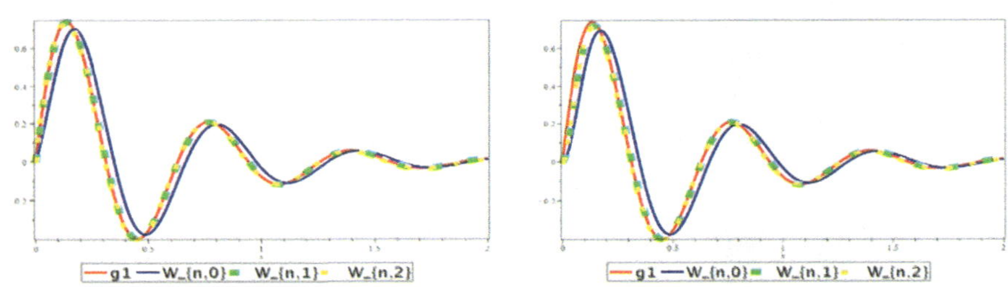

Fig. 1 Convergence of $W_{n,s}(g_1; x)$ to $g_1(x)$ when $n = 30, s = 0, 1, 2$, and $\delta_0 = 0.01, 1$, respectively

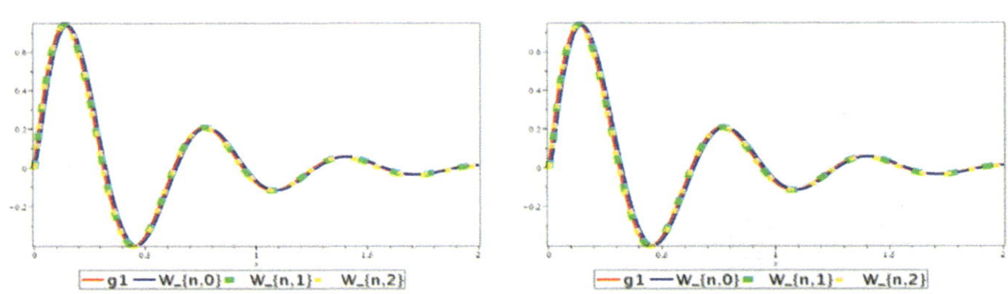

Fig. 2 Convergence error of $W_{n,s}(g_1; x)$ to $g_1(x)$ for $n = 100$, $s = 0, 1, 2$, and $\delta_o = 0.01, 1$, respectively

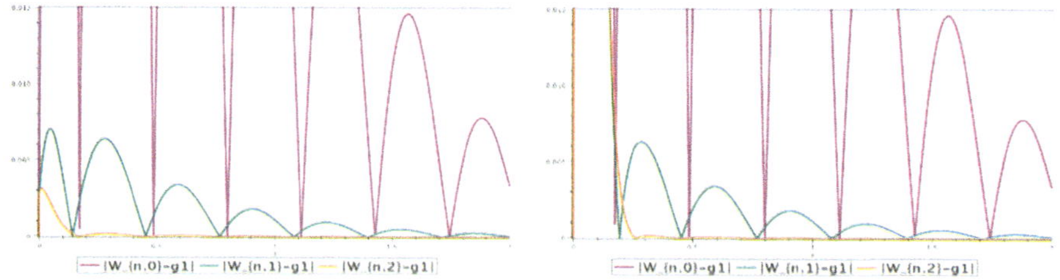

Fig. 3 Error function of $W_{n,s}(g_1; x)$ when $n = 30$, $s = 0, 1, 2$, and $\delta_0 = 0.01, 1$, respectively

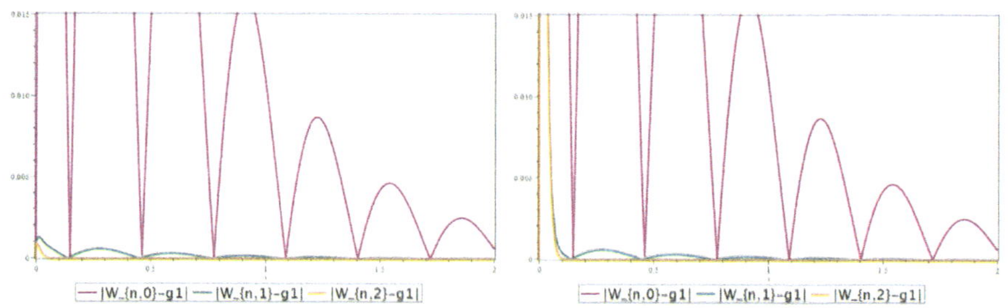

Fig. 4 Error function of $W_{n,s}(g_1; x)$ for $n = 100$, $s = 0, 1, 2$, and $\delta_o = 0.01, 1$, respectively

are tend to be zero from the right for a fixed value n. We describe this property of the sequence $W_{n,s}(.; x)$ by plotting the graphs of the test function and the approximation. We advise the researchers to use this generalization in applications.

References

1. Aral, A., Cárdenas-Morales, D., & Garrancho, P. (2018). Bernstein-Type Operators that reproduce exponential functions. *Journal of Mathematical Inequalities, 12*(3), 861–872.
2. Atakuta, C., Karatekeb, S., & Buyukyazıcıa, I. (2019). On approximation process by certain modified Dunkl generalization of Sz´asz-Beta type operators. *Journal of Mathematics and Computer Science, 19*, 9–18.
3. Bernstein, S. N. (1912/13). Démonstration du théoréme de Weierstrass fondée surle calculde probabilités. *Communication of the Mathematical Society of Kharkow, 13*, 1–2.
4. Buck, R. C. (1978). *Advanced calculus* (3rd ed.). New York: Mc Graw-Hill, Inc.
5. Cárdenas-Morales, D., Garranchoa, P., & Rasha, I. (2014). Approximation properties of Bernstein-Durrmeyer type operators. *Applied Mathematics and Computation, 232*, 1–8.
6. Costabile, F. A. (2018). Maria Italia Gualtieri, Anna Napoli, Some results on generalized Szász operators involving Sheffer polynomials. *Journal of Computational and Applied Mathematics, 337*, 244–255.
7. Khosravian-Arab,H., Dehghan, M., & Eslahchi, M. R. (2018). A new approach to improve the order of approximation of the Bernstein operators: Theory and applications. *Numerical Algorithms, 77*(1), 111–150.

8. Mohammad, A., J., Muslim, H., O. (2019). Simultaneous approximation of new sequence of integral type operators with parameter δ_0, *European Journal of Pure and Applied Mathematics, 12*(4), 1508–1523.
9. Pallini, A.. (2005). Bernstein-type approximation of smooth function. *Statistica, 65*(2).
10. Quaintance, J. (2016). *Combinatorial Identities for stirling numbers.* Singapore: World Scientific Publishing Co. Pte. Ltd.

Genetic Algorithm Based on Test Suite Prioritization for Software Testing in Neural Network

K. Hema Shankari, S. Mathivilasini, Dilli Arasu, and G. Suseendran

Abstract The paper addresses research in the area of software testing, which has a critical part of performing in software quality assurance. It becomes very inefficient for a tester to re-execute massive number of test cases again and again for small variations. The proposed artificial neural network with the genetic algorithm-based case suite prioritization (ANNGCSP) achieves better performance compared. New proposed algorithm combined the rational functional tester (RFT) tool, and new proposed algorithm combined open source tool in terms of APFD metric.

Keywords Regression testing · APFD metric · Genetic algorithm · RFT tool · Test case prioritization · Open source tool

1 Introduction

Software maintenance is one of the activities that comprises an enhancement, an error correction, optimization and the elimination of an obsolete capability. These changes in software might result in the software to be working bizarre and might impact various other portions of software, and therefore, in the sense of preventing this, regression testing has been performed [1]. Regression testing has uses in the aim of revalidating the changes in specific software. Regression testing is regarded

K. Hema Shankari
Department of Computer Application, Women's Christian College, Chennai, India
e-mail: hemashankariwcc@gmail.com

S. Mathivilasini
Department of Computer Science, Ethiraj College for Women, Chennai, India
e-mail: mathi_vilasini@yahoo.co.in

D. Arasu
Department of Computer Application, DRBCCC Hindu College, Chennai, India

G. Suseendran (✉)
Department of Information Technology, Vels Institute of Science, Technology and Advanced Studies, Chennai, India
e-mail: suseendar_1234@yahoo.co.in

S.-L. Peng et al. (eds.), *Proceedings of First International Conference on Mathematical Modeling and Computational Science*, Advances in Intelligent Systems and Computing 1292, https://doi.org/10.1007/978-981-33-4389-4_37

as costly procedures by which the test suites conduct ensuring that they have no new errors [2, 3]. Unfortunately, there might be not that much-needed resources for allowing re-executing all the test cases while carrying out the regression testing. Here, in this particular scenario, the test case prioritization approaches could achieve by enhancing the proficiency of the relapse testing by methods for sequencing the experiments accessible [4].

In this paper, it has been developed an artificial neural network with the genetic algorithm-based case suite prioritization and stated as ANNGCSP. The proposed methodology had designed for the optimization of regression testing, which is the foremost step in software maintenance [5]. In this review, the genetic algorithm is made to integrate with the Selenium tool. ANN has been used to amend the severity using without manual intervention, and that would increase the genetic algorithm.

Artificial neural network shortly stated as ANN, one among the machine learning areas, which can perform the human brain modelling and may consist of several artificial neurons [6]. An activation function is applied for those inputs that might result in outputting the neuron's value. Knowledge regarding learning task has been provided in the instance form and being called as preparation examples [7]. The pattern is provided to an input layer, and the output layer would generate various other patterns. The amount of the neurons in the output must be related directly to kind of task, and that should be carried out by the particular neural network [8]. The hidden layer of the neural network is an intermediary layer present among the input and the output layer. The initiation function might apply in the hidden layer, in case it is accessible. The hidden layer could consist of hidden nodes. The hidden nodes or neurons refer to the neurons, which are neither in the input nor output layer [9]. For regression testing of industrial applications, we need a platform that supports decision making; the platform should include, but not limited to, business system rules, description of the application, analysis of changes based on requirements, cost assessment, risk assessment and management of test cases.

The proposed work organizes test cases abusing GA thinking about the rate of blame recognition. Blame location decides how quickly an experiment can recognize the most significant number of shortcomings inside the testing process, considering that every one of the issues is of equal seriousness. The number of inhabitants in chromosomes is populated by sequentially modifying one populace with another in light of wellness work distributed to every chromosome. The potential individual with the best wellness done for the following populace and low wellness people wiped out from every age. In this proposed research work, test suites are taken as inputs, and test case ranking (TCR) is considered as the fitness function.

2 Related Work

Suri and Singhal discussed regression testing, which is an indispensable and expensive activity being regularly implemented in a manner wherein both time and resource have restraints [10]. Therefore, methods such as test case selection and prioritization, for the selection and prioritization of a subsection from the entire tests suite, sustaining few of the selected criteria are brought into use. Ant colony optimization (ACO) is a method determined by the practical behaviour of ants. The described graph representation and example runs depict the behaviour of ACO aids in exploring the probable course and then select the optimal one among them. Inferences drawn indicate ACO results in solutions, which are in proximity to optimal solutions [9].

Kaur and Goyal present a bee colony optimization (BCO) algorithm for the prioritization of blame inclusion relapse test suites. Two kinds of working drones, like the Scout honey bees and forager honey bee, exist in the ordinary honey bee province, who have the duty of creating and keeping up the settlement. The BCO calculation fundamentally sent to encourage blame inclusion relapse test suite is subject to these two honey bees' conduct. BCO calculation is sent fundamentally to encourage blame inclusion to accomplish ideal blame inclusion in lessened units of the execution time of each experiment, making utilization of two such precedents and their outcomes that match ideal arrangement. APFD measurements and diagrams sent here exhibit effectiveness of the calculation proposed [15].

Suman introduced a novel genetic algorithm enabling prioritization of the regression test suite that would dynamically arrange the test cases based on the complete code coverage. Regression testing is the technique used to authenticate changed software to certify that the modified portions of software function as required and there has been no adverse effect on the entire software sections with changes. The regression test suite is significant, and smart technique is necessary to select test cases minimizing the test expense on an overall [10, 11].

The test case prioritization methods aim at improving the regression testing effectiveness through test cases ordering to ensure ones offering the highest benefit executed first. Meanwhile, a technique to generate new test cases introduces employing partially mapped crossover (*PMX*), and cyclic crossover and analysis are carried out based on technique and test expenditure [15]. The research work aims at reducing the test cases, which have to perform post modifications.

Jacob and Ravi discussed the genetic algorithm that considers the test case information acquired from regression testing in the form of input and generates a set of test case for execution so that there is maximum coverage on the modified code. Here also takes the benefit of choosing the test case information obtained from relapse testing. They are organized based on the adjusted lines secured by experiment. The experiment covering the maximum number of modified lines has the highest priority and carried out first, and the one with the lowest coverage of modified lines has the lowest priority and not carried out until the last deadline. Thus, even though the test is not done, the complete modified lines would be covered, and the test cases are given priority [16].

Sethi et al. clarified about the experiment determination and a prioritization technique which select and sort out the experiments in a test suite that relies upon some sort of criteria guaranteeing deficiencies are secured quickly inside an insignificant time of execution. This errand can be performed in light of the ACO strategy for swarm intelligence since it not wholly inspected yet. The critical detachment of this postulation is to decide the problem of the path: to propose getting the shortest way and resolving the time issue: inferring to diminish the time required overlooking the briefest way [12]. Because of the time and cost confinements, it is not practical to complete an intricate relapse testing. Techniques including experiment determination and prioritization are used for settling the issue of period and cost restrictions. Additionally, changing the prior way to deal with accomplished better outcomes as far as execution time and the proficiency of the methodologies is estimated utilizing APFD metric [11].

3 Proposed Methodology

The proposed research has been designed using artificial neural network with specifically genetic algorithm-based case suite prioritization and shortly represented as ANNGCSP. It possesses the usage of the tool, Selenium tool, and then, the automated testing is selected, one of the open-source tools. In this approach, genetic algorithm is integrated with the Selenium tool for reconfiguring specific test suites that comprising various test cases, in the aim of enhancing prioritization.

3.1 Selenium Tool

For different Web applications, lightweight programming named Selenium is utilized. This specific device is utilized for the account and the ensuing playback later; mainly composing the experiments does not have to learn Selenium IDE; the test-specific dialect known as Selenese needs to be taken in. The experiments are then continued against the new Web browsers.

3.2 Genetic Algorithm Based Test Case Prioritization

Genetic algorithm is a heuristic that is influenced by Charles Darwin theory of natural evolution. Genetic algorithm is well said to be a smart search approach and utilized for solving the problems of optimization.

3.3 Objective Function

Three prioritization factors were considered, which include each blame, doled out the risk esteem, which relies on a ten-point scale and characterizes as taking after. For the test case namely Tk, the RDAk calculation is done by applying the severity value called Sk, Nk refers several defects which are computed using Tk, and the time k shows the time required by Tk for finding those particular bugs. The RDA registered as beneath:

$$\text{RDAk} = (\text{Sk} * \text{Nk})/\text{time k} \quad (1)$$

3.3.1 Test Case Ranking

For the positioning of the experiments, it requires to entirety up three different components includes the RFD, the PFD and the RDA. The below equation expresses it as

$$\text{TCRk} = \text{RFDK} + \text{PFDk} + \text{RDA k} \quad (2)$$

3.4 Improved Industry-Oriented Genetic Algorithm for Regression Test

3.4.1 Case Prioritization

Input: Test suite called TK and TCR for each test case is accepted to be the algorithm inputs.

Output: Prioritized sequence of the test cases.

Algorithm: An electronic task that contains a general measure of the 244 experiments. Such experiments are listed as various sets in this, and each experiment arrangement is known as 'Test Suite'. Along these lines, when the Tn would allude to be specific n, the tjk would remain for an experiment called j which present in the named k. For the prioritization, the hereditary calculation is utilized.

1. Manually arranging test cases in the form of the sets in specific test suites
2. Performing regression testing, the defects tracking, the measurement of the test time and the assignment of a severity manually which is very high risk = 10, etc., to the least risk = 2
3. I am choosing test suites for the mutation based on a formula.

3.5 Neural Network in Regression Testing

It is an insight calculation approach which has impacted by the natural sensory system and incorporates the data preparing. A worldview is utilized for the example acknowledgement. Neural systems determine the physical cell frameworks that could procure the store and the procedure of experiential information. Like people, the neural systems learn by models and their past encounters [9, 13].

3.6 Defect Severity Automation

The underlying ANN utilized for this situation examines just of two inputs to be specific the usefulness code and the blunder code. It contains just a single concealed layer. The table given beneath additionally introduces particular after-effects of the ANN in correlation severity that is with manually assigned. For every eight faults, seven faults are used for the training of ANN. 'Every output has been defined under a category' (Table 1).

Table 2 might display four possibly available categories, whereas they could place any output. As ANN has just an approximation of

Table 1 Functionality with the number of test cases and priority

Test case	Test case Name	Description	Test case
1	Select Add News & Events—News & Events	Validating the list of News & Events that can be edited in Add News & Events page	4
2	Select Add News & Events—News & Events	Entering the text that needs to display as Description for 'News & Events' Page	6
3	Select 'News & Events' in Actual APGPCL site	Validating the updated 'News & Events' description in the APGPCL site	4
4	Select 'News & Events' in Actual APGPCL site	Validating the updated 'News & Events' description in the APGPCL site	5
5	Select Add News & Events—News & Events	Updating the text that needs to display as description for 'News & Events' page	4
6	Select 'News & Events' in Actual APGPCL site	Validating the updated 'News & Events' description in the APGPCL site	3

Table 2 Four Possible Categories

	Tested application output		
	CORRECT		WRONG
ANN output Correct	1	True positive	2 True negative
Wrong	4	False negative	3 False positive

Table 3 APFD % for RFT Tool and Selenium Tool

Prioritization technique	APFD %
IIGRTCP with RFT tool	88
IIGRTCP with selenium tool	91

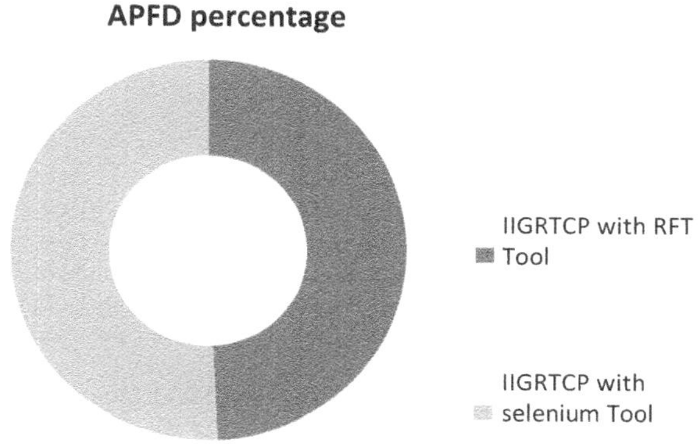

Fig. 1 APFD percentage for RFT tool and selenium tool

the original system, few outputs of it might be found erroneous. On the opposite side, the tried application without anyone else's input may produce the blunders, an essential reason for the testing process [14]. On the off chance that ANN yield has redressed while the yield of tried application has likewise wrong, assessment of correlation instrument has been named genuine negative or class.

Table 3, the proposed prioritized order is compared with previous work The newly introduced artificial neural network with the genetic algorithm based on case suite prioritization called ANNGCSP that attains superior performance in comparison with new proposed algorithm is combined with the rational functional tester and s representing as RFT tool, and the new proposed algorithm integrates with an open source tool in terms of the APFD metric (shown in Fig. 1).

4 Conclusion

The proposed research conceived as an artificial neural network with case suite prioritization based on specific genetic algorithms called ANNGCSP. Regression testing is carried out with a tool called selenium, and the genetic algorithm is used in an attempt to reduce the human intervention in the regression testing cycles below. Alter the extent of the defect in e with the aid of ANN. The proposed artificial neural network with the genetic algorithm-based case suite prioritization could achieve better performance is compared with the IIGRTCP with the rational functional tester called RFT tool and the improvised representing as IIGRTCP with Selenium tool utilizing the APFD metric.

In future, we can derive an algorithm for prioritized test cases by using various soft computing techniques such as simulated annealing (SA), anarchies society optimization (ASO) and improved ant colony optimization algorithm (ACO). In addition to this, other factors like the coverage criteria, time constraints, implementation complexity and fault impact are considered to assess the effectiveness of the regression testing.

References

1. Albert, E., Gómez-Zamalloa, M., & Puebla, G. (2010). PET: A partial evaluation-based test case generation tool for Java byte code, *ACM In Proceedings of the 2010 ACM SIGPLAN Workshop on Partial Evaluation and Program Manipulation* (pp. 25–28).
2. Amland, S. (2000). Risk-based testing: Risk analysis fundamentals and metrics for software testing, including a financial application case study. *Journal of Systems and Software, 53*(3), 287–295.
3. Anderson, J., Salem, S., & Do, H. (2014, May). Improving the effectiveness of test suite through mining historical data. In *Proceedings of the 11th Working Conference on Mining Software Repositories* (pp. 142–151).
4. Ansari, A., Khan, A., Khan, A., & Mukadam, K. (2016). Optimized regression test using test case prioritization. *Procedia Computer Science, 79,* 152–160.
5. Ashraf, E., Mahmood, K., Ahmed, T., & Ahmed, S. (2017). Value based PSO test case prioritization algorithm. *International Journal of Advanced Computer Science and Application, 8*(1), 389–394.
6. Baresi, L., & Pezze, M. (2006). An introduction to software testing. *Electronic Notes in Theoretical Computer Science, 148*(1), 89–111.
7. Basheer, I. A., & Hajmeer, M. (2000). Artificial neural networks: Fundamentals, computing, design, and application. *Journal of Microbiol Methods, 43*(1), 3–31.
8. Baudry, B., Fleurey, F., Jézéquel, J. M., & Le Traon, Y. (2002, September). Automatic test case optimization using a bacteriological adaptation model: Application to. net components. In *Proceedings 17th IEEE International Conference on Automated Software Engineering* (pp. 253–256). IEEE.
9. Suri, B., & Singhal, S. (2011). Implementing ant colony optimization for test case selection and prioritization. *International Journal of Computational Science and Engineering, 3*(5), 1924–1932.
10. Suman, S. (2012). A genetic algorithm for regression test sequence optimization. *International Journal of Advanced Research Computer and Communication Engineering, 1*(7), 478–481.
11. Sethi, N., Rani, S., & Singh, P. (2014). Ants optimization for minimal test case selection and prioritization as to reduce the cost of regression testing. *International Journal of Computer Applications, 100*(17).
12. Sathya, C., & Karthika, C. (2015). *A study on dependency optimization using machine-learning approach for test case prioritization.*
13. Bauersfeld, S. (2013, March). GUIdiff—A regression testing tool for graphical user interfaces. In *2013 IEEE Sixth International Conference on Software Testing, Verification and Validation* (pp. 499–500). IEEE.
14. Panigrahi, C. R., & Mall, R. (2010). Model-based regression test case prioritization. *ACM SIGSOFT Software Eng. Notes, 35*(6), 1–7.
15. Kaur, A., & Goyal, S. (2011). A survey on the applications of bee colony optimization techniques. *International Journal on Computer Science and Engineering, 3*(8), 3037.
16. Jacob, T. P., & Ravi, T. (2013). Optimization of test cases by prioritization.

MHD-Parabolic Flow Past an Accelerated Isothermal Vertical Plate with Variable Temperature and Uniform Mass Diffusion in the Presence of Rotation

S. Dilip Jose, A. Selvaraj, R. Muthucumaraswamy, S. Karthikeyan, and E. Jothi

Abstract A particular assessment of rotation impact of unsteady allegorical stream past of impenetrable and electrically driving liquid past a uniform quickened unbounded isothermal perpendicular plate, under the activity of diagonally workable force field has been appeared in the examination. The temperature of the plate is linearly raised with time and the concentration level close to the plate is in like way raised to C'_w. The non-dimension overseeing equation has been fathomed utilizing Laplace transform strategy. The temperature profile, concentration and quickness profile have been scrutinized for distinct physical framework like warm Grashof integer, mass Grashof integer, Schmidt number, Prandtl number and time. The study has been thought of with wavering temperature and systematic mass scattering. Finally, it is seen that the quickness increments with the developing estimations of warm Grashof or mass Grashof an integer. It is besides reviewed that the quickness increments with a decrement in magnetic field M.

Keywords Parabolic · Mass diffusion · Rotation · Isothermal vertical plate · Heat transfer · MHD · Force field

---

S. Dilip Jose · A. Selvaraj (✉) · S. Karthikeyan · E. Jothi
Department of Mathematics, Vels Institute of Science, Technology and Advanced Studies, Chennai 600117, India
e-mail: aselvaraj_ind@yahoo.co.in

S. Dilip Jose
e-mail: dilipjose.rs@gmail.com

S. Karthikeyan
e-mail: skarthikeyan492@gmail.com

E. Jothi
e-mail: jothi.1983@rediffmail.com

R. Muthucumaraswamy
Department of Applied Mathematics, Sri Venkateshwara College of Engineering, Chennai 602117, India
e-mail: msamy@svce.ac.in

S.-L. Peng et al. (eds.), *Proceedings of First International Conference on Mathematical Modeling and Computational Science*, Advances in Intelligent Systems and Computing 1292, https://doi.org/10.1007/978-981-33-4389-4_38

1 Introduction

Magnetohydrodynamic is a part of liquid elements where an attractive field is significant in the liquid stream. The liquid must be electrically directing and air protecting, on account of water, it is pitifully leading, subsequently, MHD impacts are not significant in the earth air or seas. It has wide application in astronomy, i.e., hot gases ionize. The idea of MHD stream includes both the condition of liquid elements, the Naiver equation and the condition of electrodynamics Maxwell's condition which are commonly combined with the Lorentz's power and ohms for moving electric conveyors. Let us talk about hardly any realities about the root of MHD, creator Hannes Alfren was the principal individual to present the idea and the term magnetohydrodynamic. He was the main individual depicted astrophysical marvel as a free logical control and the official birth of magnetohydrodynamic is in 1936–1937. The huge utilization of MHD generators is to be specific three-space application, mechanical and transport drive. MHD generators are utilized for driving submarines, hypersonic airstream investigations, safeguard applications and aircraft. MHD has utilized as a continuous force flexibly framework and as force plants in enterprises. Impacts of mass exchange on the stream past a consistently quickened perpendicular plate were concentrated by Soundalgekar [1] it is seen that the skin-contact increments with expanding Schmidt number, however, decline with expanding lightness parameter or Grashof number.

At that point, the examination reached out to mass exchange impacts on the stream past a quickened perpendicular plate with consistent warmth motion was thought by Singh A. K. Furthermore, Singh [2]. Then, mass exchange impacts on the stream past an exponentially quickened perpendicular plate with steady warmth motion were clarified by Jha et al. [3]. Raptis and Singh [4] clarified rotation effects on magnetohydrodynamic free convection flow past a quickened perpendicular plate. On flow past, an allegorical began unbound perpendicular plate with wavering temperature and systematic mass scattering was concentrated by R. Muthucumaraswamy and A. Neel Armstrong [5] reasons that quickness increments with expanding estimations of warm Grashof number or mass Grashof. Muthucumaraswamy and Armstrong [6] examined mass exchange impacts on stream past an explanatory began perpendicular plate with wavering temperature and mass dissemination. Hossain and Shayon [7] examined the skin grating in the flimsy free convection stream past a quickened plate. The connection of warm radiation with free convection heat move was delineated diagrammatically by Cess [8]. Muthucumaraswamy and Geetha [9] gave thought regarding effects of illustrative movement of an isothermal perpendicular plate with consistent mass motion. As such, it is proposed to dissected the rotational effects of parabolic stream past of an impenetrable viscous and electrically planning liquid past a dependably revived unbound isothermal plate in the vicinity warmth and mass exchange the nearness of magnetohydrodynamic with the limit condition wavering temperature and uniform mass dissemination. The non-dimensional controlling condition is enlightened utilizing Laplace change methodology. The derived course of action achieves as far as complementary error function and exponential outcomes.

1.1 Related Works

Rajput and Surendra Kumar [10] conveyed rotation and radiation impact on magnetohydrodynamic stream past an indiscreetly begun perpendicular plate with wavering temperature was finished up the outcome essential speed increment with increment in Grashof number and attractive field and abatement with increment in rotational parameter and time and auxiliary speed increment with increment in R and reduction within Grashof and rotational parameter. At long last temperature of the liquid increment, when R and time are expanded R.

Muthucumaraswamy et al. [11] give a brief explanation on effect of a turn on magnetohydrodynamic stream past a quickened isothermal perpendicular plate with warmth and mass dispersion finished up the outcome that speed increment estimations of mass Grashof and warm Grashof number and time yet the pattern is simply switched as for the rotational parameter or attractive field parameter. Further, the investigation reached out with wavering temperature and mass dissemination by Muthucumaraswamy et al. [12] explained rotation impact on magnetohydrodynamic stream past a quickened perpendicular plate with wavering temperature and uniform mass dispersion.

Soundalgekar [13] studied a definite investigation of the impacts of mass exchange on the Stokes issue for a boundless perpendicular plate has been introduced on considering the free convection flows. It has been seen that there is an ascent in the quickness because of the nearness of an outside mass. In any case, an expansion in Schmidt number prompts a fall in the speed. The skin-erosion increments because of the nearness of a remote mass.

Muthucumaraswamy et al. [14] noted the shut structure arrangement of pivot consequences for shaky hydromagnetic stream past a consistently quickened endless perpendicular plate within the sight of wavering temperature and mass dispersion has been considered. The impact of various criteria like warm Grashof number, mass Grashof and time are concentrated diagrammatically. It is seen that the speed increments with expanding estimations of mass Grashof, thermal Grashof and time. Be that as it may, the pattern is simply turned around regarding the revolution parameter or attractive field parameter M.

Muthucumaraswamy and Venkatesan [15] concentrated on radiative stream past an allegorical began isothermal perpendicular plate with uniform mass transition outcome thought that speed increments with diminishing estimations of the warm radiation parameter. The quickness rise with growing thermal Grashof number or mass Grashof wide variety, but the pattern is just reversed with admire to the thermal radiation criterion or Schmidt range. The temperature of the plate rise with the decline values of the radiation parameter. The plate awareness rise with decline values of the Schmidt variety. The pattern is simply turned around concerning the warm Grashof number or mass Grashof.

2 Mathematical Formulation

Consider the three-D magnetohydromagnetic allegorical take the path of least resistance past of an electrically captivating in a liquid float along with a gooey impervious liquid incited by the method of consistently revived movement of an interminable isothermal plate within the sight of revolution is thought of. The stream is flimsy, at that point x'-axis is taken nearby opposite heading and y'-axis is considered as should be expected to x'-axis and Z'-axis is typical to it. Both the liquid and the plate are in a country of rigid axis with uniform precise speed about the Z'-axis within the sight of transverse attractive order B_0. The accelerated attractive region and thick scattering are believed to be negligible. From the outset, each the liquid and plate were very still and with the indistinguishable temperature. At time t'>0, the plate begin moving with a pace $u = \left(u_0 t'\right)^2$ in its plane with the opposite way. At the equivalent time, the plate temperature is raised to T'_w, and in this manner, the fixation stage changed to C'_w and looked after consistent. Since the plate possessing the plane $Z' = 0$ is of unending degree, all the physical bit depends best Z' and t' at that point by methods for standard Boussinesq's estimation, and the flimsy stream is governed through the given condition.

$$\frac{\partial U}{\partial t} - 2\Omega V = Gr\theta + GcC + \frac{\partial^2 U}{\partial Z^2} - MU \tag{1}$$

$$\frac{\partial V}{\partial t} + 2\Omega U = \frac{\partial^2 V}{\partial Z^2} - MV \tag{2}$$

$$\frac{\partial \theta}{\partial t} = \frac{1}{pr}\frac{\partial^2 \theta}{\partial Z^2} \tag{3}$$

$$\frac{\partial C}{\partial t} = \frac{1}{sc}\frac{\partial^2 C}{\partial Z^2} \tag{4}$$

With the accompanying starting and limit condition

$$\begin{aligned} &u = 0, \quad T' = T'_\infty, \quad C' = C'_\infty \quad \text{for all } y,\ t' \leq 0 \\ &t' > 0 \quad u = \left(u_0 t'\right)^2, \quad T' = T'_w, \quad C' = C'_w \quad \text{at } y = 0 \\ &u \to 0, \quad T' \to T'_\infty, \quad C' \to C'_\infty \quad \text{at } y \to \infty \end{aligned} \tag{5}$$

On suggesting the subsequent dimensionless quantities:

$$
\begin{aligned}
U &= \frac{u}{(Vu_0)^{\frac{1}{3}}} \quad V = \frac{v}{(Vu_0)^{\frac{1}{3}}} t = t'\left(\frac{u_0^2}{v}\right)^{\frac{1}{3}} \quad Z = z\left(\frac{u_0}{v^2}\right)^{\frac{1}{3}} \\
\theta &= \frac{T - T_\infty}{T_w - T_\infty} \quad \mathrm{Gr} = \frac{g\beta(T_w - T_\infty)}{u_0} \quad C = \frac{C' - C'_\infty}{C'_w - C'_\infty} \\
\mathrm{Gc} &= \frac{g\beta^*(C'_w - C'_\infty)}{u_0} \quad M = \frac{\sigma B_0^2}{\rho}\left(\frac{v}{u_0^2}\right)^{\frac{1}{3}}, \quad pr = \frac{\mu C_p}{k}, \quad c = \frac{v}{D}
\end{aligned} \tag{6}
$$

The parabolic flow past with rotating free convective stream past a quickened perpendicular plate is depicted by differential condition (1) to (4) with the accompanying starting and limit condition (5). Since the prescribed equation is very hard to solve, therefore, we present a complex velocity $q = u + iv$ then the condition first and the second condition are understood and joined into a single condition.

$$
\frac{\partial q}{\partial t} = \mathrm{Gr}\theta + \mathrm{Gc}C + \frac{\partial^2 q}{\partial Z^2} - mq \tag{7}
$$

With the accompanying starting and limit condition in dimensionless quantities are follows

$$
\begin{aligned}
&q = 0, \quad \theta = 0, \quad C = 0 \quad \text{for all} Z, t \le 0 \\
&t > 0 \quad q = t^2, \quad \theta = t, \quad C = 1 \quad \text{at } Z = 0 \\
&q \to 0, \quad \theta \to 0, \quad C \to 0 \quad Z \to \infty
\end{aligned} \tag{8}
$$

where $m = M + 2i\Omega$

3 Coupled Equation with Solution Procedure

The dimensionless administering condition tended to in second, third and seventh equation with the accompanying starting and limit condition which is given in Eq. (8) is solved by Laplace transform technique. Finally, the inverse transform is performed and the solution are derived as follows

$$
\begin{aligned}
q = &\left[\frac{(\eta^2 + mt)t}{4m}\left[e^{2\eta\sqrt{mt}}\, erfc\left(\eta + \sqrt{mt}\right) + e^{-2\eta\sqrt{mt}}\, erfc\left(\eta - \sqrt{mt}\right)\right]\right. \\
&+ \frac{\eta\sqrt{t}(1 - 4mt)}{8m^{\frac{3}{2}}}\left[e^{-2\eta\sqrt{mt}}\, erfc\left(\eta - \sqrt{mt}\right) - e^{2\eta\sqrt{mt}}\, erfc\left(\eta + \sqrt{mt}\right)\right] \\
&\left.- \frac{\eta t}{2m\sqrt{\pi}} e^{-(\eta^2 + mt)}\right] + \left[\frac{\mathrm{Gr}}{a(1 - pr)} + \frac{\mathrm{Gc}}{b(1 - sc)}\right]
\end{aligned}
$$

$$\times\frac{1}{2}\begin{bmatrix} e^{2\eta}\sqrt{mt}\, erfc(\eta+\sqrt{mt}) \\ +e^{-2\eta}\sqrt{mt}\, erfc(\eta-\sqrt{mt}) \end{bmatrix}-\frac{\mathrm{Gr}}{a(1-pr)}$$

$$\times\left[\left(\frac{t}{2}-\frac{\eta\sqrt{t}}{2\sqrt{M}}\right)\left(e^{-2\eta\sqrt{Mt}}erfc\left(\eta-\sqrt{Mt}\right)\right)\right.$$

$$\left.+\left(\frac{t}{2}+\frac{\eta\sqrt{t}}{2\sqrt{M}}\right)\left(e^{2\eta\sqrt{Mt}}erfc\left(\eta+\sqrt{Mt}\right)\right)\right]$$

$$-\left[\frac{\mathrm{Gr}}{(1-pr)}\right]\left[\frac{e^{at}}{2}\left[e^{2\eta}\sqrt{(m+a)t}\; erfc\left(\eta+\sqrt{(m+a)t}\right)\right]\right.$$

$$\left.+\left[e^{-2\eta}\sqrt{(m+a)t}\; erfc\left(\eta-\sqrt{(ma)t}\right)\right]\right]$$

$$-\left[\frac{\mathrm{Gc}}{a(1-sc)}\right]\left[\frac{e^{bt}}{2}\left[e^{2\eta}\sqrt{(m+b)t}\; erfc\left(\eta+\sqrt{(m+b)t}\right)\right]\right.$$

$$\left.+\left[e^{-2\eta}\sqrt{(m+b)t}\; erfc\left(\eta-\sqrt{(m-b)t}\right)\right]\right]$$

$$-\frac{\mathrm{Gr}}{a^2(1-pr)}erfc\left(\eta\sqrt{pr}\right)$$

$$-\frac{G_r}{a(1-pr)}t\left[\left(1+2\eta^2\,\mathrm{Pr}\right)erfc\left(\eta\sqrt{\mathrm{Pr}}\right)-\frac{2\eta\sqrt{\mathrm{Pr}}}{\sqrt{\pi}}e^{-\eta^2\,\mathrm{Pr}}\right]$$

$$+\frac{\mathrm{Gr}}{a^2(1-pr)}\left[\frac{e^{at}}{2}(e^{2\eta\sqrt{\mathrm{Pr}\,at}}erfc\left(\eta\sqrt{\mathrm{Pr}}+\sqrt{at}\right)\right.$$

$$\left.+e^{-2\eta\sqrt{\mathrm{Pr}\,at}}erfc\left(\eta\sqrt{\mathrm{Pr}}-\sqrt{at}\right)\right]$$

$$-\frac{\mathrm{Gc}}{a(1-Sc)}erfc\;(\eta\sqrt{\mathrm{Sc}})+\frac{G_c}{b^2(1-\mathrm{Sc})}$$

$$\times\left[\frac{e^{bt}}{2}(e^{2\eta\sqrt{Scbt}}erfc\left(\eta\sqrt{Sc}+\sqrt{bt}\right)+e^{-2\eta\sqrt{Scbt}}erfc\left(\eta\sqrt{\mathrm{Sc}}-\sqrt{bt}\right)\right] \tag{9}$$

$$\theta=t\left\{\left(1+2\eta^2\,\mathrm{Pr}\right)erfc\left(\eta\sqrt{\mathrm{Pr}}\right)-\frac{2\eta\sqrt{\mathrm{Pr}}}{\sqrt{\pi}}e^{-(\eta^2\,\mathrm{Pr})}\right\} \tag{10}$$

$$C=erfc\left(\eta\sqrt{sc}\right) \tag{11}$$

where, $a=\frac{m}{pr-1}$ and$b=\frac{m}{sc-1}\eta=\frac{Z}{2\sqrt{t}}$

Condition nine shows a speed profile and (10) temperature profile and (11) concentration profile for the issue. To get physical information on the issue, numerical estimation is carried. From this equation, we can secure dominant velocity (primary) profile and auxiliary velocity profile (secondary).

$$erfc(a+ib) = erf(a) + \frac{\exp(-a^2)}{2a\pi}[1 - \cos(2ab) + i\sin(2ab)] + \frac{2\exp(-a^2)}{\pi}\sum_{n=1}^{\infty}\frac{\exp(-\eta^2/4)}{\eta^2 + 4a^2}[f_n(a,b) + ig_n(a,b)] + \in(a,b)$$

while assessing the statement of q, it delineates that the contention of the error function is perplexing, and subsequently, the articulation is separated into real and complex parts by utilizing the equation strategy above.

4 Results and Interpretation

The derived condition comprises of three temperature profiles, concentration and velocity profiles have been handled and modified in MATLAB programming, and the yield pictures have been traded recorded beneath.

For the physical perception of the issue, numerical figuring is passed on for distinct criterion enrolled for warm Grashof, mass Grashof, magnetic field, Schmidt, Prandtl number and time. The temperature profile has been determined from condition (10), the contribution of Prandtl number is picked for air is 0.71 and for water 7.0 and time 0.2 are portrayed in Fig. 1. It is seen that the temperature increase with a decrease in Prandtl number. For calculation, the estimation of Schmidt is assigned to be 2.01.

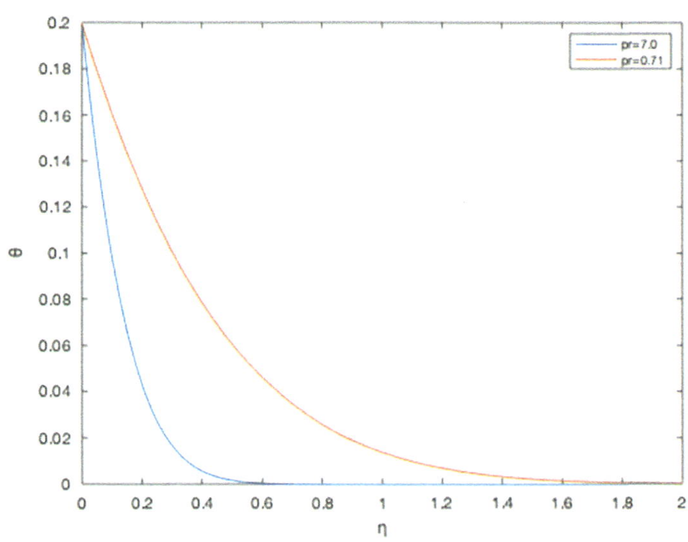

Fig. 1 Temperature profile for various values of Pr

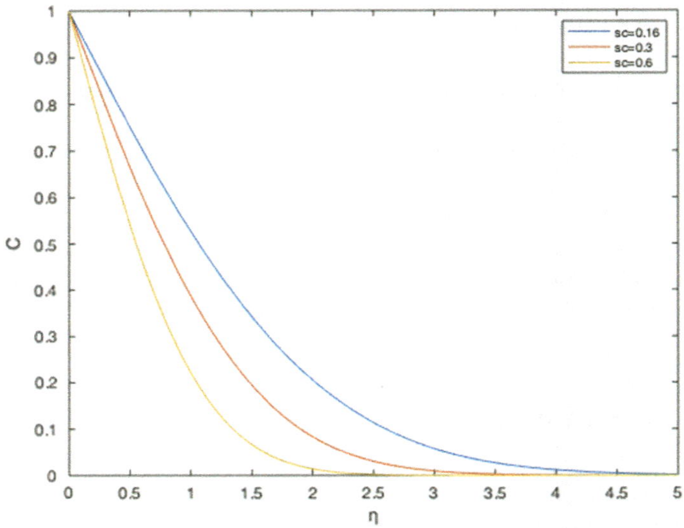

Fig. 2 Temperature profile for various values Sc

Figure 2 portrays the impact of concentration profile at time 0.2 for distinct Schmidt number is taken to be 0.16, 0.3, 0.6. It is visible that the divider concentration increments with reducing estimations of Schmidt number. The profile has the basic segment that the focus diminishes in a consistency path from the zero-regard a few separations route within the loosened stream.

Figure 3 shows the impact of dominant quickness profile for distinct warm Grashof number is consider to be 2, 2, 5 and mass Grashof number taken to be 2, 5, 5 and power field is taken to be unique for instance 1, the rotational parameter is 0.5, further Prandtl consigned as 7 and time is 0.4. It is seen that the quickness increment with the developing estimations of the warm Grashof or mass Grashof number.

Figure 4 layouts the impact of power field parameter on the speed when force field is named as 1, 5, 10 and warm Grashof is 5, mass Grashof allocated as 5 and the rotational parameter is assigned 0.5, Prandtl is 7 and time is 0.4. It is seen that the speed increment with the decrement estimations of the force field parameter. This outcome in increment in the engaging field parameter which prompts fall in speed.

The colleague discretionary speed for various estimation of warm Grashof is considered as 2, 2, 5 and mass Grashof is taken as 2, 5, 5, rotational worth held onto 0.5, Prandtl number is separate 7 and time 0.2, $M = 5$ is introduced in Fig. 5. The pattern shows that the speed increment with extending estimations of Grashof or mass Grashof whole number.

The secondary speed profile for the various rotational parameter is appropriated as 0.5, 1, 1.5, warm Grashof number is relegated as 5 and mass Grashof whole number is additionally taken as 5, further $\mathrm{Pr} = 7$, force field is 5 and time is 0.2 appear in

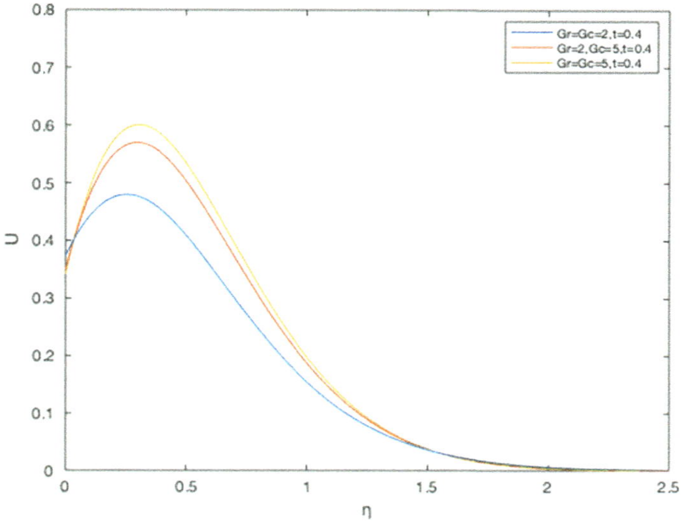

Fig. 3 Dominant velocity profile for various Gr and Gc

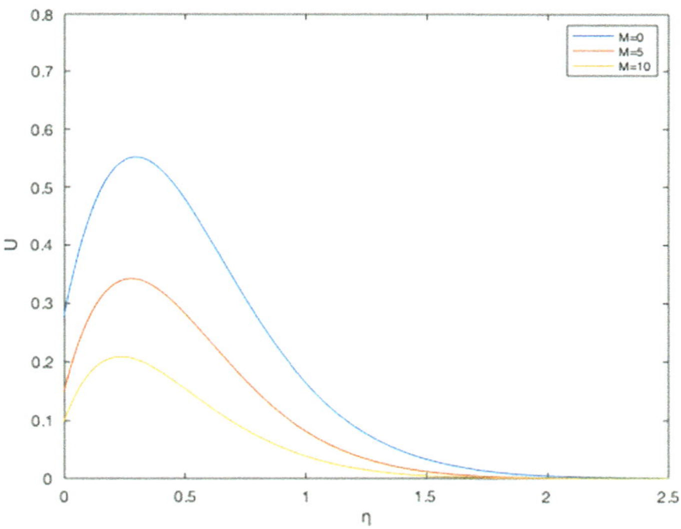

Fig. 4 Dominant velocity profile for various *M*

Fig. 6. It is seen that the auxiliary speed increment with reducing estimation of the rotation parameter.

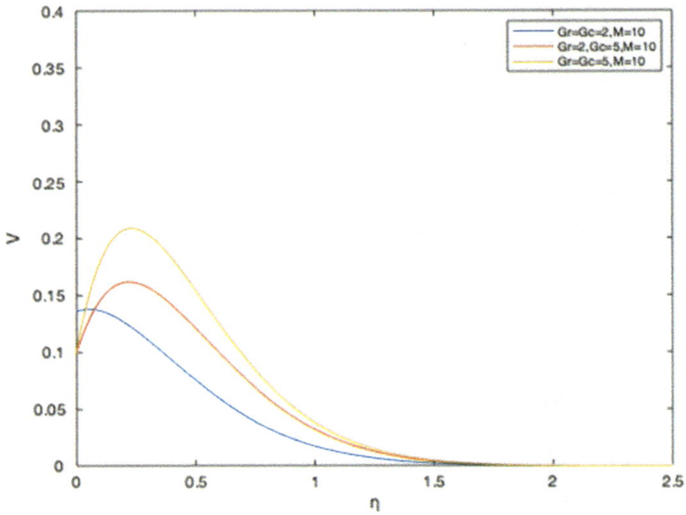

Fig. 5 Auxiliary velocity profile for various Gr & Gc

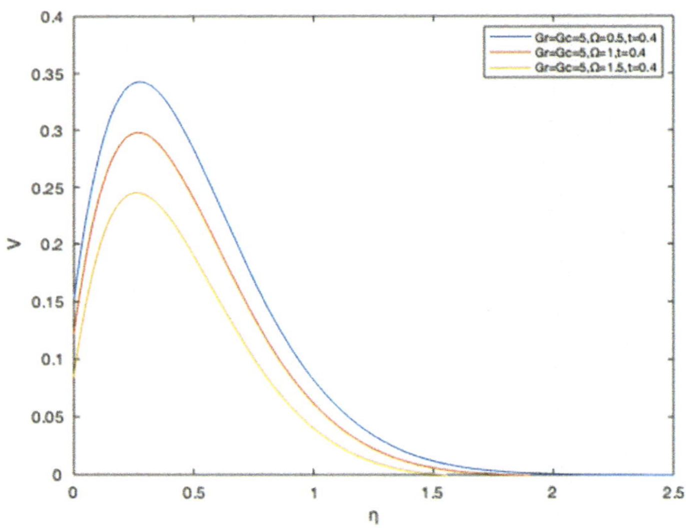

Fig. 6 Auxiliary velocity profile for various Ω

5 Conclusion

The theoretical direction of motion of parabolic circulation past a consistently quickened sizable isothermal perpendicular plate inner seeing variable mass dissemination was the concept of within the presence of wavering temperature and systematic mass scattering. The dimensionless administering circumstance is settled with the aid of Laplace exchange method. The aggregate of wonderful physical criterion like heat Grashof integer, mass Grashof integer, rotational parameter, force field and time t is inspected graphically. The conclusion of the result is enlisted below. It is far visible that the temperature increment with a decline in Prandtl number. It is far visible that the divider concentration increments with lowering estimations of Schmidt range. It is visible that the speed increment with the developing estimations of the warm Grashof or mass Grashof wide variety. It is far visible that the rate increment with the decrement estimations of the force field parameter. The instance is exchanged with appreciate to the rotational parameter or force field parameter m.

References

1. Soundalgekar, V. M. (1982). Effects of mass transfer on the flow past a uniformly accelerated vertical plate. *Letters in Heat and Mass Transfer, 9*, 65–72. https://doi.org/10.1016/0094-4548(82)90049-2.
2. Singh, A. K., & Singh, J. (1983). Mass transfer effects on the flow past an accelerated vertical plate with constant heat flux. *Astrophysics and Space Science, 97*, 57–61. https://doi.org/10.1007/BF00684609.
3. Jha, B. K., Prasad, R., & Rai, S. (1991). Mass transfer effects on the flow past an exponentially accelerated vertical plate with constant heat flux. *Astrophysics and Space Science, 181*, 125–134. https://doi.org/10.1007/BF00644119.
4. Raptis, A., & Singh, A. K. (1985). Rotation effects on MHD freecon vection flow past an accelerated vertical plate. *Mechanics Research Communications, 12*, 31–40. https://doi.org/10.1016/0093-6413(85)90032-1.
5. Muthucumaraswamy, R., & Armstrong, A. (2014). On flow past a parabolic started Infinite vertical plate with variable temperature and uniform mass diffusion. *International Journal of Engineering and Research, 3*(2), 79–82. https://doi.org/10.17950/ijer/v3s2/209.
6. Muthucumaraswamy, R., & Armstrong, A. (2014). Mass transfer effects on flow past a parabolic started vertical plate with variable temperature and mass diffusion. *International Journal of Mathematics Archive, 5*(2), 53–58.
7. Hossain, M. A., & Shayon, L. K. (1986). The skin friction in the unsteady free convection flow past an accelerated plate. *Astrophysics and Space Science, 125*, 315–324. https://doi.org/10.1007/bf00648030.
8. Cess, R. D. (1966). The interaction of thermal radiation with free convection heat transfer. *International Journal of Heat and Mass Transfer 9*, 1269–1277. https://doi.org/10.1016/0017-9310(66)90119-0.
9. Muthucumaraswamy, R., & Geetha, E. (2014). Effects of parabolic motion of an isothermal vertical plate with constant mass flux. *Ain Shams Engineering Journal, 5*, 1317–1323. https://doi.org/10.1016/j.asej.2014.05.011.
10. Rajput, U. S., Surendra Kumar. (2011). Rotation and Radiation effect on MHD flow past an impulsively started vertical plate with variable temperature. *International Journal of Mathematical Analysis, 5*, 1155–1163.
11. Muthucumaraswamy, R., Lal, T., & Ranganayakulu, D. (2010). Effect of rotation on MHD flow past an accelerated isothermal vertical plate with heat and mass diffusion. *Theory and Applied Mechanics, 27*, 289–202. https://doi.org/10.2298/tam1003189m.

12. Muthucumaraswamy, R., Lal, T., & Ranganayakulu, D. (2011). Rotation effect on MHD flow past an accelerated vertical plate with variable temperature and uniform mass diffusion. *International Journal of Engineering*.
13. Soundalgekar V. M. (1979). Effects of mass transfer and Free convection on the flow past an impulsively vertical plate. *Journal of Applied Mechanics, 46*(4), 757–760. https://doi.org/10.1115/1.3424649.
14. Muthucumaraswamy, R., Lal, T., & Ranganayakulu, D. (2013). MHD flow past an accelerated vertical plate with variable temperature and uniform mass diffusion in the presence of rotation. *International Journal of Innovative Research in science, Engineering and Technology, 2*(10).
15. Muthucumaraswamy, R., & Venkatesan, J. (2013). Radiative flow past a parabolic started isothermal Vertical plate with uniform mass flux. *International Journal of Mathematical Analysis, 7*, 2907–2921. https://doi.org/10.12988/ijma.2013.310246

Effect of MHD and Radiation Absorption Fluid Flow Past an Exponentially Accelerated Vertical Plate with Variable Temperature and Concentration

E. Jothi, A. Selvaraj, S. Dilipjose, A. Neel Armstrong, and S. Karthikeyan

Abstract The impact of magnetohydrodynamic along with diffusion retention liquid stream over an exponentially increased perpendicular plate within unsteady heat temperature and also concentration has been examined. The raised heat temperature is developing straightly among time duration and also the focus close to the plate expanded. As per outcome, results are referenced graphically and abstractly. The monetary value of velocity outline, temperature outline, and concentration outline examined as all manner of qualities. An impact uncovers that radiation includes the velocity and temperature are expanded and concentration is diminished. The set of dimensionless equations including limit conditions are determined to utilize Laplace transformation techniques. It is observed that the impact of the velocity profile is referenced as a particular physical measurement sort of thermal Grashof value, mass Grashof value, Schmidt value, and time duration.

Keywords MHD variable temperature · Exponential functional · Vertical plate · Uniform mass diffusion transform

E. Jothi · A. Selvaraj (✉) · S. Dilipjose · S. Karthikeyan
Department of Mathematics, Vels Institute of Science, Technology and Advanced Studies, Chennai 600117, India
e-mail: aselvaraj_ind@yahoo.co.in

E. Jothi
e-mail: jothi.1983@rediffmail.com

S. Dilipjose
e-mail: dilipjose.rs@gmail.com

S. Karthikeyan
e-mail: skarthikeyan492@gmail.com

A. Neel Armstrong
Department of Mathematics, Sagunthala Rand D Institute of Science and Technology, Veltech Rangarajan Dr, Chennai 600062, India
e-mail: armsambomaths@gmail.com

S.-L. Peng et al. (eds.), *Proceedings of First International Conference on Mathematical Modeling and Computational Science*, Advances in Intelligent Systems and Computing 1292, https://doi.org/10.1007/978-981-33-4389-4_39

1 Introduction

MHD is the analysis of the electro-magnetic effects and conduct of electrifying directing liquids. The basic idea driving MHD is attractive flux actuate flow in stunning conductor fluid, and the equal variation in electro-magnetic area itself. The states of MHD are a sequence of Navier–Stokes rules as liquid elements and Maxwell conditional like electromagnetism. It is far recognized with planning troubles, for instance, plasma imprisonment, liquid metallic moderate of nuclear activators, and electro-magnetic casting. MHD considers the fluid particles in the bloodstream and the external magnetic field. The dimensionless overseeing conditions are fathomed utilizing the Laplace change strategy. The arrangements are as far as exponential and correlative mistake work. Such investigation is discovered valuable in the attractive direction of liquid–solid stream within the steel business, fluid metal moderate in atomic activator, and attractive concealment of liquid remolding equipment.

1.1 Related Work

Armstrong et al. [1] assessed the hypothetical analysis of magnetohydrodynamic stream past an explanatory movement of an immense perpendicular side plate within the unstable condition and equivalent mass scattering researched [2], 3]. The raised temperature is a square within duration also focus layer closed to serving increased reliably. Such an investigation of stream past discovered helpful in attractive control of liquid iron stream inside the gird trade, fluid plate reduce in atomic activator, appealing disguise molten semi-leading elements, and weather forecasting. We infer that speed increments with expanding estimations of Gr, Gc, and time. Yet, this pattern line turned around concerning the attractive field parameter. Since the appealing progress applies a ruining power on the free-convective stream past the illustrative started vertical plate.

Muthucumarasamy et al. [4] assessed the hypothetical arrangement of stream past a consistently quickened interminable isothermic perpendicular side plate within the sight of irregular a mass circulation. This depends on whether or not they occur at an intersection and homogeneous, on the off chance that it happens in arrangement As a rule of substance responses, the response rate depends upon the centralization of the division alone. The numerical qualities of the husk grating stay organized. It is seen a certain the speed move reformed at the duration enlarge also speed is diminished over the Prandtl value enlarge. Skin contact diminishes for the expanding estimations of time or warm Grashof value and mass Grashof value. The skin contact increments an expanding estimation like Schmidt number. This pattern is simply turned around as for the substance response parameter just as attractive field parameter. Yet, the pattern is simply turned around regarding the pivot parameter Ω or attractive field parameter.

Muthucumaraswamy and. Santhana Lakshmi [5, 6] contemplated the hypothetical arrangement of the stream over an exponentially beginning flow limitless perpendicular plate within sight as irregular heat temperatures and steady a mass dispersion. Reddy et al. [7] assessed calculate numerical find out about is finalize for unstable more deportation, chemically reactive, diffusion assimilation, gooey, substantial, and electric leading liquid over an exponentially accelerated perpendicular permeable plate inside seeing sink. Unequivocally, limited contrast techniques are utilized to comprehend the conditions overseeing the stream. From the present numerical examination, the pace of warmth move raised with Prandtl value, heats, and diffusion parameter while it shows the unfavorable impact on account of diffusion retention variable [8]. The speed of a mass transmission raised within Schmidt value, cause, and effect variable. The heat diminishes as raise in the temperature assimilation variable and heat retention causes a diminishes in the kinetic vitality as well as thermal vitality of the fluid. That force and heat limit layer get more slender if there should arise an occurrence of warmth retaining liquids.

The spin impact on an MHD stream through an exponentially quickened vertical plate with steady temperature and a mass dispersion by Thamizhsudar et al. [9, 10] The impact of MHD, radiation, and the substance response in the copper water-based nano-liquid over a near unending vertical swaying plate with steady warmth motion and the mass transition is investigated by worker fourth Runge–Kutta formula within Nachtsheim–Swigert shooting procedure [11]. Therefore, it is considered hydromagnetic impacts on stream over an illustrative boundless vertical plate within the irregular temperature, uniform mass scattering subject to symbolic development inside seeing appealing range. Non-dimensional managing conditions understood utilizing Laplace change procedure. Game plans are the extent that exponential and correlative slip-up work. Such examination is found important in the appealing force of fluid cast stream in the grid business, the skill of making things out of metal in atomic power plant, alluring disguise of fluid partly leading objects, and weather forecasting.

2 Mathematical Formulation

An unstable progression, a different compressed liquid over and endless vertical plate with the irregular condition and stable a mass expansion inside the nearness of attractive transition has been thought of. Here, the precarious progression of a thick incompressible liquid is at first very still and encompasses a limitless vertical plate within the temperature T_∞ also concentration C'_∞. The x-axle brought the plate inside the horizontally above course also they-pivot seized typical into a portion. About t'$\leq$ 0, a portion also liquid other identical heat T_∞ also concentration C'_∞. About $t' > 0$, a portion is exponentially quickened within speed $u = u_0 ecp(a't')$ in particular uniform range of power and the condition against the plate is raised into T_w and in this manner, therefore, fixation level close to a portion is also embossed C'_w within time t, and a mass is dispersed consistently against a portion into liquid.

As the plate is unbounded long, all that is inside the overseeing conditions while autonomous like x and therein not stream on the system. The boundless vertical plate is furthermore exposed to a steady attractive transition of solidarity B_0 which is thought to be applied typically to the plate. At that point, the flimsy stream is represented by the free-convective progression of an electrically leading liquid during an exponentially quickened under the standard Boussinesq's and limit layer guess, which is administered by the resulting dimensionless kind of conditions:

$$\frac{\partial u}{\partial t'} = g\beta(T - T_\infty) + g\beta^*(C' - C'_\infty) + v\frac{\partial^2 u}{\partial y^2} - \frac{\sigma B_0^2}{\rho}u \tag{1}$$

$$\rho C_p \frac{\partial T}{\partial t'} = k\frac{\partial^2 T}{\partial y^2} \tag{2}$$

$$\frac{\partial C'}{\partial t'} = D\frac{\partial^2 C'}{\partial y^2} \tag{3}$$

Along the successive basic and limit constrains

$$\begin{aligned} &u = 0, \quad T = T_\infty \quad C' = C'_\infty \quad \text{as } y,\ t' \le 0 \\ &t' > 0: \quad u_0 \exp(a't'), \quad T = T_\infty + (T_w - T_\infty)At', \quad C' = C'_w \quad \text{on } y = 0 \\ &u \to 0, \quad T \to T_\infty C' \to C'_\infty \quad \text{as } y \to \infty \end{aligned} \tag{4}$$

Here, $A = \frac{u_0^2}{v}$.

On presenting the successive dimensionless measures:

$$\begin{aligned} &U = u\left(\frac{u_0}{v^2}\right)^{1/3}, \quad t = \left(\frac{u_0^2}{v}\right)^{1/3} t', \quad Y = y\left(\frac{u_0}{v^2}\right)^{1/3} \\ &\theta = y\left(\frac{u_0}{v^2}\right)^{1/3}, \quad Gr = \frac{g\beta(T_w - T_\infty)}{(vu_0)^{\frac{1}{3}}} \\ &C = \frac{C' - C'_\infty}{C'_w - C'_\infty}, \quad Gc = \frac{g\beta(C'_w - C'_\infty)}{(vu_0)^{1/3}} \\ &m = \frac{\sigma B_0^2}{\rho}\left(\frac{v}{u_0^2}\right)^{1/3}, \quad \Pr = \frac{\mu C_p}{k}, \quad Sc = \frac{v}{D} \end{aligned} \tag{5}$$

In Eqs. (1)–(4), leads to

$$\frac{\partial q}{\partial t} = Gr\theta + GcC + \frac{\partial^2 q}{\partial z^2} - mq \tag{6}$$

$$\frac{\partial \theta}{\partial t} = \frac{1}{Pr}\frac{\partial^2 \theta}{\partial z^2} \tag{7}$$

$$\frac{\partial C}{\partial t} = \frac{1}{Sc}\frac{\partial^2 C}{\partial z^2} \tag{8}$$

The basic and limit constraints in dimensionless measures exist the non-dimensional quantities are characterized in the classification.

$$\begin{aligned} &q = 0, \quad \theta = 0, \quad C = 0 \quad \text{as } Z; t \le 0 \\ &q = e^{at}, \quad \theta = e^{at}, \quad c = e^{at} \quad \text{on } z = 0 \\ &q \to 0, \quad \theta \to 0, \quad c \to 0 \quad \text{as } Z \to \infty \end{aligned} \tag{9}$$

3 Method of Solution

The non-dimensional Eqs. (6)–(8), sub to consistent starting limit constraints [12], also handled utilizing Laplace transforms strategy also result determined in this way:

$$\theta = \frac{e^{at}}{2}\left[\exp\left(-\sqrt{Pra}\right)z \operatorname{erf} c(\eta\sqrt{\Pr} - \sqrt{at}\right) + \exp\sqrt{Pra}z \operatorname{erf} c(\eta\sqrt{Pr} + \sqrt{at})\right] \tag{10}$$

$$C = \frac{e^{at}}{2}\left[\exp\left(-\sqrt{Sca}\right)z \operatorname{erf} c(\eta\sqrt{Sc} - \sqrt{at}) + \exp\sqrt{Scaz} \operatorname{erf} c(\eta\sqrt{Sc} + \sqrt{at})\right] \tag{11}$$

$$\begin{aligned} q = {} & \frac{e^{at}}{2}\begin{bmatrix} \exp(-2\eta\sqrt{(a+m)t})\operatorname{erf} c(\eta - \sqrt{(a+m)t}) \\ +\exp(2\eta\sqrt{(a+m)t})\operatorname{erf} c(\eta + \sqrt{(a+m)t}) \end{bmatrix} \\ & - \frac{Gr}{(a-b)(1-\Pr)}\frac{e^{at}}{2}\begin{bmatrix} \exp(-2\eta\sqrt{(a+m)t})\operatorname{erf} c(\eta - \sqrt{(a+m)t}) \\ +\exp(2\eta\sqrt{(a+m)t})\operatorname{erf} c(\eta + \sqrt{(a+m)t} \end{bmatrix} \\ & - \frac{Gr}{(b-a)(1-\Pr)}\frac{e^{bt}}{2}\begin{bmatrix} \exp(-2\eta\sqrt{(b+m)t})\operatorname{erf} c(\eta - \sqrt{(b+m)t}) \\ +\exp(2\eta\sqrt{(b+m)t})\operatorname{erf} c(\eta + \sqrt{(b+m)t}) \end{bmatrix} \\ & - \frac{Gc}{(a-c)(1-Sc)}\frac{e^{at}}{2}\begin{bmatrix} \exp(-2\eta\sqrt{(a+m)t})\operatorname{erf} c(\eta - \sqrt{(a+m)t}) \\ +\exp(2\eta\sqrt{(a+m)t})\operatorname{erf} c(\eta + \sqrt{(a+m)t}) \end{bmatrix} \\ & - \frac{Gc}{(c-a)(1-Sc)}\frac{e^{ct}}{2}\begin{bmatrix} \exp(-2\eta\sqrt{(c+m)t})\operatorname{erf} c(\eta - \sqrt{(c+m)t}) \\ +\exp(2\eta\sqrt{(c+m)t})\operatorname{erf} c(\eta + \sqrt{(c+m)t}) \end{bmatrix} \\ & + \frac{Gr}{(a-b)(1-\Pr)}\frac{e^{at}}{2}\begin{bmatrix} \exp\left(-2\eta\sqrt{\Pr at}\right)\operatorname{erf} c(\eta\sqrt{\Pr} - \sqrt{at}) \\ +\exp\left(2\eta\sqrt{\Pr at}\right)\operatorname{erf} c(\eta\sqrt{\Pr} + \sqrt{at}) \end{bmatrix} \\ & + \frac{Gc}{(a-c)(1-Sc)}\frac{e^{at}}{2}\begin{bmatrix} \exp\left(-2\eta\sqrt{Scat}\right)\operatorname{erf} c(\eta\sqrt{Sc} - \sqrt{at}) \\ +\exp\left(2\eta\sqrt{Scat}\right)\operatorname{erf} c(\eta\sqrt{Sc} + \sqrt{at}) \end{bmatrix} \end{aligned}$$

$$+\frac{Gc}{(c-a)(1-Sc)}\frac{e^{ct}}{2}\begin{bmatrix}\exp\left(-2\eta\sqrt{Scct}\right)\mathrm{erf}\,c(\eta\sqrt{Sc}-\sqrt{ct})\\+\exp\left(2\eta\sqrt{Scct}\right)\mathrm{erf}\,c(\eta\sqrt{Sc}+\sqrt{ct})\end{bmatrix} \quad (12)$$

where $b = \frac{m}{Pr-1}$, c $= \frac{m}{Sc-1}$ and $\eta = \frac{z}{2\sqrt{t}}$.

4 Result and Discussion

For a real understanding of the issue, the numerical arrangement is accomplished for various real constraints *Gc*, *Gr*, *Pr*, *Sc*, and *M* also time *t* possibility a stream along with transportation. An estimation of Schmidt value Sc equal to 0.6 and that identifies with liquid water steam and also the amount of the Prandtl number picked to such an extent that other exhibit atmosphere value *Pr* equal to 0.71 and liquid *Pr* equal to 7.0. Statistical evaluation speed is a display for various real constraints like the Prandtl number, warm Grashof value, mass Grashof value, Schmidt value, and time duration.

Figure 1 illustrates the impact of velocity outlines various qualities Prandtl value ($Pr = 0.30, 0.71, 7.10$), $Gr = 5$, $Gc = 5$, $Sc = 0.22$, and $M = 5$ also time t equal to 0.5. As per diagram, it exhibits the value of velocity increments with expanding Prandal number of the liquid.

Figure 2 illustrates the effects of velocity outlines various qualities of Schmidt value ($Sc = 0.16, 0.30, 0.60$), $Pr = 0.71$, $Gr = 5$, $Gc = 5$, and $M = 5$ also $t = 0.5$. As per chart, it exhibits the value of velocity increments with diminishing Schmidt value.

Fig. 1 Velocity outlines for distinct estimations of *Pr* (*Sc* = 0.22, *Gc* = 5, *Gr* = 5, *M* = 5, *t* = 0.5)

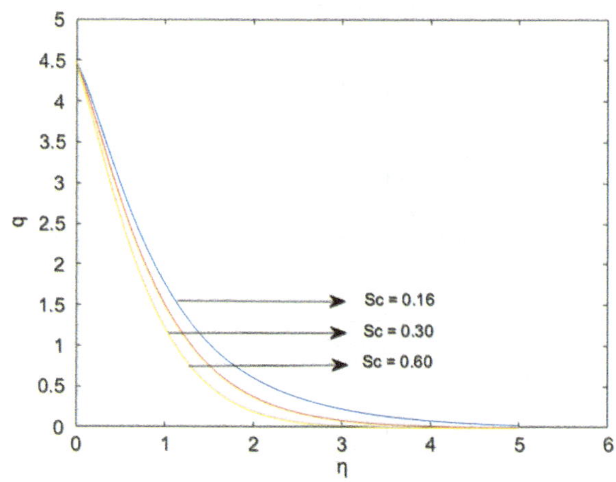

Fig. 2 Velocity outlines for distinct estimations of *Sc* ($Pr = 0.71$, $Gc = 5$, $Gr = 5$, $M = 5$, $t = 0.5$)

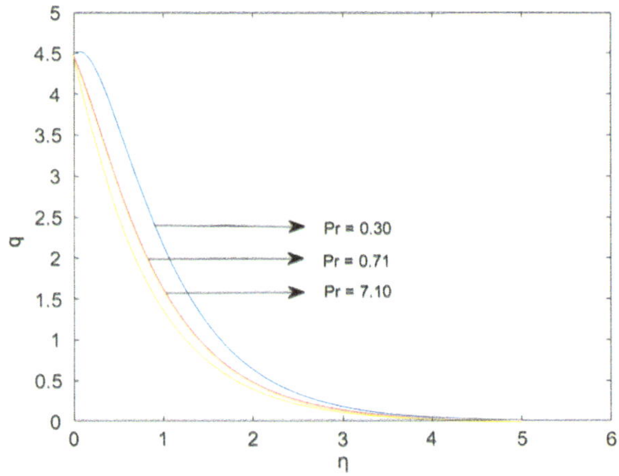

Figure 3 illustrates the impact of velocity outlines various qualities of time duration ($t = 0.2, 0.4, 0.6, 0.8$), $Pr = 0.71$, $Gr = 5$, $Gc = 5$, $Sc = 0.22$ and $M = 5$. As per diagram, it represents the velocity increments gradually as the time t increase.

Figure 4 illustrates the impact of velocity profile various qualities thermal Grashof value ($Gr = 2$ and 5), mass Grashof value ($Gc = 5$ and 10) with corresponding $Pr = 0.71$, Sc $= 0.22$, and also $M = 5$ on the velocity on time duration $t = 0.5$. As per graph, it represents the velocity rises with the extend to estimation of the thermal Grashof number just as mass Grashof number.

Figure 5 illustrates the impacts of temperature profile various qualities of Prandtl value ($Pr = 0.30, 0.71, 7.10$), $Gr = 5$, $Gc = 5$, $Sc = 0.22$, $M = 5$ also $t = 0.5$. As per diagram, it represents the growing Prandtl value pass to reduce in the heat temperature.

Fig. 3 Velocity outlines for distinct estimations of *t* ($Pr = 0.71$, $Sc = 0.22$, $Gc = 5$ $Gr = 5$, $M = 5$)

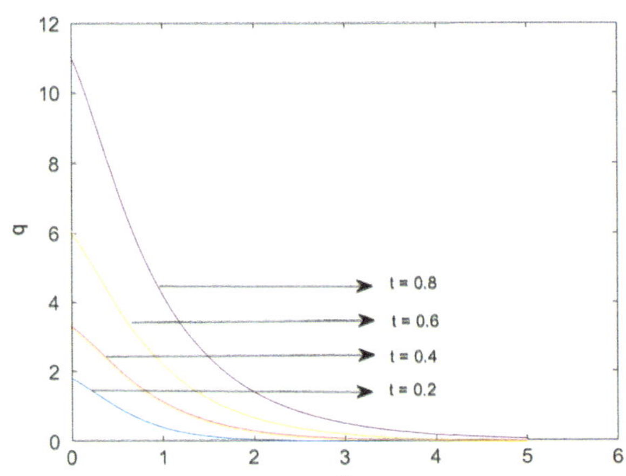

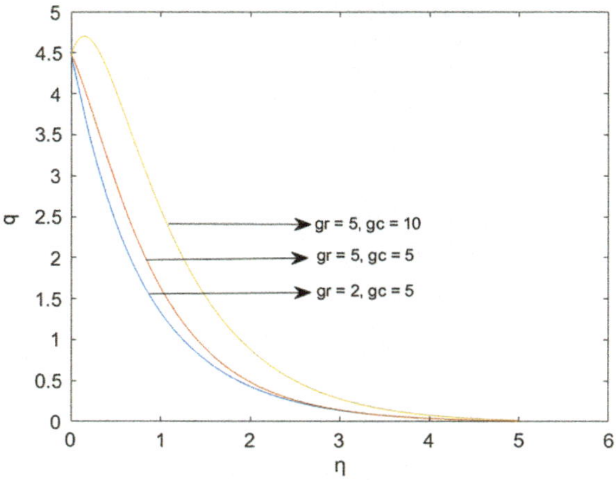

Fig. 4 Velocity outlines for distinct estimations of *Gr* and *Gc* ($Pr = 0.71$, $Sc = 0.22$, $M = 5$, $t = 0.5$)

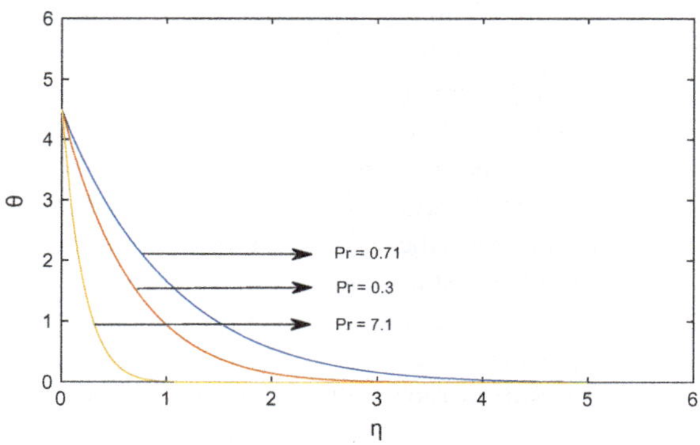

Fig. 5 Temperature outlines for distinct estimations of *Pr* ($Sc = 0.22$, $Gc = 5$, $Gr = 5$, $M = 5$, $t = 0.5$)

Figure 6 illustrates the impacts of temperature profile various qualities of the time duration ($t = 0.2, 0.4, 0.6, 0.8$), $Pr = 0.71$, $Gr = 5$, $Gc = 5$, $Sc = 0.22$, and $M = 5$. As per graph, it represents the concentration increments with expanding the estimations of time duration t.

Figure 7 illustrates the impacts of concentration profile various qualities of Schmidt value ($Sc = 0.22, 0.30, 0.50$), $t = 0.50$, $Pr = 0.71$, and $M = 5$. As per graph, it represents the Schmidt number increments, as the concentration of fluid reduce.

Figure 8 illustrates the results of concentration profile distinct estimates of the time duration ($t = 0.2, 0.4, 0.6, 0.8$), $Gr = Gc = 5$, $Sc = 0.22$, $Pr = 0.71$, and $M = 5$. As per graph, it represents concentration increments with expanding the estimations of t.

Fig. 6 Temperature outlines for distinct estimations of t ($Pr = 0.71$, $Sc = 0.22$, $Gc = 5$, $Gr = 5$, $M = 5$)

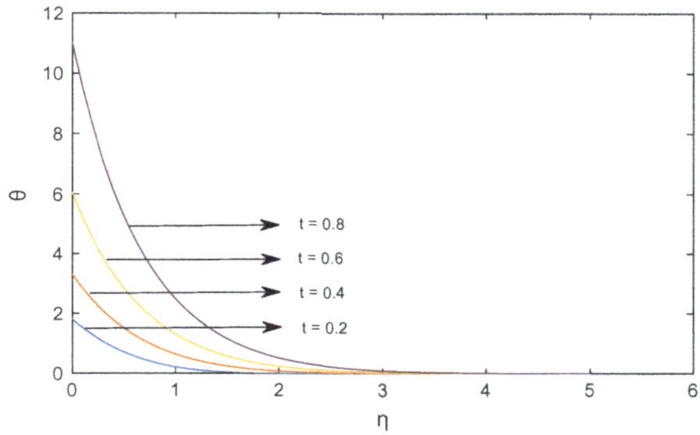

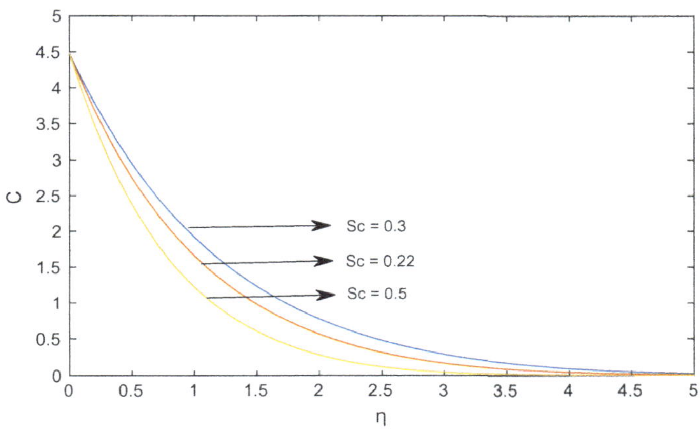

Fig. 7 Concentration outlines for distinct estimations of Sc ($Pr = 0.71$, $Gc = 5$, $Gr = 5$, $t = 0.5$, $M = 5$)

Fig. 8 Concentration outlines for distinct estimations of t ($Pr = 0.71$, $Sc = 0.22$, $Gc = 5$, $Gr = 5$, $M = 5$)

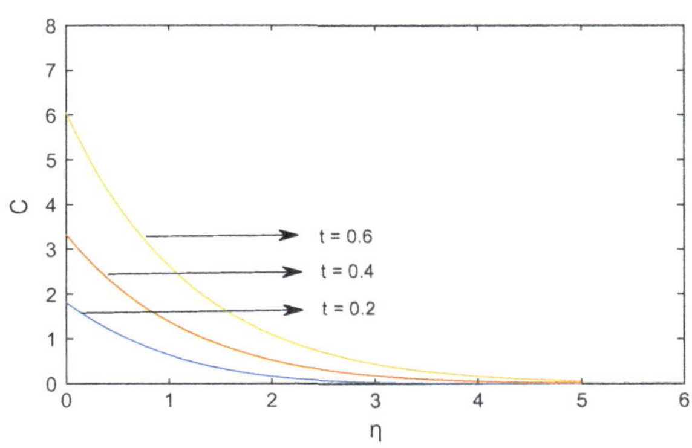

5 Conclusion

In this paper assessed, the hypothetical solution of magnetohydrodynamic along with diffusion effects retention liquid stream over an exponential increased perpendicular plate within unstable heat temperature and concentration are inspected. The non-dimensional equations including limit conditions are resolved utilizing Laplace transformation techniques. The impact of speed of velocity, heat temperature, and focus of concentration profile is referenced as a specific physical estimation sort of thermal Grashof value, a mass Grashof value, Schmidt value, and also duration. We conclude that the value of velocity increments within extending estimations of *Gr*, *Gc*, and *t*. The value of velocity reductions within expanding magnetism range value *M*. Heat temperature diminishes within expanding estimations of Prandtl value. The focus of concentration close to plate increments within diminishing estimations of Schmidt value.

References

1. Armstrong, A. N., & Muthucumaraswamy., R. (2014). MHD flow past a parabolic started vertical plate with variable temperature and mass diffusion. *Journal of Mechanical Engineering and Sciences, 7*(2014), 1251–1260.
2. Das, U. N., Deka, R., & Soundalgekar, V. M. (1994). Effects of mass transfer on flow past an impulsively started infinite vertical plate with constant heat flux and chemical reaction. *Forschung Im Ingenieurwesen, 60*(10), 284–287.
3. Muthucumaraswamy, R. (2017). Heat and mass transfer effects on MHD flow past an exponentially accelerated vertical plate in the presence of rotation and hall current.*Journal of Advanced Research in Dynamical & Control Systems, 9*(2).
4. Muthucumaraswamy, R., Dhanasekar, N., & Easwara Prasad, G. (2013). MHD and rotation effects on flow past an accelerated vertical plate with variable temperature and mass diffusion in the presence of chemical reaction. *International Journal of Applied Mechanics and Engineering, 18*(4), 1087–1097.
5. Muthucumaraswamy, R., & Santhana Lakshmi, C. (2015). Mass and Heat transfer effects on MHD fluid flow of an exponentially accelerated isothermal vertical plate with variable mass diffusion." *Adv. Appl. Fluid Mech* 18 (2015): 31–49.
6. Narayana, PV Satya, G. Ramireddy, and S. Venkataramana. "Hall current effects on free convection MHD flow past a porous plate. *International Journal of Automotive and Mechanical Engineering, 3*, 350–363.
7. Gurivi Reddy, P., Umamaheswar, M., Raju, M. C., & Varma, V. (2016). Magneto-convective and radiation absorption fluid flow past an exponentially accelerated vertical porous plate with variable temperature and concentration. *International Journal of Mathematics Trends and Technology (IJMTT), 31*(1).
8. Rajput, U. S., & Kumar, G. (2019). Effects of radiation and chemical reaction on MHD flow past a vertical plate with variable temperature and mass diffusion. *Journal of Naval Architecture and Marine Engineering, 16*(2), 99–108.
9. Sivaiah, M., Nagarajan, A. S., & Sreehari Reddy, P. (2010). Radiation effects on MHD free convection flow over a vertical plate with heat and mass flux. *Emirates Journal for Engineering Research, 15*(1), 35–40 (2010)
10. Thamizhsudar, M., Pandurangan, J., & Muthucumaraswamy, R. (2015). Hall effects and rotation effects on MHD flow past an exponentially accelerated vertical plate with combined heat

and mass transfer effects. *International Journal of Applied Mechanics and Engineering, 20*(3), 605–616.

11. Vijayalakshmi, K., Umadevi, R., Muthucumaraswamy, R. (2018). Oscillating plate in nanofluid with uniform heat and mass flux under the effect of MHD, radiation and chemical reaction is analyzed by Runge-Kutta method. 13(2018), 793–803.
12. Rajput, U. S., Kumar, G. (2018). Effect of Hall current on unsteady magneto hydrodynamic flow past an exponentially accelerated inclined plate with variable temperature and mass diffusion.*MATEMATIKA: Malaysian Journal of Industrial and Applied Mathematics, 34*(2), 433–443.

Relevant Text Extraction Using Enhanced Artificial Neural Network Fuzzy Inference System

Srikanth Bhyrapuneni and R. Anandan

Abstract Identifying the security level of a private record is an indispensable task for associations to ensure the secret data represented. Different rules and systems are being applied by human specialists. Expanding the number of private data in associations are making it hard to order all the reports cautiously with human exertion. In the present advanced period, it turns into a test for users of the web to discover explicit data on the web. Many online records are recovered, and it is difficult to process all the recovered data. The programmed text summary is a procedure that recognizes the significant focuses from all the related archives to deliver a compact outline. Enormous datasets surround covered patterns which pass on significant information about the dataset. Information mining research manages the extraction of helpful and important data from such huge datasets. The procedure of information extraction can be seen as exploration and examination of huge amounts of information, via programmed or self-loader mean, to find significant examples and rules. These days, text acknowledgement has become a significant developing zone in handling images and extraction of data. In this proposed work, an efficient text extraction method enhanced artificial neural network-based fuzzy inference system (EANN-FIS) is introduced for relevant text extraction including hidden layers. The proposed method is compared with the traditional methods and the results show that the proposed method is exhibiting better performance in terms of accuracy and speed.

Keywords Text extraction · Data handling · Clustering · Image handling procedure · Artificial neural network · Fuzzy inference system · Information mining

---

S. Bhyrapuneni (✉) · R. Anandan
Department of Computer Science and Engineering, VELS Institute of Science, Technology and Advanced Studies (VISTAS), Chennai, India
e-mail: srikanthbt3010@gmail.com

R. Anandan
e-mail: anandan.se@velsuniv.ac.in

S. Bhyrapuneni
RISE KrishnaSai Gandhi Group of Institutions, Ongole, India

S.-L. Peng et al. (eds.), *Proceedings of First International Conference on Mathematical Modeling and Computational Science*, Advances in Intelligent Systems and Computing 1292, https://doi.org/10.1007/978-981-33-4389-4_41

1 Introduction

Text extraction and text acknowledgement from the picture have been one of the most fascinating and testing research areas in the field of text acknowledgement, man-made analysis, machine vision, and image handling in the ongoing years. There are fundamentally four stages that incorporate preprocessing, extraction of relevant features, text determination, and relevant text acknowledgement to build up any of the text acknowledgement framework [1]. Text information present in pictures and video contain helpful data for programmed comment, ordering, and organizing of pictures. In general, text data extraction framework comprises of four phases: text discovery, text limitation, text extraction and improvement and acknowledgement [2]. The video comprises of two sorts of text: text on images and text as captions. Scene text is the text that in nature happens in the area of catching of video like text on flags, signs, holder, signboard, messages on vehicles [3].

Neural networks are the way towards driving a system to yield a specific reaction to particular info. The fundamental types of learning are regulated and independent [4]. In supervised learning, for every particular info, the ideal reaction of the system is known while in unsupervised learning, the ideal reaction isn't known [5]. During the process, a classifier or group of classifiers identify potential clashes. The subsequent preparing stage becomes dynamic just when a choice on the troublesome cases must be taken. The process of relevant text extraction from images is shown in Fig. 1.

In the proposed work, an enhanced artificial neural network-based fuzzy inference system (EANN-FIS) recognizer is introduced. This recognizer starts with features extricated in a littler sub-pictures of the character on each recursive pass, accordingly working with a better goal of a sub-picture each spam of time till relevant text is extracted [6]. Along these lines, a component vector is separated for each character that has more data from those sub-pictures that are considered to be a higher priority than others [7]. The component vector is produced by brushing all features removed in each sub-picture. These features depend on the histogram of inclination and minute-based projections [8].

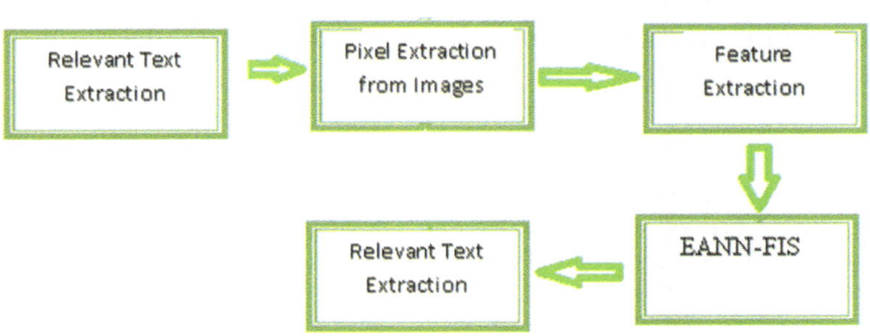

Fig. 1 Proposed text extraction model framework

2 Literature Survey

Salahuddin Unar et al. [1] proposed a novel calculation, on the grounds of hit parts and unmistakable Gabor channels and aim was to identify text regions from pictures. The authors endorsed the factual investigation on the key factor of text from preparing sets to get useful text, which is utilized as the whole portrayal of key parts. Rayleigh model is utilized for the customary run of the circulations for each image to depict appropriateness insights on positive preparing tests and the Gaussian model to portray reasonableness measurements on negative preparing tests.

Yalniz et al. [2] proposed the technique of limiting regularized varieties to remove maximally stable extremal regions (MSERs) where separation loads and the edge of the calculations on clusters are characterized for text applicants like self-preparing separation metric learning calculations that empowers the probabilities of text competitors identified with non-text borders as a character classifier; lastly, messages are set apart into a classifier.

Tsai et al. [4] proposed a binarization way to deal text certainty map with the assistance of text finder. The named parts as text or non-text are displayed in a conditional random field model identified with unary segment properties with neighbouring segments. Text parts are gathered into text lines with minimization of energy approach. This technique was proposed to guarantee the presentation for contrasting images for accurate identification of text.

3 Proposed Methodology

Artificial neural networks are used to improve the performance levels of the text extraction process. The neural system recognizes data directly from a considered image source. Each neural system separately learns the attributes that recognize pictures followed by searching for comparable qualities in the ideal picture to be distinguished. The ANN mathematical representation is depicted in Fig. 2.

The enhanced artificial neural network-based fuzzy inference system (EANN-FIS) is utilized as the classifier to classify the text and non-text parts. The linear independently neural system is utilized to order the separated text from the recordings into text and non-text classes. Preparing is through a regulated model where the info and the coordinating model are given dependent on the learning rules. The arrangement of information which encourages the preparation is known as the 'preparation set'. Once the preparation is done, now the system is prepared to proceed as the classifier where it will recognize the text (1) and non-text (0) classes [9].

EANN-FIS is prepared to order text archive sentences as an outline and non-rundown sentences. In the wake of figuring, the feature extraction records, and values for each sentence from the preparation set are modelled by the ANN. When the preparation is finished, the subsequent EANN-FIS classifier model will be capable to

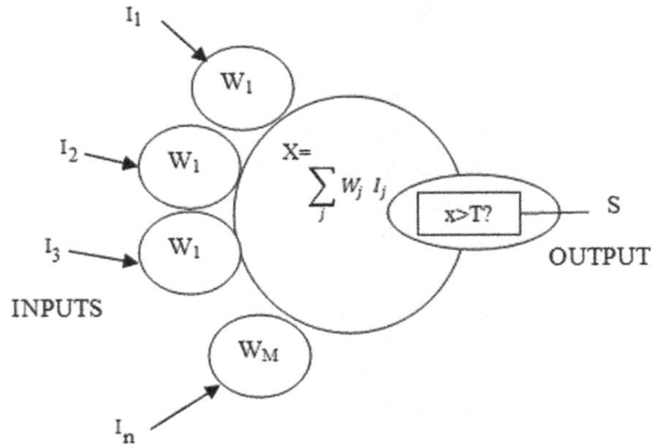

Fig. 2 EANN-FIS representation

yield the score of another report to decide its class, that is, as relevant or non-relevant text.

Algorithm EANN-FIS

Input: Image Dataset.

Output: Text (1) class or non-text (0) class.

Step 1: The image is converted with an array of fuzzy singletons represented with X * Y, each with a value of high brightness level. The fuzzy representation of the considered image is done as

$$I(x, y) = \bigcup_{p=0}^{x-1} f(x, y) U_{q=0}^{y-0} \frac{f(x, y)}{f_{\max}} \tag{1}$$

Step 2: Apply the filter using the median-based method 'λ' for noise removal that is performed as

$$\lambda_{\text{med}}^2 = \frac{1}{4In^2(p)} \approx \frac{p * \pi_i^2}{n(I) + \frac{\pi}{2} - 1} \tag{2}$$

Step 3: The considered image undergoes segmentation that results in sub-images for extracting the relevant text only and the process is performed as

$$l_\sigma\left(I\left[l_s, l_f\right]\right) = \sqrt{\sum_{l=l_s}^{l_f} \left(K - l_m\left(I\left[l_s, l_f\right]\right)\right)^2 \times P_l^{I\left[l_s, l_f\right]}} \tag{3}$$

Step 4: After partitioning, optimal segmentation partition is applied by reducing the objective function of pixels as

$$\text{Iopt} = \sum_{j=1}^{N}\sum_{i=1}^{c} U_{ij}^{m}\left\|X_j - Y_i\right\|^2 + X * Y \tag{4}$$

where K is the constant, I is the image considered and P is the pixel position. F represents the intensity level.

Step 5: A model for assigning the weights for the image pixels is used which assigns the weights for the pixels as

$$W_{i,j} = \exp\left(\frac{-(I_{\text{in}}(x) - I_{\text{in}}(y))^2}{\left(\frac{\beta^2}{2}\right) * \text{Iopt}}\right) \tag{5}$$

Step 6: Relevant features are extracted from the image pixels and from a relevant sub-image is represented using the fuzziness level that is calculated as

$$U_{i,j} = \sum_{k=1}^{c}\left\{\frac{\left\|X_j - V_i\right\|}{\left\|X_j - V_K\right\|}\right\}^{-\frac{2}{m}-1} + \text{Iopt} * I(X, Y) \tag{6}$$

Step 7: Each feature is contrasted with the neighbouring pixel feature and is grouped to form a relevant cluster as

$$CS(In) = \sum_{k} c_{j.k}\phi_{j.k}(x, y) + \sum_{j}\sum_{k} U_{j.k}(x, y) + \sum_{j=1}^{n}(U_{ij})^m X_i * Y_j \tag{7}$$

Step 8: The text values are stored in the matrix and are represented as

$$\begin{aligned} T^I(X, Y) &= CS^{(i)}(X, Y) \\ &\quad + \frac{1}{W}\sum_{k=1}^{k=N} W_k^{-1}\left(\left(\left(X_k - Y_k^{(i)}\right) + X(W)\right) + Y(W)\right) \end{aligned} \tag{8}$$

After text extraction from images, the extracted text is assembled into words, and afterwards, the words are gathered to lines by building the minimum spanning tree. The tree is based on the boundary box separation between the extracted texts. The edge cuts are utilized as the segment in the tree development which will prompt the text restriction.

4 Results and Discussions

The proposed method for relevant text extraction is implemented in Python and tested in ANACONDA SPYDER Platform [10]. The results represent that the proposed method exhibits better performance when compared to traditional methods. 500 pictures of different categories are considered from UCI machine repository for extraction of relevant text from an image by segmenting the image into sub-images haphazardly from characteristic scenes [11]. EANN-FIS is utilized and been chosen for extraction of text from images. The proposed method takes less time for clustering of relevant data. The process of identification of relevant text from the images is shown in Fig. 3. From the considered image, the relevant text is extracted by using the EANN-FIs methodology.

In an image, there is both relevant and non-relevant text [12]. The relevant text is marked with value '1' and the non-relevant text is identified and marked with '0' [13]. The time taken for relevant text recognition is shown in Fig. 4. The results represent that the proposed method takes less time in the identification of relevant text from the images.

In an image considered, initially, pixels are extracted. The borders and edges need to be calculated for accurate text identification [14]. The accuracy level of the proposed method in the identification of borders and edges is more when compared

Fig. 3 Text extraction from image

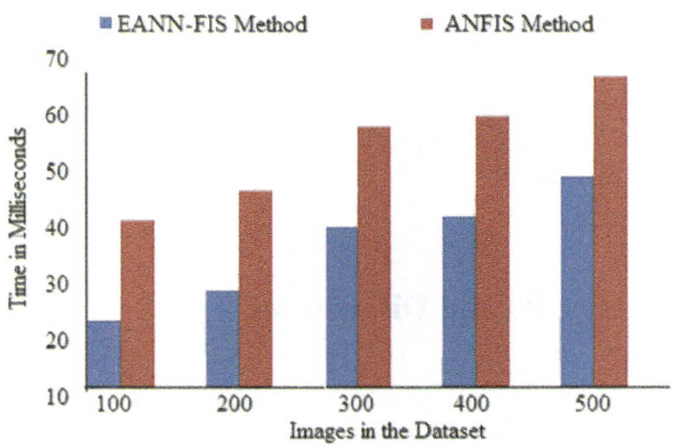

Fig. 4 Time for relevant text extraction

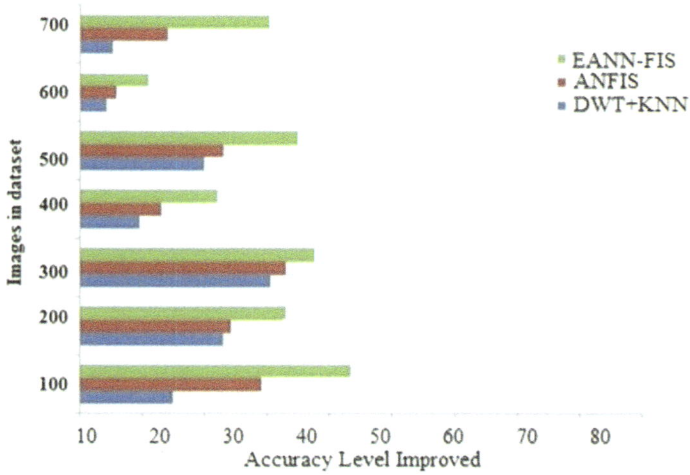

Fig. 5 Borders and edge detection accuracy level

to the traditional methods [15]. Figure 5 represents the accuracy level in the border and edge detection.

The overall accuracy rate of the proposed method is calculated by using three parameters, and they are sensitivity, specificity and accuracy level.

$$\text{Sensitivity} = \text{TP}/\,(\text{TP} + \text{FN}) * 100\%$$
$$\text{Specificity} = \text{TN}/\,(\text{TN} + \text{FP}) * 100\%$$
$$\text{Accuracy} = (\text{TP} + \text{TN})/(\text{TP} + \text{TN} + \text{FP} + \text{FN}) * 100\%$$

Table 1 specifies the ranges of the parameters which represents that the proposed method has better ranges when compared to traditional methods.

Table 1 Calculated parameters

Algorithm	Sensitivity (%)	Specificity (%)	Accuracy (%)
FCM	90.57	92.54	92.14
DWT + KNN	91.27	91.32	91.62
ANFIS	91.69	92.02	92.55
Proposed EANN-FIS	97.25	96.38	97.68

5 Conclusion

Artificial neural networks are utilized as the classifier for the extraction of relevant text from the captured images or video. Optical character recognition is utilized for character or text location identification. Text is recognized by developing the base tree with the utilization of a boundary box division. In this manuscript, an enhanced artificial neural network-based fuzzy inference system procedure is utilized to build up a framework called relevant text recognition. Since there are different procedures to build up a framework that consequently perceives the text from the images, the proposed method reduces the classification and clustering time and the accuracy level is also improved. The benefit of utilizing ANN is that they are increasingly used in various models to improve accuracy levels. The proposed model is contrasted with existing methods, and the results exhibit better performance. In future, the accuracy level of extraction of text from images can be enhanced by reducing the total hidden layers and to enhance the model to reduce the time for border and edge detection of images.

References

1. Unar, S., Wang, X., Zhang, C., & Wang, C. (2019). Detected text-based image retrieval approach for textual images. *IET Image Processing, 13*(3), 515–521. https://doi.org/10.1049/iet-ipr.2018.5277.
2. Yalniz, I. Z., & Manmatha, R. (2019). Dependence models for searching text in document images. *IEEE Transactions on Pattern Analysis and Machine Intelligence, 41*(1), 49–63. https://doi.org/10.1109/tpami.2017.2780108.
3. Wu, K., & Yu, Y. (2018). Automatic object extraction from images using deep neural networks and the level-set method. *IET Image Processing, 12*(7), 1131–1141. https://doi.org/10.1049/iet-ipr.2017.1144.
4. Tsai, C., & Tsai, S. (2018). Simultaneous 3D object recognition and pose estimation based on RGB-D images. *IEEE Access, 6,* 28859–28869. https://doi.org/10.1109/access.2018.2808225.
5. Khan, N. H., & Adnan, A. (2018). Urdu optical character recognition systems: Present contributions and future directions. *IEEE Access, 6,* 46019–46046. https://doi.org/10.1109/access.2018.2865532.
6. Baran, R., Partila, P., & Wilk, R. (2018, January). Automated text detection and character recognition in natural scenes based on local image features and contour processing techniques. In *International Conference on Intelligent Human Systems Integration* (pp. 42–48). Springer, Cham.
7. Dai, Y., Huang, Z., Gao, Y., Xu, Y., Chen, K., Guo, J., & Qiu, W. (2018, August). Fused text segmentation networks for multi-oriented scene text detection. In *2018 24th International Conference on Pattern Recognition (ICPR)* (pp. 3604–3609). IEEE.
8. V. R. Maddumala. (2020). Enhanced Morphological Operations for Improving the Pixel Intensity Level. *International Journal of Advanced Science and Technology, 29*(03), 9191–9201.
9. Manzyuk, O., Pearlmutter, B. A., Radul, A. A., Rush, D. R., & Siskind, J. M. (2019). Perturbation confusion in forward automatic differentiation of higher-order functions. *Journal of Functional Programming, 29.*
10. Yin, X., Zuo, Z., Tian, S., & Liu, C. (2016). Text detection, tracking and recognition in video: A comprehensive survey. *IEEE Transactions on Image Processing, 25*(6), 2752–2773. https://doi.org/10.1109/tip.2016.2554321.

11. Sarada, K., & Lakshman Narayana, V. (2020). Improving relevant text extraction accuracy using clustering methods. *TEST Engineering and Management, 83*, 15212–15219.
12. Yang, J., Price, B., Cohen, S., Lee, H., & Yang, M. (2016). Object contour detection with a fully Convolutional encoder-decoder network. In *2016 IEEE Conference on Computer Vision and Pattern Recognition (CVPR)*. doi:https://doi.org/10.1109/cvpr.2016.28
13. Sahare, P., & Dhok, S. B. (2016). Review of text extraction algorithms for scene-text and document images. *IETE Technical Review, 34*(2), 144–164. https://doi.org/10.1080/02564602.2016.1160805.
14. Narayana, V. L., Sudheer, B. N., Maddumala, V. R., & Anusha, P. (2020). Fuzzy base artificial neural network model for text extraction from images. *Journal of Critical Reviews, 7*(6), 350–354, doi: https://doi.org/10.31838/jcr.07.06.61.
15. Raghunandan, K. S., Shivakumara, P., Roy, S., Kumar, G. H., Pal, U., & Lu, T. (2019). Multi-script-Oriented text detection and recognition in video/Scene/Born digital images. *IEEE Transactions on Circuits and Systems for Video Technology, 29*(4), 1145–1162. https://doi.org/10.1109/tcsvt.2018.2817642.

Maximum Rate, with Joint Resource Allocation and Relay Selection for OFDM-Cognitive Network with Multiple-(Two-Way AF)-Relay

Ban Khalid Ammar, Hayder Jawad Albattat, Ahmed Ghanim Wadday, and Mohannad A. M. Al-Ja'afari

Abstract Cognitive relay network is an integrated technique that combines the cognitive radio network (CRN) with wireless relay technology. A range of challenges faces this combination and some of these are proposed in this paper. In this paper, the power allocation, subcarrier allocation, and relay selection are proposed as a problem formulation for OFDM based on cognitive radio network with assistance the two-way amplify-and-forward (AF) relay, under all interference constraints and power constraints that considered in this paper to maximize the total transmission rate for the secondary network. To solve this problem, genetic algorithm (GA) is proposed as an optimization technique to find all subcarriers allocation as matrix pairing A, the relay selection matrix B, and all powers calculated that distributed over 256 levels. The results showed the ability of this algorithm to effectively determine the frequency and power values of the system.

Keywords Cognitive relay network · Resource allocation · Relay selection · Genetic algorithm

B. K. Ammar (✉) · H. J. Albattat · A. G. Wadday
Department of Communication Techniques Engineering, Engineering Technical College/Najaf, Al-Furat Al-Awsat Technical University, Najaf, Iraq
e-mail: coj.bancom@atu.edu.iq

H. J. Albattat
e-mail: hayder.albattat@atu.edu.iq

A. G. Wadday
e-mail: ahmadghw@hotmail.com

M. A. M. Al-Ja'afari
Department of Electronics and Communication Techniques, Al-Najaf Technical Institute, Al-Furat Al-Awsat Technical University, Najaf, Iraq
e-mail: inj.muh@atu.edu.iq

S.-L. Peng et al. (eds.), *Proceedings of First International Conference on Mathematical Modeling and Computational Science*, Advances in Intelligent Systems and Computing 1292, https://doi.org/10.1007/978-981-33-4389-4_42

1 Introduction

The cognitive radio networks (CRN) have been considered the best solution for spectrum utilization efficiently, due to the scarceness of frequency band caused by the increasing requirement for high data rate services. The idea of a cognitive radio network (CRN) is in simple definition: The primary users, also known (licensed users), share its licensed spectrum band with other secondary users, also known (unlicensed users) [1, 2]. In the related literature for this work, the resources allocation (RA) works for the power allocation and/or subcarrier allocation for both secondary users and the relays, furthermore, the relay selection with and without the interference constrains from the secondary network to primary users (PUs). In [3], the dual method used as optimization technique to maximize the transmission rate under the power constraints to avoid the interference to PUs, with two activities the subcarrier pairing (SP) to all the relays and joint power allocation, and the destination received the signals from all relays not from specific relay and this is the difference with the other works. In [4], the problem of relay selection and RA for two-way relaying cognitive networks are solved, for both amplify-and-forward (AF) and decode-and-forward (DF) protocols, using dual decomposition, sub-gradient methods and proposed the genetic algorithm for sub-optimal approach. The interference management in [5] is offered with two activities for cognitive relay network, and they arranged a pairing of (subcarrier and power allocation) based on genetic algorithm. The novel optimization metric is proposed in [6] named bandwidth power product (BPP) to achieving the optimal spectrum sharing for the multi-user multi-relays with decode-and-forward (DF) relay, and using the Hungarian method for the relay selection. The RA with perfect and imperfect sensing is proposed in [7] for two-way multi-AF relays to maximization the total rates for secondary network under the interference constraints using three optimization techniques are Lagrange dual decomposition, discrete searching, and Hungarian algorithm to achieving the subcarrier allocation, relay selection, and optimal power allocation.

In this study, a perfect spectrum sensing is proposed for OFDM system based on cognitive two-way AF relay network using genetic algorithm (GA) to maximizing the total transmission rate, under all of the interference and power constraints to prevent the secondary network of harming the primary users.

2 System Model

In this paper, the system model consists of one primary user pair (sender and destination) PU_s, PU_d shares the licensed band with the other secondary users SU_s, $SU_s \in \{SU_1 \ldots SU_a\}$ where the symbol (a) refers to several secondary users. Some of the relay nodes R_k, $k \in \{1 \ldots K\}$ is proposed to help the secondary users from communicating with each other in lower transmission power, and for farther distant, where the symbol (k) refers to the number of relay nodes. The total frequency

Table 1 System variables

SU_s	$\{SU_1, SU_2\}$
R_k	$\{R_1, R_2\}$
N	$\{1, 2, 3, 4, 5, 6\}$
N_o	$\{1, 2\}$
N_v	$\{1, 2, 3, 4\}$

Fig. 1 System model

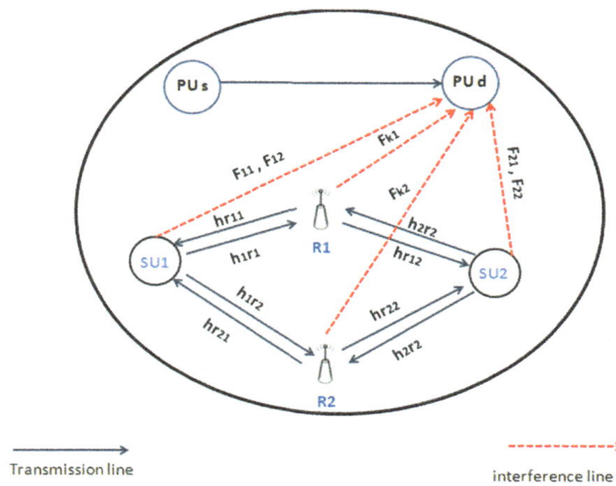

band owned by the licensed user is divided into some of the subchannels equal to N subcarriers, where the subcarriers occupied by the licensed user defined by N_o,$o \in \{1, \ldots, p, \ldots, P\}$, and the unused subcarriers or vacant are allocated to the secondary network defined by N_v, $v \in \{1, \ldots, s, \ldots S\}$. To implement this system model, the numbers for all contents of this system are illustrated in Table 1. The system model depicts in Fig. 1.

It is important to note that the numbers of N_o and N_v is represented the frequency index and not the frequency sequence itself.

2.1 Secondary Network

To analyze the rate of secondary network with the AF, two-way relay needs two phases to exchange the signals between the SUs users [8]. The system definitions are described in Table 2.

Multi-access (MA) phase: The SUs users may transmit signals at the same time to the selected relay (k) in the allocated subcarrier-i, the received signal in the k relay is:

Table 2 System model definitions

i	Subcarrier index for MA-phase from (SU_s) to relays. $i \in N_\upsilon$
J	Subcarrier index for BC-phase from relays to (SU_s) $j \in N_\upsilon$
x_{si}	Describe the transmitted symbols from SU_s through subcarrier-i
p_{Si}	The power allocated to the SU_s when are sending to kth relay on MA-phase
$h_s r_k$	Represent the channel gains when the SU_s are sending to kth relay on MA-phase
hr_{ks}	Represent the channel gains from kth-relay to the secondary users on BC-phase
p_{kj}	The allocated power to kth relay when it sending to secondary user on BC-phase
F_{sq}	Represent the interference-channel gains from secondary users on MA-phase to primary users on subcarrier q, when SU_s transmits to relay nodes ($q \in N_0$)
F_k	Represent the interference-channel gains from the kth relay in BC-phase to primary user on subcarrier, when the relays broadcast the signal to SU_s

$$y_i^k = \sqrt{p_{1i}^k} H_{1i}^k x_{1i}^k + \sqrt{p_{2i}^k} H_{2i}^k x_{2i}^k + \theta_i^k \tag{1}$$

θ_i^k is an additive white Gaussian noise(AWGN) for RF channel.

Broadcast (BC) phase: The selected relay then amplifies the received signal with an amplification factor Af_j^k and forwards this s $E\left(|x_{1i}|^2\right) = 1, E\left(|x_{2i}|^2\right) = 1$ denoted by E (.) [4].

$$p_j^k = \left(y_i^k . Af_j^k\right)^2 + \sigma^2 \tag{2}$$

$$Af_j^k = 1/\sqrt{y_i^{k2} + \sigma^2} = 1/\sqrt{p_{1i}^k H_{1i}^{k2} + p_{2i}^k H_{2i}^{k2} + \sigma^2} \tag{3}$$

where is σ^2 is the AWGN plus the noise from k relay (it is also AWGN) to SU_s. The delivered signal to SU_1 and SU_2 denotes in Eqs. (4) and (6).

$$y_{1j} = AF\sqrt{p_{kj}} HR_{1k} y_i^k + \sigma^2 \tag{4}$$

$$y_{1j} = \underbrace{Af\sqrt{p_{kj}}\, HR_{1k}\sqrt{p_{1i}}\, H_1 R_k x_{1i}}_{\text{Inter-user Interference}} + \underbrace{Af\sqrt{p_{kj}}\, HR_{1k}\sqrt{p_{2i}}\, H_2 R_k x_{2i}}_{\text{Signal}} + \underbrace{Af\sqrt{p_{kj}}\, HR_{1k}\,\sigma^2 + \sigma^2}_{\text{Noise}} \tag{5}$$

$$y_{2j} = Af\sqrt{p_{kj}} HR_{2k} y_i^k + \sigma^2 \tag{6}$$

$$y_{2j} = \underbrace{Af\sqrt{p_{kj}}\, HR_{2k}\sqrt{p_{1i}}\, H_1R_k\, x_{1i}}_{\text{Signal}} + \underbrace{Af\sqrt{p_{kj}}\, HR_{2k}\sqrt{p_{2i}}\, H_2R_k\, x_{2i}}_{\text{Inter-user Interference}} + \underbrace{Af\sqrt{p_{kj}}\, HR_{2k}\,\sigma^2 + \sigma^2}_{\text{Noise}} \tag{7}$$

$$\text{SNR1} = \frac{Af\sqrt{p_{kj}}HR_{1k}\sqrt{p_{2i}}H_2R_kx_{2i}}{Af\sqrt{p_{kj}}HR_{1k}\sigma^2 + \sigma^2} \tag{8}$$

$$\text{SNR2} = \frac{Af\sqrt{p_{kj}}HR_{2k}\sqrt{p_{2i}}H_2R_kx_{2i}}{Af\sqrt{p_{kj}}HR_{2k}\sigma^2 + \sigma^2} \tag{9}$$

Notice that as mentioned in the beginning, this system is with perfect sensing then the channel estimation can be cancelled the effectiveness of the inter-user-interference in Eqs. (5) and (7) [4]. By substitution amplification factor Af in Eq. (3) into Eqs. (8) and (9), numerator and denominator divided by σ^4, will get:

$$\text{SINR1} = \frac{P_{kj}HR_{1k}P_{2i}H_2R_k}{P_{kj}HR_{1k} + p_{1i}H_1R_k + p_{2i}H_2R_k + 1} \tag{10}$$

$$\text{SINR2} = \frac{P_{kj}HR_{2k}P_{1i}H_1R_k}{P_{kj}HR_{2k} + p_{1i}H_1R_k + p_{2i}H_2R_k + 1} \tag{11}$$

where: $HR_{sk} = \frac{|hr_{sk}|^2}{\sigma^2}$, $H_sR_k = \frac{|h_sr_k|^2}{\sigma^2}$.

The total transmission rate of SU_1 and SU_2 with assistance of two-way AF relays on subcarrier <*i*, *j*> with two time slots is [9]:

$$R_{\text{AF}}^{i,j,k} = \frac{1}{2}\log_2(1 + \text{SINR1}) + \frac{1}{2}\log_2(1 + \text{SINR2}) \tag{12}$$

where is the SINR1, SINR2 represent the signal to interference plus noise ratio of SU_1 and SU_2, respectively.

2.2 Interference Model

To analyze the interference model with assume a perfect spectrum sensing that mean the SU_s detects the spectrum perfectly without harming the PU_d, the interference from SU_s subcarrier-*i* to the PU_d on subcarrier-*o* is:

$$I_{Si} = \sum_{i\in N_v}\sum_{k\in K}\left(p_{1i}\sum_{o\in N_O}|f_{1o}|^2 + p_{2i}\sum_{o\in N_O}|f_{2o}|^2\right) \le I_{\text{th}} \tag{13}$$

The interference from relay nodes subcarrier-j to PU_d on subcarrier-o is:

$$I_{kj} = \sum_{j \in N_v} \sum_{k \in K} \left(p_{kj} \sum_{o \in N_O} |f_{ko}|^2 \right) \leq I_{\text{th}} \tag{14}$$

3 Problem Formulation

The objective function in this paper is to maximize the total system capacity that leads to maximization of the total transmission rate according to *Shannon theory*, and the resources allocation (RA) and relay selection under the interference threshold and maximum transmission power for secondary network constraints are proposed in this paper. To illustrate this problem functions, a two-dimensions binary matrix denotes ($A_{i\times j}$) that index the subcarrier pairing, where if A equal to 1 meaning that the subcarrier-i is allocated for MA-phase (when the SU_s transmit to k relay) is paired with subcarrier-j that assigned to BC-phase and allocated for k relay (when it amplify the received signal and forward it to SU_s). Three-dimension binary matrix denotes ($B_{i\times j\times k}$), if B equal to 1 that means the subcarrier pair $<i, j>$ is allocated to the k relay. All the constraints for the optimization problems to maximization the total transmission rate for the secondary network include $c1$ if the subcarrier-i is paired with subcarrier-j, c_2 the subcarrier pair $<i, j>$ is allocated to the k relay, c_3 and c_4 the transmission power from SU_s and from k relay must be less than the maximum power, respectively, and the c_5 and c_6 refer to the interference power from SU_s and from k relay must be less than the interference threshold that known by the PU_s, respectively [5, 7].

$$\mathrm{OP}: \left(\begin{matrix} \max \\ A, B, P \end{matrix} \right) \sum_{i \in N_v} \sum_{j \in N_v} \sum_{k \in K} B_{i,j}^{k} A_{i,j} R_{AF}^{i,j,k}$$

$$c1: \sum_{i \in N_v} A = 1, \sum_{j \in N_v} B = 1.$$

$$c2: \sum_{k \in K} B = 1, \quad \text{for all } \langle i, j \rangle$$

$$c3: \sum_{i \in N_v} \sum_{k \in K} p_{si} \leq p_{S\,\max} \quad \text{for} \quad s = 1, 2$$

$$c4: \sum_{j \in N_v} p_{kj} \leq p_{k\,\max}. \quad \text{for all } k$$

$$c5 : I_{Si} = \sum_{i \in N_s} \sum_{k \in K} \left(p_{1i} \sum_{o \in N_O} |f_{1o}|^2 + p_{2i} \sum_{o \in N_O} |f_{2o}|^2 \right) \leq I_{\text{th}}$$

$$c6 : I_{kj} = \sum_{j \in N_v} \sum_{k \in K} \left(p_{kj} \sum_{o \in N_O} |f_{ko}|^2 \right) \leq I_{\text{th}}$$

4 Simulation Results

To maximize the total transmission rate for secondary network and solve the optimization problems, GA is adopted in this thesis which obtains jointly the results of the resources allocation RA and relay selection. All parameters for GA and the power values used in this system model are listed in Table 3.

The adopted channel gains and interference-channel gains are independent-and-identically distributed (i.i.d) as $\mathcal{CN} \sim (\mu, \sigma^2)$ with zero means [7].

The results of this system model are:

1. The subcarrier pair matrix A and relay selection matrix B for each relay obtained from GA are:

$$A = \begin{bmatrix} 0\,1 & 0\,0 \\ 0\,0 & 0\,1 \\ 0\,0 & 1\,0 \\ 1\,0 & 0\,0 \end{bmatrix}$$

Table 3 GA parameters and power values

Number of iterations size data	10
Number of initial generations data	100
Total number of variables	$(X_T = 48)$
Power variables	$(X_P = 24)$, its range 256 levels
Subcarrier variables	$(X_F = 24)$, binary range (0 or 1)
Maximum transmission power for secondary users	$p_{Si} = 1.5$ W
Maximum transmission power for relay nodes	$p_{kj} = 1.5$ W
Interference power threshold	$I_{\text{th}} = 1 \times 10$–6 W
AWGN	Zero mean and 0.01 variance

$$B1 = \begin{bmatrix} 0\,1 & 0\,1 \\ 0\,1 & 0\,1 \\ 0\,1 & 0\,1 \\ 0\,1 & 0\,1 \end{bmatrix}, \quad B2 = \begin{bmatrix} 1\,0 & 1\,0 \\ 1\,0 & 1\,0 \\ 1\,0 & 1\,0 \\ 1\,0 & 1\,0 \end{bmatrix}$$

$$A \times B_1 = \begin{bmatrix} 0\,1 & 0\,0 \\ 0\,0 & 0\,1 \\ 0\,0 & 0\,0 \\ 0\,0 & 0\,0 \end{bmatrix}, \quad A \times B_2 = \begin{bmatrix} 0\,0 & 0\,0 \\ 0\,0 & 0\,0 \\ 0\,0 & 1\,0 \\ 1\,0 & 0\,0 \end{bmatrix}$$

In matrix A: the rows represent the subcarriers-i while the columns represent the subcarriers-j. The values of ones in matrix A mean that the subcarrier-i of the MA-phase is paired with the subcarrier-j of the BC-phase which known by subcarrier pairs <i, j> . To distinguish if the subcarrier pairs in matrix A is allocated to R_1 or R_2, the matrix A will multiply (dot product) by each of the matrices B_1 and B_2, respectively. According to that, the subcarrier pair <1, 2> is allocated for R_1, <2, 4> is allocated for R_1, <3, 3> is allocated for R_2 and <4, 1> is allocated for R_2. All of the subcarrier pairs which are allocated are shown in Fig. 2.

2. The powers allocated for each of the secondary users and relay nodes under maximum transmission power constraint ($p_{Si} = 1.5\ W$) are depicted in Figs. 3 and 4, respectively. The power values will be calculated for each subcarrier-i when the SU_s transmitted to kth relay, also, it will be calculated for each subcarrier-j when the kth the relay will AF the signal and transmitted to SU_s. Then, will find the power allocated for each subcarrier pair <i, j> according to matrix A and the multiplication of ($A \times B_1 \mathrm{and} A \times B_2$). The power values of each subcarrier pairs are listed in Table 4.

3. The signal to noise ratio for SU_s with kth the relay is shown in Fig. 5:

   Note that in Fig. 5, the value of SNR_{11} at the pair <1, 2> in Fig. 5a takes the highest value than the others SNR on the same pair. For this reason, the subcarrier pair depends on SNR when allocated to secondary users and relay nodes. Also, for the same way in SNR_{12} in Fig. 5b, the highest value is obtained at <3, 3> subcarrier pair, then it will be allocated for SU_1 with R_2, SNR of each relay according to the allocated subcarriers are listed in Table 5.

4. The total data rate from the two relays is shown in Fig. 6. From the SNR in Fig. 5, the total data rate can be calculated and shown in Fig. 6 from each node according to the Shannon theory mentioned in (12).

The calculated data rates from each relay according to the allocated subcarriers are listed in Table 6.

$$\mathrm{Rate}_1 = 0.5 \log_2(1 + \mathrm{SNR}11) + 0.5 \log_2(1 + \mathrm{SNR}21) \tag{15}$$

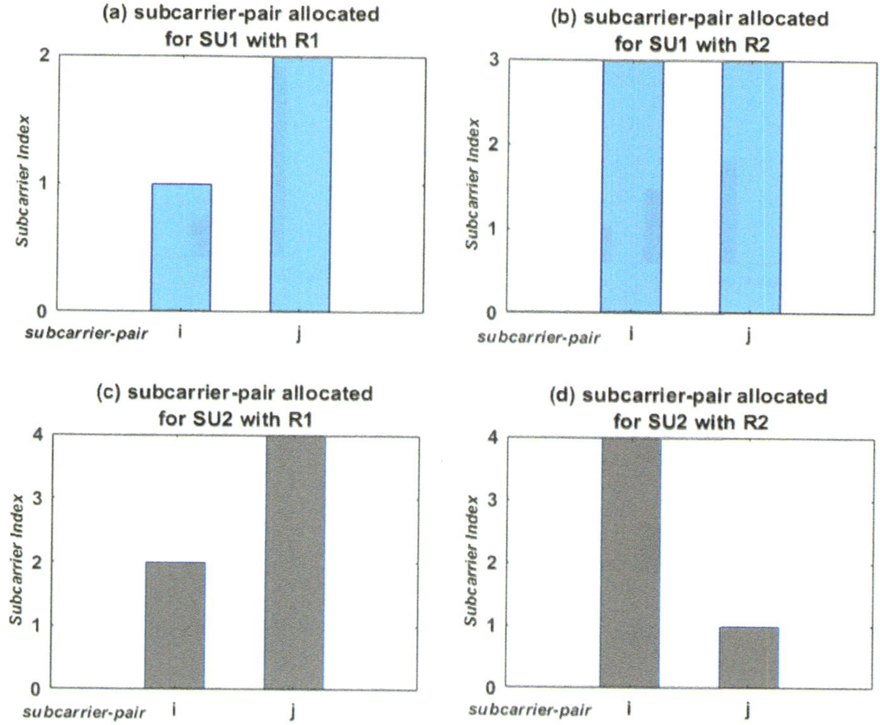

Fig. 2 Subcarrier pairs allocated for all users with each relay nodes

$$\text{Rate}_2 = 0.5\ \log_2(1 + \text{SNR12}) + 0.5\ \log_2(1 + \text{SNR22}) \tag{16}$$

Depending on subcarrier-allocation matrix A and the subcarrier pairs, the highest individual value of data rate can be calculated:

$\text{Rate}_1 = 7.525\,\text{bit/s/Hz}$, $\text{Rate}_2 = 7.006\,\text{bit/s/Hz}$, then the total data rate (bit/s/Hz) of this system model is the result of adding the two rates:

$$\text{Total data rate} = 14.5312\text{bit/s/Hz}.$$

To find the total transmission rate (bit/s) must be suggested a frequency band. This system model can respond in general for any wireless system.

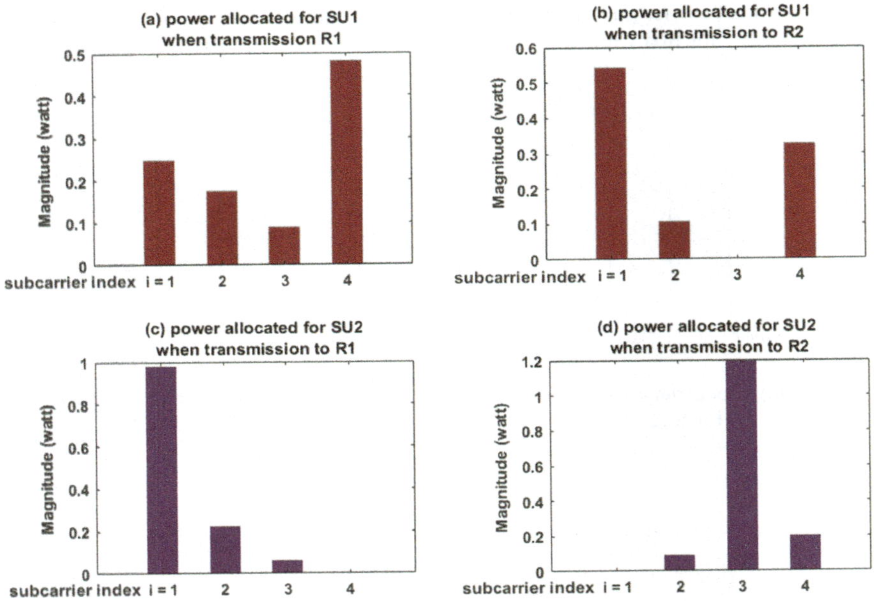

Fig. 3 Power allocated for secondary users when transmits to relay nodes on subcarrier-i for MA-phase

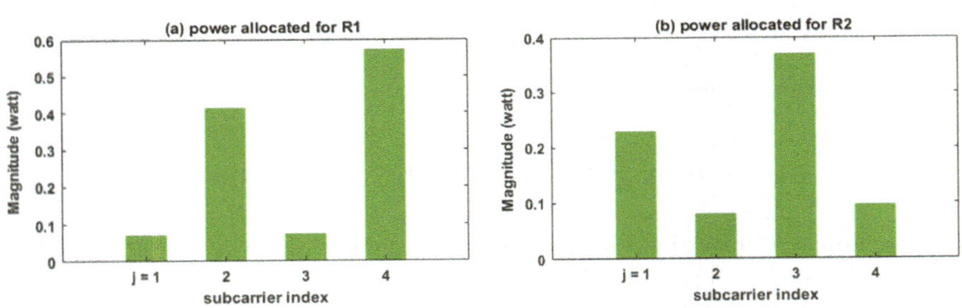

Fig. 4 Power allocated for relay nodes on subcarrier-j for BC-phase when it amplifies and forwards the signals to secondary users

Table 4 Allocated power values of subcarrier pairs

R_1		R_2	
Subcarrier pairs	Power values	Subcarrier pairs	Power values
<1, 2>	0.2477 W, 0.414 W	<3, 3>	3e-07 W, 0.3703 W
<2, 4>	0.2294 W, 0.07262 W	<4, 1>	0.2049 W, 0.229 W

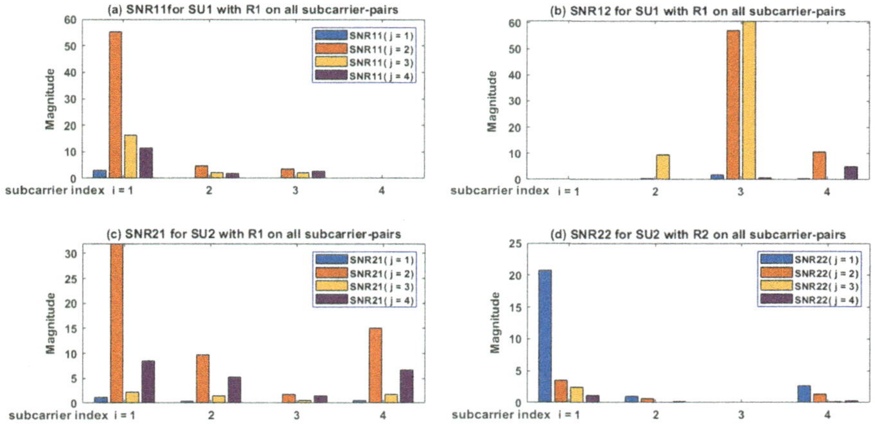

Fig. 5 SNR for all subcarrier pairs

Table 5 SNR of each relay

SNR of R_1		SNR of R_2	
Subcarrier pairs	SNR values	Subcarrier pairs	SNR values
<1, 2>	55.1365	<3, 3>	60.873
<2, 4>	5.31	<4, 1>	1.43

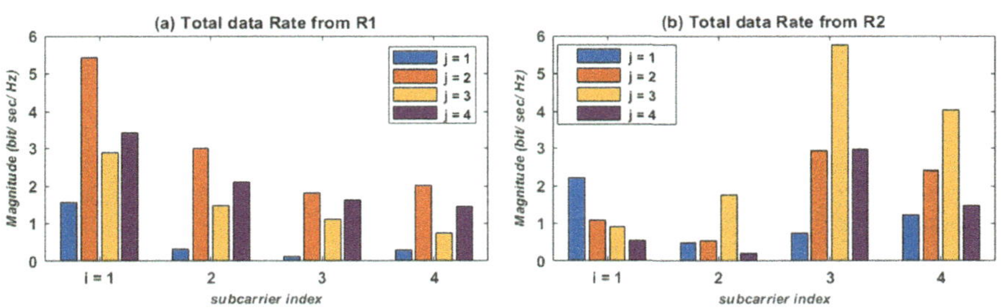

Fig. 6 Total data rate from each relay on all subcarrier pairs

Table 6 Total data rate of each relay

Rate of R_1		Rate of R_2	
Subcarrier pairs	Rate values	Subcarrier pairs	Rate values
<1, 2>	5.42672	<3, 3>	5.76676
<2, 4>	2.09849	<4, 1>	1.23926

5 Conclusion

In this paper, the genetic algorism is used for an OFDM-based cognitive two-way AF relay network with assuming perfect sensing to avoid the interference between the unlicensed network and licensed users. The obtained total system capacity has an increase in this work to (14.5312bit/s/Hz) for a secondary network with all calculations of subcarrier pairing, the powers allocated and the best relay selection. Under the interference power constraint to preventing the secondary network from harmful the primary user when sharing its spectrum band with a secondary network and under the constraints of maximum power transmitted from the secondary network.

References

1. Mitola, J., & Maguire, G. Q. (1999). Cognitive radio: Making software radios more personal. *IEEE Personal Communications, 6*(4), 13–18.
2. Doyle, L. (2009). *Essentials of cognitive radio*. Cambridge: Cambridge University Press.
3. Li, X., Zhang, Q., Zhang, G., & Qin, J. (2013). Joint power allocation and subcarrier pairing for cooperative OFDM AF multi-relay networks. *IEEE Communications Letters, 17*(5), 872–875.
4. Alsharoa, A., Bader, F., & Alouini, M. S. (2013). Relay selection and resource allocation for two-way DF-AF cognitive radio networks. *IEEE Wireless Communications Letters, 2*(4), 427–430.
5. Lang, H. S., Lin, S. C., & Fang, W. H. (2015). Subcarrier pairing and power allocation with interference management in cognitive relay networks based on genetic algorithms. *IEEE Transactions on Vehicular Technology, 65*(9), 7051–7063.
6. Chen, B., Lei, M., & Zhao, M. (2016). An optimal resource allocation method for multi-user multi-relay DF cognitive radio networks. *IEEE Communications Letters, 20*(6), 1164–1167.
7. Li, H., & Zhao, X. (2018). Joint resource allocation for OFDM-based cognitive two-way multiple AF relays networks with imperfect spectrum sensing. *IEEE Transactions on Vehicular Technology, 67*(7), 6286–6300.
8. Rodríguez, L. J., Tran, N., & Le-Ngoc, T. (2015). *Amplify-and-forward relaying in wireless communications*. New York: Springer
9. Ji, X., Zheng, B., Cai, Y., & Zou, L. (2012). On the study of half-duplex asymmetric two-way relay transmission using an amplify-and-forward relay. *IEEE Transactions on Vehicular Technology, 61*(4), 1649–1664.

Enhancement in E-Marketing Using SVM and Association Rule

Evan Madhi Hamzh Al Rubaie

Abstract Data mining has many new revolutions in E-marketing in online business. It is observed that using a set of frequently brought items, it is possible to predict the next item which user is more likely to buy. It is found that in early stages of any account this strategy fails. Reason for this failure is the lack of data about that user. This problem can be fixed by using text mining in social networks account of the user and basic information which is gathered while accounting registration. It is not a good idea to use the traditional market-basket technique on newly registered users. Therefore, this technology is developed to overcome this issue of marketing to a new user. Kernel support vector machine and Apriori algorithm are used to improve the accuracy of knowledge discovery.

Keywords E-marketing · Machine learning · Text mining · Natural language processing · Association rule mining · Kernel function · Support vector machine

1 Introduction

A cornerstone of revolution in E-marketing is data. Using the Internet and computational power to carry out promotional activities is known as E-marketing, but to use this data for user-specific marketing, hidden pattern present in the data must be found out, as the size grows, data mining and knowledge discovery becomes more reliable. The traditional method which used in marketing is steady and less robust. With small changes in products and their specifications, a marketing strategy can get changed extremely. Hence, data mining using decision tree, clustering and association rule mining is used to find business needs in marketing strategy. In such analysis, data from different sources is collected. This data is trained to find knowledge present in the data, and this process is called as knowledge discovery. Market-basket is the most popular and useful technique in determining the next product which a customer is

---

E. M. H. Al Rubaie (✉)
College of Engineering, University of Babylon, Hillah, Babylon, Iraq
e-mail: haidar_iraq77@yahoo.com

S.-L. Peng et al. (eds.), *Proceedings of First International Conference on Mathematical Modeling and Computational Science*, Advances in Intelligent Systems and Computing 1292, https://doi.org/10.1007/978-981-33-4389-4_43

most likely to buy. Using this prediction on data, advertisement marketing technique is used for that customer [7].

This approach is possible only when a user has to buy history stored in databases or cache memory. But for users with less buying history data or users who buy newly listed products, rule mining is unreliable. In such a case, the market-basket analysis produces results with less accuracy. This problem of market-basket analysis is handled by text mining on social network data and the data which is collected from the user at the time of registration [8]. Kernel's support vector machine and associate rule mining are used in this analysis. Using correct data sources and data pre-processing algorithm, prediction about marketing strategy is decided.

2 Objective of This Research

This data is collected from social media and present database and the main objective of this research is to determine the new marketing strategy, by using previous datasets. This process should be easy to handle and visualization should be there for a better understanding of the diagnosis. Few objectives of this research are:

1. Create proper marketing plan for new users.
2. Predict the next purchase for new products and new users.
3. Enhancing real-world E-marketing solutions.
4. Improve accuracy of machine learning model using SVM and Apriori algorithms.

This paper aims to find out a new marketing strategy. For better performance, it is very much important to use correct data and machine learning algorithm. In this research, it is observed that this database and data mining algorithms are useful in determining marketing strategy for new users with a high accuracy rate.

Various phases of data mining and knowledge discovery are explained in Sect. 2 and Sect. 3. Section 4 summarizes the accuracy of the algorithm. It is observed that the accuracy of the algorithm is good enough to use this system in a real-time E-marketing application.

3 Data Collection and Pre-processing

Data selection is a very first and important step in the data mining process. Authenticated and correct data are key for any data mining algorithm. Along with these, data should be in the proper format so that pre-processing and knowledge discovery becomes simple. If data is not in correct format, then the selection of the appropriate machine learning algorithm is also a necessary task. For this reason, it is very much essential to select proper data and data pre-processing techniques [2].

In this research, Twitter API is used to fetch the user's social network data. User's personal data is collected from different sources. This data contains basic information about other registered users and their interest in different products. In the later stage of this research association, data mining is used. This algorithm cannot handle numeric attribute hence sampling is required along with this. Data extracted from Twitter also need to be categorized to find important keyword from social text messages. Local data which is collected also needs to confidential and secure. Few of these data attributes are age, gender, location, loan history and many other attributes.

4 Twitter Data Extraction

R Library is used to fetch data from the Twitter source. Twitter is open-source and its data is free. Customer opinion and feedback model requires input data which contain their emotions and reviews, this data is pre-processed to normalize it for model design, and random forest classification algorithm is used to build system model [5, 6].

5 Input and Pre-processing

Data is collected from different sources which include data from Web crawler and Twitter API. RCrawler is an *R* package present in *R*, which scraps the data from different URLs. RCrawler extracts information from Internet sources and converts it into structured data, which help in the data handling part. Web crawlers usually collect the data from Web resources, but Web scraper collects the data and extracts information from it. There are many different packages presents in *R*, like ScrapeR and Rvest, but RCrawler visits all the URL in a specific domain automatically and collects information from them. RCrwaler also supports parallelization which visits multiple URLs at the same time, so that time of data collection gets reduced (Fig. 1).

RCrawler supports link extraction, hashtag-based data extraction, and user-based data extraction. User-based data extraction can be done by R command.

User_data < - get_timeline ("TwitterHandler", $n = 50$).

This command can fetch last 50 tweets from user with Twitter handler "Twitter-Handler."

5.1 Data from Dataset

Data collected from users while registration is stored in databases, this data is also used in the Apriori algorithm to create a rule for making E-marketing strategy. This data includes attributes like age, gender, location and loan history.

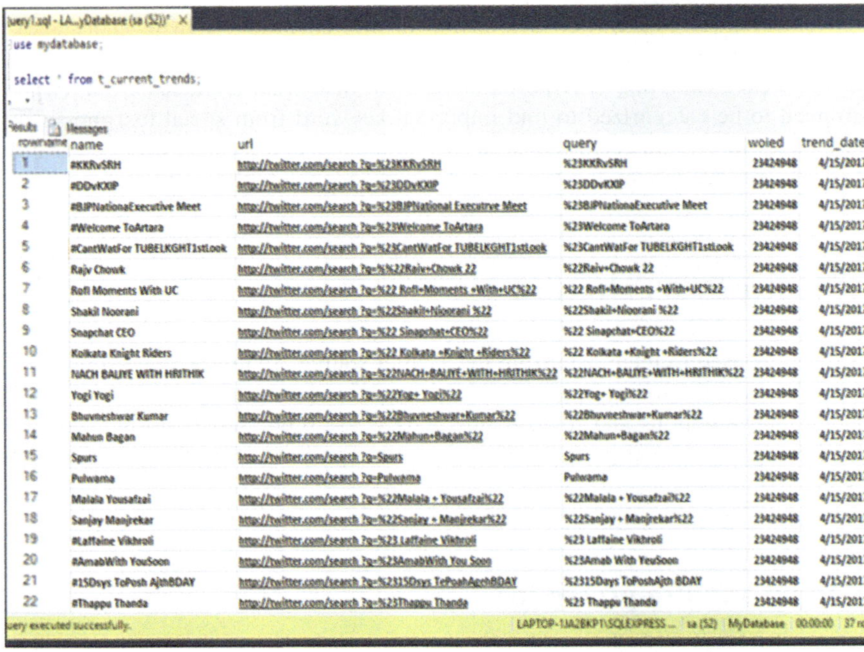

```
use mydatabase;

select * from t_current_trends;
```

rowname	name	url	query	woied	trend_date
1	#KKRvSRH	http://twitter.com/search ?q=%23KKRvSRH	%23KKRvSRH	23424948	4/15/2017
2	#DDvKXIP	http://twitter.com/search ?q=%23DDvKXIP	%23DDvKXIP	23424948	4/15/2017
3	#BJPNationaExecutive Meet	http://twitter.com/search ?q=%23BJPNational Executive Meet	%23BJPNationaExecutive Meet	23424948	4/15/2017
4	#Welcome ToArtara	http://twitter.com/search ?q=%23Welcome ToArtara	%23Welcome ToArtara	23424948	4/15/2017
5	#CantWatFor TUBELKGHT1stLook	http://twitter.com/search ?q=%23CantWatFor TUBELKGHT1stLook	%23CantWatFor TUBELKGHT1stLook	23424948	4/15/2017
6	Rajv Chowk	http://twitter.com/search ?q=%%22Raiv+Chowk 22	%22Raiv+Chowk 22	23424948	4/15/2017
7	Rofl Moments With UC	http://twitter.com/search ?q=%22 Rofl+Moments +With+UC%22	%22 Rofl+Moments +With+UC%22	23424948	4/15/2017
8	Shakil Noorani	http://twitter.com/search ?q=%22Shakil+Nioorani %22	%22Shakil+Nioorani %22	23424948	4/15/2017
9	Snapchat CEO	http://twitter.com/search ?q=%22 Sinapchat+CEO%22	%22 Sinapchat+CEO%22	23424948	4/15/2017
10	Kolkata Knight Riders	http://twitter.com/search ?q=%22 Kolkata +Knight +Riders%22	%22 Kolkata +Knight +Riders%22	23424948	4/15/2017
11	NACH BALIYE WITH HRITHIK	http://twitter.com/search ?q=%22NACH+BALIYE+WITH+HRITHIK%22	%22NACH+BALIYE+WITH+HRITHIK%22	23424948	4/15/2017
12	Yogi Yogi	http://twitter.com/search ?q=%22Yog+ Yogi%22	%22Yog+ Yogi%22	23424948	4/15/2017
13	Bhuvneshwar Kumar	http://twitter.com/search ?q=%22Bhuvneshwar+Kumar%22	%22Bhuvneshwar+Kumar%22	23424948	4/15/2017
14	Mahun Bagan	http://twitter.com/search ?q=%22Mahun+Bagan%22	%22Mahun+Bagan%22	23424948	4/15/2017
15	Spurs	http://twitter.com/search ?q=Spurs	Spurs	23424948	4/15/2017
16	Pulwama	http://twitter.com/search ?q=Pulwama	Pulwama	23424948	4/15/2017
17	Malala Yousafzai	http://twitter.com/search ?q=%22Malala + Yousafzai%22	%22Malala + Yousafzai%22	23424948	4/15/2017
18	Sanjay Manjrekar	http://twitter.com/search ?q=%22Sanjay + Manjrekar%22	%22Sanjay + Manjrekar%22	23424948	4/15/2017
19	#Laffaine Vikhroli	http://twitter.com/search ?q=%23 Laffaine Vikhroli	%23 Laffaine Vikhroli	23424948	4/15/2017
20	#AmabWith YouSoon	http://twitter.com/search ?q=%23AmabWith You Soon	%23Amab With YeuSoon	23424948	4/15/2017
21	#15Dsys ToPosh AjthBDAY	http://twitter.com/search ?q=%2315Dsys TePoahAgthBDAY	%2315Days ToPoshAjth BDAY	23424948	4/15/2017
22	#Thappu Thanda	http://twitter.com/search ?q=%23Thappu Thanda	%23 Thappu Thanda	23424948	4/15/2017

Fig. 1 Extraction of Twitter data

Data Pre-processing:

Before applying a machine learning algorithm, some pre-processing is required on the datasets, and this is a procedure where raw data is converted into a dataset for machine learning. There are few steps in pre-processing as below [3]:

- Cleaning: This is a useful step because data may contain other values which create noise in the model creation process. Removal of such noisy data comes under the cleaning step of data mining.
- Integration: In this step, different data is collected from different sources. This data is then integrated with strength factor. This process is known as data integration.
- Reduction: In data mining procedure, some data can be there which is not required or which is repeated in the dataset, this data is removed by applying different data reduction technique, and this involves dimensionality reduction, aggregation and clustering.
- Sampling: Sampling is a process in which data of a few same samples is converted under one labelled to minimize the variation present in the datasets. Age and few others are numeric dataset, these values are categorized using the sampling technique. In sampling, age is converted into nominal values, the output of these setup is like for age between 0 and 20–teenage, 20–30 it is young, 30–20 labelled as an elder. In this way, sampling is used to categorized current data.

5.2 *Natural Language Processing for Tokenization of Twitter Data*

The first step in Twitter keyword extraction is to clean and tokenize the data. It breaks data in words so that it becomes easy to handle. In the tokenization process, raw data is converted in terms, this process is known as feature generation. *R*-quanteda library is used for this purpose [7]. Next step is to reduce redundant strings from data. Natural language processing also plays an important role in pre-processing, and it helps to collect important words like an adjective, which reduces the time taken for data extraction and classification.

6 Model Design

Enchantment in E-marketing with kernel-based SVM and Apriori algorithm consists of two phases. In each phase, different algorithms were applied to generate knowledge discovery form given datasets. In the first phase, Twitter API is used to fetch data, while in the second phase, Apriori algorithm is applied to create rules for a machine learning model.

6.1 *SVM-Based Data Analysis*

This is the first phase of this research, in this, data is collected from Twitter API, as described in Sect. 2 of this paper. *R* Library is used to fetch data from Twitter API. Natural language processing is used for data pre-processing in which un-necessary keyword is removed from datasets. News contains many words which may not be useful while determining sentiments. So, the feature extraction technique extracts only use words and features from the text document. The feature is nothing but a combination of the sentiment of the text and keywords. The extracted feature is very small as compared to the whole text which increases the accuracy and speed of learning algorithm [13].

Support vector machine algorithm can handle nonlinear data hence it gives good accuracy. While dealing with text data. It is a supervised learning algorithm. It categorizes the training data into one of the classes to get the output classes for each entry record. SVM can also categorize unlabelled data for unsupervised learning approaches. SVM has hyperplane to categorize given data into two or more classes [9] (Fig. 2).

The optimal hyperplane is a line which divides the data. Other two lines which are present in the figure are used to find decision boundary or optimal hyperplane [4]. Margin is a distance between two hyperplanes. Margin is selected in such a way

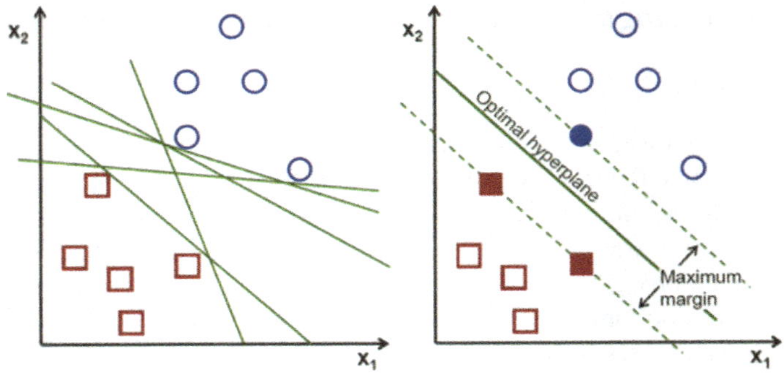

Fig. 2 Support vector machine

that all data points must be separate from the boundary, and this point is noted as support vectors.

Using these functions, hyperplane and decision boundary are selected. This equation is for simple support vector machine which can separate linear data.

6.2 *Advanced SVM*

Simple support vector machine is capable of deciding, decision boundaries of linearly separable data but when data is in nonlinear separable from then accuracy of simple support vector machine is dropped by a great margin.

An advanced version of SVM is kernel-based support vectors. In kernel-based SVM, different dimensions are used to find out nonlinearly separable data and its decision boundary. Using these dimensions, data is labelled with different classes. The kernel function is found to be very useful in many types of research as data is not always linearly separable [1]. How the kernel function improves support vector machine?

Support vector machine can handle linearly separable data, which divides data into two hyperplanes, but in case of kernel SVM, different hyperplanes are used for classification of data. The formula for support vector machine is

$$Y = W\,\mathrm{phi}(x) + b$$

In this equation, w is a weight, phi is a feature vector and b means bias.

In output calculation, if y is 0 then output is considered for class 1 and if the output is 1 then class is labelled as class 2. This is the working principle of support vector machine. This function can be improved by applying kernel function. The formula for kernel function is (Fig. 3):

$$K(x, y) = \langle f(x),\ f(y)\rangle$$

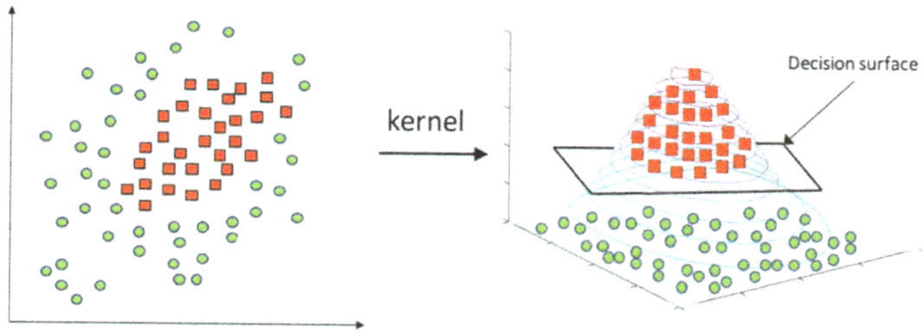

Fig. 3 SVM kernal

In this equation, k is a kernel function which is applied on two input x and y, respectively. X and Y are inputted at different dimensions, and f is a map of two different dimensions.

If we put kernel equation in support vector machine's equation in the form of $\langle f(x), f(y) \rangle$, then SVM feature vectors function can be replaced by $\langle$ phi(x), phi(y)$\rangle$, these shows overlap of x and y in the feature vector.

Advantages of kernel support vector machine.

- Support vectors can be drawn into infinite dimensions.
- Kernel function reduces the computational power required for calculation of support vector machine at a different dimension.

The output of this phase I is a labelled with the specified class which indicates whether the user is active or inactive in social media, also. These labelled are extracted from NLP as a keyword and it is assigned, and this data helps to identify user's choice. The output is labelled in such a way that it helps to determine the likes and dislikes of user. For example, it helps to understand whether the user likes mobile phones or car or travelling, reading or sports. These attribute values are also stored for that user. In later stage output of these, the phase is used in associate rule mining.

Phase II—Rule Creation using Association rule mining.

In this phase, the input is dataset and output of phase I, this input is provided to association rule mining algorithm to create rules for the given dataset, and this rule can be used for decision making in E-marketing. Data cleaning and data integration steps are used in this process to collect data from different sources. Association rule mining is a process of converting data into rules in the form of 'if and then' statement. There are many different algorithms available for association rule mining. SETM, AIS and Apriori, these are some famous association rule mining algorithms. In this research, the Apriori algorithm is used for finding association rule [12].

Apriori algorithm is used for frequent itemset identification and finding such rules in datasets. It uses a bottom up approach for finding rules. There is two important terms in Apriori algorithm which are support and resistance

$$\text{Support} = \frac{\text{Total number of transaction containing } 'X' \text{ item}}{\text{Total Number of transaction}}$$

$$\text{Confidence } (X \rightarrow Y) = \frac{\text{Support } (XUY)}{\text{Support } (X)}$$

Algorithm for Apriori Algorithm.

1. Find support '*S*' of each itemset, and it with minimum support and get the support of itemset.
2. Join all frequent item set as a *k*-item set and then use pruning to remove not frequent k-items from that dataset.
3. Check all transaction to find support '*s*' of each candidate present in the *k*-item set, in the given set, compare that value with minimum support to find out a set of *k* frequent item set.
4. If a candidate is empty then for each frequent itemset 1, a nonempty subset of 1 is generated.
5. For each subset, the output is a rule "*S* → [1 - *S*]" if confidence *C* of the rule is "*S* → [1-*S*]" minimum confidence.
6. If the candidate set is not empty, then go jump to step 2 [10].

```
Pseudo Code for Apriori Algorithm
Ck = item set with k number of candidate items
    Lk = Frequent set with k number of items
   L1 = {Frequent itemset};
For (k=1; Lk != Φ; K++) start
Ck+1 = generated candidate by Lk
Foreach Transaction p in dataset do
   Increase count of candidate by 1
Lk+1 = candidate having minimum support
  End
Return Uk Lk;
```

Figure 4 shows the flow of this research. In phase I, Twitter API and R Library are used for data collection as described in Sect. 2, natural language processing and kernel-based support vector machine is used to find an important feature from the user profile. The output of this phase is used as an input to the next phase, along with this dataset from users table is also used which contains basic information about the user like age, gender, location and loan history. Apriori algorithm is applied to this data to mine hidden rules present in it. In this way, the machine learning model is prepared with rules, so that it can serve a new request by checking existing rules present in the model. There is the working approach of this model.

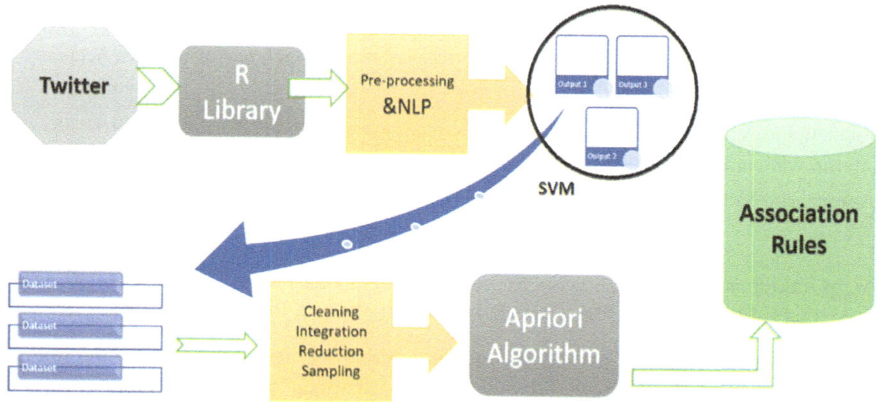

Fig. 4 Data mining process

7 Model Analysis and Result Accuracy

In this research, Apriori algorithm is used to training data for building machine learning model. This model contains a set of rules, which are called as association rule. Based on these rules, new prediction can be made on similar data [11]. If the rule is present in the dataset, then other user information can be obtained. In this research, SVM output and other data are provided to the Apriori algorithm, which generates the following rules for association.

1. Social = Sports 169 ==> Gender = Male 169 < conf:(1) > lift:(1.44) lev:(0.2) [51] conv:(51.49)
2. Social = Books Loan = No 167 ==> Gender = Female 167 < conf:(1) > lift:(1.44) lev:(0.2) [50] conv:(50.88)
3. Social = Sports 169 ==> Loan = Moderate 167 < conf:(0.99) > lift:(1.02) lev:(0.02) [3] conv:(1.98)
4. Social = High_Buyer Gender = Male 169 ==> Loan = High 167 < conf:(0.99) > lift:(1.02) lev:(0.02) [3] conv:(1.98)
5. Social = Sports 169 ==> Loan = No Gender = Female 167 < conf:(0.99) > lift:(1.45) lev:(0.2) [52] conv:(18.04)
6. Gender = Male 178 ==> Loan = Moderate 174 < conf:(0.98) > lift:(1.01) lev:(0.01) [2] conv:(1.25)
7. Loan = No Gender = Female 174 ==> Social = Sports 167 < conf:(0.96) > lift:(1.45) lev:(0.2) [52] conv:(7.39)
8. Location = Urban 182 ==> Loan = High 173 < conf:(0.95) > lift:(0.99) lev:(-0.01) [-2] conv:(0.64)
9. Gender = Male 178 ==> Social = Books 169 < conf:(0.95) > lift:(1.44) lev:(0.2) [51] conv:(6.05)

Table 1 Accuracy of different algorithms

Algorithm	Total instances	Correct instances	Incorrect result	% Accuracy
K-means clustering	1000	893	107	89.3
Random forest	1000	899	101	89.9
Apriori without SVM	1000	856	144	85.6
Apriori with SVM	1000	921	79	92.1

10. Gender = Female 178 ==> Social = Sports Loan = High 167 < conf:(0.94) > lift:(1.44) lev:(0.2) [50] conv:(5.16)

In these rules, support and confidence are major factors which help in the understanding pattern of E-marketing data. Using this analysis, we can find out the interest of the user, or the product which he most likely to buy. This algorithm can be used recursively to obtain basic information about users by rule mining.

Result Accuracy:

For the given data, different approaches were tested to find the best accuracy rate of the model. Few best possible results came with the Apriori algorithm without SVM, K-means clustering and random forest. Data is applied to each algorithm to find out the accuracy of given datasets. For testing purpose, real-time data is used on those different machine learning model. Around 1000 different samples were tested on the model. Aim of this process is to find an algorithm which is best fit to underlying data. After applying this dataset to the different algorithm, accuracy was recorded which is present in the table.

Table 1 shows the accuracy of the given dataset with *K*-means, random forest and Apriori algorithm. It is observed that the accuracy of Apriori with SVM is highest which is 92.1%. Accuracy of random forest and clustering is also considerable, as in both of them SVM pre-processing is used. For this, it is also observed that when support vector machine is not used, accuracy dropped to a great extent. This indicates that accuracy of the underlying dataset is high when support vector machine and natural language processing are used as data pre-processing and Apriori algorithm is used for association rule mining of the given dataset.

8 Conclusion and Future Work

We conclude that kernels support vector machine and association rule mining are used for enhancement in E-marketing strategy. Natural language processing and support vector machine is used on raw data. Association rules are generated with an accuracy of around 92.1% which can predict the trend for new inputs. It is found that using Twitter data and kernel-based support vector machine, accuracy is improved, and association rules have more accuracy rate than clustering and classification algorithm.

For future research: Different technical analysis techniques can be implemented which can provide great insight to data and also can improve the accuracy of the model. Along with different pre-processing algorithm and datasets with more attributes can be used in analysis. These techniques can further improve the accuracy of this model.

References

1. Demir, B., & Bruzzone, L. (2015). Fast and accurate image classification with histogram based features and additive kernel SVM. In *2015 IEEE International Geoscience and Remote Sensing Symposium (IGARSS)*. IEEE.
2. Wang, C., Wang, X. C., & Wamg, Y. Q. (2009). Research on E-shopper purchase behavior based on linguistic values of attributes of database. In *2009 International Conference on Management Science and Engineering*. IEEE.
3. Zhang, C. L., Liu, Z. F., & Yin, J. R. (2010). The Application Research on Web Log Mining in E-Marketing. In *2010 2nd International Conference on E-business and Information System Security*. IEEE.
4. Teng, G., Su, H., Liu, X., & Wang, T. (2006). Mining and sharing heterogeneous E-marketing intelligence-a universal metadata-based approach. In *1st International Conference on Innovative Computing, Information and Control-Volume I (ICICIC'06)* (Vol. 1). IEEE.
5. Zhang, G., Zhou, F., Wang, F., & Luo, J. (2008). Knowledge creation in marketing based on data mining. In *2008 International Conference on Intelligent Computation Technology and Automation*. IEEE.
6. Banati, H., & Bajaj, M. (2012). Promoting products online using firefly algorithm. In *2012 12th International Conference on Intelligent Systems Design and Applications (ISDA)*. IEEE.
7. Zeng, H., & Pan, D. (2010). A knowledge discovery and data mining process model in E-marketing. In *2010 8th World Congress on Intelligent Control and Automation*. IEEE.
8. Yang, J. Y., Myung, J., & Lee, S. G. (2009). The method for a summarization of product reviews using the user's opinion. In *2009 International Conference on Information, Process, and Knowledge Management*. IEEE.
9. Naredi, S., & Deshmukh, R. A. (2015). Improved extraction of quantitative rules using best M positive negative association rules algorithm. In *2015 IEEE International Conference on Electronics, Computing and Communication Technologies (CONECCT)*. IEEE.
10. Ouamour, S., & Sayoud, H. (2012). Authorship attribution of ancient texts written by ten arabic travelers using a SMO-SVM classifier: SihamOuamour. In *2012 International Conference on Communications and Information Technology (ICCIT)*. IEEE.
11. Wang, X. (2016). Application research on marketing data analysis using data mining technology. In *2016 International Conference on Smart Grid and Electrical Automation*. IEEE.
12. Shen, Y., Liu, J., & Shen, J. (2010). The further development of weka base on positive and negative association rules. In *2010 International Conference on Intelligent Computation Technology and Automation*. IEEE.
13. Wang, Z., & Qu, Z. (2017). Research on Web text classification algorithm based on improved CNN and SVM. In *2017 IEEE 17th International Conference on Communication Technology (ICCT)*. IEEE.

Bernoulli Polynomials and Ramanujan Summation

R. Sivaraman

Abstract The concept of Bernoulli numbers has appeared in several counting problems in mathematics. Bernoulli numbers are vastly used in many mathematical problems, especially dealing with Riemann zeta function, as Leonhard Euler in the eighteenth century, proved that the zeta function values for positive even integers are connected to even subscript Bernoulli numbers. As a natural generalization of Bernoulli numbers, we discuss Bernoulli polynomials. Bernoulli numbers occur as constant terms of the respective Bernoulli polynomials. This paper introduces the concept of Bernoulli numbers and Bernoulli polynomials and discusses few properties concerning them. In Sect. 2.1, the relation between successive Bernoulli numbers is obtained. This will help us in generating successive Bernoulli numbers from the initial Bernoulli number $B_0 = 1$. Using these Bernoulli numbers, we obtain a nice formula for generating successive Bernoulli polynomials in Sect. 3.1. In Sect. 4.1, the concept that the Bernoulli numbers are obtained by corresponding polynomials of Bernoulli at $t = 1$ has been established. Finally, in Sect. 5.1, the connection between Riemann zeta function and Bernoulli polynomials has been proved. Further, five figures are included to justify the truth of Sect. 4.1. The main focus of this paper is to establish the relation between Bernoulli polynomials and Ramanujan Summation.

Keywords Bernoulli numbers · Bernoulli polynomials · Cauchy product · Riemann zeta function · Ramanujan summation

1 Introduction

The idea of various types of summation methods has been a key point of study for nearly three centuries now. Mathematicians like Nicole Oresme, Guido Grandi, Jacob Bernoulli, Leonhard Euler, Ernesto Cesaro investigated several types of series of real numbers and had provided plenty of ways to sum a given series whose terms are real numbers. In particular, Italian mathematician Ernesto Cesaro found a clever method

---

R. Sivaraman (✉)
Department of Mathematics, D. G. Vaishnav College, Chennai, India
e-mail: rsivaraman1729@yahoo.co.in

S.-L. Peng et al. (eds.), *Proceedings of First International Conference on Mathematical Modeling and Computational Science*, Advances in Intelligent Systems and Computing 1292, https://doi.org/10.1007/978-981-33-4389-4_44

of attaining certain real numbers to particular divergent series of real numbers. Due to this concept, vast amount of interest and research in summability theory kick started and several mathematicians beginning from nineteenth century took the task of finding ways to assign finite values to divergent series. One such method was proposed by one of the greatest Indian mathematicians of twentieth century Srinivasa Ramanujan. Today, it was called as "Ramanujan Summation". In recent years, there was huge interest among everyone to know how Ramanujan could have devised such ingenious equations. Two recent papers by the corresponding author [1, 2] discuss Ramanujan Summation in detail which provides a good way of knowing how Ramanujan could have possibly thought about such a method. However, in this paper, some elementary properties of Bernoulli polynomials are proved which in turn are used to prove the Ramanujan Summation method.

For knowing more about sums of powers of natural numbers and Bernoulli numbers and Bernoulli polynomials, see [3–8]. We first begin our discussion with well known Bernoulli numbers named after Swiss mathematician Jacob Bernoulli.

2 Definition

The coefficients of $\frac{x^n}{n!}$ in Maclaurin's series expansion of $\frac{x}{e^x-1}$ are defined as Bernoulli numbers. Let the nth Bernoulli number be denoted by B_n (see [9]).

Hence, we get

$$\frac{x}{e^x - 1} = \sum_{n=0}^{\infty} B_n \frac{x^n}{n!} \tag{2.1}$$

Since the constant term $\frac{x}{e^x-1}$ is 1, from (2.1) we find that $B_0 = 1$. We first prove a theorem related with successive Bernoulli numbers (see [10]).

2.1 Theorem 1

The Bernoulli numbers B_k satisfy

$$\sum_{k=0}^{n} \binom{n+1}{k} B_k = 0 \tag{2.2}$$

where $B_0 = 1$.

Proof From Eq. (2.1), we have

$$x = \left(e^x - 1\right) \sum_{i=0}^{\infty} B_i \frac{x^i}{i!} = \sum_{j=0}^{\infty} \frac{x^{j+1}}{(j+1)!} \sum_{i=0}^{\infty} B_i \frac{x^i}{i!}$$

Using Cauchy product, we can simplify this equation (by considering i as k and j as $n-k$) as follows:

$$x = \sum_{n=0}^{\infty} \sum_{k=0}^{n} \frac{x^{n-k+1}}{(n-k+1)!} \frac{B_k x^k}{k!}$$

$$x = \sum_{n=0}^{\infty} \sum_{k=0}^{n} \frac{(n+1)! B_k}{(n-k+1)! \times k!} \cdot \frac{x^{n+1}}{(n+1)!} = \sum_{n=0}^{\infty} \sum_{k=0}^{n} \binom{n+1}{k} B_k \cdot \frac{x^{n+1}}{(n+1)!}$$

For $n = 0$, both sides co-efficient would be 1 (since $B_0 = 1$) and for all other values of n, since the left-hand side contains only first power of x, equating the powers of x (from second and higher powers) on both sides, we get (2.2) as desired.

2.2 *List of Bernoulli Numbers*

Using Eq. (2.2), we can list the Bernoulli numbers up to the required level. For example, to find B_1, considering $n = 1$, we find that $B_0 + 2B_1 = 0$. Since $B_0 = 1$, we get $B_1 = -\frac{1}{2}$.

Similarly, by substituting $n = 2, 3, 4, 5, 6, 7, 8, \ldots$ successively we find from (2.2) that the successive Bernoulli numbers are given by

$$B_0 = 1,\ B_1 = -\frac{1}{2},\ B_2 = \frac{1}{6},\ B_3 = 0,\ B_4 = -\frac{1}{30},\ B_5 = 0,$$
$$B_6 = \frac{1}{42},\ B_7 = 0,\ B_8 = -\frac{1}{30},\ B_9 = 0,\ B_{10} = \frac{5}{66}, \qquad (2.3)$$

Except B_1, all Bernoulli numbers of odd indices vanish (see [3, 10]).

3 Bernoulli Polynomials

The polynomials which occur as coefficients of $\frac{x^n}{n!}$ in Maclaurin's series expansion of $\frac{xe^{tx}}{e^x-1}$ are called Bernoulli polynomials. Let the nth Bernoulli polynomial be denoted by $B_n(t)$.

Hence, we have

$$\frac{xe^{tx}}{e^x - 1} = \sum_{n=0}^{\infty} B_n(t)\frac{x^n}{n!} \tag{3.1}$$

From (2.1) and (3.1), we find that $B_n(0) = B_n$.

The Bernoulli numbers are also values from evaluated Bernoulli polynomials at $t = 0$ (see [3, 11–13]).

We now prove an important theorem for finding Bernoulli polynomials (see [3]) through Bernoulli numbers.

3.1 Theorem 2

The nth polynomial of Bernoulli is given by the nth

$$B_n(t) = \sum_{k=0}^{n} \binom{n}{k} B_k t^{n-k} \tag{3.2}$$

Proof Using Eqs. (3.1) and (2.1), and definition of exponential series, we have

$$\sum_{n=0}^{\infty} B_n(t)\frac{x^n}{n!} = \frac{xe^{tx}}{e^x - 1} = \frac{x}{e^x - 1} \cdot e^{tx} = \sum_{i=0}^{\infty} B_i \frac{x^i}{i!} \cdot \sum_{j=0}^{\infty} \frac{(tx)^j}{j!}$$

Now, using the Cauchy product, we get

$$\sum_{n=0}^{\infty} B_n(t)\frac{x^n}{n!} = \sum_{n=0}^{\infty}\sum_{k=0}^{n} \frac{(tx)^{n-k}}{(n-k)!} \cdot \frac{B_k x^k}{k!} = \sum_{n=0}^{\infty}\sum_{k=0}^{n} \frac{n! t^{n-k} B_k}{(n-k)! \times k!} \cdot \frac{x^n}{n!}$$
$$= \sum_{n=0}^{\infty}\sum_{k=0}^{n} \binom{n}{k} B_k t^{n-k} \cdot \frac{x^n}{n!}$$

Equating the like powers, we get.

$B_n(t) = \sum_{k=0}^{n} \binom{n}{k} B_k t^{n-k}$ as desired. This completes the proof.

3.2 First Few Examples

Here, we list few examples of Bernoulli polynomials (see [3]).
For $n = 0$,

$$B_0(t) = B_0 = 1 \tag{3.3}$$

When $n = 1$,

$$B_1(t) = \sum_{k=0}^{1} \binom{1}{k} B_k t^{1-k} = B_0 t + B_1 = t - \frac{1}{2} \tag{3.4}$$

When $n = 2$,

$$B_2(t) = \sum_{k=0}^{2} \binom{2}{k} B_k t^{2-k} = B_0 t^2 + 2B_1 t + B_2 = t^2 - t + \frac{1}{6} \tag{3.5}$$

When $n = 3$,

$$B_3(t) = \sum_{k=0}^{3} \binom{3}{k} B_k t^{3-k} = B_0 t^3 + 3B_1 t^2 + 3B_2 t + B_3 = t^3 - \frac{3}{2}t^2 + \frac{1}{2}t \tag{3.6}$$

Continuing in the same fashion, if $n = 4$, 5 and 6, we obtain

$$B_4(t) = B_0 t^4 + 4B_1 t^3 + 6B_2 t^2 + 4B_3 t + B_4 = t^4 - 2t^3 + t^2 - \frac{1}{30} \tag{3.7}$$

$$\begin{aligned} B_5(t) &= B_0 t^5 + 5B_1 t^4 + 10B_2 t^3 + 10B_3 t^2 + 5B_4 t \\ &\quad + B_5 = t^5 - \frac{5}{2}t^4 + \frac{5}{3}t^3 - \frac{1}{6}t \end{aligned} \tag{3.8}$$

$$\begin{aligned} B_6(t) &= B_0 t^6 + 6B_1 t^5 + 15B_2 t^4 + 20B_3 t^3 + 15B_4 t^2 \\ &\quad + 6B_5 t + B_6 = t^6 - 3t^5 + \frac{5}{2}t^4 - \frac{1}{2}t^2 + \frac{1}{42} \end{aligned} \tag{3.9}$$

The first seven Bernoulli polynomials are (3.3)–(3.9).

4 Bernoulli Polynomials Evaluated at $t = 1$

By comparing (2.1) and (3.1), we saw that $B_n(0) = B_n$. for all natural numbers n. In the following theorem, we try to prove a similar property when evaluated at $t = 1$.

4.1 Theorem 3

For any natural number n,

$$B_{n+1}(1) = B_{n+1} \tag{4.1}$$

Proof From Eq. (3.2) in Sect. 3.1, we have $B_n(t) = \sum_{k=0}^{n} \binom{n}{k} B_k t^{n-k}$.

Equation (2.2) in Sect. 2.1, we have $\sum_{k=0}^{n} \binom{n+1}{k} B_k = 0$

Now using these two equations, we obtain.

$B_{n+1}(1) = \sum_{k=0}^{n+1} \binom{n+1}{k} B_k = \sum_{k=0}^{n} \binom{n+1}{k} B_k + B_{n+1} = 0 + B_{n+1} = B_{n+1}.$

This completes the proof.

5 Relation of Bernoulli Polynomials and Ramanujan Summation

As presented in [1] and [2], by the corresponding author regarding Ramanujan Summation, we find that

$$\zeta(-n) = 1^n + 2^n + 3^n + \cdots = -\frac{B_{n+1}}{n+1} \tag{5.1}$$

We now prove a simple theorem connecting Bernoulli polynomials and Ramanujan Summation method as presented in Eq. (5.1). Here, ζ is the Riemann zeta function.

5.1 Theorem 4

For any natural number n,

$$(n+1)\zeta(-n) = -\left[B_{n+1}(t)\right]_{t=0} = -\left[B_{n+1}(t)\right]_{t=1} \tag{5.2}$$

Proof We know that $B_{n+1}(0) = B_{n+1}$ for all natural numbers n and from Eq. (4.1) in Sect. 4.1, we have $B_{n+1}(1) = B_{n+1}$. Now, from (5.1), we have

$$(n+1)\zeta(-n) = 1^n + 2^n + 3^n + \cdots = -B_{n+1} = -B_{n+1}(0) = -\left[B_{n+1}(t)\right]_{t=0}$$
$$(n+1)\zeta(-n) = 1^n + 2^n + 3^n + \cdots = -B_{n+1} = -B_{n+1}(1) = -\left[B_{n+1}(t)\right]_{t=1}$$

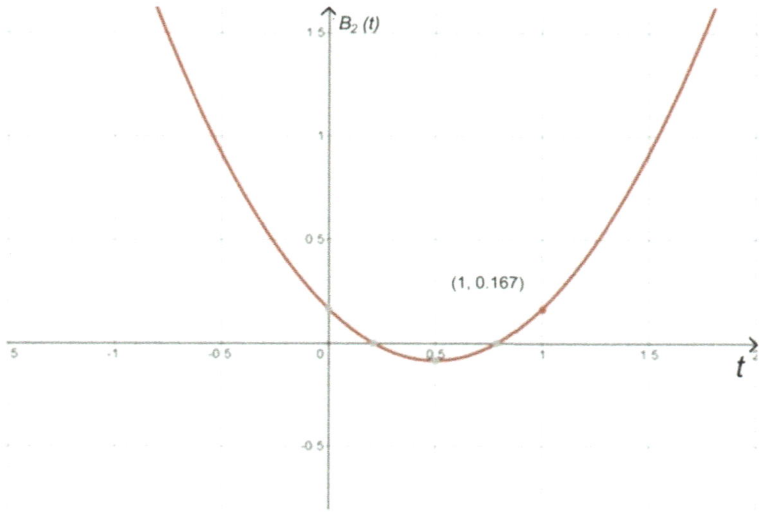

Fig. 1 Graph of $B_2(t) = t^2 - t + \frac{1}{6}$ near $t = 1$. We see that $B_2(1) = \frac{1}{6} = B_2$

This completes the proof.

6 Viewing Geometrically

The equation $B_{n+1}(0) = B_{n+1}$ is obvious in comparison with (2.1) and (3.1) through their basic definitions.

In Sect. 4.1 (Eq. (4.1)), we proved that $B_{n+1}(1) = B_{n+1}$. We will now view graphs of B_{n+1} near the neighbourhood of $t = 1$ to see visually (geometrically) that this is so.

Thus, through Figs. 1, 2, 3, 4 and 5, we have verified the fact that $B_{n+1}(1) = B_{n+1}$.

7 Conclusion

There are several ways of proving Ramanujan Summation formula as given in Eq. (5.1). For instance, we can prove it through Riemann integrals or by using generating functions. Many mathematicians had done Ramanujan Summation method using complex function theory where Euler zeta function is analytically extended from real axis to Riemann zeta function in the complex plane. The main advantage in this paper is that we have used just real-valued functions in the form of simple polynomials called Bernoulli polynomials to understand the connection between

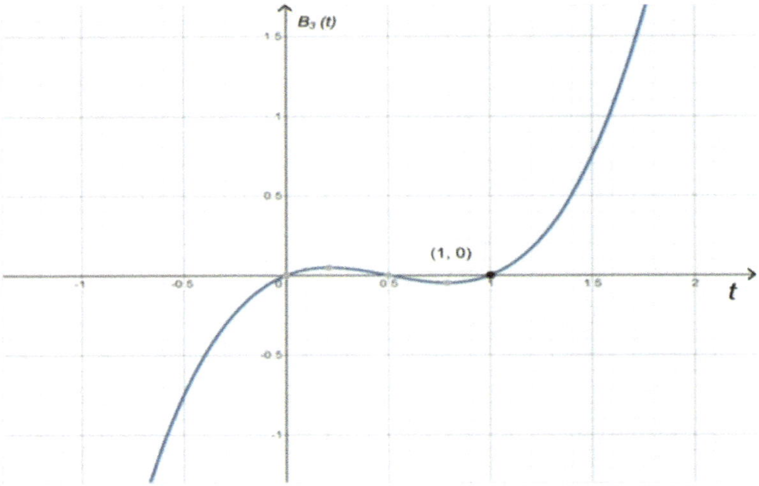

Fig. 2 Graph of $B_3(t) = t^3 - \frac{3}{2}t^2 + \frac{1}{2}t$. We see that $B_3(1) = 0 = B_3$

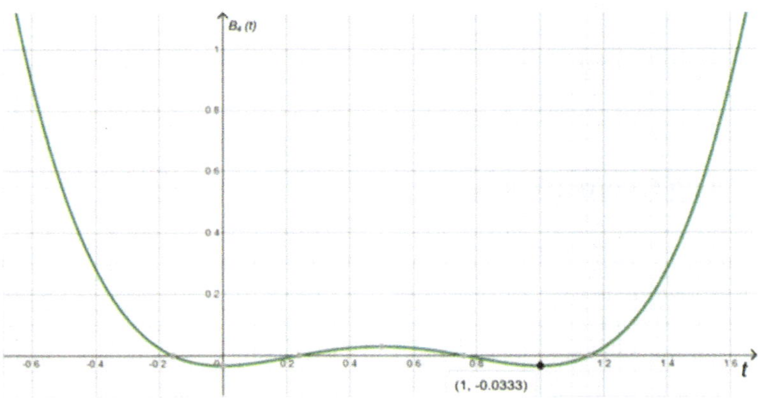

Fig. 3 Graph of $B_4(t) = t^4 - 2t^3 + t^2 - \frac{1}{30}$. We see that $B_4(1) = -\frac{1}{30} = B_4$

Bernoulli polynomials and Ramanujan Summation. Several other properties can be proved using Bernoulli polynomials discussed in Sect. 3 of this paper. Using those properties, there are ample possibilities to prove several other summation formulas. The inter-link between various concepts of mathematics is inherited in this paper. The inter-relationship between two unexpected concepts of mathematics is what makes mathematical research a fruitful experience.

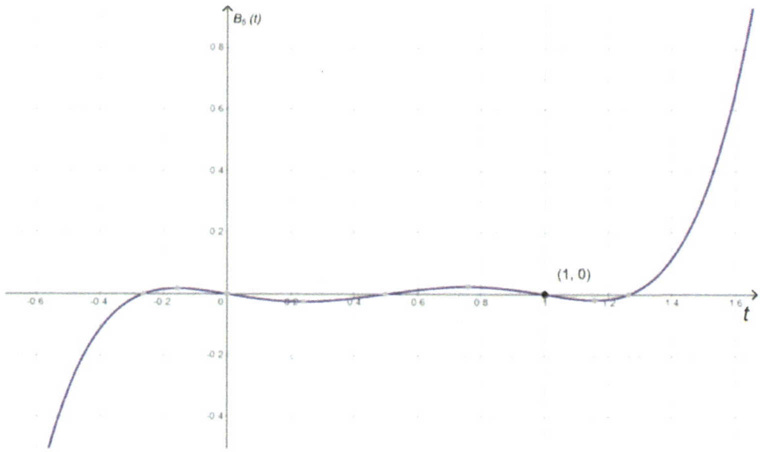

Fig. 4 Graph of $B_5(t) = t^5 - \frac{5}{2}t^4 + \frac{5}{3}t^3 - \frac{1}{6}t$. We see that $B_5(1) = 0 = B_5$

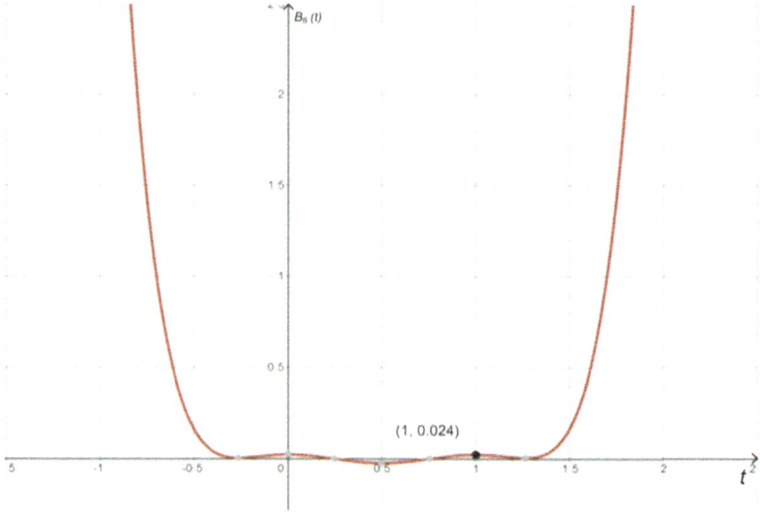

Fig. 5 Graph of $B_6(t) = t^6 - 3t^5 + \frac{5}{2}t^4 - \frac{1}{2}t^2 + \frac{1}{42}$. We see that $B_6(1) = \frac{1}{42} = B_6$

References

1. Sivaraman, R. (2020). Understanding Ramanujan Summation. *International Journal of Advanced Science and Technology, 29*(7), 1472–1485.
2. Sivaraman, R.. (2020). Remembering Ramanujan. *Advances in Mathematics: Scientific Journal*, 489–506.
3. Sivaraman, R. (2020). Properties of Bernoulli polynomials. *J. Math. Comput. Sci., 10*(5), 2025–2032.

4. Sivaraman, R. (2020). Summing through integrals. *Science Technology and Development Journal*, *9*, 267–272.
5. Sivaraman, R. (2020). Sum of powers of natural numbers.*AUT AUT Research Journal*, *11*(4), 353–359.
6. Kim, M. S., & Son, J. W. (2000). On Bernoulli numbers. *Journal of the Korean Mathematical Society*, *37*(3), 391–410.
7. Mazur, B. (2008). Bernoulli numbers and the unity of mathematics.
8. Coen, L. E. S. (1996). Sums of powers and the Bernoulli numbers.
9. Kim, M. S. (2011). A note on sums of products of Bernoulli numbers. *Applied Mathematics Letters*, *24*(1), 55–61.
10. John H. C., Richard K. G. (1998). *The book of numbers*, New York: Springer.
11. Açikgöz, M., Erdal, D., & Araci, S. (2010). A new approach to q-Bernoulli numbers and q-Bernoulli polynomials related to q-Bernstein polynomials. *Advances in Difference Equations*.
12. Momenzadeh, M., & Kakangi, I. Y. (2017). Approximation Properties of-Bernoulli Polynomials. In *Abstract and Applied Analysis* (Vol. 2017). Hindawi.
13. Bahari, H. (2006). *A Generalization of the Bernoulli Numbers*, Doctoral dissertation, Universiti Putra Malaysia.

Impact of MHD Stream Past an Exponentially Inclined Vertical Plate of First-Order Chemical Response with Variable Mass Diffusion and Thermal Radiation

V. Usharani, A. Selvaraj, R. Constance Angela, and A. Neel Armstrong

Abstract In this issue, an appraisal of temperamental MHD regular convective stream, move of mass and radiation past buoy exponentially duplicated willing plate embedded in variable temperature and consideration inside seeing an inclined attractive region has been done. The development of this issue is to dissect the effect of a slanted attractive order at the drift wonders with a heat source and dangerous reaction for exponentially extended slanted plate. The fundamental profiles are read for unmistakable real parameters comprehensive of the Prandtl assortment, warm Grashof run, mass Grashof amount, Schmidt number, and administering conditions have been perceived and understood through utilizing Laplace rework method. The estimation of float speed, mindfulness and temperature is obtained graphically. It is intriguing and found that the expense of augmentation of tendency edge of an executed attractive region bringing down speed profile.

Keywords MHD · Exponential · Transmission · Inclined standing plate · Mass · Transfer

---

V. Usharani · A. Selvaraj (✉) · R. Constance Angela · A. Neel Armstrong
Department of Mathematics, Vels Institute of Science, Technology and Advanced Studies, Chennai 600117, India
e-mail: aselvaraj_ind@yahoo.co.in

V. Usharani
e-mail: ur709185@gmail.com

R. Constance Angela
e-mail: constance_angela@hotmail.com

A. Neel Armstrong
e-mail: armsambomaths@gmail.com

A. Neel Armstrong
Department of Mathematics, Veltech Rangarajan Dr. Sagunthala R and D Institute of Science and Technology, Chennai 600062, India

S.-L. Peng et al. (eds.), *Proceedings of First International Conference on Mathematical Modeling and Computational Science*, Advances in Intelligent Systems and Computing 1292, https://doi.org/10.1007/978-981-33-4389-4_45

1 Introduction

Inflow situation magnetohydrodynamics move issues play an indispensable situation inside the recollect of innovation and period; the issue of free convection MHD float of a gooey not compressible electrically principal fluid with coordinated results of radiation action has been awesome thought in light of the reality of its tremendous outcomes on the limit layer liquid oversee[1, 2]. Differing examinations were accomplished right now various researchers practice scope of go with the drift conditions; their dimensionless overseeing conditions had been understood using way of the standard Laplace seriously change.

Approach [3]. Convection heat move takes zone among a surface and a moving liquid, while they're at remarkable temperatures. In a severe vibe, convection isn't generally an essential method of warmth move as the glow change from the floor to the liquid comprises of two components running at the same time. The essential one is the power switch because of sub-atomic development (conduction) through a liquid layer connecting to the surface [4, 5].

While liquid streams over a surface, its pace and temperature contiguous the floor are indistinguishable as that of the surface on account of the no-slip condition. The speed and temperature far away from the surface may keep on being unaffected [6].

1.1 Related Works

Chambré et al. [2] dissected a first-request concoction reaction inside the area of a level plate, and view of warmth and mass switch issues with compound reaction is of top-notch reasonable significance to designers and researchers as a result of its about conventional rate in heaps of parts of science and building. Some advisor fields of diversion wherein consolidated warmness and mass switch related to compound response assume a significant job are found in concoction strategy ventures comprising of food preparing and polymer creation [7, 8].

Soundalgekar [9] mentioned on the results of a mass switch-on go with the flow past a uniformly accelerated vertical plate and studied a particular investigation of the effects of mass exchange on the Stokes problem for a boundless perpendicular plate has been introduced on thinking about the loose-convection flows [10, 11]. It has been visible that there's an ascent in the quickness because of the nearness of an out of doors mass. Anyways, a variety in Schmidt number prompts a fall within the velocity. The pores and skin-erosion increase due to the nearness of a far-flung mass.

Muthukumaraswamy et al. [10, 12] tried compound reaction results on a vertical influencing plate with the changeable warm temperature and alluded to the effect of pivot and substance reaction of the first request on precarious hydromagnetic free convection stream of a liquid past a consistently expanded isothermal vertical plate with variable mass dissemination and explored concentrated radiative development past a symbolic initiated isothermal opposite plate with uniform mass progress conclusive outcomes imagined that speed increases with lessening estimations of the overall quite comfortable radiation parameter [12, 13]. With creating warm Grashof wide assortment or mass Grashof broad range anyway the style is essentially turned around with perceive to the warm radiation Criterion or Schmidt go. The temperature of the plate upward pushes with the decrease estimations of the warm radiation parameter. The plate mindfulness upward pushes with decrease estimations of the Schmidt range. The test is unmistakably turned around in regards to the warm Grashof amount or mass Grashof.

Rajput and Gaurav Kumar [14] examine the consequences of radiation action, concoction reaction and permeability of system on the insecure progression of a thick, not compressible and current captivating in liquid past an exponentially expanded perpendicular plate with changeable divider temperature and constant mass dispersion. Anyway, it's far proposed to find about the results of MHD accept the way things are with an exponentially beginning development of an innumerable isothermal vertical plate within the sight of warm radiation and the synthetic chemical reaction has been studied [9, 15–17]. Those procedures show up in nuclear reactor security and consuming systems, picture voltaic finders as it ought to be as substance and metallurgical planning.

Raptis et al. [16] explored MHD with the transfer heat through a fluid stream past a quickened lamina perpendicular plate with the movable area and warmth transition. Azzam [1] explained an assessment about the warm diffraction of MHD convection shaft closet a semi-bondless transferring vertical plate now the heat sorts are quite high. Nield et al. [13] consider the related to normal convection monitoring stream or mass switch over between excess regarding a couple of geometrical our bodies. Ramachandra et al. [15] considered the high-energy particle caused ionization and mass switch results on a precarious MHD unrestricted convective skim past a warmed perpendicular plate in a permeable system with thick dispersal. Mbeledogu and Ogulu [6] inspected uncertain magneto fluid dynamics regular convection movement within the fluid warmth and mass switch stream of a transmitting fluid past a perpendicular permeable plate system inside seeing warm radiation. Muthukumaraswamy et al. [8] study the interaction between the magnetic field and electrically conducting fluids which is the self-control that examines the elements of electrically directing liquids. The significant method of magnetohydrodynamics that is very fair are a possible to catch about flows in a movable conduction fluid, which makes powers on the liquid, and changes attractive subject parameter. Defeat et al. (2013) investigate the effect of compound impact and radiation effect on MHD liquid skim way a permeable model in a regular attractive control with changeable temperature and mass dissemination with warmness flexibly has been more and more growing past an exponentially quickened vertical plate. Manay et al. [5] an MHD weight or MHD propulsor is a methodology for impelling seagoing vessels the utilization of exclusively electric controlled and attractive fields with no moving parts. The working statute involves a charge of the force (gas or water), which would then be able to be coordinated through an attractive field, pushing the vehicle the opposite way [18]. MHD is utilized to balance out a stream contrary to the progress from laminar to fierce accept circumstances for what they are. Raju et al. [17] dissected shaky MHD drift underneath unique designs. Muthukumarasamy et al. (2015) contemplated the principal request substance reaction effect of more and more exponentially fast able perpendicular plate with changeable floor temperature and mass dispersion [14]. Endalew et al. [4] concentrated in the matter of precarious MHD regular convection of warmth move with a slanted plate in the permeable medium. These procedures appear in nuclear reactor security and consuming structures, photovoltaic finders as fittingly as substance and metallurgical planning.

2 Mathematical Formulation

In this problem, we don't forget and think about an unsteady go with the flow of a viscous incompressible fluid with an exponentially elevated a countless isothermal willing vertical plate in the presence of thermal temperature diffusion, and primary order chemical action has been studied. The magnetic subject $B = (B_0 \cos\phi, B_0 \sin\phi, 0)$ is corrected to make a tendency attitude angle of ϕ and a magnetic discipline of uniform energy B_0 is utilized perpendicular to the plate. The fluid is believed to be x'-axis is taken alongside the plate in the vertically upward course and the y-axis is examined perpendicular to the plate at a time $t' \leq 0$, the plate and fluid are at the same temperature T_∞ and concentration C'_∞ at a time $t' \geq 0$. The plate is started with a speed $q = q_0 e^{wt}$ in its very own airplane toward the gravitational field, the temperature from the plate is increased to T_{w} and the attention ranges close to the plate are additionally raised to C'_w with time t, the fluid is viewed a gray very interest produced electromagnetic waves, this viscous dissipation is thought to be not considered. The impact is thought to identify the region virtually within the move. Then below normal Boussinesq's estimation, the no steady initial movement is dominated with the aid of using the subsequent non-dimensional equations

$$\frac{\partial q'}{\partial t} = g\beta(T - T_\infty)\cos\phi + g\beta^*(C' - C'_\infty)\cos\phi + v\frac{\partial^2 q'}{\partial y^2} - \frac{\sigma\beta_0^2 q'}{\rho} \tag{1}$$

$$\frac{\partial \theta}{\partial t} = k\frac{\partial^2 \theta}{\partial y^2} - \frac{\partial q_{\mathrm{r}}}{\partial y} - h(T - T_\infty) \tag{2}$$

$$\frac{\partial c}{\partial t} = \frac{\partial^2 c}{\partial Y^2} - K_1(C' - C'_\infty) \tag{3}$$

With the help of preliminary and necessary boundary conditions

$$q = 0, T = T_\infty, C' = C'_\infty \text{ for all } y, t' \leq 0$$

$$t' > 0, q = q_0\left(e^{w't'}\right), T = T_\infty + (T_{\mathrm{w}} - T_\infty)\alpha t'$$

$$C' = C'_\infty + (C''_w - C'_\infty)\alpha t', C' = C'_{\mathrm{w}} \text{ at } y = 0,$$

$$q \to 0, \quad T \to T_\infty, \quad C' \to C'_\infty \quad \text{at } y \to \infty \text{ and } A = \frac{q_0^2}{v} \tag{4}$$

Initiating the suitable non-dimensionless requires quantities

$$q' = q\left(\frac{q_0}{v_2}\right)^{1/3}, t = t'\left(\frac{q_0^2}{v}\right)^{\frac{1}{3}}, Y = y\left(\frac{q_0}{v^2}\right)^{\frac{1}{3}},$$
$$\theta = \frac{T - T_\infty}{T_w - T_\infty}, G_r = \frac{g\beta(T_w - T_\infty)}{q_0}, C = \frac{C' - C'_\infty}{C'_w - C'_\infty},$$
$$G_c = \frac{g\beta^*(C'_w - C'_\infty)}{u_0} M = \frac{\sigma B_0^2}{\rho}\left(\frac{v}{q_0^2}\right)^{\frac{1}{3}},$$
$$P_r = \frac{\mu C_p}{k}, S_c = \frac{v}{D}, K_2 = K_1\left(\frac{v}{q_0^2}\right)^{1/3},$$
$$K = k_0\left(\frac{q_0}{v}\right)^2, R = \frac{16\sigma^* T_\infty^3}{3k_p k}, Y = \frac{q_0 y}{v}, N = \frac{vh}{q_0^2 \rho_c p}, \tag{5}$$

By use of the Rosseland approximation $q_r = \frac{4\sigma^*}{3k_p}\frac{\partial T^4}{\partial y}$ where σ^* and k_p Stefan–Boltzmann not changeable constant and the Rosseland mean immersion constant quantity, Taylor series fail to provide that higher-order terms we can given T^4 as a linear characteristic temperature of the format style $T^4 = 4T_\infty^3 - 3T_\infty^4$.

We get $\frac{\partial q_r}{\partial y} = -\frac{16\sigma^* T_\infty^3}{3k_p}\frac{\partial^2 T}{\partial y^2}$.

Equations (1)–(3) reduce the following non-dimensional forms

$$\frac{\partial q'}{\partial t} = G_r\theta \cos\phi + \mathrm{Gc}C \cos\phi + \frac{\partial^2 q'}{\partial Y^2} - Mq' \tag{6}$$

$$\frac{\partial \theta}{\partial t} = \left(\frac{1+R}{P_r}\right)\frac{\partial^2 \theta}{\partial Y^2} - N\theta \tag{7}$$

$$\frac{\partial q'}{\partial t} = \frac{1}{S_c}\frac{\partial^2 c}{\partial Y^2} - K_2 C \tag{8}$$

Define the error function

$$L^{-1}\left\{\frac{e^{-c\sqrt{(S+b)}}}{s-a}\right\} = \frac{e^{at}}{2}\left[e^{(-\sqrt{a+b})\mathrm{erfc}\left(\frac{c}{2\sqrt{t}} - \sqrt{(a+b)t}\right)} + e^{(\sqrt{a+b})\mathrm{erfc}\left(\frac{c}{2\sqrt{t}} + \sqrt{(a+b)t}\right)}\right] \tag{9}$$

where $\mathrm{erf}(t) = \frac{2}{\sqrt{\pi}}\int_t^\infty e^{-u^2}\mathrm{d}u$.

The initial and boundary conditions in non-dimensional forms

$$\begin{aligned} & q = 0, \theta = 0, C = 0 \text{ for all } y, t \leq 0, \\ & t > 0 : q = e^{wt}, \theta = t, C = t \text{ at } y = 0 \\ & q \to 0, \theta \to 0, \quad C \to 0 \text{ as } y \to \infty \end{aligned} \tag{10}$$

3 Solution Method

The non-dimensional governing equations (6)–(8) difficulty to corresponding preliminary and boundary prerequisites means (4) are getting by necessary usage of the Laplace transform change technique, the answer is acquired below.

$$\begin{aligned} q(t) = & \frac{e^{wt}}{2}\Big[e^{-2\eta\sqrt{(M+w)t}}\operatorname{erfc}\Big(\eta - \sqrt{(M+w)t}\Big) \\ & + e^{-2\eta\sqrt{(M+w)t}}\operatorname{erfc}\Big(\eta + \sqrt{(M+w)t}\Big)\Big] \\ & - \frac{G_r\cos\phi}{a^2(1-A)}\frac{1}{2}\Big[\Big\{e^{-2\eta\sqrt{ANt}}\operatorname{erfc}\Big(\eta\sqrt{A} - \sqrt{Nt}\Big) \\ & + e^{2\eta\sqrt{ANt}}\operatorname{erfc}\Big(\eta\sqrt{A} + \sqrt{Nt}\Big)\Big\} \\ & - \Big\{e^{-2\eta\sqrt{Mt}}\operatorname{erfc}\Big(\eta - \sqrt{Mt}\Big) + e^{2\eta\sqrt{Mt}}\operatorname{erfc}\Big(\eta + \sqrt{Mt}\Big)\Big\}\Big] \\ & - \frac{G_r\cos\phi}{a(1-A)}\Big[\frac{t}{2}\Big\{e^{-2\eta\sqrt{ANt}}\operatorname{erfc}\Big(\eta\sqrt{A} - \sqrt{Nt}\Big) \\ & + e^{2\eta\sqrt{ANt}}\operatorname{erfc}\Big(\eta\sqrt{A} + \sqrt{Nt}\Big)\Big\} \\ & - \frac{\eta\sqrt{At}}{2\sqrt{N}}\Big\{e^{-2\eta\sqrt{ANt}}\operatorname{erfc}\Big(\eta\sqrt{A} - \sqrt{Nt}\Big) \\ & - e^{2\eta\sqrt{ANt}}\operatorname{erfc}\Big(\eta\sqrt{A} + \sqrt{Nt}\Big)\Big\} \\ & - \frac{t}{2}\Big\{e^{-2\eta\sqrt{Mt}}\operatorname{erfc}\Big(\eta - \sqrt{Mt}\Big) + e^{2\eta\sqrt{Mt}}\operatorname{erfc}\Big(\eta + \sqrt{Mt}\Big)\Big\} \\ & - \frac{\eta\sqrt{t}}{2\sqrt{M}}\Big\{e^{-2\eta\sqrt{Mt}}\operatorname{erfc}\Big(\eta - \sqrt{Mt}\Big) - e^{2\eta\sqrt{Mt}}\operatorname{erfc}\Big(\eta + \sqrt{Mt}\Big)\Big\}\Big] \\ & + \frac{G_r\cos\phi}{(1-A)a^2}\frac{e^{at}}{2}\Big[\Big\{e^{-2\eta\sqrt{A(a+N)t}}\operatorname{erfc}\Big(\eta\sqrt{A} - \sqrt{(a+N)t}\Big) \\ & + e^{2\eta\sqrt{A(a+N)t}}\operatorname{erfc}\Big(\eta\sqrt{A} + \sqrt{(a+N)t}\Big)\Big\} \\ & - \Big\{e^{-2\eta\sqrt{(a+M)t}}\operatorname{erfc}\Big(\eta\sqrt{M} - \sqrt{at}\Big) \end{aligned}$$

$$
\begin{aligned}
&+ e^{2\eta\sqrt{(a+M)t}}\mathrm{erfc}\left(\eta\sqrt{M} + \sqrt{at}\right)\Big\}\Big]\\
&- \frac{G_c \cos\phi}{b^2(1-S_c)}\frac{1}{2}\Big[\Big\{e^{-2\eta\sqrt{S_c K_2 t}}\mathrm{erfc}\left(\eta\sqrt{S_c} - \sqrt{K_2 t}\right)\\
&+ e^{2\eta\sqrt{S_c K_2 t}}\mathrm{erfc}\left(\eta\sqrt{S_c} + \sqrt{K_2 t}\right)\Big\}\\
&-\Big\{e^{-2\eta\sqrt{Mt}}\mathrm{erfc}\left(\eta - \sqrt{Mt}\right) + e^{2\eta\sqrt{Mt}}\mathrm{erfc}\left(\eta + \sqrt{Mt}\right)\Big\}\Big]\\
&- \frac{G_c \cos\phi}{b^2(1-S_c)}\Big[\frac{t}{2}\Big\{e^{-2\eta\sqrt{S_c K_2 t}}\mathrm{erfc}\left(\eta\sqrt{S_c} - \sqrt{K_2 t}\right)\\
&+ e^{2\eta\sqrt{S_c K_2 t}}\mathrm{erfc}\left(\eta\sqrt{S_c} + \sqrt{K_2 t}\right)\\
&- \frac{\eta\sqrt{S_c t}}{2\sqrt{K_2}}\Big\{e^{-2\eta\sqrt{S_c K_2 t}}\mathrm{erfc}\left(\eta\sqrt{S_c} - \sqrt{K_2 t}\right)\\
&- e^{2\eta\sqrt{S_c K_2 t}}\mathrm{erfc}\left(\eta\sqrt{S_c} + \sqrt{K_2 t}\right)\Big\}\\
&- \frac{t}{2}\Big\{e^{-2\eta\sqrt{Mt}}\mathrm{erfc}\left(\eta - \sqrt{Mt}\right) + e^{2\eta\sqrt{Mt}}\mathrm{erfc}\left(\eta + \sqrt{Mt}\right)\Big\}\\
&- \frac{\eta\sqrt{t}}{2\sqrt{M}}\Big\{e^{-2\eta\sqrt{Mt}}\mathrm{erfc}\left(\eta - \sqrt{Mt}\right) - e^{2\eta\sqrt{Mt}}\mathrm{erfc}\left(\eta + \sqrt{Mt}\right)\Big\}\\
&+ \frac{G_c \cos\phi}{b^2(1-S_c)}\frac{e^{bt}}{2}\Big[-\Big\{e^{-2\eta\sqrt{S_c(b+K_2)t}}\mathrm{erfc}\left(\eta\sqrt{S_c} - \sqrt{(b+K_2)t}\right)\\
&+ e^{2\eta\sqrt{S_c(b+K_2)t}}\mathrm{erfc}\left(\eta\sqrt{S_c} + \sqrt{(b+K_2)t}\right)\Big\}\\
&- \Big\{e^{-2\eta\sqrt{(b+M)t}}\mathrm{erfc}\left(\eta - \sqrt{(b+M)t}\right)\\
&+ e^{2\eta\sqrt{(b+M)t}}\mathrm{erfc}\left(\eta + \sqrt{(b+M)t}\right)\Big\}\Big]
\end{aligned}
\tag{11}
$$

where

$$
\begin{aligned}
C = &\frac{t}{2}\Big[e^{-2\eta\sqrt{SK_2 t}}\mathrm{erfc}\left(\eta\sqrt{S_c} - \sqrt{K_2 t}\right)\\
&+ e^{2\eta\sqrt{S_c t}}\mathrm{erfc}\left(\eta\sqrt{S_c} + \sqrt{K_2 t}\right)\Big]\\
&+ \frac{\eta\sqrt{S_c}}{2\sqrt{K_2}}\Big[e^{-2\eta\sqrt{S_c K_2 t}}\mathrm{erfc}\left(\eta\sqrt{S_c} - \sqrt{K_2 t}\right)\\
&- e^{2\eta\sqrt{S_c K_2 t}}\mathrm{erfc}\left(\eta\sqrt{S_c} + \sqrt{K_2 t}\right)\Big]
\end{aligned}
\tag{12}
$$

$$
\begin{aligned}
\theta = &\frac{t}{2}\Big[e^{-2\eta\sqrt{ANt}}\mathrm{erfc}\left(\eta\sqrt{A} - \sqrt{Nt}\right)\\
&+ e^{2\eta\sqrt{ANt}}\mathrm{erfc}\left(\eta\sqrt{A} + \sqrt{Nt}\right)\Big] +
\end{aligned}
$$

$$- \frac{\eta\sqrt{A}}{2\sqrt{N}}\Big[e^{-2\eta\sqrt{ANt}}\operatorname{erfc}\Big(\eta\sqrt{A}-\sqrt{Nt}\Big)$$
$$-e^{2\eta\sqrt{ANt}}\operatorname{erfc}\Big(\eta\sqrt{A}+\sqrt{Nt}\Big)\Big] \quad (13)$$

$$A=\frac{P_r}{1+R}, a=\frac{AN-M}{1-A}, b=\frac{S_cK_2-M}{1-S_c}, \eta=\frac{y}{2\sqrt{t}}$$

4 Result and Discussions

The temperature vitiates for scarcely any estimation of the heat radiation mathematical constant parameter $R = 0, 1, 2, 3$ in the exact work sight of air at time $t = 0.4$, $N = 0.1$, $P_r = 7.1$ are constructed in Fig. 1. They have an impact on the warm absorption mathematical limitation parameter is vital in warm variate. It is resolved that the warm level will increment with diminishing radiation constant limitation.

The impact of concentration for one of a kind values of the chemical response mathematical limitation $K_2 = 2, 5, 10, 20$ at $t = 0.4$. The impact of the chemical action response mathematical limitation is essential in the attention field. The variation has the frequent function that the awareness will decrement in an uninteresting movement

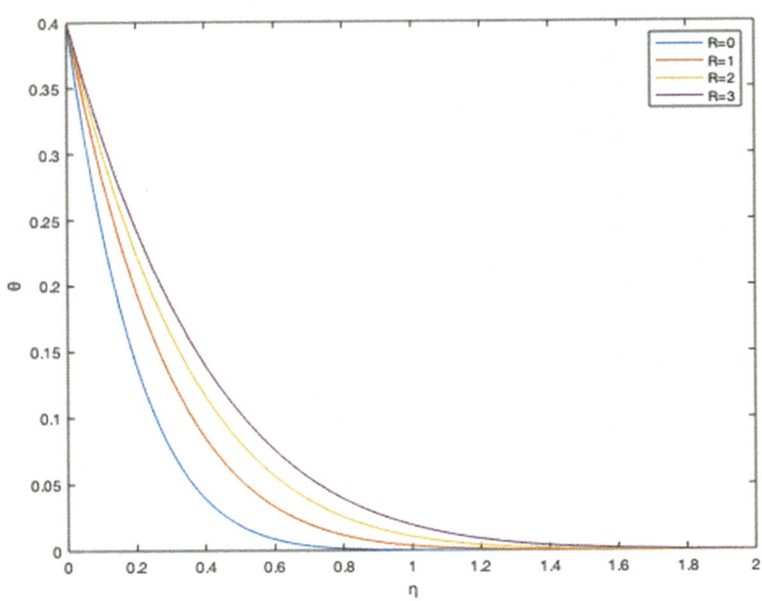

Fig. 1 Temperature variation for a few values of R

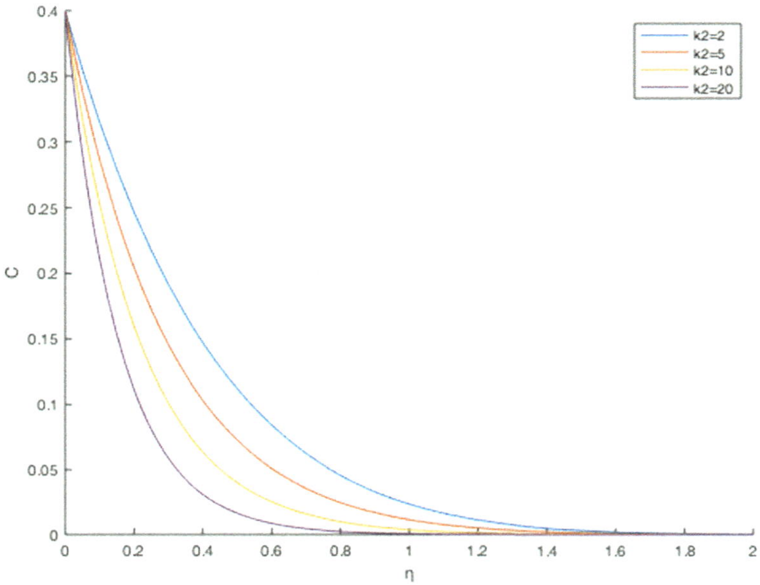

Fig. 2 Concentration profile ranges for some values of K_2

from the over surface area to zero free a long process in the system. It is determined that the focus of the study will increase with reducing values of the chemical action response constant mathematical parameter (Fig. 2).

The attention profiles variate for one kind of values of the $t = 0.2, 0.4, 0.6, 0.8, 1$ and $K_2 = 5$, Sc 0.6. The requirement of the fluid with appreciate to time is detailed in Fig. 3, and it is located that the need for profile variation will increase with making bigger in the period.

The speed profiles for explicit estimations of $M = 5$, 10, 15 and $G_r = G_c = 5$ $R = 0.2$, $t = 0.4$, $w = 1$ are considered and expressed graphically in Fig. 4, and it is the support that speed profile diminishes with reach out in attractive control parameter.

The speed profiles for explicit estimations of $w = 1$, 1.1, 1.5, 1.8 and $G_r = 5$ and $G_c = 5$ and $R = 0.2$ at $t = 0.4$ are contemplated and explained in Fig. 5. It is the support that the speed increments with developing estimations of w.

Figure 6 Effect of the speed for few estimations of $G_r = 5$, 10, 15 and $G_c = 5$, 10, 15 and $K_2 = 5$, $w = 1$, $R = 0.2$ $t = 0.4$. The speed increases with extending estimations of the warm G_r or mass G_c.

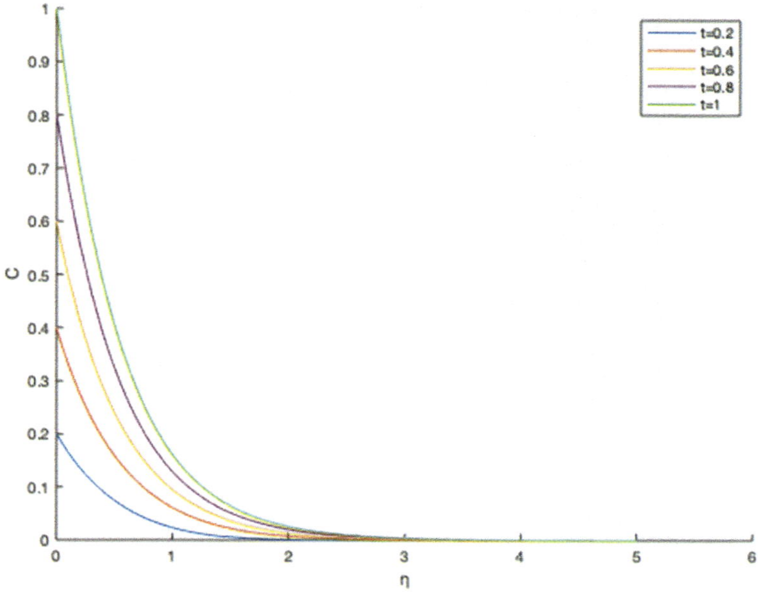

Fig. 3 Concentration variate for some values of t

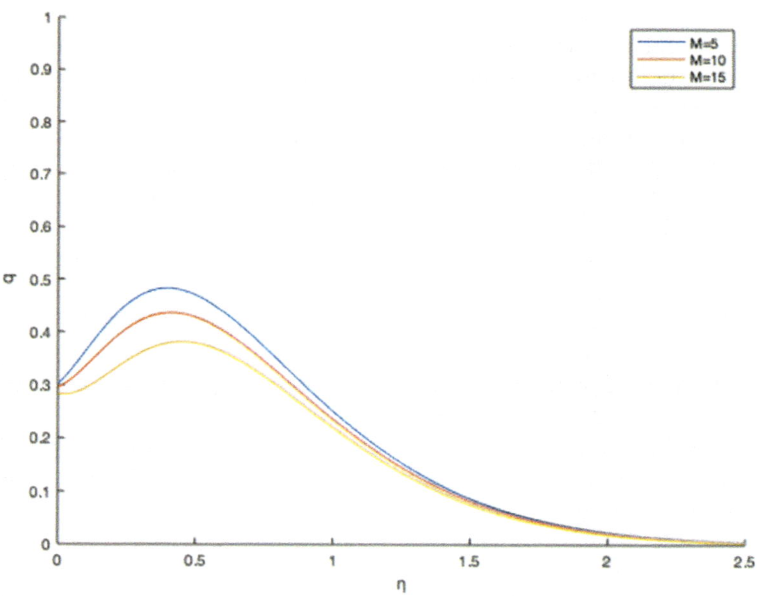

Fig. 4 Velocity study ranges variate for some values of M

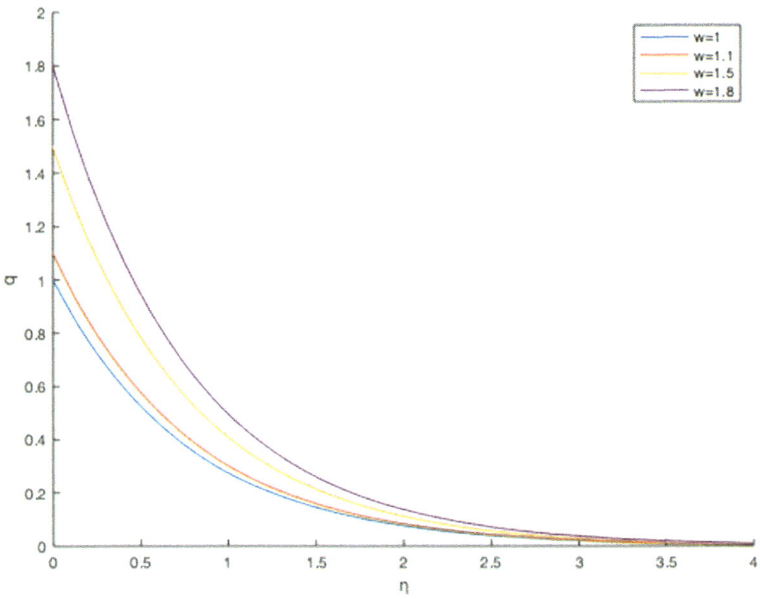

Fig. 5 Velocity study ranges variate for some values of *w*

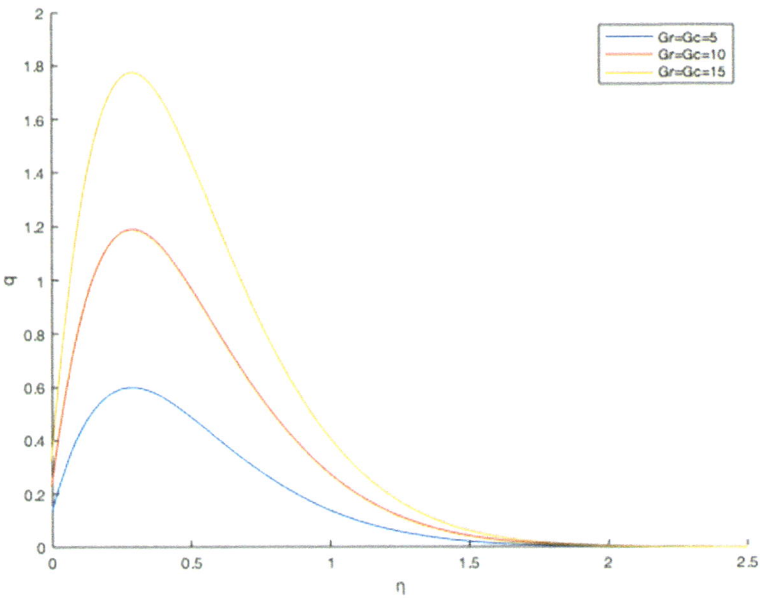

Fig. 6 Velocity study ranges variate for few values of Grand Gc

5 Conclusion

An evaluation of MHD free convection stream with an exponentially beginning improvement of a boundless isothermal inclined vertical plate in observing warm radiation and substance response The dimensionless managing conditions are applied by techniques for the customary Laplace-change methodology. The impact of remarkable limitation, for instance, the warm number G_r, mass number G_c, invention response parameter, radiation constant parameter, appealing control mathematical parameter and t are focused figural. The finishes of the find out about are temperature of the liquid stream increments as an increment in radiation parameter and intriguing seen is watched that warmth source of constant parameter and Pr whole number diminish the temperature profile. The extend in the intensity of substance responding materials as precisely correctly Schmidt number thought processes to decrease the fixation profile. The limit layer thickness and speed are augmentation with developing an assessment of mass Grashof number. The speed profile diminishes with the developing tendency point of attractive field parameter.

References

1. Azzam, G. E. D. A. (2002). Radiation effects on the MHD mixed free-forced convective flow past a semi-infinite moving vertical plate for high temperature differences. *Physica Scripta, 66*(1), 71.
2. Chambré, P. L., & Young, J. D. (1958). On the diffusion of a chemically reactive species in a laminar boundary layer flow. *the Physics of Fluids, 1*(1), 48–54.
3. Chen, S. L., Ma, H. K., & Chen, D. Y. (1993). Radiation blockage by the interaction of thermal radiation with conduction and convection in the combustion of condensed fuels. *International Communications in Heat and Mass Transfer, 20*(1), 145–157.
4. Endalew, M. F., & Nayak, A. (2019). Thermal radiation and inclined magnetic field effects on MHD flow past a linearly accelerated inclined plate in a porous medium with variable temperature. *Heat Transfer—Asian Research, 48*(1), 42–61.
5. Manay, E., Ozceyhan, V., Sahin, B., & Gunes, S. (2014). Edge length effect of bluff bodies on flow structure. *International Journal of Automotive and Mechanical Engineering, 9,* 1793–1801.
6. Mbeledogu, I. U., & Ogulu, A. (2007). Heat and mass transfer of an unsteady MHD natural convection flow of a rotating fluid past a vertical porous flat plate in the presence of radiative heat transfer. *International Journal of Heat and Mass Transfer, 50*(9–10), 1902–1908.
7. Muthukumaraswamy, R., Dhanasekar, N., & Easwara Prasad, G. (2012). Mass transfer effects on accelerated vertical plate in a rotating fluid with first order chemical reaction. *Journal of Mechanical Engineering and Sciences, 3,* 346–355.
8. Muthukumaraswamy, R., & Radhakrishnan, M. (2012). Chemical reaction effects on flow past an accelerated vertical plate with variable temperature and mass diffusion in the presence of magnetic field. *Journal of Mechanical Engineering and Sciences, 3,* 251–260.
9. Soundalgekar, V. M. (1982). Effects of mass transfer on flow past a uniformly accelerated vertical plate. *Letters in Heat and Mass Transfer, 9*(1), 65–72.
10. Muthukumaraswamy, R., Dhanasekar, N., & Prasad, G. E. (2013). Rotation effects on unsteady flow past an accelerated isothermal vertical plate with variable mass transfer in the presence of chemical reaction of first order. *Journal of Applied Fluid Mechanics, 6*(4), 485–490.

11. Muthukumaraswamy, R., & Lakshmi, C. S. (2015). First order chemical reaction effects on exponentially accelerated vertical plate with variable mass diffusion in the presence of thermal radiation. *International Journal of Applied Mechanics and Engineering, 20*(2), 329–344.
12. Muthukumaraswamy, R., & Meenakshisundaram, S. (2006). Theoretical study of chemical reaction effects on vertical oscillating plate with variable temperature. *Theoretical and Applied Mechanics, 33*(3), 245–257.
13. Nield, D. A., & Bejan, A. (2006). *Convection in porous media* (Vol. 3). New York: springer.
14. Rajput, U. S., & Kumar, G. (2019). Effects of radiation and chemical reaction on MHD flow past a vertical plate with variable temperature and mass diffusion. *Journal of Naval Architecture and Marine Engineering, 16,* 99–108.
15. Prasad, V. R., & Reddy, N. B. (2007). Radiation and mass transfer effects on an unsteady MHD free convection flow past a heated vertical plate in a porous medium with viscous dissipation. *Theoretical and Applied Mechanics, 34*(2), 135–160.
16. Raptis, A. A., Tzivanidis, G. J., & Perdikis, C. P. (1981). Hydromagnetic free convection flow past an accelerated vertical infinite plate with variable suction and heat flux. *Letters in Heat and Mass Transfer, 8*(2), 137–143.
17. Raju, M. C., Varma, S. V. K., & Seshaiah, B. (2015). Heat transfer effects on a viscous dissipative fluid flow past a vertical plate in the presence of induced magnetic field. *Ain Shams Engineering Journal, 6*(1), 333–339.
18. Rout, B. R., & Pattanayak, H. B. (2013). Chemical reaction and radiation effects on MHD flow past an exponentially accelerated vertical plate in presence of heat source with variable temperature embedded in a porous medium. *Annals of Faculty Engineering Hunedoara-Int. Jou. Of Engg, 4,* 253–259.

Eccentric Sequence of Self-centered Graphs

K. Deepika and S. Meenakshi

Abstract For a graph G, with v as any vertex, the maximum distance of v to another vertex or node is defined as the eccentricity $e(v)$ of the vertex v. Of all the eccentricities, the minimum is the radius, maximum of the eccentricities is the diameter, of graph G. If the eccentricity of any vertex is equal to the radius, then the vertex is said to be the central vertex of G. So, if for G, every vertex is the center, then G is defined as a graph that is self-centered. Here, we construct and characterize the power graph of any self-centered graph having a regular eccentric sequence. Also, the embedding of special graphs is discussed here.

Keywords Regular eccentric sequence · Self-centered graphs · Almost self centered graphs · n-pan graph · Graph power

1 Introduction

In graph theory, sequences play a major role as they have more information about graphs than a single invariant or a number [1]. Also, a sequence can be studied and dealt with ease just like a single number. But representing graphs using a sequence is preferable than a single invariant [2]. Also, the study of graphs and their properties is precise when the graphs are represented using sequences. Various types of sequences are studied in graph theory [3]. Degree sequence was the first type of sequence to be studied and developed. Many researchers have studied and developed various graph properties using degree sequence concept [4]. Next, based on the distance of a graph, its corresponding distance degree sequence and eccentric sequence concepts were developed [3].

The conditions for graphical sequences are the primary concern for any sequence under consideration [1, 2]. That is, given any sequence one must determine whether

K. Deepika (✉) · S. Meenakshi
Department of Mathematics, VISTAS, Pallavaram, Chennai, India
e-mail: maths1988@gmail.com

S. Meenakshi
e-mail: meena.sbs@velsuniv.ac.in

S.-L. Peng et al. (eds.), *Proceedings of First International Conference on Mathematical Modeling and Computational Science*, Advances in Intelligent Systems and Computing 1292, https://doi.org/10.1007/978-981-33-4389-4_46

there exists any graph realizing that sequence [2]. This is called the graph realization problem. Erdos and Gallai, Havel and Hakimi have made major contributions in the concept of degree sequence of graphs [5]. Many constructions and characterizations of graphs have been developed in this area. Researchers like Randic, Bloom and Kennedy have made a wide insight into the field of distance degree sequence of graphs [6]. Also, there are distance degree regular graphs [7] where the distance degree sequence (DDS), that is, the sequence based on the distance of the vertices of a graph is the same [8]. Similar to this concept of distance degree regular graphs is the self-centered graphs where every vertex has the same eccentricity [9–11]. Nandakumar [12] mainly dealt with the study of minimal eccentric sequences for graphs. However, for digraphs [13], the eccentric sequence was studied only in late 2000 by Gimbert and Lopez [10]. The concept of eccentric sequence of graphs provides a wide range of opportunity for researchers to analyze and develop it. The idea of the eccentric sequence of graphs leads to the development of self-centered graphs and almost self-centerd graphs [8]. In this paper, we have discussed the concept of graphs which are self-centerd and almost self-centered (analogous to distance degree injective graphs) [14] and embedding of special graphs on the former [1, 8, 9, 15].

2 Preliminaries

We now give the basic definitions required for this paper. Let G be the simple, finite graph [16–18].

Definitions.

2.1 *The Eccentricity of a Vertex*

For any graph, with "v" as any vertex, the distance that is the maximum from "v" to another node of the graph is defined to be the eccentricity $e(v)$ of the vertex v [8, 18].

$$e(v)\text{is } \{a/a \text{ is the maximum distance of node } v \text{ to any other vertex, } \forall\, v \varepsilon G\}$$

2.2 *Radius and Diameter of* G

The radius of G is the lowest value of all eccentricities. The diameter of G is the highest value of all eccentricities [3, 19].

2.3 Center of G

The enumeration of every vertex of lowest eccentricity is the center of G. That is, $e(v) =$ radius (G), then v is said to be the central vertex of G [3].

Center of G $= \{a/a = e(v),\ \text{lowest eccentricityof} G\}$.

2.4 Self-Centered Graph

For any graph, if the eccentricity of all vertices or nodes is the same, then the graph is self-centered [9].

$e(v) = a, \forall v \in G$, a is the eccentricity of vertex v.

2.5 Almost Self Centered Graph

If G has exactly two diametral vertices and the remaining all vertices are central, then G is an almost self-centered graph. Also, when the graph has only one central vertex and all the remaining vertices are diametral, then again the graph is almost self-centered [8, 9, 11].

2.6 Eccentric Sequence and Regular Eccentric Sequence of G

Enumerating or listing the eccentricity of every vertex of graph G is known as the eccentric sequence of the graph. We determine the eccentricity of any vertex of the.

graph by measuring the farthest vertex distance from that vertex. The eccentricity sequence or the eccentric sequence of a self-centered graph can be called as a regular eccentric sequence. This follows by the definition of a self-centered graph.

Eccentric Sequence $= \{a_i/a_i$ is the eccentricity of vertices $v_i, \forall v_i \in G\}$.

Regular Eccentric Sequence $= \{a/a$ is the eccentricity of vertices $v_i, \forall v_i \in G\}$.

2.7 Power Graph

For any graph G, its kth power or power graph is a graph where two vertices in the power graph are adjacent when their distance in G is at most k.

2.8 Pan Graph

A pan graph or n-pan graph is obtained by adjoining a cycle C_n with a single vertex or a path of vertices using a bridge.

2.9 Embedding of a Graph

A graph G is embedded in the topological space X if G is represented in X such that the vertices of G are distinct elements in X, and an edge in G is a simple arc connecting its two ends such that no two edges intersect except possibly at a common end[20].

3 Main Results

Theorem 3.1 For the power graph G^k of the self-centered graph $G = C_n$, the eccentric sequence is regular. In other words, the power graph G^k of the self-centered graph $G = C_n$ is again self-centered.

Proof Let $G = C_n$ be the self-centered graph as in Fig. 1. The eccentricity of all nodes or vertices of $G = C_n$ is 2. Hence, the regular eccentric sequence is of the form $\{2, 2, 2, \ldots 2\}$ for n vertices. We now construct the power graph $G^k = {C_n}^k$, that is we construct another graph with the same set of vertices, and when the distance between two vertices is at most k, then they are adjacent. So we obtain Fig. 2.

Figure 2 is a complete graph, and we note that the degree of every vertex in G^k has increased by 1 as k increases, where k is the distance between two vertices. Also once we obtain a complete graph, no further power graphs of $G = C_n$ be constructed. By

Fig. 1 Graph $G = C_n$

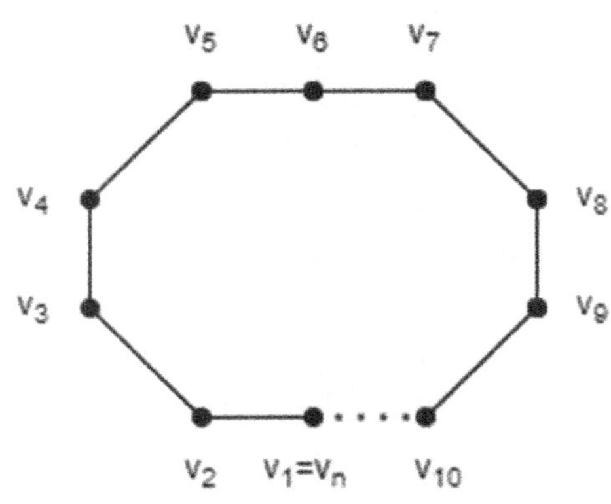

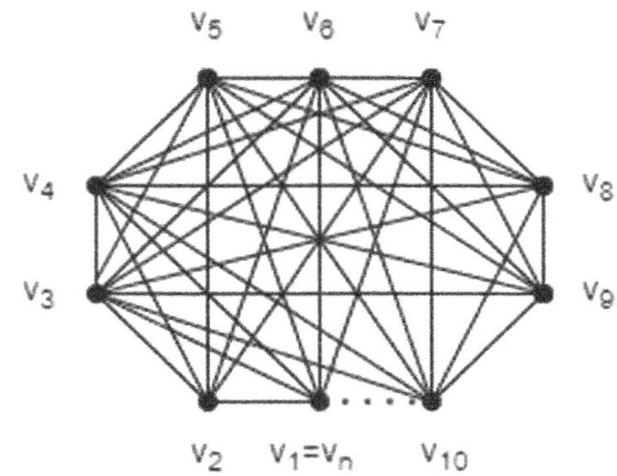

Fig. 2 Graph $G^k = C_n^k$

the definition of the regular eccentric sequence of graphs, we note that the complete graph obtained as the power graph is self-centered [3]. So the eccentric sequence is regular [3].

3.1 Corollary

The eccentric sequence of the power graph G^k of any self-centered graph is a regular eccentric sequence.

Eccentric sequence of $G^k = \{a/e(v) = a, \forall \text{ vertex } v \in G\}$. "$a$" is the eccentricity of vertex "v" in G.

Proof Suppose G is a graph which is also self-centered [8]. Then, by definition, the eccentric sequence for a self-centered graph is a regular eccentric sequence say (m, m, m, … m) where $e(v)$is m ∀ vertex m in G. As per the power graph definition, we construct another graph for the graph G in which two nodes are adjacent when the distance is at most k, k = distance 2,3,…. The degree of every vertex of the power graph increases by 1 as k increases. Finally, we obtain a complete graph as the power graph of G, and this power graph will have the same eccentricity for all its vertices, and hence, the eccentricity sequence of the power graph is regular [3]. The power graph is self-centered.

That is, eccentric sequence of $G^k = \{a/e(v) = a, \forall \text{ vertex } v \in G\}$.

We write an algorithm for the above-discussed corollary.

3.2 Algorithm

An algorithm to construct a power graph of self-centered graph G.

Constructing a power graph G^k of a self-centered graph having a regular eccentric sequence.

Step 1 Consider the self-centered graph G with vertices $v_i, i = 1, 2, \ldots n$.

Step 2 By definition, the eccentricity of all vertices or nodes of the graph G is the same, and the eccentric sequence for G is the regular eccentric sequence. We compute the eccentricity of all the vertices of G by finding the maximum distance from a vertex to another vertex.
$e(v) = a$, for all $v \in G$. a is the maximum distance from vertex v to any other vertex.
The regular eccentric sequence of G is $\{a, a, a, \ldots a\}$ for all $v \in G$.

Step 3 Construct a graph for the graph G. We do this by finding the distance from vertex v_i to vertex v_j; $i, j = 1, 2, \ldots n$. If the distance is at most two, then we make the corresponding vertices adjacent. We do this for every pair of vertices. The resultant graph is the power graph which will have same number of vertices as the graph G.

Step 4 Repeat step 3 where a graph is constructed for the graph G in which two nodes are adjacent when the distance between them is at most k, $k =$ distance 3, 4, …

Output

A complete graph is obtained as the power graph of G where all the vertices have the same eccentricity. The eccentricity sequence of the power graph is regular. So the power graph is self-centered.

Theorem 3.2 Suppose G is a graph that is regular [8]. Then, in an almost self-centered graph, G can be embedded.

Proof Consider G as a graph which is regular [8]. Let r be the regularity of G. Let v be any vertex. Let us add v to G and let every vertex of G be adjacent to v. The graph thus obtained is almost self-centered with one central vertex r, and all other vertices are diametral of the form $r + 1$. The resulting graph G_e has G as its induced subgraph. Thus, in an almost self-centered graph, any graph G can be embedded (Fig. 3).

Theorem 3.3 An n-pan graph can be embedded in a self-centered graph. Also in an almost self-centered graph, the 3-pan graph can be embedded.

Proof Suppose G is the 3-pan graph. Let us consider two vertices u, v. We make all the vertices of the 3-pan graph adjacent to the new vertices u and v. The resulting graph G_e thus obtained is an almost self-centered graph with exactly one central vertex and all other vertices as diametral vertices. We also observe that G is the induced subgraph of G_e. Hence, G_e is almost self-centered (Figs. 4 and 5).

Fig. 3 Graph G_e

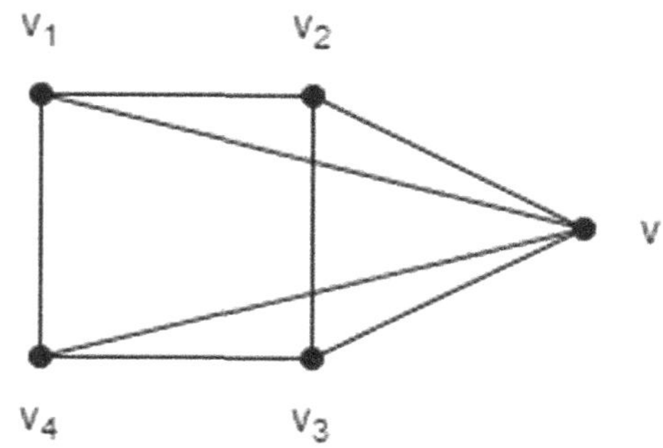

Fig. 4 Graph G

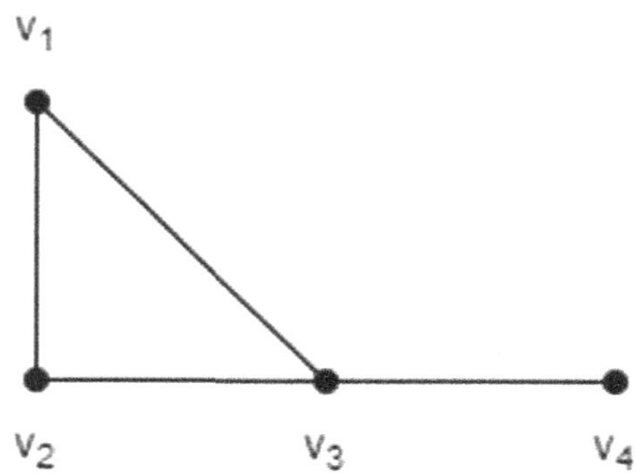

Fig. 5 Graph G_e

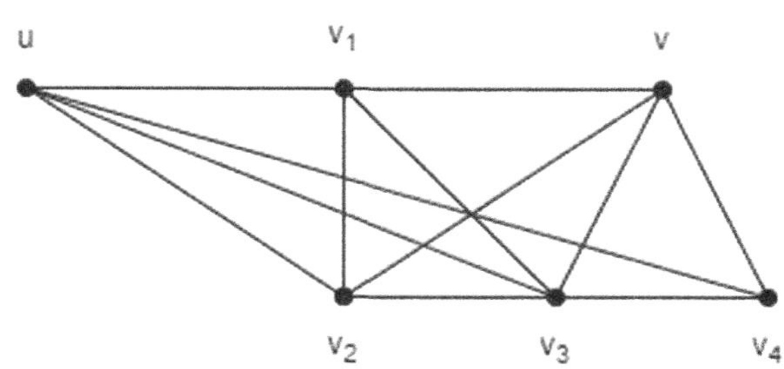

We now consider G as the n-pan graph, $n \geq 4$. Again we consider two vertices u and v. All the vertices of n-pan graph are made adjacent to the vertices u and v. We observe that the resulting graph G_e has the vertices with the same eccentricity, say some m for all vertices that is the graph is self-centered. The eccentric sequence of the graph G_e is regular. Also G_e has G as its induced subgraph. The graph G_e is thus self-centered. Hence, the embedding of n-pan graph in the self-centered graph.

4 Proposition

We also observe that G is the induced subgraph of G_e. Hence, G_e is almost self-centered. We embedded an n-pan graph in an almost self-centered graph and self-centered graph. Similarly, for an almost self-centered graph or a self-centered graph, we may determine whether a graph that is isomorphic to the pan graph can be embedded in them.

5 Conclusion

In this manuscript, we have constructed and characterized the power graph of any self-centered graph having a regular eccentric sequence. Also, the embedding of special graphs has been discussed here. Many types of research have been done in the area of self-centered and almost self-centered graphs. Also, various products of self-centered graphs have been determined earlier. Also, researches have been carried on minimal self-centered graphs. Apart from these graphs, the concept of the eccentric sequence itself provides a wide range of research idea. In the coming years, these concepts can be discussed and many properties and results can be determined. Also, algorithms of such characterisation problems can be written in the future.

References

1. Huilgol, M. I., Rajeshwari, M., & Ulla, S. S. A. (2012). Products of distance degree regular and distance degree injective graphs. *Journal of Discrete Mathematical Sciences and Cryptography, 15*(4–5), 303–314.
2. Erdös, P., & Gallai, T. (1960). Graphs with prescribed degrees of vertices. *Mat. Lapok, 11*, 264–274.
3. Meenakshi, S., Deepika, K., Abdul Saleem, R. (2019). Eccentric sequence of graphs. *International Journal of Recent Technology and Engineering, 8*(4S5), 52–54.
4. Meenakshi, S., Deepika, K., (2019) 'Some results on the eccentric sequence of graphs. *International Journal of Recent Technology and Engineering, 8*(4S5), 55–57
5. Hakimi, S. L. (1962). On realizability of a set of integers as degrees of the vertices of a linear graph. I. *Journal of the Society for Industrial and Applied Mathematics, 10*(3), 496–506.
6. Randic, M. (1979). Characterizations of atoms, molecules, and classes of molecules based on path enumerations. *Match, 7*, 5–64.
7. Bloom, G. S., Quintas, L. V., & Kennedy, J. W. (1981). Distance degree regular graphs. *The theory and applications of graphs (Kalamazoo, Mich., 1980)*, 95–108.
8. Huilgol, M. I., & Ramaprakash, C. (2014). On Some edge rotation distance graphs. *IOSR Journal of Mathematics, 10*(6), 16–25.
9. Bond, J. (1989). Graph theory and its applications. *Annals of the New York Academy of Sciences, 5*, 76.
10. Gimbert, J., & López, N. (2008). Eccentricity sequences and eccentricity sets in digraphs. *ARS Combinatoria-Waterloo then Winnipeg, 86*, 225.
11. Lesniak, L. (1975). Eccentric sequences in graphs. *Periodica Mathematica Hungarica, 6*(4), 287–293.

12. Nandakumar, R. (1986). *On some eccentricity properties of graphs* (Doctoral dissertation, Ph. D. Thesis, Indian Institute of Technology, India).
13. Chartrand, G., & Lesniak, L. Graphs & Digraphs: (Chapman k Hall, London, 1996). *Melnikov, V. Saravanov, R. Tyshkevich, V. Yemelichev and I. zverovich Exercises in Graph.*
14. Volf, J. (1987). A small distance degree injective cubic graphs, Notes from New York graph theory day XIII.
15. Huilgol, M. I., Rajeshwar, M., & Ulla, S. A. S. (2013). Embedding in distance degree regular and distance degree injective graphs. *Malaya Journal of Matematik, 4*(1), 134–141.
16. Bollobás, B., & Béla, B. (2001). *Random graphs* (No. 73). Cambridge university press.
17. Buckley, F., & Harary, F. (1990). *Distance in graphs.* Addison-Wesley.
18. Changat, M., & Das, S. (2020). *Algorithms and Discrete Applied Mathematics.* Springer International Publishing.
19. Ostrand, P. A. (1973). Graphs with specified radius and diameter. *Discrete Mathematics, 4*(1), 71–75.
20. Carsten Thomassen (1994) Embeddings of graphs. *Discrete Mathematics 124*, 217–228; North-Holland 217 Mathematisk Institut, Danmarks Tekniske Hojskoie. Bygning 303, DK-Lyngby. Denmark.

Study the Stability for Ordinary Differential Equations Using New Techniques via Numerical Methods

Huda Omran Altaie

Abstract Nonlinear differential equation stability is a very important feature of applied mathematics, as it has a wide variety of applications in both practical and physical life problems. The major object of the manuscript is to discuss and apply several techniques using modify the Krasovskii's method and the modify variable gradient method which are used to check the stability for some kinds of linear or nonlinear differential equations. Lyapunov function is constructed using the variable gradient method and Krasovskii's method to estimate the stability of nonlinear systems. If the function of Lyapunov is positive, it implies that the nonlinear system is asymptotically stable. For the nonlinear systems, stability is still difficult even though the Lyapunov methods are applied. There has to find a positive definite Lyapunov function, and its derivative function has to be negative definite. A new approach had been tested in several examples to steady the analysis of these methods. Among the results of the suggested examples, we can see the reliability and the simplicity of the method in describing the stability of differential equations.

Keywords Variable gradient method · Lyapunov functions · Stability analysis · Eigenvalue · Krasovskii's method · Gradient function

1 Introduction

Lyapunov suggested a way to study the stability for some kinds of elements without solving the models for several types of differential equations in 1982. The method includes some of algorithms and procedures that are interested with the existence of Lyapunov functions. This method has a major advantage and large applications to check stability of several systems [1, 2, 4, 6, 7]. Certain methods of Lyapunov function had been discussed in several literatures which indicate the stability. These methods are very important in this paper to discuss the most suitable methods of them via the Krasovskii method and variable gradient method [3–5, 8].

H. O. Altaie (✉)
Department of Mathematics, College of Education for Pure Science/Ibn Al-Haitham, University of Baghdad, Baghdad, Iraq
e-mail: dr.Huda_hm2019@yahoo.com

S.-L. Peng et al. (eds.), *Proceedings of First International Conference on Mathematical Modeling and Computational Science*, Advances in Intelligent Systems and Computing 1292, https://doi.org/10.1007/978-981-33-4389-4_47

2 Definitions of the Construction of Lyapunov Function [4, 5]

Consider Ø open set in D. Let S a continuous function on Ø; then, we can define Lyapunov function as follows:

1. Let $S(x)$: Ø $\rightarrow$ R is called a positive definite function on Q. If $\forall\, x \neq 0$ and $x \in$ Ø then $S(0) = 0$, and $S(x) > 0$
2. The positive function is a quadratic form: $S(x) = \sum_{k=1}^{m} \sum_{n=1}^{m} q_{kn} x_k x_n$, $k,\ n = 1, 2, \ldots, m$ When $Q = q_{kn}$ is a symmetric matrix. If $S(x)$ positive definiteness, then we apply the Sylvester's criteria.

3 Krasovskii's Method [5]

Consider $\dot{x} = N(x)$, the nonlinear system where x is a vector with n-dimensional and $N(x)$ vector consists of linear or nonlinear functions of $x_1, x_2, \ldots, x_n$. Let $\hat{J} = J^T + J$, such that J means Jacobian of function, and J^T transpose of J. So when $\hat{J}$ nonpositive definite, then the solution is asymptotically stable and Lyapunov yields: $S(x) = N_1^2 + N_2^2 + \cdots + N_n^2 = \sum_{i=1}^{n} N_i^2$.

4 A New Technique for Variable Gradient Method for Constructing Lyapunov Function Using Derivative Function

Consider the system described by the following equation:

$$\dot{x} = g(x, t) \tag{4.1}$$

Let us a Lyapunov functions as Q and we assume that Q is an explicit function of $x_1, x_2, \ldots, x_n$, that implies:

$$\dot{Q} = \frac{\partial Q}{\partial x_1}\dot{x}_1 + \frac{\partial Q}{\partial x_2}\dot{x}_2 + \cdots + \frac{\partial Q}{\partial x_n}\dot{x}_n$$

Then

$$Q^{\cdot} = (\nabla Q)^T \dot{X}$$

where $\nabla Q = \begin{bmatrix} \nabla Q_1 \\ \nabla Q_2 \\ \vdots \\ \nabla Q_n \end{bmatrix}$

where $Q = \int_0^x (\nabla Q)^T \, dx$.

That implies

$$Q = \int_0^{x_1,(x_2=x_3=\cdots=x_n=0)} \nabla Q_1 dx_1 + \int_0^{x_2,(x_1=x_3=\cdots=x_n=0)} \nabla Q_2 dx_2 + \cdots + \int_0^{x_n,(x_1=x_n=\cdots=x_{n-1}=0)} \nabla Q_n dx_n$$

where ∇X_1 is the component of ∇Q in the x_i direction, $\forall i$. The matrix F created with $\frac{\partial \nabla Q_i}{\partial x_j}$:

$$F = \begin{bmatrix} \frac{\partial \nabla Q_1}{\partial x_1} & \frac{\partial \nabla Q_1}{\partial x_2} & \cdots & \frac{\partial \nabla Q_1}{\partial x_n} \\ \frac{\partial \nabla Q_2}{\partial x_1} & \frac{\partial \nabla Q_2}{\partial x_2} & \cdots & \frac{\partial \nabla Q_2}{\partial x_n} \\ \vdots & \vdots & \ddots & \vdots \\ \frac{\partial \nabla Q_n}{\partial x_1} & \frac{\partial \nabla Q_n}{\partial x_2} & \cdots & \frac{\partial \nabla Q_n}{\partial x_n} \end{bmatrix}$$

Must be symmetric. Therefore,

$$\nabla Q = \begin{bmatrix} b_{11}x_1 + b_{12}x_2 + \cdots + b_{1n}x_n \\ b_{21}x_1 + b_{22}x_2 + \cdots + b_n x_n \\ \vdots \\ b_{n1}x_1 + b_{n2}x_2 + \cdots + b_{nn}x_n \end{bmatrix} \quad (4.2)$$

5 Application Examples

Example 5.1 Describe the stability of the following nonlinear system of differential equations:

$$\dot{x}_1 = 2x_2 - \frac{x_1}{(1+x_1^2)^2} \quad \dot{x}_2 = -x_1 - \frac{2x_2}{(1+x_1^2)}$$

Therefore,

$$\dot{Q} = (\nabla Q)^T \dot{X} = (b_{11}x_1 + b_{12}x_2)\left(\frac{-x_1}{(1+x_1^2)^2} + 2x_2\right) + (b_{21}x_1 + b_{22}x_2)\left(-x_1 - \frac{2x_2}{(1+x_1^2)}\right)$$

Hence,

$$\frac{\partial \dot{Q}}{\partial x_1} = \frac{(1+x_1^2)^2(-2b_{11}x_1 - b_{12}x_2 - 2b_{21}x_2) - (2b_{11}x_2 - 4b_{21}x_1 - 2b_{22}x_2)}{(1+x_1^2)^4}$$

which yields $\frac{\partial^2 \dot{Q}}{\partial x_1^2} = -2b_{21}$, $\frac{\partial^2 \dot{Q}}{\partial x_2^2} = 2b_{12}$, $\frac{\partial^2 \dot{Q}}{\partial x_1 \partial x_2} = 2b_{11} - 2b_{22}$, and $\frac{\partial^2 \dot{Q}}{\partial x_2 \partial x_1} = 2b_{11} - 2b_{22}$.

Setting

$$r_1 = -2b_{21}, r_2 = 2b_{11} - 2b_{22}, r_3 = 2b_{11} - 2b_{22} \text{ and } r_4 = 2b_{12}$$

Therefore, $F_1 = r_1$ and $F_2 = r_1 r_4 - r_2 r_3$. By choosing the minimization $\underline{F}_2$ with the condition that $F_1 \geq 0$ and $F_2 \geq 0$, we get: $b_{11} = 0$, $b_{12} = 0$ and $b_{22} = 2$.

We obtain:

$$Q^{\cdot} = -\frac{2x_1^2}{(1+x_1^2)^2} - \frac{x_2^2}{(1+x_1^2)^2}$$

We see that $Q^{\cdot}$ is negative definite, then

$$Q = \int_0^{x_1,(x_2=0)} 2x_1 \mathrm{d}x_1 + \int_0^{x_2,(x_1=0)} x_2 \mathrm{d}x_2 = 2x_1^2 + x_1^2 > 0.$$

Thus, the system is asymptotically stable.

By using the Krasovskii's method

$$\dot{x}_1 = \frac{-x_1}{(1+x_1^2)^2} + 2x_2 = f_1 \quad \dot{x}_2 = -x_1 - \frac{2x_2}{(1+x_1^2)}$$

Define

$$J = \begin{bmatrix} \frac{\partial f_1}{\partial x_1} & \frac{\partial f_1}{\partial x_2} \\ \frac{\partial f_2}{\partial x_1} & \frac{\partial f_2}{\partial x_2} \end{bmatrix} = \begin{bmatrix} \frac{-(1+x_1^2)^2 + 4x_1^2}{(1+x_1^2)^4} & 2 \\ -1 + \frac{4x_1x_2}{(1+x_1^2)^2} & \frac{-2}{(1+x_1^2)} \end{bmatrix}$$

Therefore:

$$J^T = \begin{bmatrix} \frac{-(1+x_1^2)^2+4x_1^2}{(1+x_1^2)^4} & -1 + \frac{4x_1x_2}{(1+x_1^2)^2} \\ 2 & \frac{-2}{(1+x_1^2)} \end{bmatrix}$$

And since $\hat{J} = J^T + J$, hence we have:

$$\hat{J} = \begin{bmatrix} \frac{-2(1+x_1^2)^2+8x_1^2}{(1+x_1^2)^4} & 1 + \frac{4x_1x_2}{(1+x_1^2)^2} \\ 1 + \frac{4x_1x_2}{(1+x_1^2)^2} & \frac{-4}{(1+x_1^2)} \end{bmatrix}$$

We obtain the solution is asymptotically stable and Lyapunov function yields:

$$Q(x) = \left(\frac{-x_1}{(1+x_1^2)^2} + 2x_2\right)^2 + \left(-x_1 - \frac{2x_2}{(1+x_1^2)}\right)^2$$

Now we use modified Krasovskii's method.
By using Euler's formula yields $x_{n+1} = x_n + h_n S(x_n)x_n$.

Define

$$S = \begin{bmatrix} h\left(1 + \frac{-(1+x_1^2)^2+4x_1^2}{(1+x_1^2)^4}\right) & 2h \\ h\left(-1 + \frac{4x_1x_2}{(1+x_1^2)^2}\right) & \frac{-2h}{(1+x_1^2)} \end{bmatrix}.$$

To find the eigenvalue of matrix S using computer programs when h = 0.0001 yields the eigenvalues n_1 = -0.5 and n_2 = 1 therefore the matrix is stable.

Example 5.2 Describe the stability of the following nonlinear differential equations:

$$\dot{x}_1 = -2x_1x_2x_3 \quad \dot{x}_2 = \frac{x_2}{t+2} + x_1^2x_3^2$$

$$\dot{x}_3 = \frac{x_3}{t+2} + x_1^2x_2^2$$

where $t \geq 0$. Begin, we choose Q so that its ∇Q is given by:

$$\nabla Q = \begin{bmatrix} b_{11}x_1 + b_{12}x_2 + b_{13}x_3 \\ b_{21}x_1 + b_{22}x_2 + b_{23}x_3 \\ b_{31}x_1 + b_{32}x_2 + b_{33}x_3 \end{bmatrix}$$

Then $Q^{\cdot} = \left(\frac{x_2}{t+2} + x_1^2x_3\right) + (b_{31}x_1 + b_{32}x_2 + b_{33}x_3)\left(\frac{x_3}{t+2} + x_1^2x_2\right)$.

Now, choosing $t = 0$, we have the following matrix:

$$F = \begin{bmatrix} 0 & \frac{b_{21}}{2} & \frac{b_{31}}{2} \\ \frac{b_{21}}{2} & \frac{b_{22}}{2} & b_{23} \\ \frac{b_{31}}{2} & b_{23} & b_{33} \end{bmatrix}$$

Hence,

$$F_1 = 0 \quad F_2 = \begin{vmatrix} 0 & \frac{b_{21}}{2} \\ \frac{b_{21}}{2} & \frac{b_{22}}{2} \end{vmatrix} = -\left(\frac{b_{21}}{2}\right)^2$$

$$F_3 = \begin{vmatrix} 0 & \frac{b_{21}}{2} & \frac{b_{31}}{2} \\ \frac{b_{21}}{2} & \frac{b_{22}}{2} & b_{23} \\ \frac{b_{31}}{2} & b_{23} & b_{33} \end{vmatrix} = b_{23}\frac{b_{21}}{2}\frac{b_{31}}{2} + \frac{b_{31}}{2}\frac{b_{21}}{2}b_{23} - \left(\frac{b_{31}}{2}\right)^2\frac{b_{22}}{2} - \left(\frac{b_{21}}{2}\right)^2 b_{33}$$

To estimate the stability of the system using a new approach, we use $f_1 > 0,\ f_2 > 0$, and $f_3 > 0$, then we have:

$$\begin{aligned} b_{11} &= 0, b_{12} = b_{21} = -0.003, \\ b_{22} &= 0.9, b_{13} = b_{31} = -0.01, \\ b_{23} &= b_{32} = 0.8 \text{ and } b_{33} = 1.1 \end{aligned}$$

with $F_1 = 0$, $F_2 = -1 \times 10^{-6}$ and $F_3 = 0.04 \times 10^{-7}$. Therefore, system is asymptotically stable.

Now, By using the Krasovskii's method

$$\begin{aligned} \dot{x}_1 &= -2x_1x_2x_3 = S_1 \\ \dot{x}_2 &= \frac{x_2}{t+2} + x_1^2x_3^2 = S_2 \\ \dot{x}_3 &= \frac{x_3}{t+2} + x_1^2x_2^2 = S_3 \end{aligned}$$

Define

$$J = \begin{bmatrix} -2x_3x_2 & -2x_1x_3 & -2x_2x_1 \\ 2x_3x_1 & \frac{1}{t+2} & x_1^2 \\ 2x_1x_2 & x_1^2 & \frac{1}{t+2} \end{bmatrix}$$

Therefore

$$J^T = \begin{bmatrix} -2x_2x_3 & 2x_1x_3 & 2x_1x_2 \\ -2x_1x_3 & \frac{1}{t+2} & x_1^2 \\ -2x_1x_2 & x_1^2 & \frac{1}{t+2} \end{bmatrix}$$

And since $\hat{J} = J^T + J$, hence we have:

$$\hat{J} = \begin{bmatrix} -4x_2x_3 & 0 & 0 \\ 0 & \frac{1}{2} & 2x_1^2 \\ 0 & 2x_1^2 & \frac{1}{2} \end{bmatrix}$$

Therefore, the solution is asymptotically stable.

Now we use modified: By using Euler's formula yields

$$x_{n+1} = x_n + h_n S(x_n)x_n$$

Define

$$S = \begin{bmatrix} -2x_2x_3h & -2x_1x_3h & -2x_1x_2h \\ 2x_1x_3h & \frac{1}{t+2}h & hx_1^2 \\ 2x_1x_2h & hx_1^2 & \frac{1}{t+2}h \end{bmatrix}$$

The eigenvalue of matrix S when $h = 0.0001$ yields $n_1 = 1$, $n_2 < 0$ and $n_3 < 0$ therefore the matrix is stable.

6 Conclusion

This paper introduced new Lyapunov function building techniques for the stability of nonlinear systems. Lyapunov functions have an enormous benefit and wide applications for the study of stability of nonlinear differential equations. In this study, new way used to check the stability of nonlinear differential equation is introduced and discussed in some examples using the two approaches proposed via the modified variable gradient method and modified Krasovskii's method. The results of the suggested examples ensure the stability of asymptotically stable for system of nonlinear differential equations, and we can see the accuracy and the simplicity of the new techniques in describing the stability of differential equations.

References

1. Rao M. R. M. (1980). Ordinary Differential Equations Theory and applications Affiliated East–West Private.
2. Adil, H. (1995). *Vector Liapunov function of large scale system* (M.Sc. thesis). University of Technology, Baghdad, Iraq.
3. Michel, A., Miller, R., & Nam, B. (1982). Stability analysis of interconnected systems using computer generated Lyapunov functions. *IEEE Transactions on Circuits and Systems, 29*(7), 431–440. https://doi.org/10.1109/tcs.1982.1085181.
4. Brayton, R., & Tong, C. (1979). Stability of dynamical systems: A constructive approach. *IEEE Transactions on Circuits and Systems, 26*(4), 224–234. https://doi.org/10.1109/tcs.1979.1084637.
5. Lapidsus, L. (1967). State space analysis of control systems, K. Ogata, Prentice-hall, Inc., Englewood Cliffs, New Jersey(1967). 596p. *AIChE Journal, 13*(3), 617–617. https://doi.org/https://doi.org/10.1002/aic.690130303
6. Brayton, R., & Tong, C. (1980). Constructive stability and asymptotic stability of dynamical systems. *IEEE Transactions on Circuits and Systems, 27*(11), 1121–1130. https://doi.org/10.1109/tcs.1980.1084749.
7. Zheng, D. (2002). *Linear System Theory* (2nd ed.). Beijing: Tsinghua University Press.
8. Mohammedali, K. H., Fadhel, F., & Ahmad, N. A. (2015). Modified variable gradient method for constructing the Lyapunov function using nonlinear programming method. *2015 10th Asian Control Conference (ASCC)*. https://doi.org/https://doi.org/10.1109/ascc.2015.7244616

Exploring a Context-Based Service Access for Trusted Pervasive Application

A. M. Hema and K. Kuppusamy

Abstract Significant advancement in the area of communication technology paves a way for a pervasive environment. Role-based access control provides fixed security in the access control mechanism. RBAC is used in combination of contextual parameters to improve the security in access control mechanism. Because of the mobility of the entities, we need to adopt new access control policies accounting the contextual parameters. In this paper, we developed a context-based service access for the trusted entity based on the analytical hierarchy process (AHP). We observed that the access control preferences are by location, time at which the request arrived; type of the network user/entity has used during their service access request and assumed initial trust value of the requested service in dynamic situations. Most preference parameter for service access as network type and then the location of the entity. It enhances security between trusted entities in the service transaction.

Keyword Pervasive · AHP · Trust · Context · Access control

1 Introduction

A most important improvement in the communication technologies has lifted the computing paradigm to pervasive nature. In the static traditional environment, network security is defined as access control policies, based on authentication and authorization of the entities. But in the pervasive computing environment, the traditional access control policies are not suitable to provide complete security to the network resources due to the device's mobility. The mobility context about the entity is the most important factor to adapt the access control policies. Therefore, we have proposed context-based access control with trust to assign the policy for access

A. M. Hema (✉)
Thiagarajar College (Research scholar at Alagappa University), Madurai, India
e-mail: jehemaeinkaran@gmail.com

K. Kuppusamy
Department of Computational Logistics, Alagappa University, Karaikudi, India
e-mail: kkdisamy@yahoo.com

S.-L. Peng et al. (eds.), *Proceedings of First International Conference on Mathematical Modeling and Computational Science*, Advances in Intelligent Systems and Computing 1292, https://doi.org/10.1007/978-981-33-4389-4_48

objects or resources in this pervasive environment. Confidentiality, integrity and availability play a significant role in adapting access control policies. Confidentiality ensures that the protected data is shared between the right users; integrity shows that request is from the authorized user and availability meets the authorized user request if the resource is in an available state. Access control policies provide authorization to the authenticated user by a set of rules as privileges to access the resources. In a pervasive environment, the entities are facilitated with dynamic access control policies based on the time, location, initial trust value and recommender or delegator. Analytical hierarchical process (AHP) selects the order of preference of the contextual and trust parameters for defining the access control policies in this environment. The access delegation is defined as the combination of the context and computed trust value. Authorized user would access the service or available resource by their predefined access privileges in a mandatory access control model. User identity is used in discretionary access control. Role of the user is signed in case of role-based access control models. All these traditional models are the static and centralized system. In context-based network access control for mobile devices, access policies were defined completely based on user's context for secured transaction [1]. So, we have explored a method to access service for the trusted entities to provide dynamic rights by analytic hierarchical process, with priority preferences of context and trust levels, to improve the security measures while accessing the services.

Remaining part of the paper is organized as follows: Sect. 2 describes literature review, Sect. 3 explains the AHP process for the trust value computation to decide the service access control and Sect. 4 disclose the conclusion and suggestions.

2 Literature Review

Trust is an important factor in Internet-based e-commerce services, in which trust is calculated using reputation and dynamic trust values are formulated mathematical [2]. In [3], trust value was calculated to fix and monitor the access rights of the authenticated nodes. In [4], trust definition was postulated for a pervasive and distributed environment. Trust description and trust evaluations are established and included. Trust level was used to decide accessing the resources between trusted entities [5]. In all cases, trust becomes a very important factor and context parameters are taken into account to adapt dynamic access control policies.

So context parameters can be viewed as value-based factors to fix the consistency in providing security for accessing the resources. Hayashi et al. [6] defined probabilistic framework for CASA, for the selection of active authentication scheme dynamically for the specified set of passive security factors. Filho and Martin [7] focus on quality factors in addition to the context, to decide and adopt the access permissions in their owner-centric QoC-aware context-based access control model. Chakraborty et al. explained a framework for privacy measure of sensor data on mobile application according to user behavior.

Table 1 Factors to be considered in decision making to provide the service or not

Parameter	1	2	3	4	5	6
Location	Anonymous	Classroom	Laboratory 1	Laboratory 2	Laboratory 3	Laboratory 4
Time	Morning	Forenoon	Noon	Afternoon	Evening	
Network	Unknown	Work Wi-Fi	Cellular	Hotspot		
Success rate	Low	Medium	High			
Initial trust value	Minimum	Average	Maximum			
Recommender trust value	Minimum	Average	Maximum			

Most significant parameters to calculate the trust value to access the requested service directly. In an indirect way of accessing, the resource would be by the recommender delegated trust to the requested entity. The following Table 1 lists the common factors which influenced doing the service transaction.

To decide whether to allow or deny the requested service by the entity/user, we need to collect various contextual parameters and trust value. As every parameter contributes in the calculation of the trust value, to allow the entity/user to access the service or to deny the request from the entity. Every parameter was consolidated as a single service priority preference value in multi-criteria decision-making method, to decide the permission to allow or deny the request for the service. Among various MCDM methods, AHP has significant important to account all direct and indirect influential factors for making a deciding preference. AHP widely used in most of the decision-making domain because of its simplified implementation. The parameters that are taken into account to decide the access permission was tabulated in Table 1.

Figure 1 shows the hierarchical level description of the parameters involved to decide the permission for the trusted entity whether to allow or deny the request

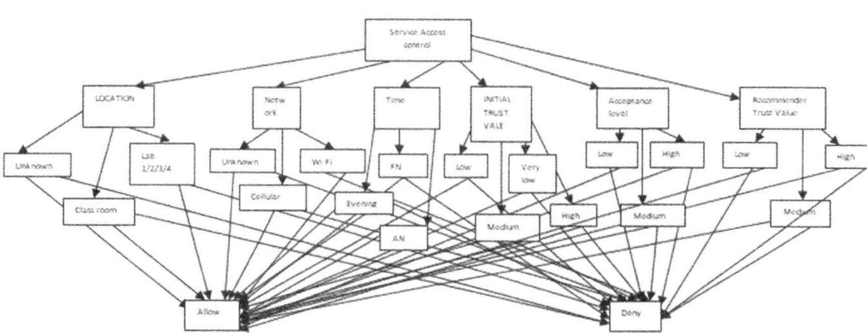

Fig. 1 Analytic hierarchical process structure

to access the service in the pervasive environment. The priority value for location context was shown in Table 2 for the trusted entity/user.

Table 3 contains the relative importance values for each parameters account to fix the service access control permission for the trusted user/entity. As stated in [1], we have calculated the value for $\lambda_{\max}$ is 6.49515, consistency index as 0.09903 using the following formula(1)

$$CI = \frac{\lambda_{\max} - n}{n - 1} \tag{1}$$

Consistency ratio is calculated by dividing the consistency index by random index. Random index (RI) is 1.24 from [9]. Consistency ratio is 0.079683, which is less than 0.1, shows that parameters are consistence and proceed with AHP, to decide the access control permission for the requested service.

From Table 3, we found that the network connectivity is the most important parametric criterion to get access for the requested service by the entity. Then, location and time parameters also had an influence on service access decision.

3 AHP Process for the Trust Value Computation

Final trust value computation is done by the context factor preferences from the AHP process. To validate the process steps, we have taken the academic learning pervasive environment. The calculation configuration comprises a list of context and trust factors to allow or deny the service request based on the importance of each context and trust values.

The context-based trust model consists of the following units:

a. Request Process Unit: This unit manages all request from the trusted entity. The global database stores information about the history of the trusted entity and the available resources in this environment. Based on the calculated trust value, this unit provides the access permission control decision for the requested service.
b. Parameter Process Unit: Which processes the preference context and trust parameters using AHP method and the decision structure shown in Fig. 1.
c. Parameter Collecting Unit: This unit collects and analyzes the context and trust states about the entities in the pervasive environment.

We have developed Web-based learning system to illustrate the above situation. Through the login page, we identify the user. The credentials are the username and their date of birth. To do authorization, we check this input with the existing global database. We had another table with permission rights for the available service in the environment. To define the decision control for service access, we have taken six different contexts and trust parameters, as listed in Table 1. Each context aspects can have a maximum of six different states. The location has six different states: like anonymous, classroom, laboratory 1, laboratory 2, laboratory 3, laboratory 4, while

Table 2 Priority value for location context to provide service access permission

Location—anonymous	Allow	Deny	Priority	Location—classroom	Allow	Deny	Priority	Location—laboratory 1	Allow	Deny	Priority
Allow	1	1/9	0.1	Allow	1	5	0.833	Allow	1	7	0.875
Deny	9	1	0.9	Deny	1/5	1	0.167	Deny	1/7	1	0.125
Location—laboratory 2	Allow	Deny	Priority	Location—laboratory 3	Allow	Deny	Priority	Location—laboratory 4	Allow	Deny	Priority
Allow	1	7	0.875	Allow	1	6	0.857	Allow	1	6	0.857
Deny	1/7	1	0.125	Deny	1/6	1	0.143	Deny	1/6	1	0.143

Table 3 Pairwise comparison of context

Context/trust	Location	Network	Time	Initial trust value	Success rate	Recommender trust value	Priority
Location	1	1/5	5	8	8	8	0.28
Network	5	1	7	8	8	8	0.48
Time	1/5	7	1	4	4	4	0.12
Initial trust value	1/8	1/8	¼	1	1	1	0.04
Success rate	1/8	1/8	¼	1	1	1	0.04
Recommender trust value	1/8	1/8	¼	1	1	1	0.04

time has five states, network has four states, trust-related parameters success rate and initial trust value (assumed), recommender trust value has three states as: low, medium and high and minimum, average, maximum.

The context and trust parameter importance to achieve the service access control level by AHP process is tabulated in Table 3.

Sum weight of the 'allow' state is 0.801435 from Table 4, and similarly, for 'deny' state, it is 0.439526 from Table 5.

From the state, we found that when the location is at classroom and laboratories user preference is high to achieve the service access. Regarding other parameters

Table 4 Relative pairwise matrix for allow state

Context/trust	State	Alternative (A)	Priority value (P)	A * P
Location	Laboratory 2	0.875	0.279272	0.244363
Network	Work	0.857	0.482524	0.413523
Time	FN	0.68	0.118007	0.080245
Initial trust value	Medium	0.2	0.040066	0.008013
Success rate	High	0.88	0.040066	0.035258
Recommender trust value	Medium	0.5	0.040066	0.020033

Table 5 Relative pairwise matrix for deny state

Context/trust	State	Alternative (A)	Priority value(P)	A * P
Location	Undefined	0.857	0.279272	0.239336
Network	Unknown	0.1	0.482524	0.048252
Time	AN	0.68	0.118007	0.080245
Initial trust value	Medium	0.3	0.040066	0.01202
Success rate	Low	0.9	0.040066	0.036059
Recommender trust value	Low	0.4	0.040066	0.016026

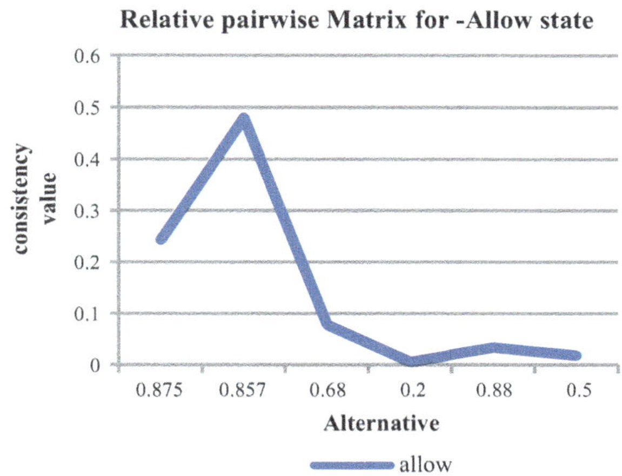

Fig. 2 Relative pairwise matrix for 'allow' state

like time, initial trust value, success rate and recommender trust value, the alternative values are calculated according to the data retrieved from the global database history and priority values. Table 4 shows an example for the user priority with the location state in laboratory 2, with the priority value 0.875 from Table 2 toward allow access alternative, with the decision goal value 0.27 from Table 3. The overall decision alternative value is obtained by multiplying the priority value and decision goal value which is 0.24. For all other context and trust aspects, we have calculated the overall decision alternative value. Allow state sum of various parameter weights is 0.801435 for a specific user. The request process unit responds to the parameter process unit with the allow access when sum weight is greater than the calculated threshold value 0.5. In our case, the overall sum weight is 0.801435, is greater than the threshold value 0.5, and decide to have access permission as 'allow' state. Another example in Table 5, where the sum weight value as 0.43, which is less than the threshold value 0.5, results with the 'deny' state.

The values are obtained through the GNU Octave software for our proposed service access control framework. Figure 2 represents the 'allow' state for the particular user whose location is at the workplace, and Fig. 3 represents the 'deny' state for the same user at an unknown place. Figure 4 represents the normalized pairwise matrix for a different contexts and trust parameters and Fig. 5 represent the relative pairwise values for allowing state for the different user who is in the workplace.

4 Conclusions

In this paper, we proposed a context-based service access control framework for the trusted entity, includes user's preferences to get access permission as 'allow' by their context parameters like location, time of accessing the service, selection of network

Fig. 3 Relative pairwise matrix for 'deny' state

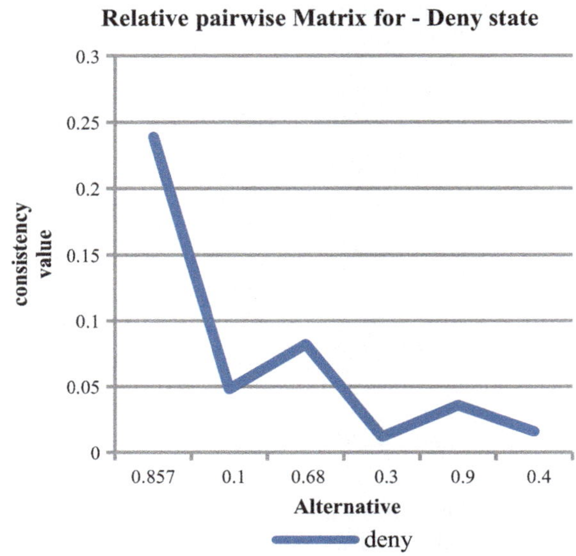

Fig. 4 Normalized pairwise matrix for the location–work

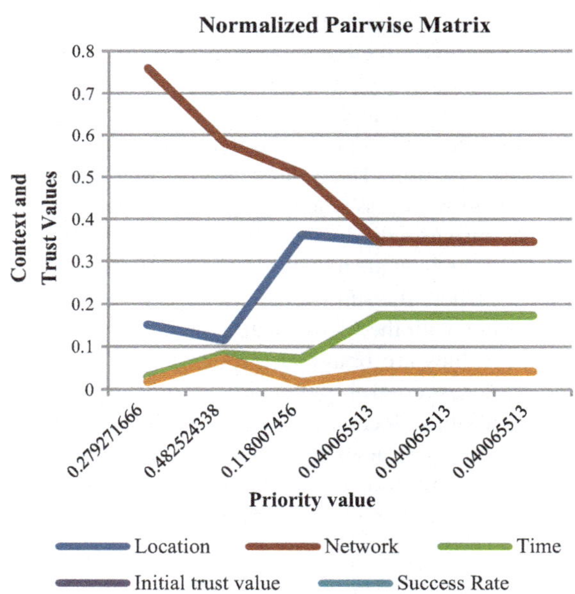

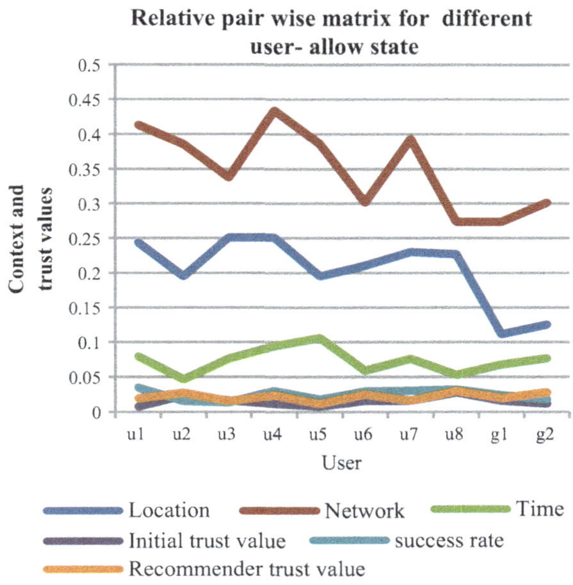

Fig. 5 Relative pairwise matrix for 'allow' state for different user

connectivity along with the trust parameters like initial trust value assigned to all the available service in that environment, the success rate of each entity/user form accessing the available services and recommender trust value for indirect access of the requested service by the trusted entity. The decision to access service by the trusted entity was defined as the AHP process. We checked the results using the GNU octave software inferred that when the user/entity was in the workplace, they have a high preference for the chance of 'allow' state, with an appropriate network selection and threshold trust value for the requested service. Further, we can improve the access preferences through fuzzy AHP method to solve the uncertain situation.

Acknowledgements This research work has been written with the financial support of Rashtriya Uchchatar Shiksha Abhiyan (RUSA–Phase 2.0) grant sanctioned vide Letter No. F.24-51/2014-U, Policy (TNMulti-Gen), Dept. of Edn. Govt. of India, Dt. 09.10.2018.

References

1. Filho, J. B., & Martin, H. (2009). A generalized context-based access control model for pervasive environments. In *Proceedings of the 2nd SIGSPATIAL ACM GIS 2009 International Workshop on Security and Privacy in GIS and LBS* (pp. 12–21).
2. Tian, B., Liu, K., & Chen, Y. (2015). Dynamical trust and reputation computation model for b2c e-commerce. *Future Internet*, *7*(4), 405–428.
3. Boukerche, A., & Ren, Y. (2008). A trust-based security system for ubiquitous and pervasive computing environments. *Computer communications*, *31*(18), 4343–4351.

4. Xiu, D., & Liu, Z. (2005). A formal definition for trust in distributed systems. In *International conference on information security* (pp. 482–489). Springer, Berlin, Heidelberg.
5. Hema, A. M., & Kuppusamy, K. (2012). Trust based access control scheme for pervasive computing environment. In *2012 International Conference on Recent Trends in Information Technology* (pp. 157–161). IEEE.
6. Hayashi, E., Das, S., Amini, S., Hong, J., & Oakley, I. (2013). CASA: context-aware scalable authentication. In *Proceedings of the Ninth Symposium on Usable Privacy and Security* (pp. 1–10).
7. Filho, J. B., & Martin, H. (2008). QACBAC: an owner-centric QoC-aware context-based access control model for pervasive environments. In *Proceedings of the SIGSPATIAL ACM GIS 2008 International Workshop on Security and Privacy in GIS and LBS* (pp. 30–38).
8. Chakraborty, S., Raghavan, K. R., Johnson, M. P., & Srivastava, M. B. (2013). A framework for context-aware privacy of sensor data on mobile systems. In *Proceedings of the 14th Workshop on Mobile Computing Systems and Applications* (pp. 1–6).
9. Taherdoost, H. (2017). Decision making using the analytic hierarchy process (AHP); A step by step approach. *International Journal of Economics and Management Systems, 2.*

Inventory Decision for Substitutable Products Under Quadratic Demand with Quantity Discounts

B. Komala Durga and E. Chandrasekaran

Abstract In this article, an inventory system consisting of two substitutable products is studied. In which, the feature of quantity discount offered to the retailer has been discussed. The inventory of both the products is depleted by their demand which is assumed to be quadratic. When one of them becomes out of stock, partial substitution has been done from the inventory of mutually substitutable product. The unmet demand is assumed to be lost. The mathematical formulation of the model is proposed. The solution procedure is brought forth to obtain the optimal order quantity by minimizing the total cost. Sensitivity analysis has been done to showcase the advantage of substitution over without substitution and hence comprehend the behaviour of the model.

Keywords Inventory control · Product substitution · Quadratic demand · Quantity discount · Substitution rate

1 Introduction

In most of the companies, supermarkets or retailers come across the situation of products being out of stock temporarily and quantity discount. Considering the behaviour of the customer in this situation is quite understandable. They make two choices, one is to wait for the same product to be in stock again, or switch over to an alternative product, which serves them in the same manner. In this case, the retailer orders two mutually substitutable products, also an additional benefit of quantity discount. We consider this scenario in this paper. Two mutually substitutable products are considered in the inventory model. When one of the product is consumed fully, the further demand for the product is partially met from the inventory of the other product. There

B. Komala Durga (✉) · E. Chandrasekaran
Vel Tech Rangarajan Dr. Sagunthala R&D Institute of Science and Technology, Avadi Chennai, India
e-mail: komalasmc20@gmail.com

E. Chandrasekaran
e-mail: e_chandrasekaran@yahoo.com

S.-L. Peng et al. (eds.), *Proceedings of First International Conference on Mathematical Modeling and Computational Science*, Advances in Intelligent Systems and Computing 1292, https://doi.org/10.1007/978-981-33-4389-4_49

are ways involved while making substitution: one may allow complete substitution or partial substitution. Many earlier types of research have worked on the demand of the product being linear and have found that partial substitution is more profitable than a complete substitution. In this article, we have considered two mutually substitutable products having quadratic demand, and partial substitution of the products is done when one stocks out.

A study on inventory model with substitutable products is firstly investigated by Mcgillivray and Silver [1], in which the effects of substitutable demand associated with the inventory and shortage cost are carried out. Salameh et al. [2] developed an inventory model for substitutable products considering the demand deterministic. Huang and Ke [3], Hsieh [4], Li et al. [5] assumed the demand stochastic in their works. Pasternack and Drezner [6] in their paper found that if substitutable does not take place, then the optimal order quantities would become larger or smaller. Zhao et al. [7] investigated the optimal ordering policy for two substitutable products having probabilistic demand whose ordering cost and holding cost are fixed. Drezner et al. [8] compared the inventory model with substitution and without substitution in his works. Gurnani and Drezner [9] extended this model for multiple products. Krommyda et al. [10], Gerchak and Grosfeld [11], Nakhai Kamalabadi and Rasouli [12] proposed inventory system for two mutually substitutable product, whose demand is assumed to be deterministic with fixed ordering and holding cost. However, many of them not considered the demand to be quadratic and quantity discounts in their works. Ernst and Kouvelis [13] suggested a numerical algorithm to find the optimal stocking levels.

In this paper, two mutually substitutable products are considered in the inventory system. The natural phenomenon in the real world of the demand being quadratic, which is studied by Durga and Chandrasekaran [14] and the feature of quantity discount is considered in this proposed model. When one product is out of stock, the demand for it is partially substituted from the leftover inventory of the other product. In the next section, the assumptions and notations used in the model are presented. In Sect. 3, we tend to propose the model featuring product substitution having quadratic demand and quantity discount. A solution procedure is presented in Sect. 4, which enables us to find the optimal ordering quantities. Sensitivity analysis has been done extensively by varying the parameters of the model, and results are presented in Sect. 5. Finally, in Sect. 6, we conclude by giving some useful remarks.

2 Assumptions

The notations and assumptions made in the proposed model are as follows.

- Let Q_1, Q_2 be the order quantity of product 1 and 2.
- Let D_1, D_2 be the demands of the products, which is assumed to be quadratic. When product 1 is fully consumed, the future demand for it is partially fulfilled by product 2.

- Let k_1, k_2 be the fixed ordering cost of product 1 and 2 and h_1, h_2 be the fixed holding cost of products.
- Let ν_1, ν_2 be the substitution rates of the products and $(1-\nu_1)$, $(1-\nu_2)$ are the rate of unmet demand, which is assumed to be lost. Let c_s be the shortage cost of a product.
- Products are ordered together in every ordering cycle. Lead time is zero.
- The quantity discount offered to the retailer by the distributor is given below.

Product 1 quantities	The unit price of product 1	Product 2 quantities	The unit price of product 2
$0 \le Q_1 \le Q_1^1$	c_1	$0 \le Q_2 \le Q_2^1$	d_1
$Q_1^1 \le Q_1 \le Q_1^2$	c_2	$Q_2^1 \le Q_2 \le Q_2^2$	d_2
$Q_1^2 \le Q_1 \le Q_1^3$	c_3	$Q_2^2 \le Q_2 \le Q_2^3$	d_3
…	…	…	…
…	…	…	…
$Q_1^{n-1} \le Q_1 \le Q_1^n$	c_n	$Q_1^{m-1} \le Q_2 \le Q_2^m$	d_m

where n, m are the maximum number of price breaks available for product 1 and 2, respectively.

3 Proposed Mathematical Model

Consider an inventory system consisting of two mutually substitutable product P_1 and P_2. The inventory of the product is depleted because of its demand, which is assumed to be quadratic. Suppose P_1 stocks are fully consumed, then further the fraction of demand of P_1 is substituted by P_2 with the rate of substitution ν_1, the unmet demand of P_1 is concluded to be lost. The rate at which it is lost is $(1-\nu_1)$.

Similarly, consider the case when P_2 stocks out, its demand is partially substituted by P_1 with substitution rate ν_2, and the unmet demand of P_2 is assumed to be lost. The rate at which it is lost is $(1-\nu_2)$. Let T_1 and T_2 be the cycle time of product 1 and product 2. The features of the quantity discount are considered in this model. The total cost equation is derived for the proposed model, based on their cycle time, i.e. when $T_1 \le T_2$ and $T_1 \ge T_2$.

Case (i): When $T_1 \le T_2$.

The total cost for both products, in this case, when $T_1 \le T_2$ as depicted in Fig. 1 comprises ordering, holding and shortage cost, which is obtained as

$$\begin{aligned} \mathrm{TC}_1 = k_1 + k_2 + \frac{2h_1 c_i Q_1^2}{3D_1} + \frac{2h_2 d_j Q_2^2}{3D_2} + \frac{D_1^2 h_2 d_j (Q_1+Q_2)^2 (1-\nu_1)^2}{2(D_2+\nu_1 D_1)(D_1+D_2)^2} \\ + \frac{c_{s_1} D_1^2 (Q_1+Q_2)(1-\nu_1)^2}{(D_2+\nu_1 D_1)(D_1+D_2)} \end{aligned} \quad (1)$$

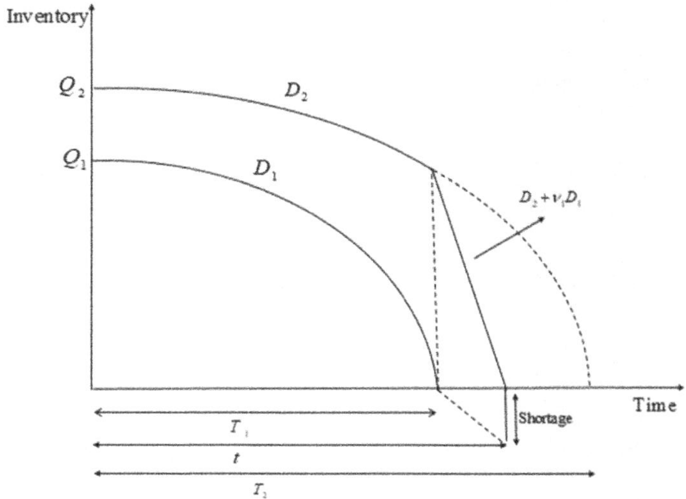

Fig. 1 Graphical representation of the inventory model of the first case when $T_1 \leq T_2$

The total cost per unit time is obtained by dividing the above equation by its cycle time, which is found as

$$\mathrm{TCU}_1 = \frac{D_2 + \nu_1 D_1}{Q_1 + Q_2}\left(\begin{array}{c} k_1 + k_2 + \frac{2h_1 c_i Q_1^2}{3D_1} + \frac{2h_2 d_j Q_2^2}{3D_2} + \frac{D_1^2 h_2 d_j (Q_1+Q_2)^2(1-\nu_1)^2}{2(D_2+\nu_1 D_1)(D_1+D_2)^2} \\ + \frac{c_{s_1} D_1^2 (Q_1+Q_2)(1-\nu_1)^2}{(D_2+\nu_1 D_1)(D_1+D_2)} \end{array}\right) \tag{2}$$

Case (ii): When $T_1 \geq T_2$.

Similarly, when $T_1 \geq T_2$ as depicted in Fig. 2, the total cost equation is obtained as

$$\mathrm{TC}_2 = k_1 + k_2 + \frac{2h_1 c_i Q_1^2}{3D_1} + \frac{2h_2 d_j Q_2^2}{3D_2} + \frac{D_2^2 h_1 c_i (Q_1 + Q_2)^2 (1 - \nu_2)^2}{2(D_1 + \nu_2 D_2)(D_1 + D_2)^2} + \frac{c_{s_2} D_2^2 (Q_1 + Q_2)(1 - \nu_2)^2}{(D_1 + \nu_2 D_2)(D_1 + D_2)} \tag{3}$$

Hence, the total cost per unit of time is obtained by dividing by its cycle time.

$$\mathrm{TCU}_2 = \frac{D_1 + \nu_2 D_2}{Q_1 + Q_2}\left(\begin{array}{c} k_1 + k_2 + \frac{2h_1 c_i Q_1^2}{3D_1} + \frac{2h_2 d_j Q_2^2}{3D_2} + \frac{D_2^2 h_1 c_i (Q_1+Q_2)^2(1-\nu_2)^2}{2(D_1+\nu_2 D_2)(D_1+D_2)^2} \\ + \frac{c_{s_2} D_2^2 (Q_1+Q_2)(1-\nu_2)^2}{(D_1+\nu_2 D_2)(D_1+D_2)} \end{array}\right) \tag{4}$$

Case (iii): When no substitution takes place.

The total cost per unit time in the case of no substitution is easily found as

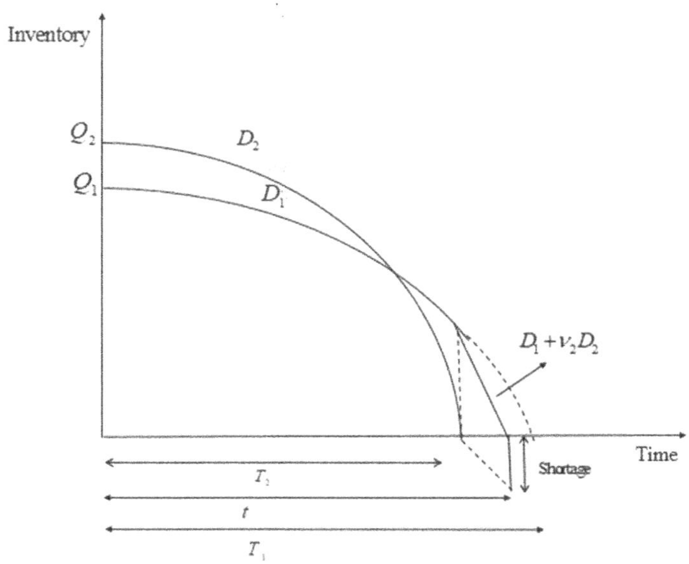

Fig. 2 Graphical representation of the inventory model of the second case when $T_1 \geq T_2$

$$\mathrm{TCU} = \left(\frac{D_1 + D_2}{Q_1 + Q_2}\right)\left[k_1 + k_2 + \frac{2h_1 c_i Q_1^2}{3D_1} + \frac{2h_2 d_j Q_2^2}{3D_2}\right] \tag{5}$$

and this is later used to comprehend the behaviour of the model, stating the advantage of substitution over no substitution.

4 Solution Procedure

A procedure for solving the proposed model is recommended in this section, in order to optimize the total cost per unit time TCU_1 and TCU_2 obtained in Eqs. (2) and (4). We first show that total costs TC_1 and TC_2 are obtained in Eqs. (1) and (3) to be a convex function by proving that it is Hessian matrix positive semi-definite. Then, applying the mathematical statement, which states, "the ratio of a positive convex function over a linear function is pseudo-convex". This guarantees a unique optimal order quantity $\left(Q_1^*, Q_2^*\right)$ by using any search technique.

To find the determinant of the Hessian matrix of TC_1, the following partial derivatives are determined.

$$\frac{\partial^2 \mathrm{TC}_1}{\partial Q_1^2} = \frac{4c_i h_1}{3D_1} + \frac{2D_1^2 h_2 d_j (1 - \nu_1)^2}{2(D_2 + D_1\nu_1)(D_1 + D_2)^2} \geq 0$$

$$\frac{\partial^2 \mathrm{TC}_1}{\partial Q_2^2} = h_2 d_j \left(\frac{4}{3D_2} + \frac{2D_1^2(1-\nu_1)^2}{2(D_2+\nu_1 D_1)(D_1+D_2)^2} \right) \geq 0$$

$$\frac{\partial^2 \mathrm{TC}_1}{\partial Q_1 \partial Q_2} = \frac{2D_1^2 h_2 d_j (1-\nu_1)^2}{2(D_2+\nu_1 D_1)(D_1+D_2)^2} \geq 0$$

Hence, the determinant of the hessian matrix is $\left(\frac{\partial^2 \mathrm{TC}_1}{\partial Q_1^2} \frac{\partial^2 \mathrm{TC}_1}{\partial Q_2^2} \right) - \left(\frac{\partial^2 \mathrm{TC}_1}{\partial Q_1 \partial Q_2} \right)^2 > 0$ which implies TC_1 is convex. Similarly, the Hessian matrix TC_2 is also positive semi-definite.

Since, $\mathrm{TCU}_1 = \frac{\mathrm{TC}_1}{t}$ where t is its cycle time given by $t = \frac{Q_1+Q_2}{D_2+\nu_1 D_1}$. By the property, TCU_1 is pseudo-convex. Similarly, TCU_2 is also pseudo-convex.

We next solve the constrained optimization problem:

$\min\limits_{Q_1, Q_2} \mathrm{TCU}_1(Q_1, Q_2)$ s.t $\frac{Q_1}{D_1} \leq \frac{Q_2}{D_2}$ and $\min\limits_{Q_1, Q_2} \mathrm{TCU}_2(Q_1, Q_2)$ s.t $\frac{Q_1}{D_1} \geq \frac{Q_2}{D_2}$.

Let the optimal solution be a minimum of $\left(\mathrm{TCU}_1^*, \mathrm{TCU}_2^*\right)$. If it is not feasible, consider all the other higher combinations of order quantity especially at price breaks and then solve the optimization problem until a reasonable order quantity is found. If it is feasible, then the retailer has to fix these quantities as the optimal value.

In the next section, an extensive illustration of numerical examples and sensitivity analysis is carried out.

5 Sensitivity Analysis

To illustrate the proposed model, a numerical example is considered. This will give an insight into the behaviour of the model. Initialize demand rates as $D_1 = 200,\ D_2 = 100$, holding cost $h_1 = 10,\ h_2 = 10$, ordering cost $k_1 = 150, k_2 = 150$, substitution rate $\nu_1 = 0.2,\ \nu_2 = 0.8$, shortage cost $c_{s_1} = 1,\ c_{s_2} = 30$ the lowest possible unit price for product 1 is $c_1 = 5,\ c_2 = 4, c_3 = 3$ and for product 2 as $d_1 = 1,\ d_2 = 2, d_3 = 3$.

MATLAB optimization toolbox is used for optimizing the total cost for case(i) and case(ii) and then compared with case(iii) to show that there is an enormous improvement in the total cost compared to substitution over without substitution.

For the initial data set-up, the optimal ordering policy for case(i) is $Q_1 = 30.93, Q_2 = 46.40$ and $\mathrm{TCU}_1 =$ Rs.1171.49. And for case (ii) $Q_1 = 50.59$, $Q_2 = 25.29$ and $\mathrm{TCU}_1 =$ Rs. 2215.07. Note that $\mathrm{TCU}_1 < \mathrm{TCU}_2$. Therefore, the optimal ordering policy is $Q_1^* = 30.93 \quad Q_2^* = 46.40 \quad \mathrm{TCU}_1^* =$ Rs. 1171.49. When no substitution takes place, the optimal ordering policy is obtained $Q_1 = 34.61 \quad Q_2 = 51.96 \quad \mathrm{TCU}_1 =$ Rs. 2078.46. An enormous improvement percentage of 43.64% is calculated for the total cost with substitution over without substitution.

Next, sensitivity analysis is extensively done by increasing the fixed ordering cost and the percentage improvement for substitution over without substitution is presented in Table 1; Fig. 3. Similarly, analysis has been done for holding cost (Table 2 and Fig. 4), shortage cost (Fig. 5) and substitution rates (Fig. 6).

Table 1 Set-up cost analysis

k_1	Unit cost of ordering quantity	Optimal order policy with substitution			Optimal order policy without substitution			% improvement
		Q_1	Q_2	TC(WS)	Q_1	Q_2	TC(WoS)	
150	$c_1 = 5, d_3 = 1$	20.59	51.47	1251.07	22.68	56.69	2267.79	44.83
	$c_1 = 5, d_2 = 2$	25.06	31.32	1575.33	28.28	35.36	2828.43	44.30
	$c_1 = 5, d_1 = 3$	27.12	22.60	1775.00	31.33	26.11	3133.40	43.35
	$c_2 = 4, d_3 = 1$	24.71	49.42	1218.59	27.39	54.77	2190.89	44.38
	$c_2 = 4, d_2 = 2$	29.36	29.36	1515.67	33.54	33.54	2683.28	43.51
	$c_2 = 4, d_1 = 3$	31.27	20.84	1697.29	36.74	24.49	2939.39	42.26
	$c_3 = 3, d_3 = 1$	30.93	46.40	1171.49	34.64	51.96	2078.46	43.64
	$c_3 = 3, d_2 = 2$	35.56	26.67	1435.10	41.40	31.05	2484.24	42.23
	$c_3 = 3, d_1 = 3$	37.05	18.52	1596.89	44.72	22.36	2683.28	40.49
175	$c_1 = 5, d_3 = 1$	21.43	53.57	1298.67	23.60	59.01	2360.39	44.98
	$c_1 = 5, d_2 = 2$	26.08	32.60	1636.17	29.44	36.80	2943.92	44.42
	$c_1 = 5, d_1 = 3$	28.22	23.52	1844.00	32.61	27.18	3261.34	43.46
	$c_2 = 4, d_3 = 1$	25.72	51.43	1264.86	28.50	57.01	2280.35	44.53
	$c_2 = 4, d_2 = 2$	30.56	30.56	1574.08	34.91	34.91	2792.85	43.64
	$c_2 = 4, d_1 = 3$	32.54	21.70	1763.11	38.24	25.50	3059.41	42.37
	$c_3 = 3, d_3 = 1$	32.20	48.30	1215.84	36.06	54.08	2163.33	43.80
	$c_3 = 3, d_2 = 2$	37.01	27.76	1490.21	43.09	32.32	2585.67	42.37
	$c_3 = 3, d_1 = 3$	38.56	19.28	1658.61	46.55	23.27	2792.85	40.61
200	$c_1 = 5, d_3 = 1$	22.24	55.59	1344.47	24.49	61.24	2449.49	45.11
	$c_1 = 5, d_2 = 2$	27.06	33.83	1694.71	30.55	38.19	3055.05	44.53
	$c_1 = 5, d_1 = 3$	29.29	24.41	1910.38	33.84	28.20	3384.46	43.55
	$c_2 = 4, d_3 = 1$	26.69	53.37	1309.39	29.58	59.16	2366.43	44.67
	$c_2 = 4, d_2 = 2$	31.72	31.72	1630.27	36.23	36.23	2898.28	43.75
	$c_2 = 4, d_1 = 3$	33.77	22.51	1826.44	39.69	26.46	3174.90	42.47
	$c_3 = 3, d_3 = 1$	33.41	50.12	1258.52	37.42	56.12	2244.99	43.94
	$c_3 = 3, d_2 = 2$	38.41	28.81	1543.25	44.72	33.54	2683.28	42.49
	$c_3 = 3, d_1 = 3$	40.02	20.01	1718.00	48.30	24.15	2898.28	40.72
225	$c_1 = 5, d_3 = 1$	23.02	57.54	1388.66	25.35	63.39	2535.46	45.23
	$c_1 = 5, d_2 = 2$	28.01	35.02	1751.20	31.62	39.53	3162.28	44.62
	$c_1 = 5, d_1 = 3$	30.32	25.26	1974.44	35.03	29.19	3503.25	43.64
	$c_2 = 4, d_3 = 1$	27.62	55.25	1352.35	30.62	61.24	2449.49	44.79
	$c_2 = 4, d_2 = 2$	32.83	32.83	1684.50	37.50	37.50	3000.00	43.85
	$c_2 = 4, d_1 = 3$	34.96	23.30	1887.55	41.08	27.39	3286.34	42.56
	$c_3 = 3, d_3 = 1$	34.59	51.88	1299.69	38.73	58.09	2323.79	44.07
	$c_3 = 3, d_2 = 2$	39.76	29.82	1594.42	46.29	34.72	2777.46	42.59
	$c_3 = 3, d_1 = 3$	41.42	20.71	1775.30	50.00	25.00	3000.00	40.82

(continued)

Table 1 (continued)

k_1	Unit cost of ordering quantity	Optimal order policy with substitution			Optimal order policy without substitution			% improvement
		Q_1	Q_2	TC(WS)	Q_1	Q_2	TC(WoS)	
300	$c_1 = 5, d_3 = 1$	25.21	63.04	1513.06	27.77	69.44	2777.46	45.52
	$c_1 = 5, d_2 = 2$	30.69	38.36	1910.20	34.64	43.30	3464.10	44.86
	$c_1 = 5, d_1 = 3$	33.21	27.68	2154.75	38.38	31.98	3837.61	43.85
	$c_2 = 4, d_3 = 1$	30.26	60.52	1473.28	33.54	67.08	2683.28	45.09
	$c_2 = 4, d_2 = 2$	35.96	35.96	1837.13	41.08	41.08	3286.34	44.10
	$c_2 = 4, d_1 = 3$	38.29	25.53	2059.57	45.00	30.00	3600.00	42.79
	$c_3 = 3, d_3 = 1$	37.89	56.83	1415.60	42.43	63.64	2545.58	44.39
	$c_3 = 3, d_2 = 2$	43.55	32.67	1738.45	50.71	38.03	3042.56	42.86
	$c_3 = 3, d_1 = 3$	45.37	22.69	1936.60	54.77	27.39	3286.34	41.07

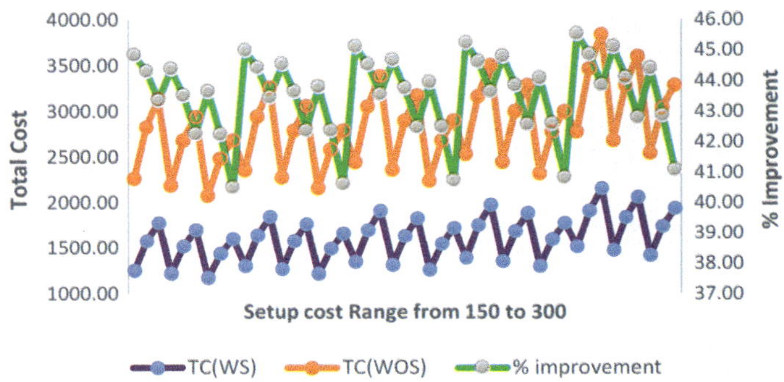

Fig. 3 Percentage improvement in total cost as set-up cost varies

Table 2 Holding cost analysis

h_1	Unit cost of ordering quantity	Optimal order policy with substitution			Optimal order policy without substitution			% improvement
		Q_1	Q_2	TC(WS)	Q_1	Q_2	TC(WoS)	
6	$c_1 = 5, d_3 = 1$	34.64	51.96	2078.46	30.93	46.40	1171.49	43.64
	$c_1 = 5, d_2 = 2$	41.4	31.05	2484.24	35.56	26.67	1435.10	42.23
	$c_1 = 5, d_1 = 3$	44.72	22.36	2683.28	37.05	18.52	1596.89	40.49
	$c_2 = 4, d_3 = 1$	41.29	49.54	1981.73	36.50	43.80	1131.38	42.91
	$c_2 = 4, d_2 = 2$	48.41	29.05	2323.79	40.82	24.49	1371.36	40.99
	$c_2 = 4, d_1 = 3$	48.04	24.02	2498.00	38.87	19.43	1526.07	38.91
	$c_3 = 3, d_3 = 1$	51.3	46.17	1846.76	44.62	40.16	1076.13	41.73
	$c_3 = 3, d_2 = 2$	56.69	28.35	2116.60	46.45	23.22	1290.95	39.01
	$c_3 = 3, d_1 = 3$	52.22	26.11	2297.83	40.99	20.49	1451.59	36.83

(continued)

Table 2 (continued)

h_1	Unit cost of ordering quantity	Optimal order policy with substitution			Optimal order policy without substitution			% improvement
		Q_1	Q_2	TC(WS)	Q_1	Q_2	TC(WoS)	
8	$c_1 = 5, d_3 = 1$	27.39	54.77	2190.89	24.71	49.42	1218.59	44.38
	$c_1 = 5, d_2 = 2$	33.54	33.54	2683.28	29.36	29.36	1515.67	43.51
	$c_1 = 5, d_1 = 3$	36.74	24.49	2939.39	31.27	20.84	1697.29	42.26
	$c_2 = 4, d_3 = 1$	32.89	52.62	2104.94	29.45	47.11	1182.54	43.82
	$c_2 = 4, d_2 = 2$	39.53	31.62	2529.82	34.11	27.29	1453.41	42.55
	$c_2 = 4, d_1 = 3$	42.83	22.84	2740.99	35.71	19.05	1619.27	40.92
	$c_3 = 3, d_3 = 1$	41.29	49.54	1981.73	36.50	43.80	1131.38	42.91
	$c_3 = 3, d_2 = 2$	48.41	29.05	2323.79	40.82	24.49	1371.36	40.99
	$c_3 = 3, d_1 = 3$	48.04	24.02	2498.00	38.87	19.43	1526.07	38.91
10	$c_1 = 5, d_3 = 1$	22.68	56.69	2267.79	20.59	51.47	1251.07	44.83
	$c_1 = 5, d_2 = 2$	28.28	35.36	2828.43	25.06	31.32	1575.33	44.30
	$c_1 = 5, d_1 = 3$	31.33	26.11	3133.40	27.12	22.60	1775.00	43.35
	$c_2 = 4, d_3 = 1$	27.39	54.77	2190.89	24.71	49.42	1218.59	44.38
	$c_2 = 4, d_2 = 2$	33.54	33.54	2683.28	29.36	29.36	1515.67	43.51
	$c_2 = 4, d_1 = 3$	36.74	24.49	2939.39	31.27	20.84	1697.29	42.26
	$c_3 = 3, d_3 = 1$	34.64	51.96	2078.46	30.93	46.40	1171.49	43.64
	$c_3 = 3, d_2 = 2$	41.4	31.05	2484.24	35.56	26.67	1435.10	42.49
	$c_3 = 3, d_1 = 3$	44.72	22.36	2683.28	37.05	18.52	1596.89	40.72
12	$c_1 = 5, d_3 = 1$	19.36	58.09	2323.79	17.65	52.96	1274.84	45.23
	$c_1 = 5, d_2 = 2$	24.49	36.74	2939.39	21.87	32.81	1621.39	44.62
	$c_1 = 5, d_1 = 3$	27.39	27.39	3286.34	23.98	23.98	1837.13	43.64
	$c_2 = 4, d_3 = 1$	23.48	56.36	2254.41	21.30	51.11	1245.40	44.79
	$c_2 = 4, d_2 = 2$	29.19	35.03	2802.60	25.81	30.97	1564.66	43.85
	$c_2 = 4, d_1 = 3$	32.27	25.82	3098.39	27.85	22.28	1760.88	42.56
	$c_3 = 3, d_3 = 1$	29.88	53.79	2151.41	26.87	48.36	1202.00	44.07
	$c_3 = 3, d_2 = 2$	36.27	32.65	2611.71	31.55	28.40	1486.53	42.59
	$c_3 = 3, d_1 = 3$	39.53	23.72	2846.05	33.33	20.00	1660.39	40.82
14	$c_1 = 5, d_3 = 1$	16.9	59.16	2366.43	15.46	54.10	1293.01	45.52
	$c_1 = 5, d_2 = 2$	21.62	37.84	3027.15	19.42	33.99	1658.07	44.86
	$c_1 = 5, d_1 = 3$	24.36	28.42	3410.39	21.51	25.09	1888.03	43.85
	$c_2 = 4, d_3 = 1$	20.57	57.58	2303.32	18.72	52.42	1266.14	45.09
	$c_2 = 4, d_2 = 2$	25.88	36.23	2898.28	23.04	32.26	1604.28	44.10
	$c_2 = 4, d_1 = 3$	28.83	26.91	3229.18	25.14	23.46	1813.83	42.79
	$c_3 = 3, d_3 = 1$	26.29	55.21	2208.49	23.76	49.89	1226.00	44.39
	$c_3 = 3, d_2 = 2$	32.33	33.95	2715.81	28.38	29.80	1528.98	42.86
	$c_3 = 3, d_1 = 3$	35.5	24.85	2982.30	30.33	21.23	1714.36	41.07

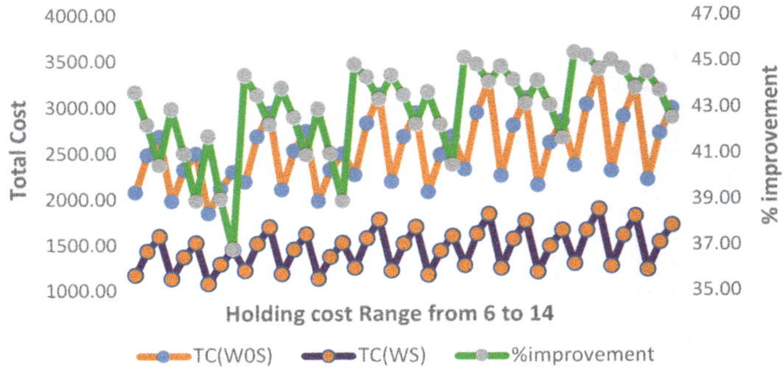

Fig. 4 Percentage improvement in total cost as holding cost varies

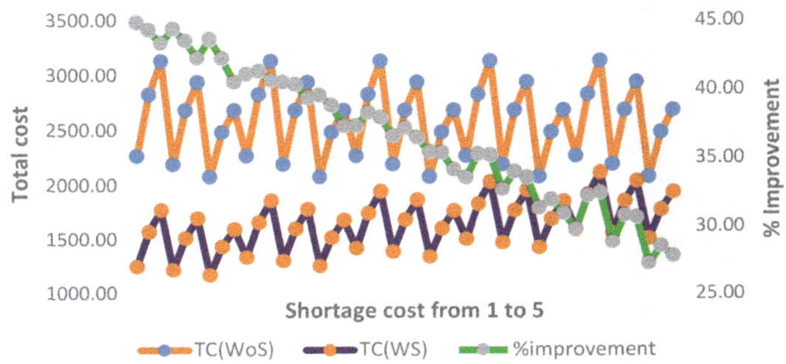

Fig. 5 Percentage improvement in total cost as shortage cost varies

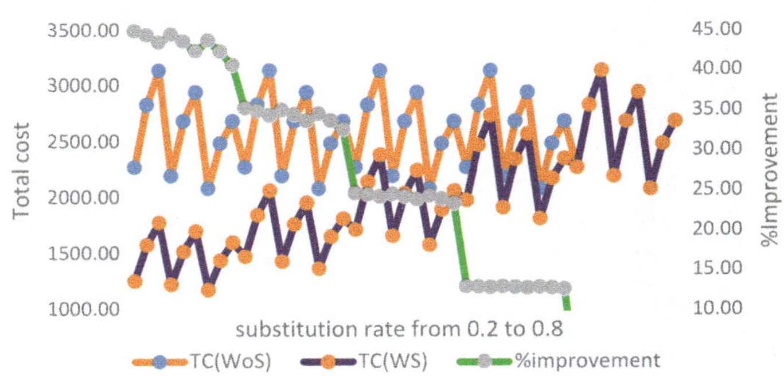

Fig. 6 Percentage improvement in total cost as the rate of substitution varies

Table 3 Summary of sensitivity analysis

Parameter	Variation	TC(WoS)	TC(WS)	% Improvement
Setup cost (k_1)	Increases	Increases	Increases	Increases
Holding cost (h_1)	Increases	Increases	Increases	Increases
Shortage cost (π_1)	Increases	Constant	Increases	Decreases
Substitution rate (υ_1)	Increases	Constant	Increases	Decreases

6 Conclusion

An inventory model with two mutually substitutable products, having quadratic demand, is proposed in this article. The aspect of quantity discount and substitution rate has been considered. A mathematical model is formulated for the proposed inventory system. A procedure for solving the model has been discussed to obtain the optimal ordering policy. Sensitivity analysis has been carried out. The summary of it is presented in Table 3. It signifies that when the ordering cost and holding cost increase, the percentage improvement for substitution over without substitution also increases, whereas when the shortage cost and substitution rate increase, the percentage improvement decreases. Further extension can be done on this model by considering multiple products including the cost of substitution.

References

1. Mcgillivray, R., & Silver, E. (1978). Some concepts for inventory control under substitutable demand. *INFOR: Information Systems and Operational Research, 16*(1), 47–63.
2. Salameh, M. K., Yassine, A. A., Maddah, B., & Ghaddar, L. (2014). Joint replenishment model with substitution. *Applied Mathematical Modelling, 38*(14), 3662–3671.
3. Huang, H., & Ke, H. (2017). Pricing decision problem for substitutable products based on uncertainty theory. *Journal of Intelligent Manufacturing, 28*(3), 503–514.
4. Hsieh, Y. J. (2011). Demand switching criteria for multiple products: An inventory cost analysis. *Omega, 39*(2), 130–137.
5. Li, X., Nukala, S., & Mohebbi, S. (2013). Game theory methodology for optimizing retailers' pricing and shelf-space allocation decisions on competing substitutable products. *The International Journal of Advanced Manufacturing Technology, 68*(1–4), 375–389.
6. Pasternack, B. A., & Drezner, Z. (1991). Optimal inventory policies for substitutable commodities with stochastic demand. *Naval Research Logistics (NRL), 38*(2), 221–240.
7. Zhao, J., Wei, J., & Li, Y. (2014). Pricing decisions for substitutable products in a two-echelon supply chain with firms' different channel powers. *International Journal of Production Economics, 153*, 243–252.
8. Drezner, Z., Gurnani, H., & Pasternack, B. A. (1995). An EOQ model with substitutions between products. *Journal of the Operational Research Society, 46*(7), 887–891.
9. Gurnani, H., & Drezner, Z. (2000). Deterministic hierarchical substitution inventory models. *Journal of the Operational Research Society, 51*(1), 129–133.
10. Krommyda, I. P., Skouri, K., & Konstantaras, I. (2015). Optimal ordering quantities for substitutable products with stock-dependent demand. *Applied Mathematical Modelling, 39*(1), 147–164.

11. Gerchak, Y., & Grosfeld-Nir, A. (1999). Lot-sizing for substitutable, production-to-order parts with random functionality yields. *International Journal of Flexible Manufacturing Systems, 11*(4), 371–377.
12. Nakhai Kamalabadi, I., & Rasouli, N. (2014). Joint pricing and inventory control for seasonal and substitutable goods mentioning the symmetrical and asymmetrical substitution. *International Journal of Engineering, 27*(9), 1385–1394.
13. Ernst, R., & Kouvelis, P. (1999). The effects of selling packaged goods on inventory decisions. *Management Science, 45*(8), 1142–1155.
14. Durga, B. K., & Chandrasekaran, E. (2020). Optimal ordering policy for substitutable products under quadratic demand with cost of substitution. *Journal of Critical Reviews, 7*(4), 200–2005.

Executing the Apriori Hybrid Algorithm in Semi-structured Mining Datasets and Comparison with HD Algorithm

D. Akila, S. Jeyalaksshmi, G. Suseendran, and D.Balaganesh

Abstract The recent development in data mining in various fields and firms, for example, hospital, medicinal research, banking, government firm attracted many users. Data examination can be done in numerous possible ways quickly. But for the data examination in semi-structured datasets like XML format, json is little tricky than standard datasets. The easy examination of data in semi-structured datasets is possible by using Apriori hybrid algorithm. Hash tree algorithm is combined with weighed Apriori to get the Apriori hybrid algorithm and when compared with hybrid distribution algorithm (HD algorithm). The pseudo-code is then created and implemented for the proposed algorithm. The systematic graph illustrates to differentiate the Apriori hybrid algorithm and other algorithms like weighted, hash tree, and hybrid distribution algorithm.

Keywords Apriori hybrid algorithm · Hash tree algorithm · Distribution algorithm

1 Introduction

Data mining had attracted much considerable attention in recent years in various fields and researchers. Data mining applies in various areas like financial time series, for example, forecasts in their financial areas, market strategy, etc. Knowledge discovery

D. Akila (✉) · S. Jeyalaksshmi · G. Suseendran
Department of Information Technology, Vels Institute of Science and Technology and Advanced Studies, Chennai, India
e-mail: akiindia@yahoo.com

S. Jeyalaksshmi
e-mail: pravija.lakshmi@gmail.com

G. Suseendran
e-mail: suseendar_1234@yahoo.co.in

D.Balaganesh
Department of Computer Science and Multimedia, Lincoln University College, Kota Bharu, Malaysia
e-mail: balaganesh@lincoln.edu.my

S.-L. Peng et al. (eds.), *Proceedings of First International Conference on Mathematical Modeling and Computational Science*, Advances in Intelligent Systems and Computing 1292, https://doi.org/10.1007/978-981-33-4389-4_50

mining also known as KDM is one of the significant steps in data mining. Knowledge discovery mining is a computational method used in large datasets for discover patterns. Various areas and techniques like AI, artificial neural network, medical research, machine learning, etc., involved in it.

In data mining complete logics, architectures devoted to the extraction of information from base datasets. Those extricated data help clients in the forecast just as the executive choices for better results. It is particularly valuable in the idea of market bushel examination. Clients, either from customary or Internet business market, may not know complete rundown of things that they need to buy. Those separated data help heads to decide the promoting conduct of individual client dependent on past designs. They may mastermind items in like manner in physical markets that are now and again chosen together by clients. Aside from that if there should arise an occurrence of the virtual market for an example Internet business, clients might be offered with beneficial items that are of clients' advantage. Besides, clients pacified with various kind of offers on specific items dependent on such removed data.

In advance, user preferences the determining, predictions are the essential tools in data mining and can be helpful in client conduct investigation also as in choice demonstrating. The Apriori calculation is the main most well-known calculation as we can as the first most popular one. This method known as the Apriori algorithm is an exceptionally mainstream design finding calculation in information mining strategy. This calculation focused on finding an item set blend which has a specific estimation of recurrence as per the ideal rules or channel. This kind of algorithm used to analyze and squish important information in data. This kind of useful tools is associated with note separate documents, find the different patterns, changing trends, etc., across multiple document data. In data mining, text mining is more and more popular—this kind of mining fewer troubles and efforts for the user. The T hash Apriori is used in this method merged with weighted Apriori to form the Apriori hybrid algorithm.

From massive data repositories, data mining aimed to extract the hidden information, for example, file archives, etc., and is used for constructing price value data patterns and some kinds of predictive models. The associative rule mining is one of the techniques in this primary data mining, and other techniques are clustering, the last technique is classification. The distribution network study is an essential content in distributed network configuration. Sometimes in the large-scale data mining, it is inevitable to deal with it becomes harder, and some business organization or research organization, there is an inevitable need for parallel and distributed computing. The combination of count distribution and intelligent data distribution is known as hybrid distribution algorithm. We used the resulted values from the Apriori hybrid algorithm and compared to the hybrid distribution algorithm for further process to get our method is better than other algorithms.

2 Related Works

Bo Wu et al. [1] illustrated that in large datasets, the association relationships founded by rule association mining. The rule of association mining has a frequent pattern as necessary. In their work, they used an Apriori growth-based algorithm to get mining frequent patterns. The FP-tree structure is used in this algorithm based on the Apriori algorithm to get the Apriori growth algorithm. The sub-conditional pattern tree generated in this Apriori growth algorithm is one of the advantages; instead, conditional patterns do not need to generate. Their method shows that Apriori growth algorithm is more significant than and performs faster when comparing the computational results with the Apriori algorithm and FP-tree structure.

Suriya et al. [2] illustrated that the large datasets are represented by acquiring data by analyzing, refining and information extraction from data mining. The proposed method called KDCL algorithm means k-direct count and intersect, and it combines with Apriori algorithm to acquire mining frequent itemsets. The hybridized KDCL algorithm is combined with the Apriori algorithm to get better performance when these two algorithm are combined; they get better results than individual results. The combined two algorithms have better mining speed and scalability than working individually. This k-direct count and intersect embrace the packed information. Presentation in count and intersect is very viable because of the embracement of advancement strategies.

Shankar et al. [3] illustrated that the knowledge discovery in data mining important aspect. This kind process that they used the Apriori algorithm. The time is one of the drawbacks in the Apriori algorithm, and also space is a disadvantage. It produces various uninteresting itemsets which lead to producing different principles which are of totally of no utilization. The two elements considered for affiliation rules age are minimum support threshold and minimum certainty threshold. Be that as it may, imperative mining decreases these two constraints of Apriori calculation to an impressive degree. Between MST also with MCT, the AND operator is used with combining constraint mining is used in this method to prune frequent itemsets [4]. It is additionally evident that it has improved the calculation if there should be an occurrence of the two parameters as existence. Future work may incorporate the blend of any man-made consciousness calculations with this proposed procedure and try for better execution [5].

Surendra Kumar Chadokar et al. [6] illustrated that to generate consecutive itemsets, a technique called associative rule mining. This application used in different areas, for example, network traffic analysis, it is based on the basics of the datasets. For the hybrid Apriori genetic algorithm is the method they proposed [7]. The hybrid Apriori genetic algorithm is the combination of genetic algorithm combined with the Apriori algorithm.

Ali Akbar et al. [8] developed a software tool to solve the shortest path in Internet network implementation. The problem that occurs commonly is to connect every node to connect network [9]. In information development, it is essential in the data network. Their exploration demonstrates that from execution time and emphasis point of view, the hereditary calculation is undeniably progressively prevalent, while Apriori calculation is better in its capacity than decide separation [10, 11]. Their method used to find the briefest way by improving their proposed calculation.

3 Proposed Method

In this proposed method, we use Apriori hybrid algorithm for association rule mining. The result we got from the Apriori hybrid algorithm, and we use hybrid distribution algorithm for the comparison process. The comparison includes consumption of memory, compare efficiency, some frequency, and also some parameters. A standard data stored in a data center might be structured and semi-structured data.

The semi-structured data mostly can be stored in the book database that implies it does not organize as there in ordinary database information. What is more, not unstructured methods total tremendous proceeds with related information.

On the off chance that we consider model, for example, novel-related content itemset, we can say that the title of the novel/books, creator ID, and so forth information is in organized structure and unique and substance of the novel shown as unorganized structure. So semi-organized information mining design examining that power information on the example thing sets is almost hard to look in the present required immense itemset. At that point apply Apriori cross-breed calculation for affiliation mining. Then having gotten the results from our Apriori hybrid algorithm, we compare with the hybrid association algorithm in the graph for better comparison.

A. *Hash function*

The hash table is used in this hash tree Apriori algorithm to achieve uses in the data structure. In the hash tree calculation, the order to improve drawback in some algorithm like Apriori like improving quality in k itemset. The key to improve some type of presentation the key method is the two itemsets.

This calculation utilizes a hash tree-based strategy to consolidate the quantity of up-and-comer thing sets delivered in the main pass. It attested that the number of thing sets in C-2 delivered utilizing hashing can be minor so that the sweep necessary to set up L-2 is progressively productive.

Algorithm

1. Create two itemsets from all transaction by the scan.
2. Input value 8 for size in the hash table.
3. The candidate pairs are assigned in every bucket using ASCII itemset values.
4. The count in the hash table taken in each bucket, in the bucket the hashed itemset
5. Bucket count if it is equal to or greater than the bit vector = one or else zero.
6. In detached bit, the vector is not set in candidate sets location that hash.
7. The competitor matches that hash to areas where the bit vector bit isn't set are disconnected.

8. Candidate sets only.

Drawbacks:

- The requirement in the hash tree algorithm for calculations is high.
- Consumption of memory is high.
- High time is needed to compute in knots dispensation.

B *Weighted Apriori*

In this algorithm, a record is formed whenever an operation in the dataset is conducted in this Apriori algorithm. The item is each dataset resulted in the record.

Given the created affiliation rules, routinely happened Item sets were mined effortlessly. Given in the equation below

$$W_s(AB) = \frac{\sum_{t=1}^{|W_{st}|\&(A\cup B)\subseteq K_t} \text{wieght}(K_t)}{\sum_{t=1}^{W_{st}} \text{weight}(K_t)} \quad (1)$$

Drawbacks.

- The candidate itemset amplified because the arrangement is generated recurrently in itemset.
- The more time it took for each transaction in the weight algorithm.
- Accurateness is a small amount in weight algorithm.
- The data divergence in this algorithm is in depend.

From Table 1; Fig. 1, we can say that the memory compared with the weight Apriori algorithm and hash tree algorithm the memory needed for the weighted algorithm is large. And the time compared with the weight Apriori algorithm and hash tree algorithm, the time needed for a weighted algorithm is larger than a hash tree.

C. *Proposed algorithm*

The proposed algorithm we used is named as Ah Algorithm A is Apriori and h-hybrid, which is explained in Fig. 2. This algorithm is a combination of a hash tree with weighted Apriori calculation.

By thinking about both the calculations, gainful focuses on the semi-structured dataset to sort that itemset to mine. This method proposed to improve the disadvantages faced in the other algorithm faced above like time and space.

Table 1 Table of weighted, hash tree and hybrid distribution

	Weighted	Hash tree
Time	29	3
Memory	18	15

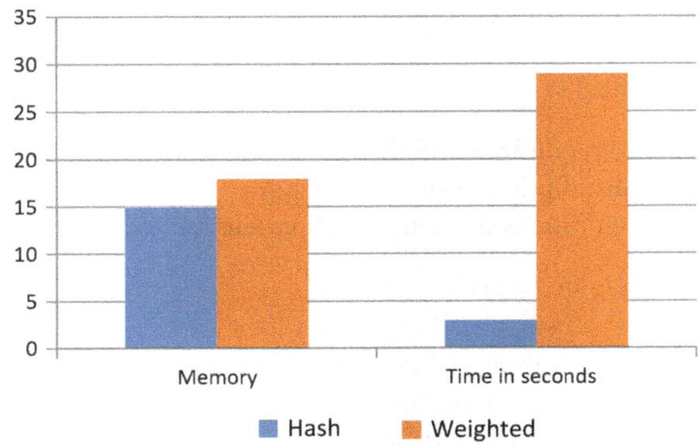

Fig. 1 Memory and time comparison

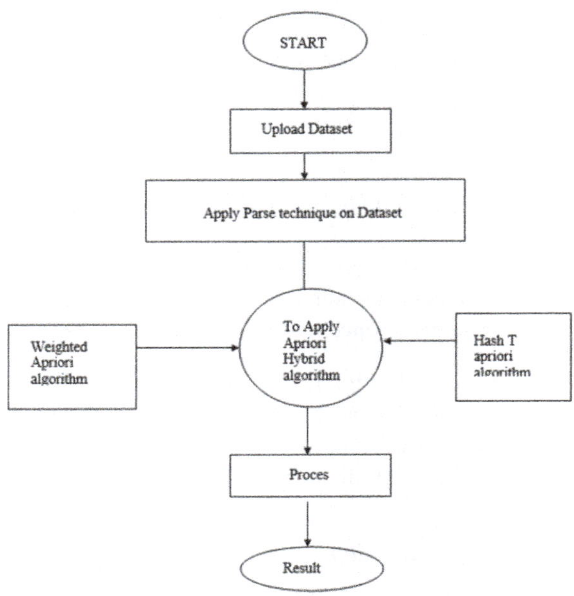

Fig. 2 Proposed method

$$\text{Hybrid}(K_t) = \frac{\sum_{i=1}^{|H_{st}|} \text{Hybrid}(\text{item}(i))}{H_{st}(K_t)} \tag{2}$$

The proposed algorithm is depicted below

```
T_D= Dataset of input items
D_o=output
D_T= Transactional Database
For t=1…..n
L_i=n.getitems();
For i=1:n
If (F_i==I_i)
X_yi=1;
Else
            I_x=Index of (I_i);
End if;
End for;
End for;
T_Tree=Tree(F_i);
For t=1…..n
     If (T_Tree!=0) then
     {
     CT=N_i;
     SC_i=N_i.SC_i;
RT=CT;
Repeat;
}
End if;
End for;
```

4 Result and Discussion

The supremacy of our method

Table 2; Fig. 3 explain the time and memory comparison with other models. The time needed for the calculation is small, and it is accurately predicting the numerous items than other weight and hash tree algorithm. The memory needed in Apriori hybrid algorithm for the process its consumption is lesser when compared with the other weight and hash tree algorithm. The candidate set production is avoided in this method.

For each knot, the width, height and every reachability examined, and tree structure configuration produced.

From Table 3; Fig. 4, it is proved that the memory and time consumption is lesser in Apriori Hybrid algorithm when compared to the hybrid distribution algorithm.

Table 2 Table of weighted, hash tree and hybrid distribution

	Weighted	Hash tree	Hybrid
Time	29	3	1
Memory	18	15	7

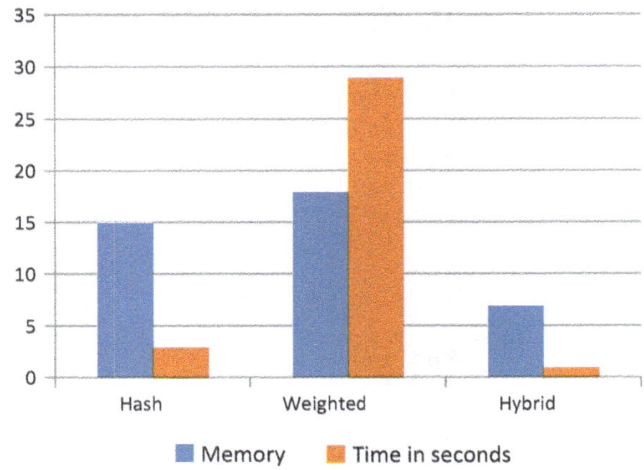

Fig. 3 Time and memory Comparison with other models

Table 3 Table of Apriori hybrid versus hybrid distribution

	Memory	Time
Hybrid Distribution	3900	6000
Apriori Hybrid	2500	2300

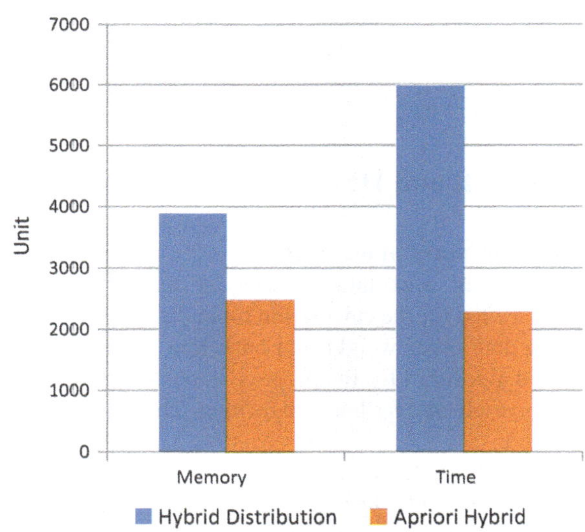

Fig. 4 Apriori hybrid versus hybrid distribution

5 Conclusion

The objective of removing the fascinating examples from the a significant extent substance of the datasets in the executed system accomplished for the motivation behind finding the information. From the graphical results shown above, we can conclude that Apriori hybrid algorithm, when compared with either hybrid distribution algorithm or with other weight or hash tree algorithm the time consumption and memory utilization is so much lesser. The CPU utilization is also lesser when gathering information from the dataset. So we conclude that our proposed algorithm is better than a lot of conventional methods.

References

1. Wu, B., Zhang, D., Lan, Q., & Zheng, J. (2008). An efficient frequent patterns mining algorithm based on apriori algorithm and the FP-tree structure. In *2008 Third International Conference on Convergence and Hybrid Information Technology* (Vol. 1, pp. 1099–1102). IEEE.
2. Suriya, S., & Shantharajah, S. P. (2013). A hybrid k-DCI and Apriori algorithm for mining frequent itemsets. In *2013 International Conference on Circuits, Power and Computing Technologies (ICCPCT)* (pp. 1059–1064). IEEE.
3. Shankar, S. K., & Kaur, A. (2016). Constraint data mining using apriori algorithm with AND operation. In *2016 IEEE International Conference on Recent Trends in Electronics, Information & Communication Technology (RTEICT)* (pp. 1025–1029). IEEE.
4. Kesarwani, S., Goel, A., & Sardana, N. (2017). MSD-Apriori: Discovering borderline-rare items using association mining. In *2017 Tenth International Conference on Contemporary Computing (IC3)* (pp. 1–4). IEEE.
5. Kalyanathaya, K. P., Akila, D., & Rajesh, P. (2019). Advances in natural language processing–a survey of current research trends, development tools and industry applications. *International Journal of Recent Technology and Engineering, 7*(5C), 199–202.
6. Ni, Y., Du, X., Ye, P., Xiao, R., Yuan, Y., & Li, W. (2019). Frequent pattern mining assisted energy consumption evolutionary optimization approach based on surrogate model at GCC compile time. *Swarm and Evolutionary Computation,* Vol. 50, p. 100574, 2019/11/01, doi: https://doi.org/https://doi.org/10.1016/j.swevo.2019.100574.
7. Du, J., Zhang, X., Zhang, H., & Chen, L. (2016). Research and improvement of Apriori algorithm. In *2016 Sixth International Conference on Information Science and Technology (ICIST)* (pp. 117–121). IEEE.
8. Han, E. H., Karypis, G., & Kumar, V. (1997). Scalable parallel data mining for association rules. *ACM SIGMOD Record, 26*(2), 277–288.
9. Saxena, P., Pant, B., & Goudar, R. H. (2014). Inter—Transactional pattern discovery applying comparative apriori and modified reverse apriori approach. In *2014 IEEE 8th International Conference on Intelligent Systems and Control (ISCO)* (pp. 300–305). IEEE.
10. Sumithra, R., & Paul, S. (2010). Using distributed apriori association rule and classical apriori mining algorithms for grid based knowledge discovery. In *2010 Second International conference on Computing, Communication and Networking Technologies* (pp. 1–5). IEEE.
11. Chun-Sheng, Z., & Yan, L. (2014). Extension of local association rules mining algorithm based on apriori algorithm. In *2014 IEEE 5th International Conference on Software Engineering and Service Science* (pp. 340–343). IEEE.

Grouping and Scheduling the Examinations Using Visual Basic and Alpha Cut of Fuzzy Graphs

R. Shanmugapriya and D. Mary Jiny

Abstract Examination scheduling problem can be solved in different ways. The main idea to solve the problem is to group the examinations into independent sets so that no student writes two or more examinations in the same schedule. The method which is being followed is graph colouring using a backtracking algorithm but which is an NP-hard problem for scheduling more number of examinations. In this paper, we suggest a simple heuristic method to do the grouping using visual basic. But for scheduling the examination in the different slot we need to consider many constraints like no two examinations is conducted in consecutive days for the same student, no two examinations can be conducted in a consecutive slots morning and evening for the same student, etc. Depend on the necessity, the scheduling will change. Here in this paper, we have used α − cut of a fuzzy graph to schedule the examination which is more effective than the existing method to reduce the number of days of examinations.

Keywords Fuzzy subset · Fuzzy graph · α Cut of a fuzzy graph · Fuzzy graph colouring

1 Introduction

Fuzzy mathematics was introduced by Lotfi Asker Zadeh's in the year 1965. Later in the year, 1975 Rosenfield defined fuzzy graphs, it has many applications in real life, such as networking, colouring [1], cluster analysis [8], telecommunication [9] and traffic network [7]. Examination scheduling problem has mainly two important constraints: the first one is, scheduling the examination date without any conflict, that is, no student should have more than one examination in the same slot. Secondly,

R. Shanmugapriya (✉) · D. Mary Jiny
Department of Mathematics, Sagunthala R&D Institute of Science and Technology, Tamil Nadu, Vel Tech Rangarajan DrAvadi, Chennai 600062, India
e-mail: spriyasathish11@gmail.com

D. Mary Jiny
e-mail: djstudy.jiny94@gmail.com

S.-L. Peng et al. (eds.), *Proceedings of First International Conference on Mathematical Modeling and Computational Science*, Advances in Intelligent Systems and Computing 1292, https://doi.org/10.1007/978-981-33-4389-4_51

students must be given enough number of holidays before each examination and minimizing the total number of days to conduct all the examination. Graph colouring using backtracking algorithm is a well-known NP-hard problem, m-partition of the vertices into independent subset is known as m-colouring problem and the chromatic number of a graph is the minimum colour or minimum independent partition of the vertices, which is used to avoid conflict in examination schedule, but to fix the examination date, the different university has different priority or constraints. In this paper, we have particularly considered the percentage of student writing the consecutive examinations for scheduling. That is, different α − cut levels will give different scheduling or colouring based on the membership value of the edges between the two set of examination.

2 Definition

Definition 2.1 Consider a nonempty set X. A **fuzzy subset** A of X is an ordered pair $A = \{(x, \mu_A(x))/x \in X\}$ where $\mu_A : X \to [0, 1]$ and $\mu_A(x)$ are the membership value of element x in fuzzy set A for each $x \in X$ [2].

Definition 2.2 Graph $G(V(G), E(G))$ is said to be an ordered pair which has a set $V(G)$ of vertices, edges $E(G)$, together with an incidence map $\Psi(G)$ that relates every edge of G an unordered tuple of vertices of G [3].

Definition 2.3 Chromatic number of a crisp graph is the lowest number of colours necessary to colour the vertices of a graph so that no two adjacent vertexes can have the same colour and which is denoted as $\chi(G)$ [3].

Definition 2.4 Fuzzy graph is an ordered triple $G(V, \sigma, \mu)$, where V is a nonempty set of vertices together with the functions $\sigma : V \to [0, 1]$ and $\mu : V \times V \to [0, 1]$ such that for all $x, y \in V$, $\mu(x, y) \leq \sigma(x) \wedge \sigma(y)$. The support of μ and σ is represented as $\mu^* = \{(x, y)/\mu(x, y) > 0\}$ and $\sigma^* = \{x/\sigma(x) > 0\}$. The fuzzy subgraph induced by S is defined as $H = (S, \tau, \nu)$ where $S \subseteq V$, $\tau(x) \subseteq \sigma(x) \forall x \in S$, $\nu(x, y) = \mu(x, y) \forall x, y \in S$, and it is denoted as $\langle S \rangle$ [2].

Definition 2.5 In a fuzzy graph $G(V, \sigma, \mu)$, the sequence of different vertices $v_1, v_2, ...v_n$ with $\mu(v_{i-1}, v_i) > 0, i = 1, 2, ...n$ is called a fuzzy path P of length $'n'$. The consecutive pairs are named us the edges of the path. The length of the longest path between v_1 and v_2 is called the diameter of $v_1 - v_2$ denoted as diam (v_1, v_n). The strength or weight of the path P is defined as the weight of the membership value of the weakest edge in the path. The weight of connectedness between v_1 to v_2 is the maximum of the weight of all the paths between v_1 to v_2 and is denoted by $\mu^\infty(v_1, v_n)$ [3]. The fuzzy path P is called a fuzzy cycle if $v_1 = v_n$ and $n \geq 3$.

Definition 2.6 Let $G(V, \sigma, \mu)$ be a fuzzy graph and $\alpha_1 < \alpha_2 < \alpha_3... < \alpha_n$ is the levels of edge membership values. Then G_α denote the crisp graph $G(V, E_\alpha)$

where $E_\alpha = \{uv/\mu(uv) \geq \alpha\}$. $\chi(G_\alpha)$ denotes the chromatic number of the graph G_α. $L = \{\alpha_1, \alpha_2, ...\alpha_n\} \cup \{0\}$ is called the **level set** of G.

Definition 2.7 A family $K = \{k_1, k_2, ...k_n\}$ of subsets of σ is called ***n*-fuzzy colouring** of the fuzzy graph G if,

(i) $\cup K = \sigma$
(ii) $k_i \cap k_j = 0\ [i \neq j]$
(iii) for all strongly adjacent vertices x and y of G, min $\{k_i(x), k_i(y)\} = 0 (1 \leq i \leq n)$

3 A Survey on Related Works

Malkawi et al. [1] in the paper "A New Exam Scheduling Algorithm Using Graph Coloring" defined a weighted matrix using the number of students common in any two courses and the colouring of courses is done using an algorithm by pre-assuming the number of colours available. But it does not say the least number of colours needed and the possible way of colouring the subjects. The paper by Dey and Pradhan [4] "The Fuzzy Robust Graph Coloring Problem" which handle uncertainty in the process of decision making and makes fuzzy graph colouring using robust optimization. They used minimal colouring problem to reduce the duration of examinations and the probability is checked using statistical independence since the algorithm is not explained which may not suitable for large data. Pillary [5], "Evolving construction heuristics for the curriculum-based University Course Time table problem", discuss certain constrains in scheduling courses time tables such as room capacity, the minimum number of days, curriculum density and room strength. He uses a heuristic method to achieve the constraints. Komaki [6] is also discussed course time table problem using the heuristic method in their paper "Interactive Optimization techniques based on Column generation model for timetabling problems of university makeup courses". Abayomi-alli [7], "Automatic Examination Timetable Scheduling Using Particle Swarm Optimization and Local Search Algorithm", they have considered (i) each examination must be assigned to a room at a certain time slot, (ii) students strength should not go beyond the room capacity, (iii) the maximum time in the day must not exceed the limit, (iv) students choice and preference, (v) preference to have their examination on the first two weeks as constrains and used particle swarm optimization algorithm with local search method to find the optimal solution Fahmy [2] "A fuzzy algorithm for scheduling non-periodic jobs on soft real-time single processor system", he explains a fuzzy algorithm to test the main concern of the jobs comes in the queue to be executed first and the performance analysis is done with an example.

4 A Simple Heuristic Technique for Grouping the Subjects into Independent Sets

In this section, we will see how the adjacency matrix of the subjects is formed using VBA. A small sample data of the students registered in 11 courses with a maximum of 65 students in the computer science department of the University A with their register number and paper code is taken for grouping. Table 1 shows how the raw data will look like.

The adjacency matrix A_{ij} gives the relation between any two subjects i and j. That is, if a student is writing the paper i and the paper j, then $a_{ij} = 1$, else $'0'$. The algorithm to find the adjacency matrix from the given datasheet in visual basic for applications (VBA) is given below.

Step 1: Let "nc" be the number of courses and "nr" be the maximum number of students writing the same examination.

Step 2: Now check each pair of the column for $c = 2$ to nc + 1 has any student in common. If any pair say *i*th column and *j*th column have a common student, then print the *i*th row and *j*th column of the adjacency matrix AM as "1".

Step 3: Now For $r = 1$ To nc + 1.

For $c = 1$ To nc + 1.

Worksheets ("AM"). Cells (c, r) = Worksheets ("AM"). Cells (r, c).

Next c.

Next r.

The output of the program is given in Table 2. This adjacency matrix can be represented graphically to find the colouring or partitioning of the subjects into disjoint sets. To find the chromatic number of the graph and colouring, more number of courses is difficult and it is considered as an NP-hard problem. Therefore, we suggest a simple heuristic method to find one such grouping of subjects. For that, we give preference to the first subject code 1151CS103 and start collecting its friends in the sense that, two subjects are said to be "friends", if they don't have any common student writing both examinations. We collect the friends of 1151CS103 from the left-hand side to the right-hand side. We can begin from the second course based on our preference. The grouping will defer based on the subject preference we give for first place. Here we analyse the grouping done by taking 1141CS103 as the first subject in the first group. The algorithm for the VB program to collect all the friends of each subject is as follows and the output is given in Table 3.

Step 1: If *i*th row and *j*th column of the adjacent matrix AM is empty, then print the *j*th course in the *i*th row and *j*th column of the matrix "friends".

Step 2: If *i*th row and *j*th column of the adjacent matrix AM is nonempty, then empty the *i*th row and *j*th column of the matrix "friends".

The first group is formed by collecting the friends of the first course say $C_1 =$ 1151CS103 from the left-hand side. If there exist any two friends of C_1 (other than C_1), then check whether the second friend of C_1 is a friend of the first friend of C_1, if they are friends those three can be put in one group, else delete the second friend of C_1 from the group. Continue this process for all of his friends and form a

Table 1 List of courses with the roll number of the students registered (datasheet)

Course code/S. No.	1151CS103	1151CS113	1151CS116	1152CS101	1152CS102	1152CS104	1152CS106	1152CS107	1152CS108	1152CS110	1152CS111
1	REG006026	REG010037	REG005872	REG006119	REG010256	REG010139	REG010233	REG010139	REG010010	REG006119	REG006768
2	REG006119	REG010344	REG005987	REG006123	REG010339	REG010162	REG010264	REG010179	REG010136	REG006604	REG006938
3	REG006128	REG010426	REG006087	REG006363	REG010340	REG010223	REG010474	REG010188	REG006069	REG007110	REG006986

Table 2 Adjacency matrix "AM" sheet showing the common students in each pair of courses

Courses	1151CS103	1151CS113	1151CS116	1152CS101	1152CS102	1152CS104	1152CS106	1152CS107	1152CS108	1152CS110	1152CS111
1151CS103		1	1	1	1	1	1	1	1	1	1
1151CS113	1		1	1	1	1	1	1	1	1	1
1151CS116	1	1		1	1	1	1	1	1	1	1
1152CS101	1	1	1		1	1	1	1	1	1	
1152CS102	1	1	1	1		1	1	1	1	1	1
1152CS104	1	1	1	1	1		1	1			
1152CS106	1	1	1	1	1	1		1	1		
1152CS107	1	1	1	1	1	1	1		1	1	1
1152CS108	1	1	1	1	1		1	1		1	1
1152CS110	1	1	1	1	1			1	1		1
1152CS111	1	1	1		1			1	1	1	

Table 3 Friends sheet represents the friends of each subject

Courses											
1151CS103	1151CS103										
1151CS113		1151CS113									
1151CS116			1151CS116								
1152CS101				1152CS101							1152CS111
1152CS102					1152CS102						
1152CS104						1152CS104			1152CS108	1152CS110	1152CS111
1152CS106							1152CS106			1152CS110	1152CS111
1152CS107								1152CS107			
1152CS108						1152CS104			1152CS108		
1152CS110						1152CS104	1152CS106			1152CS110	
1152CS111				1152CS101		1152CS104	1152CS106				1152CS111

Table 4 Grouping sheet represents grouping by deletion of rows and columns

Courses											
1151CS103	1151CS103										
1151CS113		1151CS113									
1151CS116			1151CS116								
1152CS101				1152CS101							1152CS111
1152CS102					1152CS102						
1152CS104						1152CS104			1152CS108	1152CS110	1152CS111
1152CS106							1152CS106			1152CS110	1152CS111
1152CS107								1152CS107			
1152CS108						1152CS104					
1152CS110						1152CS104	1152CS106			1152CS110	
1152CS111				1152CS101		1152CS104	1152CS106				1152CS111

group in such a way that no two of its friends are not friends; that is, no two of its friends have a common student writing both examinations. Once the first friend list is formed delete the entire row elements of C_1's friends (except C_1)and delete the column element below C_1's friends. Continue this process for all nonempty rows.

we can see that in Table 4, the course 1152CS110 is not a friend of 1152CS108, therefore which is deleted from the friend list of 1152CS104 and is denoted in red colour. The row deletion and column deletion are represented by green and yellow colours of Table 4. Arrangement of the courses in independent groups is shown in Table 5.

The grouping of the courses obtained by the above method is one among the different solution of colouring. Now, we are given only two rooms with strength 40 each, and then we could accommodate only 80 students in one slot. We have to check any of our group exceeds the strength 80. The sixth group has total strength 104, and hence, we split the 6th group into two groups. The final grouping is given in Table 5, and we can draw a fuzzy graph G indicating the number of students common in any two groups say u and v and the membership values are calculated as follows.

$\sigma(u)$ = number of students in the group u/80.

$\mu(uv)$ = number of students present both in u and v/total number of students in u and v.

$$\mu(v_1v_2) = 36/104 = 0.35; \quad \mu(v_1v_3) = 10/93 = 0.1\mu(v_1v_4) = 4/78 = 0.09$$
$$\mu(v_1v_5) = 0.08; \quad \mu(v_1v_6) = 0.03; \quad \mu(\mathrm{v}_1\mathrm{v}_7) = 0.18; \quad \mu(v_1v_8) = 0.07;$$
$$\mu(v_1v_9) = 0.04; \quad \mu(v_2v_3) = 0.09; \quad \mu(v_2v_4) = 0.1; \quad \mu(v_2v_5) = 0.08;$$
$$\mu(v_2v_6) = 0.04; \quad \mu(v_2v_7) = 0.1; \quad \mu(v_2v_8) = 0.07; \quad \mu(v_2v_9) = 0.1;$$
$$\mu(v_3v_4) = 0.1; \quad \mu(v_3v_7) = 0.03; \quad \mu(v_3v_8) = 0.03; \quad \mu(v_3v_5) = 0.04;$$
$$\mu(v_3v_6) = 0.02; \quad \mu(v_3v_9) = 0.06; \quad \mu(v_4v_5) = 0.03;$$
$$\mu(v_4v_6) = 0.03; \quad \mu(v_4v_7) = 0.1; \quad \mu(v_4v_8) = 0.02;$$
$$\mu(v_4v_9) = 0.05; \quad \mu(v_5v_6) = 0.06; \quad \mu(v_5v_7) = 0.02;$$
$$\mu(v_5v_8) = 0.1; \quad \mu(v_5v_9) = 0.04; \quad \mu(v_6v_7) = 0;$$
$$\mu(v_6v_8) = 0.04; \quad \mu(v_6v_9) = 0.02; \quad \mu(v_7v_8) = 0.03;$$
$$\mu(v_7v_9) = 0.07; \quad \mu(v_8v_9) = 0.06$$

Table 5 Final grouping

Disjoint sets	1	2	3	4	5	6	7	8	9
Course code	1151CS103	1151CS113	1151CS116	1152CS101	1152CS102	1152CS104	1152CS108	1152CS106	1152CS107
Course code				1152CS111				1152CS110	
Vertices	V1	V2	V3	V4	V5	V6	V7	V8	V9
$\sigma(\boldsymbol{u})$	0.812	0.812	0.36	0.38	0.812	0.812	0.48	0.88	0.21

5 A Fuzzy Graphical Approach of Finding the Examination Schedule

We can see that except v_6 and v_7 all other vertices are interrelated to each other. If our concern is to give at least one day left before each examination of all students, then the total scheduling will be done in 17 days. But problem comes if we have a more number of subjects and the days given for the total scheduling of the examination are less than the required number of days. In such a case, we give preference to the two examinations with more number of common students and have at least one day holiday between the examinations. This can be done using alpha cut of the fuzzy graph. The edge level set of G is $L = \{0.01, 0.02, 0.03, 0.04, 0.05, 0.06, 0.07, 0.08, 0.09, 0.1, 0.18, 0.35\}$. We draw G_α for $\alpha \leq 0.05$:

$$\chi_{0.05} = \chi(G_{0.05}) = 4 = \chi_{0.06} = \chi_{0.07} = \chi_{0.08} = \chi_{0.09};\ \chi_{0.1} = 3;\ \chi_{0.18} = 2 = \chi_{0.35}$$

To conduct the examinations less than 17 days, we can fix one day holiday only between the examinations having greater than or equal to 0.05 edge weight. That is, the two vertexes with less than 0.05 edge weight between then can be conducted in consecutive days. The fuzzy colouring for $G_{0.05}$ is:

$$K = \{k_1, k_2, k_3, k_4\}$$

$$k_1(u_i) = \begin{cases} 0.8 & \text{if } i = 2 \\ 0.8 & \text{if } i = 6 \\ 0 & \text{otherwise} \end{cases}$$

$$k_2(u_i) = \begin{cases} 0.8 & \text{if } i = 1 \\ 0.21 & \text{if } i = 9 \\ 0 & \text{otherwise} \end{cases}$$

$$k_3(u_i) = \begin{cases} 0.36 & \text{if } i = 3 \\ 0.48 & \text{if } i = 7 \\ 0.88 & \text{if } i = 8 \\ 0 & \text{otherwise} \end{cases}$$

$$k_4(u_i) = \begin{cases} 0.38 & \text{if } i = 4 \\ 0.8 & \text{if } i = 5 \\ 0 & \text{otherwise} \end{cases}$$

Total number of days of scheduling $= 2 + 1 + 2 + 1 + 3 + 1 + 2 = 12$ days.

Similarly, we can find the number of days of scheduling for alpha cuts 0.06, 0.07, 0.08, 0.09, 0.1, 0.18 and 0.35 from Fig. 1 is given in Table 6.

If we allow 5–9 percentage of common students in consecutive examinations, the examinations can be conducted in 12 days. And if we allow 10 and 18 percentage of

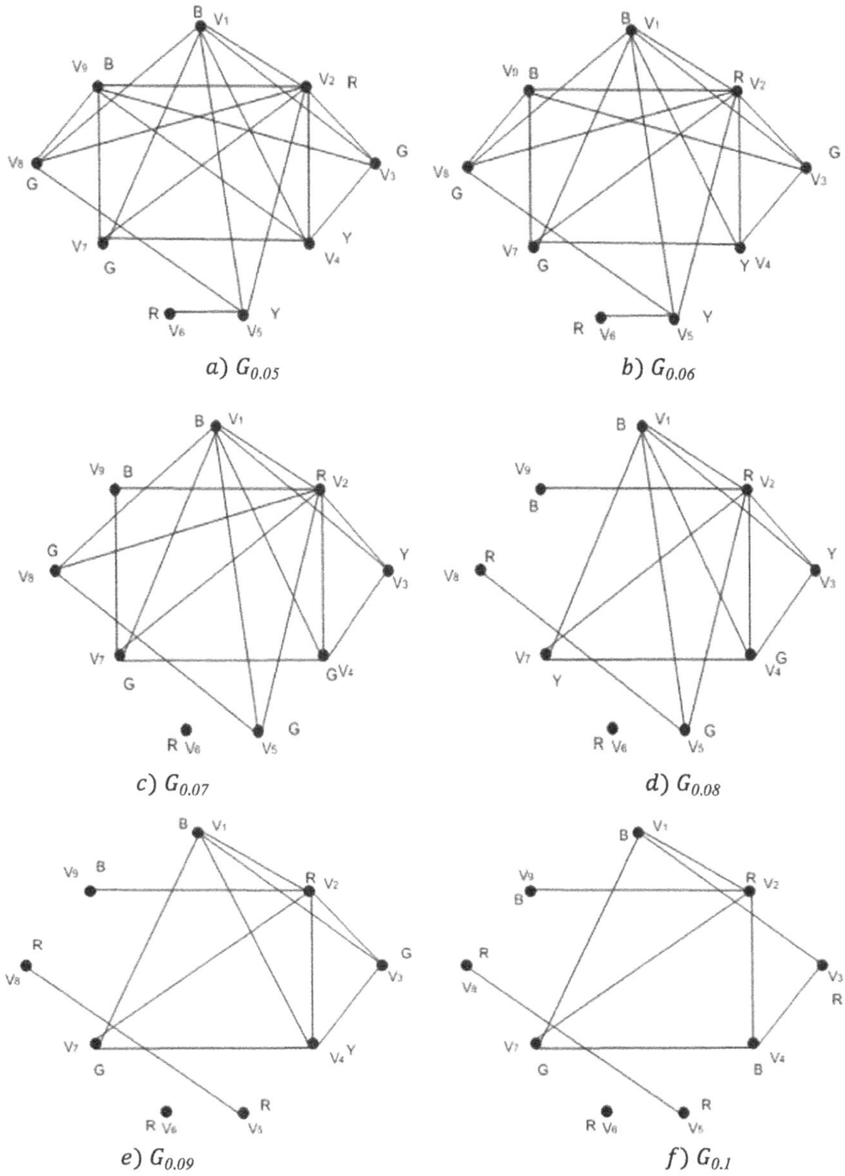

Fig. 1 Different alpha cuts

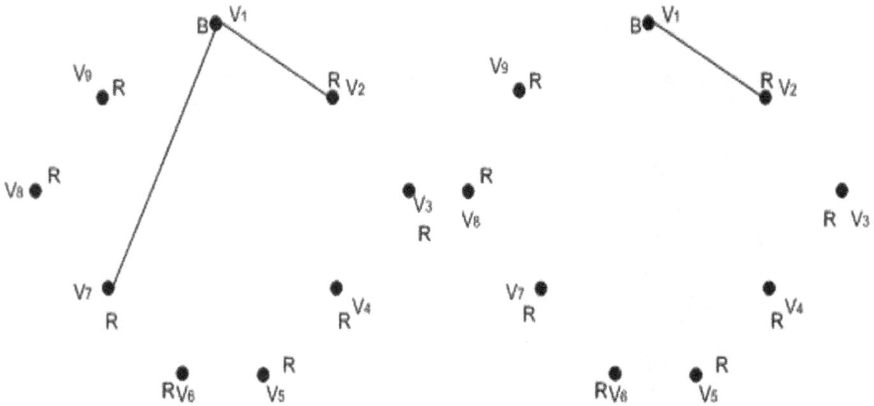

g) $G_{0.18}$ h) $G_{0.35}$

Fig. 1 (continued)

Table 6 Total number of days of schedule

$\alpha - cuts$	0.05	0.06	0.07	0.08	0.09	0.1	0.18	0.35
Chromatic number of G_α	4	4	4	4	4	3	2	2
Total number of days of scheduling	12	12	12	12	12	11	10	10

common students in consecutive examination the scheduling can be done in 11 and 10 days, respectively. Throughout this paper, we have considered any disjoint set of examination is conducted in a single day. But if we allow two slots in a day, morning and evening, the grouping can be done in less number of days. This can be done in considering the consecutive two slots is one day in the grouping process.

6 Conclusion

We have discussed a simple heuristic technique to do a grouping of the subjects into independent sets such that no two courses in the same group have a common student. After grouping is done, the total number of days of examination is calculated based on the two constraints, the number of holidays between each group and α-cut of the fuzzy graph representing the relation between each pair of groups as vertices. For grouping, we have suggested a VBA (macro in excel) programming which will help the reader to apply in the real situation directly and will get benefited out of it.

References

1. Malkawi, M., Hassan, M. A. H., & Hassan, O. A. H. (2008). A New Exam Scheduling Algorithm Using Graph Coloring. *International Arab Journal of Information Technology (IAJIT), 5*(1).
2. Fahmy, M. M. M. (2010). A fuzzy algorithm for scheduling non-periodic jobs on soft real-time single processor system. *Ain Shams Engineering Journal, 1*(1), 31–38.
3. Mordeson, J. N., Mathew, S., & Malik, D. S. (2018). Human trafficking: Source, transit, destination designations. In *Fuzzy Graph Theory with Applications to Human Trafficking* (pp. 181–208). Cham: Springer.
4. Dey, A., Pradhan, R., Pal, A., & Pal, T. (2015). The fuzzy robust graph coloring problem. In *Proceedings of the 3rd International Conference on Frontiers of Intelligent Computing: theory and applications (FICTA) 2014* (pp. 805–813). Cham: Springer.
5. Pillay, N. (2016, July). Evolving construction heuristics for the curriculum based university course timetabling problem. In *2016 IEEE Congress on Evolutionary Computation (CEC)* (pp. 4437–4443). IEEE.
6. Komaki, H., Shimazaki, S., Sakakibara, K., & Matsumoto, T. (2015, November). Interactive optimization techniques based on a column generation model for timetabling problems of university makeup courses. In *2015 IEEE 8th International Workshop on Computational Intelligence and Applications (IWCIA)* (pp. 127–130). IEEE.
7. Abayomi-Alli, O., Abayomi-Alli, A., Misra, S., Damasevicius, R., &Maskeliunas, R. (2019). Automatic examination timetable scheduling using particle swarm optimization and local search algorithm. In *Data, Engineering and Applications* (pp. 119–130). Singapore: Springer.
8. Samanta, S., Pramanik, T., & Pal, M. (2016). Fuzzy colouring of fuzzy graphs. *AfrikaMatematika, 27*(1–2), 37–50.

Simulation of the Dynamic Voltage Restorer Using Two Different Types of the Controller in a Distribution System

Basim Mohammed Al Kadhimi and Ali Basim Mohammed

Abstract This study addresses and utilizes the dynamic voltage restorer to mitigate the power quality problems such as voltage sag/swell and harmonics to keep the distribution system work effectively with minimum losses. The most important issues related to the quality problems in the distribution system are the voltage sag/swell and harmonics. Many devices are available for solving power quality problems like static synchronous compensator (STATCOM), dynamic voltage restorer (DVR), and uninterruptible power supply (UPS). The DVR is a simple and effective device to mitigate the disturbances when occur in the distribution system. DVR with two control techniques exhibits in this paper; the first one is a proportional–integral controller, and the second controller is the hysteresis controller. The proposed system was implemented by utilized MATLAB/Simulink to investigate and improve the check the ability of the proposed controller. The simulation results show the ability of the dynamic voltage restorer to mitigate the power quality problems in the distribution system.

Keywords Power quality · Compensation methods · Dynamic voltage restorer · Voltage compensation voltage sag/swell

1 Introduction

Nowadays, power quality is a very important issue related to the distribution system. It's a problem effected to the equipment of the consumer and reduced their life's and get major losses for industries. A good power quality needed for commercial buildings and manufacture industries. Many techniques are available to solve the power quality problems like (D-STATCOM, UPQC, and DVR). It called also custom power

B. M. Al Kadhimi (✉)
Wasit University, Wasit, Iraq
e-mail: alaoibasim2013@gmail.com

A. B. Mohammed
Faculty of Electrical and Electronic Engineering, University Tun Hussein Onn, Batu Pahat, Johor, Malaysia
e-mail: alaoibasim2013@gmail.com

S.-L. Peng et al. (eds.), *Proceedings of First International Conference on Mathematical Modeling and Computational Science*, Advances in Intelligent Systems and Computing 1292, https://doi.org/10.1007/978-981-33-4389-4_52

device. They worked on the controller concept to mitigate the power quality problems like sag/swells and harmonics. The power received by industries and consumer is not continuous; the power quality problem is a flicker, sag, swell, transient, harmonics, and notching. Some of these problems are in a short time natures, and the other's natures are along time [1]. The first one is the voltage sagging and swelling, the voltage sag means a reduction in the range of the RMS voltage 0.9–0.1 PU at the period of a half cycles to 1 min when a large size load or a large size motor are connected. While voltage swell means an in the range of the RMS voltage 1.1–1.8 PU at the period of a half cycles to 1 min. It happens when capacitor bank energization, switch off the larger size motor or large size of the load, etc. [2].

This paper deals with the DVR in the distribution system is located between a series transformer and the consumer. From the side of the distribution system, it is connected through a series transformer. The first installation of the DVR was in a 12.14 kv substation duke power company in Anderson South Carolina of America in 1996 [3]. This device is providing a reliable and better quality power to the customer. Model of the DVR with RL is discussed in [4, 5]. For power quality problem, there are different types of controller techniques available like untreatable power supply and generator motor set. These devices have disadvantages like maintenance required frequent, noise produces, the bulk of the system and reduce efficiency. The hysteresis controller is discussed in [6].

Two controls are used; one is the hysteresis controller, and the second one is the PI controller utilized to control the DVR for different types of unbalance condition. The DVR model performance in MATLAB simulation with the nonlinear load is when the voltage swell, sag and harmonic occurred in the system.

2 The Structure of the DVR

The general structure of the DVR illustrated as below (Fig. 1):

Voltage source converter: The generation of the proper injection voltage is done by the VSC to compensate the voltage disturbances. It contains six pulses linked in series with the secondary winding of the injection transformer. The VSC rating is high in current and low in voltage [7].

Injection transformer: It is utilized between the DVRs and the distribution system. The purpose of using it to isolate VSC, limit the noise produces and isolate the control circuit from system.

Pulse width modulation: Employing the inverter to convert the voltage supplied by the energy storage into AC voltage to be inject to the system.

Line filter: There are different types of the filter available RL, RC, double-tuned filter and RLC, etc. It is used to suppress the short-circuit current into the system and also to mitigate the order of the high-frequency harmonics in the output voltage.

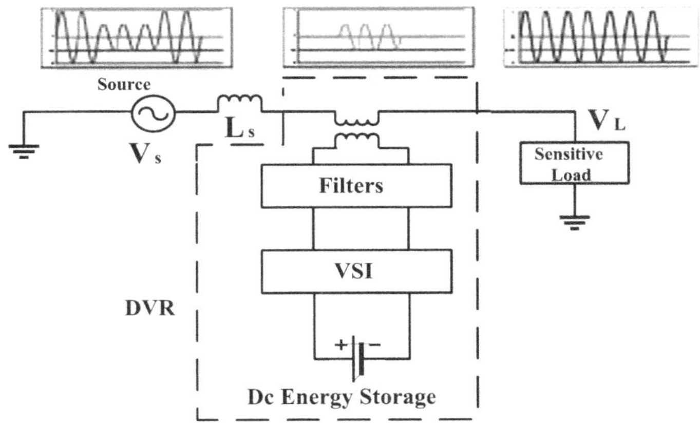

Fig. 1 General structure of the DVR

Storage device: It is a device uses to store a DC voltage when the sag event is appearing and maintain the DC voltage at a nominal voltage with providing a discharge circuit.

2.1 The DVR Operation Modes

Dynamic voltage restorer operates in three modes:

The bypass mode: It is also called (protection mode). As a protective tool, the bypass switch can safeguard the DVR. So, the current of the system will be very high. In this type of mode, the DVR operates as a protection device.

The active mode: In this case, the voltage sag/swell was detected and inject a proper magnitude of voltage by the DVR.

Standby mode: In this case, the supply voltage in a balance condition. The DVR will not need to supply voltage in the system. The DVR will measure the sag/swell voltages when appear and provide this signal to the controller.

3 Voltage Injection/the Compensation Technique

Via pre-sag compensation method: The DVR in this method is compensated for the magnitude and the angle of the voltages. The DVR will inject a proper amount of the voltage measured by DVR and making the system balanced.

The In-phase compensation method: The DVR in this method, the inject voltage in phase with the voltage supply. It is done only for the linear load [9].

In the phase advanced method: This method is used to reduce the injection of active power to the system. It is done by decreasing the demand voltage injected and reducing the rating of the storage device [9].

4 Hysteresis Controllers

The object of the control circuit is to detect and compare the voltage sag/swell with a reference voltage, reference voltage refers to a three-phase sine wave.

$$\begin{aligned} f_{(a)} &= V_m \sin(wt) \\ f_{(b)} &= V_m \sin(wt - 120) \\ f_{(c)} &= V_m \sin(wt + 120) \end{aligned} \tag{1}$$

The comparison between reference voltage with the output voltage and the error generated was send to the relay of the hysteresis. This relay will set the band of the hysteresis with the tolerance value. It selects a proper gate pulse to IGBTs switches of the VSC; the VSC will generate a proper amount of voltages to mitigate the disturbances related with the power quality problems [8] (Fig. 2).

Hysteresis controller operates for 1 leg of VSC. The dotted lines show the upper and the lower limit of the band of hysteresis. Middle line pointing to the reference voltage, the triangular line shows the voltages of the VSC. The two-level VSC switching operation as below:

$$\text{if } u\ (k) > h \text{ then } sw = 1$$

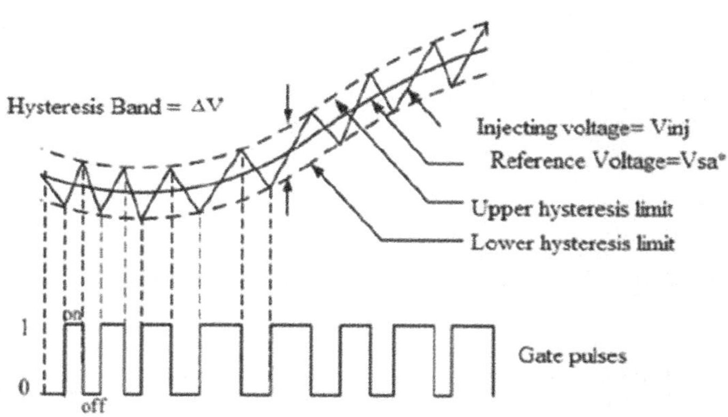

Fig. 2 Principle operation of the hysteresis controller [8]

$$\text{else if } u\ (k) < h \text{ then } sw = -1 \qquad (2)$$

The hysteresis band (h) and the switch state (SW). The obtained error will go from the lower to the upper limit. The switch (1) indicates to ON, while the switch (2) indicate to OFF mode. The error goes from the upper to lower limit. The switch (2) indicates to ON, while the switch (1) indicate to OFF [6, 8].

5 MATLAB Simulation and the Results

Many unbalanced conditions are applied and checked the DVR performance. Implement a small distribution system with the DVR and simulated it in MATLAB/Simulink. Creation a voltage sag or swell by supplying grid voltage. Connecting the DVR with the system by a series transformer to inject the amount of voltage needed. The simulation is made utilizing the PI and hysteresis controller [10, 11]. Where the reference value considered 1 P.U., the magnitude of sag condition will reduce to 0.5 P.U, and in voltage swell condition magnitude increases 1.5 P.U. and 5th and 7th harmonic injected using the programmable source in the MATLAB/Simulink, the parameter shown in Table 1 below:

a. Simulation without DVR (Fig. 3)

b. The controller of DVR in MATLAB (Figs. 4 and 5)

Table 1 System parameters

System parameters	Rating components
Grid voltage	380 v, 50 Hz
Nonlinear load	6-pulse diode with RL load $R = 60\ \Omega$, $L = 0.15$ mH
DC voltage source	700 v
Transformer rating	4000 VA, 50 Hz

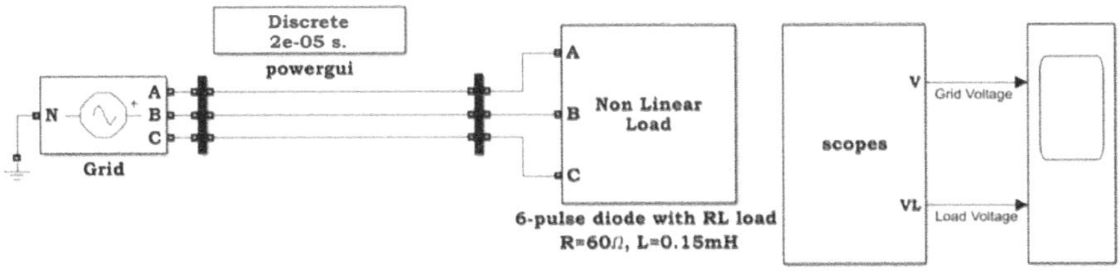

Fig. 3 Simulation without DVR connected with nonlinear load

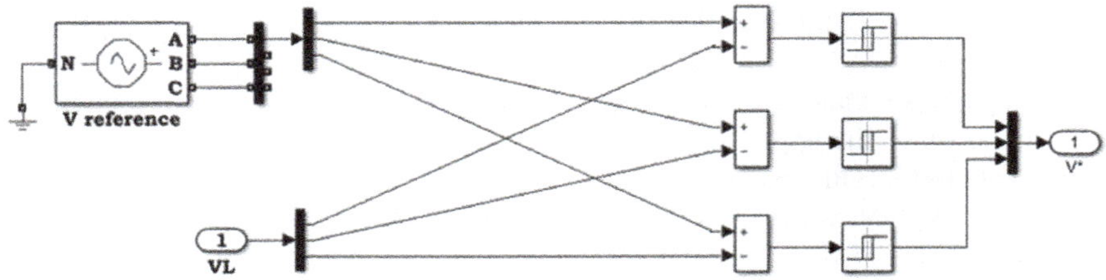

Fig. 4 Hysteresis controller

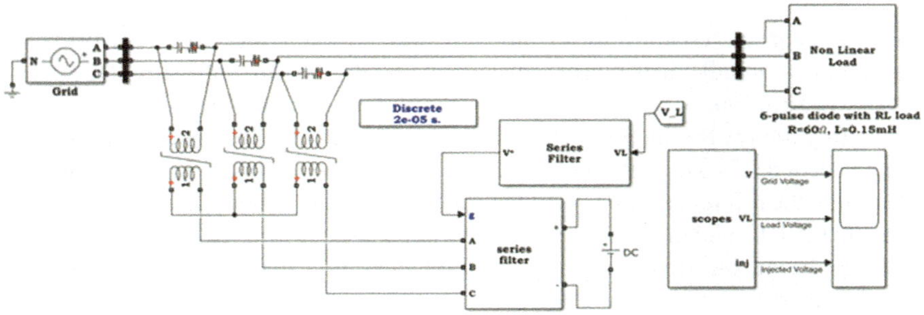

Fig. 5 DVR in simulation with the hysteresis controller

The case 1: Fig. 6 shows a 5th, 7th harmonic injected into a distribution system during time period 0.1–0.2 s. Figure 6a shows a balance load voltage in a distribution system.

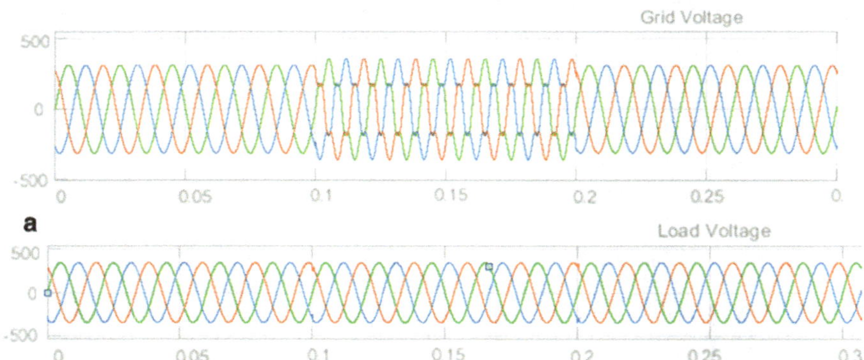

Fig. 6 5th, 7th orders of harmonic injected in the grid voltage. **a** 5th, 7th orders of harmonic in the balance load voltage

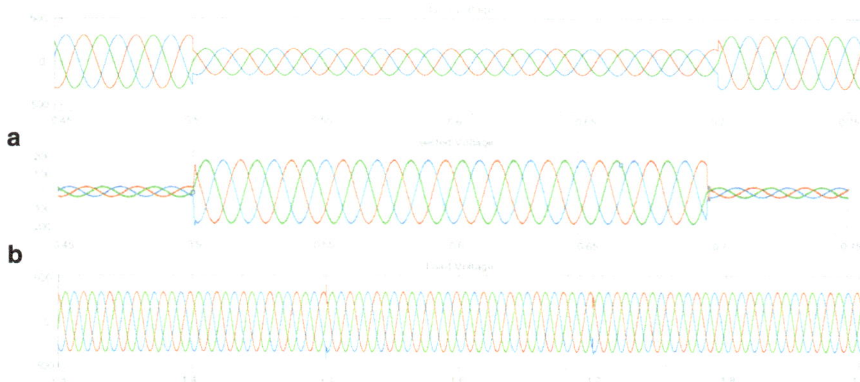

Fig. 7 Voltage sag magnitude 0.5 P. U during period 0.5–0.7 s. **a** Injected voltage by the DVR, **b** Balance voltage in the load side

The case 2: Fig. 7 creates a sag reduce voltage magnitude 0.5 P. U during period 0.5–0.7 s. Figure 7a shows the injected voltage by the DVR. Figure 7b shows a balance load voltage of the distribution system.

The case 3: Fig. 8 creates a swell increase in voltage magnitude 1.5 P. U during period 1.5–1.7 s. Figure 8a shows the injected voltage by the DVR. Figure 8b shows a balance load voltage of the system.

c. The simulation of the DVR with the hysteresis controller
d. The 5th,7th orders of harmonic injected in the grid voltage
e. The voltage sag condition
f. The voltage swell condition.

6 The Simulation of the DVR with the PI Controller

Considering the same parameters given in Table 1, with the same conditions, the reference voltage compared with the measured load voltage exhibits an error. The error enters to the PI controller after that delivers to the PWM. The parameters of the PI parameter are calculated analytically by utilizing the system transfer function [10, 12].

The DVR in simulation with the PI controller.

a. The 5th, 7th orders of harmonic injected in the grid voltage (Figs. 9 and 10)
b. The voltage sag condition (Figs. 11 and 12)
c. The voltage swell condition (Figs. 13 and 14)

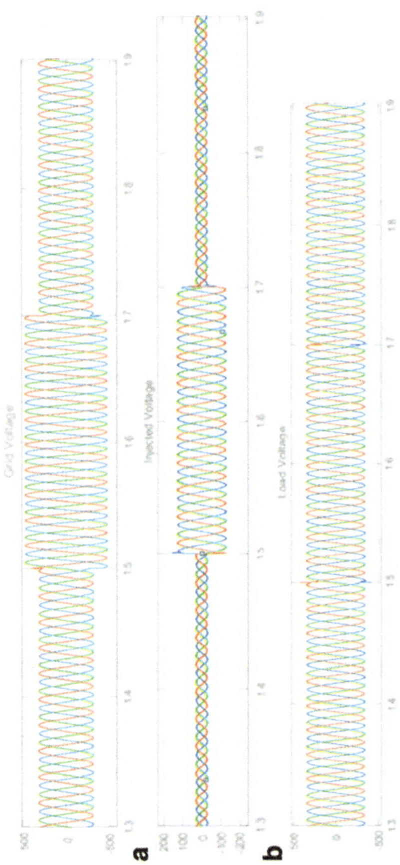

Fig. 8 Voltage swells magnitude 1.5 P. U during period 1.5–1.7 s. **a** Injected voltage by the DVR, **b** Balance voltage in the load side

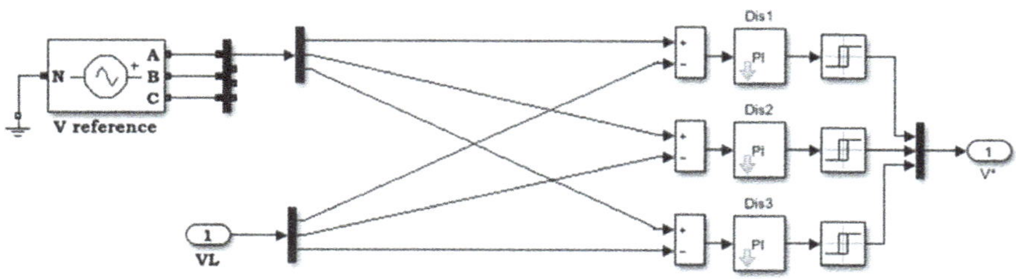

Fig. 9 DVR in simulation with the hysteresis controller

Fig. 10 5th, 7th orders of harmonic injected in the grid voltage

Fig. 11 5th, 7th orders of harmonic in the balance load voltage

Fig. 12 Voltage sag magnitude 0.5 P. U during period 0.5–0.7 s. **a** Injected voltage by the DVR **b** Balance voltage in the load side

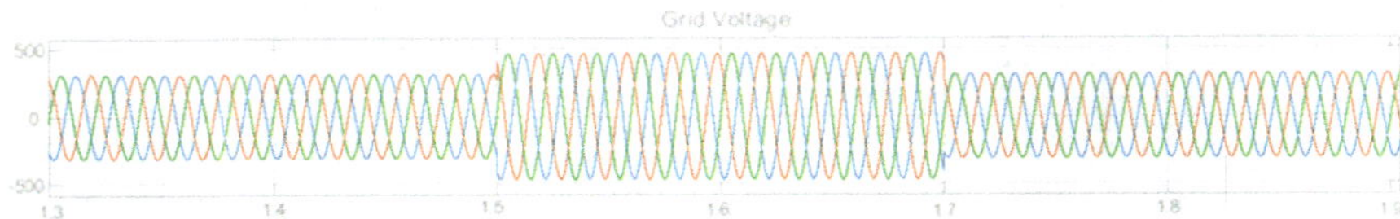

Fig. 13 Voltage swells magnitude 1.5 P. U during period 1.5–1.7 s.

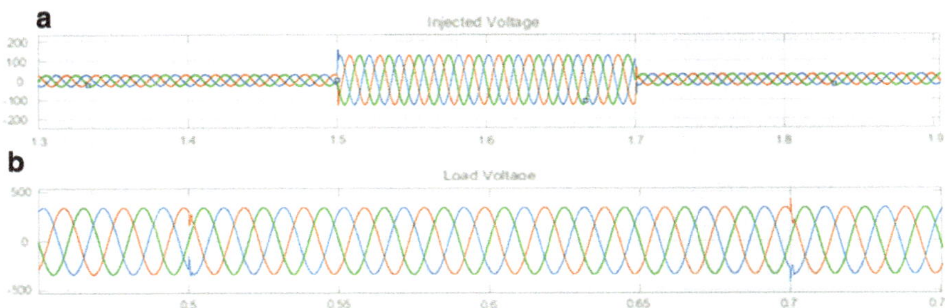

Fig. 14 **a** Injected voltage by the DVR. **b** Balance voltage in the load side

7 FFT Analysis

In this simulation, two controller's techniques are used with FFT analysis of the DVR (Figs. 15, 16 and 17).

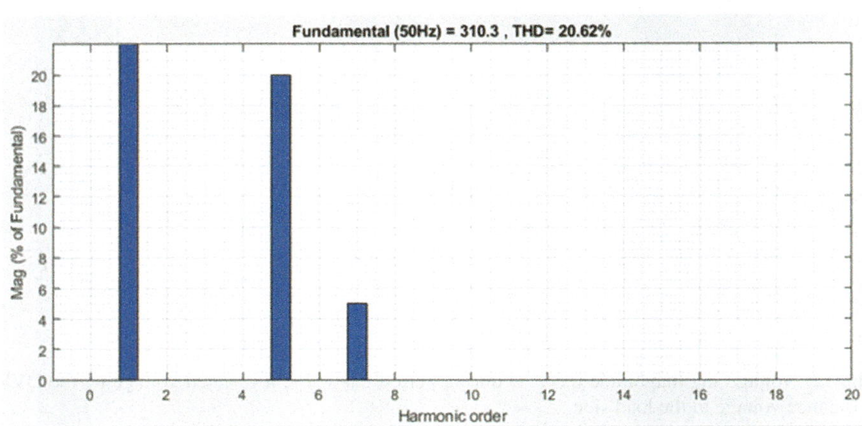

Fig. 15 THD without DVR

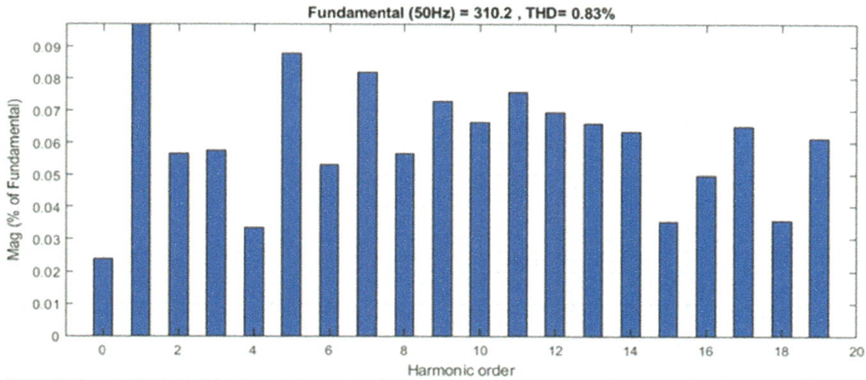

Fig. 16 THD with DVR (hysteresis controller)

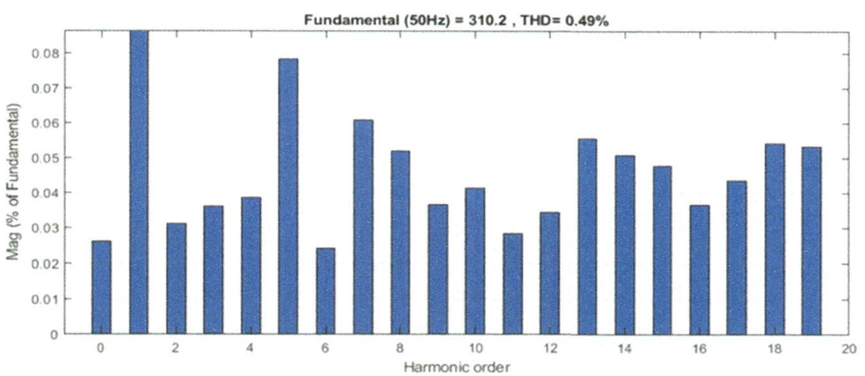

Fig. 17 THD with DVR (PI controller)

8 Conclusion

It presents the structure, operation principles and control circuity of the DVR using two different controllers PI and the hysteresis controller. Evaluation of the DVR performance is by creating different types of disturbances (voltage sag, voltage swell, and harmonics). This performance was verified in the MATLAB/Simulink. The results of simulation and the analysis are done for the load waveform and find out the difference between the PI and the hysteresis controller. The PI controller has a simple operation, fast response and easy to implement compared to hysteresis controller in terms of reduction in harmonics as illustrated in Table 2.

Table 2 Controller with reducing THD

Controller	Without DVR	With the DVR
PI controller	20.62	0.49%
Hysteresis controller	20.62	0.83%

References

1. IEEE.std. (1995). *Recommended Practice for Monitoring Electric Power Quality*, 1159–1995.
2. Dugan, R. C. (1996). *Electrical Power Systems Quality*. Mc-Graw Hill.
3. Woodley, N. H., Member, S., Morgan, L., & Sundaram, A. (1999). Experience with an inverter-based dynamic voltage restorer. *IEEE Transactions on Power Delivery, 14*(3), 1181–1186.
4. Jowder, F. A. L. (2007). Modeling and simulation of dynamic voltage restorer (DVR) based on hysteresis voltage control. In *IECON conference IEEE* (pp. 1726–1731).
5. Pal, Y., Swarup, A., Member, S., Singh, B., & Member, S. (2008). *A Review of Compensating Type Custom Power Devices for Power Quality Improvement*.
6. Ghosh, A., & Ledwich, G. (2002). *Power Quality Enhancement Using Custom Power Devices*. Norwell: Mc-Graw Hill, Kluwer International Series.
7. Nielsen, J.G., & Blaabjerg, F. (2005). A detailed comparison of system topologies for dynamic voltage restorers. *41*(5), 1272–1280.
8. Pathan, A. I. (2014). *Different Control Techniques of Dynamic Voltage Restorer for Power Quality Problems*.
9. Vilathgamuwa, K. T. M. & Perera, R. (1999). Control of energy optimized dynamic voltage restorer. In *Proceedings of IEEE IECON* (Vol. 2, pp. 873–878).
10. Nielsen, J.G., Blaabjerg, F., & Technology, E. (2001). *Control Strategies for Dynamic Voltage Restorer Compensating Voltage Sags with Phase Jump*.
11. Ghosh, A., Member, S., Ledwich, G., & Member, S. (2002). *Compensation of Distribution System Voltage Using DVR* (Vol. 17, no. 4, pp. 1030–1036).
12. Ali, M. T., Abbas, F., Nadeem, A., Yaqoob, M., Malik, S. M., & Iqbal, M. J. (2014). *Effect of PI Controller on Efficiency of Dynamic Voltage Restorer for Compensation of Voltage Quality Problems* (pp. 1134–1139).

Application of Fuzzy Vertex Magic Graph

R. Shanmugapriya and P. K. Hemalatha

Abstract Fuzzy labeling of graph $G = (V, E)$ which have p nodes and q links is a one to one map on to a fuzzy set V. We consider the independent set of vertices which satisfies a fuzzy vertex magic total labeling such that adding the value of individual node label and its connected edges labels must be uniform, and it should satisfy for every node $\sigma(vi) = \sigma(vi) + \sigma(viej) = K$, where $ej's$ are edges incident to V. We show generalized Peterson P (m, n) graph is FVMT labeling. By FVMT, the sum of the degree of the odd nodes is equivalent to double the degree of all edge membership and the difference of the sum of degrees of even nodes on P (m, n). An application about current passing through the transformers by P (4, 1) is demonstrated under specific conditions.

Keywords Labeling · Graph labeling · Magic labeling · Vertex labeling · Vertex magic total (VMT) labeling · Fuzzy labeling graph · Fuzzy vertex magic total (FVMT) labeling · Peterson graph

1 Introduction

Let G of (V, E) be a simple and undirected graph with v nodes and e links representing details required between objects. It is Zadeh's [1] contribution that he provides us with a formal framework that the membership function of a fuzzy set σ of V, called fuzzy set σ. The fusion of a fuzzy set with a graph is known as a fuzzy graph. Rosenfeld [2] developed a fuzzy graph after Zadeh's introduced fuzzy set. In [3], the concepts of eccentricity and center have been discussed in fuzzy graphs.

A valuation (or labeling) of a graph is a map that carries graph components (elements) to a special number. Graph labeling gave birth to a different form of

R. Shanmugapriya (✉) · P. K. Hemalatha
Vel Tech Rangarajan Dr. Sangunthala R&D Institute of Science and Technology, Chennai 600062, India
e-mail: spriyasathish11@gmail.com

P. K. Hemalatha
e-mail: pkhemalathamsc@gmail.com

S.-L. Peng et al. (eds.), *Proceedings of First International Conference on Mathematical Modeling and Computational Science*, Advances in Intelligent Systems and Computing 1292, https://doi.org/10.1007/978-981-33-4389-4_53

labeling such as graceful, harmonious, magic, vertex magic, edge magic, prime magic, anti-magic, bi-magic, and so on. In 1963, the theory of labeling was initiated and explained about total magic labeling by Rosa. In [5], double fan graphs and D ω_m of fuzzy graceful labeling have been discussed. The VMT labeling notation is established by MacDougall et al. [4] in 1999. Gallion [6] is defined by disparate kind of magic labeling. The fuzzy labeling graph is an emerging topic in the fuzzy graph. The idea of labeling to fuzzy graphs and features was suggested by Gani et al.

2 Preliminaries

A graph G is a finite nonempty set v nodes, and e links are shown with G of (V, E) which is called to be a vertex magic graph if a vertex label and an edges label connected to that individual vertex added to get a fixed constant.

A fuzzy graph G of (V, μ,σ) is a nonempty set V together with a pair of functions where μ is a vertex set of G and σ is an edge set of G then μ: $V \to [0,1]$ and $\sigma: V \times V \to [0,1]$ such that $\sigma(x, y) \leq \mu(x) \wedge \mu(y)$. A graph if $\mu(x) + \sigma(x, y) + \mu(y) = k$, has a uniform magic value then the graph G be a fuzzy magic graph for all $x,y \in V$.

2.1 Definition

A fuzzy vertex magic total labeling [FVMT] is assign of a label to nodes and to link of a graph G which is always mapped to positive integers that is f: $\sigma \cup \mu$ to the finite subset of membership function [0,1] then$\sigma(vi) = \sigma(vi) + \sigma(viej)$, where $e's$ are edges incident to V.

2.2 Example

Consider the path graph $P3$ with 2 links and 3 nodes as shown in Example 2.2. At the vertex labeled 0.05, adding of the node label and link labels connecting to that individual node shows us the total of all labels related to a vertex is the same $k = 0.06$ [9, 10] (Fig. 1).

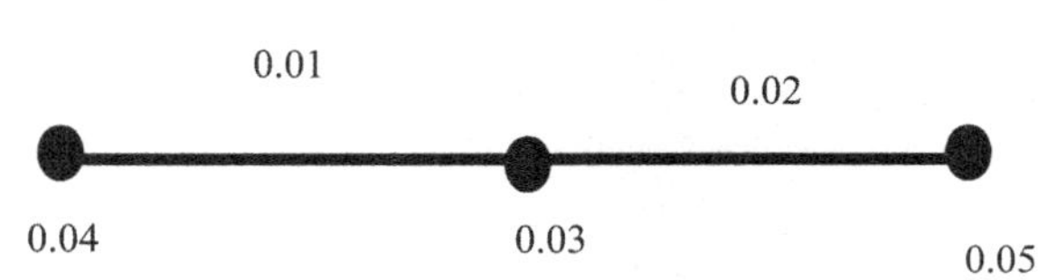

Fig. 1 Path graph P_3

2.3 Peterson Graph

A generalized Peterson graph P (m, n) consists of an outer p-cycles $u_1, u_2, .., u_{p-1}, u_p$ a set of spokes $u_i v_i$, $1 \leq i \leq m$, the q-inner edges $vivi + n$ with indices taken modulo m which can also be described as a cartesian product $P2 \times Cn$ of a path graph on two nodes with a cycle on n nodes[7] 8.

3 Theorem

Let G of (V, E) be a generalized Peterson P (m, n) graph. Then the graph G can be an FVMT labeling.

Proof Consider a generalized Peterson P $(m$, n) graph, where $m \geq 3$. Here an outer cycle consists of p vertices $V = \{u_1, u_2, .., u_{p-1}, u_p\}$ and $\{v_1, v_2, .., v_{q-1}, v_q\}$ are vertices of the inner cycle. $upvq$ are spokes $1 \leq i \leq m$ of a graph.(u_i, u_{i+1}),$\mu(v_i, v_{i+n})$ are the outer and inner edges of a graph such that $\mu(u_i, u_{i+1}) > 0$ where $i = 1$ to m and $\mu(v_i, v_{i+n}) > 0$ where $i = 1$ to m. αm- labeling is required for graph P (m, n). To prove FVM labeling of P (m, n).

Case (i): To prove Peterson P (m, n) graph is a fuzzy Peterson graph, then we have to prove $\mu(e_j) \leq \sigma(v_i) \wedge \sigma(v_j)$.

In the graph P (m, n), the values of the vertex set and the values of edge set are.$\mu(u_i, u_{i+1})$ and $\mu(v_i, v_{i+n})$ starts from $\frac{1}{n^i}$, where n, $i = 1, 2 \ldots$ etc.

Then continue the values to set of vertices (u_p) *and* extended to (v_q) to the set of edges. By fuzzy definition, all the values of nodes and links are distinct and the link takes the minimum label then the vertex label (i.e.) $(e_j) \leq (v_i) \wedge (v_j)$.

Hence, Peterson graph is a fuzzy Peterson graph P (m, n) Type equation here.

Case (ii): To prove fuzzy Peterson graph FP (m, n) is an FVMT labeling graph.

Consider the fuzzy Peterson graph if $m \geq 3$, we consider the total of all labels connected to that individual node is the same, at every vertex V, then $\sigma(vi) = \sigma(vi) + \sigma(viej) = K$, where $ej's$ are edges incident to V and K is fuzzy vertex magic constant. Hence, fuzzy Peterson graph is an FVMTL graph as shown in Fig. 2.

3.1 Example

Consider a graph P (6, 1) with 12 vertices and 18 edges and there are 30 labelings are required to label a graph, the value starts from $\frac{1}{n^i}$ where n, $i = 1, 2, \ldots$, etc., are used such that.

For the outer cycle,

(i) At the vertex v1, $\sigma(v_1) = \sigma(v_1) + \sigma(v_i e_j)$
$= 0.27 + (0.16 + 0.09 + 0.01) = 0.53$.

Fig. 2 Peterson graph (3, 1)

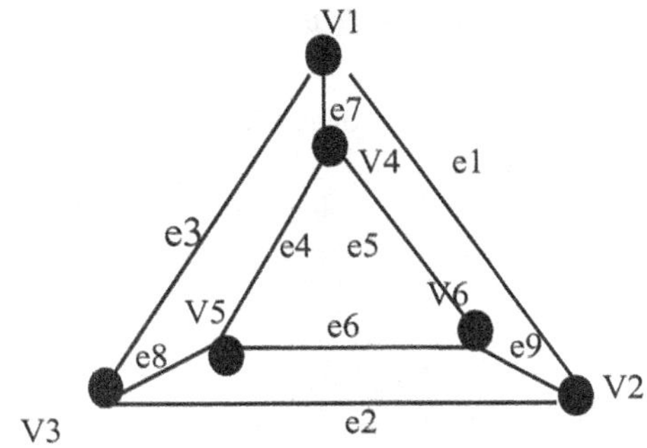

(ii) At the vertex v2, $\sigma(v_2) = \sigma(v_2) + \sigma(v_i e_j)$
$= 0.30 + (0.04 + 0.01 + 0.18) = 0.53.$

(iii) At the vertex v3, $\sigma(v_3) = \sigma(v_3) + \sigma(v_i e_j)$
$= 0.26 + (0.07 + 0.18 + 0.02) = 0.53.$

(iv) At the vertex v4, $\sigma(v_4) = \sigma(v_4) + \sigma(v_i e_j)$
$= 0.28 + (0.02 + 0.06 + 0.17) = 0.53.$

(v) At the vertex v5, $\sigma(v_5) = \sigma(v_5) + \sigma(v_i e_j)$
$= 0.25 + (0.03 + 0.08 + 0.17) = 0.53.$

(vi) At the vertex v6, $\sigma(v_6) = \sigma(v_6) + \sigma(v_i e_j)$

$= 0.39 + (0.03 + 0.05 + 0.16) = 0.53.$

Similarly, we can prove for an inner cycle. Clearly, we understood from the above example *P*(6, 1) the total of all labels connected to a vertex is uniform using FVMT labeling as shown in Fig. 3

4 Lemma

In FVMT labeling of a Peterson graph, the sum of the degree of the odd nodes is equivalent to double the degree of all edge membership and the difference of the sum of degrees of even nodes.

Proof Let *G* of (, μ) is FVMT labeling of a Peterson graph *P*(*m*, *n*). If the edge membership values are incident on any degree of node (v_i) are added then the total of corresponding node membership values are found.

The degree of vertices has been split into an odd vertex degree and even vertex degree. If the edge membership values are added, then we find the degree of edges $\sum_{i=1}^{n} \mu(u_i, v_{i+1})$. We know that the sum total of the degree of nodes is double, the sum of the degree of the membership of (*ui*, *vi* + 1) then

$$\sum_{i=1}^{n} \deg(v_i) = 2\sum_{i=1}^{n} \mu(u_i, v_{i+1})$$

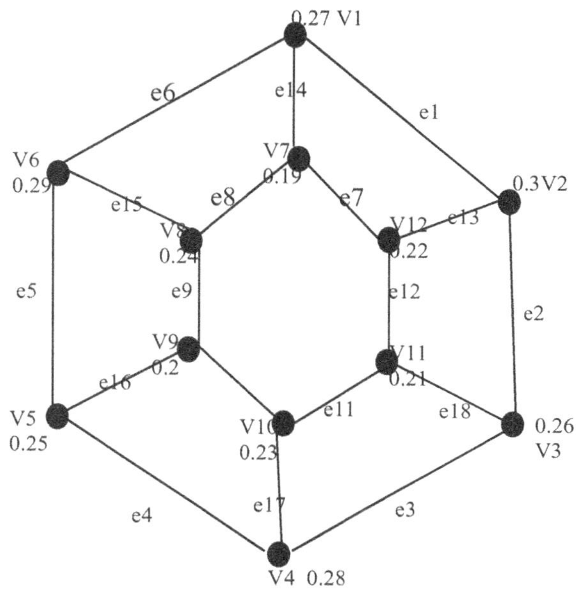

Fig. 3 Fuzzy vertex magic total labeling on Peterson graph P(6,1)

From above we get,

$$\sum_{i=1}^{n} \text{even deg } \sigma(v_i) + \sum_{i=1}^{n} \text{odd deg } \sigma(v_i) = 2\sum_{i=1}^{n} \mu(u_i, v_{i+1})$$

Then,

$$\sum_{i=1}^{n} \text{odd deg } \sigma(v_i) = 2\sum_{i=1}^{n} \mu(u_i, v_{i+1}) - \sum_{i=1}^{n} \text{even deg } \sigma\ (v_i)$$

Thus, the FVMT labeling of Peterson graph the sum of the degree of the odd node is equivalent to double the degree of all edge membership and the difference of the sum of degrees of even nodes.

4.1 Lemma

A fuzzy graph be a generalized Peterson $P\ (n, k)$ when n is even and k is either odd or even such that a fuzzy set V is specified by membership function σ, σ: $V \rightarrow [0,1]$ by assigning the values anticlockwise and clockwise direction to the outer cycle and inner cycle then G is an FVMTL graph.

4.2 Lemma

A Peterson graph $P(n, k)$ has fuzzy vertex magic constant, then $\delta(G^*) \leq \Delta(G^*) \leq S(G^*) \leq O(G^*)$.

Proof Consider $G^* = (\sigma, \mu)$ be an FVMT graph since $\delta(G^*)$ and $\Delta(G^*)$ are the smallest and largest degree of $G^*, \delta(G^*) = \vee \{d(v)/v \in V\}$ and $\Delta(G^*) = \vee \{d(v)/v \in V\}$.

then.

$$\delta(G^*) \leq \Delta(G^*) \tag{1}$$

The degree of a node v is defined by $(v) = \sum_{x \neq y} \mu\ (x, y)$.

We denote order and size of G^* has $O(G^*) = \sum_{x \in V}{'}\sigma(x)$ and $S(G^*) = \sum_{x \neq y} \mu\ (x, y)$.

By degree of a vertex,

$\sum x \neq y\ (x, y) \geq \vee \{(v)/v \in V\}$, then

$$\Delta(G^*) \leq S(G^*) \tag{2}$$

An FVMT labeling graph, an edge membership value is smaller than or equal to the minimum value of the edge connecting vertices.

Therefore,

$$(G^*) \leq (G^*) \tag{3}$$

From (1), (2), and (3), we get, $(G^*) \leq \Delta(G^*) \leq S(G^*) \leq O(G^*)$.

5 An Application of Fuzzy Vertex Magic Total Labeling Graph

Described the current passing through the transformer shown in Fig. 4. Consider the transformer $\tau_i$where $i = 1, 2, \ldots, n$ represents the vertex. The transformer sharing the current from τ_i to τ_j is taken as an edge between the vertex (Table 1).

The above diagonal entry(voltage) shows the current use by its transformer, not shared by the other transformers $\tau_i \neq \tau_j$ where $i, j = 1, 2, \ldots, n$. The other diagonal entries (voltage) are sharing the current from τ_i to $\tau_j$respective pair of transformers. The entry zero indicates the current is not passing through the path.

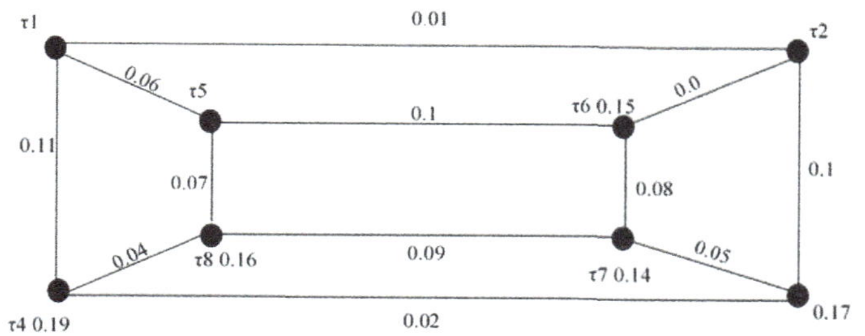

Fig. 4 Fuzzy vertex magic total labeling on Peterson graph (4, 1)

Table 1 Current shared between the transformers

Trans	τ1	τ2	τ3	τ4	τ5	τ6	τ7	τ8
τ 1	0.18	0.01	0	0.11	0.06	0	0	0
τ 2	0.01	0.2	0.12	0	0	0.03	0	0
τ 3	0	0.12	0.17	0.02	0	0	0.05	0
τ 4	0.11	0	0.02	0.19	0	0	0	0.04
τ 5	0.06	0	0	0	0.13	0.10	0	0.07
τ 6	0	0.03	0	0	0.1	0.15	0.08	0
τ 7	0	0	0.05	0	0	0.08	0.14	0.09
τ 8	0	0	0	0.04	0.07	0	0.09	0.16

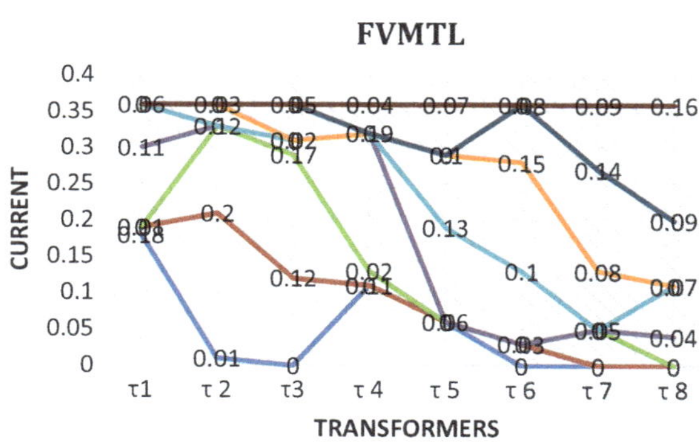

Graph.5.1

This FVMT labeling gives a magic constant $k = 0.36$. Thus, all transformers distributed the current equally by using FVMT labeling shown in the graph.

6 Conclusion

We proposed the idea of FVMT labeling of a graph by considering the concept of fuzzy magic labeling. Here we choose Peterson graph $P\ (m, n)$ is FVMT labeling under some specific condition has been investigated. We interpreted some results related to FVMT labeling using $P\ (m, n)$. Then we introduced an application using $P\ (4, 1)$ how all transformers distributed the current equally using FVMT labeling. Further, we are planning to enlarge how to construct FVMT labeling for some classes of graphs. Also, we are extended our field to FEMT labeling on some more graphs for jellyfish, tadpole, unicycle, bi-star, friendship graphs.

References

1. Zadeh, L. A. (1971). Similarity relations and fuzzy orderings. *Information Sciences, 3*(2), 177–200.
2. Rosenfeld, A. (1975). Fuzzy graphs. In *Fuzzy Sets and Their Applications to Cognitive and Decision Processes* (pp. 77–95). Academic Press.
3. Bhattacharya, P. (1987). Some remarks on fuzzy graphs. *Pattern Recognition Letters, 6*(5), 297–302.
4. MacDougall, J. A., Miller, M., & Bača, M. (2006). Vertex-magic total labeling of generalized Petersen graphs and convex polytopes. *Journal of Combinatorial Mathematics and Combinatorial Computing, 58,* 89–99.
5. Shanmugapriya, R., Hemalatha, P. K., & Suba, M. (2019, July). Fuzzy graceful labeling on the double fan graphs and the double wheel graphs. *IJITEE, 8*(9).
6. Dey, A., Ghosh, D., & Pal, A. (2013, December). An application of fuzzy graph in traffic light control. In *International Conference on Mathematics* (pp. 1–8).
7. Prasanna, N. L., Sravanthi, K., & Sudhakar, N. (2014). Applications of graph labeling in communication networks. *Oriental Journal of Computer Science and Technology, 7*(1), 139–145.
8. Bibi, K. A., & Devi, M. (2017). Fuzzy Bi-Magic labeling on Cycle Graph and Star Graph. *Global Journal of Pure and Applied Mathematics, 13*(6), 1975–1985.
9. Sumathi, P., & Fathima, B. (2017). Mod(k) vertex magic labeling of some non hamiltonian graphs-paper-1. *International Journal of Pure and Applied Mathematics*, 115(9), 279–290.
10. Getzimah, H. V., & Ramesh, D. S. T. *Vertex Magic Pyramidal Graphs.*

Metric Measures of Optic Nerve Head in Screening Glaucoma with Machine Learning

M. Antony Ammal, D. Gladis, and Atheek Shaik

Abstract Glaucoma is an irreversible disease of the eyes that prevails over the optic nerve head of the retinal image due to upsurge in intra-ocular pressure. Due to the asymptomatic and slow progressive nature of glaucoma, many people are unaware of this visual upheaval until it stretches an intricate stage. As of now prevention is the only mean to avoid vision loss and hence periodical screening of retinal image is essential. Creating awareness among the people and reasonably priced diagnosing techniques are the need of the hour. The aim of this analysis is to design an algorithmic method to identify primary open-angle glaucoma in a cost effective way. The vertical and horizontal optic cup-to-disc ratio, neuroretinal rim area and ISNT rule analysis are the prime factors of detecting the glaucomatous eye. Optic disc and cup feature extraction is prerequisite to analyze, interpret and predict the abnormality. The proposed method is a multivariate approach that use K-means clustering concept in segregating the optic disc and cup from retina, measures the cup-to-disc ratio vertically and horizontally, estimate neuro retinal rim area and ISNT rule analysis automatically. It is tested on imaging dataset Bin Rushed and Magrabi Eye center file of Retinal Fundus Images for Glaucoma Analysis, to qualify the images as either normal or glaucomatous. The results have been compared with the ophthalmologist's prediction. A graphical user interface has been designed and implemented using MATLAB.

Keywords Machine learning · K-means clustering · Neuroretinal rim · ISNT rule · VCDR · Glaucoma

M. Antony Ammal (✉)
Research Scholar, Presidency College (A), University of Madras, Chennai 600005, India
e-mail: manto2k4@yahoo.co.in

D. Gladis
Principal, Bharathi Women's College (A), University of Madras, Chennai 600108, India
e-mail: gladischristopher@gmail.com

A. Shaik
Ophthalmologist, Fathima Eye Clinic, Chennai 600004, India
e-mail: dratheek@gmail.com

S.-L. Peng et al. (eds.), *Proceedings of First International Conference on Mathematical Modeling and Computational Science*, Advances in Intelligent Systems and Computing 1292, https://doi.org/10.1007/978-981-33-4389-4_54

1 Introduction

Glaucoma, macular degeneration, and retinal detachment are certain retinal diseases that cause damage to the eye and leads to blindness. Primary-open-angle glaucoma is the maximum prevalent type of glaucoma that makes up the majority of 90% of all and other than that congenital, angle-closure and normal-tension glaucoma [1] are few more types. Glaucoma is a disease of eye which is insidious and usually asymptomatic until the visual field has lost significantly. It is characterized by slowly progressive with elevated intraocular pressure (IOP) [2]. It is the key contributing threat aspect for the advance of primary open-angle glaucoma also known as chronic glaucoma. In the preliminary stage, the IOP may not be raised permanently, but in due course its raise is stable. The main cause for the rise in IOP is due to the disproportion between the aqueous fluid formed and outflow [3]. An eye requires a significant amount of pressure to retinal eyeball to preserve its correct form, so it can function properly. The foremost task of the retinal optic nerve is to relay visual data from the light-sensitive layer existing in the eye, the retina to the brain [4], where it is acknowledged as an object. If the optic nerve in the retinal tract is concealed by elevated pressure (IOP), it can be impaired and is shown in Fig. 1. The deformation of the optic nerve head (ONH) results in adverse effect. Minimal awareness towards the cause of it and the high cost associated with diagnosing are the factors that cause glaucoma. Significant loss of vision and blindness is the end result of untreated case of glaucoma [4]. Creating alertness for screening periodically after middle age and assessment are the demanding course of action desired for treating glaucoma.

Uncovering glaucomatous vicissitudes earlier is the strategy to contribute to the reduction of severe sight depletion. In general, glaucoma is detected by a blend of multiple experiments. The objective of this effort is to use retinal fundus photograph as it suits well for diagnosing retinal diseases and relatively simple procedure with minimal cost compared to other methods. Retina, the innermost tunic of the eyeball, is a fragile and translucent sheath and this is the finely formed eye tissue [5]. The fundus of the eye is the back portion of the interior of the eye ball and the fundus photography of the retina is depicted in Fig. 2. It is imperative to comprehend that IOP is not only one risk aspect for the augmentation of glaucoma [6], and it needs to be evaluated with a numeral clinical dynamics to decide if someone has glaucoma.

The medical term for the clinically evident region of the optic nerve head's neural and connective tissues is called the optic disc (OD) [7] and it is the region in which

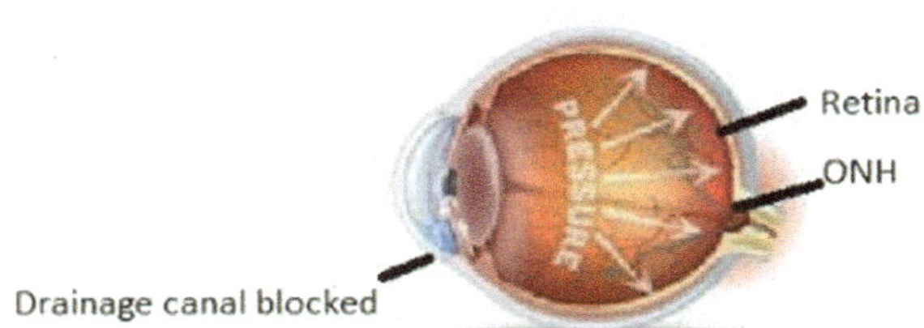

Fig. 1 Anatomy of human eye [3]

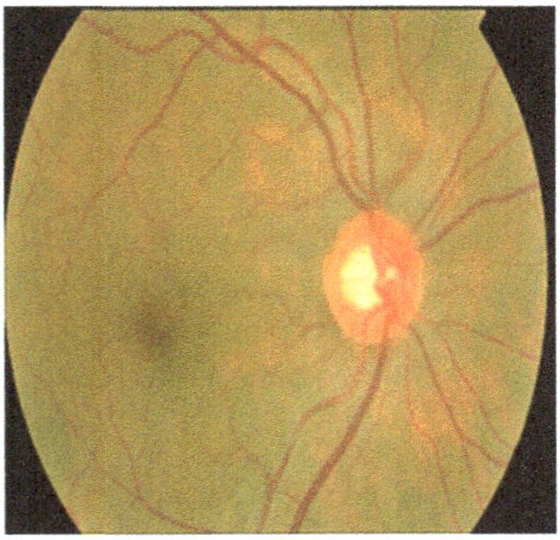

Fig. 2 Fundus image of Retina

optic nerve affix to the retina with a color ranging from orange to pink and comprises of two distinct regions of which the brightest center region called optic cup (OC) and a marginal area named neuroretinal rim (NRR). The parameters used widely for diagnosis of glaucoma are vertical and horizontal cup-to-disc ratio and neuroretinal rim area. Segmenting the OD and OC from the fundus image of the retina, is the prerequisite for the above and it is a complex task due to the shape and size of OD and OC and also due to the passage of the central retinal artery and vein through the ONH. The quality of image acquisition process also disturbs the image segmentation process. The ratio between OC and OD increases as the disease progress. The progression and severity of the retinal diseases is such that timely medical interference is indispensable to preclude visual deterioration. Based on the ophthalmologists registered with All India Ophthalmological Society (AIOS), 1: 1000 is the ratio of ophthalmologist and people and it is essential to reassure quicker and precision-based coordination of medical treatment [8]. Computerized technique is significant for computing quantitatively the multitudinal diagnostic parameters for persistent assessment for the continuous observations [7]. Computer-assisted analysis of medical imaging generates precise results more swiftly compared with manual recognition. This paper presented an overview of computing vertical cup-to-disc ratio (VCDR), horizontal cup-to-disc ratio (HCDR), neuroretinal rim area (NRR), and inferior superior nasal temporal (ISNT) rule analysis in Sect. 2, details about dataset source in Sect. 3, discussion of result analysis of data processed by machine learning technique is detailed in Sect. 4 followed by conclusion in Sect. 5.

2 Methodology

Segmentation of image is the cataloging of an image into disparate factions that is more worthwhile and apparent to explore. It is the procedure of apportioning a label to every single pixel in an image such that pixels with the identical label stake positive features. The proposed methodology is represented in the flowchart in

Fig. 3 and it applies K-means clustering algorithm which is a self-learning exercise. It is an unsupervised machine learning algorithm that is used to itemize the field of interest from the context by partitioning the image into group of clusters with high inter-cluster and low intra-cluster property [9]. Machine learning is the study of computer algorithms that showed potentially improved performance over other by automatically learning through experience together with some statistical techniques that permit better adaptation to the data and thus well suited for automatic detection of glaucoma. The application of machine learning techniques in healthcare industry to pinpoint the abnormalities at the early stage is phenomenal.

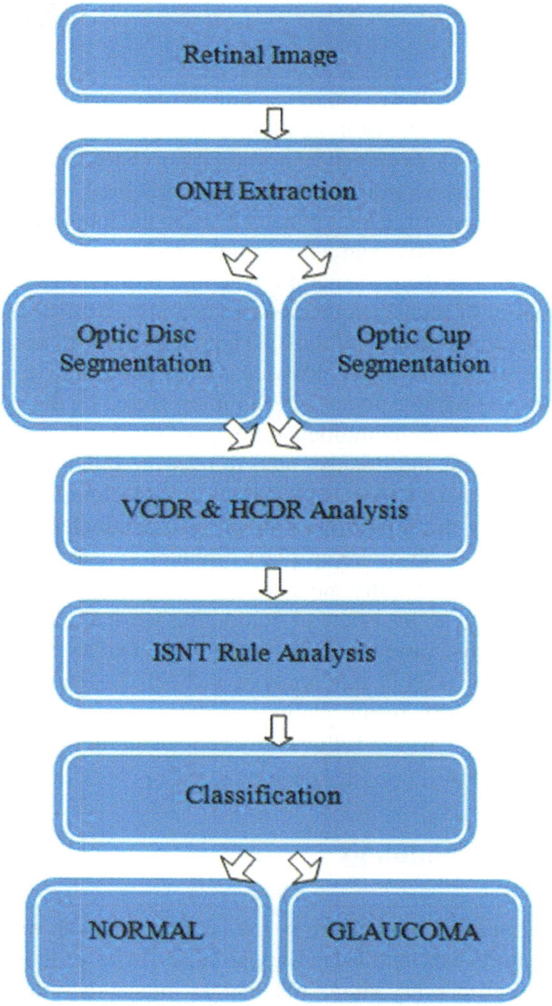

Fig. 3 Flowchart of the proposed algorithm

2.1 Segmentation of Optic Disc and Optic Cup

The 2D retinal fundus image is an RGB image of size 1584*2376 or 1936*2743 pixels and if the image is processed as such the computational complexity will be high. To make the process easier and to reduce the examining space, the preprocessing task is applied as follows. The region with the maximum brightness, the ONH is extracted as the region of interest (ROI) as represented in Fig. 4a. Hence, there is a considerable reduction of processing time and the speed of computation is evident.

K-means clustering algorithm bundles the entities adjoining by spatially gathering the K points. Iteratively, the centroid of groups is reorganized one after another till the preeminent bundling outcomes are achieved. It is a prototypical grouping style created on the function of model with distance measure as the target of optimization. The algorithm makes use of Euclidean distance as the metric of similarity to catch the optimum grouping of preliminary group midpoint, thus minimizing the assessment indicator. The clustering criterion applied is based on the concept of error square sum criterion function [9]. The input to the K-means clustering algorithm is a set of points $X_1 \ldots X_n$, with the intention to partition of the points into clusters $C_1 \ldots C_k$, along with a center μj for each cluster as in (1), so as to minimize

$$\sum_{j=1}^{K} \sum_{i \varepsilon C_j} \| X_i - \mu_j \|^2 \tag{1}$$

where $\|xi - \mu_j\|^2$is the Euclidean distance from the pixel to the cluster center, and K is the number of clusters. It can be checked that in any optimal solution, μ_j is the mean of the points in C_j.

$$\sum_{i \varepsilon C_j} \| X_i - \mu_j \|^2 = \frac{1}{2|C_j|} \sum_{i,i' \varepsilon C_j} \| X_i - X_i' \|^2 \tag{2}$$

Therefore, the K-means cost function can be equivalently written as in (3) [10].

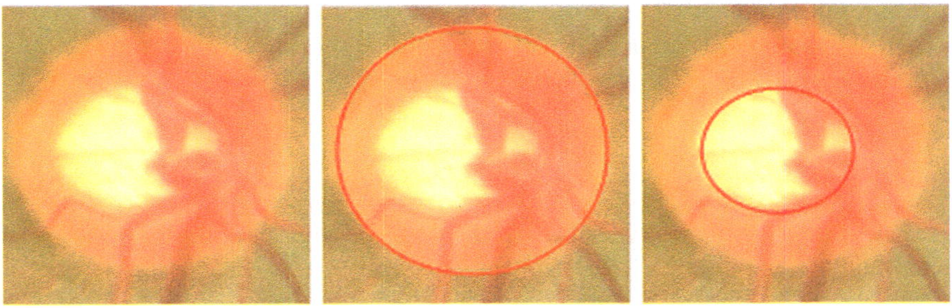

Fig. 4 **a** ROI **b** Optic disc **c** Optic cup

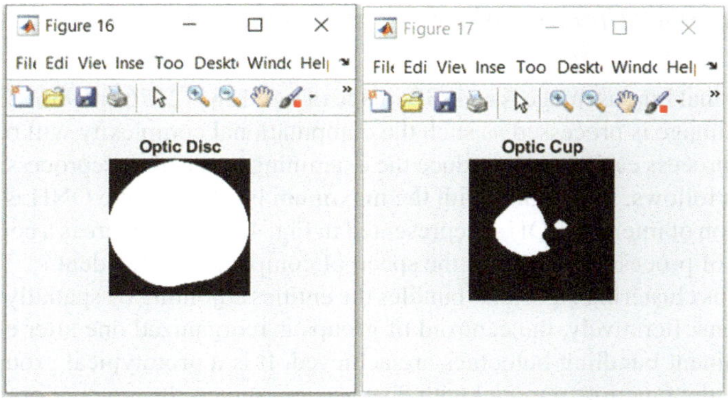

Fig. 5 **a** Optic disc **b** Optic cup

$$\sum_{j=1}^{K} \frac{1}{2|C_i|} \sum_{1,i' \varepsilon C_j} \| X_i - X_i' \|^2 \tag{3}$$

The dotted line in Fig. 4b and c shows elliptically fitted area of optic disk and cup images which are of interest to the analysis and they are well segregated using K-means algorithm and the binary images encompass inadequacies. In order to guarantee the succeeding stage accuracy, the segmented images are subjected to morphological operations and to smooth the boundaries, median filter is applied [11] as post-processing and the resultant picture is shown in Fig. 5.

2.2 *VCDR and HCDR Analysis*

Cup-to-disc ratio (CDR) computation is one among the other measure indicators used to detect glaucoma and is the most valuable method adopted by professionals. It is the ratio of size of optic cup-to-optic disc region. Larger cup size and low rim region result in higher CDR. CDR can be computed by three procedures namely VCDR, HCDR, and average CDR. VCDR is the ratio of measurement of the optic cup and optic disc diameter vertically as expressed in (4).

$$\text{VCDR} = \frac{\text{VCD}}{\text{VDD}} \tag{4}$$

HCDR is the ratio of measurement of the optic cup and optic disc diameter horizontally as expressed in (5).

$$\text{HCDR} = \frac{\text{HCD}}{\text{HDD}} \tag{5}$$

Average CDR is the ratio of optic cup area and optic disc area as shown in (6).

$$\text{Average CDR} = \frac{\text{Cup Area}}{\text{Disc Area}} \tag{6}$$

2.3 *NRR and ISNT Rule Analysis*

Neuroretinal rim area calculation is the other measure indicator used to detect glaucoma. Neuroretinal rim area is the section in the midst of optic cup and disc and is represented in Fig. 6a. As the optic cup size enlarges, proportionately the CDR increases and consequently the thickness of NRR area decreases which leads to glaucoma.

The NRR area is calculated by finding the difference of the disc area and cup area. Neuroretinal rim ratio is the ratio between rim area-to-disc area as in (7).

$$\text{NRR ratio} = \frac{\text{Rim Area}}{\text{Disc Area}} = \frac{\text{Disc Area - Cup Area}}{\text{Disc Area}} \tag{7}$$

Neuroretinal rim can be partitioned into four quadrants, namely inferior (I), superior (S), nasal (N), and temporal (T) as shown in Fig. 6b. Most healthy optic nerve head rim tissue follows the ISNT rule, where the rim tissue is usually the thickest inferiorly, then superior, then nasally and then temporally and it can be expressed as (8):

$$\text{Inferior} > \text{Superior} > \text{Nasal} > \text{Temporal} \tag{8}$$

If this ratio is not followed, and there is significant thinning of the rim tissue in any of these regions, the patient may have glaucomatous damage [12]. The optic nerve fibers are damaged in vertical direction earlier than horizontally, thus leading to trimming of upper and lower frames and the outcome is infringement of the rule

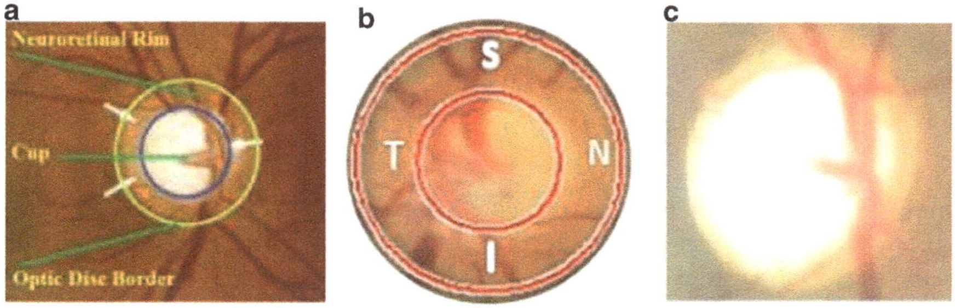

Fig. 6 **a** NRR [8] **b** ISNT quadrants **c** ISNT

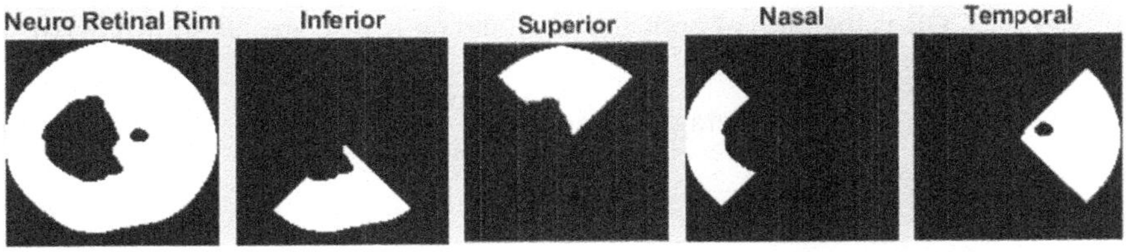

Fig. 7 **a** NRR **b–e** ISNT quadrants

[12], that is, the thickness of N&T is more than I&S. Thus, infraction of the order is evident in detecting glaucoma [13]. NRR image of Fig. 5 is shown in Fig. 7a. The NRR area inside the ISNT quadrants can be figured by assigning dissimilar mask to the NRR field extracted [12]. Figure 7b–e shows the extracted regions of the NRR area from Fig. 7a in the corresponding ISNT quadrants.

2.4 Classification

A decision tree classifier is applied for automatic classification of image as either healthy or glaucoma. The factors considered in discriminating normal eye and glaucomatous eye in the fundus image are marked cupping, focal notch, ISNT rule analysis and asymmetry between both eyes [5]. Cupping ratio is identified by depth of cup-to-disc ratio vertically, horizontally and average CDR, and focal notch is measured by NRR area and ISNT rule analysis by measuring ISNT quadrants. Notching of the rim especially up to disc margin is pathognomic [5]. CDR and the ISNT rule analysis are dominant in predicting and classifying the eye as healthy or glaucomatous [12].

The examination procedures performed are VCDR, HCDR, average CDR, NRR area and ISNT rule analysis. The usage of the above parameters enriches the accurateness of taxonomy. The classification process is as follows: if the value of the VCDR is lesser than 0.5 and complies by the ISNT rule, it is categorized as normal whereas if it is greater than 0.5 and do not complies by the ISNT rule, it is categorized as glaucoma. Also, if the asymmetry of the right eye to the left eye differs by 0.2 CDR, then it is also classified as glaucoma.

2.5 Category of Glaucoma Stages

Optic nerve with healthy tissue is considered as normal. Based on the significant loss of nerve tissue, the stages can be categorized as mild, moderate and advanced. Based on the stages of glaucoma the perceivable nature of vision also changes.

2.6 *Performance Evaluation Metrics*

The performance metrics of any algorithm can be evaluated by a confusion matrix known as an error matrix. For the proposed algorithm, performance metrics is evaluated using (9)–(12).

$$\text{Sensitivity (SN)} = \frac{\text{TP}}{\text{TP} + \text{FN}} \tag{9}$$

$$\text{Specificity (SP)} = \frac{\text{TN}}{\text{TN} + \text{FP}} \tag{10}$$

$$\text{Accuracy (Acc)} = \frac{\text{TP} + \text{TN}}{\text{TP} + \text{FN} + \text{FP} + \text{TN}} \tag{11}$$

$$\text{Precision} = \frac{\text{TP}}{\text{TP} + \text{FP}} \tag{12}$$

Sensitivity and specificity represent the most widely used metrics in medical research [14]. The pursuance of the technique is equated with ground truth provided by the ophthalmologists prediction which is considered as the gold standard.

3 Dataset for Investigation

The algorithm is tested on images of Bin Rushed and Magrabi Eye center file of retinal fundus images for glaucoma analysis (RIGA) dataset to investigate and classify the imaging data as normal eye or glaucomatous eye. The optic cup and disc boundaries of these images were marked and interpreted physically by six skillful ophthalmologists independently [15]. Bin Rushed and Magrabi Eye center file has been updated on January 17, 2018 and December 18, 2018, and the datasets included a number of testing images existing in 8 bits per color plane with resolution of 1584*2376 and 1936*2743 pixels and stored in JPG and TIFF format [15] respectively, and consist of both normal and glaucoma types. To assess the proposed method, a total of 90 images were considered, of which 63 images from Bin Rushed Ophthalmic center file and 27 images from Magrabi Eye center file. Hence for each original image, the six ophthalmologist manual markings were compared, so a total of 630 images were analyzed.

4 Results and Discussion

A graphical user interface (GUI) has been designed and implemented using image processing toolbox in MATLAB 2018a [16]. The absolute program code developed extracts the ocular structures, measures the values of the parameters, and classifies the image of the fundus photograph as normal or glaucomatous status.

The proposed methodology tried to design and implement a diagnosing tool to simulate most of the parameters analyzed in the OCT scan report and they are VCDR, HCDR, average CDR, NRR, *I*, *S*, *N* and *T* values. The estimation of distinctive parameters contributes in testing the proposed methodology and appraised outcomes are trustworthy if settled nearly equal to clinical outcomes marked by the ophthalmologists.

The algorithm constructed with the theory of K-means clustering procedure was applied to excerpt the OD and OC and eventually equated by automatic elliptical approximation. The results have been worked out by means of metric measures and visualized in Fig. 8a and b for a healthy and a glaucoma image respectively. OD segmentation is more accurate as compared with OC segmentation and it is confronted by the solidity of blood vessels wrapping portions of the cup and the plodding variation in luminosity between the rim and cup. The system was verified with image datasets which express encouraging effects for both healthy and glaucoma detection. The metric measures are automatically calculated without user intervention and the elliptical marking of the ROI and NRR images are also automatically marked based on the area calculated from the segmented regions.

VCDR, HCDR, average CDR, IS, and NT values of the test images and the ophthalmologists manual marked images which are calculated using the tool are tabulated in Table 1, which includes both healthy and glaucoma image. Figure 9a and b shows the evaluation chart of the results of k-means algorithm with the ground truth.

It can be perceived that the result achieved by the suggested system is amenable to ophthalmologist's diagnosis. From Table 1, it is evident that in the Magrabi dataset, ophthalmologist 4 manual marking deviates from others. The algorithm showed better accuracy for both datasets in disc segmentation, but in case of cup segmentation, Bin Rushed dataset performed better than Magrabi dataset.

As per Table 1, the impacts of VCDR and HCDR factors and ISNT rule analysis are evident in predicting the given retinal image as healthy or glaucomatous eye. If the CDR value is less than 0.5 and IS > NT, it is recognized as healthy image. Factors like broadening of OC and diminishing of NRR leads to glaucoma and hence glaucoma is suspected centered on the CDR and NRR measures. The practice of these parameters reinforces the cataloguing accuracy, that is, if the CDR is greater than 0.5 and the IS area is less than NT area. Table 2 displays the sample results obtained from the images analyzed and the calculations are carried out successfully and tabulated.

From Table 2, it is robust that, ISNT rule may not be always up to the mark in predicting the results. Its reliability depends on the position of the expansion of the

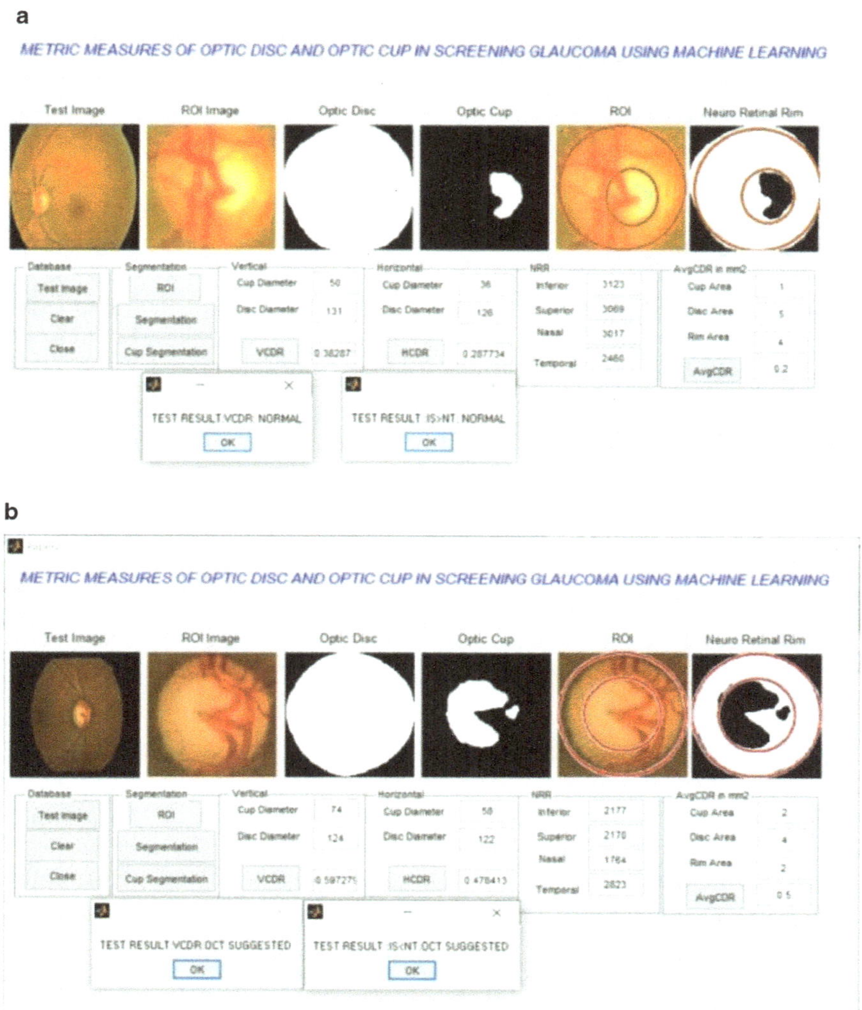

Fig. 8 **a** Healthy image, **b** Glaucoma image

cup region and the volume of the cup. In the following Fig. 6c the enlargement of cup is tilted toward temporal area and its CDR value is greater than 0.5 but it does not follow the ISNT rule. Similarly, ISNT rule differs in some cases when CDR value is nearer to 0.5. This results in lack of effectiveness of ISNT rule in screening glaucomatous eyes [17]. Hence it is insignificant in predicting glaucoma. Also, NRR area does not allow robust difference of normal and glaucoma eye and clinically not appreciable [18]. Thus NRR and ISNT rules have lesser effect in distinguishing glaucomatous from non-glaucomatous eye clinically [17]. Hence NRR and ISNT

Table 1 Result analysis of algorithm and ophthalmologists

Image	VCD	VDD	VCDR	HCD	HDD	HCDR	Avg. CDR	IS	NT
K-means	42	111	0.37	33	106	0.31	0.33	4320	3827
Ophthalmologist 1	44	108	0.4	36	108	0.34	0.36	4022	3867
Ophthalmologist 2	31	117	0.27	31	110	0.28	0.25	4869	4440
Ophthalmologist 3	31	115	0.27	21	105	0.2	0.2	4788	4117
Ophthalmologist 4	71	112	0.63	60	110	0.55	0.4	5540	6381
Ophthalmologist 5	31	115	0.27	29	109	0.26	0.21	4856	4236
Ophthalmologist 6	39	114	0.34	34	103	0.33	0.25	4279	3834
K-means	33	111	0.29	30	111	0.27	0.25	4451	4398
Ophthalmologist 1	31	107	0.29	29	106	0.28	0.25	4084	4066
Ophthalmologist 2	37	113	0.33	35	103	0.34	0.28	4308	3740
Ophthalmologist 3	41	117	0.35	31	109	0.28	0.25	4695	4300
Ophthalmologist 4	58	110	0.53	47	109	0.43	0.36	3490	3740
Ophthalmologist 5	35	109	0.32	32	105	0.31	0.26	4162	3888
Ophthalmologist 6	34	112	0.3	30	111	0.27	0.23	4468	4463
K-means	190	238	0.8	170	219	0.78	0.64	2041	2147
Ophthalmologist 1	161	209	0.77	159	208	0.76	0.59	2480	2527
Ophthalmologist 2	165	207	0.8	162	197	0.82	0.64	2008	2528
Ophthalmologist 3	178	218	0.82	166	206	0.81	0.67	1951	2547
Ophthalmologist 4	161	209	0.77	157	201	0.77	0.59	2658	2884
Ophthalmologist 5	161	200	0.8	160	199	0.8	0.65	2288	2172
Ophthalmologist 6	169	205	0.82	152	184	0.83	0.68	1841	2158
K-means	166	253	0.66	152	250	0.61	0.43	3572	3866

(continued)

Table 1 (continued)

Image	VCD	VDD	VCDR	HCD	HDD	HCDR	Avg. CDR	IS	NT
Ophthalmologist 1	145	227	0.64	143	222	0.64	0.41	3151	3225
Ophthalmologist 2	146	226	0.65	144	223	0.65	0.42	3142	3176
Ophthalmologist 3	157	227	0.69	146	226	0.65	0.48	3138	3429
Ophthalmologist 4	163	230	0.71	163	224	0.73	0.5	2789	2853
Ophthalmologist 5	147	224	0.66	144	224	0.64	0.43	3244	3178
Ophthalmologist 6	149	217	0.69	134	211	0.64	0.47	3000	3135

rule may not be considered as the stand-alone decision factor for glaucoma, but its impact is considerable when considered with other clinical parameters like VCDR and HCDR.

As specified in the performance metrics, the sensitivity, specificity, accuracy, and precision are calculated for the tabulated data in Table 3. True positive (TP) means a number of observations the model predicted were positive that were actually positive, false negative (FN) means a number of observations the model predicted were negative that were actually positive, false positive (FP) says a number of observations the model predicted were positive that were actually negative, and true negative (TN) gives a number of observations the model predicted were negative that were actually negative. Among the sixty-three images from Bin Rushed dataset eleven images and among twenty-seven images from Magrabi dataset five images were noted to have deviation in the midst of the markings by the six ophthalmologists due to the perceptibility and eliminated as lack of agreement (LA) among ophthalmologist. Generally manual markings lead to inter-observer variations. Magrabi dataset had the highest percentage of eliminated images compared to Bin Rushed.

The proposed method observed the performance metrics sensitivity, specificity, accuracy, and precision as 80%, 86%, 82%, and 90%, respectively, for both the datasets. As can be seen in Table 3, the results of Bin Rushed dataset achieved 89% of sensitivity, 82% specificity, 87% accuracy, and 91% precision which is best compared with the Magrabi dataset. Figure 10 display the percentage of performance metrics comparison chart of Bin Rushed and both datasets. It stands that greater the sample size, the more accurate the effects of the learning. This tool is effective and encompasses less computational complexity to facilitate the analyst to take well-timed assessment for factual treatment and step as a feasible alternative in mass screening places since it may be expensive and time consuming tasks.

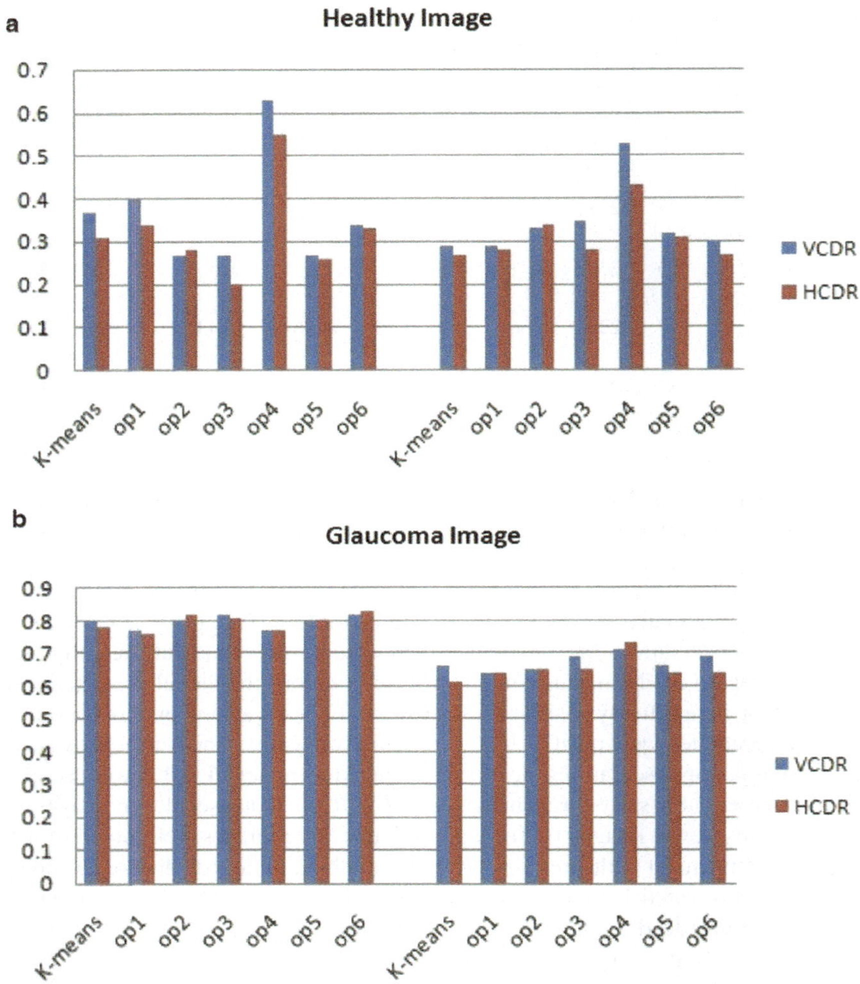

Fig. 9 **a** VCDR and HCDR comparison of healthy image, **b** VCDR and HCDR comparison of glaucoma image

5 Conclusion

Careful and routine observation of optic nerve head through fundus image examination throws light on the status of glaucoma which helps to prevent it. To speed up the processing and to increase the accuracy, a GUI has been developed for automatic segmentation of optic disc and cup to measure the CDR factors vertically and horizontally, to compute average CDR, to find NRR area and to analyze ISNT rule. Though glaucoma can be detected by using automated machines in a precise manner,

Table 2 Result analysis of dataset images

Image	VCDR	HCDR	IS	NT	IS > NT	Inference
1	0.44	0.35	5926	5285	Yes	Normal
2	0.21	0.2	5822	5792	Yes	Normal
3	0.36	0.36	5838	5060	Yes	Normal
4	0.48	0.45	5508	5392	Yes	Normal
5	0.37	0.27	6361	5713	Yes	Normal
6	0.48	0.44	5473	4564	Yes	Normal
7	0.43	0.44	5343	4511	Yes	Normal
8	0.36	0.37	5067	4924	Yes	Normal
9	0.39	0.34	6086	5666	Yes	Normal
10	0.42	0.39	5581	4907	Yes	Normal
11	0.5	0.49	4640	4957	No	Glaucoma
12	0.51	0.5	4733	4106	Yes	Glaucoma
13	0.64	0.61	3938	4158	No	Glaucoma
14	0.6	0.57	3631	4037	No	Glaucoma
15	0.53	0.47	4035	4147	No	Glaucoma
16	0.53	0.48	3334	3392	No	Glaucoma
17	0.56	0.52	3259	4078	No	Glaucoma
18	0.67	0.55	3672	4130	No	Glaucoma
19	0.57	0.48	3744	4107	No	Glaucoma
20	0.8	0.78	2147	2041	Yes	Glaucoma

Table 3 Prediction rate of datasets

Dataset	Total	LA	TP	TN	FP	FN
Bin Rushed	63	11	31	14	3	4
Magrabi	27	5	5	11	1	5
Total	90	16	36	25	4	9

the cost of diagnosing is an aspect to be reflected which are not affordable for all. This GUI tool cannot replace an expert's role but used as paramedical assistance in a primary health center and used as screening tool for large population in eye awareness camps so that the first level identification can be done with the help of optometrists with nominal fee. When the abnormality is predicted with the tool with the trained technicians, it facilitates the ophthalmologist to investigate further using OCT and diagnose rapidly, which considerably delay the progression of glaucoma and decrease the cost involved in finding. Hence, automated image classification system may be helpful and treatment becomes more affordable for the community. The diagnosis of the presence of a hemorrhage on the rim tissue of the optic nerve can be considered as the future enhancement techniques.

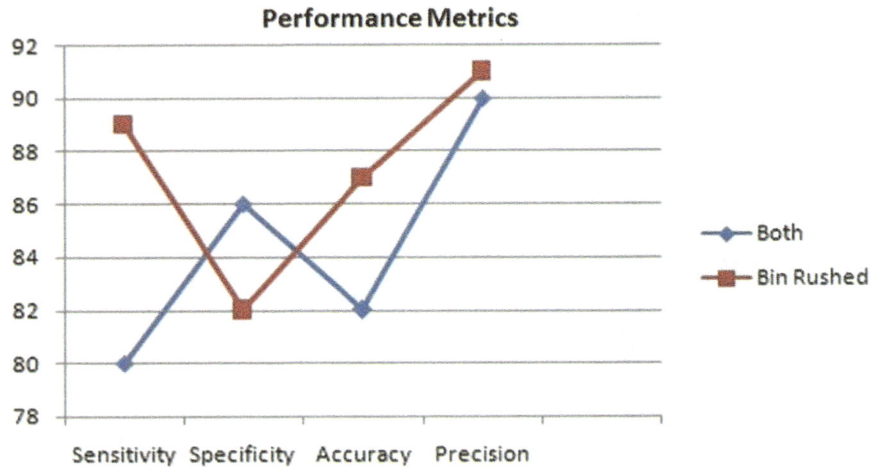

Fig. 10 Performance metrics

References

1. Kausu, T. R., Gopi, V. P., Wahid, K. A., Doma, W., & Niwas, S. I. (2018). Combination of clinical and multiresolution features for glaucoma detection and its classification using fundus images. *Biocybernetics and Biomedical Engineering, 38*(2), 329–341.
2. Naveed, M., Ramzan, A., & Akram, M. U. (2017, July). Clinical and technical perspective of glaucoma detection using OCT and fundus images: a review. In *2017 1st International Conference on Next Generation Computing Applications (NextComp)* (pp. 157–162). IEEE.
3. Salam, A. A., Khalil, T., Akram, M. U., Jameel, A., & Basit, I. (2016). Automated detection of glaucoma using structural and non structural features. *Springerplus, 5*(1), 1519.
4. Dhanachandra, N., Manglem, K., & Chanu, Y. J. (2015). Image segmentation using K-means clustering algorithm and subtractive clustering algorithm. *Procedia Computer Science, 54,* 764–771.
5. Khurana, A. K. (2015).*Comprehensive Ophthalmology* (6th ed.). Jaypee Brothers Medical publishing.
6. Jiang, Y., Duan, L., Cheng, J., Gu, Z., Xia, H., Fu, H., … Liu, J. (2019). Jointrcnn: A region-based convolutional neural network for optic disc and cup segmentation *IEEE Transactions on Biomedical Engineering, 67*(2), 335–343
7. Hong, S. W., Koenigsman, H., Ren, R., Yang, H., Gardiner, S. K., Reynaud, J., … Burgoyne, C. F. (2018). Glaucoma specialist optic disc margin, rim margin, and rim width discordance in glaucoma and glaucoma suspect eyes *American Journal of Ophthalmology, 192*, 65–76.
8. https://www.thehindu.com/news/cities/chennai/india-needs-more-eye-doctors/article26906320.ece.
9. Zheng, X., Lei, Q., Yao, R., Gong, Y., & Yin, Q. (2018). Image segmentation based on adaptive K-means algorithm. *EURASIP Journal on Image and Video Processing, 2018*(1), 68.
10. Dasgupta, S. (2008). *The hardness of k-means clustering.* Department of Computer Science and Engineering: University of California, San Diego.
11. Das, P., Nirmala, S. R., & Medhi, J. P. (2016). Diagnosis of glaucoma using CDR and NRR area in retina images. *Network Modeling Analysis in Health Informatics and Bioinformatics, 5*(1), 3.

12. Qiu, K., Wang, G., Lu, X., Zhang, R., Sun, L., & Zhang, M. (2018). Application of the ISNT rules on retinal nerve fibre layer thickness and neuroretinal rim area in healthy myopic eyes. *Acta Ophthalmologica, 96*(2), 161–167.
13. Park, D. Y., Lee, E. J., Han, J. C., & Kee, C. (2018). Applicability of ISNT rule using BMO-MRW to differentiate between healthy and glaucomatous eyes. *Journal of Glaucoma, 27*(7), 610–616.
14. Fatima Bokhari, S. T., Sharif, M., Yasmin, M., & Fernandes, S. L. (2018). Fundus image segmentation and feature extraction for the detection of glaucoma: A new approach. *Current Medical Imaging, 14*(1), 77–87.
15. Almazroa, A., Alodhayb, S., Osman, E., Ramadan, E., Hummadi, M., Dlaim, M., ... Lakshminarayanan, V. (2018, March). Retinal fundus images for glaucoma analysis: the RIGA dataset. In *Medical Imaging 2018: Imaging Informatics for Healthcare, Research, and Applications* (Vol. 10579, p. 105790B). International Society for Optics and Photonics.
16. https://www.mathworks.com.
17. Sihota, R., Srinivasan, G., Dada, T., Gupta, V., Ghate, D., & Sharma, A. (2008). Is the ISNT rule violated in early primary open-angle glaucoma—a scanning laser tomography study. *Eye, 22*(6), 819–824.
18. Rao, H. L., Yadav, R. K., Addepalli, U. K., Begum, V. U., Senthil, S., Choudhari, N. S., & Garudadri, C. S. (2015). The ISNT rule in glaucoma: Revisiting with spectral domain optical coherence tomography. *Acta Ophthalmologica, 93*(3), e208–e213.

Certain Separation Axioms and R_i Spaces with α_{γ_o}-open sets

N. Kalaivani and D. Saravanakumar

Abstract A new class of open sets, namely α_{γ_o}-open sets, was introduced by the authors, and its basic properties through the corresponding τ_{γ_o}-interior and τ_{γ_o}-closure operators are studied in their earlier work. Also, it has been proved the concept of γ_o-open sets and α_{γ_o}-open sets are independent and every γ_o-open set is an α_{γ_o}-open set, but the converse need not be true. The concepts of $\alpha_{\gamma_o} T_i$ spaces are introduced and established the relationship between these spaces by providing numerous theorems, tantamount statements, outcomes and corollaries. The theory of $\alpha_{\gamma_o} R_i$ spaces has been introduced, and their characterizations have been studied by exemplifying $\tau_{\alpha_{\gamma_o}}$-kernel and many dictums comprehend to that notion. Besides, the communication between these $\alpha_{\gamma_o} R_i$ spaces have been formed and analyzed using various axioms and obverse affidavits.

Keywords α_{γ_o}-open set · τ_{γ_o}-int · τ_{γ_o}-cl · $\alpha_{\gamma_o} T_i$ spaces ($i = 0$, 1/2 , 1, 2), · $\alpha_{\gamma_o} R_i$ spaces ($i = 0$, 1, 2) Mathematics Subject Classification: AMS(2000)54A05,54A10

1 Introduction

This paper is devoted to introduce new type of separation axioms which build on α_{γ_o}-open sets and explore their essential properties, communication and characterizations. At the end of this manuscript, we extend and examine the synergy between R_0 and R_1 spaces.

The various motifs which are used in this article are o-open, c-closed, os-open set, oss-open sets, cs-closed set, css-closed sets, T-topology, op-operation, ma-mapping, pos-power set, st-such that, sus-subset, TS-topological space, TSS-topological spaces, γ_0-γ, X_T denotes X.

N. Kalaivani (✉)
Department of Mathematics, Vel Tech Rangarajan Dr. Sagunthala R & D Institute of Science and Technology, Chennai, India
e-mail: kalaivani.rajam@gmail.com; drkalaivani@veltech.edu.in

D. Saravanakumar
Department of Mathematics, M S S Wakf Board College, Madurai, India
e-mail: drdstop@gmail.com

S.-L. Peng et al. (eds.), *Proceedings of First International Conference on Mathematical Modeling and Computational Science*, Advances in Intelligent Systems and Computing 1292, https://doi.org/10.1007/978-981-33-4389-4_55

Njastad [1] introduced α-oss in a TS. The semi-oss and pre-oss were investigated, respectively, by Levine [2], Corson and Michael [3]. Andrijevic [4] found a new class of T generated by pre-oss. Kasahara [5] expounded the idea of op on TSS and instigated α-closed graphs of an os. Ogata [6] labeled the op α as γ_o op and imported the insight of τ_{γ_o} that is the heap of all γ_o-oss in a TS (X_T, τ). G.Sai Sundara Krishnan and K.Balachandran [7, 8] announced γ-semi-oss and γ-pre-oss. Thenceforth Kalaivani and Sai Sundara Krishnan [9] proposed α_{γ_o}-oss that are generated through α-oss and investigated some of its basic properties. The notion of $\tau_{\alpha_{\gamma_o}}$- which is the collection of α_{γ_o}-oss in a TS was introduced. Further, the concepts of $\tau_{\alpha_{\gamma_o}}$-interior and $\tau_{\alpha_{\gamma_o}}$-closure operators were introduced, and some of their properties were studied. The concepts of α-os and α_{γ_o}-oss were independent was proved. Also, every γ-os was an α_{γ_o}-oss, but the converse need not be true was proved. Also, some properties of the T generated by α_{γ_o}-oss was studied.

In division 3, the concepts of $\alpha_{\gamma_o} T_i$ spaces (i = 0, 1/2, 1, 2) are introduced and characterized using the notion of α_{γ_o}-css or α_{γ_o}-oss. Some of their properties are investigated, and the relationship between them are studied. In division 4, the concepts of $\alpha_{\gamma_o} R_i$ spaces (i = 0, 1, 2) are introduced and some of their properties and the relationship between them are analyzed.

2 Preface

We present the fundamental terminologies and axioms related to this paper.

Let (X_T, τ) be a TS and D be a sus of X_T. Then D is forenamed as α- os [1] if $D \subseteq$ int(cl(int(D))). An op γ_o [6] on the T τ is a ma from τ on to the pos $P(X)$ of X_T st $F \subseteq F^{\gamma_o}$ for each $F \in \tau$,where F^{γ_o} stand for the value of γ_o at F. Let E be a sus of X_T and γ_o be an op on τ. Then E is said to be: a γ_o- os [6] if for each $e \in E$ there exists an os W such that $e \in W$ and $W^{\gamma_o} \subseteq E$. τ_{γ_o} stand for the set of all γ_o-oss in (X_T, τ). A γ_o- semi-os [7] if and only if $E \subseteq \tau_{\gamma_o}$-cl($\tau_{\gamma_o}$-int($E$)). A γ_o- preos [9] if and only if $E \subseteq \tau_{\gamma_o}$-int($\tau_{\gamma_o}$-cl($E$)). A γ_o- semi pre os [8] if and only if $E \subseteq \tau_{\gamma_o}$-cl($\tau_{\gamma_o}$-int($\tau_{\gamma_o}$-cl($E$))). The τ_{γ_o}- interior of E [8] is interpreted as an amalgamation of entire γ_o- oss encompassed in E and it is symbolized as τ_{γ_o}- int(E). Specifically τ_{γ_o}- int(E) $= \cup\{W : W$ is a $\gamma_o$os and $W \subseteq E\}$. The τ_{γ_o}-closure of A [6] is delineated as the intersection of the entire γ_o- cssencompassing E and it is symbolized by τ_{γ_o}-cl(E). Specifically τ_{γ_o}-cl(E) $= \cap\{Q : Q$ is a $\gamma_o$cs and $E \subseteq Q\}$. *A* sus D is γ_o- precs [8] if and only if τ_{γ_o}-cl(τ_{γ_o}- int(D))$\subseteq D$. *A* sus D is γ_o-semi pre-closed [8] if and only if τ_{γ_o}-int(τ_{γ_o}-cl(τ_{γ_o}-int(D)))$\subseteq A$. *A* sus D of X_T is said to be an α_{γ_o}-os [9] if and only if $D \subseteq \tau_{\gamma_o}$-int($\tau_{\gamma_o}$-cl($\tau_{\gamma_o}$-int($D$))). The $\tau_{\alpha_{\gamma_o}}$- interior of D [9] is the union of the entire α_{γ_o}-oss enclosed in D, and it is expressed by $\tau_{\alpha_{\gamma_o}}$-int(D). Specifically $\tau_{\alpha_{\gamma_o}}$-int(D) $=\{W : W$ is an α_{γ_o}os and $W \subseteq D\}$. The $\tau_{\alpha_{\gamma_o}}$- closure of D [9] is the intersection of the entire α_{γ_o}-css subsisting of D, and it is connoted by $\tau_{\alpha_{\gamma_o}}$-cl(D). Specifically $\tau_{\alpha_{\gamma_o}}$-cl(D) $= \{Q : Q\, is\, an\, \alpha_{\gamma_o}\, cs\, and\, D \subseteq Q\}$.

3 $\alpha_{\gamma_o} T_i$ Spaces

We present the general operation approaches on T_i spaces where $i = 0, 1/2, 1, 2$. Here γ_o be an operation on τ.

Definition 3.1 A TS (X_T, τ) is called an $\alpha_{\gamma_o} T_0$ space if for each distinct points a, $b \in X_T$ there exists an α_{γ_o}- os M such that a $\in M$ and $b \notin M$ or $b \in M$ and a $\notin M$.

Definition 3.2 A TS (X_T, τ) is called an $\alpha_{\gamma_o} T_1$ space if for each distinct points a, $b \in X_T$ there exists α_{γ_o}- oss M, N accommodating a and b, respectively, such that $b \notin M$ and a $\notin M$.

Definition 3.3 A TS (X_T,τ) is called an $\alpha_{\gamma_o} T_2$ space if for each distinct points a, $b \in X_T$ there exists α_{γ_o}-oss M, N such that a $\in M$, $b \in N$ and $M \cap N = \phi$.

Definition 3.4 A sus P of X_T is forenamed as α_{γ_0} g-cs if $\tau_{\alpha_{\gamma_0}}$-cl(P) $\subseteq Q$ whenever $P \subseteq Q$ and Q is an α_{γ_O}- os in (X_T, τ).

Remark 3.1 From the above Definition 3.3, every α_{γ_0}-cs in $(X_T, \tau)$is α_{γ_o} g-cs. However, the converse need not be true.

Definition 3.5 A TS $(X_T, \tau)$is called an $\alpha_{\gamma_0} T_{\frac{1}{2}}$ space if every α_{γ_o} g-cs in $(X_T, \tau)$is an α_{γ_o}-cs.

Theorem 3.1 *A sus S of X_T is aforesaid to be an α_{γ_o} g-cs if and only if $\tau_{\alpha_{\gamma_0}}$-cl ($\{a\}) \cap S \neq \phi$ holds for every $a \in \tau_{\alpha_{\gamma_0}}$-cl(S).*

Proof Let M be an α_{γ_o}-os such that $S \subseteq M$. Let a $\in \tau_{\alpha_{\gamma_0}}$-cl(S). Then by assumption, there exists a $w \in \tau_{\alpha_{\gamma_0}}$-cl ($\{a\}$) and $w \in S \subseteq M$. Hence, by Theorem 3.21 [10] $M \cap\{a\} \neq \phi$, this implies that an $\in M$. Therefore, $\tau_{\alpha_{\gamma_0}}$-cl(S) $\subseteq M$. Hence, S is a α_{γ_0} g-cs in (X_T, τ).

Conversely, suppose a $\in \tau_{\alpha_{\gamma_0}}$-cl($S$) such that $\tau_{\alpha_{\gamma_0}}$-cl ($\{a\}) \cap S = \phi$,then $S \subseteq (X_T - \tau_{\alpha_{\gamma_0}}$-cl ($\{a\}$). Then by Remark 3.19 (i) [10] and the assumption, it ensures that $\tau_{\alpha_{\gamma_0}}$-cl(S) $\subseteq (X_T - \tau_{\alpha_{\gamma_0}}$-cl ($\{a\}$). Hence a $\notin \tau_{\alpha_{\gamma_0}}$-cl ($\{a\}$). Hence, a $\notin \tau_{\alpha_{\gamma_0}}$-cl($S$). This is a contradiction. Therefore, $\tau_{\alpha_{\gamma_0}}$-cl ($\{a\}) \cap S \neq \phi$.

Theorem 3.2 *A sus P of X_T is forenamed to be an α_{γ_o} g- cs, then $\tau_{\alpha_{\gamma_0}}$-cl(P) P does not contain a non-empty α_{γ_o}-cs.*

Proof Suppose that there exists a non-empty α_{γ_o}- cs E such that $E \subseteq \alpha_{\gamma_o}$-cl($B$)- B ($\subseteq \tau_{\alpha_{\gamma_0}}$-cl ($B$)). Let e $\in E$. Then $e \in \tau_{\alpha_{\gamma_0}}$-cl ($B$); hence, it follows that from Remark 3.19 (i) [10] that $E \cap B = \tau_{\alpha_{\gamma_0}}$- cl $(E) \cap B \supseteq \tau_{\alpha_{\gamma_0}} cl(\{e\}) \cap B \neq \phi$. This implies that $E \cap B \neq \phi$. This is a contradiction. Hence, $\tau_{\alpha_{\gamma_0}}$-cl(B)-B does not contain a nonempty α_{γ_o}-cs.

Theorem 3.3 *For each $c \in X_T$, $\{c\}$ is an α_{γ_o}-cs or $X_T - \{c\}$ is an α_{γ_o} g-cs in (X_T, τ).*

Proof Suppose that $\{a\}$ is not an α_{γ_o}- cs then $X_T - \{a\}$ is not an α_{γ_o}-os. This implies that X_T is the only α_{γ_o}-os containing $X_T - \{a\}$. Therefore $X_T - \{a\}$ is an α_{γ_o} g- cs.

Theorem 3.4 *A TS (X_T, τ) is an α_{γ_o} $T_{\frac{1}{2}}$ space if and only if $\{a\}$ is an α_{γ_o}-cs or an α_{γ_o}-os in (X_T, τ).*

Proof Suppose $\{a\}$ is not an α_{γ_o}-cs. Then, it chases from the assumption and Theorem 3.3 that $\{a\}$ is an α_{γ_o}-os. Conversely, let H be an α_{γ_o} g-cs in (X_T, τ) and $a \in \tau_{\alpha_{\gamma_o}}$-cl $(H)(\subseteq X_T)$. Then by assumption $\{a\}$ is an α_{γ_o}-os or an α_{γ_o}-cs.

Case (i) suppose $\{a\}$ is an α_{γ_o}-os. Then by Theorem 3.21 [10], $\{a\} \cap H \neq \phi$, hence a $\in H$.

Houses (ii) Suppose $\{a\}$ is an α_{γ_o}-cs. Assume an $\notin H$, then $\{a\} \in \tau_{\alpha_{\gamma_o}}$- cl (H) - H . This is not possible by Theorem 3.2. Thus, we have a $\in H$. Therefore, $\tau_{\alpha_{\gamma_o}}$- cl $(H) = H$ and hence H is an α_{γ_o}-cs.

Theorem 3.5 *A TS (X_T, τ) an $\alpha_{\gamma_o} T_1$ space if and only if any singleton of X_T is an α_{γ_o}-cs.*

Proof Follows from Definition 3.2.

Remark 3.2

(a) Let $X_T = \{e, f, g\}$, $\tau = \{\emptyset, X_T, \{e\}, \{g\}, \{e, f\}, \{e, g\}\}$ exemplify an op γ_o on τ st $B^{\gamma_o} = \begin{cases} cl(B)\text{iff} \in B \\ B \cup \{c\}\text{iff} \notin B \end{cases}$. Then $\tau_{\gamma_o} = \{\emptyset, X_T, \{e\}, \{g\}, \{e, f\}, \{e, g\}\}$, and $\tau_{\alpha_{\gamma_o}} = \{\emptyset, X_T, \{e\}, \{g\}, \{e, f\}, \{e, g\}\}$. (X_T, τ) is an α_{γ_o} T_0 space but not an α_{γ_o} $T_{\frac{1}{2}}$ space.

(b) Let $X_T = \{e, f, g\}$, $\tau = \{\emptyset, X_T, \{f\}, \{g\}, \{e, g\}, \{f, g\}\}$, construe an op γ_o on τ st $C^{\gamma_o} = \begin{cases} C \cup \{e\} if\ C = \{e\}, \{f\}\, or\, \{g\} \\ C\text{ otherwise} \end{cases}$. Then, $\tau_{\gamma_o} = \{\emptyset, X_T, \{e, g\}, \{f, g\}\}$, and $\tau_{\alpha_{\gamma_o}} = \{\emptyset, X_T, \{e, g\}, \{f, g\}\}$. Then (X_T, τ) is an α_{γ_o} $T_{\frac{1}{2}}$ space but not an α_{γ_o} T_1 space.

(c) Let $X_T = \{e, f, g\}$, $\tau = \{\emptyset, X_T, \{f\}, \{g\}, \{e, g\}, \{f, g\}\}$ construe an op γ_o on τ st $D^{\gamma_o} = \begin{cases} cl(D)\text{ iff} \notin D \\ D\text{ iff} \in D \end{cases}$
Then $\tau_{\gamma_o} = \{\emptyset, X_T, \{f\}, \{e, f\}, \{e, g\}\}$, and $\tau_{\alpha_{\gamma_o}} = \{\emptyset, X_T, \{f\}, \{e, g\}, \{f, g\}\}$. Then $(X_T, \tau)$is an $\alpha_{\gamma_o} T_1$ space but not an $\alpha_{\gamma_o} T_2$ space.

(d) Let $X_T = \{e, f, g\}$, $\tau = \{\emptyset, X_T, \{f\}, \{g\}, \{e, g\}, \{f, g\}\}$ construe an op γ_o on τ st $E^{\gamma_o} = \begin{cases} cl(E)\text{ iff} \notin E \\ E\text{ iff} \in E \end{cases}$
Then $\tau_{\gamma_o} = \{\emptyset, X_T, \{f\}, \{e, f\}, \{e, g\}\}$, and $\tau_{\alpha_{\gamma_o}} = \{\emptyset, X_T, \{f\}, \{e, g\}, \{f, g\}\}$. Then (X_T, τ) is an $\alpha_{\gamma_o} T_1$ space but not an $\gamma_o T_1$ space

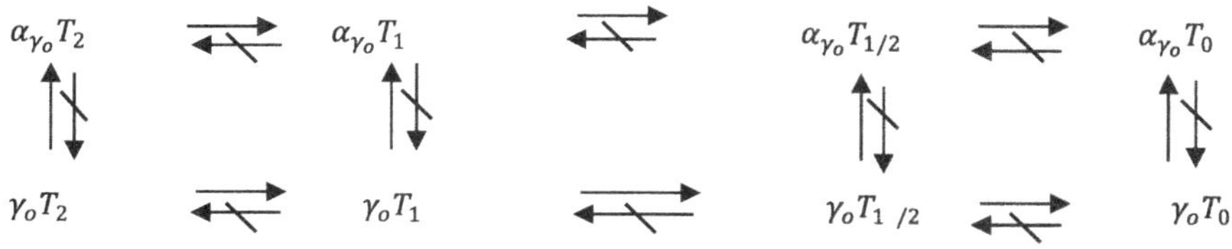

A→ B represents A implies B, A ↛ B represents A does not imply B and γ_o is a regular op on τ .

Fig. 1 Relationship between T_i spaces

(e) Let $X_T = \{e, f, g\}, \tau = P(X)$ construe an op γ_o on τ st $F^{\gamma_o} = \begin{cases} F \cup \{g\} \; if \; F = \{e\} \, or \, \{f\} \\ F \cup \{e\} \; if \; F = \{g\} \\ F \qquad if \; F \neq \{e\}, \{f\} \, and \, \{g\} \end{cases}$.
Then $\tau_{\gamma_o} = \{\emptyset, X_T, \{e, f\}, \{e, g\}, \{f, g\}\}$, and $\tau_{\alpha_{\gamma_o}} = \{\emptyset, X_T, \{e, f\}, \{e, g\}, \{f, g\}\}$. Then $(X_T, \tau)$is an $\alpha_{\gamma_o} T_{\frac{1}{2}}$ space but not an $\gamma_o T_{\frac{1}{2}}$ space.

(f) Let $X_T = \{e, f, g\}, \tau = \{\emptyset, X_T, \{e\}, \{f\}, \{g\}, \{e, f\}, \{e, g\}, \{f, g\}\}$ detail an op γ_o on τ st $G^{\gamma_o} = \begin{cases} G \cup \{g\} if \; G = \{e\}, \{f\}, \{f\} \\ \qquad G \qquad \text{otherwise} \end{cases}$.
Then $\tau_{\gamma_o} = \{\emptyset, X_T, \{e\}, \{f\}, \{g\}, \{e, g\}, \{f, g\}\}$, and $\tau_{\alpha_{\gamma_o}} = \{\emptyset, X_T, \{e\}, \{f\}, \{g\}, \{e, g\}, \{f, g\}\}$. Then $(X_T, \tau)$is an $\alpha_{\gamma_o} T_2$ space but not a $\gamma_o T_2$ space.

Remark 3.3 From the Theorems 3.3, 3.4, 2.3.5, Remark 3.2, Propositions 4.11 (Ogata 1991) and Remark 3.14 (Ogata 1991), we have the following implications as mentioned in Fig. 1.

4 Characterizations of $\alpha_{\gamma_o} R_0$ and $\alpha_{\gamma_o} R_1$ Spaces

Definition 4.1 Let (X_T, τ) be a TS and B be a sus of X_T. The α_{γ_o}-kernel of B is defined as the intersection of the entire α_{γ_o}-oss that contains B, and it is expressed as $\tau_{\alpha_{\gamma_o}}$-ker(B). That is $\tau_{\alpha_{\gamma_o}}$ -ker(B)= $\cap\{F : F \in \tau_{\alpha_{\gamma_o}} : B \subseteq F\}$

Theorem 4.1 *Let (X_T, τ) be a TS, C and D be subsets of X_T, then*

(i) $z \in \tau_{\alpha_{\gamma_o}}$*-ker(C) if and only if $C \cap F \neq \phi$ for any α_{γ_o}-cs F that contains z.*
(ii) $C \subseteq \tau_{\alpha_{\gamma_o}}$*-ker(C) and $C = \tau_{\alpha_{\gamma_o}}$-ker(C) if C is an α_{γ_o}-os.*
(iii) *If $C \subseteq D$, then $\tau_{\alpha_{\gamma_o}}$-ker(C) $\subseteq \tau_{\alpha_{\gamma_o}}$-ker(D).*

Proof (i) Let F_0 be the set of all $z \in X_T$ such that $W \cap C \neq \phi$ for every α_{γ_o}-cs W and $z \in W$. To demonstrate the dictum, it is enough to justify that $F_0 = \tau_{\alpha_{\gamma_o}}$- ker($C$).

Let $z \in \tau_{\alpha_{\gamma_0}}$- ker$(C)$. Let us assume that $z \notin F_0$ then there exists a α_{γ_0}- cs U_c of z such that $U_c \cap C = \phi$. This implies that $C \subseteq X_T - U_c$ and hence $\tau_{\alpha_{\gamma_0}}$-ker$(C) \subseteq X_T - U_c$. Therefore, $z \notin \tau_{\alpha_{\gamma_0}}$-ker$(C)$ which is a contradiction and hence $\tau_{\alpha_{\gamma_0}}$- ker$(C) \subseteq F_0$.

Conversely, let F_1 be a set such that $C \subseteq F_1$ and $(X_T - F_1)$is an α_{γ_0}-cs. Let $z \notin F_1$ then $z \in (X_T - F_1)$and $(X_T - F_1) \cap C = \phi$. This implies $z \notin F_0$. Therefore, $F_0 \subseteq F_1$. Hence, $F_0 \subseteq \tau_{\alpha_{\gamma_0}}$- ker$(C)$. Hence, the proof.

(ii) and (iii) The proof follows from Definition 4.1.

Theorem 4.2 *Let (X_T, τ)be a TS and e,f be any two distinct points in X_T.Then, $y \in \tau_{\alpha_{\gamma_0}}$-ker($\{e\}$) if and only if $e \in \tau_{\alpha_{\gamma_0}}$-cl ($\{f\}$).*

Proof Let $f \in \tau_{\alpha_{\gamma_0}}$-ker ($\{e\}$). If $e \notin \tau_{\alpha_{\gamma_0}}$-cl($\{f\}$), then $e \notin \cap\{F_1 : X - F_1 \in \tau_{\alpha_{\gamma_0}}\}$ and $\{f\} \subseteq F_1$ implies that e does not belong to at least one F_1, say F_0 such that $e \in X_T - F_0$ and $f \notin X_T - F_0$. Therefore $f \notin \cap\{X_T - F_1 : X - F_1 \in \tau_{\alpha_{\gamma_0}}\}$ and $\{e\} \subseteq X_T - F_1$ and hence $f \notin \tau_{\alpha_{\gamma_0}}$-ker($\{e\}$), which is a contradiction. Thus $e \in \tau_{\alpha_{\gamma_0}}$-cl ($\{f\}$).

Conversely, Let $e \in \tau_{\alpha_{\gamma_0}}$-cl ($\{f\}$). If $f \notin \tau_{\alpha_{\gamma_0}}$-ker ($\{e\}$), then $f \notin \cap\{U_c : U_c \in \tau_{\alpha_{\gamma_0}}\}$ and $\{e\} \subseteq U_c$ implies that $f \in X_T - U_c$ and $e \notin X_T - U_c$. Therefore $e \notin \cap\{X_T - U_c : U_c \in \tau_{\alpha_{\gamma_0}}\}\}$ and $\{f\} \subseteq X_T - U_c$and hence $\{e\} \notin \tau_{\alpha_{\gamma_0}}$-cl($\{f\}$), a contradiction. Thus, $f \in \tau_{\alpha_{\gamma_0}}$-ker($\{e\}$).

Theorem 4.3 *Let H be a subset of X_T. Then $\tau_{\alpha_{\gamma_0}}$-ker(H)= $\{e \in X : \tau_{\alpha_{\gamma_0}}$-cl($\{e\}) \cap H \neq \phi\}$.*

Proof Let e $\in \tau_{\alpha_{\gamma_0}}$-ker$(H)$. If $\tau_{\alpha_{\gamma_0}}$-ker($\{e\}) \cap H = \phi$, then $e \notin X_T - \tau_{\alpha_{\gamma_0}}$-cl($\{e\}$), which is an α_{γ_0}-os containing H. Thus $e \notin \tau_{\alpha_{\gamma_0}}$-ker$(H)$,a contradiction. Hence $\tau_{\alpha_{\gamma_0}}$-cl($\{e\}) \cap H \neq \phi$. Conversely, let $e \in X_T$ be such that $\tau_{\alpha_{\gamma_0}}$-cl($\{e\}) \cap H \neq \phi$. If possible, let $e \notin \tau_{\alpha_{\gamma_0}}$-ker(H). Then there exists a $U_c \in \tau_{\alpha_{\gamma_0}}$ such that $e \notin U_c$ and $H \subseteq U_c$. Let $f \in \tau_{\alpha_{\gamma_0}}$-cl($\{e\}) \cap H$. Then $f \in \tau_{\alpha_{\gamma_0}}$-cl($\{e\}$) and $f \in U_c$, which gives $e \in U_c$, a contradiction. Hence $e \in \tau_{\alpha_{\gamma_0}}$-ker$(H)$.

Definition 4.2 A TS (X_T, τ) is said to be an $\alpha_{\gamma_0} R_0$ space if for each α_{γ_0}-os U, $c \in H$ implies that $\tau_{\alpha_{\gamma_0}}$-cl($\{c\}) \subseteq H$.

Example 4.1 Let $X_T = \{l, m, m\}$, $\tau = P(X)$, exemplify an op γ_o on τ

$$D^{\gamma_o} = \begin{cases} D \cup \{c\} if\ D = \{l\}\ or\ \{m\} \\ D \cup \{a\} if\ D = \{n\} \\ D if\ D \neq \{l\}, \{m\} and\ \{n\} \end{cases} \quad \text{Then } \tau_{\gamma_o} = \{\phi, X_T, \{l, m\}, \{l, n\}, \{m, n\}\},$$

and$\tau_{\alpha_{\gamma_0}} = \{\phi, X_T, \{l, m\}, \{l, n\}, \{m, n\}\}$. Then (X_T, τ) is an $\alpha_{\gamma_0} R_0$ space.

Theorem 4.4 *Let (X_T, τ) be a TS. Then the affidavits indicated beneath are obverse:*

(i) *(X_T, τ) is an $\alpha_{\gamma_0} R_0$ space;*
(ii) *let M be an α_{γ_0}-cs and a point $c \notin M$. There endures an $N \in \tau_{\alpha_{\gamma_0}}$ such that $c \notin N$ and $M \subseteq N$;*

(iii) *If M is an α_{γ_o}-cs and a point $c \notin M$, $\tau_{\alpha_{\gamma_0}}$-cl$\{c\} \cap M = \phi$.*

Proof $(i) \Rightarrow (ii)$ Let M be an α_{γ_o}-cs and $c \notin M$. By (i), $\tau_{\alpha_{\gamma_0}}$-cl$(\{c\}) \subseteq X_T - M$. Let $N = X_T - \tau_{\alpha_{\gamma_0}}$-cl$(\{c\})$. Then $N \in \tau_{\alpha_{\gamma_0}}$ and also $M \subseteq N$ and $c \notin N$. $(ii) \Rightarrow (iii)$ Let M be an α_{γ_o}-cs and a point $c \notin F$. Suppose the given condition holds. Then by (ii), there exists an $N \in \tau_{\alpha_{\gamma_0}}$ such that $c \notin N$ and $M \subseteq N$. Thus, $c \in X_T - N \subseteq X_T - M$. Hence $X_T - N$ is an α_{γ_o}-cs containing c, we have $\tau_{\alpha_{\gamma_0}}$-cl$(\{c\}) \cap (X_T - N) \neq \phi$ implies that $\tau_{\alpha_{\gamma_0}}$-cl$(\{c\}) \subseteq X_T - N$. Therefore $N \cap \tau_{\alpha_{\gamma_0}}$-cl$(\{c\}) = \phi$ and hence $M \cap \tau_{\alpha_{\gamma_0}}$-cl$(\{c\}) = \phi$.

$(iii) \Rightarrow (i)$ Let $P \in \tau_{\alpha_{\gamma_0}}$ and $c \in P$. Then $X_T - P$ is an α_{γ_o}-cs and $c \notin X_T - P$. By (iii), $\tau_{\alpha_{\gamma_0}}$-cl$(\{c\}) \cap (X_T - P) = \phi$ implies that $\tau_{\alpha_{\gamma_0}}$-cl$(\{c\}) \subseteq P$. Hence (X_T, τ) is an $\alpha_{\gamma_o} R_0$ space.

Theorem 4.5 *Let (X_T, τ) be a TS and c,d be any two points in X_T. Then the statements mentioned below are equivalent:*

{i} (X_T, τ) *is an* $\alpha_{\gamma_o} R_0$ *space*
{ii} $\tau_{\alpha_{\gamma_0}}$*-cl*$(\{c\}) \subseteq \tau_{\alpha_{\gamma_0}}$*-ker*$(\{c\})$
{iii} d $\in \tau_{\alpha_{\gamma_0}}$*-ker*$(\{c\})$ *if and only if* $c \in \tau_{\alpha_{\gamma_0}}$*-ker* $(\{d\})$
{iv} d $\in \tau_{\alpha_{\gamma_0}}$*-cl*$(\{c\})$ *if and only if* $c \in \tau_{\alpha_{\gamma_0}}$*-cl*$(\{d\})$
{v} *Let E be an* α_{γ_o}*-cs and a point* $c \notin E$*. Then there exists a* $P \in \tau_{\alpha_{\gamma_0}}$ *and* $E \subseteq P$
{vi} Each α_{γ_o}*-cs E can be expressed as* $E = \cap\{P : P \in \tau_{\alpha_{\gamma_0}}$ *and* $E \subseteq P\}$.
{vii} Each α_{γ_o}*- os P,* $P = \cup\{E : X_T - E \in \tau_{\alpha_{\gamma_0}}$ *and* $E \subseteq P\}$
{viii} Let E be an α_{γ_o}*-cs and* $c \notin E$ *implies that* $\tau_{\alpha_{\gamma_0}}$*-cl*$(\{c\}) \cap E = \phi$.

Proof $\{i\} \Rightarrow \{ii\}$ By Definition 4.1, $\tau_{\alpha_{\gamma_0}}$-ker$(\{c\}) = \cap\{P : P \in \tau_{\alpha_{\gamma_0}} and \{x\} \subseteq U\}$.Then by (i), $\tau_{\alpha_{\gamma_0}}$-cl$(\{c\}) \subseteq P$.Hence $\tau_{\alpha_{\gamma_0}}$-cl$(\{c\}) \subseteq \tau_{\alpha_{\gamma_0}}$-ker$(\{c\})$.

$\{ii\} \Rightarrow \{iii\}$ If $d \in \tau_{\alpha_{\gamma_0}}$-ker$(\{c\})$, then by Theorem 4.2, $c \in \tau_{\alpha_{\gamma_0}}$-cl$(\{d\})$.By (ii) $c \in \tau_{\alpha_{\gamma_0}}$-ker$(\{d\})$. Similarly, if $c \in \tau_{\alpha_{\gamma_0}}$-ker$(\{d\})$,then $d \in \tau_{\alpha_{\gamma_0}}$-ker$(\{c\})$.

$\{iii\} \Rightarrow \{iv\}$ If $d \in \tau_{\alpha_{\gamma_0}}$-cl$(\{c\})$, then by Theorem 4.2, $c \in \tau_{\alpha_{\gamma_0}}$-cl$(\{d\})$.By (iii) $d \in \tau_{\alpha_{\gamma_0}}$-ker$(\{c\})$.Again by Theorem 4.2, $d \in \tau_{\alpha_{\gamma_0}}$-cl$(\{d\})$. Similarly, if $c \in \tau_{\alpha_{\gamma_0}}$-cl$(\{d\})$, then $d \in \tau_{\alpha_{\gamma_0}}$-cl$(\{c\})$.

$\{iv\} \Rightarrow \{v\}$ Let E be an α_{γ_o}-cs and a point $c \notin E$. Then for any $d \in E$, $\tau_{\alpha_{\gamma_0}}$-cl$(\{d\}) \subseteq E$ and so $c \notin \tau_{\alpha_{\gamma_0}}$-cl$(\{d\})$. By (iv), $c \notin \tau_{\alpha_{\gamma_0}}$-cl$(\{d\})$ implies $d \notin \tau_{\alpha_{\gamma_0}}$-cl$(\{c\})$, that is there exists a $P_d \in \tau_{\alpha_{\gamma_0}}$ such that $d \in P_d$ and $c \notin P_d$. Let $P = \cup_{d\in E}\{P_d : P_d \in \tau_{\alpha_{\gamma_0}}, d \in P_d \, and \, c \notin P_d\}$.Then by Theorem 2.1 [10], P is an α_{γ_o}-os such that $c \notin P$ and $E \subseteq P$.

$\{v\} \Rightarrow \{vi\}$ Let E be a α_{γ_o}-cs and $H = \cap\{P : P \in \tau_{\alpha_{\gamma_0}}$ *and* $E \subseteq P\}$. $E \subseteq H$ and it remains to show that $H \subseteq E$. Let $c \notin E$. Then by (v), there exists an α_{γ_o}-os P such that $c \notin P$ and $E \subseteq P$ and hence $c \notin H$. Therefore, each α_{γ_o}-cs E can be expressed as $E = \cap\{P : P \in \tau_{\alpha_{\gamma_0}}$ *and* $E \subseteq P\}$.

$\{vi\} \Rightarrow \{vii\}$ Follows from the Definitions 3.16 and 3.18 [10].

$\{vii\} \Rightarrow \{viii\}$ Let E be an α_{γ_o}-cs and $c \notin E$. Then $X_T - E = P$ is an α_{γ_o}-os containing c. Then by $\{vii\}$, P can be written as the union of α_{γ_o}-css and so there is an α_{γ_o}-cs H such that $c \in E \subseteq P$ and $\tau_{\alpha_{\gamma_o}}$-cl($\{c\}$) $\subseteq P$. Thus $\tau_{\alpha_{\gamma_o}}$-cl($\{c\}$) $\cap E = \phi$.

$\{viii\} \Rightarrow \{i\}$ Let P be an α_{γ_o}-os and $c \in P$. Take $E = X_T - P$. Then by (viii), there exists an α_{γ_o}-cs E such that $c \in E \subseteq P$ and $\tau_{\alpha_{\gamma_o}}$-cl($\{c\}$) $\cap E \neq \phi$.Therefore $\tau_{\alpha_{\gamma_o}}$-cl($\{c\}$) $\subseteq E$ and hence $\tau_{\alpha_{\gamma_o}}$-cl($\{c\}$) $\subseteq P$. Thus X_T is an $\alpha_{\gamma_o} R_0$ space.

Theorem 4.6 *For a TS the given statements are equivalent:*

(i) *(X_T, τ) is an $\alpha_{\gamma_o} R_0$ space*
(ii) *For any two distinct points c, d of X_T, $\tau_{\alpha_{\gamma_o}}$-cl($\{c\}$) $=$ $\tau_{\alpha_{\gamma_o}}$-cl($\{d\}$) or $\tau_{\alpha_{\gamma_o}}$-cl($\{c\}$) $\cap$ $\tau_{\alpha_{\gamma_o}}$-cl($\{d\}$) $= \phi$.*

Proof (i) $\Rightarrow$ (ii) Let X_T be an $\alpha_{\gamma_o} R_0$ space, $\tau_{\alpha_{\gamma_o}}$-cl($\{c\}$) $\neq$ $\tau_{\alpha_{\gamma_o}}$-cl($\{d\}$) and $\tau_{\alpha_{\gamma_o}}$-cl($\{c\}$) $\cap$ $\tau_{\alpha_{\gamma_o}}$-cl($\{d\}$) $\neq \phi$. Let $s \in \tau_{\alpha_{\gamma_o}}$-cl($\{c\}$) $\cap$ $\tau_{\alpha_{\gamma_o}}$-cl($\{d\}$) and c $\notin \tau_{\alpha_{\gamma_o}}$-cl($\{d\}$). Then c $\in X_T - \tau_{\alpha_{\gamma_o}}$-cl($\{d\}$) $\in \tau_{\alpha_{\gamma_o}}$. But $\tau_{\alpha_{\gamma_o}}$-cl($\{c\}$) $\notin X_T - \tau_{\alpha_{\gamma_o}}$-cl($\{d\}$), since $s \in \tau_{\alpha_{\gamma_o}}$-cl($\{c\}$) $\cap$ $\tau_{\alpha_{\gamma_o}}$-cl($\{d\}$), which is a contradiction to the hypothesis that X_T is an $\alpha_{\gamma_o} R_0$ space. Hence either $\tau_{\alpha_{\gamma_o}}$-cl($\{c\}$) $=$ $\tau_{\alpha_{\gamma_o}}$-cl($\{d\}$) or $\tau_{\alpha_{\gamma_o}}$-cl($\{c\}$) $\cap$ $\tau_{\alpha_{\gamma_o}}$-cl($\{d\}$) $= \phi$.

(ii) $\Rightarrow$ (i) Let $W \in \tau_{\alpha_{\gamma_o}}$ with $c \in W$. To show that $\tau_{\alpha_{\gamma_o}}$-cl($\{c\}$) $\subseteq W$. Let $d \notin W$. Then $d \neq c$ and $c \notin \tau_{\alpha_{\gamma_o}}$-cl($\{d\}$). Thus $\tau_{\alpha_{\gamma_o}}$-cl($\{c\}$) $\neq$ $\tau_{\alpha_{\gamma_o}}$-cl($\{d\}$). Hence by (ii), $\tau_{\alpha_{\gamma_o}}$-cl($\{c\}$) $\cap$ $\tau_{\alpha_{\gamma_o}}$-cl($\{d\}$) $= \phi$ for each $d \notin W$ and hence $\tau_{\alpha_{\gamma_o}}$-cl($\{c\}$) $\cap [\bigcup_{d \in X_T - W} \tau_{\alpha_{\gamma_o}}$-cl($\{d\}$)$] = \phi$. Again, since $W \in \tau_{\alpha_{\gamma_o}}$ and $d \in X_T - W$, $d \in \tau_{\alpha_{\gamma_o}}$-cl($\{d\}$) $\subseteq X_T - W$. Thus $X_T - W = \bigcup_{d \in X_T - W} \tau_{\alpha_{\gamma_o}}$-cl($\{d\}$). Therefore $(X_T - W) \bigcap \tau_{\alpha_{\gamma_o}}$-cl($\{c\}$) $= \phi$ and hence $\tau_{\alpha_{\gamma_o}}$-cl($\{c\}$) $\subseteq W$.

Corollary 4.1 *A TS (X_T, τ) is an $\alpha_{\gamma_o} R_0$ space if and only if for any c,d $\in X_T$, $\tau_{\alpha_{\gamma_o}}$-cl($\{c\}$) $\neq$ $\tau_{\alpha_{\gamma_o}}$-cl($\{d\}$) implies that $\tau_{\alpha_{\gamma_o}}$-cl($\{c\}$) $\cap$ $\tau_{\alpha_{\gamma_o}}$-cl($\{d\}$) $= \phi$.*

Proof Follows from the Theorem 4.6.

Theorem 4.7 *Let c, d be any two distinct points in a TS X_T. Then these statements are equivalent:*

{i} *$\tau_{\alpha_{\gamma_o}}$-ker($\{c\}$) $\neq$ $\tau_{\alpha_{\gamma_o}}$-ker($\{d\}$);*
{ii} *$\tau_{\alpha_{\gamma_o}}$-cl($\{c\}$) $\neq$ $\tau_{\alpha_{\gamma_o}}$-cl($\{d\}$).*

Proof $\{i\} \Rightarrow \{ii\}$ Let $\tau_{\alpha_{\gamma_o}}$-ker($\{c\}$) $\neq$ $\tau_{\alpha-\gamma_o}$-ker($\{d\}$). Then there prevails $z \notin \tau_{\alpha_{\gamma_o}}$-ker($\{d\}$) (say). By Theorem 4.2, $c \in \tau_{\alpha_{\gamma_o}}$-cl($\{z\}$) and $d \notin \tau_{\alpha_{\gamma_o}}$-cl($\{z\}$). As $\tau_{\alpha_{\gamma_o}}$-cl($\{c\}$) $\subseteq$ $\tau_{\alpha_{\gamma_o}}$-cl($\{z\}$), we have d $\notin \tau_{\alpha_{\gamma_o}}$-cl($\{c\}$). Hence $\tau_{\alpha_{\gamma_o}}$-cl($\{c\}$) $\neq$ $\tau_{\alpha_{\gamma_o}}$-cl($\{d\}$).

$\{ii\} \Rightarrow \{i\}$ Let $\tau_{\alpha_{\gamma_o}}$-cl($\{c\}$) $\neq$ $\tau_{\alpha_{\gamma_o}}$-cl($\{d\}$). Then there exists $z \in X_T$, and $z \notin \tau_{\alpha_{\gamma_o}}$-cl($\{d\}$)(say), which signifies that there occurs a $M \in \tau_{\alpha_{\gamma_o}}$ such that $z \in M$, $d \notin M$ and $c \in M$ implies that $d \notin \tau_{\alpha_{\gamma_o}}$-ker($\{c\}$). Hence $\tau_{\alpha_{\gamma_o}}$-cl($\{c\}$) $\neq$ $\tau_{\alpha_{\gamma_o}}$-cl($\{d\}$).

Theorem 4.8 *A TS (X_T, τ) is an $\alpha_{\gamma_o} R_0$ space if and only if for each p,q $\in X_T$, $\tau_{\alpha_{\gamma_o}}$-ker($\{p\}$) $\neq$ $\tau_{\alpha_{\gamma_o}}$-ker($\{q\}$) implies that $\tau_{\alpha_{\gamma_o}}$-ker ($\{p\}$) $\cap$ $\tau_{\alpha_{\gamma_o}}$-ker ($\{q\}$) $= \phi$.*

Proof Let X_T be an $\alpha_{\gamma_o}R_0$ space and $\tau_{\alpha_{\gamma_o}}$-ker $(\{p\}) \neq \tau_{\alpha_{\gamma_o}}$-ker $(\{q\})$ where $p,q \in X_T$. Suppose that $\tau_{\alpha_{\gamma_o}}$-ker $(\{p\}) \cap \tau_{\alpha_{\gamma_o}}$-ker $(\{d\}) \neq \phi$. Let $s \in \tau_{\alpha_{\gamma_o}}$-ker $(\{p\}) \cap \tau_{\alpha_{\gamma_o}}$-ker $(\{q\})$.Then $s \in \tau_{\alpha_{\gamma_o}}$-ker $(\{p\})$ and $s \in \tau_{\alpha_{\gamma_o}}$-ker $(\{q\})$. By Theorem 4.5 (iii), $p \in \tau_{\alpha_{\gamma_o}}$-ker $(\{s\})$ and $q \in \tau_{\alpha_{\gamma_o}}$-ker $(\{s\})$. Hence $\tau_{\alpha_{\gamma_o}}$-ker $(\{p\}) \subseteq \tau_{\alpha_{\gamma_o}}$-ker $(\{s\}) \subseteq \tau_{\alpha_{\gamma_o}}$-ker $(\{q\})$ and $\tau_{\alpha_{\gamma_o}}$-ker $(\{q\}) \subseteq \tau_{\alpha_{\gamma_o}}$-ker $(\{s\}) \subseteq \tau_{\alpha_{\gamma_o}}$-ker $(\{p\})$ implies that $\tau_{\alpha_{\gamma_o}}$-ker $(\{p\}) = \tau_{\alpha_{\gamma_o}}$-ker $(\{q\})$, a contradiction. Hence $\tau_{\alpha_{\gamma_o}}$-ker $(\{p\}) \cap \tau_{\alpha_{\gamma_o}}$-ker $(\{q\}) = \phi$.

Conversely, suppose that for each $p,q \in X$, $\tau_{\alpha_{\gamma_o}}$-ker $(\{p\}) \neq \tau_{\alpha_{\gamma_o}}$-ker $(\{q\})$ implies that $\tau_{\alpha_{\gamma_o}}$-ker $(\{p\}) \cap \tau_{\alpha_{\gamma_o}}$-ker $(\{q\}) = \phi$. If $\tau_{\alpha_{\gamma_o}}$-cl$(\{p\}) \neq \tau_{\alpha_{\gamma_o}}$-cl$(\{q\})$ then by Theorem 4.7, $\tau_{\alpha_{\gamma_o}}$-ker $(\{p\}) \neq \tau_{\alpha_{\gamma_o}}$-ker $(\{q\})$. Then by the given condition $\tau_{\alpha_{\gamma_o}}$-ker $(\{p\}) \cap \tau_{\alpha_{\gamma_o}}$-ker $(\{q\}) = \phi$ which implies $\tau_{\alpha_{\gamma_o}}$-cl$(\{p\}) \neq \tau_{\alpha_{\gamma_o}}$-cl$(\{q\}) = \phi$. Because if $z \in \tau_{\alpha_{\gamma_o}}$-cl$(\{p\}) \cap \tau_{\alpha_{\gamma_o}}$-cl$(\{q\})$ by Theorem 4.2, $p \in \tau_{\alpha_{\gamma_o}}$-ker $(\{z\})$,implies $\tau_{\alpha_{\gamma_o}}$-ker $(\{p\}) \cap \tau_{\alpha_{\gamma_o}}$-ker $(\{z\}) \neq \phi$ and similarly $\tau_{\alpha_{\gamma_o}}$-ker $(\{p\}) \cap \tau_{\alpha_{\gamma_o}}$-ker $(\{z\}) \neq \phi$.So by hypothesis $\tau_{\alpha_{\gamma_o}}$-ker $(\{p\}) = \tau_{\alpha_{\gamma_o}}$-ker $(\{z\}) = \tau_{\alpha_{\gamma_o}}$-ker $(\{q\})$, which is a contradiction. Thus, $\tau_{\alpha_{\gamma_o}}$-cl$(\{p\}) \cap \tau_{\alpha_{\gamma_o}}$-cl$(\{q\}) = \phi$ and by Corollary 4.1, the TS X_T is an $\alpha_{\gamma_o}R_0$ space.

Corollary 4.2 *Let c, d be any two distinct points in a TS (X_T, τ). Then statements mentioned below are equivalent:*

(i) *(X_T, τ) is an $\alpha_{\gamma_o}R_0$ space*
(ii) *If R is an α_{γ_o}- cs in X_T, then $R = \tau_{\alpha_{\gamma_o}}$-ker$(\{F\})$*
(iii) *If R is an α_{γ_o}- cs in X_T and $c \in R$, then $\tau_{\alpha_{\gamma_o}}$-ker$(\{c\}) \subseteq R$*
(iv) *If $c \in X_T$, then $\tau_{\alpha_{\gamma_o}}$-ker$(\{c\}) \subseteq \tau_{\alpha_{\gamma_o}}$-cl$(\{c\})$*

Proof $(i) \Rightarrow (ii)$ Let R be an α_{γ_o}- cs and $c \notin R$. Then $X_T - R$ is a α_{γ_o}-os containing c. Since X_T is $\alpha_{\gamma_o}R_0$, $\tau_{\alpha_{\gamma_o}}$-cl$(\{c\}) \subseteq X_T - R$. Therefore $\tau_{\alpha_{\gamma_o}}$-cl$(\{c\}) \cap R = \phi$ and by Theorem 4.3, $c \notin \tau_{\alpha_{\gamma_o}}$-ker$(R)$. Hence $\tau_{\alpha_{\gamma_o}}$-ker(R) $\subseteq R$. By Definition 3.1, $R \subseteq \tau_{\alpha_{\gamma_o}}$-ker$(R)$. Thus $R = \tau_{\alpha_{\gamma_o}}$-ker$(R)$.

$(ii) \Rightarrow (iii)$ Let R be an α_{γ_o}-cs containing c. Then $\{c\} \subseteq R$ and $\tau_{\alpha_{\gamma_o}}$-ker$(\{c\}) \subseteq \tau_{\alpha_{\gamma_o}}$-ker$(R)$. By (ii), $\tau_{\alpha_{\gamma_o}}$-ker$(\{c\}) \subseteq R$.

$(iii) \Rightarrow (iv)$ Since $c \in \tau_{\alpha_{\gamma_o}}$-cl$(\{c\})$ and $\tau_{\alpha_{\gamma_o}}$-cl$(\{c\})$ is an α_{γ_o}-cs in X_T. Then by (iii), $\tau_{\alpha_{\gamma_o}}$-ker$(\{c\}) \subseteq \tau_{\alpha_{\gamma_o}}$-cl$(\{c\})$.

$(iv) \Rightarrow (i)$ Let $c \in \tau_{\alpha_{\gamma_o}}$-cl$(\{d\})$. Then by Theorem 4.6 (iv), we have X_T is $\alpha_{\gamma_o}R_0$.

Theorem 4.9 *In a TS (X_T, τ), the statements given below are equivalent:*

(i) *(X_T, τ) is an $\alpha_{\gamma_o}T_1$ space;*
(ii) *Let c be any point in X_T. Then $\tau_{\alpha_{\gamma_o}}$-cl$(\{c\}) = \{c\}$;*
(iii) *X_T is an $\alpha_{\gamma_o}R_0$ and$\alpha_{\gamma_o}T_0$ space.*

Proof $(i) \Rightarrow (ii)$ If $d \notin \{c\}$, then there exists $\in \tau_{\alpha_{\gamma_o}}$ such that $d \in P$, $c \notin P$. Therefore $P \cap \{c\} = \phi$ and hence $d \notin \tau_{\alpha_{\gamma_o}}$-cl$(\{c\})$.

$(ii) \Rightarrow (iii)$ Let $c,d \in X_T$ with $c \neq d$. Then $\{c\}$ and $\{d\}$ are α_{γ_o}-cs and hence $X_T - \{c\}$ is an α_{γ_o}-os containing d but not c showing X_T to be $\alpha_{\gamma_o}T_0$. Suppose that

P is an α_{γ_o}-os and $c \in P$. Then by (ii), $\tau_{\alpha_{\gamma_o}}$-cl$(\{c\}) = \{c\} \subseteq P$. Hence (X_T, τ) is an $\alpha_{\gamma_o} R_0$ space.

$(iii) \Rightarrow (i)$ Let $c,d \in X_T$ with $c \neq d$. Then there exists $P \in \tau_{\alpha_{\gamma_o}}$ such that $c \in P$ and $d \notin P$ (say),which implies that $\tau_{\alpha_{\gamma_o}}$-cl$(\{c\}) \subseteq P$ and so $d \notin \tau_{\alpha_{\gamma_o}}$-cl$(\{c\})$.Hence $c \in P \in \tau_{\alpha_{\gamma_o}}$, $d \notin P$ and $d \in X_T - \tau_{\alpha_{\gamma_o}}$-cl$(\{c\}) \notin \tau_{\alpha_{\gamma_o}}$, $c \notin X_T - \tau_{\alpha_{\gamma_o}}$-cl$(\{d\})$. So (X_T, τ) is an $\alpha_{\gamma_o} T_1$ space.

Definition 4.3 Let (X_T, τ) be a TS and $c \in X_T$. A net $\{c_\alpha : \alpha \in D\}$ in (X_T, τ) is said to be α_{γ_o}-convergent to c in X_T if for every $\tau_{\alpha_{\gamma_o}}$-os P containing c there exists $\alpha_0 \in D$ such that $c_\alpha \in P$, for each $\alpha \geq \alpha_0$.

Lemma 4.1 *Let p and q be any two points of a TS (X_T, τ) such that every point in (X_T, τ) converging to q also α_{γ_o}-converging to p. Then $p \in \tau_{\alpha_{\gamma_o}}$-cl$(\{q\})$.*

Proof Suppose that $p_n = q$ for each $n \in N$ (N is the set of natural numbers). Then $\{p_n\}$ is a net in $\tau_{\alpha_{\gamma_o}}$-cl$(\{y\})$. By the fact that $\{x_n\}$, α_{γ_o}-converges to y, $\{x_n\}$, α_{γ_o} converging to p. Thus $p \in \tau_{\alpha_{\gamma_o}}$-cl$(\{q\})$.

Theorem 4.10 *For a TS the statements stated below are equivalent:*

(i) *(X_T, τ) is an $\alpha_{\gamma_o} R_0$ space*
(ii) *If c,d $\in (X_T, \tau)$, then $d \in \tau_{\alpha_{\gamma_o}}$-cl$(\{c\})$ if and only if every net in (X_T, τ) α_{γ_o}-converging to d also α_{γ_o}-converging to c.*

Proof

(i) $\Rightarrow$ (ii) Let c,d $\in X_T$ such that d $\in \tau_{\alpha_{\gamma_o}}$-cl$(\{c\})$. Let $\{c_\alpha\}_{\alpha \in D}$ be a net in (X_T, τ) such that $\{c_\alpha\}_{\alpha \in D} \alpha_{\gamma_o}$-converges to d. Since d $\in \tau_{\alpha_{\gamma_o}}$-cl$(\{c\})$ by Theorem 4.6., $\tau_{\alpha_{\gamma_o}}$-cl$(\{c\}) = \tau_{\alpha_{\gamma_o}}$-cl$(\{d\})$. Therefore $c \in \tau_{\alpha_{\gamma_o}}$-cl$(\{d\})$. Thus $\{c_\alpha\}_{\alpha \in D} \alpha_{\gamma_o}$-converging to c. Conversely, let c, $d \in (X_T, \tau)$ such that every net in (X_T, τ) α_{γ_o}-converging to d also α_{γ_o}-converging to c.Then by Lemma 4.1, $c \in \tau_{\alpha_{\gamma_o}}$-cl$(\{d\})$. Thus by Theorem 4.6, $\tau_{\alpha_{\gamma_o}}$-cl$(\{c\}) = \tau_{\alpha_{\gamma_o}}$-cl$(\{d\})$.Thus $d \in \tau_{\alpha_{\gamma_o}}$-cl$(\{c\})$.
(ii) $\Rightarrow$ (i) Let c and d be any two points of (X_T, τ) such that $\tau_{\alpha_{\gamma_o}}$-cl$(\{c\}) \cap \tau_{\alpha_{\gamma_o}}$-cl$(\{d\}) \neq \phi$. Given Theorem 4.6, it is sufficient to show that $\tau_{\alpha_{\gamma_o}}$-cl$(\{c\}) = \tau_{\alpha_{\gamma_o}}$-cl$(\{d\})$. Let $z \in \tau_{\alpha_{\gamma_o}}$-cl$(\{c\}) \cap \tau_{\alpha_{\gamma_o}}$-cl$(\{d\})$. So there exists a net $\{c_\alpha\}_{\alpha \in D}$ in $\tau_{\alpha_{\gamma_o}}$-cl$(\{c\})$ such that $\{c_\alpha\}_{\alpha \in D}$ converges to z. Since $z \in \tau_{\alpha_{\gamma_o}}$-cl$(\{d\})$, $\{c_\alpha\}_{\alpha \in D}$ converges to d. Hence by (ii) $z \in \tau_{\alpha_{\gamma_o}}$-cl$(\{d\}) \Rightarrow \tau_{\alpha_{\gamma_o}}$-cl$(\{z\}) \subseteq \tau_{\alpha_{\gamma_o}}$-cl$(\{d\})$ (1). By Lemma 4.1, $d \in \tau_{\alpha_{\gamma_o}}$-cl$(\{z\})$ implies $\tau_{\alpha_{\gamma_o}}$-cl$(\{d\}) \subseteq \tau_{\alpha_{\gamma_o}}$-cl$(\{z\})$ (2). Hence by (1) and (2), $\tau_{\alpha_{\gamma_o}}$-cl$(\{d\}) = \tau_{\alpha_{\gamma_o}}$-cl$(\{z\})$. Similarly, it can be shown that $\tau_{\alpha_{\gamma_o}}$-cl$(\{c\}) = \tau_{\alpha_{\gamma_o}}$-cl$(\{z\})$ by taking the net in $\tau_{\alpha_{\gamma_o}}$-cl$(\{d\})$. So $\tau_{\alpha_{\gamma_o}}$-cl$(\{c\}) = \tau_{\alpha_{\gamma_o}}$-cl$(\{d\})$.

Definition 4.4 A TS (X_T, τ) is said to be an $\alpha_{\gamma_o} R_1$ space if for$c,d \in X_T$,$\tau_{\alpha_{\gamma_o}}$-cl$(\{c\}) \neq \tau_{\alpha_{\gamma_o}}$-cl$(\{d\})$, there exists P, $Q \in \tau_{\alpha_{\gamma_o}}$ such that $\tau_{\alpha_{\gamma_o}}$-cl$(\{c\}) \subseteq P$ and $\tau_{\alpha_{\gamma_o}}$-cl$(\{d\}) \subseteq Q$ and $P \cap Q = \phi$.

Example 4.2 Let $X_T = \{f, g, h\}$, $\tau = \{\phi, X_T, \{f\}, \{g\}, \{f, g\}\}$, define γ_o on τ such that γ_o $(D) =$ cl(D) for every $D \in \tau$. Then (X_T, τ) is an $\alpha_{\gamma_o} R_1$ space.

Theorem 4.11 *If (X_T, τ) is an $\alpha_{\gamma_o} R_1$ space, then it is an $\alpha_{\gamma_o} R_0$ space.*

Proof Let f,g be distinct points in X_T. Then there exists an α_{γ_o}-os M such that $f \in M$ and $g \notin M$. This hints that $\tau_{\alpha_{\gamma_o}}$-cl$(\{f\}) \neq \tau_{\alpha_{\gamma_o}}$-cl$(\{g\})$. Since X_T is an $\alpha_{\gamma_o} R_1$ space signifies that $\tau_{\alpha_{\gamma_o}}$-cl$(\{f\}) \cap \tau_{\alpha_{\gamma_o}}$-cl$(\{g\})$ = void set. Thence by 4.6 doctrines, (X_T, τ) is an $\alpha_{\gamma_o} R_0$ space.

Remark 4.1 The overturn of the above doctrine 4.10 need not be accurate generally.

Theorem 4.12 *In a TS (X_T, τ), these statements are equivalent:*

(i) *(X_T, τ) is an $\alpha_{\gamma_o} T_2$ space*
(ii) *(X_T, τ) is an $\alpha_{\gamma_o} R_1$ and $\alpha_{\gamma_o} T_1$ space*
(iii) *(X_T, τ) is an $\alpha_{\gamma_o} R_1$ and $\alpha_{\gamma_o} T_0$ space*

Proof $(i) \Rightarrow (ii)$ Let (X_T, τ) be an $\alpha_{\gamma_o} T_2$ space. Then (X_T, τ) is an $\alpha_{\gamma_o} T_1$ space. If $c,\ d \in (X_T, \tau)$ with $\tau_{\alpha_{\gamma_o}}$-cl$(\{c\}) \neq \tau_{\alpha_{\gamma_o}}$-cl$(\{d\})$, then $c \neq d$ and so there exists p, $Q \in \tau_{\alpha_{\gamma_o}}$ such that $c \in P$, $d \in Q$ and $P \cap Q = \phi$. Hence by Theorem 4.8, $\tau_{\alpha_{\gamma_o}}$-cl$(\{x\}) = \{x\} \subseteq U$ and $\tau_{\alpha_{\gamma_o}}$-cl$(\{d\}) = \{d\} \subseteq Q$ and $P \cap Q = \phi$. Then (X_T, τ) is an $\alpha_{\gamma_o} R_1$ space.

$(ii) \Rightarrow (iii)$ The proof is immediate from the Definitions 3.1 and 3.2.

$(iii) \Rightarrow (i)$ Let (X_T, τ) be an $\alpha_{\gamma_o} R_1$ and $\alpha_{\gamma_o} T_0$ sace. Then by Theorem 4.9, (X_T, τ) is an $\alpha_{\gamma_o} R_0$ and $\alpha_{\gamma_o} T_0$ space. By Theorem 4.9, (X_T, τ) is an $\alpha_{\gamma_o} T_1$ space. Let $c,d \in X_T$ with $c \neq d$. Then $\tau_{\alpha_{\gamma_o}}$-cl$(\{c\}) = \{c\} \neq \{d\} = \tau_{\alpha_{\gamma_o}}$-cl$(\{d\})$. As X_T is an $\alpha_{\gamma_o} R_1$ space, there exists $P, Q \in \tau_{\alpha_{\gamma_o}}$ such that $\tau_{\alpha_{\gamma_o}}$-cl$(\{c\}) = \{c\} \subseteq P$, $\tau_{\alpha_{\gamma_o}}$-cl$(\{d\}) = \{d\} \subseteq Q$ and $P \cap Q = \phi$. Thus X_T is an $\alpha_{\gamma_o} T_2$ space.

Theorem 4.13 *In a TS (X_T, τ), the statements stated below are equivalent:*

(i) *$(X_T, \tau))$ is an $\alpha_{\gamma_o} R_1$ space;*
(ii) *If c,d are any two distinct points $\in (X_T, \tau)$, then, one of the following holds;*

(a) For $P \in \tau_{\alpha_{\gamma_o}}$, *$c \in P$ if and only if $d \in P$;*
(b) *There exists P,Q $\in \tau_{\alpha_{\gamma_o}}$ such that $c \in P$ and $d \in Q$ and $P \cap Q = \phi$;*

(iii) *If c,d $\in$ such that $\tau_{\alpha_{\gamma_o}}$-cl$(\{c\}) \neq \tau_{\alpha_{\gamma_o}}$-cl$(\{d\})$, then there exists α_{γ_o}-css E_1 and E_2 such that $c \in E_1$, $d \notin E_1$, $d \in E_2$, $c \notin E_2$ and $X_T = E_1 \cup E_2$.*

Proof $(i) \Rightarrow (ii)$ Let $c,d \in X_T$. Then $\tau_{\alpha_{\gamma_o}}$-cl$(\{c\}) = \tau_{\alpha_{\gamma_o}}$-cl$(\{d\})$ or $\tau_{\alpha_{\gamma_o}}$-cl$(\{c\}) \neq \tau_{\alpha_{\gamma_o}}$-cl$(\{d\})$. Suppose $\tau_{\alpha_{\gamma_o}}$-cl$(\{c\}) = \tau_{\alpha_{\gamma_o}}$-cl$(\{d\})$ and $c \in P \in \tau_{\alpha_{\gamma_o}}$. This implies that $d \in \tau_{\alpha_{\gamma_o}}$-cl$(\{d\}) = \tau_{\alpha_{\gamma_o}}$-cl$(\{c\}) \subseteq P$. Hence $d \in P$. Similarly, it can be proved that if $d \in P$, then $C \in P$. Suppose $\tau_{\alpha_{\gamma_o}}$-cl$(\{c\}) \neq \tau_{\alpha_{\gamma_o}}$-cl$(\{d\})$. Then there exists $P, Q \in \tau_{\alpha_{\gamma_o}}$ such that $c \in \tau_{\alpha_{\gamma_o}}$-cl$(\{c\}) \subseteq P$ and $d \in \tau_{\alpha_{\gamma_o}}$-cl$(\{d\}) \subseteq Q$ and $P \cap Q = \phi$.

$(ii) \Rightarrow (iii)$ Let $c,d \in X_T$ such that $\tau_{\alpha_{\gamma_o}}$-cl$(\{c\}) \neq \tau_{\alpha_{\gamma_o}}$-cl$(\{d\})$. Then $c \notin \tau_{\alpha_{\gamma_o}}$-cl$(\{d\})$, so that there exists $M \in \tau_{\alpha_{\gamma_o}}$ such that $c \in M$ and $d \notin M$. Thus by (ii) there exists $P, Q \in \tau_{\alpha_{\gamma_o}}$ such that $c \in P$ and $d \in Q$ and $P \cap Q = \phi$. Let $E_1 = X_T - Q$ and

$E_2 = X_T - P$. Then E_1 and E_2 are α_{γ_o}-css and $c \in E_1, d \notin E_1, d \in E_2, c \notin E_2$ and $X_T = E_1 \cup E_2$.

$(iii) \Rightarrow (i)$ Let $P \in \tau_{\alpha_{\gamma_o}}$ and $c \in P$. Then $\tau_{\alpha_{\gamma_o}}$-cl($\{c\}$) $\subseteq P$. Otherwise, there exists $d \in \tau_{\alpha_{\gamma_o}}$-cl($\{c\}$)$\cap(X_T - P)$. Then $\tau_{\alpha_{\gamma_o}}$-cl($\{c\}$) $\neq \tau_{\alpha_{\gamma_o}}$-cl($\{d\}$) and by (iii), there exists α_{γ_o}-css E_1 and E_2 such that $c \in E_1, d \notin E_1, d \in E_2, c \notin E_2$ and $X_T = E_1 \cup E_2$. Then $d \in E_2 - E_1 = X_T - E_1$ and $c \notin X_T - E_1$, where $X_T - E_1 \in \tau_{\alpha_{\gamma_o}}$,which is a contradiction to the fact that $d \in \tau_{\alpha_{\gamma_o}}$-cl($\{c\}$). Hence $\tau_{\alpha_{\gamma_o}}$-cl($\{c\}$) $\subseteq P$. Thus X_T is $\alpha_{\gamma_o} R_0$. To show X_T to be an $\alpha_{\gamma_o} R_1$ space assume that $a,b \in X_T$ with $\tau_{\alpha_{\gamma_o}}$-cl($\{a\}$) $\neq \tau_{\alpha_{\gamma_o}}$-cl($\{b\}$). Then, there exists α_{γ_o}-css A_1 and A_2 such that $c \in A_1, d \notin A_1$, $d \in A_2, c \notin A_2$ and $X_T = A_1 \cup A_2$. Thus $a \in A_1 - A_2 \in \tau_{\alpha_{\gamma_o}}, b \in A_2 - A_1 \in \tau_{\alpha_{\gamma_o}}$. So $\tau_{\alpha_{\gamma_o}}$-cl($\{a\}$) $\subseteq A_1 - A_2$, $\tau_{\alpha_{\gamma_o}}$-cl($\{b\}$) $\subseteq A_2 - A_1$. Thus $(X_T, \tau)$is an $\alpha_{\gamma_o} R_1$ space.

Theorem 4.14

(i) *space (X_T, τ) is an $\alpha_{\gamma_o} T_2$ space if and only if for c,d $\in X_T$ with $c \neq d$, there exists α_{γ_o}-css E_1 and E_2 such that $c \in E_1$, $d \notin E_1$, $d \in E_2$, $c \notin E_2$ and X_T $=E_1 \cup E_2$.*

(ii) Space X_T *is an $\alpha_{\gamma_o} R_1$ space if and only if c,d $\in X_T$ with $\tau_{\alpha_{\gamma_o}}$-ker($\{c\}$) $\neq \tau_{\alpha_{\gamma_o}}$-ker($\{d\}$), there exists P, Q $\in \tau_{\alpha_{\gamma_o}}$ such that $\tau_{\alpha_{\gamma_o}}$-cl($\{c\}$) $\subseteq P$ and $\tau_{\alpha_{\gamma_o}}$-cl($\{d\}$) $\subseteq$ Q and $P \cap Q = \phi$.*

Proof

(i) The proof is evident from the doctrine 4.12 and 4.13.
(ii) The proof is prompt from doctrine 4.7 and Definition 4.4.

5 Conclusion

In this document, the concept of $\alpha_\gamma T_i$ spaces that are generated through α_{γ_o}, γ_o and α-open sets is reviewed, and some of their essential properties are explored. Moreover, the authors proposed the perception and documentation of $\alpha_{\gamma_o} R_i$ spaces. The α_{γ_o} generalized-closed set(α_{γ_o}-gcs) has been defined, and dictums associated are established. The philosophy of α_{γ_o} the convergence of net has been detailed. The notions of $\alpha_{\gamma_o} R_0$, $\alpha_{\gamma_o} R_1$ are elaborated using α_{γ_o}-closed sets, α_{γ_o}-open sets and their corresponding $\tau_{\alpha_{\gamma_o}}$-closure operator. Besides they build the relationship among them and learned their nature.

References

1. Njåstad, O. (1965). On some classes of nearly open sets. *Pacific Journal of Mathematics, 15*(3), 961–970.
2. Levine, N. (1963). Semi-open sets and semi-continuity in topological spaces. *The American Mathematical Monthly, 70*(1), 36–41.

3. Corson, H. H., & Michael, E. (1964). Metrizability of certain countable unions. *Illinois Journal of Mathematics, 8*(2), 351–360.
4. Andrijević, D. (1987). On the topology generated by preopen sets. *MatematičkiVesnik, 39*(100), 367–376.
5. Kasahara, S. (1979). Operation-compact spaces. *Mathematica Japonica, 24,* 97–105.
6. Ogata, H. (1991). Operations on topological spaces and associated topology. *Mathematica Japonica, 36,* 175–184.
7. Krishnan, G. S. S. (2006).On γ-semi-open sets in topological spaces. *Bulletin of the Calcutta Mathematical Society, 98*, 517-530.
8. Krishnan, G., & Balachandran, K. (2006). ON A CLASS OF of γ-PREOPEN SETS IN A TOPOLOGICAL SPACE. *East Asian Mathematical Journal, 22*(2), 131–149.
9. Kalaivani, N., & Krishnan, G. S. S. (2013). Operation approaches on α-γ-open sets in topological spaces. *International Journal of Mathematical Analysis, 7*(10), 491–498.

Time-Sensitive Probabilities of an $M/M/1$ Queueing Model with Semi-Vacation and Threshold Policy

B. Janani and M. Lakshmi Priya

Abstract An $M/M/1$ queueing model with multiple exponential working vacation and threshold policy is considered. Here, the server will go for a working vacation after completing the service for all waiting customers in the system. During the working vacation, the service rate in which customers are served is slow compared to normal service rate. The server will move to a busy state from working vacation state when he finds N or more number of customers in the system. In this paper, we derived explicit analytical expressions of time-dependent probabilities of the model explained above.

Keywords Multiple working vacations · Threshold · Generating function · Continued fraction · Time-dependent probabilities

1 Introduction

Queueing models with vacation have been studied by many researchers due to their applications in power saving mechanism in communication networks, production management system, and computer system. The period during which the server is unavailable to the customers is called vacation of the server. Vacations of the server are of three types namely single vacation, multiple vacations, and semi-vacation. In a single vacation policy, the server will go for a vacation after providing service to all waiting customers and return after some time. If he finds customer, he will provide service otherwise he will wait for the customer. But in multiple vacations, if the server does not find any waiting customer after returning from vacation, then the server will take one more vacation. In semi-vacation policy, the server continues the

B. Janani (✉)
Rajalakshmi Institute of Technology, Chennai, India
e-mail: jananisrini2009@gmail.com

M. Lakshmi Priya
Department of Mathematics, Vel Tech Rangarajan Dr. Sagunthala R&D Institute of Science and Technology, Chennai, India
e-mail: vishnuvcstrust@gmail.com

S.-L. Peng et al. (eds.), *Proceedings of First International Conference on Mathematical Modeling and Computational Science*, Advances in Intelligent Systems and Computing 1292, https://doi.org/10.1007/978-981-33-4389-4_56

service to the customer at a slower rate instead of stopping service during a vacation. Also, this type of policy is seen to prevail in a bank, ticket counter, and so on.

1.1 Related Works

The readers may have a look on the monographs of Takagi [10], Tian and Zhang [12] and also the survey paper of Doshi [4] and Teghem [11] to know about various vacation policies and various vacation models.

Usually, in a classical vacation queue, the server stops working during the vacation period. Lately, the semi-vacation policy was introduced by Servi and Finn [8]. Liu et al. [7] extended the same work and found out the explicit expressions for the stationary queue length and waiting time. Also, they derived stochastic decompositions structures and busy period. $M/G/1$ queue with working vacation was analyzed by Kim [5] and Wu and Takagi [14]. $GI/\text{Geom}/1$ queue with working vacation and service interruption was analyzed by Li and Tian [6]. $GI/M/1$ queue with working vacation was investigated by Baba [1] and Banik et al. [2] extended the same for the finite capacity system. Semi-vacation policy with threshold policy for efficient usage of energy in WSN was analyzed by Boutoumi and Gharbi [3].

There are lots of papers related to queueing models with a vacation in literature. But most of the authors analyzed the system under the steady-state regime. However, the steady-state analysis will not make any sense for the system which never approaches equilibrium. Hence, transient analysis is important to know about the behavior of the system concerning time. Readers may refer papers by Sudhesh [9], Vijayashree and Janani [13] for the better understanding of transient analysis of queueing models with vacation.

1.2 Importance of the Considered Model

To optimize economic cost and energy consumption, it is better to operate the system at a lower speed when the number of customers in the system is relatively few. Also to minimize the switching cost, it is necessary to fix a threshold N. Hence, by keeping these factors in consideration in this paper, an $M/M/1$ queue with multiple working vacation and N-policy is considered and we derived time-dependent probabilities explicitly.

The remaining sections are arranged as follows.

Section 2—Model Description.

Section 3—Time-Sensitive Probabilities.

Section 4—Conclusion and Future Scope.

2 Model Description

A single server queueing model is considered. The pattern in which arrivals are joining the system is distributed as Poisson with parameter λ and the pattern in which service take place is distributed exponentially with parameter μ. Also, the server will go for a working vacation where he serves at a slow rate $\mu_v(<\mu)$, only if there are no customers in the system and return back when he finds N or more number of customers waiting for service. The vacation time is also exponentially distributed with parameter θ.

Let

$$\Lambda\ (t);\ \text{denotes count of users in the system}$$
$$\Omega\ (t);\ \text{represents the current state of the system.}$$

The state

$$\Omega(t) = \begin{cases} 1; & \text{functional state} \\ 0; & \text{semi-functional state} \end{cases}$$

Then $(\Omega(t), \Lambda(t))$ defines a Markov process. The pictorial depiction of the model is given in Fig. 1.

Let $P_{l,m}(t)$ denote the time-sensitive probabilities. The governing equations are,

$$P'_{0,0}(t) = -\lambda P_{0,0}(t) + \mu_v P_{0,1}(t) + \mu P_{1,1}(t), \tag{2.1}$$

$$P'_{0,n}(t) = -(\lambda + \mu_v) P_{0,n}(t) + \mu_v P_{0,n+1}(t) + \lambda P_{0,n-1}(t); n = 1, 2, .., N-1, \tag{2.2}$$

$$P'_{0,n}(t) = -(\lambda + \mu_v + \theta) P_{0,n}(t) + \mu_v P_{0,n+1}(t) + \lambda P_{0,n-1}(t); n = N, N+1..., \tag{2.3}$$

$$P'_{1,1}(t) = -(\lambda + \mu) P_{1,1}(t) + \mu P_{1,2}(t), \tag{2.4}$$

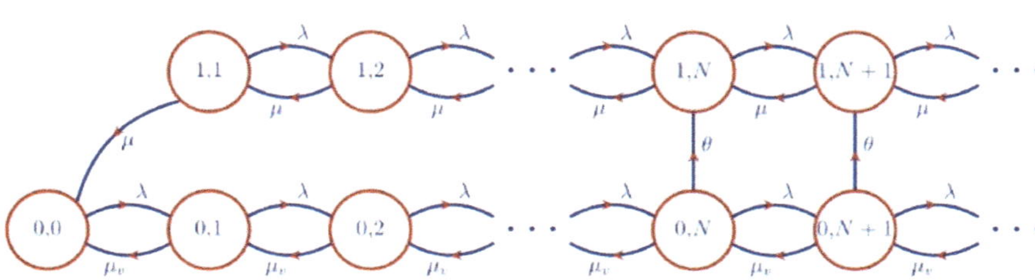

Fig. 1 Pictorial depiction of the model

$$P'_{1,n}(t) = -(\lambda + \mu)P_{1,n}(t) + \mu P_{1,n+1}(t) + \lambda P_{1,n-1}(t); n = 2, 3 \ldots N - 1, \quad (2.5)$$

and

$$P'_{1,n}(t) = -(\lambda + \mu)P_{1,n}(t) + \mu P_{1,n+1}(t) + \lambda P_{1,n-1}(t) + \theta P_{0,n}(t); \quad n = N, N+1, \ldots. \quad (2.6)$$

with the initial condition $P_{0,0}(0) = 1$ and $P_{j,n}(0) = 0$ for $n = 1, 2, 3...$ and $j = 0, 1$.

3 Time-Sensitive Probabilities

In this section, the time-sensitive probabilities of the model are obtained by applying generating function technique and continued fraction approach.

Evaluation of $P_{1,n}(t)$.

Define

$$\mathcal{M}(z,t) = \sum_{n=1}^{N-1} P_{1,n}(t)z^n + \sum_{n=N}^{\infty} P_{1,n}(t)z^n.$$

Then differentiating partially with respect to 't' yields,

$$\frac{\partial \mathcal{M}(z,t)}{\partial t} = \sum_{n=1}^{N-1} P'_{1,n}(t)z^n + \sum_{n=N}^{\infty} P'_{1,n}(t)z^n.$$

From (2.5) and (2.6), after some algebra, we have

$$\frac{\partial \mathcal{M}(z,t)}{\partial t} - \left(-(\lambda + \mu) + \frac{\mu}{z} + \lambda z\right)\mathcal{M}(z,t) = \theta \sum_{n=N}^{\infty} P_{0,n}(t)z^n - \mu P_{1,1}(t).$$

Integrating with respect to 't' yields,

$$\mathcal{M}(z,t) = \theta \int_0^t \left(\sum_{n=N}^{\infty} P_{0,n}(y)z^n\right) e^{-(\lambda+\mu)(t-y)} e^{\left(\frac{\mu}{z}+\lambda z\right)(t-y)} dy$$
$$- \mu \int_0^t P_{1,1}(y) e^{-(\lambda+\mu)(t-y)} e^{\left(\frac{\mu}{z}+\lambda z\right)(t-y)} dy. \quad (3.1)$$

Using the fact that

$$\exp\left(\frac{\mu}{z} + \lambda z\right)t = \sum_{n=-\infty}^{\infty} (\beta z)^n I_n(\alpha t),$$

where $\alpha = 2\sqrt{\lambda\mu}$, $\beta = \sqrt{\frac{\lambda}{\mu}}$ and relating the coefficients of z^n in (3.1) leads to,

$$P_{1,n}(t) = \theta \int_0^t \sum_{m=N}^{\infty} P_{0,m}(y)\beta^{n-m} I_{n-m}(\alpha(t-y))e^{-(\lambda+\mu)(t-y)}dy$$
$$- \mu \int_0^t P_{1,1}(y)e^{-(\lambda+\mu)(t-y)}\beta^n I_n(\alpha(t-y))dy. \tag{3.2}$$

Using $I_{-n}(t) = I_n(t)$ and relating the coefficients of z^{-n} in (3.1) yields,

$$0 = \theta \int_0^t \sum_{m=N}^{\infty} P_{0,m}(y)\beta^{-n-m} I_{n+m}(\alpha(t-y))e^{-(\lambda+\mu)(t-y)}dy$$
$$- \mu \int_0^t P_{1,1}(y)e^{-(\lambda+\mu)(t-y)}\beta^{-n} I_n(\alpha(t-y))dy. \tag{3.3}$$

From (3.2) and (3.3) for $n = 1, 2, 3....$, we get

$$P_{1,n}(t) = \theta \int_0^t \sum_{m=N}^{\infty} P_{0,m}(y)\beta^{n-m} \Big\{ I_{n-m}(\alpha(t-y))$$
$$- I_{n+m}(\alpha(t-y))\Big\} e^{-(\lambda+\mu)(t-y)}dy \tag{3.4}$$

Hence, $P_{1,n}(t)$ for $n = 1, 2..$ are obtained in terms of $P_{0,n}(t)$.
Estimation of $P_{0,n}(t)$ for $n = 1, 2...N - 1$.
Define

$$\mathbb{R}(z, t) = \sum_{n=1}^{N-1} P_{0,n}(t)z^n.$$

Then

$$\frac{\partial \mathbb{R}(z, t)}{\partial t} = \sum_{n=1}^{N-1} P'_{0,n}(t)z^n.$$

On substituting Eq. (2.2), we get

$$\frac{\partial \mathbb{R}(z,t)}{\partial t} - \left(-(\lambda+\mu_v) + \frac{\mu_v}{z} + \lambda z\right)\mathbb{R}(z,t) = \lambda z P_{0,0}(t) - \mu_v P_{0,1}(t).$$

Integrating the above equation with respect to 't' yields,

$$\mathbb{R}(z,t) = \int_0^t \lambda z P_{0,0}(y) \mathrm{e}^{-(\lambda+\mu_v)(t-y)} \mathrm{e}^{\left(\frac{\mu_v}{z}+\lambda z\right)(t-y)} dy$$
$$- \int_0^t \mu_v P_{0,1}(y) \mathrm{e}^{-(\lambda+\mu_v)(t-y)} \mathrm{e}^{\left(\frac{\mu_v}{z}+\lambda z\right)(t-y)} dy. \tag{3.5}$$

Using the fact that

$$\exp\left(\frac{\mu_v}{z} + \lambda z\right)t = \sum_{n=-\infty}^{\infty} (\beta_1 z)^n I_n(\alpha_1 t),$$

where $\alpha_1 = 2\sqrt{\lambda\mu_v}$, $\beta_1 = \sqrt{\frac{\lambda}{\mu_v}}$ and relating the coefficients of z^n on both sides of Eq. (3.5), we get

$$P_{0,n}(t) = \int_0^t \lambda P_{0,0}(y) e^{-(\lambda+\mu_v)(t-y)} \beta_1^{n-1} I_{n-1}(\alpha_1(t-y)) dy$$
$$- \int_0^t \mu_v P_{0,1}(y) e^{-(\lambda+\mu_v)(t-y)} \beta_1^n I_n(\alpha_1(t-y)) dy. \tag{3.6}$$

Similarly collecting the terms containing z^{-n} on both sides of Eq. (3.5) and multiplying the resulting equation by β_1^{2n}, we get

$$0 = \int_0^t \lambda P_{0,0}(y) e^{-(\lambda+\mu_v)(t-y)} \beta_1^{n-1} I_{n+1}(\alpha_1(t-y)) dy$$
$$- \int_0^t \mu_v P_{0,1}(y) e^{-(\lambda+\mu_v)(t-y)} \beta_1^n I_n(\alpha_1(t-y)) dy. \tag{3.7}$$

Subtracting Eq. (3.7) from Eq. (3.6), we get

$$P_{0,n}(t) = \int_0^t \lambda P_{0,0}(y) e^{-(\lambda+\mu_v)(t-y)} \beta_1^{n-1} (I_{n-1}(\alpha_1(t-y)) - I_{n+1}(\alpha_1(t-y)))dy,$$
$$n = 1, 2 \ldots N-1. \tag{3.8}$$

Hence, $P_{0,n}(t)$ for $n = 1, 2 \ldots N-1$ are obtained in terms of $P_{0,0}(t)$.

Estimation of $P_{0,n}(t)$ for $n = N, N+1 \ldots$

Using the Laplace transform technique for the Eq. (2.3), we get

$$(s + \lambda + \mu_v + \theta)\hat{P}_{0,n}(s) = \lambda \hat{P}_{0,n-1}(s) + \mu_v \hat{P}_{0,n+1}(s); n = N, N+1, \ldots$$

After rearranging the terms, we have

$$\frac{\hat{P}_{0,n}(s)}{\hat{P}_{0,n-1}(s)} = \frac{\lambda}{(s + \lambda + \mu_v + \theta) - \mu_v \frac{\hat{P}_{0,n+1}(s)}{\hat{P}_{0,n}(s)}}.$$

Using continued fraction approach, we get

$$\hat{P}_{0,n}(s) = \left(\frac{1}{\mu_v} \frac{P_1 - \sqrt{P_1^2 - \alpha_1^2}}{2}\right)^{n-N+1} \hat{P}_{0,N-1}(s); n = N, N+1, \ldots, \tag{3.9}$$

where $P_1 = s + \lambda + \mu_v + \theta$.

Equation (3.9) can be rewritten as,

$$\hat{P}_{0,n}(s) = \beta_1^{n-N+1} \left(\frac{P_1 - \sqrt{P_1^2 - \alpha_1^2}}{\alpha_1}\right)^{n-N+1} \hat{P}_{0,N-1}(s); \quad n = N, N+1, \ldots \tag{3.10}$$

On inverting the Eq. (3.10) yields,

$$P_{0,n}(t) = \int_0^t \lambda \beta_1^{n-N} P_{0,N-1}(y) e^{-(\lambda+\mu_v+\theta)(t-y)} \big(I_{n-N}(\alpha_1(t-y)) - I_{n-N+2}(\alpha_1(t-y))\big)dy;$$
$$n = N, N+1, \ldots \tag{3.11}$$

Hence, $P_{0,n}(t)$ for $n = N, N+1 \ldots$ are obtained in terms of $P_{0,N-1}(t)$.

Estimation of $\hat{P}_{0,N-1}(s)$ in terms of $\hat{P}_{0,0}(s)$.

Using the Laplace transform technique for Eq. (3.8), we get

$$\hat{P}_{0,n}(s) = \lambda\beta_1^{n-1}\hat{P}_{0,0}(s)\left[\frac{1}{\sqrt{P_2^2-\alpha_1^2}}\left(\frac{P_2-\sqrt{P_2^2-\alpha_1^2}}{\alpha_1}\right)^{n-1} - \frac{1}{\sqrt{P_2^2-\alpha_1^2}}\left(\frac{P_2-\sqrt{P_2^2-\alpha_1^2}}{\alpha_1}\right)^{n+1}\right]; \quad n = 1, 2..N-1 \tag{3.12}$$

where $P_2 = s + \lambda + \mu_v$.
Substituting $n = N - 1$ in Eq. (3.12) gives,

$$\hat{P}_{0,N-1}(s) = \lambda\beta_1^{N-2}\hat{P}_{0,0}(s)\left[\left(\frac{P_2-\sqrt{P_2^2-\alpha_1^2}}{\alpha_1}\right)^{N-2}\frac{1}{\sqrt{P_2^2-\alpha_1^2}} - \left(\frac{P_2-\sqrt{P_2^2-\alpha_1^2}}{\alpha_1}\right)^{N}\frac{1}{\sqrt{P_2^2-\alpha_1^2}}\right]. \tag{3.13}$$

Estimation of $P_{0,n}(t)$ for $n = N, N+1, \ldots$ in terms of $P_{0,0}(t)$.
Substituting Eqs. (3.13) in (3.10) gives,

$$\hat{P}_{0,n}(s) = \lambda\beta_1^{n-1}\hat{P}_{0,0}(s)\left(\frac{P_1-\sqrt{P_1^2-\alpha_1^2}}{\alpha_1}\right)^{n-N+1}\left[\left(\frac{P_2-\sqrt{P_2^2-\alpha_1^2}}{\alpha_1}\right)^{N-2} \times \frac{1}{\sqrt{P_2^2-\alpha_1^2}} - \left(\frac{P_2-\sqrt{P_2^2-\alpha_1^2}}{\alpha_1}\right)^{N}\frac{1}{\sqrt{P_2^2-\alpha_1^2}}\right]; \quad n = N, N+1, \ldots. \tag{3.14}$$

On inverting the Eq. (3.14), we get

$$P_{0,n}(t) = \lambda\beta_1^{n-1}P_{0,0}(t) * \left[e^{-(\lambda+\mu_v)t}I_{N-2}(\alpha_1 t) - e^{-(\lambda+\mu_v)t}I_N(\alpha_1 t)\right] * e^{-(\lambda+\mu_v+\theta)t}\frac{(n-N+1)I_{n-N+1}(\alpha_1 t)}{t} \tag{3.15}$$

Hence, $P_{0,n}(t)$ for $n = N, N+1\ldots$ are expressed in terms of $P_{0,0}(t)$. Now vacation state probabilities, $P_{0,n}(t)(n = 1, 2\ldots)$ and busy state probabilities $P_{1,n}(t)(n = 1, 2...)$ are expressed explicitly in terms of $P_{0,0}(t)$. Hence, it remains to find $P_{0,0}(t)$.

Estimation of $P_{0,0}(t)$.
Using the Laplace transform technique for Eq. (3.4), we get

$$\hat{P}_{1,n}(s) = \theta \sum_{m=N}^{\infty} \hat{P}_{0,m}(s)\beta^{n-m}\left[\left(\frac{P_0 - \sqrt{P_0^2 - \alpha^2}}{\alpha}\right)^{n-m} \times \frac{1}{\sqrt{P_0^2 - \alpha^2}} - \left(\frac{P_0 - \sqrt{P_0^2 - \alpha^2}}{\alpha}\right)^{n+m} \frac{1}{\sqrt{P_0^2 - \alpha^2}}\right]; \quad n = 1, 2, \ldots. \tag{3.16}$$

where $P_0 = s + \lambda + \mu_v + \theta$. Substituting $n = 1$ in Eq. (3.16), we get

$$\hat{P}_{1,1}(s) = \theta \sum_{m=N}^{\infty} \hat{P}_{0,m}(s)\beta^{1-m}\left[\left(\frac{P_0 - \sqrt{P_0^2 - \alpha^2}}{\alpha}\right)^{1-m} \frac{1}{\sqrt{P_0^2 - \alpha^2}} - \left(\frac{P_0 - \sqrt{P_0^2 - \alpha^2}}{\alpha}\right)^{1+m} \frac{1}{\sqrt{P_0^2 - \alpha^2}}\right]. \tag{3.17}$$

Substituting Eq. (3.14) in Eq. (3.17), we get

$$\hat{P}_{1,1}(s) = \theta \sum_{m=N}^{\infty} \hat{P}_{0,0}(s)\lambda\beta^{1-m}\beta_1^{m-1}\left(\frac{P_1 - \sqrt{P_1^2 - \alpha_1^2}}{\alpha_1}\right)^{m-N+1} \left[\hat{\psi}_{N-2}(s) - \hat{\psi}_N(s)\right]\left[\hat{\Omega}_{\mathrm{n-m}}(\mathrm{s}) - \hat{\Omega}_{\mathrm{n+m}}(\mathrm{s})\right], \tag{3.18}$$

where

$$\hat{\psi}_i(s) = \left(\frac{P_2 - \sqrt{P_2^2 - \alpha_1^2}}{\alpha_1}\right)^i \frac{1}{\sqrt{P_2^2 - \alpha_1^2}},$$

and

$$\hat{\Omega}_{\mathrm{i}}(\mathrm{s}) = \left(\frac{P_0 - \sqrt{P_0^2 - \alpha^2}}{\alpha}\right)^i \frac{1}{\sqrt{P_0^2 - \alpha^2}}.$$

Using Laplace transform technique for Eq. (3.8) yields,

$$\hat{P}_{0,n}(s) = \lambda\beta_1^{n-1}\hat{P}_{0,0}(s)\left[\left(\frac{P_2-\sqrt{P_2^2-\alpha_1^2}}{\alpha_1}\right)^{n-1}\frac{1}{\sqrt{P_2^2-\alpha_1^2}}\right.$$
$$\left.-\left(\frac{P_2-\sqrt{P_2^2-\alpha_1^2}}{\alpha_1}\right)^{n+1}\frac{1}{\sqrt{P_2^2-\alpha_1^2}}\right]; n = 1, 2 \ldots N-1 \quad (3.19)$$

Substituting $n = 1$ in Eq. (3.19), we get

$$\hat{P}_{0,1}(s) = \lambda\hat{P}_{0,0}(s)\left[\frac{1}{\sqrt{P_2^2-\alpha_1^2}} - \hat{\psi}_2(s)\right]. \quad (3.20)$$

Using the Laplace transform technique for Eq. (2.1) gives,

$$\hat{P}_{0,0}(s) = \frac{1}{s+\lambda} + \frac{\mu_v}{s+\lambda}\hat{P}_{0,1}(s) + \frac{\mu}{s+\lambda}\hat{P}_{1,1}(s). \quad (3.21)$$

By substituting Eqs. (3.18) and (3.20) in Eq. (3.21) and after some simplification, we get

$$\hat{P}_{0,0}(s) = \frac{1}{(1-F(s))(s+\lambda)}, \quad (3.22)$$

where

$$F(s) = \frac{\lambda\mu_v}{s+\lambda}\left(\frac{1}{\sqrt{P_2^2-\alpha_1^2}} - \hat{\psi}_2(s)\right)$$
$$+\frac{\mu\theta}{s+\lambda}\sum_{m=N}^{\infty}\left\{\lambda\beta^{1-m}\beta_1^{m-1}\left(\frac{P_1-\sqrt{P_1^2-\alpha_1^2}}{\alpha_1}\right)^{m-N+1}\right.$$
$$\left.\times\left[\hat{\psi}_{N-2}(s) - \hat{\psi}_N(s)\right]\left[\hat{\Omega}_{n-m}(s) - \hat{\Omega}_{n+m}(s)\right]\right\}.$$

On inverting Eq. (3.22), we get

$$P_{0,0}(t) = \mathrm{e}^{-\lambda t} * \sum_{\mathrm{j}=0}^{\infty}(\mathrm{F(t)})^{*\mathrm{j}},$$

where

$$F(t) = \mathrm{e}^{-\lambda t}\mu_v\lambda * \left(\mathrm{e}^{-(\lambda+\mu_v)t}I_0(\alpha_1 t) - \psi_2(t)\right) + \mu\theta\mathrm{e}^{-\lambda t} * \sum_{m=N}^{\infty}\left\{\begin{array}{l}\lambda\beta^{1-m}\beta_1^{m-1}\left(e^{-(\lambda+\mu_v+\theta)t}\dfrac{(m-N+1)I_{m-N+1}(\alpha_1 t)}{t}\right)\\ *(\psi_{N-2}(t) - \psi_N(t)) * (\Omega_{n-m}(t) - \Omega_{n+m}(t))\end{array}\right\},$$

$$\psi_i(t) = e^{-(\lambda+\mu_v)t}I_i(\alpha_1 t),$$

and

$$\Omega_i(t) = e^{-(\lambda+\mu)t}I_i(\alpha t).$$

Hence, $P_{0,0}(t)$ is explicitly determined.
Special case:
As $\mu_v = 0$ and $N = 1$ then,

$$P_{1n}(t) = \theta\int_0^t \sum_{m=1}^{\infty} P_{0m}(y)\beta^{n-m}\{I_{n-m}(\alpha(t-y)) - I_{n+m}(\alpha(t-y))\}e^{-(\lambda+\mu)(t-y)}dy,$$

and

$$P_{0,n}(t) = \int_0^t \lambda\beta_1^{n-1}P_{0,\ 0}(y)e^{-(\lambda+\mu_v+\theta)(t-y)}(I_{n-1}(\alpha_1(t-y)) - I_{n+1}(\alpha_1(t-y)))dy.$$

The above results are seen similar to the results given by Sudhesh (2012).

4 Conclusion and Future Scope

A massive task in network design is minimizing energy consumption. To minimize energy consumption, the vacation queueing model plays a vital role in modeling and analyzing the sensor nodes. By considering the need for efficient energy-saving schemes in WSN, an $M/M/1$ queueing model with multiple working vacation and N-policy is considered and we derived the transient state probabilities of the model explicitly. Further, this work will be extended by allowing disaster to occur in a busy state and it will be interesting to generalize the current model to multi-server model.

References

1. Baba, Y. (2005). Analysis of a GI/M/1 queue with multiple working vacations. *Operations Research Letters, 33*(2), 201–209.
2. Banik, A. D., Gupta, U. C., & Pathak, S. S. (2007). On the GI/M/1/N queue with multiple working vacations—analytic analysis and computation. *Applied Mathematical Modelling, 31*(9), 1701–1710.
3. Boutoumi, B., & Gharbi, N. (2018, July). Two thresholds working vacation policy for improving energy consumption and latency in WSNs. In *International Conference on Queueing Theory and Network Applications* (pp. 168–181). Cham: Springer.
4. Doshi, B. T. (1986). Queueing systems with vacations—a survey. *Queueing Systems, 1*(1), 29–66.
5. Kim, J. D. (2003). Analysis of queue-length distribution of the M/G/1 queue with working vacations (M/G/1/WV). In *Proceedings of Hawaii International Conference on Statistics and Related Fields, 2003* (pp. 1191–1200).
6. Li, J. H., & Tian, N. S. (2007). The discrete-time GI/Geo/1 queue with working vacations and vacation interruption. *Applied Mathematics and Computation, 185*(1), 1–10.
7. Liu, W., Xu, X., & Tian, N. (2007). Some results on the M/M/1 queue with working vacations. *Operations Research Letters, 35*(5), 595–600.
8. Servi, L. D., & Finn, S. G. (2002). M/M/1 queues with working vacations (m/m/1/wv). *Performance Evaluation, 50*(1), 41–52.
9. Sudhesh, R., & Raj, L. F. (2011, December). Computational analysis of stationary and transient distribution of single server queue with working vacation. In *International Conference on Computing and Communication Systems* (pp. 480–489). Berlin, Heidelberg: Springer.
10. Takagi, H., & Boguslavsky, L. B. (1991). A supplementary bibliography of books on queueing analysis and performance evaluation. *Queueing Systems, 8*(1), 313–322.
11. TeghemJr, J. (1986). Control of the service process in a queueing system. *European Journal of Operational Research, 23*(2), 141–158.
12. Tian, N., & Zhang, Z. G. (2006). *Vacation queueing models: theory and applications* (Vol. 93). Springer Science & Business Media.
13. Vijayashree, K. V., & Janani, B. (2015). Transient analysis of an queue with multiple exponential vacation and N-policy. *Pakistan Journal of Statistics and Operations Research, 11*(4), 587–600.
14. Wu, D. A., & Takagi, H. (2006). M/G/1 queue with multiple working vacations. *Performance Evaluation, 63*(7), 654–681.

Estimating the Optimum Completion Time of Project Using Binomial Distribution and Probabilistic PERT Network

Zainab Al-Baldawi and Iftikhar Ali Hussein

Abstract Using an efficient and scientific planning tool for planning and scheduling project can be considered a crucial process. Two approaches have been applied to find the total completion time of project, namely program evaluation and review technique (probabilistic PERT network) and binomial distribution cumulative density function (CDF). Cumulative density function is assumed that time is a random variable that followed the discrete distribution (binomial distribution). The coefficient of variation that depends on (S, X-) has been calculated to determine uncertainty of activity completion (c.v) at each stage of project where it is between (0, 0.103), and it is a very weak value so this illustrates that most activities are worked as it planned. The final results show that the cumulative function method is more accurate than the traditional method (PERT) where the wasted time was decreased around 4 days. The total project completion time by using PERT is 33 days where it is 29 days by using the cumulative function method.

Keywords Binomial distribution · Cumulative density function (CDF) · Discrete random variable · Project completion time · PERT technique

1 Introduction

Program evaluation and review technique (PERT) is a planning and scheduling method that uses to evaluate and estimate the completion time of project in addition to control the project resources. Project is a combination of activities when all activities and events that formed the project network are connected logically and sequentially. Event is represented by a node (a circle), and it represents either starting or ending of an activity also not consume any resource. Activity is represented by an arrow

Z. Al-Baldawi (✉)
University of Technology, Baghdad, Iraq
e-mail: engzainab@yahoo.com

I. Ali Hussein
Middle Technical University, Baghdad, Iraq
e-mail: iahk1@yahoo.com

S.-L. Peng et al. (eds.), *Proceedings of First International Conference on Mathematical Modeling and Computational Science*, Advances in Intelligent Systems and Computing 1292, https://doi.org/10.1007/978-981-33-4389-4_57

and consumes resources like money, time, and materials [1]. Basic steps to form project network are included, list all the activities in a project, determine the logical relationship between activities and finally identify the activity time and resources.

PERT technique has been used as a project planning technique in various fields, like [2] Research and development for new product, The new plant construction, The ship building, Project of NASA space exploration, The production of films, Conducting an advertisement campaign, Relocation of a major facility, Developing a new weapons system for government-sponsored project, Maintenance and repair process for nuclear reactor, and Installation of a management information system.

The probabilistic PERT network studies were beginning since the latest 1950s. The researchers were interested about this topic. Malcolm et al. [3] researches earlier in probabilistic PERT network where he illustrated that all PERT analyses depend on extracting the parameters according to the two following formulas [3]

$$\overline{y}_{(\text{pert})} = (\mathrm{T_O} + 4\mathrm{T_L} + \mathrm{T_P})/6 \quad (1)$$

$$\sigma^2 = (Tp - To)2/36 \quad (2)$$

PERT activities are probabilistic in nature, so activities time cannot be specified correctly, because of uncertainties in carrying out the activity. The PERT network is used when the times of activity are probabilistic.

Three times are used in PERT calculations [4]: **Optimistic time** (T_O) is a symbol that refers to the optimistic time, and it is lowest time to complete any activity, **pessimistic time** (T_P) is the highest time to complete activity; and **likely time** (T_L) is in between pessimistic times and optimistic time.

These times are identified according experiences and the parameters that have been estimated according propose that the random variable (t) follow the **Beta distribution** that knowing by the probability function (Fig. 1):

$$f_{T(t)} = 1\ /(tP - tO)^{\alpha+\beta-1}\beta(\alpha,\beta)\frac{1}{(tP - tO)^{\alpha+\beta-1}\beta(\alpha,\beta)}(\mathrm{t} - \mathrm{To})^{\beta-1}$$

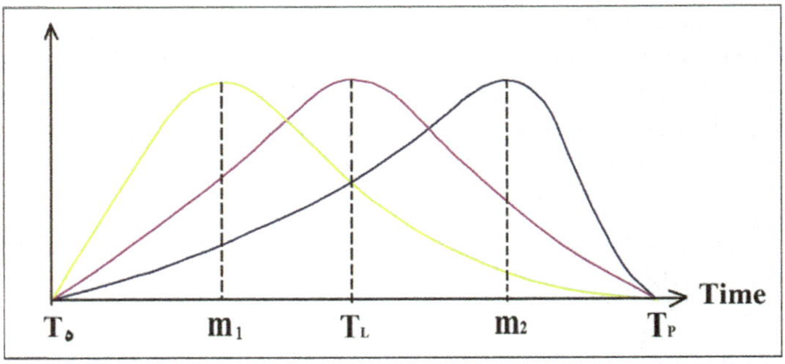

Fig. 1 Three times of PERT

$$T_o \leq t \leq T_p, \alpha, \beta > 0 \quad (3)$$

Two types of random variable can be recognized [5]—discrete random variable and ccontinuous random variable.

Most researches have studied the probabilistic PERT network by assumption that completion time of any activity of project is a continuous random variable, so many researchers have studied the probabilistic PERT network for continuous variables by using different distributions;

Maidamisa [6] compares beta distribution against triangular distribution to determine the efficiency of each one in scheduling the project activity duration that lead the optimum completion time. Triangular distribution was selected and got minimum completion time.

Chin et al. [7] used the normal distributions and the beta distribution to calculate the probability in completing the project using the commonly methods to schedule the project activities time, namely critical path method (CPM) and program evaluation review technique (PERT).

Ballesteros-Pérez [8] suggested M-PERT to avoid the drawbacks of the classic PERT where it is quite accurate when estimating real project duration. The M-PERT allows manual calculation.

Doubravský and Doskocil [9] studied the probability of the project completion time by comparison two different methods: Monte Carlo method and PERT method. The statistical hypothesis test was used to compare calculated probabilities of project completion. The hypothesis was rejected for all simulations and showed that from application's point of view there is difference between the methods.

In the case of closed large-scale infrastructure projects, Hajdu and Bokor (2014) [10] applied different distributions (uniform, triangular, β). The potential impact of such distributions on the project completion period was analyzed and examined. The result showed that using such distributions with the same three-point estimate has a smaller effect on the completion period of the project than a 10 percent difference in the three-point estimate values.

Ahmad [11] combined three methods, namely genetic algorithm (GA), program evaluation and review (PERT) and Gantt chart (GC) as integrated approach for many reasons, to review the existing scheduling techniques, to solve the scheduling problem in the construction industry, and finally to compare the applicable of the proposed approach with the existing techniques.

Yakhchali [12] proposed a novel approach to schedule in networks the project with stochastic activity durations for random variables with beta distribution to deal with uncertainty in the scheduling process.

In the PERT estimation technique, Edward et al. [13] used Weibull probability distribution to estimate the task time. The results showed that it is a better choice and for this application it has the same benefits as the conventional beta distribution.

2 Methodology of the Research

In this work, a comparative study is done by using probabilistic PERT network and cumulative density function binomial distribution to find the optimum total completion time of project, and Fig. 2 illustrates methodology of this research.

The binomial distribution is considered as one of the most important discrete distributions that is increasingly used in various fields like engineering, medical, administration, scientific, and any applications that frequently face failure and success trial, where it refers to the probability that a binomial experiment results in exactly "*x*" successes.

Three conditions should satisfy to consider any probability of an event as a binomial probability; there are a fixed number of trials (the number of trials is denoted by *n*). The trials are independent (the outcome on one trial does not affect the outcome on other trials) and finally, each trial has only two possible outcomes; a failure or a success, and the probability of success, denoted by p and for any trial is the same.

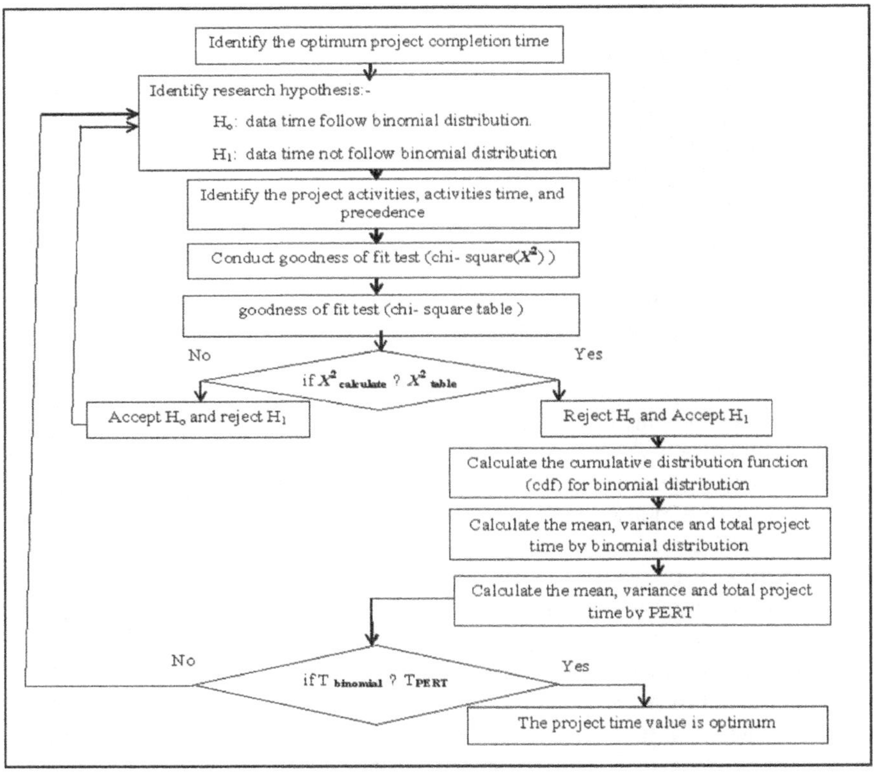

Fig. 2 Methodology of the research

So, for the binomial experiment that consists of "n" trails and results in exactly "x" successes, the general formula to find the binomial probability is:

$$\mathrm{P}\{\mathrm{x}=\mathrm{k}\} = C_k^n P^k (1-P)^{n-k}, k = 0, 1, 2, 3 \ldots, n \tag{4}$$

where.
x: The number of successes that results from the binomial experiment.
n: The number of trials in the binomial experiment.
p: The probability of success on an individual trial.
q: Theprobabilityoffailureonanindividualtrial,
The mean and variance are

$$\mathrm{E}\{\mathrm{x}\} = \mathrm{np} \tag{5}$$

$$\mathrm{ar}\{\mathrm{x}\} = \mathrm{np}(1-\mathrm{p}) \tag{6}$$

$$\sigma = \sqrt{npq} \tag{7}$$

$$\text{Variant coefficient(c.v)} = (\text{sqrt(var)/mean} * 100\%) \tag{8}$$

To estimate (P) for the binomial distribution, we can use maximum likelihood method (MLM), where this method may reduce the errors and increase and make the parameter in high value.

$$\begin{aligned}
P(x, n, p) &= c_x^n p^x (1-p)^{n-x} \\
LP(x, n, p) &= c_x^n p^{\sum xi} (1-p)^{n-\sum xi} c_x^n \prod_{i=1}^{n} p^x (1-P)^{n-x} \\
&= \prod \left[\left(\frac{ni}{xi(n-x)!} \right) p^{xi} (1-p)^{n-xi} \right] \\
&= \left(\frac{ni}{xi(n-x)!} \right)^n p^{\sum xi} (1-p)^{n-\sum xi} \\
LnL &= nLn \left(\frac{ni}{xi(n-x)!} \right) + \sum xiLnP + (n - \sum xi) Ln(1-P) \\
\frac{\partial LnL}{\partial P} &= 0 + \frac{\sum xi}{P} - \frac{n - \sum xi}{1-P} \\
0 \Rightarrow \frac{\sum xi}{\hat{p}} &= \frac{(n - \sum xi)}{1-\hat{p}} \\
\sum xi - \hat{p} \sum xi &= n\hat{p} - \hat{p} \sum xi \\
\therefore \hat{p} &= \frac{\sum xi}{n} = \overline{x}
\end{aligned} \tag{9}$$

The practical application of the research was applied at sawing factory for men suits where factory consists of many various departments.

Design department is responsible for new suit design. Control department is responsible for quality control of suits during each stage. Preparation department is responsible about processes and operation planning. Programming department is responsible to release the production orders Sawing department is responsible for sawing all men suits models. Maintenance department is responsible for maintaining all factory machines. The processes technology of the men suit is illustrated in Table 1, where the three times of each activity were determined with activity description and relationship with other activities. Figure 3 showed the network of the men suit activities.

3 Results and Discussion

The statistical analysis for the data of time, Table 1 was depended on matching the data of time for the production stages of the men suits with the binomial distribution where data that represent the most likely time (TL) was depended.

After divvied each time value by the sum of time, $x = 1, 2, \ldots, 10$ will be got, then the value that result by $k = 6$ periods, where the period length = (0.03),

where:

$n = 10$, the number of activities.

$\overline{x} = 0.0322.$

$S_x^2 = 0.0098.$

$x =$ the activity completion time.

$\overline{x} =$ the mean of x.

$S_x^2 =$ the variance of x.

After that, Qi- test (×2 square tests) has been established to determine whether the xi values subject the binomial distribution or not.

The null hypothesis (Ho) = x's ~ Bin(n ,p) is subjected to binomial distribution.

The alternative hypothesis (H_1) = x's ~ has not been subjected to binomial distribution.

This value is depended on the fit of goodness criteria between the frequency of the actual observing and the frequency of the expecting by using statistical × 2 test, the formula of test is:

$$x^2 = \sum \frac{(Oi - Ei)}{(Ei)} \tag{10}$$

where:

$K =$ the classes number of observing frequency.

$O_i =$ the observing frequency.

$E_i =$ the expecting frequency.

Table 1 Description and time of design and process planning the men suit activities

Seq	Activity symbol	Activities description for new design for men suit	Precedence	Optimistic time T_o	Likely time T_L	Optimistic time T_o
1	A	Designing the suit model (model no.)	–	13	6	5
2	B	Approval the suit model by skilled and experts committee	A	7	5	3
3	C	Preparing the preliminary model for various sizes	B	14	10	6
4	D	Process planning for suit sawing by programming department	C	5	3	2
5	E	Operations planning for suit sawing for every model in the preparation department	D	8	6	4
6	F	Scheduling the suit sawing operation for every model	E	6	4	2
7	G	Release the production order to the sawing department	F	4	3	2
8	H	Workers training for sawing the various models by technology and control department	G	8	5	3
9	I	Inspection the suit sawing quality for every model	H	5	3	2

(continued)

Table 1 (continued)

Seq	Activity symbol	Activities description for new design for men suit	Precedence	Optimistic time T_o	Likely time T_L	Optimistic time T_o
10	J	Purchasing and providing all the suit sawing requirements	I	10	8	5

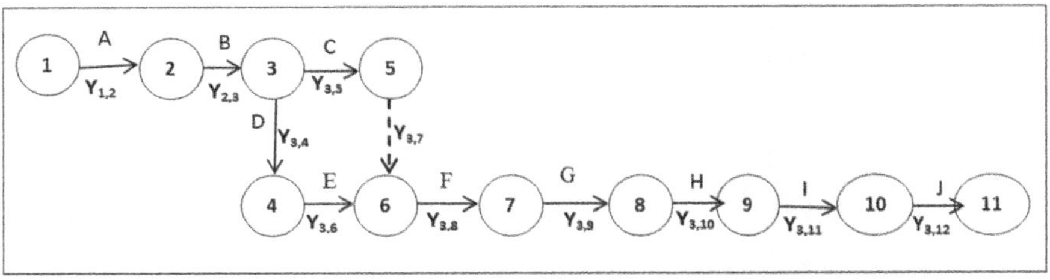

Fig. 3 Network of the men suit activities

The freedom degree is $v = \text{k m}^{-1}$.

where:

$M =$ the number of the estimated parameters.

So,x^2 the tabular value will be contained quality level α and freedom degree v, where The null hypothesis (H_o) may be refused if the calculated value x^2 greater than the tabular value and will accepted reverse that will accepted reverse that (Table 2).

The formula (10) has been used to calculated the calculated value x^2.

$$x^2_{\text{cal}} = \sum_{i=1}^{5} \frac{(Oi - Ei)2}{Ei}$$

Table 2 Calculation of goodness-of-fit test

k	Class	Observing frequency Oi	Class probability Pi	Expecting frequency Ei
1	0.03–0.06	3	0.1698	1.698
2	0.06–0.09	1	0.0755	0.755
3	0.09–0.12	4	0.415	4.15
4	0.12–0.15	0	0	0
5	0.15–0.18	1	0.1509	1.509
6	0.18–0.21	1	0.1888	1.88

Table 3 PERT calculations

Activity	Variance	Mean
A	1.777778	7
B	0.444444	5
C	1.777778	10
D	0.25	3.166667
E	0.444444	6
F	0.444444	4
G	0.111111	3
H	0.694444	5.166667
I	0.694444	7.833333
J	1	3.666667

$$= \frac{(3-1.698)2}{1.698} + \frac{(1-0.755)2}{0.755} + \frac{(4-4.15)2}{4.15} + \frac{(1-1.509)2}{1.509} = \frac{(1-1.888)2}{1.888} x^2_{\text{cal}} = 1.67263$$

So, the calculated value is less than the tabular value. The null Hypothesis (H_o) has been accepted because it considered that the time data is followed by the binomial distribution and in the same time refused by the alternative hypothesis (H_1).

PERT calculations have been done in steps, the three times of activities have been estimated previously, namely optimistic time (TO), pessimistic time (TP), likely time (TL) as shown in Table 1. The formulas (1) and (2) have been used to determine the mean and variance as shown in Table 3.

The critical path of the activities for the PERT network has been identify and it is equal to (33) days

$$\text{The critical path} = \text{A} \rightarrow \text{B} \rightarrow \text{D} \rightarrow \text{E} \rightarrow \text{F} \rightarrow \text{G} \rightarrow \text{H} \rightarrow \text{I} \rightarrow \text{J}$$

The cumulative function of the binomial distribution was used to calculate the total time to complete the project, and Table 4 shows the results of these calculations.

The calculation of the mean, variance, and standard division in Table. 5 has been calculated by using the formulas (5), (6), (7) and (8) where:

Formula (5) was applied to find the mean of activity time, formula (6) was used to find the variance, the standard deviation (SD) was found by using the formula (7) where the formula (8) (variant coefficient c.v%) is used to estimate the uncertainty the completion time of activities.

Table 4 Calculations of the probability cumulative function for binomial distribution

R.V	Representation	Probability density function (p.d.f)							
T_1	$Y_{1,2}$	t	6						
		p	1						
T_2	$Y_{2,3}$	t	2						
		p	1						
T_3	$Y_{3,5}+Y_{3,7}$	t	3	8	11				
		p	0.25	0.75	1				
T_4	$Y_{3,4}+Y_{3,6}$	t	3	8	11				
		p	0.25	0.75	1				
W_1	$Max(T_3, T_4)$	t	3	8	11	16			
		p	0.125	0.25	0.75	1			
W_2	$Y_{3,8}+W_1$	t	3	3	8	11	16	19	
		p	0.0625	0.125	0.4375	0.5	0.875	1	
W_3	$Y_{3,9}+W_2$	t	6	8	11	19	25	31	
		p	0.015625	0.09375	0.328125	0.5	0.875	1	
W_4	$Y_{3,10}+W_3$	6	8	11	19	25	31	34	
		0.015625	0.03125	0.0625	0.125	0.25	0.875	1	
W_5	$Y_{3,11}+W_4$	t	8	11	19	25	31	34	36
		p	0.015625	0.03125	0.0625	0.125	0.25	0.875	1
W_6	$Y_{3,12}+W_5$	11	19	25	31	34	36	38	
		0.015625	0.03125	0.0625	0.125	0.25	0.875	1	

Table 5 Calculations of the standard deviation, variance, mean, and variant coefficient%

Mean	Variance	Standard division (S.D.)	Variant coefficient c.v%
10	0	0	0
10	0	0	0
15	2.5	1.581139	0.105409
15	2.5	1.581139	0.105409
25	6.25	2.5	0.1
26.25	7.34375	2.709935	0.103236
26.875	7.9296875	2.81597	0.10478
28.4375	7.451171875	2.729683	0.095989
28.59375	7.604980469	2.757713	0.096445
28.67188	7.682495117	2.771731	0.096671

4 Conclusions

Two approaches have been applied to find the total completion time of project, namely probabilistic PERT network and cumulative function under assumption that the time is a random variable that followed the discrete distribution (binomial distribution). The final results illustrate that the cumulative function method is more accurate than the traditional method (PERT) where the wasted time was decreased around 4 days where the total project completion time was concluded using PERT = 33 days where the total project completion time approximately = 29 days was identified using the cumulative function method. The value of uncertainty (c.v) for each activity is between (0, 0.103) and it is a very weak value and this illustrates that most activities are worked as planned. Researchers can use all the discrete and continuous distributions but this can do after conduct the fit of goodness for distribution with the time under study.

References

1. Taha., H. A. (1989). *Operation research: An introduction* (4h ed.). New York: Macmillan Company.
2. Hillier, L. (2001). *Introduction to operation research*. McGraw-Hill.
3. Malcolm, D. G., Rosboon, H., Clark, C. E., & Fazar, W. (1959). Application of technique for research and development program evaluation. *Journal of operation Research, 17*, 646–649.
4. Rama Murthy, P. (2007). *Operation Research*, New Age International (P) Ltd.
5. Soong, T. (2004). Fundamentals of probability and statistic in engineering. (Wiley).
6. Maidamisa, A. A. (2017). Comparative analysis of activity duration in project management. *The International Journal of Business &Management, 5*(12).
7. Chin, M. K., Kek, S. L., Sim, S. Y., & Seow, T. W. (2017). Probablistic completion time in project scheduling. *International Journal of Engineering Research & Science (IJOER), 3*(4).
8. Ballesteros Perez, P. (2017). M-PERT manual project duration estimation technique for teaching scheduling basics. *Journal of Construction Engineering and Management, 9*(143).
9. Doubravsky, K., & Doskocil, R. (2015). Comparision of approach's for calculating the probability of a project completion. *Journal of Eastern Europe Research in Business & Economics*.
10. Haidu, M., & Bokor, O. (2014). The effect of different activity distribution on project duration in pert networks. In *Procedia-Social and Behavioral Science 119(2014), 27th IPMA World Congress*.
11. Ahmad, S. H. B. (2014). *Intergration of Program Evaluation and Review Technique (PERT), Gantt Chart and Genetic Algorthim/ Master thesis*. Malaysia: Faculty of Technology Management and Business, University Tun Hussien On Malaysia.
12. Yakhchali, S. H. (2011). A simple approach to project scheduling in networks with stochastic duration. In *Proceeding of the World Congress on Engineering, V(1), WCE 2011*. London, UK.
13. McCombs, E. L., Elam, M. E., & Pratt, D. B. (2009). Estimating Task duration in PERT using the Weibull probability distribution. *Journal of Modern Applied Statistical Methods, 8*(26, 1).

Auto-Adaptive Learning for Machine Perception of Native Accent Using Deep Learning

Mekala Srinivasa Rao, P. S. V. Srinivasa Rao, and S. Ranga Swamy

Abstract One of the solutions to artificial intelligence is machine learning. It enables us to create machines that can learn from experience rather than be programmed explicitly. Current formulations of machine learning are mostly designed with the help of data available for learning and performing specific tasks from neural networks. Deep learning is an effective machine learning approach that can solve multiple and specific tasks with mini mother change. Deep learning extends machine learning to multi-level distributed representations with the necessary mapping functions into a single composite function, and in particular neural networks. Along with their capability to learn dynamic hierarchical representations, the advent of deep learning and neural networks has opened up the way for continuous training. The main objective of this thesis is to research and establish a systematic approach to continuous learning that facilitates the success of profound education and neural networks.

Keywords Deep learning · Machine learning · Artificial intelligence · Neural networks

1 Introduction

Based on current artificial intelligence (AI) research, the imitation of a typical human brain. Computers are faster than humans but not smarter than human brain, because the human brain has much intelligence than a computer. That is.

- Recognizing
- Accepting
- Listing
- Feeling
- Intellectual

M. Srinivasa Rao (✉)
Lakireddy Bali Reddy College of Engineering, Mylavaram, Andhra Pradesh, India
e-mail: srinu.mekala@gmail.com

P. S. V. Srinivasa Rao · S. Ranga Swamy
Vignan's Institute of Management and Technology for Women, Kondapur, Telangana, India

S.-L. Peng et al. (eds.), *Proceedings of First International Conference on Mathematical Modeling and Computational Science*, Advances in Intelligent Systems and Computing 1292, https://doi.org/10.1007/978-981-33-4389-4_58

AI's purpose is to make the human mind's actions clever enough to computers and computer programs. It will be nice to describe what human understanding means before we move further into artificial intelligence [1]. Present days, most of the computers and smartphones, Google, Alexa and so on use voice recognition to simplify the system operation. AI refers only to static computers. Predictions are templates that intelligent machines differentiate. A prediction is to produce knowledge based on something we don't know. Intelligent machinery with a wide range of algorithms. The goal of artificial intelligence is to comprehend and develop smart organizations. Thinking machines are technically only constrained by equations and physics. Several approaches to AI, for example, symbolic reasoning and artificial intelligence are possible. It can lead one day to thought machines that have more analytical skills than humans. Deep learning is an overall machine learning approach, capable of solving several specific tasks with minimal changes. This offers hierarchical support for learning representations and concepts. Deep learning is state of the art and is popular in many areas, including computer vision, speech recognition and the processing of natural languages. The most prevalent example of deep learning is a deep neural network (DNN), a network with many hierarchical layers.

2 Deep Learning in Machine Learning

Since the 1950s there has been artificial intelligence as a term. However, the computing capacity constraints and prohibitive costs of data storage have not been a big feature. In the past few decades, AI has become viable for large-scale usages due to dropping costs and the storage capacity and the exponential rise in processing speeds of computers [2]. Even though there is a comprehensive list of different AI industries, a brief version of some important divisions is as shown in Fig. 1 as a layered architecture of artificial intelligence. Although it is good to concentrate on industry while learning the diverse aspects of AI, it is good to remember that you could require skills which are classified under different branches, in this work

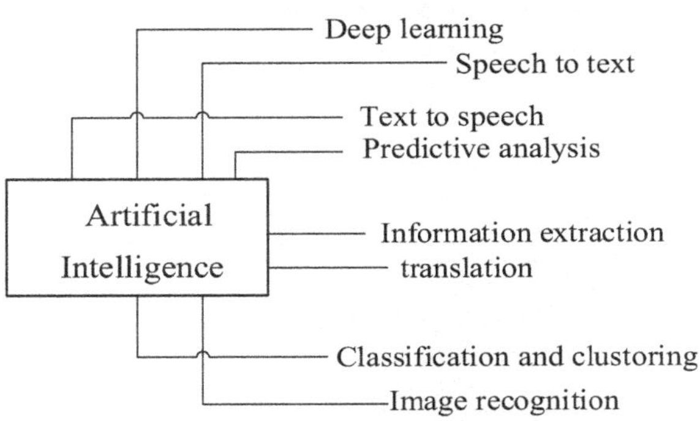

Fig. 1 Branches of artificial intelligence

author can concentrate on few branch, which are deep learning, speech to text and information extraction.

Machine learning is an artificial intelligence (AI) technology that enables systems to learn and develop automatically without needing to be specifically programmed. Machine learning focuses on the creation of computer programs that can access and use data for their purposes [3]. Within a fundamental research model, data with a logical framework close to the way that a person can conclude are continuously analyzed, for this purpose, a deep learning program uses an artificial neural network, a complex system of algorithms. The creation of an artificial neural network is informed by the human brain's biological neural network which contributes to a learning process much more capable than standard machine learning models [4].

Machine learning requires large data analysis. While the results are typically faster and more precise to identify productivity or risk, additional time and resources may be required for appropriate education. The analysis processing large volumes processing knowledge can be much more effective using AI learning and cognitive technology.

Profound learning is a form of master learning through a multi-layered neural network [5]. It is one of the many methods for the machine learning of statistical synthesis; it is an algorithmic approach to the structure and function of the brain, called the artificial neural networks. The inputs (and even the output) are analogous to the problem domains. In other words, it is not only in the tabular format where there are few amounts but also images of pixel data, text data documents or audio files. This is a difficult prospect of making sure that a profound learning model will not draw erroneous conclusions—it needs lots of training like other AI instances, to correct the learning process [6]. Most popular AI technologies use machine learn algorithms in customer service. We help to encourage self-service, increasing the efficiency of agents and boost the reliability of workflows. The data incorporated into such algorithms is a continuous stream of incoming consumer inquiries, which provides an important background in the issues facing consumers. The introduction of this background in an AI application leads, in effect, to reliable and better predictions [7]. This has turned artificial intelligence into an exciting prospect for several businesses and industry leaders speculate that customer service will be the most practical application for AI (Fig. 2).

3 Accent Recognition

Present days, most of the devices made up with virtual assistance like, Amazon Alexa, Google Assistance, iPhone Siri [8]. In comparison, virtual supports let us communicate in a natural spoken language with computers with keys and mouse, tablets and phones with touch screen. Audio interfaces simplify our relationship with technology significantly. Figure 3 shows the simple form of voice recognition.

This is the ultimate goal of deep learning recognition of speech, but we are not yet. At least as soon as I authored that—I bet we're going to be there a few years from

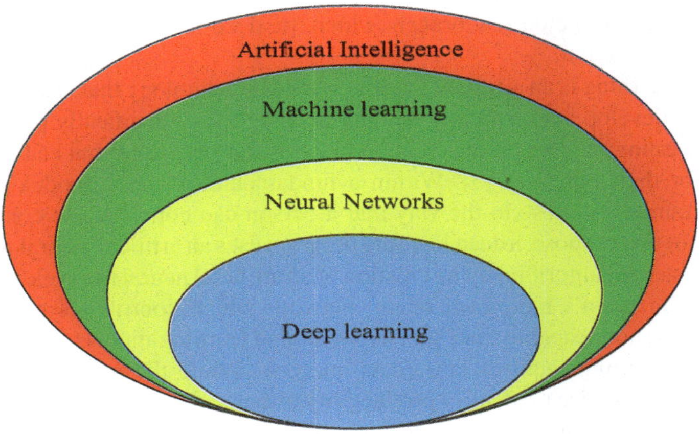

Fig. 2 Layered architecture of AI

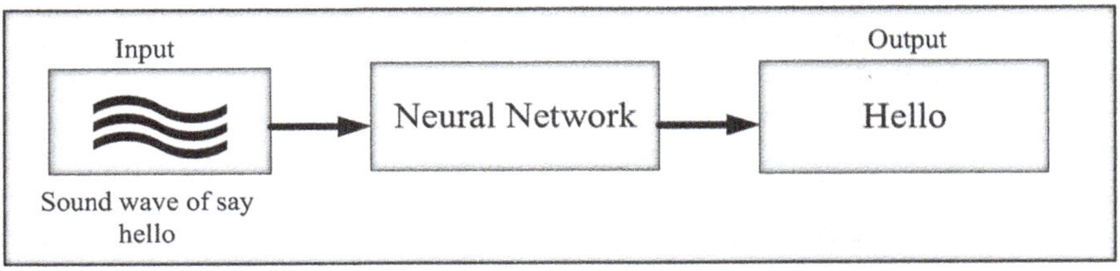

Fig. 3 Simple voice recognition

now. The major problem is that the speed of accent varies. You could say "hello!" So fast and somebody may say, "Heeeellllllllllllooooooo!". There's a lot longer sound file with much more data very slowly. The same text should be remembered as both sound files—"hi!" Aligning audio files with a fixed-length text piece in different lengths is very difficult. To get around this, in addition to the deep neural network we need to use some special technique and supplementary processing. The first step in the detection of speech is obvious—sound waves must be fed into a device [9]. However, the sound is passed on like waves. How are sound waves converted into numbers? It's just using this "Hello" sound clip of me. Sound waves are one-dynamic. We have such a single value at each moment in a time heavily reliant on wave height (Fig. 4).

A neural network is a group of nodes or units interconnected by adaptive weights, creating a weighted diagram that helps you to know distributed representation and finally the task involved [10]. Originally, neural networks were developed as a simplified model of biological neural brain network; however, it is now clear that neural networks and biological neural systems have no resemblance. Neural networks, however, tend to be one of the most important methods for learning [11]. An arbitrary precision of any practical Borel function is achievable by a neural network with at least one nonlinear layer of hidden elements comprised of an arbitrary number of

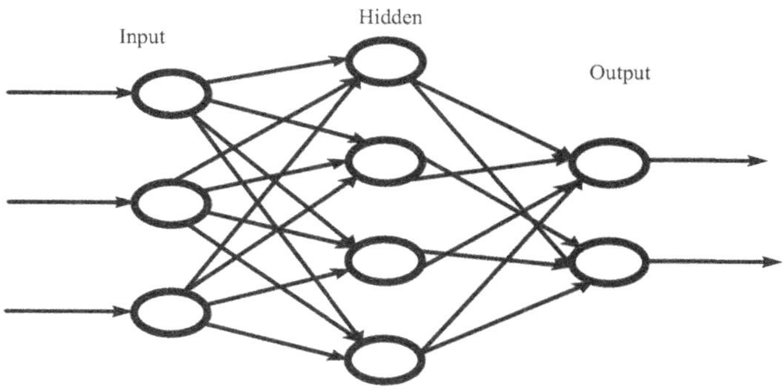

Fig. 4 Simple accent recognition model

units and linear output layer, it is considered as a more unit in a buried layer like as shown in Fig. 4.

4 Analysis

Virtual assistants are designed on a complex AI technology system to address requests like as shown in Fig. 5. In this article, the model for speech recognition systems imitates the model of human perception. The identification of acoustic processing, extraction and identification features by computer is based on three levels of processes. Samples, digitization, and filters will be made for the analog speech. By these steps, the signal is confidential. The signal is then set for the beginning and end

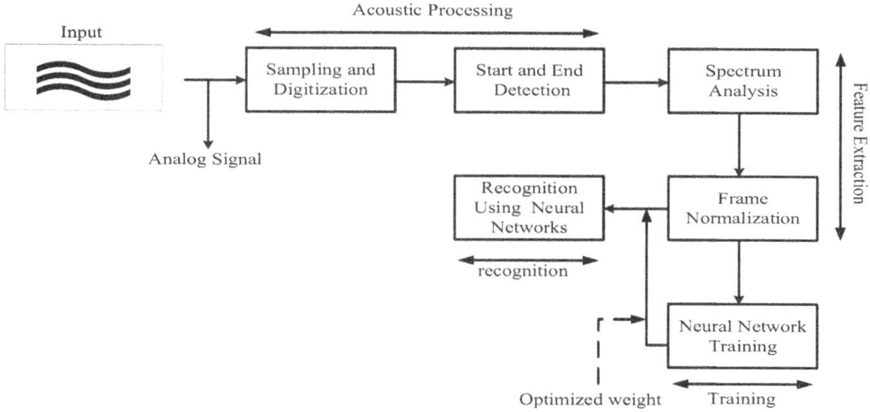

Fig. 5 Accent recognition using deep learning

points to only analyze the signal with information. The next move is to remove functionality in the beginning and end points from the signal. Linear Predictive Coding (LPC) for spectral analysis and voice representation was used in this experiment. The number of frames from the resulting features varies from word to word. Frame normalization must, therefore, be applied to equate to the number of input nodes used in this experiment in the neural network [12]. The testing and training data are handled in the same way. The training data will be incorporated into the neural network to improve the learning process by modifying the weights between network layers on an iterative basis. This is used to check the training patterns after having optimized weights. The understanding of human speech begins with the pulse of speech through the mouth. The speaker enters the basilar membrane located in the middle of the inner ear, and the signal waveform is analyzed and a spectrum signal produced [13]. The neural transducer, which translates the signal to a nerve operation in the ear nerve, must join the spectrum signal. The operation of the neural impulses is converted into language code, and the message is transmitted to the brain for insight. Figure 6 illustrates the process schematic [14] Virtual assistants draw on a deep AI technology pipeline to meet requests like as shown in Fig. 7.

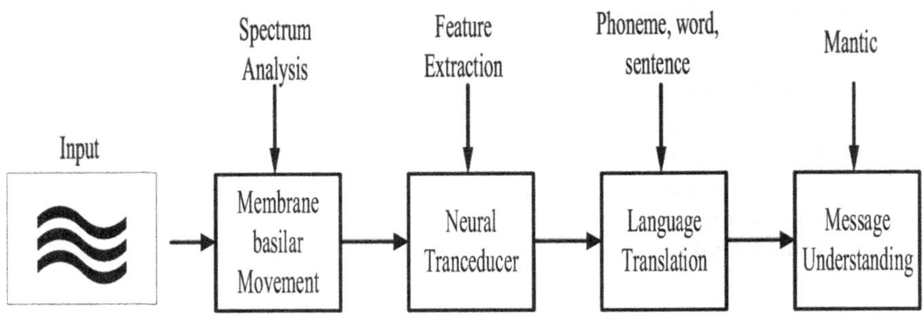

Fig. 6 Human perception method

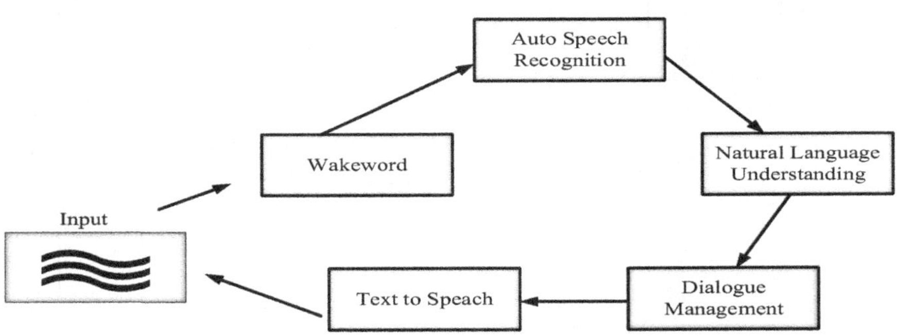

Fig. 7 Complex pipeline method for accent recognition

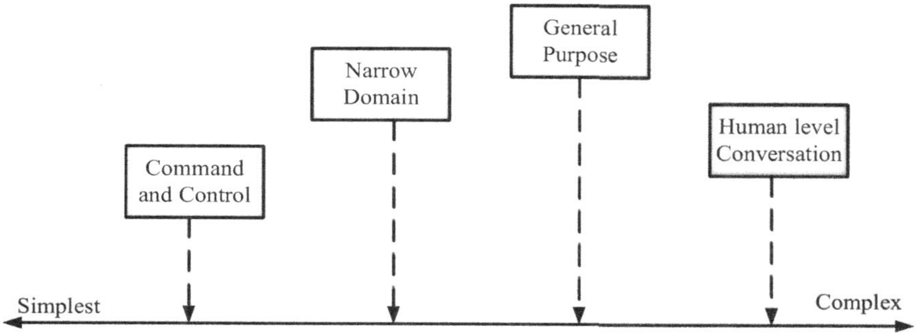

Fig. 8 Complexity for native accent recognition

The technology in this pipeline must deal with natural language's scale and ambiguity. Therefore, it is based on machine learning along with manually defined rules—a group of AI algorithms that learn their behavior from data rather than being explicitly programmed [15]. This enables helpers to learn how people speak and to make new speakers or requests widespread. There are several ways AI assistants can be used, e.g., via a smartphone application, a phone call, or on a specific device like a creative speaker. Digital support staff has proven useful in many ways, and new technologies are constantly being developed. A monitoring and control system is the easiest configuration. Here the user has just a few commands to control a computer, which are not very much self-organizing. For situations where hands-free operation increases performance, simple assistants are also used, e.g., giving operators more voice control on the production floor.

Usually, these assistants are employed in restricted areas such as finance or customer support and need a constant conversation with the user to carry out the task. Figure 8 shows the complexity of the recognition and implements the commands.

More general, Amazon's Alexa or Apple's Siri virtual helpers manage several different queries in many regions. You can play music, ask for the weather, test your smart home appliances, ask for jokes and much more. Their communications remain task-oriented, although they typically have some flimsy responses.

The accent recognition rates for various approaches such as the deep learning method, the artificial neural network, the convolutional neural network, and the neural structured learning methods are given in Fig. 9, and the deep learning method has its enhanced accent recognition rates compared to other approaches.

5 Conclusion

Academic research extends beyond task-based dialog into new forms of interaction. Fully conversational agents are not designed and used on a scale, but current work focuses on social modes of communication between human and computer systems.

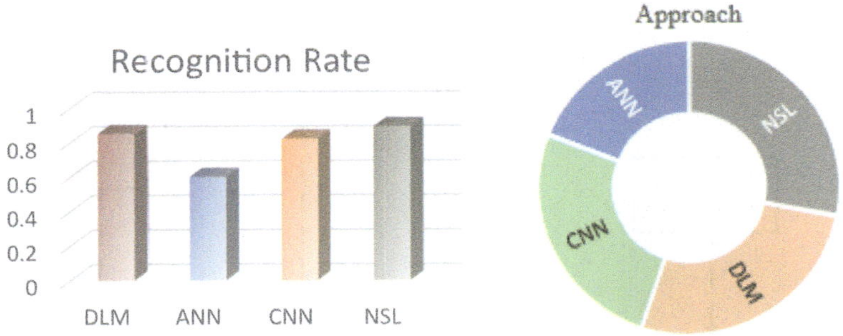

Fig. 9 Accent recognition rate using different approaches

Any of these outcomes are provided by competition such as the Alexa Prize—a university competition to create assistants who connect regularly and profoundly with people. AI assistants are still in their development despite their widespread adoption. Siri Apple debuted in 2011, and Amazon debuted Alexa in 2014 on the iPhone. The technology behind it is constantly improving. We expect AI assistants to become in the next few years. The future scope of this work is concentrated on a few categories such as customizable, contextual and conversational.

In as customizable, customized experiences are simpler for organizations to create. The first tools to make the voice assistants easy to configure are already being used. In contextual, context from various sources will be used in the assistants. Related contexts may emerge from real-world experience, customized user information or the current conversation history. In conversational, while people's communication rates are still a very long way to go, AI assistants will develop a more primitive capacity for communication in the coming years.

References

1. Ostrowski, D. (2018). Artificial intelligence with big data. In *First International Conference on Artificial Intelligence for Industries (AI4I)*. Laguna Hills, CA, USA.
2. Pleshkova, S., Bekyarski, A., & Zahariev, Z. (2019). Reduced database for voice commands recognition using cloud technologies, artificial intelligence and deep learning. In *Conference on Electrical Machines, Drives and Power Systems (ELMA)*. Varna, Bulgaria.
3. Dat, P. T (2019). Application of convolutional neural network for gender and age group recognition from speech. In *th NAFOSTED Conference on Information and Computer Science (NICS)*. Hanoi, Vietnam.
4. Qayyum, A. B. A., Arefeen, A., & Shahnaz, C. (2019). Convolutional neural network (CNN) based speech-emotion recognition. In *IEEE International Conference on Signal Processing, Information, Communication & Systems (SPICSCON)*. Dhaka, Bangladesh.
5. Guiming, D., Xia, W., Guangyan, W., Yan, Z., & Dan, L. (2016). Speech recognition based on convolutional neural networks. In *IEEE International Conference on Signal and Image Processing (ICSIP)*. https://doi.org/10.1109/SIPROCESS.2016.7888355.
6. Becerra, A., De La Rosa, J. I., & González, E. (2017). Speech recognition in a dialog system: from conventional to deep processing. In *Multimed Tools* (pp. 15875–15911). Springer.

7. Jothi, K. R., Mamatha, V. L., & Yawalkar, P. (2019). Speech intelligence using machine learning for aphasia individual. In *International Conference on Computational Intelligence and Knowledge Economy (ICCIKE)*. Dubai, United Arab Emirates.
8. Tzirakis, P., Zhang, J., & Schuller, B. W. (2018). End-to-end speech emotion recognition using deep neural networks. In *IEEE International Conference on Acoustics, Speech and Signal Processing (ICASSP)*. Calgary, AB.
9. Kamath, U., Liu, J., & Whitaker, J. (2019). *Deep learning for NLP and speech recognition*. Switzerland AG: Springer International Publishing.
10. SParthasarathy, S., & Tashev, I. (2018). Convolutional neural network techniques for speech emotion recognition. In *16th International Workshop on Acoustic Signal Enhancement (IWAENC)*. Tokyo.
11. Ahmad, A. S., & Sumari, A. D. W. (2017). Cognitive artificial intelligence: Brain-inspired intelligent computation in artificial intelligence. In *Computing Conference*. London.
12. Sen, S., Dutta, A., & Dey, N. (2019). *Audio processing and speech recognition*. Singapore: Springer Singapore.
13. Dehghani, A., & Seyyedsalehi, S. A. (2018). Performance evaluation of deep convolutional maxout neural network in speech recognition. In *25th National and 3rd International Iranian Conference on Biomedical Engineering (ICBME)*. Qom, Iran.
14. Park, S., Jeong, Y., Kim, M. S., & Kim, H. S. (2018). Linear prediction-based dereverberation with very deep convolutional neural networks for reverberant speech recognition. In *International Conference on Electronics, Information, and Communication (ICEIC)*. Honolulu, HI.
15. Tatulli, E., & Hueber, T. (2017). Feature extraction using multimodal convolutional neural networks for visual speech recognition. In *IEEE International Conference on Acoustics, Speech and Signal Processing (ICASSP)* New Orleans, LA.

An Approach on Data Visualization and Data Mining with Regression Analysis

G. Suseendran, Srinath Doss, Souvik Pal, Niti Dey, and Ton Quang Cuong

Abstract Right off the bat, the ideas and arrangement about imagined information mining innovation expounds in this paper, data mining at that point the innovation and regression analysis technology and the visualization data, what's more, are talked about; finally, the use of data mining addressed in combination with sliding bearing wear coefficient testing, regression analysis technology and data visualization. In this paper structure, a framework to actualize data mining technology for visualized data mining under the web condition. Data visualization helps to explore data with some satisfactory results in a powerful way. Also, as a part of the data reduction process by combining data, a significant role has been played by data visualization. The linear and nonlinear regression analyses are used in this method, and application research progresses to proving the factors in our methods.

Keywords Data mining · Data visualization mining technology · Linear and nonlinear regression analysis

---

G. Suseendran (✉)
Department of Information Technology, Vels Institute of Science, Technology & Advanced Studies, Chennai, India
e-mail: suseendar_1234@yahoo.co.in

S. Doss
Faculty of Computing, Botho University, Gaborone, Botswana
e-mail: srinath.doss@bothouniversity.ac.bw

S. Pal · N. Dey
Department of Computer Science and Engineering, Global Institute of Management and Technology, Krishnanagar, India
e-mail: souvikpal22@gmail.com

N. Dey
e-mail: dey.niti@gmail.com

T. Quang Cuong
VNU University of Education, National University Vietnam, Hanoi, Vietnam
e-mail: tonquangcuong@gmail.com

S.-L. Peng et al. (eds.), *Proceedings of First International Conference on Mathematical Modeling and Computational Science*, Advances in Intelligent Systems and Computing 1292, https://doi.org/10.1007/978-981-33-4389-4_59

1 Introduction

There has been a necessity for demonstrating massive proportions of data in a way that is adequately accessible and reasonable. Affiliations make data reliable. In this manner, the proportion of data open on the Web has extended substantially. It is difficult for customers to envision, research, and use this titanic data. The ability to information imagine is crucial to the consistent investigation. Today, PCs use to process a ton of data. Imagined information stresses over the structure, improvement, and usage of PC made the graphical depiction of the data. It gives a convincing data depiction of data starting from different sources. This enables the boss to see examination in visual structure and makes it straightforward for them to understand the data. It energizes them to discover plans, comprehends information, and structure an end.

Information mining is productively utilized in a considerable lot of areas, including showcasing, medication, fund, building and bioinformatics. There still are, in any case, various elements that militate against the great reception and utilization of this innovation. This is principal because of the way that the aftereffects of numerous information mining procedures are regularly hard to comprehend. For instance, the aftereffects of an information mining exertion creating 300 pages of rules would be hard to break down. The visual portrayal of the information implanted in such standards would assist with increasing the conceivability of the outcomes. The representation of the information itself, just as the information mining procedure, ought to go far towards expanding the client's comprehension of and confidence in the information mining process. That is, information and data perception give clients the capacity to get new bits of knowledge into the information, as found from enormous vaults.

Pictured information is seen as information portrayal or intelligent observation. Individuals have reliably used observations to make messages or information around the end in time. What can't be reached, smelled, or tasted can be addressed visual like apparently.

Perception information is the usage of PC supported, visual depiction of data. As opposed to static data on visualized information, natural data portrayal licenses customers to demonstrate the association used in indicating data. The typical portrayal of visualized information is shown in Fig. 1.

- Line diagram: The connection between things is shown here. Changes over a timeframe are thought by utilizing it.
- Bar diagram: Various classification amounts are calculated using it.
- Scatter plot: This is a two-dimensional plot indicating a variety of two things.
- Pie outline: This utilizes to look at the pieces of an entirety.

With the constant development of the size of present-day creation tasks, data trade progressively visits, enormous databases are developing. The ventures, for example, preparing assembling, discount and retail, exchange account, and data innovation

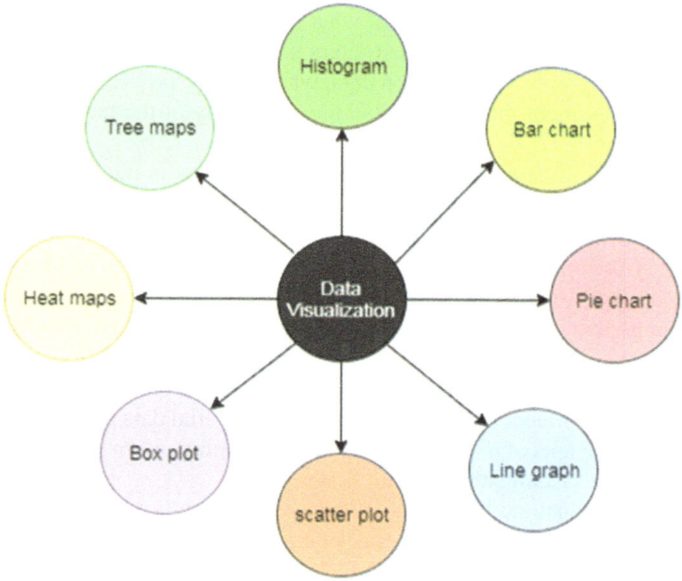

Fig. 1 Types of data visualization

confronts with the issues which are data blast brought about by as well much information, or helpless exactness and comprehensiveness of the choice results they are brought about by generally little information. So mining data innovation rises as the occasions require. Also, database question results, for the most part, appear as the report structure, it can't outwardly show information connections, information pattern and correlation, and the connection with clients isn't straightforwardly, which presented the information representation in applications in data mining.

Visualized data innovation implies speculation strategies. Furthermore, procedures that apply PC designs and picture preparing procedure change information to designs or pictures furthermore show them on screens and interface with clients.

2 Related Works

Qi et al. [1] introduced visualization methods in the field of big data analysis. Their paper represents some media-type visualization methods and along with media data developments. Data visualization improves individuals' familiarity with conceptual data, however supporting by PC intelligently and the visual portrayal of conceptual information. As an issue of first significance, we present essential principal substance and techniques for text portrayal; additionally, we present little scope blog data portrayal limits and applications under the upgrade of text portrayal improvement, and discussion about its characteristics what's increasing, social centrality; finally, according to the little scope blog customers sending hotspots well-known notion, we present the geographic information data, which can be progressively through and through data assessment moreover, show the possibility of the law of the things by recognition.

Qiang et al. [2] illustrate spatial data mining concepts in the visualization technology applications. There are two kinds of technical method; one is spatial data mining and two is spatial data visualization in recent years; it is so prevalent in comparatively technical methods. Both reasons for existing are to find geography wonders what spatial data express and find diverse data and laws irrefutable in geology component. So it is essential to join both regularly and structure another examination bearing—Visualization Spatial Data Mining (VSDM). This paper discusses the critical associations of portrayal and spatial data mining, the essential application of discernment theories and advances in spatial data mining, the standard strategies and occasions of recognition spatial data mining, we in like manner present a reference model Visualization Spatial Information.

Chen et al. [3] illustrate that visualized data representation is a compelling technique for individuals to complete further perception and investigation with regards to colossal information. Understanding the lack of communication with static information, the paper led research dependent on D3 information representation innovation. It proposed a D3 information-driven representation model and applied the model in the school gear to support the executive's framework with examination diagrams of support bunch remaining task at hand. The paper talked about the whole systems of dynamic and intuitive information perception from a few points: information preparing, application, and information perception. The outcomes bolstered the clients' need for factual graphs and affirmed the upside of the D3 procedure in unique furthermore, intelligent information perception.

Ali et al. [4] represent tools and challenges in big data visualization methods. In our daily life, where we see everything digitally and also recorded in digital like patterns in web surfs in our medical records. In every circle of daily life, there is a big data formed by just analyzing the process in the big data and resulted as digital data cannot be enough we have to make the resulted data in a visualized manner to get a better view so our human brains would get the patterns in a secure, effective manner. Visualization data, what's more, Analytics expect noteworthy employment in decision making in various divisions. It is like manner prompts new possibilities in the discernment space addressing the creative ideation for handling the huge data issue by methods for visual techniques. It is with everything taken into account a test to imagine such a mammoth proportion of data in certifiable time or a static structure. In this paper, we inspect why colossal data recognition is of most extraordinary essentialness, what are the challenges related to it and review some immense data portrayal gadgets.

Li et al. [5] illustrate about to meet the prerequisites of the high-dimensional information preparing in the data field; this paper endeavours to investigate the strategies and procedures of bunching information perception for general information assets.

Through dimensionality decrease and visual mapping of high-dimensional information; this paper sets up a perception framework model that would envision the affecting elements. The visual framework model methodology tried by utilizing the IRIS dataset from the UCL database means the University of California, Irvine. The results exhibit that this model can adequately break down the informational collection, envision the highlights of IRIS information progressively and accomplish the typical outcomes.

3 Proposed Method

Visualized data innovation implies speculation strategies; furthermore, procedures that apply PC designs and picture preparing procedure change information to designs or pictures furthermore show them on screens and interface with clients.

3.1 *Visualization of Data and Data Mining*

Data mining uses a broad range of techniques in large datasets to make predictions and to find patterns anomalies and correlation, for increasing revenue, large organizations use this extracted information also for increasing customer relationship and reducing risk [6]. Graphically representing data has been represented as data visualization. As a visualization tool, they use visual elements such as graphs, chart maps, etc. A massive amount of information are analyzed by this method, and decisions made on top of it [7] (Table 1).

A. *Linear Analysis*

The study of multivariate linear regression is a key analytical methodology. The function of this linear analysis is

$$Z = \alpha_0 + \alpha_1 v_1 + C + D_r \alpha_r \tag{1}$$

It is 1 + r hyper-dimensional plane.

The hyperplane calculated by least square method, $\alpha_e (e = 0, 1, C, r)$.

The empirical regression function is

$$\hat{Z} = \hat{\alpha}_0 + \hat{\alpha}_1 v_1 + C + \hat{D}_r \alpha_r \tag{2}$$

B. *Nonlinear Analysis*

For the most part, the nonlinear models used in anticipation are divided into three classifications: a unitary capacity bend model, a polynomial relapse model and a multivariate capacity bend model [8].

Table 1 Difference between data mining and data visualization

Basis for comparison	Data mining	Data visualization
Definition	Searches and generates an acceptable result from the large chunks of data	Gives a quick look at complex data
Preference	This has various uses, and web-search engines favoured	Data modelling and forecasts favoured
Area	Covered by data science	Covered by data science area
Platform	Web software is the major platform used	Complex data analyses support for better work
Generality	A recent underdeveloped technology	More efficient when predicting real-time data
Algorithm	There are several algorithms to the use of data mining	Algorithms are not required
Integration	Several algorithms exist for using data mining	It gives visual details irrespective of the hardware or software

As a rule, the relationship between considered factors is nonlinear in a viable application and should express by multivariate curve

$$Z = Fu(v_1, v_2, C, v_k) \tag{3}$$

Between them, the particular form of f driven by the genuine component relationship in the specific questions, the typically used models: hyperbolic model, logarithmic twist model, exponential twist model, list twist model as well, trigonometric limit model and any mix between them [9]. The treatment of various backslide models combined is logically disconcerted, so the direct twist model is in use, and its nonlinear model is as follows:

A multivariate model for hyperbolic relapse:

$$Z = g_0 + \frac{g_1}{v_1} + \frac{g_2}{v_2} + C \tag{4}$$

multivariate log bend relapse model:

$$Z = g_0 + g_1 \ln v_1 + g_2 \ln v_2 + C \tag{5}$$

multivariate power work bend relapse model:

$$Z = g_0 v_1^{g1} v_2^{g2} C \tag{6}$$

multivariate exponential capacity relapse model:

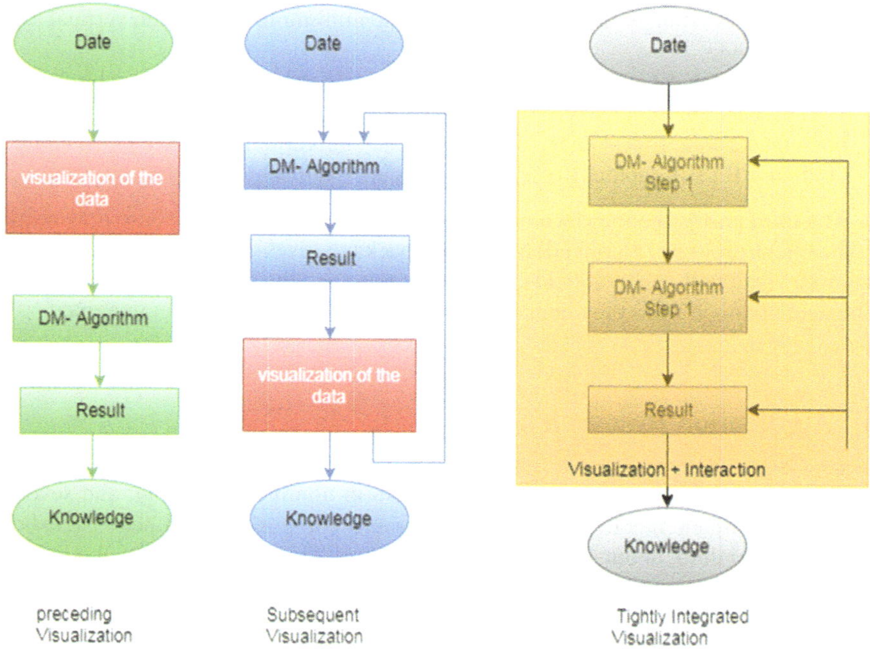

Fig. 2 Overview of different approaches to data visualization

$$Z = g_0 g_1^{v1} g_2^{v2} C \tag{7}$$

The above models can't utilize legitimately; it should be changed into various direct structures by numerous straight relapse examination technique; at last, its reverse change has done likewise.

Figure 2 visualizes different approaches, and the following roles and pieces include:

The initial section is the UI: The web page connects to JSP and ASP is the UI, can bind to clients and obtain details [10].

Notwithstanding giving the contribution of the discrete information source, the customer bundle download administration, site page can likewise show the diagram show brings about the client program.

The subsequent part is information show and investigation: information investigated by utilizing data mining and information perception, relapse investigation, and the outcomes can appear in Fig. 1. Java Applet, which can complete information figuring, diagram age is inserted into website pages in user dynamic interaction to realize the function.

The column chart example:

void setUpperRange (double d){…}// Set the maximum value of the vertical axis display range.

void setLowerRange (double d){…}// Set the minimum value of the vertical axis display range.
void setBarLabelsOn (boolean flag){…}// If the coordinate axis should be shown.
void setBarLabelsOn (String as[]){…}// Set the Coordinate Axis value.
boolean setValueLinesOn(){…}// If to display numeric line.
void paintBars (Graphic, Rectangle rectangle) {…}// Draw a column chart.

The third part is the interface of the database: The customer database may relate to the client program. Application software package gives decisions on the form of database, data table as such fields; customers must see the information shown [11]. The data obtained from the client system is submitted to a web page, imagined.

4 Application Research

Sliding wear of bearings is a common problem in nonlinear planning, and several factors are influencing it. Subsequently, a multi-component nonlinear backslide model built to think about sliding wear on bearings. This paper explores the impact of the material and weight of bearings on wear.

Table 2 displays test data of specific bearing/Lubricant compound wear coefficient at the standard temperature setting, tolerating mechanical oil as oil.

The information in Table 2 utilizes as the perception information source, three-run of the mill outline types is given below:

Figures 3, 4 and 5 shows the results in line, bar and pie chart. As demonstrated by the aftereffects of a few nonlinear relapse models, it is best to mirror the relationship between the load and wear coefficient. Table 3 indicates the aftereffects of the appropriate curve for strength function.

Table 2 Wear coefficient test data

Lubricants/ Bearing	Load				
	30	40	50	60	70
PAO4	13.9	12.8	11.2	10.1	9.8
Surfactant (SMO)	13.5	12.3	10.8	9.4	8.5
Carbon black	11.9	10.8	9.7	8.3	6.9
ZChPbSb	9.15	7.34	6.10	5.28	4.72
ZChSnSb	9.22	8.58	8.05	7.73	7.52
MCMB	7.4	6.3	5.1	3.9	3.2
CS-700	6.7	5.2	4.3	2.8	2.1
CS-2400	5.5	4.2	3.3	2.8	1.9
CS-700 + SMO	4.9	3.6	2.4	1.8	1.3
ZQSn	1.71	1.59	1.55	1.35	1.26

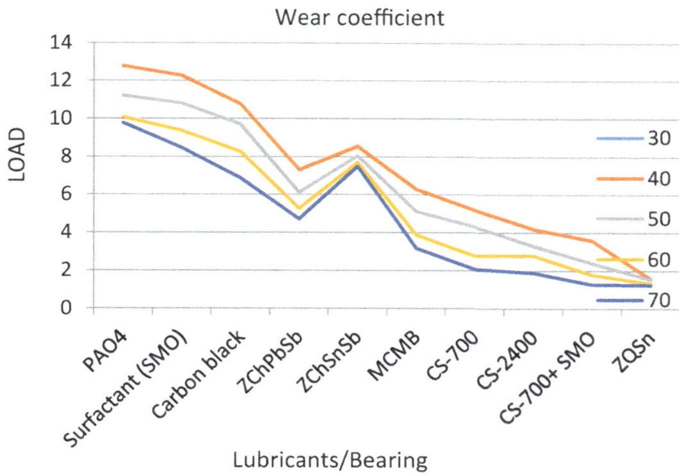

Fig. 3 Line graph

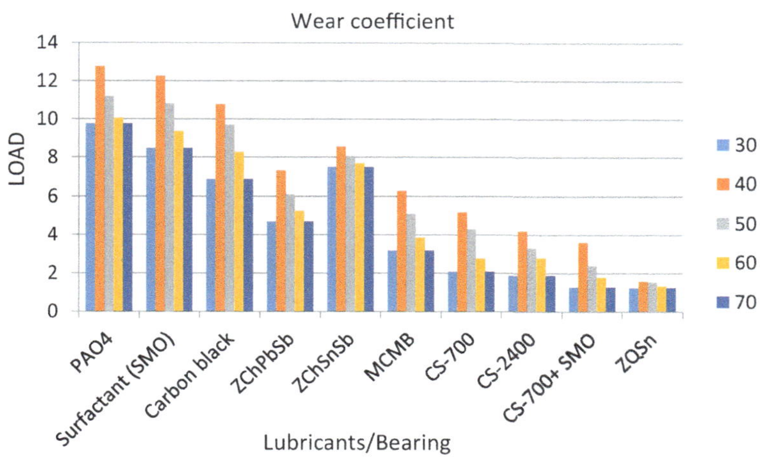

Fig. 4 Bar chart

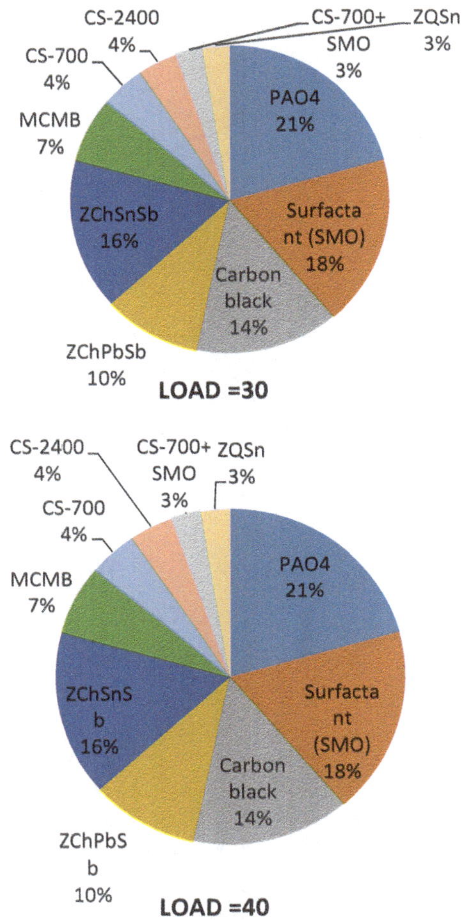

Fig. 5 Pie chart

Fig. 5 (continued)

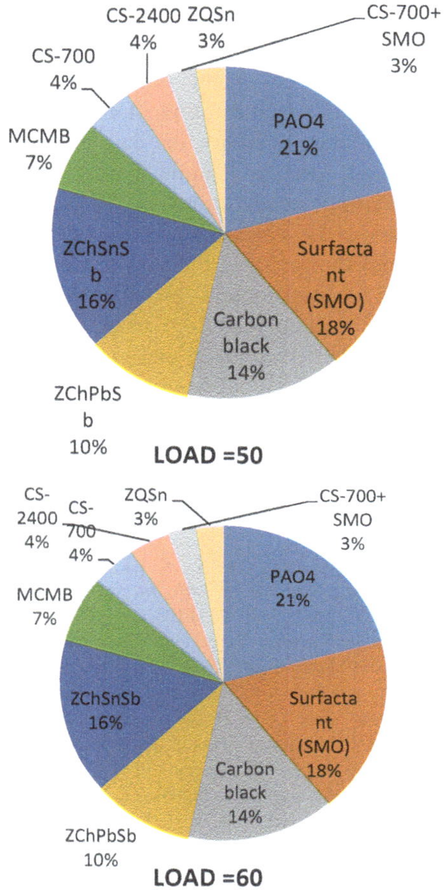

Figure 5. pie chart

Table 3 Power function results in match curve

Regression model	Model	Mean value of absolute values of relative residuals	Mean variance of relative residuals
Power function curve	$K_1 = 47.9314\ F^{-0.8055}\ V^{-0.3875}$	0.0236	0.0100
	$K_2 = 3.0314\ F^{-0.205}\ V^{-0.6750}$	0.0412	0.0458
	$K_1 = 0.00028766\ F^{-0.445}\ V^{-2.4175}$	0.3724	2.2148

5 Conclusion

It has been essential and complicated in the exploration of large datasets. In solving the problem, visualization techniques would be beneficial. This method has applied to various applications like data mining and fraud detection; also, its exploration has a high potential. Unmistakably, the information perception mining development is incredibly steady to improve the accessibility and precision of data. Web service operation, the system can be exchanged by customers; the mechanism running is in satisfactory shape; the structure's transparency and improved awareness would satisfy most customers' prerequisites. The ultimate aim is to introduce the power of visualization technology to every device so that extensive data resources can be explored better, quicker, and more intuitively. In an economic sense, this will not only be beneficial but would motivate and delight the consumer.

References

1. Qi, Y., Shi, G., Yu, X., & Li, Y. (2015, June). Visualization in media big data analysis. In *2015 IEEE/ACIS 14th International Conference on Computer and Information Science (ICIS)* (pp. 571–574). IEEE.
2. Qiang, X., Wei, Y., & Hanfei, Z. (2010, June). Application of visualization technology in spatial data mining. In *2010 International Conference on Computing, Control and Industrial Engineering* (Vol. 2, pp. 153–157). IEEE.
3. Chen, L., & Zhou, H. (2016, July). Research and application of dynamic and interactive data visualization based on D3. In *2016 International Conference on Audio, Language and Image Processing (ICALIP)* (pp. 150–155). IEEE.
4. Ali, S. M., Gupta, N., Nayak, G. K., & Lenka, R. K. (2016, December). Big data visualization: Tools and challenges. In *2016 2nd International Conference on Contemporary Computing and Informatics (IC3I)* (pp. 656–660). IEEE.
5. Li, Y., & Hou, S. (2017, December). Methods and Techniques in Data Visualization Model. In *2017 International Conference on Computer Technology, Electronics and Communication (ICCTEC)* (pp. 71–74). IEEE.
6. Ebert, D. S., Favre, J. M., & Peikert, R. (2002). Data visualization. *Computers & Graphics, 2*(26), 207–208.
7. Qin, X., Luo, Y., Tang, N., & Li, G. (2018). Deepeye: An automatic big data visualization framework. *Big Data Mining and Analytics, 1*(1), 75–82.
8. Hooi-Ten Wong, D., & Manickam, S. (2010, August). Intelligent expertise classification approach: An innovative artificial intelligence approach to accelerate network data visualization. In *2010 3rd International Conference on Advanced Computer Theory and Engineering (ICACTE)* (Vol. 6, pp. V6–437). IEEE.
9. Yan, H., Wang, J., & Xia, C. (2017, June). Research and Application of the Test Data Visualization. In *2017 IEEE Second International Conference on Data Science in Cyberspace (DSC)* (pp. 661–665). IEEE.
10. Rubel, O., Weber, G. H., Huang, M. Y., Bethel, E.W., Biggin, M.D., Fowlkes, C.C., … Malik, J. (2008). Integrating data clustering and visualization for the analysis of 3d gene expression data. *IEEE/ACM Transactions on Computational Biology and Bioinformatics, 7*(1), 64–79.
11. Roshdy, A., Sharaf, N., Saad, M., & Abdennadher, S. (2018, July). Generic Data Visualization Platform. In *2018 22nd International Conference Information Visualisation (IV)* (pp. 56–57). IEEE.

Indoor Wearable Navigation System Using 2D SLAM Based on RGB-D Camera for Visually Impaired People

Heba Hakim and Ali Fadhil

Abstract To guide the visually impaired and blind person to his/her final goal safely and efficiently in unknown indoor environment, a wearable navigation system is presented in this paper. The essential components of the proposed navigation system are mapping, path planning, path following and obstacle avoiding. The system uses a wearable RGB-D camera mounted on a head to construct a 2D map for the surrounding environment which is implemented by a well-known simultaneously and location mapping (SLAM) in mobile robotics. Once a 2D map is creating, the optimal path is generated by path planning method to guide the visually impaired person from start location to his/her destination. Also, an ultrasonic sensor is used in order to detect the temporary obstacles during the following of the optimal path. All data collection and processing is done on Raspberry Pi 3 B+ board due to its low cost, small size and lower power consumption. The visually impaired person receives the continuous guiding commands as voice commands through earphone. The performance of the proposed system has been successfully evaluated in various real-time scenarios.

Keywords Obstacle avoidance · Visually impaired people · Indoor navigation system · Path following

1 Introduction

There are about 258 million visually impaired people according to the statement of the World Health Organization (WHO) in their last report. The visually impaired people (VI) face many difficulties in perceiving and navigating independently in an unknown environment. This represents a main problem in their daily life. Many

H. Hakim (✉)
Department of Computers Engineering, University of Basrah, Basrah, Iraq
e-mail: hebah.hakem@gmail.com

A. Fadhil
Department of Electrical Engineering, University of Basrah, Basrah, Iraq
e-mail: alifm60@gmail.com

S.-L. Peng et al. (eds.), *Proceedings of First International Conference on Mathematical Modeling and Computational Science*, Advances in Intelligent Systems and Computing 1292, https://doi.org/10.1007/978-981-33-4389-4_60

navigation systems are available in the market in the recent years to assist the visually impaired and blind people in their daily life activities such as Microsoft glasses and Horus wearable device. Since these assistive devices are expensive and 90% of VI people with low income, it will be difficult to the most visually impaired people to use these technologies. The vast majority of visually impaired people rely on the traditional white cane in their navigation. However, white cane can't specify the destination of the visually impaired, nor how to reach there. The guide dog is a good choice for assisting the VI person in navigation but it required costly training, which the VI person can't affordable. Recently, many researchers developed and implemented the navigation systems for VI people with different types of technology and different degrees of performance. Most of these systems are only intended for detecting and avoiding the obstacles existence in the way of visually impaired person without providing the path planning and path following functions [1–3]. While other researchers focused on implementing a path finding and path following functions with ignoring the obstacle detection and avoidance [4, 5]. As the navigation system based on GPS technology [6] can't be used due to the performance limitation of GPS sensor in the indoor environment, Simultaneous localization and mapping (SLAM) is adopted in [7–11]. SLAM is an approach of the mobile robot that can be utilized to develop a system by building a map of unknown environment according to the acquired information from the sensor and simultaneously computes the user position within a map. Many SLAM methods have been implemented with different sensors, including laser scanners, sonar sensors and cameras. These SLAM techniques use an expensive sensor or a sensor with limited capabilities. Nowadays, most researchers adopt an RGB-D camera for SLAM research as a good sensor with low cost that provides both RGB image and depth information to utilize it for 3D mapping. 3D mapping requires a device with the higher performance such as an extensive memory amount and higher processing frequency. On other hand, two-dimensional laser scanner used in 2D SLAM techniques is simple, fast and not required a lot amount of memory. Thus, 2D SLAM methods are widely used and preferred in creating of indoor mobile robot map. Table 1 shows the specification comparison between Asus Xtion Pro depth sensor and 2D laser scanners. As shown in this table, the laser scanner is a better than depth camera in range terms. However,

Table 1 Specification comparison between Asus Xtion Pro depth sensor and 2D laser scanners

Component	Asus Xtion Pro	2D laser scanner
Distance range	0.8–3.5 m	~0–250 m
Horizontal viewing angle	58°	180°–360°
Vertical viewing angle	45°	–
# Measurement points	640 .480 = 307,200	Up to 6000
Cost	~150 USD	1000–15,000 USD

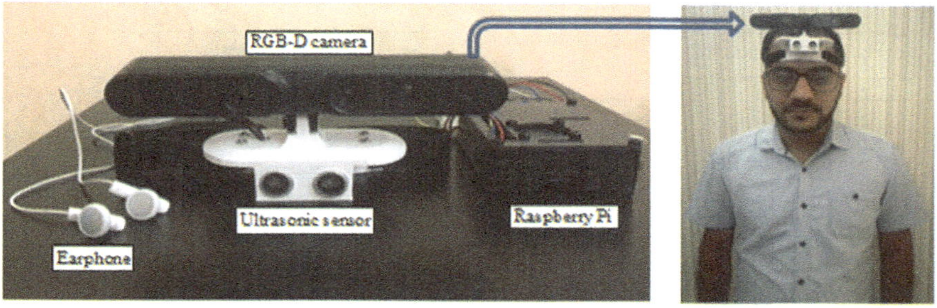

Fig. 1 Design of the proposed system

RGB-D camera has advantages: (1) It is lower cost (2) and it has an extra dimensionality (3D) that makes it more effective in performing SLAM methods in a real environment with obstacles of variable size and shape. Otherwise in 2D laser scanner, we can't see the obstacle that locates below or above its visible range. Therefore, 2D SLAM-based RGB-D sensor is adopted in our proposed navigation system.

2 The Proposed System

The wearable proposed navigation system consists of Raspberry Pi B+ , RGB-D camera (Asus Xtion Pro live), an ultrasonic sensor, power source and earphone. The sensors in this system are of low cost and small size. The RGB-D camera is used for building a map and locating the visually impaired person on a map. Besides, the RGB-D camera is also used to detect obstacle [12]. To overcome the limitation of RGB-D camera in detecting the transparent objects, the ultrasonic sensor is used. The integration of the depth camera and ultrasonic sensor for object detection can ensure a safety of VI person. All the algorithms such as Hector SLAM, object detection, path planning, path following and sound synthesis are performed on Raspberry Pi. The earphone is used for guiding the VI person with sound commands. The prototype of the proposed navigation system is described in Fig. 1.

3 Implementation of Navigation System

The framework of the proposed navigation system in indoor environment is illustrated in Fig. 2 which consists of four phases: (1) object detection, (2) hector SLAM, (3) path planning and (4) path following. The guiding information has been sent to VI person as audio commands through the earphone. The object detection algorithm has been illustrated in the previous work [12]. The hector SLAM method for 2D SLAM [13] is used to create an environment map and localize the visually impaired

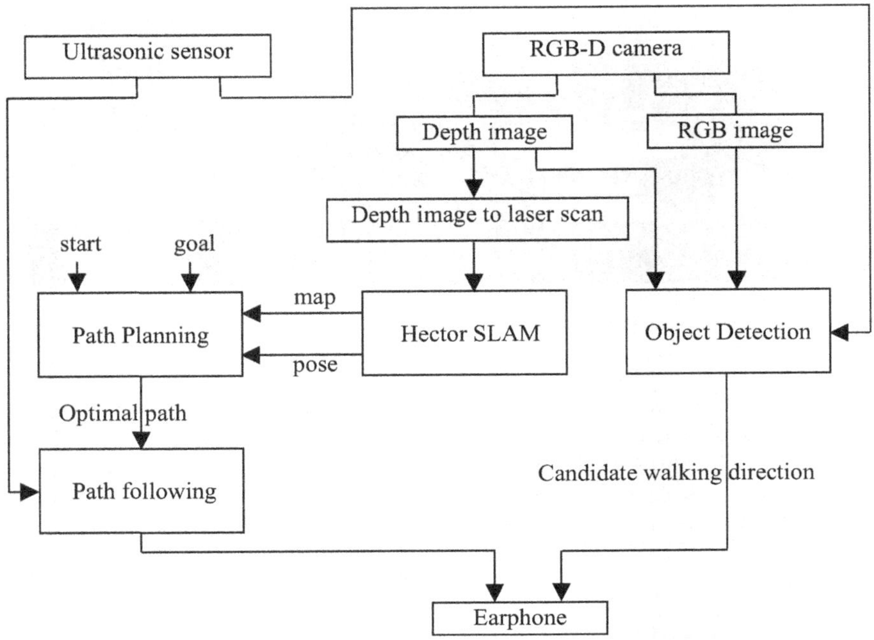

Fig. 2 Framework of the proposed

person within this map. When the VI person is moving, the sensor data is used to update the map. In order to use RGB-D camera as a sensor for the 2D Hector SLAM, the 3D depth information is converted into 2D laser data [14]. The Hector SLAM algorithm is performed using robot operating system (ROS), a robotics middleware which provided many libraries and tools for developing the robot system. Since ROS is a software framework for controlling robot, it does not have any function to guide visually impaired person to the destination. Therefore, a function is implemented to export the environment map and the position provided by hector SLAM in ROS into Python programming language to perform the path planning and path follower to guide the VI people. The path planning module based on A^* algorithm [15] is adopted to generate a global shortest path from the current position of visually impaired person to his/her destination. The global shortest path with current position (pose) of the VI person and object distance calculated by ultrasonic sensor is taken as inputs to the path following module to produce the guiding instruction. This makes the VI person to follow the path and avoid the obstacles in front of him/her at the same time. The following subsection described the algorithms which are used in the proposed navigation system.

3.1 Hector SLAM

Hector SLAM is one of the most popular 2D SLAM algorithm designed for the laser sensor, and it is used in many mobile robot projects. By comparing the Hector SLAM with other most 2D SLAM algorithms [13], it does not require the odometer information or inertial measurement unit (IMU) sensor to work well and localization. So, the 2D pose estimation is based on the robust scan matching process in Hector SLAM only. The scan matching process is based on the Gaussian–Newton method, which optimally aligns the laser scan's endpoints with the map by finding the rigid transformation. The rigid transformation $\mathfrak{z} = (p_x, p_y, \psi)^T$ that is expressed in Eq. (1):

$$\mathfrak{z}^* = argmin_{\mathfrak{z}} \sum_{i=1}^{n} [1 - M(S_i(\mathfrak{z}))]^2 \quad (1)$$

In this case, $M(S_i(\mathfrak{z}))$ is a function returns the map value at $S_i(\mathfrak{z})$ which is the coordinates of laser scan end points. Hector SLAM is also based on 3D navigation approach by extended Kalman filter when an IMU is used. In this work, IMU will not be used. A map of the surrounding environment is creating by Hector SLAM in the form of 2D occupancy grid with a map resolution 0.05 m/pixel. Each cell of 2D map has a probability of being occupied (white color for a free cell, black color for an occupied cell, while dark grey color for cell that was not scanned).

As mentioned before, RGB-D camera was utilized as a replacement for the laser sensor to get a 2D laser data in order to create a 2D map by Hector SLAM. After that, all cells in a 2D map must be classified into two categories: obstacle (occupied/undefined cell) and free. The undefined cell is remapped to the same value of the occupied cell. This is done by converting a map to a binary image using Otsu's thresholding algorithm [16] as shown in Fig. 3. Before applying a path planning algorithm, the map is reduced to match the distance between the two pixels that will represent one walking step which leads to reduce the computation time for path planning algorithm. Furthermore, the path planning algorithm does not take into account the VI person's width during computing an obstacle free path on a map. Hence, to ensure that the visually impaired person avoids collision with any obstacle, the proposed algorithm increases the size of obstacles before providing a map to $A*$ path planning algorithm. This is done by sliding the kernel (matrix of ones 3×3) through the map such as in 2D convolution. Since the obstacle with value 1 after applying Otsu method, the process of scanning kernel over the map makes a pixel value is 1 if there is one pixel at least with value 1 under the kernel.

3.2 Path Planning

The next part of an indoor navigation system is finding the optimal path to the target point on the environment map. There are many approaches of path planning. The

(a)

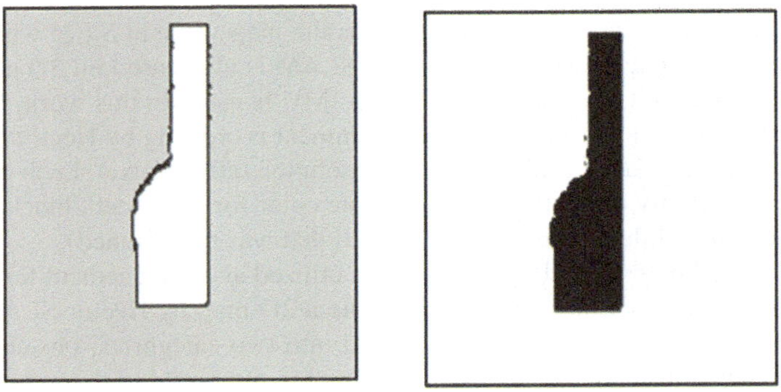

(b) (c)

Fig. 3 Process on a real-time scenario: **a** image of a real-time scenario, **b** hector map and **c** Otsu thresholding image

main role of the optimal path planning is finding a shortest path with free collision from starting point to the goal point. Two popular path planning algorithms for computation the shortest path named: Dijkstra and A^* algorithm.

These algorithms have been successfully used by various researches on path planning problem to give the optimal path. Since the visually impaired people prefer to walk aside the walls in straight lines, the path planning based on A^* algorithm was adopted in this model.

A^* finding algorithm is the Dijkstra's algorithm extension, which uses best first search. A^* algorithm runs faster than Dijkstra algorithm because it tries to search for the shortest path by using the heuristic search which gives priority to nodes that are supposed to be better than others while all possible path are explored in Dijkstra algorithm. This is done by selecting the path with minimum $f(\nu)$:

$$f(\nu) = g(\nu) + h(\nu) \tag{2}$$

where $g(\nu)$ is the total cost of path from the start node to node (ν) and $h(\nu)$ is the heuristic estimated distance between node (ν) to the goal. In this work, the heuristic is used for four-way movement (up, down, right and left) without taking the diagonal movement into consideration, and Manhattan distance is used to calculate $h(\nu)$:

$$h(\nu) = \text{abs}(x_1 - x_2) + \text{abs}(y_1 - y_2) \tag{3}$$

where x_1, y_1 are the coordinates of the current node and x_2, y_2 are the coordinates of the goal node.

This stage as described in Fig. 2 is taken the current position, goal (by using recognition with deep learning algorithms) and Hector map as inputs to provide the final optimal path.

3.3 Path Following

The path following module can be illustrated as:

Inputs:

(1) The optimal path that is provided by path finding module.
(2) The current position of visually impaired person is produced by the Hector SLAM.
(3) The distance of obstacle is measured by an ultrasonic sensor.

Outputs:

The guiding commands that guide the VI person to follow the optimal path and avoid the transit obstacle in his/her way at the same time.

The procedure of proposed path following approach is as follows:

The optimal path (P): represented as $P = \{N_1(x_1,y_1), N_2(x_2,y_2), N_3(x_3,y_3), \ldots, N_n(x_n,y_n)\}$ is a sequence of nodes (positions) which is used to guide the VI person individually from his/her position to the destination.

Navigation node $N_n(x_n,y_n)$: each node N in the optimal path has its own coordinates (x_n, y_n). This node is considered as the sub-goal by a VI person to reach and will guide a VI person towards his/her next sub-goal or final goal.

Distance (d): is the distance between two nodes in the optimal path which represented one step in this work as mentioned in previous subsection.

The path following algorithm of VI person depends on the segment of three successive nodes in the optimal path at each time to determine the direction of his/her next step. This segment has a start point N_n and an end point N_{n+2}. The x and y coordinates of three successive nodes are checked to know the kind of the segment (straight segment or turning segment). When all three successive nodes have different x and y coordinates, this segment can be considered as a turning segment (i.e., $N_1(x_n,y_n)$, $N_2(x_n,y_{n+1})$, $N_3(x_{n+1},y_{n+1})$ will be "turn right"). While the segment represents a straight line if all three successive nodes have share the same

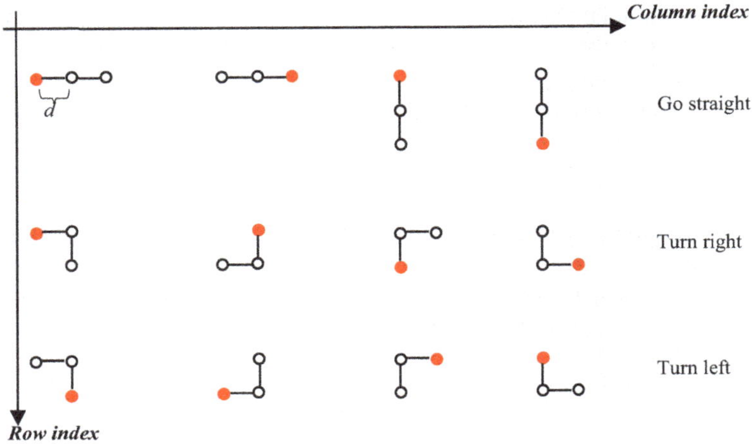

Fig. 4 Model of walking pattern

x or *y* coordinates (i.e., $N_1(x_n,y_n)$, $N_2(x_n,y_{n+1})$, $N_3(x_n,y_{n+2})$ will be "go straight"). The walking pattern model of VI person is shown in Fig. (4), and a red node in this figure represents the current position of VI person. This walking pattern model is continuously executed for all nodes in the path until the VI person reaches his/her final destination.

At each step during movement, the ultrasonic sensor will be used to detect the temporary obstacle in front of VI person. The detection range of ultrasonic sensor is 2–400 cm and the scanning field of 15°. A guiding command at each step will be converted to audio command to guide the visually impaired person till reach to his/her destination. The flowchart of the overall proposed indoor navigation system is shown in Fig. (5).

4 Demonstration

To test the effectiveness of the proposed navigation system to guide VI person to his/her final destination, many experimental environments in real time without/with obstacles along the path are conducted. The boxes and chairs in the laboratory were used as obstacles with different start positions and destinations. These experiments have shown that the proposed navigation system is suitable for guiding a human in a building. The information of movement for a VI person is to be incorporated into path following algorithm. The step length of VI person was assumed approximately 35 cm. Figure 6 shows the path following in the real-time scenarios to reach goal point.

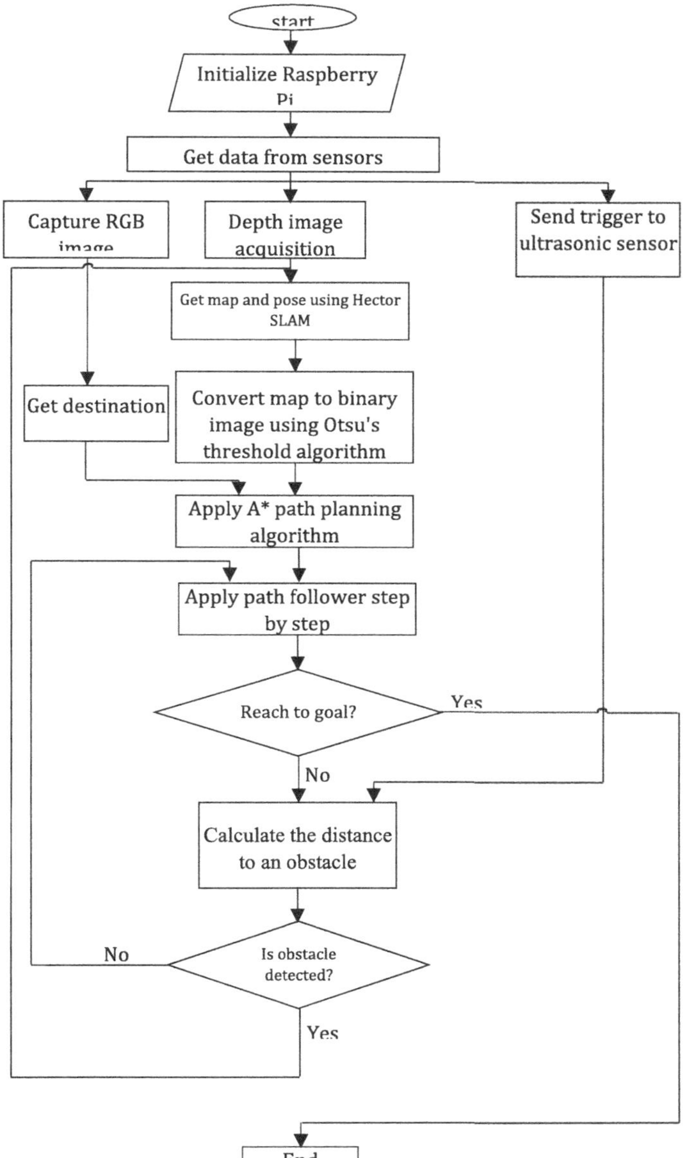

Fig. 5 Flowchart of the proposed system

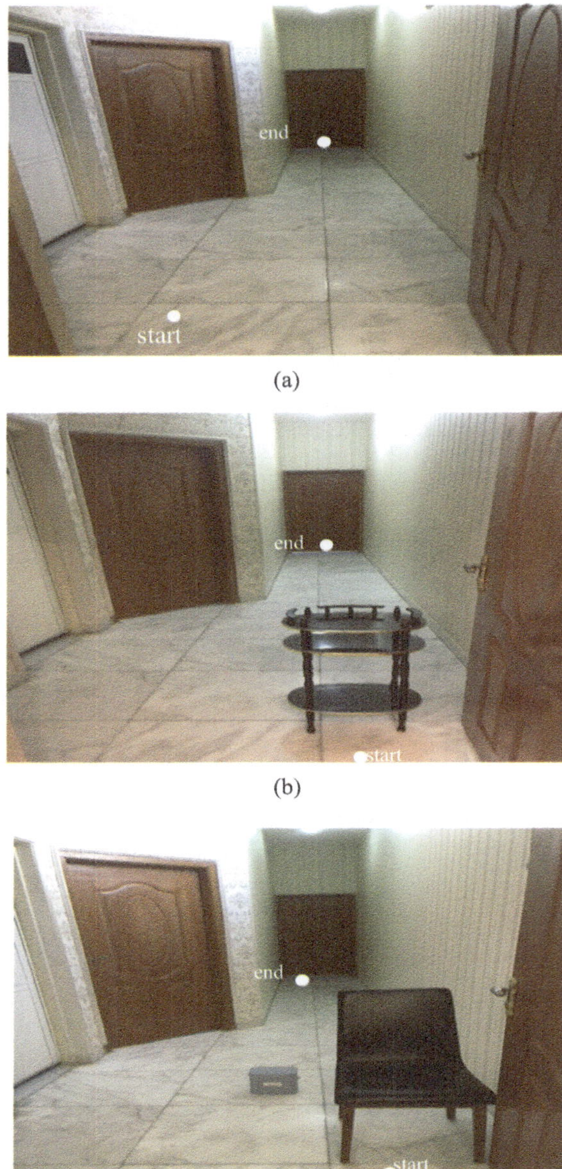

(a) (b) (c)

Fig. 6 Path following in the real-time scenarios

5 Conclusion

An efficient wearable indoor navigation system to guide the visually impaired person with voice commands to his/her final destination safely is presented in this paper. Hector SLAM method is selected to create a global map because it estimates the pose of the visually impaired person by relying on scan registration without need odometry information; in addition, 2D SLAM techniques are faster than 3D SLAM and they are not required a platform with high performances. The RGB-D camera is used instead of 2D laser scanners to solve the problem of detection obstacle with variable sizes and shapes. All adopted sensors in this work are of low cost and suitable size to be mounted on a head that will make the proposed system is comfortable and low cost. The experimental results showed that the performance of the proposed navigation system was a reliable in different real-time scenarios on assisting the visually impaired person navigates from one place to another.

References

1. Hoang, V., Nguyen, T., Le, T., Tran, T., Vuong, T., & Vuilerme, N. (2017). Obstacle visualy impaired people based on electrode matrix and mobile kinect Springer, Vietnam. *Journal of Computer Science, 4*(2), 71–83.
2. Simos, W., & de Lucena, V. (2016). Blind user wearable audio assistance for indoor nanigation based on visual markers and ultrasonic obstacle detection. In *IEEE International Conference on Consumer Electronics (ICCE)* (pp. 60–63).
3. Kang, M., Chae, S., Sun, J., Lee, S., & Ko, S. (2017). An enhanced obstacle avoidance method for the visually impaired using deformable grid. *IEEE Transactions on Consumer Electronics, 63*(2), 169–177.
4. Wu, H., Marshall, A., & Yu, W. (2007). Path planning and following algorithms in an indoor navigation model for visually impaired. In *Conference Internet Monitoring and Protection (ICIMP)* (pp. 38–38).
5. Yusof, T., Toha, A. S., & Yusof, H. (2015). Path Planning for visually impaired people in an unfamiliar environment using particle swarm optimization Elseiver. In *IEEE International Symposium on Robotics and Intelligent Sensors (IRIS)* (pp. 80–86)
6. W. Balachandran, Cecelja F., & Ptasinski, P. (2003). A GPS based navigation aid for the blind. In IEEE International conference on Applied Electromagnetics Communication (pp. 34–36).
7. Bai, J., Lian, Sh., Liu, Z., Wang, K., & Liu, D. (2018). Virtual-blind-road following based wearable navigation device for blind people. *IEEE Transactions on Consumer Electronics, 64*(11), 136–143.
8. Chaudhari, G., & Deshpande, A. (2017). Robotic assistant for visually impaired using sensor fusion. In *IEEE Conference (SmartWorld/SCALCOM/UIC/ATC/CBDCom/IOP/SCI)* (pp. 1–4).
9. Endo, Y., Sato, K., Yamashita, A., Matsubayashi, K. (2017). Indoor positioning and obstacle detection for visually impaired navigation system based on LSD-SLAM. In *IEEE International Conference on Biometrics and Kansei Engineering (ICBAKE)* (pp. 158–162).
10. Hakim, H., & Fadhil, A. (2019). Indoor low-cost assistive device using 2D SLAM based on LiDAR for visually impaired people. *Iraqi Journal of Electrical and Electronic Engineering, 15*(2), 115–121.
11. Ramesh, K., Naganada, S., Ramasangu, H., Deshpande, R. (2018). Real-time localization and navigation in an indoor environment using monocular camera for visually impaired. In *IEEE International Conference on Industrial Engineering and Applications (ICIEA)* (pp. 122–128).

12. Hakim, H., & Fadil, A. (2019). Navigation system for visually impaired people based on RGB-D camera and ultrasonic sensor. In *ACM International Conference on Information and Communication Technology* (pp. 172–177).
13. Kohlbrecher, S., Stryk, O., Meyer, J., & Klingauf, U. (2011). A flexible and scalable SLAM system with full 3D motion estimation. In *IEEE International Symposium on Safety, Security, and Rescue Robotics* (pp. 155–160).
14. Kamarud, K., Mamduh, S., Shakaff, A., Saad, S., Zakaria, A., Abdullah, A., & Kamarudin, L. (2013). Method to Convert Kinect's 3D Depth Data to a 2D Map for indoor SLAM. In *IEEE International Colloquium on Signal Processing and its Applications (CSPA)* (pp. 247–251).
15. Hart, P., Nilsson, N., & Raphael, B. (1968). A formal basis for the heuristic determination of minimum cost paths. *IEEE Transactions on Systems Science and Cybernetics (SSC), 4*(2), 100–107.
16. Otsu, N. (1979). A threshold selection method from gray-level histograms. *IEEE Transaction on Systems, Man and Cybernetics, 9*(1), 62–66.

Author Index

S.-L. Peng et al. (eds.), *Proceedings of First International Conference on Mathematical Modeling and Computational Science*, Advances in Intelligent Systems and Computing 1292, https://doi.org/10.1007/978-981-33-4389-4

The manufacturer's authorised representative in the EU is Springer Nature Customer Service Centre GmbH, Europaplatz 3, 69115 Heidelberg, Germany. If you have any concerns regarding our products, please contact ProductSafety@springernature.com

Printed and bound by CPI Group (UK) Ltd, Croydon, CR0 4YY
30/09/2026
02225292-0017